T0189245

Lecture Notes in Computer Science

Lecture Notes in Artificial Intelligence   13715

Founding Editor

Jörg Siekmann

Series Editors

Randy Goebel, *University of Alberta, Edmonton, Canada*
Wolfgang Wahlster, *DFKI, Berlin, Germany*
Zhi-Hua Zhou, *Nanjing University, Nanjing, China*

The series Lecture Notes in Artificial Intelligence (LNAI) was established in 1988 as a topical subseries of LNCS devoted to artificial intelligence.

The series publishes state-of-the-art research results at a high level. As with the LNCS mother series, the mission of the series is to serve the international R & D community by providing an invaluable service, mainly focused on the publication of conference and workshop proceedings and postproceedings.

Massih-Reza Amini · Stéphane Canu ·
Asja Fischer · Tias Guns · Petra Kralj Novak ·
Grigorios Tsoumakas
Editors

Machine Learning and Knowledge Discovery in Databases

European Conference, ECML PKDD 2022
Grenoble, France, September 19–23, 2022
Proceedings, Part III

 Springer

Editors
Massih-Reza Amini
Grenoble Alpes University
Saint Martin d'Hères, France

Stéphane Canu
INSA Rouen Normandy
Saint Etienne du Rouvray, France

Asja Fischer
Ruhr-Universität Bochum
Bochum, Germany

Tias Guns
KU Leuven
Leuven, Belgium

Petra Kralj Novak
Central European University
Vienna, Austria

Grigorios Tsoumakas
Aristotle University of Thessaloniki
Thessaloniki, Greece

ISSN 0302-9743   ISSN 1611-3349 (electronic)
Lecture Notes in Artificial Intelligence
ISBN 978-3-031-26408-5   ISBN 978-3-031-26409-2 (eBook)
https://doi.org/10.1007/978-3-031-26409-2

LNCS Sublibrary: SL7 – Artificial Intelligence

Preface

The European Conference on Machine Learning and Principles and Practice of Knowledge Discovery in Databases (ECML–PKDD 2022) in Grenoble, France, was once again a place for in-person gathering and the exchange of ideas after two years of completely virtual conferences due to the SARS-CoV-2 pandemic. This year the conference was hosted for the first time in hybrid format, and we are honored and delighted to offer you these proceedings as a result.

The annual ECML–PKDD conference serves as a global venue for the most recent research in all fields of machine learning and knowledge discovery in databases, including cutting-edge applications. It builds on a highly successful run of ECML–PKDD conferences which has made it the premier European machine learning and data mining conference.

This year, the conference drew over 1080 participants (762 in-person and 318 online) from 37 countries, including 23 European nations. This wealth of interest considerably exceeded our expectations, and we were both excited and under pressure to plan a special event. Overall, the conference attracted a lot of interest from industry thanks to sponsorship, participation, and the conference's industrial day.

The main conference program consisted of presentations of 242 accepted papers and four keynote talks (in order of appearance):

- Francis Bach (Inria), Information Theory with Kernel Methods
- Danai Koutra (University of Michigan), Mining & Learning [Compact] Representations for Structured Data
- Fosca Gianotti (Scuola Normale Superiore di Pisa), Explainable Machine Learning for Trustworthy AI
- Yann Le Cun (Facebook AI Research), From Machine Learning to Autonomous Intelligence

In addition, there were respectively twenty three in-person and three online workshops; five in-person and three online tutorials; two combined in-person and one combined online workshop-tutorials, together with a PhD Forum, a discovery challenge and demonstrations.

Papers presented during the three main conference days were organized in 4 tracks, within 54 sessions:

- Research Track: articles on research or methodology from all branches of machine learning, data mining, and knowledge discovery;
- Applied Data Science Track: articles on cutting-edge uses of machine learning, data mining, and knowledge discovery to resolve practical use cases and close the gap between current theory and practice;
- Journal Track: articles that were published in special issues of the journals *Machine Learning* and *Data Mining and Knowledge Discovery*;

– Demo Track: short articles that propose a novel system that advances the state of the art and include a demonstration video.

We received a record number of 1238 abstract submissions, and for the Research and Applied Data Science Tracks, 932 papers made it through the review process (the remaining papers were withdrawn, with the bulk being desk rejected). We accepted 189 (27.3%) Research papers and 53 (22.2%) Applied Data science articles. 47 papers from the Journal Track and 17 demo papers were also included in the program. We were able to put together an extraordinarily rich and engaging program because of the high quality submissions.

Research articles that were judged to be of exceptional quality and deserving of special distinction were chosen by the awards committee:

– Machine Learning Best Paper Award: "*Bounding the Family-Wise Error Rate in Local Causal Discovery Using Rademacher Averages*", by Dario Simionato (University of Padova) and Fabio Vandin (University of Padova)
– Data-Mining Best Paper Award: "*Transforming PageRank into an Infinite-Depth Graph Neural Network*", by Andreas Roth (TU Dortmund), and Thomas Liebig (TU Dortmund)
– Test of Time Award for highest impact paper from ECML–PKDD 2012: "*Fairness-Aware Classifier with Prejudice Remover Regularizer*", by Toshihiro Kamishima (National Institute of Advanced Industrial Science and Technology AIST), Shotaro Akashi (National Institute of Advanced Industrial Science and Technology AIST), Hideki Asoh (National Institute of Advanced Industrial Science and Technology AIST), and Jun Sakuma (University of Tsukuba)

We sincerely thank the contributions of all participants, authors, PC members, area chairs, session chairs, volunteers, and co-organizers who made ECML–PKDD 2022 a huge success. We would especially like to thank Julie from the Grenoble World Trade Center for all her help and Titouan from Insight-outside, who worked so hard to make the online event possible. We also like to express our gratitude to Thierry for the design of the conference logo representing the three mountain chains surrounding the Grenoble city, as well as the sponsors and the ECML–PKDD Steering Committee.

October 2022

Massih-Reza Amini
Stéphane Canu
Asja Fischer
Petra Kralj Novak
Tias Guns
Grigorios Tsoumakas
Georgios Balikas
Fragkiskos Malliaros

Organization

General Chairs

Massih-Reza Amini   University Grenoble Alpes, France
Stéphane Canu   INSA Rouen, France

Program Chairs

Asja Fischer   Ruhr University Bochum, Germany
Tias Guns   KU Leuven, Belgium
Petra Kralj Novak   Central European University, Austria
Grigorios Tsoumakas   Aristotle University of Thessaloniki, Greece

Journal Track Chairs

Peggy Cellier   INSA Rennes, IRISA, France
Krzysztof Dembczyński   Yahoo Research, USA
Emilie Devijver   CNRS, France
Albrecht Zimmermann   University of Caen Normandie, France

Workshop and Tutorial Chairs

Bruno Crémilleux   University of Caen Normandie, France
Charlotte Laclau   Telecom Paris, France

Local Chairs

Latifa Boudiba   University Grenoble Alpes, France
Franck Iutzeler   University Grenoble Alpes, France

Proceedings Chairs

Wouter Duivesteijn   Technische Universiteit Eindhoven,
                              the Netherlands
Sibylle Hess   Technische Universiteit Eindhoven,
                              the Netherlands

Industry Track Chairs

Rohit Babbar   Aalto University, Finland
Françoise Fogelmann   Hub France IA, France

Discovery Challenge Chairs

Ioannis Katakis   University of Nicosia, Cyprus
Ioannis Partalas   Expedia, Switzerland

Demonstration Chairs

Georgios Balikas   Salesforce, France
Fragkiskos Malliaros   CentraleSupélec, France

PhD Forum Chairs

Esther Galbrun   University of Eastern Finland, Finland
Justine Reynaud   University of Caen Normandie, France

Awards Chairs

Francesca Lisi   Università degli Studi di Bari, Italy
Michalis Vlachos   University of Lausanne, Switzerland

Sponsorship Chairs

Patrice Aknin   IRT SystemX, France
Gilles Gasso   INSA Rouen, France

Web Chairs

Martine Harshé   Laboratoire d'Informatique de Grenoble, France
Marta Soare   University Grenoble Alpes, France

Publicity Chair

Emilie Morvant   Université Jean Monnet, France

ECML PKDD Steering Committee

Annalisa Appice   University of Bari Aldo Moro, Italy
Ira Assent   Aarhus University, Denmark
Albert Bifet   Télécom ParisTech, France
Francesco Bonchi   ISI Foundation, Italy
Tania Cerquitelli   Politecnico di Torino, Italy
Sašo Džeroski   Jožef Stefan Institute, Slovenia
Elisa Fromont   Université de Rennes, France
Andreas Hotho   Julius-Maximilians-Universität Würzburg,
                                 Germany
Alípio Jorge   University of Porto, Portugal
Kristian Kersting   TU Darmstadt, Germany
Jefrey Lijffijt   Ghent University, Belgium
Luís Moreira-Matias   University of Porto, Portugal
Katharina Morik   TU Dortmund, Germany
Siegfried Nijssen   Université catholique de Louvain, Belgium
Andrea Passerini   University of Trento, Italy
Fernando Perez-Cruz   ETH Zurich, Switzerland
Alessandra Sala   Shutterstock Ireland Limited, Ireland
Arno Siebes   Utrecht University, the Netherlands
Isabel Valera   Universität des Saarlandes, Germany

Program Committees

Guest Editorial Board, Journal Track

Richard Allmendinger   University of Manchester, UK
Marie Anastacio   Universiteit Leiden, the Netherlands
Ira Assent   Aarhus University, Denmark
Martin Atzmueller   Universität Osnabrück, Germany
Rohit Babbar   Aalto University, Finland

Jaume Bacardit	Newcastle University, UK
Anthony Bagnall	University of East Anglia, UK
Mitra Baratchi	Universiteit Leiden, the Netherlands
Francesco Bariatti	IRISA, France
German Barquero	Universität de Barcelona, Spain
Alessio Benavoli	Trinity College Dublin, Ireland
Viktor Bengs	Ludwig-Maximilians-Universität München, Germany
Massimo Bilancia	Università degli Studi di Bari Aldo Moro, Italy
Ilaria Bordino	Unicredit R&D, Italy
Jakob Bossek	University of Münster, Germany
Ulf Brefeld	Leuphana University of Lüneburg, Germany
Ricardo Campello	University of Newcastle, UK
Michelangelo Ceci	University of Bari, Italy
Loic Cerf	Universidade Federal de Minas Gerais, Brazil
Vitor Cerqueira	Universidade do Porto, Portugal
Laetitia Chapel	IRISA, France
Jinghui Chen	Pennsylvania State University, USA
Silvia Chiusano	Politecnico di Torino, Italy
Roberto Corizzo	Università degli Studi di Bari Aldo Moro, Italy
Bruno Cremilleux	Université de Caen Normandie, France
Marco de Gemmis	University of Bari Aldo Moro, Italy
Sebastien Destercke	Centre National de la Recherche Scientifique, France
Shridhar Devamane	Global Academy of Technology, India
Benjamin Doerr	Ecole Polytechnique, France
Wouter Duivesteijn	Technische Universiteit Eindhoven, the Netherlands
Thomas Dyhre Nielsen	Aalborg University, Denmark
Tapio Elomaa	Tampere University, Finland
Remi Emonet	Université Jean Monnet Saint-Etienne, France
Nicola Fanizzi	Università degli Studi di Bari Aldo Moro, Italy
Pedro Ferreira	University of Lisbon, Portugal
Cesar Ferri	Universität Politecnica de Valencia, Spain
Julia Flores	University of Castilla-La Mancha, Spain
Ionut Florescu	Stevens Institute of Technology, USA
Germain Forestier	Université de Haute-Alsace, France
Joel Frank	Ruhr-Universität Bochum, Germany
Marco Frasca	Università degli Studi di Milano, Italy
Jose A. Gomez	Universidad de Castilla-La Mancha, Spain
Stephan Günnemann	Institute for Advanced Study, Germany
Luis Galarraga	Inria, France

Esther Galbrun	University of Eastern Finland, Finland
Joao Gama	University of Porto, Portugal
Paolo Garza	Politecnico di Torino, Italy
Pascal Germain	Université Laval, Canada
Fabian Gieseke	Westfälische Wilhelms-Universität Münster, Germany
Riccardo Guidotti	Università degli Studi di Pisa, Italy
Francesco Gullo	UniCredit, Italy
Antonella Guzzo	University of Calabria, Italy
Isabel Haasler	KTH Royal Institute of Technology, Sweden
Alexander Hagg	Bonn-Rhein-Sieg University, Germany
Daniel Hernandez-Lobato	Universidad Autónoma de Madrid, Spain
Jose Hernandez-Orallo	Universidad Politecnica de Valencia, Spain
Martin Holena	Neznámá organizace, Czechia
Jaakko Hollmen	Stockholm University, Sweden
Dino Ienco	IRSTEA, France
Georgiana Ifrim	University College Dublin, Ireland
Felix Iglesias	Technische Universität Wien, Austria
Angelo Impedovo	Università degli Studi di Bari Aldo Moro, Italy
Frank Iutzeler	Université Grenoble Alpes, France
Mahdi Jalili	RMIT University, Australia
Szymon Jaroszewicz	Polish Academy of Sciences, Poland
Mehdi Kaytoue	INSA Lyon, France
Raouf Kerkouche	Helmholtz Center for Information Security, Germany
Pascal Kerschke	Westfälische Wilhelms-Universität Münster, Germany
Dragi Kocev	Jožef Stefan Institute, Slovenia
Wojciech Kotlowski	Poznan University of Technology, Poland
Lars Kotthoff	University of Wyoming, USA
Peer Kroger	Ludwig-Maximilians-Universität München, Germany
Tipaluck Krityakierne	Mahidol University, Thailand
Peer Kroger	Christian-Albrechts-University Kiel, Germany
Meelis Kull	Tartu Ulikool, Estonia
Charlotte Laclau	Laboratoire Hubert Curien, France
Mark Last	Ben-Gurion University of the Negev, Israel
Matthijs van Leeuwen	Universiteit Leiden, the Netherlands
Thomas Liebig	TU Dortmund, Germany
Hsuan-Tien Lin	National Taiwan University, Taiwan
Marco Lippi	University of Modena and Reggio Emilia, Italy
Daniel Lobato	Universidad Autonoma de Madrid, Spain

Corrado Loglisci	Università degli Studi di Bari Aldo Moro, Italy
Nuno Lourenço	University of Coimbra, Portugal
Claudio Lucchese	Ca'Foscari University of Venice, Italy
Brian MacNamee	University College Dublin, Ireland
Davide Maiorca	University of Cagliari, Italy
Giuseppe Manco	National Research Council, Italy
Elio Masciari	University of Naples Federico II, Italy
Andres Masegosa	University of Aalborg, Denmark
Ernestina Menasalvas	Universidad Politecnica de Madrid, Spain
Lien Michiels	Universiteit Antwerpen, Belgium
Jan Mielniczuk	Polish Academy of Sciences, Poland
Paolo Mignone	Università degli Studi di Bari Aldo Moro, Italy
Anna Monreale	University of Pisa, Italy
Giovanni Montana	University of Warwick, UK
Gregoire Montavon	Technische Universität Berlin, Germany
Amedeo Napoli	LORIA, France
Frank Neumann	University of Adelaide, Australia
Thomas Nielsen	Aalborg Universitet, Denmark
Bruno Ordozgoiti	Aalto-yliopisto, Finland
Panagiotis Papapetrou	Stockholms Universitet, Sweden
Andrea Passerini	University of Trento, Italy
Mykola Pechenizkiy	Technische Universiteit Eindhoven, the Netherlands
Charlotte Pelletier	IRISA, France
Ruggero Pensa	University of Turin, Italy
Nico Piatkowski	Technische Universität Dortmund, Germany
Gianvito Pio	Università degli Studi di Bari Aldo Moro, Italy
Marc Plantevit	Université Claude Bernard Lyon 1, France
Jose M. Puerta	Universidad de Castilla-La Mancha, Spain
Kai Puolamaki	Helsingin Yliopisto, Finland
Michael Rabbat	Meta Platforms Inc, USA
Jan Ramon	Inria Lille Nord Europe, France
Rita Ribeiro	Universidade do Porto, Portugal
Kaspar Riesen	University of Bern, Switzerland
Matteo Riondato	Amherst College, USA
Celine Robardet	INSA Lyon, France
Pieter Robberechts	KU Leuven, Belgium
Antonio Salmeron	University of Almería, Spain
Jorg Sander	University of Alberta, Canada
Roberto Santana	University of the Basque Country, Spain
Michael Schaub	Rheinisch-Westfälische Technische Hochschule, Germany

Erik Schultheis   Aalto-yliopisto, Finland
Thomas Seidl   Ludwig-Maximilians-Universität München,
                                        Germany
Moritz Seiler   University of Münster, Germany
Kijung Shin   KAIST, South Korea
Shinichi Shirakawa   Yokohama National University, Japan
Marek Smieja   Jagiellonian University, Poland
James Edward Smith   University of the West of England, UK
Carlos Soares   Universidade do Porto, Portugal
Arnaud Soulet   Université de Tours, France
Gerasimos Spanakis   Maastricht University, the Netherlands
Giancarlo Sperli   University of Campania Luigi Vanvitelli, Italy
Myra Spiliopoulou   Otto von Guericke Universität Magdeburg,
                                        Germany
Jerzy Stefanowski   Poznan University of Technology, Poland
Giovanni Stilo   Università degli Studi dell'Aquila, Italy
Catalin Stoean   University of Craiova, Romania
Mahito Sugiyama   National Institute of Informatics, Japan
Nikolaj Tatti   Helsingin Yliopisto, Finland
Alexandre Termier   Université de Rennes 1, France
Luis Torgo   Dalhousie University, Canada
Leonardo Trujillo   Tecnologico Nacional de Mexico, Mexico
Wei-Wei Tu   4Paradigm Inc., China
Steffen Udluft   Siemens AG Corporate Technology, Germany
Arnaud Vandaele   Université de Mons, Belgium
Celine Vens   KU Leuven, Belgium
Herna Viktor   University of Ottawa, Canada
Marco Virgolin   Centrum Wiskunde en Informatica,
                                        the Netherlands
Jordi Vitria   Universität de Barcelona, Spain
Jilles Vreeken   CISPA Helmholtz Center for Information
                                        Security, Germany
Willem Waegeman   Universiteit Gent, Belgium
Markus Wagner   University of Adelaide, Australia
Elizabeth Wanner   Centro Federal de Educacao Tecnologica de
                                        Minas, Brazil
Marcel Wever   Universität Paderborn, Germany
Ngai Wong   University of Hong Kong, Hong Kong, China
Man Leung Wong   Lingnan University, Hong Kong, China
Marek Wydmuch   Poznan University of Technology, Poland
Guoxian Yu   Shandong University, China
Xiang Zhang   University of Hong Kong, Hong Kong, China

Ye Zhu   Deakin University, USA
Arthur Zimek   Syddansk Universitet, Denmark
Albrecht Zimmermann   Université de Caen Normandie, France

Area Chairs

Fabrizio Angiulli   DIMES, University of Calabria, Italy
Annalisa Appice   University of Bari, Italy
Ira Assent   Aarhus University, Denmark
Martin Atzmueller   Osnabrück University, Germany
Michael Berthold   Universität Konstanz, Germany
Albert Bifet   Université Paris-Saclay, France
Hendrik Blockeel   KU Leuven, Belgium
Christian Böhm   LMU Munich, Germany
Francesco Bonchi   ISI Foundation, Turin, Italy
Ulf Brefeld   Leuphana, Germany
Francesco Calabrese   Richemont, USA
Toon Calders   Universiteit Antwerpen, Belgium
Michelangelo Ceci   University of Bari, Italy
Peggy Cellier   IRISA, France
Duen Horng Chau   Georgia Institute of Technology, USA
Nicolas Courty   IRISA, Université Bretagne-Sud, France
Bruno Cremilleux   Université de Caen Normandie, France
Jesse Davis   KU Leuven, Belgium
Gianmarco De Francisci Morales   CentAI, Italy
Tom Diethe   Amazon, UK
Carlotta Domeniconi   George Mason University, USA
Yuxiao Dong   Tsinghua University, China
Kurt Driessens   Maastricht University, the Netherlands
Tapio Elomaa   Tampere University, Finland
Sergio Escalera   CVC and University of Barcelona, Spain
Faisal Farooq   Qatar Computing Research Institute, Qatar
Asja Fischer   Ruhr University Bochum, Germany
Peter Flach   University of Bristol, UK
Eibe Frank   University of Waikato, New Zealand
Paolo Frasconi   Università degli Studi di Firenze, Italy
Elisa Fromont   Université Rennes 1, IRISA/Inria, France
Johannes Fürnkranz   JKU Linz, Austria
Patrick Gallinari   Sorbonne Université, Criteo AI Lab, France
Joao Gama   INESC TEC - LIAAD, Portugal
Jose Gamez   Universidad de Castilla-La Mancha, Spain
Roman Garnett   Washington University in St. Louis, USA
Thomas Gärtner   TU Wien, Austria

Aristides Gionis	KTH Royal Institute of Technology, Sweden
Francesco Gullo	UniCredit, Italy
Stephan Günnemann	Technical University of Munich, Germany
Xiangnan He	University of Science and Technology of China, China
Daniel Hernandez-Lobato	Universidad Autonoma de Madrid, Spain
José Hernández-Orallo	Universität Politècnica de València, Spain
Jaakko Hollmén	Aalto University, Finland
Andreas Hotho	Universität Würzburg, Germany
Eyke Hüllermeier	University of Munich, Germany
Neil Hurley	University College Dublin, Ireland
Georgiana Ifrim	University College Dublin, Ireland
Alipio Jorge	INESC TEC/University of Porto, Portugal
Ross King	Chalmers University of Technology, Sweden
Arno Knobbe	Leiden University, the Netherlands
Yun Sing Koh	University of Auckland, New Zealand
Parisa Kordjamshidi	Michigan State University, USA
Lars Kotthoff	University of Wyoming, USA
Nicolas Kourtellis	Telefonica Research, Spain
Danai Koutra	University of Michigan, USA
Danica Kragic	KTH Royal Institute of Technology, Sweden
Stefan Kramer	Johannes Gutenberg University Mainz, Germany
Niklas Lavesson	Blekinge Institute of Technology, Sweden
Sébastien Lefèvre	Université de Bretagne Sud/IRISA, France
Jefrey Lijffijt	Ghent University, Belgium
Marius Lindauer	Leibniz University Hannover, Germany
Patrick Loiseau	Inria, France
Jose Lozano	UPV/EHU, Spain
Jörg Lücke	Universität Oldenburg, Germany
Donato Malerba	Università degli Studi di Bari Aldo Moro, Italy
Fragkiskos Malliaros	CentraleSupelec, France
Giuseppe Manco	ICAR-CNR, Italy
Wannes Meert	KU Leuven, Belgium
Pauli Miettinen	University of Eastern Finland, Finland
Dunja Mladenic	Jožef Stefan Institute, Slovenia
Anna Monreale	Università di Pisa, Italy
Luis Moreira-Matias	Finiata, Germany
Emilie Morvant	University Jean Monnet, St-Etienne, France
Sriraam Natarajan	UT Dallas, USA
Nuria Oliver	Vodafone Research, USA
Panagiotis Papapetrou	Stockholm University, Sweden
Laurence Park	WSU, Australia

Program Committee Members

Amos Abbott	Virginia Tech, USA
Pedro Abreu	CISUC, Portugal
Maribel Acosta	Ruhr University Bochum, Germany
Timilehin Aderinola	Insight Centre, University College Dublin, Ireland
Linara Adilova	Ruhr University Bochum, Fraunhofer IAIS, Germany
Florian Adriaens	KTH, Sweden
Azim Ahmadzadeh	Georgia State University, USA
Nourhan Ahmed	University of Hildesheim, Germany
Deepak Ajwani	University College Dublin, Ireland
Amir Hossein Akhavan Rahnama	KTH Royal Institute of Technology, Sweden
Aymen Al Marjani	ENS Lyon, France
Mehwish Alam	Leibniz Institute for Information Infrastructure, Germany
Francesco Alesiani	NEC Laboratories Europe, Germany
Omar Alfarisi	ADNOC, Canada
Pegah Alizadeh	Ericsson Research, France
Reem Alotaibi	King Abdulaziz University, Saudi Arabia
Jumanah Alshehri	Temple University, USA
Bakhtiar Amen	University of Huddersfield, UK
Evelin Amorim	Inesc tec, Portugal
Shin Ando	Tokyo University of Science, Japan
Thiago Andrade	INESC TEC - LIAAD, Portugal
Jean-Marc Andreoli	Naverlabs Europe, France
Giuseppina Andresini	University of Bari Aldo Moro, Italy
Alessandro Antonucci	IDSIA, Switzerland
Xiang Ao	Institute of Computing Technology, CAS, China
Siddharth Aravindan	National University of Singapore, Singapore
Héber H. Arcolezi	Inria and École Polytechnique, France
Adrián Arnaiz-Rodríguez	ELLIS Unit Alicante, Spain
Yusuf Arslan	University of Luxembourg, Luxembourg
André Artelt	Bielefeld University, Germany
Sunil Aryal	Deakin University, Australia
Charles Assaad	Easyvista, France
Matthias Aßenmacher	Ludwig-Maxmilians-Universität München, Germany
Zeyar Aung	Masdar Institute, UAE
Serge Autexier	DFKI Bremen, Germany
Rohit Babbar	Aalto University, Finland
Housam Babiker	University of Alberta, Canada

Antonio Bahamonde	University of Oviedo, Spain
Maroua Bahri	Inria Paris, France
Georgios Balikas	Salesforce, France
Maria Bampa	Stockholm University, Sweden
Hubert Baniecki	Warsaw University of Technology, Poland
Elena Baralis	Politecnico di Torino, Italy
Mitra Baratchi	LIACS - University of Leiden, the Netherlands
Kalliopi Basioti	Rutgers University, USA
Martin Becker	Stanford University, USA
Diana Benavides Prado	University of Auckland, New Zealand
Anes Bendimerad	LIRIS, France
Idir Benouaret	Université Grenoble Alpes, France
Isacco Beretta	Università di Pisa, Italy
Victor Berger	CEA, France
Christoph Bergmeir	Monash University, Australia
Cuissart Bertrand	University of Caen, France
Antonio Bevilacqua	University College Dublin, Ireland
Yaxin Bi	Ulster University, UK
Ranran Bian	University of Auckland, New Zealand
Adrien Bibal	University of Louvain, Belgium
Subhodip Biswas	Virginia Tech, USA
Patrick Blöbaum	Amazon AWS, USA
Carlos Bobed	University of Zaragoza, Spain
Paul Bogdan	USC, USA
Chiara Boldrini	CNR, Italy
Clément Bonet	Université Bretagne Sud, France
Andrea Bontempelli	University of Trento, Italy
Ludovico Boratto	University of Cagliari, Italy
Stefano Bortoli	Huawei Research Center, Germany
Diana-Laura Borza	Babes Bolyai University, Romania
Ahcene Boubekki	UiT, Norway
Sabri Boughorbel	QCRI, Qatar
Paula Branco	University of Ottawa, Canada
Jure Brence	Jožef Stefan Institute, Slovenia
Martin Breskvar	Jožef Stefan Institute, Slovenia
Marco Bressan	University of Milan, Italy
Dariusz Brzezinski	Poznan University of Technology, Poland
Florian Buettner	German Cancer Research Center, Germany
Julian Busch	Siemens Technology, Germany
Sebastian Buschjäger	TU Dortmund Artificial Intelligence Unit, Germany
Ali Butt	Virginia Tech, USA

Narayanan C. Krishnan   IIT Palakkad, India
Xiangrui Cai   Nankai University, China
Xiongcai Cai   UNSW Sydney, Australia
Zekun Cai   University of Tokyo, Japan
Andrea Campagner   Università degli Studi di Milano-Bicocca, Italy
Seyit Camtepe   CSIRO Data61, Australia
Jiangxia Cao   Chinese Academy of Sciences, China
Pengfei Cao   Chinese Academy of Sciences, China
Yongcan Cao   University of Texas at San Antonio, USA
Cécile Capponi   Aix-Marseille University, France
Axel Carlier   Institut National Polytechnique de Toulouse,
                               France
Paula Carroll   University College Dublin, Ireland
John Cartlidge   University of Bristol, UK
Simon Caton   University College Dublin, Ireland
Bogdan Cautis   University of Paris-Saclay, France
Mustafa Cavus   Warsaw University of Technology, Poland
Remy Cazabet   Université Lyon 1, France
Josu Ceberio   University of the Basque Country, Spain
David Cechák   CEITEC Masaryk University, Czechia
Abdulkadir Celikkanat   Technical University of Denmark, Denmark
Dumitru-Clementin Cercel   University Politehnica of Bucharest, Romania
Christophe Cerisara   CNRS, France
Vítor Cerqueira   Dalhousie University, Canada
Mattia Cerrato   JGU Mainz, Germany
Ricardo Cerri   Federal University of São Carlos, Brazil
Hubert Chan   University of Hong Kong, Hong Kong, China
Vaggos Chatziafratis   Stanford University, USA
Siu Lun Chau   University of Oxford, UK
Chaochao Chen   Zhejiang University, China
Chuan Chen   Sun Yat-sen University, China
Hechang Chen   Jilin University, China
Jia Chen   Beihang University, China
Jiaoyan Chen   University of Oxford, UK
Jiawei Chen   Zhejiang University, China
Jin Chen   University of Electronic Science and Technology,
                               China
Kuan-Hsun Chen   University of Twente, the Netherlands
Lingwei Chen   Wright State University, USA
Tianyi Chen   Boston University, USA
Wang Chen   Google, USA
Xinyuan Chen   Universiti Kuala Lumpur, Malaysia

Yuqiao Chen	UT Dallas, USA
Yuzhou Chen	Princeton University, USA
Zhennan Chen	Xiamen University, China
Zhiyu Chen	UCSB, USA
Zhqian Chen	Mississippi State University, USA
Ziheng Chen	Stony Brook University, USA
Zhiyong Cheng	Shandong Academy of Sciences, China
Noëlie Cherrier	CITiO, France
Anshuman Chhabra	UC Davis, USA
Zhixuan Chu	Ant Group, China
Guillaume Cleuziou	LIFO, France
Ciaran Cooney	AflacNI, UK
Robson Cordeiro	University of São Paulo, Brazil
Roberto Corizzo	American University, USA
Antoine Cornuéjols	AgroParisTech, France
Fabrizio Costa	Exeter University, UK
Gustavo Costa	Instituto Federal de Goiás - Campus Jataí, Brazil
Luís Cruz	Delft University of Technology, the Netherlands
Tianyu Cui	Institute of Information Engineering, China
Wang-Zhou Dai	Imperial College London, UK
Tanmoy Dam	University of New South Wales Canberra, Australia
Thi-Bich-Hanh Dao	University of Orleans, France
Adrian Sergiu Darabant	Babes Bolyai University, Romania
Mrinal Das	IIT Palakaad, India
Sina Däubener	Ruhr University, Bochum, Germany
Padraig Davidson	University of Würzburg, Germany
Paul Davidsson	Malmö University, Sweden
Andre de Carvalho	USP, Brazil
Antoine de Mathelin	ENS Paris-Saclay, France
Tom De Schepper	University of Antwerp, Belgium
Marcilio de Souto	LIFO/Univ. Orleans, France
Gaetan De Waele	Ghent University, Belgium
Pieter Delobelle	KU Leuven, Belgium
Alper Demir	Izmir University of Economics, Turkey
Ambra Demontis	University of Cagliari, Italy
Difan Deng	Leibniz Universität Hannover, Germany
Guillaume Derval	UCLouvain - ICTEAM, Belgium
Maunendra Sankar Desarkar	IIT Hyderabad, India
Chris Develder	University of Ghent - iMec, Belgium
Arnout Devos	Swiss Federal Institute of Technology Lausanne, Switzerland

Laurens Devos	KU Leuven, Belgium
Bhaskar Dhariyal	University College Dublin, Ireland
Nicola Di Mauro	University of Bari, Italy
Aissatou Diallo	University College London, UK
Christos Dimitrakakis	University of Neuchatel, Switzerland
Jiahao Ding	University of Houston, USA
Kaize Ding	Arizona State University, USA
Yao-Xiang Ding	Nanjing University, China
Guilherme Dinis Junior	Stockholm University, Sweden
Nikolaos Dionelis	University of Edinburgh, UK
Christos Diou	Harokopio University of Athens, Greece
Sonia Djebali	Léonard de Vinci Pôle Universitaire, France
Nicolas Dobigeon	University of Toulouse, France
Carola Doerr	Sorbonne University, France
Ruihai Dong	University College Dublin, Ireland
Shuyu Dong	Inria, Université Paris-Saclay, France
Yixiang Dong	Xi'an Jiaotong University, China
Xin Du	University of Edinburgh, UK
Yuntao Du	Nanjing University, China
Stefan Duffner	University of Lyon, France
Rahul Duggal	Georgia Tech, USA
Wouter Duivesteijn	TU Eindhoven, the Netherlands
Sebastijan Dumancic	TU Delft, the Netherlands
Inês Dutra	University of Porto, Portugal
Thomas Dyhre Nielsen	AAU, Denmark
Saso Dzeroski	Jožef Stefan Institute, Ljubljana, Slovenia
Tome Eftimov	Jožef Stefan Institute, Ljubljana, Slovenia
Hamid Eghbal-zadeh	LIT AI Lab, Johannes Kepler University, Austria
Theresa Eimer	Leibniz University Hannover, Germany
Radwa El Shawi	Tartu University, Estonia
Dominik Endres	Philipps-Universität Marburg, Germany
Roberto Esposito	Università di Torino, Italy
Georgios Evangelidis	University of Macedonia, Greece
Samuel Fadel	Leuphana University, Germany
Stephan Fahrenkrog-Petersen	Humboldt-Universität zu Berlin, Germany
Xiaomao Fan	Shenzhen Technology University, China
Zipei Fan	University of Tokyo, Japan
Hadi Fanaee	Halmstad University, Sweden
Meng Fang	TU/e, the Netherlands
Elaine Faria	UFU, Brazil
Ad Feelders	Universiteit Utrecht, the Netherlands
Sophie Fellenz	TU Kaiserslautern, Germany

Stefano Ferilli	University of Bari, Italy
Daniel Fernández-Sánchez	Universidad Autónoma de Madrid, Spain
Pedro Ferreira	Faculty of Sciences University of Porto, Portugal
Cèsar Ferri	Universität Politècnica València, Spain
Flavio Figueiredo	UFMG, Brazil
Soukaina Filali Boubrahimi	Utah State University, USA
Raphael Fischer	TU Dortmund, Germany
Germain Forestier	University of Haute Alsace, France
Edouard Fouché	Karlsruhe Institute of Technology, Germany
Philippe Fournier-Viger	Shenzhen University, China
Kary Framling	Umeå University, Sweden
Jérôme François	Inria Nancy Grand-Est, France
Fabio Fumarola	Prometeia, Italy
Pratik Gajane	Eindhoven University of Technology, the Netherlands
Esther Galbrun	University of Eastern Finland, Finland
Laura Galindez Olascoaga	KU Leuven, Belgium
Sunanda Gamage	University of Western Ontario, Canada
Chen Gao	Tsinghua University, China
Wei Gao	Nanjing University, China
Xiaofeng Gao	Shanghai Jiaotong University, China
Yuan Gao	University of Science and Technology of China, China
Jochen Garcke	University of Bonn, Germany
Clement Gautrais	Brightclue, France
Benoit Gauzere	INSA Rouen, France
Dominique Gay	Université de La Réunion, France
Xiou Ge	University of Southern California, USA
Bernhard Geiger	Know-Center GmbH, Germany
Jiahui Geng	University of Stavanger, Norway
Yangliao Geng	Tsinghua University, China
Konstantin Genin	University of Tübingen, Germany
Firas Gerges	New Jersey Institute of Technology, USA
Pierre Geurts	University of Liège, Belgium
Gizem Gezici	Sabanci University, Turkey
Amirata Ghorbani	Stanford, USA
Biraja Ghoshal	TCS, UK
Anna Giabelli	Università degli studi di Milano Bicocca, Italy
George Giannakopoulos	IIT Demokritos, Greece
Tobias Glasmachers	Ruhr-University Bochum, Germany
Heitor Murilo Gomes	University of Waikato, New Zealand
Anastasios Gounaris	Aristotle University of Thessaloniki, Greece

Yupeng Hou	Renmin University of China, China
Binbin Hu	Ant Financial Services Group, China
Jian Hu	Queen Mary University of London, UK
Liang Hu	Tongji University, China
Wen Hu	Ant Group, China
Wenbin Hu	Wuhan University, China
Wenbo Hu	Tsinghua University, China
Yaowei Hu	University of Arkansas, USA
Chao Huang	University of Hong Kong, China
Gang Huang	Zhejiang Lab, China
Guanjie Huang	Penn State University, USA
Hong Huang	HUST, China
Jin Huang	University of Amsterdam, the Netherlands
Junjie Huang	Chinese Academy of Sciences, China
Qiang Huang	Jilin University, China
Shangrong Huang	Hunan University, China
Weitian Huang	South China University of Technology, China
Yan Huang	Huazhong University of Science and Technology, China
Yiran Huang	Karlsruhe Institute of Technology, Germany
Angelo Impedovo	University of Bari, Italy
Roberto Interdonato	CIRAD, France
Iñaki Inza	University of the Basque Country, Spain
Stratis Ioannidis	Northeastern University, USA
Rakib Islam	Facebook, USA
Tobias Jacobs	NEC Laboratories Europe GmbH, Germany
Priyank Jaini	Google, Canada
Johannes Jakubik	Karlsruhe Institute of Technology, Germany
Nathalie Japkowicz	American University, USA
Szymon Jaroszewicz	Polish Academy of Sciences, Poland
Shayan Jawed	University of Hildesheim, Germany
Rathinaraja Jeyaraj	Kyungpook National University, South Korea
Shaoxiong Ji	Aalto University, Finland
Taoran Ji	Virginia Tech, USA
Bin-Bin Jia	Southeast University, China
Yuheng Jia	Southeast University, China
Ziyu Jia	Beijing Jiaotong University, China
Nan Jiang	Purdue University, USA
Renhe Jiang	University of Tokyo, Japan
Siyang Jiang	National Taiwan University, Taiwan
Song Jiang	University of California, Los Angeles, USA
Wenyu Jiang	Nanjing University, China

Zhen Jiang	Jiangsu University, China
Yuncheng Jiang	South China Normal University, China
François-Xavier Jollois	Université de Paris Cité, France
Adan Jose-Garcia	Université de Lille, France
Ferdian Jovan	University of Bristol, UK
Steffen Jung	MPII, Germany
Thorsten Jungeblut	Bielefeld University of Applied Sciences, Germany
Hachem Kadri	Aix-Marseille University, France
Vana Kalogeraki	Athens University of Economics and Business, Greece
Vinayaka Kamath	Microsoft Research India, India
Toshihiro Kamishima	National Institute of Advanced Industrial Science, Japan
Bo Kang	Ghent University, Belgium
Alexandros Karakasidis	University of Macedonia, Greece
Mansooreh Karami	Arizona State University, USA
Panagiotis Karras	Aarhus University, Denmark
Ioannis Katakis	University of Nicosia, Cyprus
Koki Kawabata	Osaka University, Tokyo
Klemen Kenda	Jožef Stefan Institute, Slovenia
Patrik Joslin Kenfack	Innopolis University, Russia
Mahsa Keramati	Simon Fraser University, Canada
Hamidreza Keshavarz	Tarbiat Modares University, Iran
Adil Khan	Innopolis University, Russia
Jihed Khiari	Johannes Kepler University, Austria
Mi-Young Kim	University of Alberta, Canada
Arto Klami	University of Helsinki, Finland
Jiri Klema	Czech Technical University, Czechia
Tomas Kliegr	University of Economics Prague, Czechia
Christian Knoll	Graz, University of Technology, Austria
Dmitry Kobak	University of Tübingen, Germany
Vladimer Kobayashi	University of the Philippines Mindanao, Philippines
Dragi Kocev	Jožef Stefan Institute, Slovenia
Adrian Kochsiek	University of Mannheim, Germany
Masahiro Kohjima	NTT Corporation, Japan
Georgia Koloniari	University of Macedonia, Greece
Nikos Konofaos	Aristotle University of Thessaloniki, Greece
Irena Koprinska	University of Sydney, Australia
Lars Kotthoff	University of Wyoming, USA
Daniel Kottke	University of Kassel, Germany

Anna Krause   University of Würzburg, Germany
Alexander Kravberg   KTH Royal Institute of Technology, Sweden
Anastasia Krithara   NCSR Demokritos, Greece
Meelis Kull   University of Tartu, Estonia
Pawan Kumar   IIIT, Hyderabad, India
Suresh Kirthi Kumaraswamy   InterDigital, France
Gautam Kunapuli   Verisk Inc, USA
Marcin Kurdziel   AGH University of Science and Technology,
                                     Poland
Vladimir Kuzmanovski   Aalto University, Finland
Ariel Kwiatkowski   École Polytechnique, France
Firas Laakom   Tampere University, Finland
Harri Lähdesmäki   Aalto University, Finland
Stefanos Laskaridis   Samsung AI, UK
Alberto Lavelli   FBK-ict, Italy
Aonghus Lawlor   University College Dublin, Ireland
Thai Le   University of Mississippi, USA
Hoàng-Ân Lê   IRISA, University of South Brittany, France
Hoel Le Capitaine   University of Nantes, France
Thach Le Nguyen   Insight Centre, Ireland
Tai Le Quy   L3S Research Center - Leibniz University
                                     Hannover, Germany
Mustapha Lebbah   Sorbonne Paris Nord University, France
Dongman Lee   KAIST, South Korea
John Lee   Université catholique de Louvain, Belgium
Minwoo Lee   University of North Carolina at Charlotte, USA
Zed Lee   Stockholm University, Sweden
Yunwen Lei   University of Birmingham, UK
Douglas Leith   Trinity College Dublin, Ireland
Florian Lemmerich   RWTH Aachen, Germany
Carson Leung   University of Manitoba, Canada
Chaozhuo Li   Microsoft Research Asia, China
Jian Li   Institute of Information Engineering, China
Lei Li   Peking University, China
Li Li   Southwest University, China
Rui Li   Inspur Group, China
Shiyang Li   UCSB, USA
Shuokai Li   Chinese Academy of Sciences, China
Tianyu Li   Alibaba Group, China
Wenye Li   The Chinese University of Hong Kong, Shenzhen,
                                     China
Wenzhong Li   Nanjing University, China

Xiaoting Li	Pennsylvania State University, USA
Yang Li	University of North Carolina at Chapel Hill, USA
Zejian Li	Zhejiang University, China
Zhidong Li	UTS, Australia
Zhixin Li	Guangxi Normal University, China
Defu Lian	University of Science and Technology of China, China
Bin Liang	UTS, Australia
Yuchen Liang	RPI, USA
Yiwen Liao	University of Stuttgart, Germany
Pieter Libin	VUB, Belgium
Thomas Liebig	TU Dortmund, Germany
Seng Pei Liew	LINE Corporation, Japan
Beiyu Lin	University of Nevada - Las Vegas, USA
Chen Lin	Xiamen University, China
Tony Lindgren	Stockholm University, Sweden
Chen Ling	Emory University, USA
Jiajing Ling	Singapore Management University, Singapore
Marco Lippi	University of Modena and Reggio Emilia, Italy
Bin Liu	Chongqing University, China
Bowen Liu	Stanford University, USA
Chang Liu	Institute of Information Engineering, CAS, China
Chien-Liang Liu	National Chiao Tung University, Taiwan
Feng Liu	East China Normal University, China
Jiacheng Liu	Chinese University of Hong Kong, China
Li Liu	Chongqing University, China
Shengcai Liu	Southern University of Science and Technology, China
Shenghua Liu	Institute of Computing Technology, CAS, China
Tingwen Liu	Institute of Information Engineering, CAS, China
Xiangyu Liu	Tencent, China
Yong Liu	Renmin University of China, China
Yuansan Liu	University of Melbourne, Australia
Zhiwei Liu	Salesforce, USA
Tuwe Löfström	Jönköping University, Sweden
Corrado Loglisci	Università degli Studi di Bari Aldo Moro, Italy
Ting Long	Shanghai Jiao Tong University, China
Beatriz López	University of Girona, Spain
Yin Lou	Ant Group, USA
Samir Loudni	TASC (LS2N-CNRS), IMT Atlantique, France
Yang Lu	Xiamen University, China
Yuxun Lu	National Institute of Informatics, Japan

Massimiliano Luca	Bruno Kessler Foundation, Italy
Stefan Lüdtke	University of Mannheim, Germany
Jovita Lukasik	University of Mannheim, Germany
Denis Lukovnikov	University of Bonn, Germany
Pedro Henrique Luz de Araujo	University of Brasília, Brazil
Fenglong Ma	Pennsylvania State University, USA
Jing Ma	University of Virginia, USA
Meng Ma	Peking University, China
Muyang Ma	Shandong University, China
Ruizhe Ma	University of Massachusetts Lowell, USA
Xingkong Ma	National University of Defense Technology, China
Xueqi Ma	Tsinghua University, China
Zichen Ma	The Chinese University of Hong Kong, Shenzhen, China
Luis Macedo	University of Coimbra, Portugal
Harshitha Machiraju	EPFL, Switzerland
Manchit Madan	Delivery Hero, Germany
Seiji Maekawa	Osaka University, Japan
Sindri Magnusson	Stockholm University, Sweden
Pathum Chamikara Mahawaga	CSIRO Data61, Australia
Saket Maheshwary	Amazon, India
Ajay Mahimkar	AT&T, USA
Pierre Maillot	Inria, France
Lorenzo Malandri	Unimib, Italy
Rammohan Mallipeddi	Kyungpook National University, South Korea
Sahil Manchanda	IIT Delhi, India
Domenico Mandaglio	DIMES-UNICAL, Italy
Panagiotis Mandros	Harvard University, USA
Robin Manhaeve	KU Leuven, Belgium
Silviu Maniu	Université Paris-Saclay, France
Cinmayii Manliguez	National Sun Yat-Sen University, Taiwan
Naresh Manwani	International Institute of Information Technology, India
Jiali Mao	East China Normal University, China
Alexandru Mara	Ghent University, Belgium
Radu Marculescu	University of Texas at Austin, USA
Roger Mark	Massachusetts Institute of Technology, USA
Fernando Martínez-Plume	Joint Research Centre - European Commission, Belgium
Koji Maruhashi	Fujitsu Research, Fujitsu Limited, Japan
Simone Marullo	University of Siena, Italy

Elio Masciari	University of Naples, Italy
Florent Masseglia	Inria, France
Michael Mathioudakis	University of Helsinki, Finland
Takashi Matsubara	Osaka University, Japan
Tetsu Matsukawa	Kyushu University, Japan
Santiago Mazuelas	BCAM-Basque Center for Applied Mathematics, Spain
Ryan McConville	University of Bristol, UK
Hardik Meisheri	TCS Research, India
Panagiotis Meletis	Eindhoven University of Technology, the Netherlands
Gabor Melli	Medable, USA
Joao Mendes-Moreira	INESC TEC, Portugal
Chuan Meng	University of Amsterdam, the Netherlands
Cristina Menghini	Brown University, USA
Engelbert Mephu Nguifo	Université Clermont Auvergne, CNRS, LIMOS, France
Fabio Mercorio	University of Milan-Bicocca, Italy
Guillaume Metzler	Laboratoire ERIC, France
Hao Miao	Aalborg University, Denmark
Alessio Micheli	Università di Pisa, Italy
Paolo Mignone	University of Bari Aldo Moro, Italy
Matej Mihelcic	University of Zagreb, Croatia
Ioanna Miliou	Stockholm University, Sweden
Bamdev Mishra	Microsoft, India
Rishabh Misra	Twitter, Inc, USA
Dixant Mittal	National University of Singapore, Singapore
Zhaobin Mo	Columbia University, USA
Daichi Mochihashi	Institute of Statistical Mathematics, Japan
Armin Moharrer	Northeastern University, USA
Ioannis Mollas	Aristotle University of Thessaloniki, Greece
Carlos Monserrat-Aranda	Universität Politècnica de València, Spain
Konda Reddy Mopuri	Indian Institute of Technology Guwahati, India
Raha Moraffah	Arizona State University, USA
Pawel Morawiecki	Polish Academy of Sciences, Poland
Ahmadreza Mosallanezhad	Arizona State University, USA
Davide Mottin	Aarhus University, Denmark
Koyel Mukherjee	Adobe Research, India
Maximilian Münch	University of Applied Sciences Würzburg, Germany
Fabricio Murai	Universidade Federal de Minas Gerais, Brazil
Taichi Murayama	NAIST, Japan

Stéphane Mussard	CHROME, France
Mohamed Nadif	Centre Borelli - Université Paris Cité, France
Cian Naik	University of Oxford, UK
Felipe Kenji Nakano	KU Leuven, Belgium
Mirco Nanni	ISTI-CNR Pisa, Italy
Apurva Narayan	University of Waterloo, Canada
Usman Naseem	University of Sydney, Australia
Gergely Nemeth	ELLIS Unit Alicante, Spain
Stefan Neumann	KTH Royal Institute of Technology, Sweden
Anna Nguyen	Karlsruhe Institute of Technology, Germany
Quan Nguyen	Washington University in St. Louis, USA
Thi Phuong Quyen Nguyen	University of Da Nang, Vietnam
Thu Nguyen	SimulaMet, Norway
Thu Trang Nguyen	University College Dublin, Ireland
Prajakta Nimbhorkar	Chennai Mathematical Institute, Chennai, India
Xuefei Ning	Tsinghua University, China
Ikuko Nishikawa	Ritsumeikan University, Japan
Hao Niu	KDDI Research, Inc., Japan
Paraskevi Nousi	Aristotle University of Thessaloniki, Greece
Erik Novak	Jožef Stefan Institute, Slovenia
Slawomir Nowaczyk	Halmstad University, Sweden
Aleksandra Nowak	Jagiellonian University, Poland
Eirini Ntoutsi	Freie Universität Berlin, Germany
Andreas Nürnberger	Magdeburg University, Germany
James O'Neill	University of Liverpool, UK
Lutz Oettershagen	University of Bonn, Germany
Tsuyoshi Okita	Kyushu Institute of Technology, Japan
Makoto Onizuka	Osaka University, Japan
Subba Reddy Oota	IIIT Hyderabad, India
María Óskarsdóttir	University of Reykjavík, Iceland
Aomar Osmani	PRES Sorbonne Paris Cité, France
Aljaz Osojnik	JSI, Slovenia
Shuichi Otake	National Institute of Informatics, Japan
Greger Ottosson	IBM, France
Zijing Ou	Sun Yat-sen University, China
Abdelkader Ouali	University of Caen Normandy, France
Latifa Oukhellou	IFSTTAR, France
Kai Ouyang	Tsinghua University, France
Andrei Paleyes	University of Cambridge, UK
Pankaj Pandey	Indian Institute of Technology Gandhinagar, India
Guansong Pang	Singapore Management University, Singapore
Pance Panov	Jožef Stefan Institute, Slovenia

Apostolos Papadopoulos	Aristotle University of Thessaloniki, Greece
Evangelos Papalexakis	UC Riverside, USA
Anna Pappa	Université Paris 8, France
Chanyoung Park	UIUC, USA
Haekyu Park	Georgia Institute of Technology, USA
Sanghyun Park	Yonsei University, South Korea
Luca Pasa	University of Padova, Italy
Kevin Pasini	IRT SystemX, France
Vincenzo Pasquadibisceglie	University of Bari Aldo Moro, Italy
Nikolaos Passalis	Aristotle University of Thessaloniki, Greece
Javier Pastorino	University of Colorado, Denver, USA
Kitsuchart Pasupa	King Mongkut's Institute of Technology, Thailand
Andrea Paudice	University of Milan, Italy
Anand Paul	Kyungpook National University, South Korea
Yulong Pei	TU Eindhoven, the Netherlands
Charlotte Pelletier	Université de Bretagne du Sud, France
Jaakko Peltonen	Tampere University, Finland
Ruggero Pensa	University of Torino, Italy
Fabiola Pereira	Federal University of Uberlandia, Brazil
Lucas Pereira	ITI, LARSyS, Técnico Lisboa, Portugal
Aritz Pérez	Basque Center for Applied Mathematics, Spain
Lorenzo Perini	KU Leuven, Belgium
Alan Perotti	CENTAI Institute, Italy
Michaël Perrot	Inria Lille, France
Matej Petkovic	Institute Jožef Stefan, Slovenia
Lukas Pfahler	TU Dortmund University, Germany
Nico Piatkowski	Fraunhofer IAIS, Germany
Francesco Piccialli	University of Naples Federico II, Italy
Gianvito Pio	University of Bari, Italy
Giuseppe Pirrò	Sapienza University of Rome, Italy
Marc Plantevit	EPITA, France
Konstantinos Pliakos	KU Leuven, Belgium
Matthias Pohl	Otto von Guericke University, Germany
Nicolas Posocco	EURA NOVA, Belgium
Cedric Pradalier	GeorgiaTech Lorraine, France
Paul Prasse	University of Potsdam, Germany
Mahardhika Pratama	University of South Australia, Australia
Francesca Pratesi	ISTI - CNR, Italy
Steven Prestwich	University College Cork, Ireland
Giulia Preti	CentAI, Italy
Philippe Preux	Inria, France
Shalini Priya	Oak Ridge National Laboratory, USA

Ricardo Prudencio	Universidade Federal de Pernambuco, Brazil
Luca Putelli	Università degli Studi di Brescia, Italy
Peter van der Putten	Leiden University, the Netherlands
Chuan Qin	Baidu, China
Jixiang Qing	Ghent University, Belgium
Jolin Qu	Western Sydney University, Australia
Nicolas Quesada	Polytechnique Montreal, Canada
Teeradaj Racharak	Japan Advanced Institute of Science and Technology, Japan
Krystian Radlak	Warsaw University of Technology, Poland
Sandro Radovanovic	University of Belgrade, Serbia
Md Masudur Rahman	Purdue University, USA
Ankita Raj	Indian Institute of Technology Delhi, India
Herilalaina Rakotoarison	Inria, France
Alexander Rakowski	Hasso Plattner Institute, Germany
Jan Ramon	Inria, France
Sascha Ranftl	Graz University of Technology, Austria
Aleksandra Rashkovska Koceva	Jožef Stefan Institute, Slovenia
S. Ravi	Biocomplexity Institute, USA
Jesse Read	Ecole Polytechnique, France
David Reich	Universität Potsdam, Germany
Marina Reyboz	CEA, LIST, France
Pedro Ribeiro	University of Porto, Portugal
Rita P. Ribeiro	University of Porto, Portugal
Piera Riccio	ELLIS Unit Alicante Foundation, Spain
Christophe Rigotti	INSA Lyon, France
Matteo Riondato	Amherst College, USA
Mateus Riva	Telecom ParisTech, France
Kit Rodolfa	CMU, USA
Christophe Rodrigues	DVRC Pôle Universitaire Léonard de Vinci, France
Simon Rodríguez-Santana	ICMAT, Spain
Gaetano Rossiello	IBM Research, USA
Mohammad Rostami	University of Southern California, USA
Franz Rothlauf	Mainz Universität, Germany
Celine Rouveirol	Université Paris-Nord, France
Arjun Roy	Freie Universität Berlin, Germany
Joze Rozanec	Josef Stefan International Postgraduate School, Slovenia
Salvatore Ruggieri	University of Pisa, Italy
Marko Ruman	UTIA, AV CR, Czechia
Ellen Rushe	University College Dublin, Ireland

Dawid Rymarczyk	Jagiellonian University, Poland
Amal Saadallah	TU Dortmund, Germany
Khaled Mohammed Saifuddin	Georgia State University, USA
Hajer Salem	AUDENSIEL, France
Francesco Salvetti	Politecnico di Torino, Italy
Roberto Santana	University of the Basque Country (UPV/EHU), Spain
KC Santosh	University of South Dakota, USA
Somdeb Sarkhel	Adobe, USA
Yuya Sasaki	Osaka University, Japan
Yücel Saygın	Sabancı Universitesi, Turkey
Patrick Schäfer	Humboldt-Universität zu Berlin, Germany
Alexander Schiendorfer	Technische Hochschule Ingolstadt, Germany
Peter Schlicht	Volkswagen Group Research, Germany
Daniel Schmidt	Monash University, Australia
Johannes Schneider	University of Liechtenstein, Liechtenstein
Steven Schockaert	Cardiff University, UK
Jens Schreiber	University of Kassel, Germany
Matthias Schubert	Ludwig-Maximilians-Universität München, Germany
Alexander Schulz	CITEC, Bielefeld University, Germany
Jan-Philipp Schulze	Fraunhofer AISEC, Germany
Andreas Schwung	Fachhochschule Südwestfalen, Germany
Vasile-Marian Scuturici	LIRIS, France
Raquel Sebastião	IEETA/DETI-UA, Portugal
Stanislav Selitskiy	University of Bedfordshire, UK
Edoardo Serra	Boise State University, USA
Lorenzo Severini	UniCredit, R&D Dept., Italy
Tapan Shah	GE, USA
Ammar Shaker	NEC Laboratories Europe, Germany
Shiv Shankar	University of Massachusetts, USA
Junming Shao	University of Electronic Science and Technology, China
Kartik Sharma	Georgia Institute of Technology, USA
Manali Sharma	Samsung, USA
Ariona Shashaj	Network Contacts, Italy
Betty Shea	University of British Columbia, Canada
Chengchao Shen	Central South University, China
Hailan Shen	Central South University, China
Jiawei Sheng	Chinese Academy of Sciences, China
Yongpan Sheng	Southwest University, China
Chongyang Shi	Beijing Institute of Technology, China

Zhengxiang Shi	University College London, UK
Naman Shukla	Deepair LLC, USA
Pablo Silva	Dell Technologies, Brazil
Simeon Simoff	Western Sydney University, Australia
Maneesh Singh	Motive Technologies, USA
Nikhil Singh	MIT Media Lab, USA
Sarath Sivaprasad	IIIT Hyderabad, India
Elena Sizikova	NYU, USA
Andrzej Skowron	University of Warsaw, Poland
Blaz Skrlj	Institute Jožef Stefan, Slovenia
Oliver Snow	Simon Fraser University, Canada
Jonas Soenen	KU Leuven, Belgium
Nataliya Sokolovska	Sorbonne University, France
K. M. A. Solaiman	Purdue University, USA
Shuangyong Song	Jing Dong, China
Zixing Song	The Chinese University of Hong Kong, China
Tiberiu Sosea	University of Illinois at Chicago, USA
Arnaud Soulet	University of Tours, France
Lucas Souza	UFRJ, Brazil
Jens Sparsø	Technical University of Denmark, Denmark
Vivek Srivastava	TCS Research, USA
Marija Stanojevic	Temple University, USA
Jerzy Stefanowski	Poznan University of Technology, Poland
Simon Stieber	University of Augsburg, Germany
Jinyan Su	University of Electronic Science and Technology, China
Yongduo Sui	University of Science and Technology of China, China
Huiyan Sun	Jilin University, China
Yuwei Sun	University of Tokyo/RIKEN AIP, Japan
Gokul Swamy	Amazon, USA
Maryam Tabar	Pennsylvania State University, USA
Anika Tabassum	Virginia Tech, USA
Shazia Tabassum	INESCTEC, Portugal
Koji Tabata	Hokkaido University, Japan
Andrea Tagarelli	DIMES, University of Calabria, Italy
Etienne Tajeuna	Université de Laval, Canada
Acar Tamersoy	NortonLifeLock Research Group, USA
Chang Wei Tan	Monash University, Australia
Cheng Tan	Westlake University, China
Feilong Tang	Shanghai Jiao Tong University, China
Feng Tao	Volvo Cars, USA

Youming Tao	Shandong University, China
Martin Tappler	Graz University of Technology, Austria
Garth Tarr	University of Sydney, Australia
Mohammad Tayebi	Simon Fraser University, Canada
Anastasios Tefas	Aristotle University of Thessaloniki, Greece
Maguelonne Teisseire	INRAE - UMR Tetis, France
Stefano Teso	University of Trento, Italy
Olivier Teste	IRIT, University of Toulouse, France
Maximilian Thiessen	TU Wien, Austria
Eleftherios Tiakas	Aristotle University of Thessaloniki, Greece
Hongda Tian	University of Technology Sydney, Australia
Alessandro Tibo	Aalborg University, Denmark
Aditya Srinivas Timmaraju	Facebook, USA
Christos Tjortjis	International Hellenic University, Greece
Ljupco Todorovski	University of Ljubljana, Slovenia
Laszlo Toka	BME, Hungary
Ancy Tom	University of Minnesota, Twin Cities, USA
Panagiotis Traganitis	Michigan State University, USA
Cuong Tran	Syracuse University, USA
Minh-Tuan Tran	KAIST, South Korea
Giovanni Trappolini	Sapienza University of Rome, Italy
Volker Tresp	LMU, Germany
Yu-Chee Tseng	National Yang Ming Chiao Tung University, Taiwan
Maria Tzelepi	Aristotle University of Thessaloniki, Greece
Willy Ugarte	University of Applied Sciences (UPC), Peru
Antti Ukkonen	University of Helsinki, Finland
Abhishek Kumar Umrawal	Purdue University, USA
Athena Vakal	Aristotle University, Greece
Matias Valdenegro Toro	University of Groningen, the Netherlands
Maaike Van Roy	KU Leuven, Belgium
Dinh Van Tran	University of Freiburg, Germany
Fabio Vandin	University of Padova, Italy
Valerie Vaquet	CITEC, Bielefeld University, Germany
Iraklis Varlamis	Harokopio University of Athens, Greece
Santiago Velasco-Forero	MINES ParisTech, France
Bruno Veloso	Porto, Portugal
Dmytro Velychko	Carl von Ossietzky Universität Oldenburg, Germany
Sreekanth Vempati	Myntra, India
Sebastián Ventura Soto	University of Cordoba, Portugal
Rosana Veroneze	LBiC, Brazil

Jan Verwaeren	Ghent University, Belgium
Vassilios Verykios	Hellenic Open University, Greece
Herna Viktor	University of Ottawa, Canada
João Vinagre	LIAAD - INESC TEC, Portugal
Fabio Vitale	Centai Institute, Italy
Vasiliki Voukelatou	ISTI - CNR, Italy
Dong Quan Vu	Safran Tech, France
Maxime Wabartha	McGill University, Canada
Tomasz Walkowiak	Wroclaw University of Science and Technology, Poland
Vijay Walunj	University of Missouri-Kansas City, USA
Michael Wand	University of Mainz, Germany
Beilun Wang	Southeast University, China
Chang-Dong Wang	Sun Yat-sen University, China
Daheng Wang	Amazon, USA
Deng-Bao Wang	Southeast University, China
Di Wang	Nanyang Technological University, Singapore
Di Wang	KAUST, Saudi Arabia
Fu Wang	University of Exeter, UK
Hao Wang	Nanyang Technological University, Singapore
Hao Wang	Louisiana State University, USA
Hao Wang	University of Science and Technology of China, China
Hongwei Wang	University of Illinois Urbana-Champaign, USA
Hui Wang	SKLSDE, China
Hui (Wendy) Wang	Stevens Institute of Technology, USA
Jia Wang	Xi'an Jiaotong-Liverpool University, China
Jing Wang	Beijing Jiaotong University, China
Junxiang Wang	Emory University, USA
Qing Wang	IBM Research, USA
Rongguang Wang	University of Pennsylvania, USA
Ruoyu Wang	Shanghai Jiao Tong University, China
Ruxin Wang	Shenzhen Institutes of Advanced Technology, China
Senzhang Wang	Central South University, China
Shoujin Wang	Macquarie University, Australia
Xi Wang	Chinese Academy of Sciences, China
Yanchen Wang	Georgetown University, USA
Ye Wang	Chongqing University, China
Ye Wang	National University of Singapore, Singapore
Yifei Wang	Peking University, China
Yongqing Wang	Chinese Academy of Sciences, China

Yuandong Wang	Tsinghua University, China
Yue Wang	Microsoft Research, USA
Yun Cheng Wang	University of Southern California, USA
Zhaonan Wang	University of Tokyo, Japan
Zhaoxia Wang	SMU, Singapore
Zhiwei Wang	University of Chinese Academy of Sciences, China
Zihan Wang	Shandong University, China
Zijie J. Wang	Georgia Tech, USA
Dilusha Weeraddana	CSIRO, Australia
Pascal Welke	University of Bonn, Germany
Tobias Weller	University of Mannheim, Germany
Jörg Wicker	University of Auckland, New Zealand
Lena Wiese	Goethe University Frankfurt, Germany
Michael Wilbur	Vanderbilt University, USA
Moritz Wolter	Bonn University, Germany
Bin Wu	Beijing University of Posts and Telecommunications, China
Bo Wu	Renmin University of China, China
Jiancan Wu	University of Science and Technology of China, China
Jiantao Wu	University of Jinan, China
Ou Wu	Tianjin University, China
Yang Wu	Chinese Academy of Sciences, China
Yiqing Wu	University of Chinese Academic of Science, China
Yuejia Wu	Inner Mongolia University, China
Bin Xiao	University of Ottawa, Canada
Zhiwen Xiao	Southwest Jiaotong University, China
Ruobing Xie	WeChat, Tencent, China
Zikang Xiong	Purdue University, USA
Depeng Xu	University of North Carolina at Charlotte, USA
Jian Xu	Citadel, USA
Jiarong Xu	Fudan University, China
Kunpeng Xu	University of Sherbrooke, Canada
Ning Xu	Southeast University, China
Xianghong Xu	Tsinghua University, China
Sangeeta Yadav	Indian Institute of Science, India
Mehrdad Yaghoobi	University of Edinburgh, UK
Makoto Yamada	RIKEN AIP/Kyoto University, Japan
Akihiro Yamaguchi	Toshiba Corporation, Japan
Anil Yaman	Vrije Universiteit Amsterdam, the Netherlands

Hao Yan	Washington University in St Louis, USA
Qiao Yan	Shenzhen University, China
Chuang Yang	University of Tokyo, Japan
Deqing Yang	Fudan University, China
Haitian Yang	Chinese Academy of Sciences, China
Renchi Yang	National University of Singapore, Singapore
Shaofu Yang	Southeast University, China
Yang Yang	Nanjing University of Science and Technology, China
Yang Yang	Northwestern University, USA
Yiyang Yang	Guangdong University of Technology, China
Yu Yang	The Hong Kong Polytechnic University, China
Peng Yao	University of Science and Technology of China, China
Vithya Yogarajan	University of Auckland, New Zealand
Tetsuya Yoshida	Nara Women's University, Japan
Hong Yu	Chongqing Laboratory of Comput. Intelligence, China
Wenjian Yu	Tsinghua University, China
Yanwei Yu	Ocean University of China, China
Ziqiang Yu	Yantai University, China
Sha Yuan	Beijing Academy of Artificial Intelligence, China
Shuhan Yuan	Utah State University, USA
Mingxuan Yue	Google, USA
Aras Yurtman	KU Leuven, Belgium
Nayyar Zaidi	Deakin University, Australia
Zelin Zang	Zhejiang University & Westlake University, China
Masoumeh Zareapoor	Shanghai Jiao Tong University, China
Hanqing Zeng	USC, USA
Tieyong Zeng	The Chinese University of Hong Kong, China
Bin Zhang	South China University of Technology, China
Bob Zhang	University of Macau, Macao, China
Hang Zhang	National University of Defense Technology, China
Huaizheng Zhang	Nanyang Technological University, Singapore
Jiangwei Zhang	Tencent, China
Jinwei Zhang	Cornell University, USA
Jun Zhang	Tsinghua University, China
Lei Zhang	Virginia Tech, USA
Luxin Zhang	Worldline/Inria, France
Mimi Zhang	Trinity College Dublin, Ireland
Qi Zhang	University of Technology Sydney, Australia

Qiyiwen Zhang	University of Pennsylvania, USA
Teng Zhang	Huazhong University of Science and Technology, China
Tianle Zhang	University of Exeter, UK
Xuan Zhang	Renmin University of China, China
Yang Zhang	University of Science and Technology of China, China
Yaqian Zhang	University of Waikato, New Zealand
Yu Zhang	University of Illinois at Urbana-Champaign, USA
Zhengbo Zhang	Beihang University, China
Zhiyuan Zhang	Peking University, China
Heng Zhao	Shenzhen Technology University, China
Mia Zhao	Airbnb, USA
Tong Zhao	Snap Inc., USA
Qinkai Zheng	Tsinghua University, China
Xiangping Zheng	Renmin University of China, China
Bingxin Zhou	University of Sydney, Australia
Bo Zhou	Baidu, Inc., China
Min Zhou	Huawei Technologies, China
Zhipeng Zhou	University of Science and Technology of China, China
Hui Zhu	Chinese Academy of Sciences, China
Kenny Zhu	SJTU, China
Lingwei Zhu	Nara Institute of Science and Technology, Japan
Mengying Zhu	Zhejiang University, China
Renbo Zhu	Peking University, China
Yanmin Zhu	Shanghai Jiao Tong University, China
Yifan Zhu	Tsinghua University, China
Bartosz Zieliński	Jagiellonian University, Poland
Sebastian Ziesche	Bosch Center for Artificial Intelligence, Germany
Indre Zliobaite	University of Helsinki, Finland
Gianlucca Zuin	UFM, Brazil

Program Committee Members, Demo Track

Hesam Amoualian	WholeSoft Market, France
Georgios Balikas	Salesforce, France
Giannis Bekoulis	Vrije Universiteit Brussel, Belgium
Ludovico Boratto	University of Cagliari, Italy
Michelangelo Ceci	University of Bari, Italy
Abdulkadir Celikkanat	Technical University of Denmark, Denmark

Tania Cerquitelli	Informatica Politecnico di Torino, Italy
Mel Chekol	Utrecht University, the Netherlands
Charalampos Chelmis	University at Albany, USA
Yagmur Gizem Cinar	Amazon, France
Eustache Diemert	Criteo AI Lab, France
Sophie Fellenz	TU Kaiserslautern, Germany
James Foulds	University of Maryland, Baltimore County, USA
Jhony H. Giraldo	Télécom Paris, France
Parantapa Goswami	Rakuten Institute of Technology, Rakuten Group, Japan
Derek Greene	University College Dublin, Ireland
Lili Jiang	Umeå University, Sweden
Bikash Joshi	Elsevier, the Netherlands
Alexander Jung	Aalto University, Finland
Zekarias Kefato	KTH Royal Institute of Technology, Sweden
Ilkcan Keles	Aalborg University, Denmark
Sammy Khalife	Johns Hopkins University, USA
Tuan Le	New Mexico State University, USA
Ye Liu	Salesforce, USA
Fragkiskos Malliaros	CentraleSupelec, France
Hamid Mirisaee	AMLRightSource, France
Robert Moro	Kempelen Institute of Intelligent Technologies, Slovakia
Iosif Mporas	University of Hertfordshire, UK
Giannis Nikolentzos	Ecole Polytechnique, France
Eirini Ntoutsi	Freie Universität Berlin, Germany
Frans Oliehoek	Delft University of Technology, the Netherlands
Nora Ouzir	CentraleSupélec, France
Özlem Özgöbek	Norwegian University of Science and Technology, Norway
Manos Papagelis	York University, UK
Shichao Pei	University of Notre Dame, USA
Botao Peng	Chinese Academy of Sciences, China
Antonia Saravanou	National and Kapodistrian University of Athens, Greece
Rik Sarkar	University of Edinburgh, UK
Vera Shalaeva	Inria Lille-Nord, France
Kostas Stefanidis	Tampere University, Finland
Nikolaos Tziortziotis	Jellyfish, France
Davide Vega	Uppsala University, Sweden
Sagar Verma	CentraleSupelec, France
Yanhao Wang	East China Normal University, China

Zhirong Yang   Norwegian University of Science and Technology, Norway

Xiangyu Zhao   City University of Hong Kong, Hong Kong, China

Sponsors

Contents – Part III

Generative Models

Computer Vision

Meta-learning, Neural Architecture Search

Deep Learning

DialCSP: A Two-Stage Attention-Based Model for Customer Satisfaction Prediction in E-commerce Customer Service

Zhenhe Wu[1,2,3], Liangqing Wu[3], Shuangyong Song[3], Jiahao Ji[1], Bo Zou[3], Zhoujun Li[1,2(✉)], and Xiaodong He[3]

[1] School of Computer Science and Engineering, Beihang University, Beijing, China
{wuzhenhe,jiahaoji,lizj}@buaa.edu.cn
[2] State Key Lab of Software Development Environment, Beihang University, Beijing, China
[3] JD AI Research, Beijing, China
{wuliangqing,songshuangyong,cdzoubo,hexiaodong}@jd.com

Abstract. Nowadays, customer satisfaction prediction (CSP) on e-commerce platforms has become a hot research topic for intelligent customer service. CSP aims to discover customer satisfaction according to the dialogue content of customer and intelligent customer service, for the purpose of improving service quality and customer experience. Previous works have made some progress in many aspects, but they mostly ignore the huge expressional differences between customer questions and customer service answers, and fail to adequately consider the internal relations of these two kinds of personalized expressions. In this paper, we propose a two-stage dialogue-level classification model containing an intra-stage and an inter-stage, to emphasize the importance of modeling customer part (content of customer questions) and service part (content of customer service answers) separately. In the intra-stage, we model customer part and service part separately by using attention mechanism combined with personalized context to obtain a *customer state* and a *service state*. Then we interact these two states with each other in the inter-stage to obtain the final satisfaction representation of the whole dialogue. Experiment results demonstrate that our model achieves better performance than several competitive baselines on our in-house dataset and four public datasets.

Keywords: Customer satisfaction prediction · Intelligent customer service · Attention-based model

1 Introduction

With the development of e-commerce platforms in recent years, a large number of companies use customer service chatbots, for the reasons that they could answer to customers' questions quickly and save labor cost. Customer satisfaction

M.-R. Amini et al. (Eds.): ECML PKDD 2022, LNAI 13715, pp. 3–18, 2023.
https://doi.org/10.1007/978-3-031-26409-2_1

prediction (CSP) for the dialogues in customer service chatbots has become an important problem in industry. For one thing, customers' satisfaction is a crucial indicator to evaluate the quality of service, which can help improve the ability of chatbots. For another, predicting customers' satisfaction in real time helps platforms handle problematic dialogues by transferring customer service chatbots to staffs timely, which can improve the customers' experience.

Fig. 1. A dialogue of customer and chatbot on e-commerce platform.

CSP is a multi-class classification task. Existing researches on CSP is mainly divided into two directions, one is the turn-level CSP, the other is the dialogue-level CSP. The former direction concerns satisfaction prediction in every customer-service turn [19, 20, 24], while the latter one predicts satisfaction level of the whole dialogue [9, 13, 14, 21, 22]. Turn-level CSP can only capture temporary user's satisfaction results which may have certain contingency, while dialogue-level CSP is the key point to evaluate the quality of the service and whether the customer's problem has been solved. Thus, in this study, we concentrate on dialogue-level CSP with five satisfaction levels (*strongly satisfied, satisfied, neutral, dissatisfied,* or *strongly dissatisfied*). As shown in Fig. 1, the customer expresses his anxiety and displeasure at the beginning, then turns into satisfied after the good answers of the chatbot.

To address the dialogue-level CSP task, many approaches extracted features from dialogue content and built models to fully utilize the interaction between customer questions and customer service answers. Some earlier studies used manual features to represent conversational context [21, 22], while recent studies concerned more on how well the questions and answers match each other [13, 14]. Although these works have made great progress in CSP task, two issues still remain: 1) existing studies ignore the huge expressional differences between customers and customer service chatbots, in terms of the emotion intensity, speaking habits, language richness, sentence length and etc. 2) most prior studies fail to adequately consider the internal relations of personalized expressions for customers and staffs/chatbots respectively.

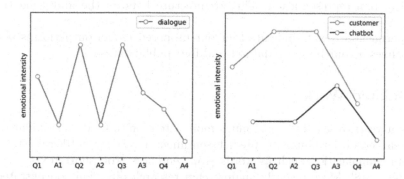

Fig. 2. The emotional intensity trends are obviously after split.

According to the above analysis, we figure that besides handling the interaction of customer questions and customer service answers, modeling customer part and service part separately should also be taken into consideration due to their expressional differences in many aspects. For example, customers' questioning emotion is volatile and the emotional intensity is usually high, while the answering emotion of service is relatively stable and the emotional intensity is low. Figure 2 shows the emotional intensity trend of the case in Fig. 1, in which the customer's emotional intensity is higher with greater fluctuation, while the chatbot is the opposite. After splitting the dialogue into customer part and service part, we are able to catch their emotional intensity trends intuitively. For the similar reason, other aspects of expressional differences also matter.

Thus, we propose a classification model for CSP in E-commerce customer service dialogues which is called DialCSP. Besides an encoder and a decoder, our model contains an intra-stage and an inter-stage as core structures. Firstly, we adopt an encoding module to extract features of the dialogue content. Next comes the intra-stage, which consists of a customer part and a service part. We split customer questions and customer service answers as two independent sequences and send them into these two parts separately. Specifically, each part is designed to fully extract the internal relations of the sequence. In the end of the intra-stage, we get *customer state* and *service state* as the results of the customer part and service part. Then, the inter-stage apply an interactive attention mechanism to capture satisfaction representations of the whole dialogue from *customer state* and *service state*. In the end, a decoder module is used to predict the final satisfaction level.

To summarize, our contributions are as follows:

– We propose a dialogue-level classification model DialCSP, for CSP in E-commerce customer service chatbots.

- By bringing forward a two-stage architecture, we split the dialogue content into customer part and service part to model them separately. With the results of *customer state* and *service state*, we construct interaction to capture final satisfaction representation. This architecture handles the above-mentioned two issues well.
- Experimental results indicate that our proposed model outperforms several baselines on our in-house dataset and four public datasets.

2   Related Work

In recent years, researchers paid much more attention to CSP and similar tasks. Some earlier works aimed to predict sentiment levels for subjective texts in different granularities, such as words [15], sentences [16], short texts [17] and documents [18]. More recently, mainstream research direction concentrated on turn-level and dialogue-level CSP.

Some researchers explored the turn-level structure, such as modeling dialogues via a hierarchical RNN [19], keeping track of satisfaction states of dialogue participants [20], exploring contrastive learning [24] and so on. But, due to the labels of turn-level satisfaction is difficult to obtain and dialogue-level CSP appears to reflect service quality more realistically, we focus on dialogue-level CSP in this paper.

To study the dialogue-level CSP, earlier methods used manual features [21, 22], while recent studies preferred deep neural networks and attention mechanism to explore how questions and answers interact with each other. Some researchers adopted a Bi-directional LSTM network to capture the contextual information of conversational services and use the hidden vector of the last utterance for satisfaction prediction [9], some researchers used each question to capture information from all answers to model customer-service interaction [13], while another study focusing on dialogue-level CSP used LSTM networks to capture contextual features and computed the semantic similarity scores between customer questions and customer service answers across different turns to model customer-service interaction [14]. However, these works didn't consider the expressional differences between customer and customer service. Morever, they failed to excavate the internal relations of personalized expression sequences. In this work, we work on addressing the two existing issues above, thus proposing DialCSP model for dialogue-level CSP.

3   Methodology

3.1   Problem Definition

In the real scenario, a customer asks questions and the chatbot provides the corresponding answers in turn, so a customer service dialogue is defined as a sequence of utterances $C = \{q_1, a_1, q_2, a_2, ..., q_n, a_n\}$. Each question q_i is followed by an answer a_i, and the length of dialogue is $2n$. The goal of our task is to predict

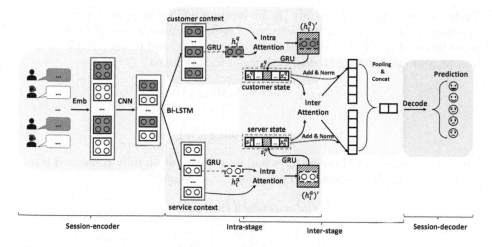

Fig. 3. Framework of the DialCSP model.

the satisfaction level y based on the dialogue content C, while the satisfaction level is divided into five classes: *strongly satisfied, satisfied, neutral, dissatisfied, strongly dissatisfied.*

3.2 Proposed Model

As shown in Fig. 3, we propose DialCSP, a two-stage classification model for dialogue-level CSP. Besides a session-encoder and a session-decoder, the core part of our model contains an intra-stage and an inter-stage. The session-encoder is a dialogue encoding module to process the raw conversation content. Intra-stage is comprised of a customer part and a service part, which helps extract sufficient internal features of question sequences and answer sequences separately. For both parts in inter-stage, we utilize attention mechanisms to adequately discover the sentence characteristics at each time step from their personalized context, served as *customer state* and *service state*. Next, inter-stage applies an interactive attention mechanism to fully capture satisfaction representations of the whole dialogue from *customer state* and *service state*. Finally, the session-decoder contributes to predict the final satisfaction level. In the following sections, we will introduce the details of the model structure in order.

3.3 Session-Encoder

Session-encoder aims to encode natural language dialogues into semantic representations. Our input is the whole dialogue text, in which words are separately transformed into 300 dimensional vectors by using pre-trained GloVe model [23]:

$$E = \text{GloVe}(C) \tag{1}$$

Then, inspired by previous study [1], we leverage a CNN layer with max-pooling to extract context independent features of each utterance. Concretely, we apply three filters of size 1, 2, 3 with 50 feature maps each, and employ ReLU activation [2] and max-pooling to deal with these feature maps:

$$fm_{1,2,3} = \text{ReLU}\left(\text{CNN}_{1,2,3}\left(E\right)\right) \tag{2}$$

$$fm'_{1,2,3} = \text{max-pooling}\left(fm_{1,2,3}\right) \tag{3}$$

Then, we concatenate these features and send them into a fully connected layer, which produces the context representations cr as follow:

$$fm' = \text{concat}\left(fm'_{1,2,3}\right) \tag{4}$$

$$cr = \text{ReLU}\left(W_0 fm' + b_0\right) \tag{5}$$

3.4   Intra-stage

Intra-stage is a core module of our two-stage model, which consists of the customer part and service part. We can alternately divide cr into question representations $qr = \{qr_1, qr_2, ..., qr_n\}$ and answer representations $ar = \{ar_1, ar_2, ..., ar_n\}$ as the input of customer part and service part. In the following, we will illustrate how these two parts of intra-stage adequately exploit the inside relations of their own utterance sequences.

Customer Part. LSTM has a special unit called memory cell, which is similar to an accumulator or a gated neuron. We adapt a Bi-directional LSTM to capture long-term dependencies of qr:

$$m_i^q = \text{BiLSTM}^q\left(m_{i\pm1}^q, qr_i\right) \tag{6}$$

where $i = 1, 2, ..., n$. m_i is the output of Bi-directional LSTM at time step i, the whole context representation of question sequence is $m^q = \{m_1^q, m_2^q, ..., m_n^q\}$.

To better explore the internal relations of question sequence, we capture the satisfaction representation of each time step iteratively by adequately interacting current features with context information. Firstly, an GRU encoder is used to process the sequence:

$$h_i^q = \text{GRU}_{\text{encode}}^q\left(m_i^q, h_{i-1}^q\right) \tag{7}$$

where $h^q = \{h_1^q, h_2^q, ..., h_n^q\}$, h^q is the hidden state of GRU. Secondly, we use an attention mechanism to match h_i^q with the masked personalized context:

$$masked_i\left(m^q\right) = \begin{cases} m_j^q, & j \in \{1, 2, ..., i\} \\ 0, & \text{Otherwise} \end{cases} \tag{8}$$

$$q, k, v = h_i^q, masked_i \left(m^q \right), masked_i \left(m^q \right) \tag{9}$$

$$h_i^{q\prime} = \text{IntraAtt}^\text{q} \left(q, k, v \right) \tag{10}$$

where $h^{q\prime} = \{h_1^{q\prime}, h_2^{q\prime}, ..., h_n^{q\prime}\}$, $h^{q\prime}$ is the result of this attention layer.

Up to now, we have adequately obtained the internal relations of question sequence. Then, a GRU is used to decode the result from the intra attention layer:

$$s_i^q = \text{GRU}_{\text{decode}}^\text{q} \left(h_i^{q\prime}, s_{i-1}^q \right) \tag{11}$$

$s^q = \{s_1^q, s_2^q, ..., s_n^q\}$, where s^q is *customer state* after the complete process of customer part.

Service Part. Service part is the other part in intra-stage, which contributes to the satisfaction state of service. The whole structure of service part is similar to customer part:

$$m_i^a = \text{BiLSTM}^\text{a} \left(m_{i\pm1}^a, ar_i \right) \tag{12}$$

$$h_i^a = \text{GRU}_{\text{encode}}^\text{a} \left(m_i^a, h_{i-1}^a \right) \tag{13}$$

$$masked_i \left(m^a \right) = \begin{cases} m_j^a, & j \in \{1, 2, ..., i\} \\ 0, & \text{Otherwise} \end{cases} \tag{14}$$

$$q, k, v = h_i^a, masked_i \left(m^a \right), masked_i \left(m^a \right) \tag{15}$$

$$h_i^{a\prime} = \text{IntraAtt}^\text{a} \left(q, k, v \right) \tag{16}$$

$$s_i^a = \text{GRU}_{\text{decode}}^\text{a} \left(h_i^{a\prime}, s_{i-1}^a \right) \tag{17}$$

where s^a is the *service state*.

3.5   Inter-stage

Inter-stage aims to fully interact s^q with s^a. Some researchers utilize attention mechanisms to capture the most relevant information and construct interaction between two sequences [3]. Inspired by those works, we use an attention mechanism to interact s^q with s^a:

$$\tilde{s}^q = \text{InterAtt}^\text{q} \left(s^q, s^a, s^a \right) \tag{18}$$

$$\tilde{s}^a = \text{InterAtt}^a \left(s^a, s^q, s^q \right) \tag{19}$$

In order to make the learning process smoother, we adopt a layer of add & normalization [4]:

$$\tilde{s}^{q'} = \text{Normalization} \left(\text{Add} \left(\tilde{s}^q, s^q \right) \right) \tag{20}$$

$$\tilde{s}^{a'} = \text{Normalization} \left(\text{Add} \left(\tilde{s}^a, s^a \right) \right) \tag{21}$$

In the end of the inter-stage, by using average pooling, we transform $\tilde{s}^{q'}$ and $\tilde{s}^{a'}$ into vectors and concatenate them together as follow:

$$s = \text{concat} \left(\text{pooling} \left(\tilde{s}^{q'} \right), \text{pooling} \left(\tilde{s}^{a'} \right) \right) \tag{22}$$

where s is the final satisfaction representation of the whole dialogue.

3.6   Session-Decoder

The session-decoder module is used to decode the satisfaction representation s to predict the customer satisfaction. We use two layers of fully connected network with ReLU activation and softmax, then get the probability distribution of classification P. $\hat{y}$ is the final prediction of satisfaction level:

$$H = \text{ReLU} \left(W_1 s + b_1 \right) \tag{23}$$

$$P = \text{softmax} \left(W_2 H + b_2 \right) \tag{24}$$

$$\hat{y} = \underset{k}{\text{argmax}} \left(P[k] \right) \tag{25}$$

As for the loss function, we choose cross-entropy:

$$\mathcal{L}(\theta) = - \sum_{v \in y_\mathcal{V}} \sum_{z=1}^{Z} Y_{vz} \ln P_{vz} \tag{26}$$

where $y_\mathcal{V}$ is the set of dialogues that have real labels. Y is the label indicator matrix, and θ is the collection of trainable parameters in DialCSP.

4   Experimental Settings

This section mainly introduces datasets, hyper parameters and baselines used in our experiments (Table 1).

Table 1. The statistics of the five datasets. While **CECSP** is our constructed Chinese E-commerce CSP dataset, **Clothes** and **Makeup** are two released corpora in different domains. **MELD** and **EmoryNLP** are two CER datasets.

Datasets	Train	Val	Test	Avg-turns
CECSP	22576	2822	2801	3.67
Clothes	8000	1000	1000	8.14
Makeup	2832	354	354	8.01
MELD	1037	113	279	3.19
EmoryNLP	685	88	78	3.86

4.1 Datasets

We evaluate DialCSP on our in-house dataset (*a Five-classification task*) and four released public datasets (*Three-classification tasks*).

– **CECSP.** This is our in-house Chinese E-commerce CSP dataset collected from one of the largest E-commerce platforms. We use real customer feedback as the dialogue-level satisfaction labels which include *strongly satisfied, satisfied, neutral, dissatisfied* and *strongly dissatisfied*.

– **Clothes & Makeup.** These are two CSP datasets in clothes and makeup domain collected from a top E-commerce platform [13]. Each dialogue is annotated as one of the three satisfaction classes: *satisfied, neutral* and *dissatisfied*.

– **MELD.** This is a multi-party conversation corpus collected from the TV show Friends [5]. Each utterance is annotated as one of the three sentiment classes: *negative, neutral* and *positive*. While *negative* and *positive* are considered as *dissatisfied* and *satisfied* respectively, *neutral* is kept unchanged.

– **EmoryNLP.** This is also a multi-party conversation corpus collected from Friends, but varies from MELD in the choice of scenes and emotion labels [6]. The emotion labels include *neutral, joyful, peaceful, powerful, scared, mad* and *sad*. To create three satisfaction classes: *joyful, peaceful* and *powerful* are positive emotion so we group them together to form the *satisfied* class; *scared, mad* and *sad* are negative emotion so we group them together to form the *dissatisfied* class; and *neutral* is kept unchanged.

– **Transforming rules for MELD & EmoryNLP.** Original MELD and EmoryNLP are two released conversational emotion recognition (CER) datasets. We transform them into the conversational service scenario following four rules: (1) We consider the first speaker of a dialogue as the customer (all other speakers as the customer service) and map all the emotion labels into turn-level satisfaction labels; (2) We concatenate consecutive utterances from the same person as a long utterance; (3) If a dialogue is ended by the first speaker, we use utterance "NULL" as the answer of the last turn; (4) We set the dialogue-level satisfaction as the average of turn-level satisfaction.

4.2   Baselines

We compared DialCSP with the following baselines in our experiments:

- **LSTMCSP** [9]: This model uses a Bi-directional LSTM network to capture the user's intent and identify user's satisfaction.
- **CMN** [10]: It is an end-to-end memory network which updates contextual memories in a multi-hop fashion for conversational emotion recognition.
- **DialogueGCN** [11]: It is a graph-based approach which leverages inter-speakers' dependency of the interlocutors to model conversational context for emotion recognition.
- **CAMIL** [13]: This Context-Assisted Multiple Instance Learning model predicts the sentiments of all the customer utterances and then aggregates those sentiments into service satisfaction polarity.
- **LSTM-Cross** [14]: This model uses LSTM networks to capture contextual features. Then, these features are concatenated with the cross matching scores to predict the satisfaction.
- **DialogueDAG** [12]: This model uses directed graphs to collect nearby and distant historical informative cues. We aggregate the node representations to capture dialogue-level representations for CSP.
- **BERT** [25] & **Dialog-BERT** [26]: We use pre-trained language models and linear layers with softmax on CSP problem. For each dialogue, we use [sep] to concat utterances as input of the models. We use pre-trained *bert-base-chinese* and *dialog-bert-chinese* on CECSP, and pre-trained *bert-base-uncased* and *dialog-bert-english* on MELD & EmoryNLP (In Clothes & Makeup datasets, words are replaced with ids for the data-safety, so we can not use pre-trained model on them).

4.3   Hyper Parameters

We reproduce all baselines with their original experimental settings. In our two-stage model, the batch sizes are set to be 64 for CECSP, Clothes, Makeup, MELD and EmoryNLP. We adopt Adam [8] as the optimizer with initial learning rates of $1e-3$ and L2 weight decay rates of $1e-4$, respectively. The dropout is set to be 0.5 [7]. We train all models for a maximum of 200 epochs and stop training if the validation loss does not decrease for 30 consecutive epochs. The total number of parameters in this model is 59.84 million. We use a piece of Tesla P40 24 GB. Each epoch of these experiments costs around 400 s.

5   Results and Analysis

5.1   Overall Results

The overall results of all the models on five datasets are shown in Table 2. We can learn from the results that DialCSP achieves better performance than all the baselines on five datasets.

Table 2. Overall performance on the five datasets. We use the accuracy and the weighted F1 score to evaluate each model. Scores marked by "#" are reported results in authors' paper, while others are based on our re-implementation.

Model	CECSP		Clothes		Makeup		MELD		EmoryNLP	
	Acc.	F1	Acc.	F1	Acc.	F1	Acc.	F1	Acc.	F1
LSTMCSP	51.55	50.10	75.59	75.78	76.31	76.56	42.29	43.08	50.01	47.56
CMN	52.09	50.32	78.5	78.1	81.07	80.88	45.52	44.08	52.56	48.52
DialogueGCN	52.69	50.25	76.89	76.82	77.72	77.78	46.39	44.99	52.72	48.78
CAMIL	55.43	52.92	78.30#	78.40	78.50#	78.64	44.44	39.02	55.13	49.52
LSTM-Cross	55.51	53.11	78.91	79.33	79.88	79.58	46.70	45.41	55.28	51.00
DialogueDAG	55.12	51.97	75.4	75.04	73.73	73.73	48.03	47.28	59.26	54.82
BERT	55.57	52.86	-	-	-	-	50.18	49.79	58.97	57.31
Dialog-BERT	56.44	51.72	-	-	-	-	47.31	46.42	**64.10**	**60.35**
DialCSP	**57.34**	**54.69**	**81.2**	**80.71**	**82.2**	**82.07**	**50.9**	**50.35**	61.54	57.88

LSTMCSP, CMN, and DialogueGCN achieve similar performance on CECSP, MELD and EmoryNLP. CMN is capable of capturing the emotional cues in context, thus achieving better F1 scores than LSTMCSP on Clothes and Makeup. However, chatbot answers are always neutral in conversational service, which narrow the gap between CMN and LSTMCSP on CECSP. In our scene, customer questions and chatbot answers are alternating, so the related positions between them cannot provide additional information. Thus, the position method in DialogueGCN does not have better performance here.

CAMIL takes turn-level sentiment information into account and outperforms previous strategies on four datasets except MELD. Due to the customer-service interaction modeling method, LSTM-Cross has made further improvement on all datasets, which implies the importance of interactions in single turn. DialogueDAG uses graphical structure to effectively collect nearby and distant information, so it performs well on datasets with shorter average turns, such as MELD and EmoryNLP. But when the average turns become longer, it doesn't work well.

BERT is one of the strongest baslines in multiple NLP tasks. We use pre-trained *bert-base-chinese* on CECSP and *bert-base-uncased* on MELD & EmoryNLP. Dialog-BERT is further designed to focus on dialogue tasks, we use pre-trained *dialog-bert-chinese* on CECSP and *dialog-bert-english* on MELD & EmoryNLP. The superiority of pre-training makes BERT and Dialog-BERT achieve better weighted F1 score over other baselines on small datasets MELD & EmoryNLP, but the performance is mediocre on big dataset CECSP.

DialCSP reaches the new state of the art on four datasets except EmoryNLP. On the one hand, intra-stage extracts internal correlation features of question sequence and answer sequence in customer part and service part separately. Using attention mechanism with the personalized context of both sequences makes feature extraction sufficient at each time step. On the other hand, we think each customer question is not only associated with the answer behind, but

Table 3. Results of ablation study on the two representative datasets.

Method	Weighted F1 score	
	CECSP	Clothes
(1) Two-stage model	**54.69**	**80.71**
(2) - Inter-stage	53.68(↓ 1.01)	78.64(↓ 2.07)
(3) - Intra-stage	54.07(↓ 0.62)	78.23(↓ 2.48)
(4) - Intra-stage & Inter-stage	53.53(↓ 1.16)	78.44(↓ 2.27)
(5) + context part	53.61(↓ 1.08)	79.37(↓ 1.34)
(6) + context part & attention	54.31(↓ 0.38)	79.16(↓ 1.55)

also the answers in other turns, so inter-stage conducts fully interaction between *customer state* and *service state*, which is different from the turn-level approaches in earlier researches. As the results, our proposed model has improved by at least 1%–3% on F1 score over five datasets, compared with non-pretrained baseline models. It also performs better than BERT and Dialog-BERT on CECSP and MELD, meanwhile, taking the advantage of light and fast. Only on the smallest dataset EmoryNLP, Dial-BERT performs better than DialCSP. However, the test set of EmoryNLP contains only 78 samples, so the results may have certain contingency.

5.2   Ablation Study

To study the impact of the modules in our two-stage model, we evaluate it by removing 1) inter-stage 2) intra-stage 3) intra-stage and inter-stage together. Removing the inter-stage means the we only retain the intra-stage (The output of the Bi-LSTM is taken as the input of the intra-stage). Removing the intra-stage means only the inter-stage remains (The output of the inter-stage is taken as the input of session-decoder after pooling & concatenate). Removing both intra-stage & inter-stage means we no longer separate the dialogue into two parts and only retain the customer part of the intra-stage (The output of session-encoder is taken as the input of customer part). We use CECSP and Clothes as the representatives in this study because they are larger datasets with short and long average turns. The results are shown in Table 3.

Here are two sets of comparative experiments. Firstly, let's pay attention to the comparison of rows (1) (2) (3). Without inter-stage, the weighted F1 score drops by 1.01% on CECSP and 2.07% on Clothes. Without intra-stage, the weighted F1 score drops by 0.62% on CECSP and 2.48% on Clothes. The results imply the importance of both two stages, none of them can be removed. Secondly, experiments on rows (2) (4) illustrate the advantage of intra attention. Both of them don't have inter-stage, and the only difference between them is whether to split the dialogue into question sequence and answer sequence. As shown in the table, the weighted F1 score drops by 0.15% and 0.20% if we don't apply intra method. Thus we can draw a conclusion, the intra method helps extract

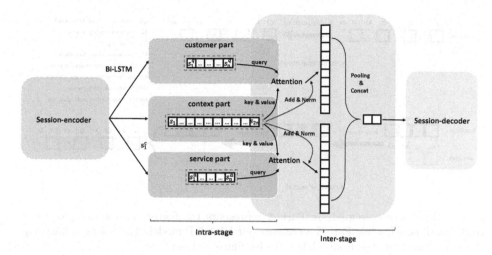

Fig. 4. Framework of DialCSP model with extra context part in the intra-stage.

the internal correlation of customer context and service context respectively, and indeed improves the performance of our model.

Furthermore, to verify if full context features may improve model performance, we conduct two groups of experiments. 1) We add context part in the intra-stage (The full output of session-encoder is taken as the input of context part, the structure of context part is the same as customer part and service part) and concatenate its output with $\tilde{s}^q$ and $\tilde{s}^a$ after pooling. 2) We add context part and get context state s as the output, then we use attention mechanism to interact s with s^q and s^a separately, as shown in Fig. 4. The results are shown in Table 3.

In the experiments on row (5), the only difference between this model and DialCSP is an extra context part. As shown in the Table 3, the weighted F1 score drops by 1.08% and 1.34%, which proves that simply increasing extra fully context information would not improve the performance of DialCSP. In the experiments on row (6), we conduct interaction of context part with customer part and service part by using attention mechanism, as shown in Fig. 4. The weighted F1 score drops by 0.38% and 1.55%, which shows the effect of interaction compared to row (5), while it still can't improve DialCSP model.

In conclusion, the ablation study proves that both intra-stage and inter-stage play important roles. In particular, the intra method of separating context representations into questions and answers contributes to the improvement of our model. Furthermore, we find extra fully context features extraction can't improve the performance of DialCSP model, which signifies the completeness and rationality of our model.

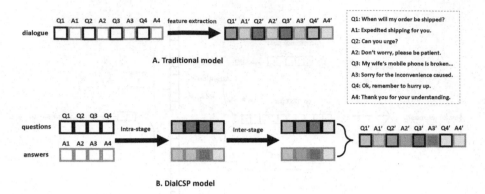

Fig. 5. Results of case analysis. Part A represents the feature extraction process of traditional model, while Part B represents our DialCSP model. The colors of heatmap show the values of attention weights. (Color figure online)

5.3 Case Analysis

In order to better understand the advantages of the DialCSP model, we analyse the case in Fig. 1. The result shows in Fig. 5. The heatmap is used to represent the values of attention weights, where darker colors mean larger weights.

Part A illustrates the feature extraction process in traditional models, in which the dialogue would not be separated into questions and answers. As customer's expression contains richer information ("when", "urge", "broken") of his problem and emotion, the model will pay more attention to Q1, Q2 and Q3. So, it is likely to ignore the importance of answers, which are critical to deciding whether these questions are solved, thus affecting customer's satisfaction deeply too.

By contrast, Part B illustrates our DialCSP model. The dialogue is split into customer questions and chatbot answers, so the model can better learn the inside relations of the two sequences separately in the intra-stage, which ensures the expressions of customers would not attract much more attention than chatbots. In this case, the customer expresses his anxiety and tells the mobile phone is broken in Q2 and Q3, so the weights of those two are larger in the question sequence. Similarly, A3 have larger weights in the answer sequence due to its obvious comforting expression. Then, the inter-stage conducts the interaction to adjust the attention weights of the two parts. In the end, we concatenate two parts and find Q2, A2, Q3, A3 are important utterances of this dialogue. In this dialogue situation, although the customer mainly shows his bad emotion and unsolved problem in Q2 and Q3, the chatbot comforts him in A2 and A3, which leads to a satisfied result. The result of part B appears to be more reasonable.

By comparing the two results, we find the intra-stage of our DialCSP model can balance the expressional differences of customer questions and chatbot answers, while the traditional model pays more attention to customer questions. What's more, the inter-stage interacts *customer state* with *service state* to adjust the weights of attention, which can help capture the characteristics of dialogue more smoothly.

6   Conclusion

In this paper, we propose a two-stage model for dialogue-level CSP task. We first introduce an intra-stage to discover the relations inside customer part and service part respectively, in which an attention mechanism with masked personalized context is used to fully capture the *customer state* and *service state*. Then, we use an inter attention mechanism to combine those two states in inter-stage and predict the customer satisfaction of the whole dialogue. Experimental results on our in-house dataset and four public datasets indicate our model outperforms all the baseline models on the dialogue-level CSP task.

In the future work, we will further improve our two-stage model by constructing more specific structures. For example, we can make differentiated design on customer part and service part in intra-stage. Moreover, we will try DialCSP or its variants on other interesting tasks in customer service dialogues, such as good dialogue mining or dialogue-level use intent detection.

Acknowledgments. This work was supported in part by the National Natural Science Foundation of China (Grant Nos. U1636211, 61672081, 61370126), and the Fund of the State Key Laboratory of Software Development Environment (Grant No. SKLSDE-2021ZX-18).

References

1. Kim, Y.: Convolutional neural networks for sentence classification. In: EMNLP (2014)
2. Nair, V., Hinton, G.E.: Rectified linear units improve restricted Boltzmann machines. In: ICML (2010)
3. Zhou, X., et al.: Multi-turn response selection for chatbots with deep attention matching network. In: ACL (2018)
4. Vaswani, A., et al.: Attention is all you need. In: NIPS (2017)
5. Poria, S., Hazarika, D., Majumder, N., Naik, G., Cambria, E., Mihalcea, R.: MELD: a multimodal multi-party dataset for emotion recognition in conversations. In: ACL (2019)
6. Zahiri, S.M., Choi, J.D.: Emotion detection on TV show transcripts with sequence-based convolutional neural networks. In: The Workshops of AAAI (2018)
7. Srivastava, N., Hinton, G.E., Krizhevsky, A., Sutskever, I., Salakhutdinov, R.: Dropout: a simple way to prevent neural networks from overfitting. J. Mach. Learn. Res. **15**(1), 1929–1958 (2014)
8. Kingma, D.P., Ba, J.: Adam: a method for stochastic optimization. In: ICLR (2015)

9. Hashemi, S.H., Williams, K., Kholy, A.E., Zitouni, I., Crook, P.A.: Measuring user satisfaction on smart speaker intelligent assistants using intent sensitive query embeddings. In: CIKM (2018)
10. Hazarika, D., Poria, S., Zadeh, A., Cambria, E., Morency, L., Zimmermann, R.: Conversational memory network for emotion recognition in dyadic dialogue videos. In: NAACL-HLT (2018)
11. Ghosal, D., Majumder, N., Poria, S., Chhaya, N., Gelbukh, A.F.: DialogueGCN: a graph convolutional neural network for emotion recognition in conversation. In: EMNLP-IJCNLP (2019)
12. Shen, W., Wu, S., Yang, Y., Quan, X.: Directed acyclic graph network for conversational emotion recognition. In: ACL/IJCNLP (2021)
13. Song, K., et al.: Using customer service dialogues for satisfaction analysis with context-assisted multiple instance learning. In: EMNLP-IJCNLP (2019)
14. Yao, R., et al.: Session-level user satisfaction prediction for customer service chatbot in e-commerce (student abstract). In: AAAI (2020)
15. Song, K., Gao, W., Chen, L., Feng, S., Wang, D., Zhang, C.: Build emotion lexicon from the mood of crowd via topic-assisted joint non-negative matrix factorization. In: SIGIR (2016)
16. Ma, D., Li, S., Zhang, X., Wang, H.: Interactive attention networks for aspect-level sentiment classification. In: IJCAI (2017)
17. Song, K., Feng, S., Gao, W., Wang, D., Yu, G., Wong, K.: Personalized sentiment classification based on latent individuality of microblog users. In: IJCAI (2015)
18. Yang, J., Yang, R., Wang, C., Xie, J.: Multi-entity aspect-based sentiment analysis with context, entity and aspect memory. In: AAAI (2018)
19. Cerisara, C., Jafaritazehjani, S., Oluokun, A., Le, H.T.: Multi-task dialog act and sentiment recognition on mastodon. In: COLING (2018)
20. Majumder, N., Poria, S., Hazarika, D., Mihalcea, R., Gelbukh, A.F., Cambria, E.: DialogueRNN: an attentive RNN for emotion detection in conversations. In: EAAI (2019)
21. Yang, Z., Li, B., Zhu, Y., King, I., Levow, G., Meng, H.: Collaborative filtering model for user satisfaction prediction in spoken dialog system evaluation. In: SLT (2010)
22. Jiang, J., et al.: Automatic online evaluation of intelligent assistants. In: WWW (2015)
23. Pennington, J., Socher, R., Manning, C.D.: Glove: global vectors for word representation. In: EMNLP (2014)
24. Kachuee, M., Yuan, H., Kim, Y.B., Lee, S.: Self-supervised contrastive learning for efficient user satisfaction prediction in conversational agents. In: NAACL-HLT (2021)
25. Devlin, J., Chang, M., Lee, K., Toutanova, K.: BERT: pre-training of deep bidirectional transformers for language understanding. In: NAACL-HLT (2019)
26. Xu, Y., Zhao, H.: Dialogue-oriented pre-training. In: Findings of the Association for Computational Linguistics: ACL/IJCNLP (2021)

Foveated Neural Computation

Matteo Tiezzi[1]([✉]), Simone Marullo[1,2], Alessandro Betti[3], Enrico Meloni[1,2], Lapo Faggi[1,2], Marco Gori[1,3], and Stefano Melacci[1]

[1] DIISM, University of Siena, Siena, Italy
{mtiezzi,marco,mela}@diism.unisi.it
[2] DINFO, University of Florence, Florence, Italy
{simone.marullo,enrico.meloni}@unifi.it
[3] Inria, Lab I3S, MAASAI, Universitè Côte d'Azur, Nice, France
alessandro.betti@inria.fr

Abstract. The classic computational scheme of convolutional layers leverages filter banks that are shared over all the spatial coordinates of the input, independently on external information on what is specifically under observation and without any distinctions between what is closer to the observed area and what is peripheral. In this paper we propose to go beyond such a scheme, introducing the notion of Foveated Convolutional Layer (FCL), that formalizes the idea of location-dependent convolutions with foveated processing, i.e., fine-grained processing in a given-focused area and coarser processing in the peripheral regions. We show how the idea of foveated computations can be exploited not only as a filtering mechanism, but also as a mean to speed-up inference with respect to classic convolutional layers, allowing the user to select the appropriate trade-off between level of detail and computational burden. FCLs can be stacked into neural architectures and we evaluate them in several tasks, showing how they efficiently handle the information in the peripheral regions, eventually avoiding the development of misleading biases. When integrated with a model of human attention, FCL-based networks naturally implement a foveated visual system that guides the attention toward the locations of interest, as we experimentally analyze on a stream of visual stimuli.

Keywords: Foveated convolutional layers · Convolutional neural networks · Visual attention

1 Introduction

In several visual tasks the salient information is distributed in regions of limited size. Objects of interest do not typically occupy the whole visual field, while peripheral areas could contain both relevant or redundant (if not misleading) information. Processing the whole visual scene in a uniform manner can lead to the development of learning machines which inherit spurious correlations from

Supplementary Information The online version contains supplementary material available at https://doi.org/10.1007/978-3-031-26409-2_2.

the training data [18,23], that might behave in an unexpected manner at inference time, when exposed to out-of-distribution inputs. A *foveated* artificial vision system is characterized by a high-acuity fovea, at the center of gaze and a lower resolution in the periphery. Several recent approaches implement this principle by transforming the input [4,22], for example blurring the image periphery [16] or sub-selecting a portion of it. Many other works introduced foveation patterns into a variety of tasks [8], architectures [24] and rendering operations [11]. Foveation in machine vision systems has been investigated both for its computational advantages [15] and for its representational and perceptual consequences, it can play a relevant role in terms of reducing undesirable correlations, noise dependence and weakness to adversarial attacks [10,17].

The importance of foveated vision systems clashes with how classic 2D convolutional layers are designed, where all the input locations are treated the same way, exploiting and sharing the same bank of filters over all the image plane [1]. This entails an architectural prior, implicitly assuming that all the input locations equally contribute to the learning process of the layer filters. From the perspective of the computational costs, extracting convolutional features requires the same computational budget over all the spatial locations. Transformer architectures and related models [5,20,21] marked a paradigm shift towards the removal of the inductive bias induced by convolutional layers, thanks to the self-attention mechanism which basically gives different importance to sub-portions of the vision field. However, this actually takes place due to further operations that are applied to predict the importance of the convolutional features extracted on image patches, and not due to an inherently foveated computational scheme, with low computational efficiency. Similar considerations hold for the efficiency of Locally Connected Layers (LCL) [6,13], that implements different filters for each local receptive field. Moreover, LCLs hinder the generalization capability of the network, losing interesting properties (invariances) and not capturing some correlations due to their strong locality [14].

Recent activity in the context of modeling human attention [26] has shown that it is possible to predict human-like scanpaths to tell deep networks what are the important locations to "observe/focus", thus filtering out non-relevant information [19]. When paired with the aforementioned properties and benefits of foveated visual systems, this calls to the need of developing neural models that can naturally and efficiently exploit the information on what is focused. Inspired by this intuition, we introduce a novel kind of foveated neural layer for computer vision, named Foveated Convolutional Layer (FCL), that rethinks the role of classic 2D convolutions. FCLs go beyond the idea of exploiting the same filters over all the spatial coordinates of the input stimuli, formalizing the idea of location-dependent foveated convolutions. Given the coordinates of a point of interest, also referred to as focus of attention (FOA), either coming from external knowledge on the task at hand or generated by a scanpath predictor [26], an FCL extracts feature maps that depend on the FOA coordinates and on where convolution is evaluated, giving a different emphasis to what is closer and farther from the FOA. In particular, FCLs perform a finer-grained processing in the focused areas (*foveal* region), and a progressively coarser processing when moving far away (*peripheral* regions). We propose several variants of FCL, which differ in the way this principle is implemented. One of the proposed instances of FCLs easily leads to faster processing with respect to classic convolutional layers, allowing the user to select the appropriate trade-off between the processing capability and computation streamlining, where the reduction of per-pixel floating point operations (faster processing) is due to the coarser feature extraction in non-focused areas. We show how some instances of FCLs can loosen the weight sharing constraint of classic 2D convolutions [13], limiting it to portions of the image at similar distances from the FOA, thus introducing a FOA-based form of locality in the connections [6]. When integrated with the dynamic model of human attention of [26], FCL-based networks naturally implement an efficient foveated visual system that guides the attention toward plausible locations of interest, leveraging peripheral low-budget computations, as we experiment in the context of a continual stream of visual stimuli.

The scope of our work is different from the one of Recurrent Attention Models [12] (and related work), that iteratively process the input, focussing on different portions of it, and learning to identify what is more relevant for the task at hand [9]. In these models, the way attention behaves is intrinsically interleaved with the task-related predictor, either by means of non-differentiable components [12] or differentiable approximations, while what we study is indeed completely agnostic to the source of the attention coordinates. Moreover, this paper is not oriented toward designing systems that make predictions as the outcome of a dynamic exploration of the input, being potentially complementary to the aforementioned approaches and other dynamic models [7]. The idea of re-structuring the kernel function is also present in Locally Smoothed Neural Networks (LSNNs) [14], that, however, are based on the idea of factorizing the weight matrix to determine the importance of different local receptive fields.

In detail, the contributions of this paper are the following: (i) We propose Foveated Convolutional Layers (FCLs) to implement location-dependent foveated processing, investigating several out-of-the-box FCLs. (ii) We study how FCLs can be stacked or injected into neural architectures, reducing the overall number of floating point operations and running times, as experimentally investigated in multiple tasks. (iii) Thanks to faster processing on the peripheral areas, we implement an all-in-one foveated visual system that can be used to drive the gaze patterns of a focus-of-attention trajectory predictor, extending a state-of-the-art scanpath model [26]. (iv) We evaluate the foveated visual system in continual learning, manipulating attention at a symbolic level, coherently with the skills that are progressively gained by the network.

2   Foveated Convolutional Layers

Let us consider an input image/tensor $I \colon \Omega \to \mathbb{R}^c$ with c channels, where $I(x)$ is the c-element vector at coordinates $x = (x_1, x_2) \in \Omega$, being Ω the domain to which the spatial coordinates belong. Let us also introduce a 2D convolutional layer composed of a bank of F kernels/filters. Without any loss of generality, and for the sake of simplicity, we restrict the following analysis to the case of

$c = 1$. For each kernel $k^j : \mathbb{R}^2 \to \mathbb{R}$, $j = 1, \ldots, F$, the convolution between I and k_j is defined as follows,

$$o^j(x) = \left(k^j * I \right)(x) = \int_\Omega k^j(\tau) I(x - \tau) d\tau = \int_\Omega I(\tau) k^j(x - \tau) d\tau. \qquad (1)$$

where $o^j(x)$ denotes the j-th output feature map computed at coordinates x. Notice that, as usual, the same filter is shared over all the spatial locations. This implies that all $o^j(x)$'s, $\forall x$, are the outcome of having applied the exact same filter function, after having centered it in x. However, from a very qualitative standpoint, this clashes with the fact that humans do not process the visual scene in such a uniform manner. The extraction of visual information depends on what the gaze is specifically observing, $a \in \Omega$. What is closer to a, the *foveal* region, is not processed the same way as what is far from it. Usually, a finer-grain processing is applied when close to a, while a coarser visual representation is modeled as long as we depart from a. It is convenient to think that the former is related to a larger usage of the computational resources, while the latter is associated to less expensive processing.

We propose a novel class of convolutional layers, named Foveated Convolutional Layers (FCLs), that make convolution dependent on a given location of interest a, and that do not exploit the exact same filter over all the x's. The information on a might come from additional knowledge (e.g., knowing the location of an object or simply focussing on the center of the image in Image Classification) or from a model of human attention that predicts where to focus, both in static images and videos [2, 26]. In FCLs, the kernel exploited for the convolution operation in Eq. 1 becomes a function of a, in order to model the dependency on the location focused by the gaze, and also function of x, to differentiate the way convolution is performed in different locations of the image plane. For example, the notion of foveated processing implies that a coarser computation is performed when x is far from a. We propose to implement this behaviour by transforming the original kernel k^j through a spatial convolution with a newly introduced function μ, that depends both on a and x, and, in particular, on the relative location of x with respect to a,

$$\tilde{k}^j_{x,a}(z) = \left(\mu_{\theta, x-a} * k^j \right)(z) := \int_\Omega \mu(\theta, x - a, z - \xi) k^j(\xi) d\xi. \qquad (2)$$

We refer to μ as the modulating function, while θ are its structural parameters. Notice that when $c > 1$, k^j includes c 2D spatial components, and the spatial convolution of Eq. 2 is intended to be applied to each of them. Features $o(x)$ are obtained as in Eq. 1, replacing k^j with $\tilde{k}^j_{x,a}$ from Eq. 2.

This definition paves the road to a broad range of instances of FCLs that differ in the way in which the modulating function $\mu(\theta, \cdot, \cdot)$ is defined, and in how we make operations less costly when departing from the FOA, that will be the subject of the rest of this section, and that are briefly summarized in Fig. 1 (top-left). In 2D convolutional layers, k^j is assumed to be defined on a limited region that, in the discrete case, is $(c\times)$ $K \times K$. The portion of I (resolution $w \times h$)

that is covered by k^j when computing such a discrete convolution at a certain location x is what is usually referred to as receptive input. The major assumption that we follow in designing out-of-the-box FCLs is that features extracted in the peripheral regions (far from a) or in the focused regions (close to a) should be about input portions of the same size, to avoid introducing strong biases in the nature of the features extracted when varying x or a. In other words, all the filters $\tilde{k}^j_{x,a}$, regardless of x and a, must cover a receptive input of the same size. We term this condition the *uniform spatial coverage assumption*.

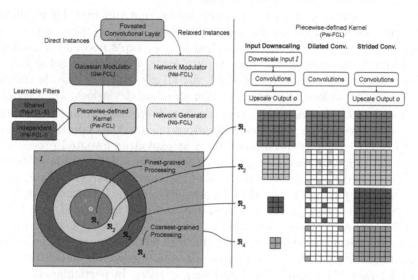

Fig. 1. Top-left: out-of-the-box FCLs. Bottom-left: example of $R = 4$ regions in the piecewise-defined kernel case, when the attention a is given. Right: three strategies (one-per-column) to implement a piecewise-defined kernel, with examples of spatial coverage of the 4 region-wise kernels (coordinates not covered due to dilation are blank). We report right after the strategy name further operations needed to fulfil the uniform spatial coverage assumption.

The most basic instance of FCLs that directly applies the idea of a finer-grade feature extraction around a and a coarser processing in the periphery, can be obtained by blurring kernel k^j with increasing intensity of the blurring operation as long as we move farther from a. We can achieve this behaviour by selecting $\mu(\theta, x - a, \cdot)$ to be a Gaussian function $g(\sigma, x - a, \cdot)$, or, more compactly, $g_{\sigma, x-a}$, characterized by a standard deviation $\sigma(x - a)$ that depends on the distance between x and a. FCLs that exploit such Gaussian modulator are referred to as GM-FCLs, and are based on σ defined as

$$\sigma(x - a) = \hat{\sigma}_a \cdot \ell \left(\|x - a\|^2 \right) + \hat{\sigma}_A \cdot \left(1 - \ell \left(\|x - a\|^2 \right) \right), \tag{3}$$

being $\cdot$ the classic multiplication, ℓ a function that is 1 when $x = a$ and it is 0 when $\|x - a\|$ reaches the maximum possible distance considering the image resolution, while σ_a and σ_A are the standard deviations on a and on the farthest location from it.[1] Due to the commutative property of the convolution (when Ω is $\mathbb{R}^2$), putting together Eq. 2 and Eq. 1 and replacing $\mu_{\theta,x-a}$ with the $g_{\sigma,x-a}$, we have $o^j(x) = (\tilde{k}^j_{x,a} * I)(x) = (g_{\sigma,x-a} * k^j_{x,a} * I)(x) = (k^j_{x,a} * (g_{\sigma,x-a} * I))(x)$. This shows that we can implement this type of FCLs by blurring I with a location-dependent Gaussian, filling the gap between FCLs and the common idea of blurring the visual scene with a progressively increasing levels of intensity [15,16]. However, the computational burden is larger than classic 2D convolutional layers, due to the additional convolution with $g_{\sigma,x-a}$.

If σ of Eq. 3 is modeled with a piecewise-constant function defined in non-overlapping ranges involving its argument, such as $\rho_i \leq \|x - a\| < \rho_{i+1}$, $i = 1, \ldots, R - 1$, $\rho_1 = 0$, $\rho_R = \infty$, then $\tilde{k}^j_{x,a}$ in Eq. 2 becomes a piecewise-defined kernel. In other words, once we are given a, the exact same kernel is used to compute features $o^j(x)$ of Eq. 1, for all x's which fall inside the same range. Moreover, as long as σ returns larger standard deviations, $\tilde{k}^j_{x,a}$ becomes a blurrier copy of k^j, leaving room to approximated representations that reduce its $K \times K$ spatial resolution. These considerations open to the definition of another instance of FCLs that is specifically aimed at exploiting foveated processing to speedup the computations, giving the user full control on the trade-off between computational cost and the level of detail of the features extracted when moving away from a. In piecewise-defined FCLs (Pw-FCLs), the input image is divided into $R \geq 2$ regions $\mathcal{R}_1, \ldots, \mathcal{R}_R$, centered in a and with no overlap, e.g., in function of $\|x - a\|$ as previously described, naturally introducing a dependency on the focused location, as shown in Fig. 1 (bottom-left).[2] The cost of the convolution operation is controlled by a user-customizable *reduction factor* $0 < r_i \leq 1$, defined for each $\mathcal{R}_i$, where $r_1 = 1$ and $r_{i+1} < r_i, \forall i$. In particular, the cost of computing a convolution in $x \in \mathcal{R}_i$ is forced to be $r_i \mathcal{C}_1$, where $\mathcal{C}_1$ is the cost of a convolution for $x \in \mathcal{R}_1$. This means that convolutions in peripheral regions are performed in a faster way than those in regions closer to a. As we will describe in the following, there are multiple ways of fulfilling this computational constraints by introducing a coarser processing. In turn, coarser processing makes Pw-FCLs exposed to less details and less data variability when far away from a, that can result in a more data-efficient learning in non-focused areas.

We propose three different strategies for enforcing the cost requirement imposed by the reduction factor, summarized in Fig. 1 (right). The first two ones are based on the fact that the cost of convolution is directly proportional to the number of spatial components of the kernel, thus the computational burden can be controlled by reducing the size of the kernel defined in each region in function of r_i. However, a smaller kernel size implies covering smaller receptive inputs, thus violating the previously introduced uniform spatial coverage assumption.

---

[1] In our experiments we used an exponential law, with $\hat{\sigma}_a$ almost zero and $\hat{\sigma}_A = 10$. Function g is computed on a discrete grid of fixed-size 7×7.

[2] The innermost region $\mathcal{R}_1$ is then a circle, and the other regions are circular crowns with increasing radii. The outermost region $\mathcal{R}_R$ is simply the complementary area. We also tested the case of a squared $\mathcal{R}_1$ and frame-like $\mathcal{R}_i$, $i > 1$.

For this reasons, further actions are needed in order to ensure such a condition is fulfilled, leading to two variants of Pw-FCLs, based on *input downscaling* and *dilated convolutions* [25], respectively. The former appropriately downscales the input to compensate the kernel reduction, and it requires then to upscale the feature maps to match their expected resolution (Fig. 1, top-right). The third strategy consists in not reducing the kernel and relying on *strided convolutions*, thus skipping some x's, that still requires to upscale the resulting feature maps (more details in Appendix A.1).

The piecewise defined kernel is the outcome of adapting the same base kernel k^j over the different regions, thus we denote what we described so far as Pw-FCL with *shared* weights, or, more compactly, Pw-FCL-S (Fig. 1, top-left). This implies that all the region-defined filters share related semantics across the image plane, due to the shared nature of the learnable k^j. A natural alternative to this model consists in using *independent* learnable filters in each region. In this case, referred to as Pw-FCL-I, features extracted in different areas could be fully different (or not) and associated to different (or same) meanings. Of course, if the information stored in the foveal and peripheral areas share some properties, then a Pw-FCL-S might be more appropriate.

It is interesting to show how the framework of FCLs can be further extended along directions that depart a little bit from the idea of foveated processing, but that are still oriented toward location-dependent processing in function of the FOA coordinates a. Of course, a detailed analysis of them goes beyond the scope of this paper. In particular, the degrees of freedom of FCLs can be extended when $\mu(\theta, x - a, z)$ is implemented with a neural network modulator, naming these layers NM-FCLs. It is convenient to think about such modulator as a multi-layer feed-forward network with input $x - a$, and that yields $N \times N$ real numbers as output, which is the filter that modulates k^j in the discrete counterpart of Eq. 1. Output values are normalized to ensure the filter sums to 1 in the $N \times N$ grid, and we set $N = 7$ in our experiments, in analogy with the way $g_{\sigma, x-a}$ was discretized. Another step in relaxing the formulation of FCLs consists in fully re-defining $\tilde{k}_{x,a}$ of Eq. 2 as a discrete filter whose values are generated by a neural network. The net acts as a neural generator, thus NG-FCLs, that learns to produce a bank of F distinct $(c\times)$ $K \times K$ filters given $x - a$, to be used when computing convolution in coordinates x. It is easy to see that the computation/memory burden of generating a different kernel for (almost) every location x in the image plane, by means of a multi-layer network, makes it less practical than the other described types of FCLs, even if it opens to further investigations into this direction (details in Appendix A.2/A.3 suppl. material).

2.1 Learning with Attention in Foveated Neural Networks

From the point of view of the input-output, FCLs are equivalent to classic 2D convolutional layers, with the exception of the additional input signal a. As a result, they can be straightforwardly stacked into deep architectures, learning the kernel components by Backpropagation. It is pretty straightforward to exploit (single or stacked) FCLs to extract features for each pixel of the $w \times h$ network input. Whenever FCLs are used after pooling layers or, in any case, after having reduced the resolution of the latent representations, a must be rescaled accordingly. Depending on the properties of the considered task, it might be convenient to use FCLs in the last portions of a deep architecture, at the beginning of it, or in other configurations, for example stacking FCLs in a way that the lower layers are mostly specialized in fine-grained processing over large areas around a, that progressively get smaller in the upper layers.

In this paper, we study the case in which a is given, either due to specific external knowledge or when it is estimated by a human-like model of visual attention. A very promising model of human attention was recently proposed in [26], well suited for generic free-viewing conditions too. Such a model estimates attention a at time t, i.e., $a(t)$, as a dynamic process driven by the following law

$$\ddot{a}(t) + \gamma a(t) - E(t, a(t), \{m_j(x,t), \ j = 1, \ldots, M\}) = 0, \tag{4}$$

where E is a gravitational field that depends on a distribution of masses, each of them indicated with m_j, and γ is a customizable weight that controls dissipation. Each mass attracts the attention in a way that is proportional to the value of $m_j(x,t)$, eventually tuned by a customizable scalar. The authors of [26] considered the case of $M = 2$, with m_1 and m_2 that yield high values when x includes strong variations of brightness and motion, respectively. However, other masses could be added over time, as briefly mentioned in [26] but never investigated. Let us introduce a stream of visual information, being I_t the frame at time t, and a neural network $f(I_t, \omega_t)$ with weights ω_t. In a C-class semantic labeling problem, f returns a vector of C class membership scores for each image coordinate x, i.e., $f(I_t, \omega_t)(x)$. The notation $f_{i,j,z,\ldots}(I_t, \omega_t)(x)$ is used to consider only the scores of the classes listed in the subscript. Let us assume that the user is interested in forcing the model to focus on specific object classes h and z. For example, in a stadium-like scene during a soccer match, the model should be attracted by the players, by the ball, and not by all the people in the bleachers or by the sky. We can pool the class membership scores to simulate a novel mass function, such as $m_q(x,t) = (\max f_{h,z}(I_t, \omega_t)(x))$, so that the attention model is automatically attracted by those pixels that are predicted as belonging to classes h or z (or both—it holds for any number of classes). If f is based on Pw-FCLs, then peripheral areas will be characterized by faster processing and slightly lower prediction quality (due to the coarser feature extraction), as it happens in the human visual system, and attention will be also influenced by such predictions.

3   Experiments

We implemented FCLs using PyTorch,[3] performing experiments in a Linux environment, using a NVIDIA GeForce RTX 3090 GPU (24 GB). We performed four types of experiments on four different datasets, aimed at comparing FCLs

---

[3] https://github.com/sailab-code/foveated_neural_computation.

with classic convolutional layer in a variety of settings. Before going into further details, we showcase the speedup obtained when using Pw-FCLs ($R = 4$ regions, $r_2 = 0.7$, $r_3 = 0.4$, $r_4 = 0.1$), comparing the 3 different strategies of Sect. 2 and Fig. 1 (right), also including a baseline with a classic convolutional layer, and a vanilla configuration of Pw-FCLs without any cost reduction (i.e., $r_i = 1$, $\forall i$). The bounding box of each region $\mathcal{R}_i$ (up to $i = R - 1$) linearly increases from 10% of w up to 70% of it. Figure 2 (left) shows the average time required to process an input I with $c = 64$ channels, whose resolution varies in $\{256 \times 256, 512 \times 512, 1024 \times 1024, 1920 \times 1080\}$, using $K = 11$, while Fig. 2 (right) is about 512×512 and variable base kernel size K. Overall, the proposed reduction strategies achieve big improvements with respect to classic convolutional layers, with better scaling capabilities for larger resolutions and kernels. The region-management overhead, that is evident in the vanilla case, is completely compensated by all the reduction mechanisms. The downscaling strategy is the faster solution (of course, reducing the number of regions or r_i's yields even better times–see Appendix C.1, supplementary material), that is what we will use in the following experiments.

The first task we consider is referred to as DUAL-INTENTION, and it consists in classifying 200×200 images in one out of 10 classes. In each image two different digits are present (each of them covering $\approx 28 \times 28$ pixel, MNIST dataset), one in the middle of it and one closer to the image border. A special *intention sign* is also present, far from the middle (one of $\{\square, \bullet\}$)). If the sign is $\square$, then the image class is the one of the digit placed close to the border. When $\bullet$ is used the

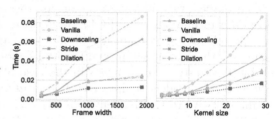

Fig. 2. Inference time of the 3 Pw-FCLs strategies, of a convolutional layer (baseline), of a fixed-kernel-size Pw-FCLs (vanilla). Left: changing input resolution. Right: varying K.

target class is the one of the digit in the center of the image—Fig. 3 (left). The challenging nature of the task comes from fact that the training data only include intention signs randomly located in locations close to the bottom border of the image, while in the test data the sign is randomly located closer to the top border (the peripheral digit stands on the opposite side with respect to the intention sign). In order to gain generalization skills, the network must be able to learn representations that are pretty much location-independent in the peripheral area, and to differentiate them from the ones developed in the central area. We compared multiple convolutional feature extractors, each of them followed by pooling operation and a classification head (128 hidden neurons, ReLU). The first extractor is a CNN with 4 layers (denoted with CNN*), with final *global-max* pooling. Then, we focused on a single classic convolutional layer (CL) followed by *max* pooling (stride 10) or *global-max* pooling. We also introduced another type of pooling that is aware of the existence of two main image regions, that we simulated with a 33 × 33 box surrounding the image center and the rest of the image, performing max pooling in each of them and concatenating the results. We refer to it as region-wise max pooling (*reg-max*). Finally, we extracted features with FCLs, leveraging Pw-FCL-S and Pw-FCL-I with FOA positioned at the image center and $R = 2$ regions (a 33 × 33 box region and the complementary one—in this case, region-wise max pooling is always exploited). We report the average test accuracy over three runs (and standard deviations) in Table 1, considering a 1K samples dataset and a 10K samples one (see Appendix C.2, supplementary materials, for the model validation procedure; r_2 is always < 1). Models based on *global-max* pooling, being them deeper (CNN*) or shallower (CL+*global-max*), loose all the location-related information, thus they do not distinguish among what is in the middle of the image and what is in the peripheral area. CL-MAX only aggregates a few spatial coordinates, thus yielding (relaxed) location-related features that let the classifier learn that the intention sign that is expected to be always at the bottom. Thanks to region-wise max pooling, the CL+*reg-max* network achieves good results, even if at the same computational cost of CL. Interestingly, when a Pw-FCL is used, strongly reducing the computational burden, we still achieve similar performances to CL+*reg-max*, and in the low-data regime both Pw-FCLs outperform it. The foveated computations in the peripheral region implicitly filters the image, reducing noise and smoothing samples, that turns out in making FCLs more data efficient (1K dataset). In Table 2 we restrict our analysis in the case of $K = 15$ and $F = 64$, comparing the two best models equipped with *reg-max* pooling, and showing the performance relative to the different values of the reduction factor r_2 (the one of the peripheral region), along with the number of performed floating point operations (GFLOPs).[4] Interestingly enough, the proposed Pw-FCL-S can achieve a very similar performance with respect to CL+*reg-max*, at a fraction of its cost.

Fig. 3. Left: train and test samples from DUAL-INTENTION dataset (class 0, and class 6, respectively). Middle: training sample from STOCK-FASHION (class is "in stock shoes"). Right: sample from the data by Xiao et al. [23] in its ORIGINAL, MIXED-SAME (1st row), MIXED-RAND and MIXED-NEXT (2nd row) versions.

---

[4] We measured the number of floating point operations using the PyTorch profiling utilities https://pytorch.org/docs/stable/profiler.html.

Table 1. DUAL-INTENTION. Average test accuracy (and std) in low (1K) and large (10K) data regimes.

Model	1K	10K
CNN*	54.3 ± 3.3	57.3 ± 1.0
CL+*global-max*	53.0 ± 1.4	53.1 ± 0.5
CL+*max*	24.7 ± 1.9	47.6 ± 1.1
CL+*reg-max*	71.0 ± 2.4	$\mathbf{96.5 \pm 0.0}$
Pw-FCL-S	$\mathbf{73.7 \pm 0.5}$	95.4 ± 0.3
Pw-FCL-I	72.3 ± 0.5	95.7 ± 0.1

Table 2. DUAL-INTENTION. The case of $K = 15$ and $F = 64$, varying the reduction factor r_2 (peripheral region).

Model	r_2	GFLOPs	Accuracy	
			1K	10K
CL+*reg-max*	-	0.996	71.0	**96.5**
Pw-FCL-S	0.1	0.078	56.7	87.7
	0.25	0.159	69.7	94.7
	0.5	0.398	**73.7**	95.4

Our next experimental activity is about a task that is based on what we refer to as STOCK-FASHION dataset, that is somewhat related to the previous one, since we still have an entity in the middle of the image and another one closer to the border. The middle area contains patterns that are harder to classify, simulated with samples from Fashion-MNIST dataset, paired with MNIST digits placed in the upper image portion at training time, and in the bottom image portion at test time. The goal is to recognize the class of the fashion item in the middle (10 types) and also it is largely available *in-stock* or with *limited* availability, in function of the value of the peripheral digit (0 to 4: limited; 5 to 9: in-stock)—Fig. 3 (middle). Hence, the total number of classes is 20. The peripheral information is still crucial for the final purpose of classifying the image, but it is of different type with respect to what is in the middle. We follow the same experimental setup of the previous experiment (10K samples) and we report in Table 3 the test accuracy of the compared models. Once again, region-wise pooling-based models are the best performing ones. The independent filters of Pw-FCL-I are able to learn dedicated properties for the foveal region and for the peripheral region, that in this task do not share any semantic similarities, leading to better performance (recall that $r_2 < 1$). Recognizing the peripheral digits is relatively simpler with respect to the fashion items, so that the foveated computational scheme perfectly balances the computational resources over the image. In Fig. 4 we report the test accuracy obtained by the foveated models, restricting our analysis to the case of $K = 15$ and $F = 128$ and varying the reduction factor r_2. We remark that even with an evident reduction of their computational capabilities ($r_2 \in \{0.1, 0.25\}$), the models yield a robust representational quality that allows the classifiers to reach large accuracies, with a preference for Pw-FCLs-I.

Table 3. STOCK-FASHION, 10K, average test accuracy (and std).

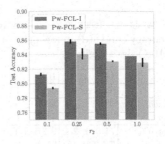

Fig. 4. STOCK-FASHION, $F = 128$, $K = 15$, varying the reduction factor r_2.

Model	1K	Accuracy
CNN*	61.7 ± 1.4	83.4 ± 1.6
CL+global-max	59.7 ± 3.3	79.8 ± 1.4
CL+max	37.3 ± 4.7	42.4 ± 0.7
CL+reg-max	$\mathbf{68.3} \pm 0.5$	85.9 ± 0.9
Pw-FCL-S	66.7 ± 0.0	83.1 ± 0.1
Pw-FCL-I	64.0 ± 0.9	$\mathbf{86.0} \pm 0.3$

In our next experimental activity, we consider a context in which the information in the background of an image might or might-not help in gaining robust generalization skills, thus ending up in learning *spurious correlations*. An out-of-distribution context change in the background usually leads to poor performances in the classification accuracy of deep neural models, as studied in the benchmark of Xiao et al. [23], based on ImageNet. Multiple test sets are made available, ORIGINAL, MIXED-SAME, MIXED-NEXT, MIXED-RAND, where the background of the images is left untouched, replaced with the one from another example of the "same" class, of the "next" class, of a "random" class, as shown Fig. 3 (right). We argue that the injection of FCLs into neural architectures, even if very shallow or simple, favors the model robustness to issues related to background correlations. In particular, several advantages come from the fact that FCLs reserve the finest-grained processing solely to the focused area, whilst the peripheral portion of the frame is processed via coarser resources/reduced resolution. Thanks to this architectural prior, the variability of features extracted from peripheral regions is reduced and they are harder to be tightly correlated with a certain class. We considered the already described CNN* as reference, replacing the *first* or *last* convolutional layer with a FCL, evaluating all the types of FCLs of Sect. 2 (see Appendix C.4 for more details). In this case, the FOA a can be either located at the center of the picture, or on the barycenter of the main object of each picture (the one that yields the class label). Figure 5 shows our experimental findings in the latter case, analyzing the first three classes of the dataset (dog, bird, vehicle)–similar results with the whole dataset (Appendix C.4). These results support the intuition that a standard CNN* suffers from the background correlation issues, when comparing accuracy in ORIGINAL (90.8%) and MIXED-NEXT (57.1%), where the background is always extrapolated from a different class. As expected, thanks to the introduction of FCLs, the performance drop is remarkably reduced in most of the cases. In Pw-FCL-I (*all*) such drop is almost halved (accuracy goes from 86.3% to 68.1%—remind that this model features a reduced number of parameters with respect to CNN*). Using neural modulators or generators (NM-FCL, NG-FCL) was less effective in the *all* and *first* architectures, while in the *last* setting they yielded outstanding performances in terms of generalization both in the ORIGINAL and MIXED-NEXT test sets. This is due to the fact that these models are not explicitly foveated, since they can freely learn how to perform convolutions at different coordinates with no restrictions, thus they can still learn to specialize in background features. However, when used in the last stages of a hierarchy, they actually learn to further refine the learned representations to better focus on the main object.

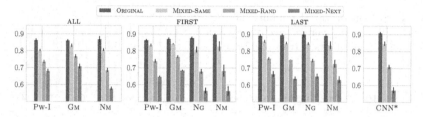

Fig. 5. Background correlations. Accuracy in the first three classes of the dataset in [23]. We dropped the string FCL from all the model names (except CNN*).

Going beyond experiments on static images, we studied an incremental learning setting in which a video stream is presented, frame-by-frame, to an FCL-based foveated network, with the goal of learning to classify each single pixel as belonging to one of 20 classes or as being part of the background. The attention model of [26] is exploited to explore the visual scene, yielding $a(t)$ for each frame (Eq. 4). The visual stream is composed of a sequence of what we refer to as *visual stimuli*, each of them involving 20 unique pictures of handwritten digits and letters from the EMNIST dataset, also generically referred to as "entities" (10 digits and 10 letters–A to K, excluding I that is too similar to digit one). A single entity covers $\approx 25 \times 25$ pixels. For each stimulus (2260 frames) an entity enters the scene from top, slowly moving down until it reaches the bottom of the frame, and it stands still. A different entity does the same, until the scene is completely populated by 20 entities with no overlap. Finally, the entities leave the scene moving down (in reverse order, sequentially). The network processes two sequences of training stimuli, learning in a supervised manner, initially receiving supervisions on digits only (first training sequence) and then on letters only (second training sequence). We implemented a simple rehearsal-based continual learning scheme to store a small subset of frames ($\approx 20 - 40$) for learning purposes [3], that are selected whenever the attention $a(t)$ performs a saccadic movement (details in Appendix C.5, supplementary materials). After each training sequence, a test sequence is presented, generating end evaluating multiple foveated-network-dependent masses that attract the attention toward custom salient elements, as described in Sect. 2.1. Masses are initially about all the *digits*, *even* digits, *odd* digits. Then, in the second test sequence, they are about *letters*, the first 3 digits and first 3 letters (*mix1*), the last 3 digits and last 3 letters (*mix2*) respectively. Masses are activated in mutually exclusive manner in evenly partitioned portions of each test sequence. All the types of FCLs of this paper are evaluated, excluding the network-generated ones that resulted to be too memory demanding. The foveated network is composed of 2 convolutional layers and a final FCL (*last*), or of 3 FCLs (*all*), $R = 2$, $r_2 = 0.25$, compared with a reference network with classic convolutional layers only (CNN). The $w \times h$ pixel-wise feature vectors of the last layer are processed by a classifying head that marks as "background" those coordinates with too small prediction confidence.

Table 4. Stream of visual stimuli, 200 × 200. GFLOPs and average accuracy (3 runs) in classifying digits, letters and background.

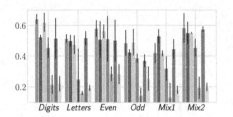

	Model	GFLOPs	Ex. Time (ms)	Digits	Letters	Back.
	▪ CNN	6.13	1.3	71.9	23.7	84.0
Last	▪ Pw-FCLs	2.72	3.1	69.2	25.3	80.6
	▪ Gм-FCLs	6.19	4.3	70.4	22.7	79.8
	▪ Nм-FCLs	6.23	4.5	81.0	28.4	67.7
All	▪ Pw-FCLs	1.16	5.8	70.8	24.7	80.0
	▪ Gм-FCLs	6.25	6.7	75.2	24.2	77.6
	▪ Nм-FCLs	6.37	7.4	58.0	25.3	52.0

Fig. 6. Stream of visual stimuli, 200 × 200. Fraction of time spent on the salient elements of the test stimuli (colors are about the models of Table 4). (Color figure online)

Table 4 reports our results for a stream at the resolution of 200 × 200, and it shows that foveated models are able to perform similarly to (or even better than) CNN, even if in Pw-FCLs the number of floating point operations is approximately 3 to 5 times smaller. It is interesting to see that Nм-FCLs (*last*) and Gм-FCLs yield significantly better results in the digit/letter pixels, compared to CNN, but they have more difficulties in classifying background, since the modulated kernel tends to respond in a less precise manner closer to the digit/letter boundaries. In Fig. 6 we report the fraction of time spent in those pixels that are about elements for which an apposite mass was created (colors are about different models, see Table 4). Results show that foveated networks based on Pw-FCLs (*last*/blue), Gм-FCLs (*last*/light-green, *all*/red) explore the expected elements in a similar manner to what happens in CNN (light-blue). The other models frequently return a small margin between the winning prediction and the other ones, thus the resulting mass is noisy. Figure 7 shows the saliency map generated by the attention model exploiting foveated networks with Pw-FCLs (*last*), from which we can qualitatively observe that the expected salient elements are indeed explored by $a(t)$.

In order to emphasize the computational gains when using Pw-FCLs, we performed the same experiment considering a stream with resolution 1000 × 1000. The original size of the digits/letters is left untouched, so that frames are almost composed of background pixels. One might be tempted to simply downscale each frame and provide it to the net, that actually turns out to be a complete failure, since the size of digits/letters become significantly small, as shown in Table 5. Differently, the performance of CNN processing the high-resolution stream is almost matched by Pw-FCLs (*last, all*), with a running

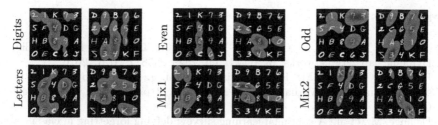

Fig. 7. Attention model of [26] with masses predicted by Pw-FCLs (*last*). For each salient element (all *digits*, *even* digits, *odd* digits, *letters*, *mix1*, *mix2*), saliency maps are shown for 2 visual stimuli, when the frame is fully populated (brighter-red: frequently visited by $a(t)$; lighter-red: sporadically visited by $a(t)$).

time that is approximately 70%–60% smaller than CNN (the time spent on the salient elements is still comparable to the one of CNN–Appendix C.5). This shows how foveated processing can let the network implement a good trade-off between speed and computational burden, being able to focus and recognize small digits/letters on a large resolution image.

Table 5. Stream of visual stimuli, 1000×1000. Average test accuracy ($\pm$ std, 3 runs), GFLOPs, inference time in classifying digits, letters and background.

Model	GFLOPs	Time (ms)	Digits	Letters	Background
CNN (resize to 200×200)	6.13	1.3 ± 0.04	27.1 ± 1.1	6.0 ± 0.8	94.4 ± 0.6
CNN (full resolution)	153.15	13.2 ± 0.01	77.6 ± 5.4	30.3 ± 1.4	99.5 ± 0.0
Pw-FCLs (last)	63.72	9.1 ± 0.13	76.0 ± 1.5	30.6 ± 1.8	99.4 ± 0.1
Pw-FCLs (all)	18.84	8.6 ± 0.04	74.4 ± 2.3	28.5 ± 1.1	99.4 ± 0.0

4   Conclusions and Future Work

We presented Foveated Convolutional Layers to extract features in function of a focused location with foveated processing. We proposed several instances of this model, emphasizing the one that yields a significantly faster processing than plain 2D convolutions. When injected into a human-like attention model, FCL-based networks naturally implement a user-customizable visual system with fast inference and coarser processing in the peripheral areas. Future work includes the analysis of foveated networks in problems driven by motion invariance principles.

Acknowledgements. This work was partly supported by the PRIN 2017 project RexLearn, funded by the Italian Ministry of Education, University and Research (grant no. 2017TWNMH2), and also by the French government, through the 3IA Côte d'Azur, Investment in the Future, project managed by the National Research Agency (ANR) with the reference number ANR-19-P3IA-0002.

References

1. Almahairi, A., Ballas, N., Cooijmans, T., Zheng, Y., Larochelle, H., Courville, A.: Dynamic capacity networks. In: ICML, pp. 2549–2558. PMLR (2016)
2. Borji, A., Cheng, M.-M., Hou, Q., Jiang, H., Li, J.: Salient object detection: a survey. Comput. Vis. Media **5**(2), 117–150 (2019). https://doi.org/10.1007/s41095-019-0149-9
3. Delange, M., et al.: A continual learning survey: defying forgetting in classification tasks. IEEE TPAMI **44**(7), 3366–3385 (2021)
4. Deza, A., Konkle, T.: Emergent properties of foveated perceptual systems. arXiv:2006.07991 (2020)
5. Dosovitskiy, A., et al.: An image is worth 16x16 words: transformers for image recognition at scale. In: ICLR (2020)
6. Gregor, K., LeCun, Y.: Emergence of complex-like cells in a temporal product network with local receptive fields. arXiv:1006.0448 (2010)
7. Han, Y., Huang, G., Song, S., Yang, L., Wang, H., Wang, Y.: Dynamic neural networks: a survey. IEEE TPAMI **44**(11), 7436–7456 (2021)
8. Kong, T., Sun, F., Liu, H., Jiang, Y., Li, L., Shi, J.: Foveabox: beyound anchor-based object detection. IEEE TIP **29**, 7389–7398 (2020)
9. Larochelle, H., Hinton, G.E.: Learning to combine foveal glimpses with a third-order Boltzmann machine. In: Advances in NeurIPS, vol. 23 (2010)
10. Luo, Y., Boix, X., Roig, G., Poggio, T., Zhao, Q.: Foveation-based mechanisms alleviate adversarial examples. arXiv:1511.06292 (2015)
11. Malkin, E., Deza, A., et al.: CUDA-optimized real-time rendering of a foveated visual system. In: NeurIPS 2020 Workshop SVRHM (2020)
12. Mnih, V., Heess, N., Graves, A., et al.: Recurrent models of visual attention. In: Advances in Neural Information Processing Systems, vol. 27 (2014)
13. Ott, J., Linstead, E., LaHaye, N., Baldi, P.: Learning in the machine: to share or not to share? Neural Netw. **126**, 235–249 (2020)
14. Pang, L., Lan, Y., Xu, J., Guo, J., Cheng, X.: Locally smoothed neural networks. In: Asian Conference on Machine Learning, pp. 177–191. PMLR (2017)
15. Poggio, T., Mutch, J., Isik, L.: Computational role of eccentricity dependent cortical magnification. arXiv:1406.1770 (2014)
16. Pramod, R., Katti, H., Arun, S.: Human peripheral blur is optimal for object recognition. arXiv:1807.08476 (2018)
17. Reddy, M.V., Banburski, A., Pant, N., Poggio, T.: Biologically inspired mechanisms for adversarial robustness. In: Advances in NeurIPS, vol. 33 (2020)
18. Sagawa, S., Koh, P.W., Hashimoto, T.B., Liang, P.: Distributionally robust neural networks. In: ICLR (2019)
19. Tiezzi, M., Melacci, S., Betti, A., Maggini, M., Gori, M.: Focus of attention improves information transfer in visual features. In: Advances in NeurIPS, vol. 33, pp. 22194–22204 (2020)
20. Tolstikhin, I.O., et al.: MLP-mixer: an all-MLP architecture for vision. In: NeurIPS, vol. 34 (2021)
21. Trockman, A., Kolter, J.Z.: Patches are all you need? arXiv:2201.09792 (2022)
22. Wang, P., Cottrell, G.W.: Central and peripheral vision for scene recognition: a neurocomputational modeling exploration. J. Vis. **17**(4), 9–9 (2017)

23. Xiao, K.Y., Engstrom, L., Ilyas, A., Madry, A.: Noise or signal: the role of image backgrounds in object recognition. In: ICLR (2020)
24. Yang, J., Li, C., Gao, J.: Focal modulation networks. arXiv:2203.11926 (2022)
25. Yu, F., Koltun, V.: Multi-scale context aggregation by dilated convolutions (2015). https://doi.org/10.48550/ARXIV.1511.07122
26. Zanca, D., Melacci, S., Gori, M.: Gravitational laws of focus of attention. IEEE TPAMI **42**(12), 2983–2995 (2020)

Class-Incremental Learning
via Knowledge Amalgamation

Marcus de Carvalho[1(✉)], Mahardhika Pratama[2], Jie Zhang[1], and Yajuan Sun[3]

[1] Nanyang Technological University, Singapore, Singapore
{marcus.decarvalho,zhangj}@ntu.edu.sg
[2] University of South Australia, Adelaide, Australia
dhika.pratama@unisa.edu.au
[3] A*Star SIMTech, Singapore, Singapore
sun_yajuan@simtech.a-star.edu.sg

Abstract. Catastrophic forgetting has been a significant problem hindering the deployment of deep learning algorithms in the continual learning setting. Numerous methods have been proposed to address the catastrophic forgetting problem where an agent loses its generalization power of old tasks while learning new tasks. We put forward an alternative strategy to handle the catastrophic forgetting with knowledge amalgamation (CFA), which learns a student network from multiple heterogeneous teacher models specializing in previous tasks and can be applied to current offline methods. The knowledge amalgamation process is carried out in a single-head manner with only a selected number of memorized samples and no annotations. The teachers and students do not need to share the same network structure, allowing heterogeneous tasks to be adapted to a compact or sparse data representation. We compare our method with competitive baselines from different strategies, demonstrating our approach's advantages. Source-code: github.com/Ivsucram/CFA.

Keywords: Continual learning · Transfer learning · Knowledge distillation

1 Introduction

Computational learning systems driven by the success of deep learning have obtained great success in several computational data mining and learning system as computer vision, natural language processing, clustering, and many more [1]. However, although deep models have demonstrated promising results on unvarying data, they are susceptible to *catastrophic forgetting* when applied to dynamic settings, i.e., new information overwrites past experiences, leading to a significant drop in performance of previous tasks.

In other words, current learning systems depend on batch setting training, where the tasks are known in advance, and the training data of all classes are accessible. When new knowledge is introduced, an entire retraining process of the network parameters is required to adapt to changes. This becomes impractical in terms of time and computational power requirements with the continual introduction of new tasks.

© The Author(s), under exclusive license to Springer Nature Switzerland AG 2023
M.-R. Amini et al. (Eds.): ECML PKDD 2022, LNAI 13715, pp. 36–50, 2023.
https://doi.org/10.1007/978-3-031-26409-2_3

To overcome catastrophic forgetting, learning agents must integrate continuous new information to enrich the existing knowledge. The model must then prevent the new information from significantly obstructing the acquired knowledge by preserving all or most of it. A learning system that continuously learns about incoming new knowledge consisting of new classes is called a *class-incremental learning* agent.

A class-incremental solution showcases three properties:

1. It should learn from a data stream that introduces different classes at different times,
2. It should provide a multi-class inference for the learned classes at any requested time,
3. Its computational requirements should be bounded, or grow slowly, to the number of classes learned.

Many strategies and approaches in the continual learning field attempt to solve the catastrophic forgetting problem in the class-incremental scenario. Regularization techniques [5] identify essential parameters for inference of previous tasks and avoid perturbing them when learning new tasks. Knowledge distillation methods have also been used [6], where knowledge from previous tasks and incoming tasks are jointly optimized. Inspired by work in reinforcement learning, memory replay has also been an important direction explored by researchers [18], where essential knowledge acquired from previous experiences is re-used for faster training, or retraining, of a learning agent.

In this paper, we propose a catastrophic forgetting solution based on knowledge amalgamation (CFA). Given multiple trained teacher models - each on a previous task - knowledge amalgamation aims to suppress catastrophic forgetting by learning a student model that handles all previous tasks in a single-head manner with only a selected number of memorized samples and no annotations. Furthermore, the teachers and the students do not need to share the same structure so that the student can be a compact or sparse representation of the teachers' models.

A catastrophic forgetting solution based on the knowledge amalgamation approach is helpful because it allows heterogeneous tasks to be adapted to a single-head final model. At the same time, knowledge amalgamation explores the relationship between the tasks without the need for any identifier during the amalgamation process, being smoothly integrated into already existing learning pipelines. This approach can be perceived as a post-processing continual learning solution, where a teacher model is developed for each task and flexibly combined into a single compact model when inference is required. As a result, it does not need to maintain specific network architectures for each task.

Contributions:

- A novel class-incremental learning approach via knowledge amalgamation which:
    - Allows teachers and students to present different structures and tasks;
    - Integrable into existing learning pipelines (including non-continual ones);

2   Related Work

A neural network model needs to learn a series of tasks in sequences in the continual learning setting. Thus, only data from the current task is available during training. Furthermore, classes are assumed to be clearly separated. As a result, catastrophic forgetting occurs when a new task is introduced and the model loses its generalization power of old tasks through learning.

Currently, there are three scenarios in which a continual learning experiment can be configured. *Task-incremental learning* is the easiest of the scenarios, as a model receives knowledge about which task needs to be processed. Models with task-specific components are the standard in this scenario, where the multi-headed output layer network represents the most common solution.

The second scenario, referred to as *domain-incremental learning*, does not have task identity available during inference, and models only need to solve the given task without inferring its task.

Finally, **class-incremental learning**, the third scenario, requires that the models must solve each task seen so far while at the same time inferring its task. The currently proposed method falls into this scenario. Furthermore, most real-world problems of incrementally learning new classes of objects also belong to this scenario.

Existing works to handle the continual learning problem are mainly divided into three categories:

- **Structure-based approach**: One reason for catastrophic forgetting to occur is that the parameters of a neural network are optimized for new tasks and no longer for previous ones. This suggests that not optimizing the entire network or expanding the internal model structure to deal with the new tasks while isolating old network parameters could attenuate catastrophic forgetting. PNN [8] pioneered this approach by adding new components to the network and freezing old task parameters during training. Context-dependent gating (XdG) [16] is a simple but popular approach that randomly assigns nodes to tasks. However, these approaches are limited to the *task-incremental learning* scenario by design, as task identity is required to select the correct task-specific components during training.
- **Regularization-based approach**: When task knowledge is only available during training time, training a different part of the network for each task can still happen, but then the whole network is used through inference. Standard methods in this approach estimate the importance of the network parameters for the previously learned tasks and penalize future changes accordingly. Elastic Weight Consolidation (EWC) [7] and its online counterpart (EWCo) [31] adopt the Fisher information matrix to estimate the importance of the network synapses. Synaptic intelligence (SI) [17] utilizes an accumulated gradient to quantify the significance of the network parameters.
- **Memory-based approach**: This strategy replays old, or augmented, samples stored in memory when learning a new task. Learning without forgetting (LWF) [6] uses a pseudo-data strategy where it labels the samples of the current tasks using the model trained on the previous tasks, resulting in training

that mixes hard-target (likely category according to previous tasks) with soft-target (predicted probabilities within all classes). Gradient episodic memory (GEM) [20] and Averaged GEM (A-GEM) [21] successfully boost continual learning performance by the usage of exact samples stored in memory to estimate the forgetting case and to constraint the parameter updates accordingly. Gradient-based Sample Selection (GSS) [26] focuses on optimizing the selection of samples to be replayed. Dark Experience Replay (DER/DER++) [27] and Function Distance Regularization (FDR) [28] use past samples and soft outputs to align past and current outputs. Hindsight Anchor Learning (HAL) [25] adds additional objectives into replaying, aiming to reduce forgetting of key learned data points.

Alternatively, methods can also take advantage of generative models for pseudo-rehearsal. For example, Deep Generative Replay (DGR) [18] utilizes a separated generative model that is sequentially trained on all tasks to generate samples from their data distribution. Additionally, knowledge distillation can be combined with DGR (DGR+distill) [19] to pair generated samples with soft target knowledge. Alternatively, methods can also take advantage of generative models for pseudo-rehearsal. For example, Deep Generative Replay (DGR) [18] utilizes a separated generative model that is sequentially trained on all tasks to generate samples from their data distribution. Additionally, knowledge distillation can be combined to DGR (DGR+distill) [19] to pair generated samples with soft target knowledge.

The proposed CFA is a memory-based approach that presents a novel way to perform continual learning using knowledge amalgamation, a derivation of knowledge distillation, and domain adaptation to merge several teacher models into a single student model.

2.1 Domain Adaptation

Transfer learning (TL) [2] is defined by the reuse of a model developed for a task to improve the learning of another task. Neural networks have been applied to TL because of their power in representing high-level features.

While there are many sub-topics of TL, we are deeply interested in domain adaptation (DA) [10]. While there are many approaches to measure and reduce the disparity between the distributions of these two domains, Maximum Mean Discrepancy (MMD) [11] and Kullback-Leibler divergence (KL) [12] are widely used in the literature. Our approach uses KL to approximate the representation of learning distributions between the teachers and the student, differing itself from the original knowledge amalgamation method [3], where the MMD approach is applied.

2.2 Knowledge Distillation

Knowledge distillation (KD) [9] is a method of transferring learning from one model to the other, usually by compression, where a larger teacher model supervises the training of a smaller student model. One of the benefits of KD is that it can handle heterogeneous structures, i.e., the teacher and the student do not need to share the same network structure. Instead, the teacher supervises the student training via its logits, also called the soft target. In other words, KD minimizes the distance between the student network output $\hat{z}$ and the logits z from a teacher network, generated from an arbitrary input sample:

$$\mathcal{L}_{KD} = ||\hat{z} - z||_2^2 \tag{1}$$

Although KD has become a field itself in the machine learning community, many approaches are still performed under a single teacher-student relationship, with a sharing task [9]. Contrary to these constraints, our method can process multiple and heterogeneous teachers, condensing their knowledge into a single student model covering all tasks.

2.3   Knowledge Amalgamation

Knowledge Amalgamation (KA) [3,4] aims to acquire a compact student model capable of handling the comprehensive joint objective of multiple teacher models, each specialized in their task. Our approach extends the concept of knowledge amalgamation in [3,23] to the continual learning environment.

3   Problem Formulation

In continual learning, within the class-incremental learning scenario, we experience a stream of data tuples (x_i, y_i) that satisfies $(x_i, y_i) \overset{iid}{\sim} P_{t_i}(X, Y)$, containing an input x_i and a target y_i organized into sequential tasks $t_i \in \mathcal{T} = 1, ...T$, where the total number of tasks T is unknown a priori. The goal is to learn a predictor $f : \mathcal{X} \times \mathcal{T} \rightarrow \mathcal{Y}$, which can be queried *at any time* to predict the target vector y associated to a test sample x, where $(x, y) \sim P_t$. Such test pair can belong to a task that we have observed in the past or the current task.

We define the knowledge amalgamation task as follows. Assume that we are given N teacher models $t_{i=1}^N$ trained a priori, each of which implements a specific task T. Let $\mathcal{D}_i$ denote the set of classes handled by model t_i. Without loss of generality, we assume $\mathcal{D}_i \neq \mathcal{D}_j, \forall i \neq j$. In other words, for any pair of models t_i and t_j, we assume they classify different tasks. The goal of knowledge amalgamation is to derive a compact single-head student model that can infer all tasks, in other words, to be able to simultaneously classify all the classes in $\mathcal{D} = \cup_{i=1}^N \mathcal{D}_i$. In other words, the knowledge amalgamation mechanism is done in the post-processing manner where all teacher models trained to a specific task are combined into a single model to perform comprehensive classification as per its teacher models. This approach provides flexibility over existing continual learning approaches because a teacher model can be independently built for a specific task. Their knowledge can later be amalgamated into a student model without loss of generalization power.

Algorithm 1: CFA

Input: Teacher models $\mathcal{T}_N$, Task Memory $\mathcal{M}$, Student model $\mathcal{S}$, number of epochs

Output: Amalgamated Model $\mathcal{S}$

1 **for** *epoch in epochs* **do**
2   M ← shuffle(M); # Optional
3   **for** *sample m in $\mathcal{M}$* **do**
4   **begin** Joint Representation Learning:
5              $\mathcal{L}_M = \mathcal{L}_R = 0$; # Loss initialization
6              $\hat{f}_S \leftarrow f_S =\leftarrow F_S$; # Student encoder
7   **for** T_i *in* $\mathcal{T}_N$ **do**
8                  $\hat{f}_{Ti} \leftarrow f_{Ti} \leftarrow F_{Ti}$; # Teacher encoder
9                  $F'_{Ti} \leftarrow f_{Ti} \leftarrow \hat{f}_{Ti}$; # Teacher decoder
10                 $\mathcal{L}_M=\mathcal{L}_M + H(\hat{f}_S, \hat{f}_{Ti}) - H(\hat{f}_S)$; # Eq. 3
11                 $\mathcal{L}_R = \mathcal{L}_R + ||F'_{Ti} - F_{Ti}||^2_2$; # Eq. 4

12   **begin** Soft Domain Adaptation:
13             $y_T \leftarrow \mathcal{T}_N(m)$; # Stacked teachers' soft output
14             $D_{\mathrm{KL}_{\mathrm{soft}}} = H(\hat{y}_S, y_T) - H(\hat{y}_S)$; # Eq. 5
15         $\mathcal{L} = \alpha D_{\mathrm{KL}_{\mathrm{soft}}} + (1 - \alpha)(\mathcal{L}_M + \mathcal{L}_R)$; # Eq. 6
16         $S_\theta = S_\theta - \lambda \nabla \mathcal{L}$; # Parameter learning

---

4 Proposed Method

In this section, we introduce the proposed CFA and its details. The knowledge amalgamation element is an extension of [3,4] and consists of two parts: a joint representational learning and a soft domain adaptation.

4.1 Joint Representation Learning

The joint representation learning scheme is depicted in Fig. 1. The features of the teachers and those to be learned from the students are first transformed into a common feature space, and then two loss terms are minimized. First, a feature ensemble loss $\mathcal{L}_M$ encourages the features of the student to approximate those of the teachers in the joint space. Then a reconstruction loss $\mathcal{L}_R$ ensures the transformed features can be mapped back to the original space with minimum possible errors.

Adaptation Layer. The adaptation layer aligns the output feature dimension of the teachers and students via a 1×1 convolution kernel [13] that generates a predefined length of output with different input sizes. Let F_S and F_{T_i} be respectively the original features of the student and teacher T_i, and f_S and f_{T_i} their respective aligned features. In our implementation, f_S and f_{T_i} have the same size of F_S and F_{T_i}.

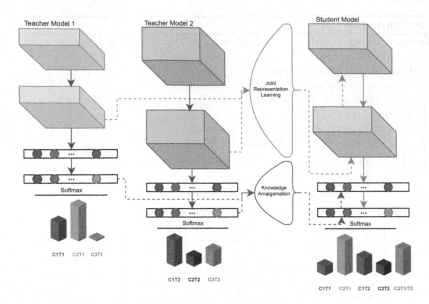

Fig. 1. A summarized workflow of the proposed CFA. It consists of two parts: A joint representation learning and knowledge amalgamation. In the joint representation learning, the features of the teachers (showing two here) and those to be learned by the students are first transformed into a joint space. Later on, knowledge amalgamation enforces a domain invariant feature space between the student and teachers via KL.

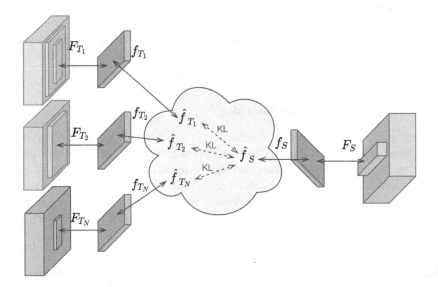

Fig. 2. An illustration of the shared extractor sub-network. A sub-network is shared between the teachers and the student during the joint representation learning procedure. This shared extractor aims to create a domain-invariant space via KL, which is then decompressed back into the student model.

Shared Extractor. Once the aligned features are derived, a naive approach would be to directly average the features of the teachers f_{T_i} as that of the student f_S. However, due to domain discrepancy of the training data and architectural differences of the teacher networks, the roughly aligned features may remain heterogeneous. To this end, the teachers and students share the parameters of a small learnable sub-network, illustrated in Fig. 2. This shared extractor consists of three consecutive residual blocks of 1 stride. It converts f_{T_i} and f_S into the common representation spaces $\hat{f}_{T_i}$ and $\hat{f}_S$. In our implementation, $\hat{f}_{T_i}$ and $\hat{f}_S$ is half the size of f_{T_i} and f_S.

Knowledge Amalgamation. To amalgamate knowledge from heterogeneous teachers, we enforce a domain invariant feature space between the student and teachers via the KL divergence, computed as follows:

$$D_{KL_i}(\hat{f}_S||\hat{f}_{T_i}) = H(\hat{f}_S, \hat{f}_{T_i}) - H(\hat{f}_S), \tag{2}$$

where $H(\hat{f}_S, \hat{f}_{T_i})$ is the cross entropy of $\hat{f}_{T_i}$ and $\hat{f}_S$ and $H(\hat{f}_S)$ is the entropy of $\hat{f}_S$.

We then aggregate all such pairwise KL losses between each teacher and the student, as shown in Fig. 2, and write the overall discrepancy $\mathcal{L}_M$ in the shared space as:

$$\mathcal{L}_M = \sum_{i=1}^{N} D_{KLi}, \tag{3}$$

To further enhance the joint representation learning, we add an autoencoder [22] reconstruction loss between the original teachers' feature space. Let F'_{Ti} denote the reconstructed feature of teacher T_i, the reconstruction loss $\mathcal{L}_R$ is defined as

$$\mathcal{L}_R = \sum_{i=1}^{N} ||F'_{Ti} - F_{Ti}||_2, \tag{4}$$

4.2 Soft Domain Adaptation

Apart from learning the teacher's features, the student is also expected to produce identical or similar inferences as the teachers do. We thus also take the teachers' predictions by feeding unlabelled input samples to them and then supervise the student's training.

We assume that all teacher models handle non-overlapping classes, then directly stack their score vectors and use them as the student's target. A similar strategy can be used for teachers with overlapping classes, where the logits of repeating classes can be summed or averaged, but we do not explore it here. Instead of directly applying a cross-entropy loss between the student output and the teachers' soft output, as most knowledge distillation solutions do, we also enforce a domain invariant space into the discriminative (fully-connected) layers of the student by applying the KL between the student output and the stacked teachers' soft output.

Let y_T denote the stacked teachers' soft output and $\hat{y}_S$ denote the corresponding student soft output, then KL is applied as:

$$D_{KL_{soft}}(\hat{y}_S \| y_T) = H(\hat{y}_S, y_T) - H(\hat{y}_S) \tag{5}$$

4.3   Final Loss

We incorporate the loss terms in Eqs. 3, 4 and 5 into our final loss function. The whole framework is trained end-to-end by optimizing the following objective:

$$\mathcal{L} = \alpha D_{KL_{soft}}(\hat{y}_S \| y_T) + (1 - \alpha)(\mathcal{L}_M + \mathcal{L}_R) \tag{6}$$

where $\alpha \in [0,1]$ is a hyper-parameter to balance the three terms of the loss function. By optimizing this loss function, the student network is trained from the amalgamation of its teachers without annotations.

5   Experiments

We evaluate CFA and its baselines under four benchmarks. Then, an ablation study gives further insight regarding CFA memory usage and internal procedures. Finally, we executed all CFA experiments using the same structure for the teachers and students; a ResNet18 backbone [24] as a feature extractor and two fully-connected layers ahead of it.

5.1   Replay Memory

To retrieve *proper*[1] logits from the teachers, CFA uses the replay memory strategy, where it records some previous samples to be replayed during the amalgamation process. The nearest-mean-of-exemplars strategy was used to build the replay memory, but any other sample selection strategy can be used.

Nearest-Mean-of-Exemplars Strategy. Consider $t_i(x)$ the logits of a teacher t_i on a specific task i. We compute the mean exemplar for each class in class y as $\mu_y = \frac{1}{\|\mathcal{D}_i\|} \sum_{x \in \mathcal{D}_i} t_i(x)$. A sample x is then added to the memory if there is free space or by descending sorting out the memory and x by their L_2 distance.

---

[1] Meaning, related to the original data distribution.

5.2   Baselines Setup

We set up two different configurations of CFA. $\text{CFA}_{\text{fixed}}$ uses the nearest-mean-of-exemplars replay memory strategy with a fixed memory footprint of 1000 samples. Meanwhile, CFA_{grow} uses the teachers' confidence replay memory strategy with a growing memory allowing 1000 samples per task. Hence, $\text{CFA}_{\text{fixed}}$ memory footprint maintains the same, independent of the number of classes learned so far[2], while CFA_{grow} memory footprint slowly grows are more classes are introduced.

All teachers and students have the same architecture, a pre-trained ResNet18 feature-extractor followed by two fully-connected layers, a $\mathbb{R}^{1000 \times 500}$ followed by a $\mathbb{R}^{500 \times \text{output}}$. CFA is optimized under Adam with learning rate $\lambda = 10^{-4}$, hyper-parameter $\alpha = 0.5$, and 100 training epochs.

The other baselines are based on the source-code release by [27]. Their configuration is also detailed in the supplemental document. The ones which are memory-based contains a memory budget of 1000 samples per task, making them similar to CFA_{grow}.

All methods have been evaluated using the same computation environment, a Windows machine with an Intel Core i9-9900K 5.0 GHz with 32 GB of main memory and an Nvidia GeForce 2080 Ti.

5.3   Metrics

The continual learning protocol is followed, where we observe three metrics:

$$\textbf{Average Accuracy: } \text{ACC} = \frac{1}{T} \sum_{i=1}^{T} R_{T,i} \tag{7}$$

$$\textbf{Backward Transfer: } \text{BWT} = \frac{1}{T-1} \sum_{i=1}^{T-1} R_{T,i} - R_{i,i} \tag{8}$$

$$\textbf{Forward Transfer: } \text{FWT} = \frac{1}{T-1} \sum_{i=2}^{T} R_{i-1,i} - \bar{b}_i \tag{9}$$

where $R \in \mathbf{R}^{T x T}$ is a test classification matrix, where $R_{i,j}$ represents the test accuracy in task t_j after completely learn t_i. The details are given by [20].

5.4   Benchmarks

SplitMNIST is a standard continual learning benchmark that adapts the entire MNIST problem [15] into five sequential tasks, with a total of 10 classes.

SplitCIFAR10 features the incremental class problem where the full CIFAR10 problem [14] is divided into five sequential tasks, with a total of 10 classes.

---

[2] Storage of the original teacher models parameters is still required, usually in secondary memory, as HDD or SSD.

Table 1. Numerical results over five execution runs.

Baseline	Metric (%)	SplitMNIST	SplitCIFAR10	SplitCIFAR100	SplitTinyImageNet
EWCo [7,31]	ACC	19.13 ± 0.02	18.57 ± 2.04	7.91 ± 0.79	7.39 ± 0.03
	BWT	-97.37 ± 0.26	-88.51 ± 5.18	-83.80 ± 1.56	-71.79 ± 0.62
	FWT	-13.36 ± 1.38	-10.09 ± 5.64	-1.02 ± 0.15	-0.44 ± 0.16
LWF [6]	ACC	19.20 ± 0.06	16.27 ± 4.55	9.13 ± 0.43	0.41 ± 0.20
	BWT	-96.52 ± 0.94	-89.24 ± 9.60	-83.34 ± 5.57	-50.33 ± 3.25
	FWT	-11.92 ± 2.01	-11.05 ± 1.88	0.16 ± 0.71	-0.25 ± 0.03
ER [29]	ACC	23.41 ± 0.60	69.07 ± 3.31	27.41 ± 2.94	12.33 ± 1.23
	BWT	-93.83 ± 0.92	-24.15 ± 14.17	-66.35 ± 1.22	-71.07 ± 2.35
	FWT	-8.83 ± 3.14	-11.84 ± 0.40	-0.98 ± 0.08	-0.5 ± 0.05
AGEM [21]	ACC	9.19 ± 0.65	13.49 ± 4.12	0.94 ± 0.32	1.55 ± 0.32
	BWT	-40.52 ± 46.02	-47.26 ± -47.26	2.99 ± 5.74	-14.93 ± 2.12
	FWT	-8.97 ± 3.54	-6.06 ± -6.06	0.74 ± 1.74	-0.55 ± 0.20
DER [27]	ACC	60.85 ± 2.87	72.99 ± 6.43	$\mathbf{32.60 \pm 9.77}$	23.62 ± 3.30
	BWT	-42.80 ± 3.81	-22.71 ± 5.00	-44.64 ± 8.54	-52.19 ± 3.66
	FWT	-12.25 ± 2.71	-9.36 ± 8.93	-0.93 ± 0.09	-0.46 ± 2.12
DER++ [27]	ACC	72.86 ± 0.95	$\mathbf{77.86 \pm 7.59}$	$\mathbf{38.82 \pm 8.28}$	23.94 ± 2.52
	BWT	-24.64 ± 1.21	-16.27 ± 5.71	-49.03 ± 7.60	-43.82 ± 5.95
	FWT	-12.59 ± 0.48	-6.26 ± 8.81	-0.91 ± 0.07	-0.26 ± 2.16
FDR [28]	ACC	78.08 ± 3.41	48.00 ± 5.36	$\mathbf{32.26 \pm 5.51}$	13.30 ± 1.64
	BWT	-21.73 ± 4.36	-86.58 ± 4.37	-62.87 ± 5.83	-67.08 ± 1.69
	FWT	-10.10 ± 1.13	-11.41 ± 2.95	-0.87 ± 7.29	-0.67 ± 0.22
GSS [26]	ACC	24.69 ± 0.80	43.96 ± 2.86	13.94 ± 0.30	9.60 ± 0.84
	BWT	-91.69 ± 1.04	-55.71 ± 2.57	-78.21 ± 0.32	-69.36 ± 0.28
	FWT	-10.31 ± 1.96	-10.56 ± 3.30	-0.39 ± 0.55	-0.53 ± 0.05
HAL [25]	ACC	$\mathbf{88.25 \pm 0.46}$	50.11 ± 1.18	11.00 ± 2.87	3.23 ± 0.11
	BWT	-13.61 ± 0.62	-47.01 ± 2.14	-44.74 ± 1.84	-32.68 ± 4.10
	FWT	-8.81 ± 3.28	-11.70 ± 1.69	-0.97 ± 0.26	-0.24 ± 0.31
$\mathbf{CFA_{fixed}}$ (Ours)	ACC	83.51 ± 1.35	74.96 ± 0.46	27.76 ± 2.28	23.44 ± 2.55
	BWT	-7.95 ± 1.53	-14.25 ± 1.76	-16.41 ± 1.49	-17.58 ± 1.89
	FWT	69.46 ± 9.41	54.28 ± 6.57	26.91 ± 3.17	20.49 ± 5.50
$\mathbf{CFA_{grow}}$ (Ours)	ACC	$\mathbf{89.25 \pm 3.66}$	$\mathbf{79.40 \pm 1.15}$	$\mathbf{38.74 \pm 3.26}$	$\mathbf{32.50 \pm 3.35}$
	BWT	69.77 ± 1.31	49.00 ± 2.78	11.49 ± 2.65	23.33 ± 3.45
	FWT	91.77 ± 5.18	65.67 ± 8.22	21.84 ± 4.26	32.58 ± 5.12

SplitCIFAR100 features the incremental class problem where the complete CIFAR100 problem [14] is divided into 10 sequential subsets, totalling 100 classes.

SplitTinyImageNet features the incremental class problem where 200 classes from the full ImageNet [30] are resized to 64×64 colored pixels and divided into 10 sequential tasks.

5.5   Numerical Results

We compare CFA_{fixed} and CFA_{grow} against one regularization-based approaches (EWCo), one knowledge distillation approaches (LWF[3]), and six memory-based approaches (AGEM, DER, DER++, FDR, GSS, HAL).

---

[3] A multi-class implementation was put forward to deal with class-incremental learning, as in [27].

Table 1 presents a metric summary between the chosen baselines and benchmarks. It demonstrates that CFA_{fixed} and CFA_{grow} are comparable, or even stronger, in comparison with the current state-of-the art methods, specially when we take in consideration that CFA_{fixed} presents a fixed memory footprint. Furthermore, both CFA_{fixed} and CFA_{grow} have great BWT and FWT metrics, with CFA_{grow} being the only model providing positive values to all metrics. This means that CFA signalizes some zero-shot learning [20], although not explicitly focused here.

Furthermore, the most outstanding achievement of CFA is being able to achieve good continual learning performance when applied to an offline environment while maintaining competitive results. In other words, all other methods are fully continual learning procedures, which would require an organization to shift its entire learning pipeline from scratch. In contrast, CFA leverages the power of individual teachers trained on the tasks, be it in an online or offline environment. This scenario is expected in current organization pipelines, saving costs in an inevitable paradigm shift from offline to online learning agent technologies.

5.6   Ablation Study

Memory Analysis. Table 2 put the strongest baselines face to face to compare how their accuracies change over different memory budgets. CFA_{fixed} maintains a competitive performance, even though it presents a fixed memory footprint. So, even though it has a performance similar to DER, DER++, and FDR, it benefits from using less memory and being applied to current offline learning pipelines.

Joint Representation Learning Analysis. As shown in Table 3, Joint Representation Learning (JTL) is the main adaptation driver, responsible for driving the student's latent space to represent different tasks. Furthermore, as we are using the same architecture for the teachers and students, the difference between $\alpha = 1.0$ and $\alpha = 0.5$ is not that significant here but immensely important when

Table 2. ACC(%) metrics over different budget memories.

Benchmark	Memory budget	CFA_{fixed}	CFA_{grow}	DER	DER++	FDR
Split CIFAR10	100	48.87 ± 5.26	**61.36 ± 2.02**	46.77 ± 3.12	51.91 ± 4.21	39.60 ± 4.54
	200	61.20 ± 4.35	**69.33 ± 1.54**	58.41 ± 3.23	64.92 ± 6.15	44.49 ± 4.31
	500	69.53 ± 3.30	**74.63 ± 1.20**	65.63 ± 5.95	72.45 ± 6.85	48.20 ± 5.30
	1000	74.96 ± 0.46	**79.40 ± 1.15**	72.99 ± 6.43	**77.86 ± 7.59**	41.91 ± 6.42
	2000	76.45 ± 1.90	**82.21 ± 2.52**	73.81 ± 5.12	77.44 ± 8.90	47.39 ± 7.01
Split CIFAR100	100	7.75 ± 1.20	**21.34 ± 2.30**	13.23 ± 0.00	**22.88 ± 4.90**	12.23 ± 4.50
	200	12.87 ± 2.12	**26.01 ± 2.53**	19.98 ± 0.00	23.78 ± 5.20	14.74 ± 2.24
	500	21.23 ± 2.23	**30.59 ± 3.54**	26.53 ± 5.23	**31.45 ± 6.43**	22.26 ± 4.21
	1000	27.76 ± 2.28	**38.74 ± 3.26**	32.60 ± 9.77	**38.82 ± 8.28**	32.26 ± 5.51
	2000	41.50 ± 3.45	**47.54 ± 3.18**	36.78 ± 9.88	43.45 ± 8.54	33.12 ± 5.40

Table 3. ACC(%) results with varying hyper-parameter α of the $\text{CFA}_{\text{fixed}}$ with 1000 of memory budget, controlling the influence of the Joint Representation Learning (JTL) and Soft Domain Adaptation (SDA) into its main loss.

Description	Split CIFAR10
$\alpha = 1.0 \mid \text{JTL}(\times)\text{SDA}(\checkmark)$	35.78 ± 10.78
$\alpha = 0.5 \mid \text{JTL}(\checkmark)\text{SDA}(\checkmark)$	74.96 ± 0.46
$\alpha = 0.0 \mid \text{JTL}(\checkmark)\text{SDA}(\times)$	69.68 ± 2.23

dealing with entire heterogeneous structures, as noted by [3]. When JTL is disabled, the model has difficulties learning high-feature representations only with the soft domain adaptation (SDA), resulting in a tremendous catastrophic forgetting.

6   Conclusion

This paper proposes CFA, an approach to handle catastrophic forgetting for the class-incremental environment with knowledge amalgamation. CFA can amalgamate the knowledge of multiple heterogeneous trained teacher models, each for a previous task, into a single-headed student model capable of handling all tasks altogether.

We compared CFA with a set of competitive baselines under the class-incremental learning scenario, yielding positive generalization with excellent average accuracy and knowledge transfer capabilities, backed by backward and forward knowledge transfer metrics. At the same time, CFA demonstrated some zero-shot learning aptitude and handled an enormous number of classes simultaneously.

CFA presents a novel approach towards continual learning using knowledge amalgamation, enabling easy integration to current learning pipelines, enabling the shift from offline to online learning with a performance similar to or superior to the best of the only-online existing methods. Our approach is perceived as a post-processing approach of continual learning, distinguishing itself from existing approaches. Our future work is directed to explore the continual learning problem in multi-stream environments.

Acknowledgement. This work is financially supported by National Research Foundation, Republic of Singapore under IAFPP in the AME domain (contract no.: A19C1A0018).

References

1. Gama, J.: Knowledge Discovery from Data Streams. CRC Press, Boca Raton (2010)

2. Pan, S.J., Yang, Q.: A survey on transfer learning. IEEE Trans. Knowl. Data Eng. **22**(10), 1345–1359 (2010)

3. Luo, S., Wang, X., Fang, G., Hu, Y., Tao, D., Song, M.: Knowledge amalgamation from heterogeneous networks by common feature learning. In: Kraus, S. (ed.) Proceedings of the Twenty-Eighth International Joint Conference on Artificial Intelligence, IJCAI 2019, Macao, China, 10–16 August 2019, pp. 3087–3093. ijcai.org (2019). https://doi.org/10.24963/ijcai.2019/428

4. Shen, C., Wang, X., Song, J., Sun, L., Song, M.: Amalgamating knowledge towards comprehensive classification. In: Proceedings of the AAAI Conference on Artificial Intelligence, vol. 33, no. 01, pp. 3068–3075 (2019). https://doi.org/10.1609/aaai.v33i01.33013068

5. Lee, S.W., Kim, J.H., Jun, J., Ha, J.W., Zhang, B.T.: Overcoming catastrophic forgetting by incremental moment matching. In: Proceedings of the 31st International Conference on Neural Information Processing Systems, NIPS 2017, pp. 4655–4665. Curran Associates Inc., Red Hook (2017)

6. Li, Z., Hoiem, D.: Learning without forgetting. IEEE Trans. Pattern Anal. Mach. Intell. **40**(12), 2935–2947 (2018). https://doi.org/10.1109/TPAMI.2017.2773081

7. Kirkpatrick, J., et al.: Overcoming catastrophic forgetting in neural networks. Proc. Natl. Acad. Sci. **114**(13), 3521–3526 (2017). https://doi.org/10.1073/pnas.1611835114. https://www.pnas.org/content/114/13/3521

8. Rusu, A.A., et al.: Progressive neural networks. arXiv abs/1606.04671 (2016)

9. Hinton, G., Dean, J., Vinyals, O.: Distilling the knowledge in a neural network, pp. 1–9 (2014)

10. Ben-David, S., Blitzer, J., Crammer, K., Kulesza, A., Pereira, F., Vaughan, J.W.: A theory of learning from different domains. Mach. Learn. **79**(1–2), 151–175 (2010)

11. Gretton, A., Borgwardt, K.M., Rasch, M.J., Schölkopf, B., Smola, A.: A kernel two-sample test. J. Mach. Learn. Res. **13**(25), 723–773 (2012). http://jmlr.org/papers/v13/gretton12a.html

12. Joyce, J.M.: Kullback-Leibler Divergence. In: Lovric, M. (ed.) International Encyclopedia of Statistical Science, pp. 720–722. Springer, Heidelberg (2011). https://doi.org/10.1007/978-3-642-04898-2_327

13. Szegedy, C., et al.: Going deeper with convolutions. In: 2015 IEEE Conference on Computer Vision and Pattern Recognition (CVPR), pp. 1–9 (2015). https://doi.org/10.1109/CVPR.2015.7298594

14. Krizhevsky, A., Hinton, G.: Learning multiple layers of features from tiny images. Master's thesis, Department of Computer Science, University of Toronto (2009)

15. LeCun, Y., Cortes, C.: MNIST handwritten digit database. ATT Labs (2010)

16. Masse, N.Y., Grant, G.D., Freedman, D.J.: Alleviating catastrophic forgetting using context-dependent gating and synaptic stabilization. Proc. Natl. Acad. Sci. **115**(44), E10467–E10475 (2018). https://doi.org/10.1073/pnas.1803839115. https://www.pnas.org/content/115/44/E10467

17. Zenke, F., Poole, B., Ganguli, S.: Continual learning through synaptic intelligence. In: Precup, D., Teh, Y.W. (eds.) Proceedings of the 34th International Conference on Machine Learning. Proceedings of Machine Learning Research, vol. 70, pp. 3987–3995. PMLR (2017). https://proceedings.mlr.press/v70/zenke17a.html

18. Shin, H., Lee, J.K., Kim, J., Kim, J.: Continual learning with deep generative replay. In: Guyon, I., et al. (eds.) Advances in Neural Information Processing Systems, vol. 30. Curran Associates, Inc. (2017). https://proceedings.neurips.cc/paper/2017/file/0efbe98067c6c73dba1250d2beaa81f9-Paper.pdf

19. van de Ven, G.M., Tolias, A.S.: Three scenarios for continual learning. arXiv abs/1904.07734 (2019)

20. Lopez-Paz, D., Ranzato, M.A.: Gradient episodic memory for continual learning. In: Guyon, I., et al. (eds.) Advances in Neural Information Processing Systems, vol. 30. Curran Associates, Inc. (2017). https://proceedings.neurips.cc/paper/2017/file/f87522788a2be2d171666752f97ddebb-Paper.pdf
21. Chaudhry, A., Ranzato, M., Rohrbach, M., Elhoseiny, M.: Efficient lifelong learning with a-GEM. arXiv abs/1812.00420 (2019)
22. Rumelhart, D.E., McClelland, J.L.: Learning Internal Representations by Error Propagation, pp. 318–362 (1987)
23. Shen, C., Wang, X., Song, J., Sun, L., Song, M.: Amalgamating knowledge towards comprehensive classification. arXiv abs/1811.02796 (2019)
24. He, K., Zhang, X., Ren, S., Sun, J.: Deep residual learning for image recognition. In: 2016 IEEE Conference on Computer Vision and Pattern Recognition (CVPR), pp. 770–778 (2016). https://doi.org/10.1109/CVPR.2016.90
25. Chaudhry, A., Gordo, A., Dokania, P.K., Torr, P., Lopez-Paz, D.: Using hindsight to anchor past knowledge in continual learning (2020). https://doi.org/10.48550/ARXIV.2002.08165. https://arxiv.org/abs/2002.08165
26. Aljundi, R., Lin, M., Goujaud, B., Bengio, Y.: Gradient Based Sample Selection for Online Continual Learning. Curran Associates Inc., Red Hook (2019)
27. Buzzega, P., Boschini, M., Porrello, A., Abati, D., Calderara, S.: Dark experience for general continual learning: a strong, simple baseline. In: Larochelle, H., Ranzato, M., Hadsell, R., Balcan, M.F., Lin, H. (eds.) Advances in Neural Information Processing Systems, vol. 33, pp. 15920–15930. Curran Associates, Inc. (2020)
28. Benjamin, A.S., Rolnick, D., Kording, K.: Measuring and regularizing networks in function space (2018). https://doi.org/10.48550/ARXIV.1805.08289. https://arxiv.org/abs/1805.08289
29. Ratcliff, R.: Connectionist models of recognition memory: constraints imposed by learning and forgetting functions. Psychol. Rev. **97**(2), 285–308 (1990)
30. Tavanaei, A.: Embedded encoder-decoder in convolutional networks towards explainable AI. arXiv abs/2007.06712 (2020)
31. Schwarz, J., et al.: Progress & compress: a scalable framework for continual learning. In: Dy, J., Krause, A. (eds.) Proceedings of the 35th International Conference on Machine Learning. Proceedings of Machine Learning Research, vol. 80, pp. 4528–4537. PMLR (2018). https://proceedings.mlr.press/v80/schwarz18a.html

Trigger Detection for the sPHENIX Experiment via Bipartite Graph Networks with Set Transformer

Tingting Xuan[1], Giorgian Borca-Tasciuc[1], Yimin Zhu[1], Yu Sun[2], Cameron Dean[3], Zhaozhong Shi[3], and Dantong Yu[4(✉)]

[1] Stony Brook University, Stony Brook, NY, USA
{tingting.xuan,giorgian.borca-tasciuc,yimin.zhu}@stonybrook.edu
[2] Sunrise Technology Inc., Stony Brook, NY, USA
yu.sun@sunriseaitech.com
[3] Los Alamos National Laboratory, Los Alamos, NM, USA
{ctdean,zhaozhongshi}@lanl.gov
[4] New Jersey Institute of Technology, Newark, NJ, USA
dantong.yu@njit.edu

Abstract. Trigger (interesting events) detection is crucial to high-energy and nuclear physics experiments because it improves data acquisition efficiency. It also plays a vital role in facilitating the downstream offline data analysis process. The sPHENIX detector, located at the Relativistic Heavy Ion Collider in Brookhaven National Laboratory, is one of the largest nuclear physics experiments on a world scale and is optimized to detect physics processes involving charm and beauty quarks. These particles are produced in collisions involving two proton beams, two gold nuclei beams, or a combination of the two and give critical insights into the formation of the early universe. This paper presents a model architecture for trigger detection with geometric information from two fast silicon detectors. Transverse momentum is introduced as an intermediate feature from physics heuristics. We also prove its importance through our training experiments. Each event consists of tracks and can be viewed as a graph. A bipartite graph neural network is integrated with the attention mechanism to design a binary classification model. Compared with the state-of-the-art algorithm for trigger detection, our model is parsimonious and increases the accuracy and the AUC score by more than 15%.

Keywords: Graph neural networks · Event detection · Physics-aware machine learning

1 Introduction

sPHENIX is a high-energy nuclear physics experiment under construction at Brookhaven National Laboratory and situated on the Relativistic Heavy Ion Collider (RHIC). The goal of sPHENIX is to probe the initial moments after

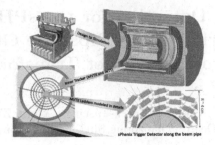

Fig. 1. sPHENIX track detector and trigger design.

the Big Bang by studying quark-gluon plasma, a state of matter where atomic nuclei melt under extremely hot and dense conditions.

The sPHENIX detector shown in Fig. 1 consists of several subdetectors for collecting a wide range of patterns of physics events. The two subsystems closest to the collision point are of most interest to this study. They are the MAPS vertex detector (MVTX) and intermediate tracker (INTT). The MVTX detector provides vertexing and tracking with high precision, while the INTT provides tracking with a resolution capable of determining the individual beam crossings at RHIC. sPHENIX also uses an outer calorimetry system to measure the energy of particles in the detector at a low speed of 15 kHz, limited by the readout electronics. As the collision rate for protons at RHIC is approximately 2 MHz, the calorimeter system does not work with the online setting and is not considered in this paper.

Heavy Flavor decays that we attempt to detect exhibit several prominent characteristics with a wide value range that overlaps with background events. The complex pattern and non-trivial decision boundary between heavy flavor delay and background events provide an ideal playground to apply ML techniques. The particles of interest decay on short time scales, typically a few nanoseconds or less. These particles may travel several millimeters at the speed of light before they decay. Physicists often extrapolate Particle tracks in the detector space to determine whether the tracks coincide with the beam collision point.

A tremendous volume of data is produced during collider experiments, but only a tiny fraction of the data needs to be selected due to the rarity of the targeted events. For example, an event that includes a charm quark typically occurs once for every 50 background events [1,21] while a beauty quark occurs once for roughly 1000 background events [2]. Collider experiments require a trigger system to reduce data in real-time and resolve the big data problem that is impractical for any data processing facility [11]. The triggers make decisions to keep or discard an event in situ and significantly reduce the data volume that needs to be retained for physics experiments. Our paper brings forth significant impacts to physics experiments by shifting many offline analysis tasks into an online setting and significantly shortening the latency between experiment and scientific discovery.

Before experiments start, we rely on simulation data to train and test our model. The simulated data is expected to match real data at a high level and has been extensively validated from previous physics experiments.

Contributions

- In this paper, we design a highly effective graph pooling/distributing mechanism for graph-level classification and prediction.
- Our model does not demand any pre-existing graph topology. Instead, it employs a set transformer to combine local and global features into each network node, enables effective knowledge exchange among network nodes, and supports local and global-scale graph learning.
- We incorporate well-known physics analysis into the multi-task neural network architecture and explicitly inference a crucial physics property in nuclear physics experiments, i.e., transverse momentum. This physics-driven learning improves our model accuracy and AUC score (Area Under Curve) by about 15%.

2   Related Work

The domain of experimental physics has a history of utilizing Machine Learning (ML) in physics-related tasks like particle identification, event selection, and reconstruction. Neural Networks have been used in these experiments [3,16,35] at first and were replaced by Boosted Decision Trees [40]. Recently Neural Networks and their modern implementation of Deep Learning (DL) regain their popularity in physics because of their superior ability to automatically learn effective features for many tasks and their outstanding performance [26].

Convolutional Neural Network (CNN) [25] is the most commonly used architecture with DL in the field of particle physics. It has proven its success in many tasks such as jet identification [9,20,23], particle identification [12,19,24], energy regression [12,15,36], and fast simulation [12,31,33]. The majority of CNN architecture models take a fixed-grid input tensor to represent the detector of an array of sensors. More efforts attempt to explore other alternative DL architectures for a better representation of physics: Recurrent Neural Network [14], recursive networks [28], Graph Neural Network (GNN) [37], and DeepSets [24] for the particle jet identification tasks. GNN is also applied to the classification tasks in neutrino physics [13]. Garnet [36] is distance-weighted graph network that can efficiently detect irregular pattern of sparse data. Transformer model architecture [43] shows its success in many applications [10,17,39]. Taking the data cloud as a set, Set Transformer [27] utilizes attention mechanism to learn interactions between elements with EdgeConv that is permutation equivariant and fits the set property.

Our previous works using ML to solve the same trigger detection problem appear in [44,45]. Beyond these efforts, no other studies are addressing the same issue. Jet taggers address a similar situation of tagging tracks with particles. Nevertheless, Jet taggers rely on the different physics properties collected from

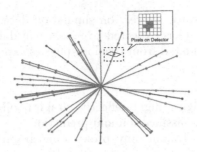

Fig. 2. An example of trigger event. In trigger events, particles decay into two or more different particles soon after the collision. Lines represent the trajectories of the particles. Green ones come from the center of the collision. Blue, red, and orange tracks start from the position where the decay happens. (Color figure online)

calorimeter detectors and they are interested in different target particles. Jet tagger belongs to offline analysis, and the readout speed from the calorimeter is much slower than the event rate, invalidating their applications in the online use case. Several ML methods for top tagging are discussed in [22]. We gain insights from the model design of existing tagging algorithms, especially those supporting the particle cloud [36–38]. Our final goal is to apply the algorithm to an online data-driven trigger system in an end-to-end fashion [29,44].

3 Problem Definition

Figure 2 schematically illustrates the trigger problem. The input of the physics event consists of a set of tracks and is represented as a matrix $X \in \mathcal{R}^{n \times d}$, where n is the total number of tracks, and d is the dimensionality of the features of each track. Tracks are treated as vertices in a graph. The goal is to determine whether this graph corresponds to a trigger event and triggers the data acquisition system to retain the readouts from the entire detector for future studies.

The commonly adopted GNN-based trigger prediction model attempts to perform end-to-end prediction from the raw hits that are the coordinates of the detector pixels where a particle traverses the detector. This domain-agnostic approach does not offer any insight into why an event becomes a trigger and results in inferior performance. Domain scientists require physics-aware reasoning and interpretation by explicitly incorporating physics models and properties. Since collecting advanced physics properties by detector requires sophisticated detectors and incurs considerable latency compared to the fast geometric detector, it is not feasible to use advanced physics properties in an online data acquisition environment. To incorporate the advanced physics properties, we must regress them onto the available geometric data. Our ultimate goal is to predict triggers while retaining the interpretability and rationality of intermediate tasks by replicating offline physics analysis workflows.

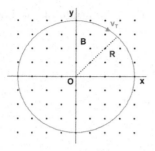 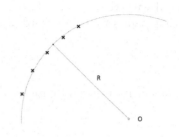

Fig. 3. The left figure shows that a positively charged particle will undergo a circular motion clockwise with a radius R in the uniform magnetic field B along the $+z$ direction. The right figure shows an example track with a fitted circle. The black cross markers represent five hits on the example track; the red dashed curve approximate a particle track and is the fitted circle with a radius R. (Color figure online)

4   Transverse Momentum Estimation

Transverse momentum, as part of the kinematics of particles, is crucial for studying particle dynamics in high-energy and nuclear physics experiments. The transverse momentum of a charged particle can be estimated with the knowledge of the magnetic field where it travels and the radius of its curved path in the magnetic field. The magnetic field in the sPhenix detector is fixed and can be precomputed; once we redefine the triggering task on the graph of tracks instead of hits, this transverse momentum is accessible for individual tracks and correlated with the radius of the particle's curved path (tracks) calculated using the geometry information of the hits on the track.

4.1   Physics Relation Between Transverse Momentum and Track Curvature

In particle collider experiments, we typically choose the cylindrical coordinates to describe the particle momentum $\vec{p} = (p_T, p_z)$ for simplicity. It is conventional to choose the beam direction as the z-axis. Here, p_T is called the transverse momentum, an analog of track radius R in the cylindrical coordinates in the transverse direction. p_z is called the longitudinal momentum, an analog of z in the cylindrical coordinates in the transverse direction.

The sPHENIX experiment uses a solenoid magnet with the field aligning with the beam direction in the z-axis. The left figure in Fig. 3 shows a positively charged particle moving clockwise under a magnetic field.

For a charged particle with charge q traveling across the magnetic field, a Lorentz force acts on the charged particle. In our case, B, defined as the magnetic field strength of the sPHENIX solenoidal magnet, is along the z direction. The Lorentz force maximizes in the transverse direction. The velocity vector $\vec{v}$ is decomposed like the momentum $\vec{v} = (v_T, v_z)$. When the charge of a particle is e, combined with the equation for circular motion, the magnitude of the Lorentz force that points radially inward, is given by:

$$F = m\frac{v_T^2}{R} = ev_T B.$$

The momentum is given by $p_T = mv_T$. Hence, we get

$$p_T = eBR. \tag{1}$$

All particles tracked in the detector have a unit charge of one electron volt (eV) and so, if we change units so that the momentum is measured in GeV/c where c is the speed of light, then Eq. 1 becomes

$$p_T = 0.3BR, \tag{2}$$

where the magnetic field is measured in Tesla (T). The detailed derivation is beyond this paper. Equation 2 shows that the crucial physics property of particle momentum is proportional to the track radius and guides us to integrate this physics insight into the ML-based detection.

4.2 Track Curvature Fitting

A track consists of a list of hits tracking particles traversing detector layers. The transverse momentum is highly correlated with the detected curvature of particle tracks. We fit a circle to those hits to approximate the momentum and calculate its radius. Here, regarding the direction of the magnetic field, we only need to consider the x-axis and y-axis. The image on the right in Fig. 3 shows an example track with its fitted circle.

A circle is represented by the following formula: $x^2 + y^2 + \beta_1 x + \beta_2 y + \beta_3 = 0$. Given a track of k_T hits $T = \{(x_1, y_1), (x_2, y_2), ..., (x_{k_T}, y_{k_T})\}$, we define a linear system that consists of k_T equations for these hits and attempt to derive the circle's coefficients $\beta = [\beta_1, \beta_2, \beta_3]^T$. To get the best circle approximation, we use the least-squares (LS) optimization to solve the linear regression equation and extract the β coefficients:

$$\beta = (A^T A)^{-1} A^T B.$$

Here $A = \begin{bmatrix} x_1 & y_1 & 1 \\ x_2 & y_2 & 1 \\ ... & & \\ x_{k_T} & y_{k_T} & 1 \end{bmatrix}$, $B = [-x_1^2 - y_1^2, -x_2^2 - y_2^2, ..., -x_{k_T}^2 - y_{k_T}^2]^T$. With the optimized coefficients for the fitted circle, the circle radius is as follows:

$$R = \frac{1}{2}\sqrt{\beta_1^2 + \beta_2^2 - 4\beta_3}. \tag{3}$$

4.3   Momentum Estimation

The momentum can be calculated by Eq. 2 using the estimated radius. However, this has the drawback that at least three hits are required for the track to estimate the radius. We propose the second method based on a feed-forward (FF) neural network to predict the transverse momentum of the given track and its LS-estimated radius R. We compare the estimation accuracy in Sect. 6.2 and evaluate their effect on trigger detection in Sect. 6.3.

5   Bipartite Graph Networks with Set Transformer for Trigger Detection

To resolve the trigger detection problem, we attempt to label events based on their entire tracks. The previous work [45] builds an affinity matrix of tracks and forms a graph where each node represents a track. Given a track graph, our algorithm first applies GNN to learn local embeddings and uses various pooling methods to aggregate and label the event. In this paper, we discard the common practice of building the fine-scale affinity matrix (graph) among hits and directly apply physical analysis to guide our neural network architecture. Determining whether an event is a trigger event involves three types of interactions:

1. Local track-to-track interactions, such as determining whether two tracks share the same origin vertex.
2. Track-to-global interactions, such as determining the collision vertex of the event and potential secondary vertices of decaying.
3. Global-to-track interactions, such as comparing each track's origin with the collision vertex of the event.

To incorporate various interactions, our neural network model uses set attention mechanisms to facilitate local track-to-track information flow, accumulate track information into aggregators to facilitate local track-to-global information flow, and updating the track embeddings based on the aggregators to facilitate global-to-local information flow. Thereby, the bipartite GNN architecture performs the local track-to-global and global-to-local information flow.

Our model consists of several Set Encoder with Bipartite Aggregator (SEBA) Blocks. The SEBA Blocks update our track embeddings and exchange information between tracks. SEBA Blocks contain a Set Attention Block, a Bipartite Aggregation module, and a Feed-Forward (FF) network for transformation, as shown in Fig. 5. After several SEBA Blocks, we use some aggregation functions as the readout functions to obtain the global representation for the whole set and feed the representation into a FF network to get the final output. Figure 4 shows the entire architecture of our model.

5.1   Set Attention Blocks

We use the Set Attention Blocks designed by Lee in [27]. The module applies a self-attention mechanism to every element in the input set to enable the model

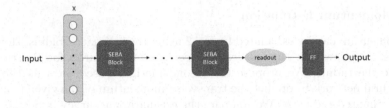

Fig. 4. Bipartite graph networks with set transformer model architectures.

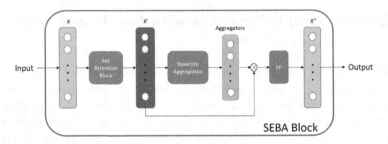

Fig. 5. Set encoder with bipartite aggregator (SEBA) blocks.

to encode pairwise- or higher-order interactions between the elements in the set. From the physics perspective, physicists often need to analyze the interactions between tracks and which tracks are from the same origin points. The self-attention mechanism neatly follows this physics practice.

Set Attention Blocks use Multi-head attention that Vaswani introduced originally in [43]. It packs a set of queries into a matrix Q. The keys and values are also stacked into matrices K and V.

For a single attention function, We compute the matrix of outputs as:

$$\text{Attention}(Q, K, V) = \text{softmax}(\frac{QK^T}{\sqrt{d_k}})V$$

Multi-head attention projects each of the Q, K, V matrices onto h different matrices separately, and applies an attention function to each of these h projections. The multiple heads allow the model to jointly attend to information from different representations subspaces at different positions.

$$\text{Multihead}(Q, K, V) = \text{Concat}(O_1, ..., O_h)W^O,$$

where $O_i = \text{Attention}(QW_i^Q, KW_i^K, VW_i^V)$ and W_i^Q, W_i^K, W_i^V, W^O are learnable parameters.

Given the matrix $X \in \mathcal{R}^{n \times d}$ which represents a set of d-dimensional vectors. Set Attention Blocks applies a multi-head attention function with $Q = K = V = X$ and serves as a mapping function using the following equation:

$$SAB(X) = \text{LayerNorm}(H + rFF(H)),$$

where $H = \text{LayerNorm}(X + \text{Multihead}(X, X, X))$, rFF is a row-wise FF layer applied to the input matrix and LayerNorm is the layer normalization [8].

5.2 Bipartite Aggregation

The idea of bipartite aggregation comes from GarNet architecture in [36]. Gar-Net is a distance-weighted graph network that partitions nodes into two groups: regular nodes of input elements and k aggregators, with both groups sharing the same node space. The original GarNet uses FF networks to measure the 'distance' between the input elements and these aggregators and aggregates information from elements to aggregators using a distance-weighted potential function. Instead of using the potential function and physical positions to interpret the relationship between track nodes and aggregators, we use the neural network to learn the affinity scores dynamically among network nodes.

The trigger decision is a graph-level prediction and requires local and global pooling. Experiments in [30] show that sophisticated pooling algorithms have no significant advantages over simple global pooling. The problem arises from the one-way rigid pooling mechanism that aggregates limited, sometimes biased information from the lower layer to the higher layer while ignoring the reverse information flow for distributing the aggregator information back to the low-level graph nodes for adjustment and adaptation. We propose an iterative message passing between two sides of networks, i.e., two-way gathering (from track nodes to aggregators)/scattering (from aggregators to track nodes) operations to resolve the missing link and inflexibility caused by the standard pooling algorithm.

Figure 6 shows a complete bipartite graph that consists of two type of nodes: track nodes $X = \{\vec{x}_1, ..., \vec{x}_n\}$ and aggregators $A = \{\vec{a}_1, ..., \vec{a}_k\}$. Our track graph uses aggregators to gather/disseminate information, where each track node is connected to each aggregator. The aggregators in our problem setting are closely related to two physics notions: the primary vertex where a physics event of collision happens and the secondary vertices where particles decay.

Each aggregator gathers information from all elements via the edges. The embedding of the edge between the ith element and the jth aggregator is

$$\vec{e}_{ij} = s_{ij}\vec{x}_i, \text{where } S = \sigma(\text{FF}_{\text{agg}}(X))$$

In the above equation, $E = \{e_{ij}\} \in \mathcal{R}^{n \times k \times d}$ represents the message from track nodes to aggregators, and $S = \{s_{ij}\} \in \mathcal{R}_+^{n \times k}$ is the score matrix between track nodes and aggregations nodes. FF_{agg} is a feed-forward neural network that assigns k weights to each element in X. The FF_{agg} acts as a gate function to the k aggregators, with each output weigh s_{ij} determining node i's contribution to the j-th aggregators, and σ is an activation function. After all information flows through the neural network gates in FF_{agg}, each aggregator performs a readout

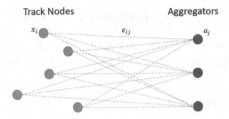

Fig. 6. Bipartite graph formed by elements and aggregators.

function on its gathered node information from tracks. We define the aggregator function as follows:

$$A = \text{readout}(E), \text{where } A \in \mathcal{R}^{k \times 2d}.$$

In our model, we concatenate the result of the mean and maximum readout functions and assign them to each aggregator.

Given the fixed number of aggregators, once we aggregate the information from elements to aggregators, we treat the k aggregators as a global graph embedding vector $g \in R^{k \times 2d}$. Then we concatenate this global vector with each track node vector to implement the by-pass connection in Fig. 5 and then apply another feed-forward network to update each node vector. This by-pass design allows us to scale the network stack into deep layers for complex event detections.

$$x_i' = \text{FF}_{\text{Node}}(\text{Concat}(x_i, g))$$

The bipartite aggregators first extract the local information to a global vector. Then the global information is fed back to each element by concatenation. It generates an information cycle from local to global, then back to local. Considering also the Set Attention Block that allows the pairwise information exchange between elements, by repeating our SEBA block several times, the intra-set relationship is well explored by our model.

6   Experiment Results

6.1   Dataset and Experiment Settings

Our experiment utilizes a simulation dataset[1]. We simulate the physics process with the PYTHIA8.3 package [41] and the GEANT4 simulation toolkit [5,7]. PYTHIA8 is a software package simulating QCD processes in small collisions systems and has been widely validated at many colliders [6,32,42]. GEANT4 is a package that simulates the passage of particles and radiation through matter. For this project, only the three-layer MVTX and two-layer INTT detectors are

---

[1] Some example data files can be found at https://github.com/sPHENIX-Collaborati on/HFMLTrigger.

Table 1. Performance of Momentum Regression Models. MLP-n indicates an MLP with 4 layers and hidden dimensions of size n. Denote Pearson's R as P's R and Spearman's R as S's R.

Method	All tracks			Tracks with at least 3 hits		
	R^2	P's R	S's R	R^2	P's R	S's R
LS	-116.18	0.0997	0.9376	-117.40	0.0999	0.9618
MLP-8	0.9071	0.9524	0.9534	0.9071	0.9524	0.9534
MLP-256	0.9100	0.9540	0.9564	0.9100	0.9540	0.9564

simulated because they are the only detectors capable of operation fast enough to achieve the goal of extremely high speed online data analysis and decision making. It has been extensively validated from previous physics experiments that a high-level agreement exists between the simulation data and real data [4]. Our model will be used to filter data in the real experiment that will start taking data in February 2023. The simulation data supporting the findings of this study are available from the corresponding author upon reasonable request[2].

The input vector for each track consists of coordinates of five hits in each detector layer, the length of each track segment, the angle between segments sequentially, and the total length of the track. The number of tracks per event varies from several to dozens. The coordinates of the geometric center of all the hits in the graph are calculated as complementary features to ease the downstream learning task. We also use the LS-estimated radius as another feature. All of the experiments, unless otherwise noted, use $1,000,000$ training samples, $400,000$ validation samples, and $400,000$ test samples. All models use the same pre-split training, validation, and test data sets, to ensure no information leakage and fair comparison. We adopt the Adam optimizer with a decayed learning rate from $1e-4$ to $1e-5$ in 50 epochs for all the training experiments. Experiments are run on various GPU architectures, including NVIDIA Titan RTX, A5000, and A6000. All baselines and our model are implemented using PyTorch [34] and PyTorch Geometric [18]. The code is publicly available on GitHub[3].

6.2   Transverse Momentum Estimation

We compare two methods for estimating the transverse momentum. The transverse momentum p_T is linearly proportional to the radius of a track R, as shown in Eq. 2. Here, we use a constant magnetic field with $B = 1.4\,\mathrm{T}$ in our dataset settings. We estimate the radius R using the LS fitting described in Sect. 4. If there are not enough hits to estimate the radius using LS fitting, we set the estimated radius to be zero.

---

[2] Please contact yu.sun@sunriseaitech.com for data access.
[3] https://github.com/Sunrise-AI-Tech/ECML2022-TriggerDetectionForTheSPHENI XExperimentViaBipartiteGraphNetworksWithSetTransformer.

Table 2. Comparison to Baseline Models with Estimated Radius.

Model	With LS-radius			Without radius		
	#Parameters	Accuracy	AUC	#Parameters	Accuracy	AUC
Set Transformer	300,802	84.17%	90.61%	300,418	69.80%	76.25%
GarNet	284,210	90.14%	96.56%	284,066	75.06%	82.03%
PN+SAGPool	780,934	86.25%	92.91%	780,678	69.22%	77.18%
BGN-ST	355,042	**92.18%**	**97.68%**	354,786	**76.45%**	**83.61%**

We make several observations on the p_T estimation results in Table 1. The LS estimated p_T achieves a high degree of correlation with the true p_T as measured by Spearman's R. However, the low value of Pearson's R and the highly negative coefficient value of determination indicate a poor linear correlation between the two. We hypothesize that this is due to outliers because some tracks occasionally produce an impossibly large estimated p_T. Using an MLP on the track to refine the LS p_T results seems to be highly effective, with all three correlation coefficients indicating that the models are highly predictive of the true p_T even with a small neural network. Noticeably, however, for tracks with at least three hits, the LS method outperforms the MLP method for Spearman's R. This might explain similar performance on triggering when using the LS-estimated p_T and the MLP-estimated p_T shown in Table 3.

6.3   Trigger Detection

Baselines. We compare our model with the ParticleNet(PN)+SAGPool method proposed in [45]. We also use Set Transformer and GarNet as our baselines because they are also well-suited to the problem of classifying a set of tracks. For Set Transformer, we use hidden dimension of 128 and four attention heads. For GarNet, we set hidden dimension of 64 and sixteen aggregators. For BGN-ST, we also use hidden dimension of 64 and sixteen aggregators. We use two-layer neural network architecture for all three models. The PN+SAGPool model has two stages. The first stage uses PN to generate an affinity matrix and track embeddings. Three edge-convolutional layers are used, with the hidden dimensions of 64, 128, and 64, respectively. All three edge-convolutional layers use 15 nearest neighbors when updating the node embeddings. The second stage includes the Sagpool layer aggregating the embeddings to perform trigger prediction. Sagpool uses three hierarchical pooling layers with a pooling ratio of 0.75.

Table 2 compares the performance between BGN-ST and the baseline models. From Table 2, we observe that BGN-ST outperforms all other methods by a significant margin. It usually takes no more than one day to train the Set Transformer, GarNet and BGN-ST on a single GPU card. The baseline PN+SAGPool has more parameters than ours. It takes two to three days to train the baseline model PN+SAGPool.

Table 3. Comparison of BGN-ST with LS-Estimated Radius and MLP-Refined Radius. A three-layer model with sixteen aggregators is used in the experiment.

Hidden dim	LS		MLP	
	Accuracy	AUC	Accuracy	AUC
32	91.52%	97.33%	91.48%	97.31%
64	92.18%	97.68%	92.23%	97.73%
128	**92.44%**	**97.82%**	**92.49%**	**97.86%**

Table 4. Hyperparameter grid search for BGN-ST

Hyperparameter	Range
Hidden dim	32, 64, 128, 256
#Aggregators	8, 16, 32
#Layers	2, 3
Activations	ReLU, Tanh, Potential, Softmax

Effect of Transverse Momentum Estimation. Table 2 shows the performance comparison between different models with or without radius. From the table, we observe both accuracy and AUC jump by 15% when the radius is added to all these four models under the same model setting.

Effect of Refining Transverse Momentum Estimate with an MLP. From Table 3, it is clear that further refining the momentum with an MLP trained to predict the momentum from the track and the LS-estimated radius does not yield any tangible improvement in the model performance. This applies to both the smaller and larger models.

Ablation Study of Hyperparameters. We perform a grid search on hyperparameters in Table 4 to find the best setting for trigger detection with BGN-ST.

We compare activation functions for aggregators in Table 4 in a two-layer BGN-ST with 64 hidden dimensions and 16 aggregators. The table shows that Softmax has the highest accuracy and AUC score among all choices.

We also undertook some other ablation studies. Figure 7 shows the accuracy comparison for different hidden dimensions, the number of aggregators, and layers. A larger model tends to perform better, but the number of parameters also increases exponentially. Our best performance is using a three-layer model with 256 hidden dimensions and 32 aggregators. The best accuracy for the test dataset is 92.52% (Fig. 7), and the best AUC score is 97.86% (Table 3).

Table 5. Ablation study of activations

Activation	Accuracy	AUC
ReLU	90.74%	96.87%
Tanh	90.19%	96.58%
Potential	90.41%	96.75%
Softmax	**92.18%**	**97.68%**

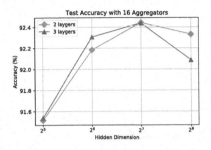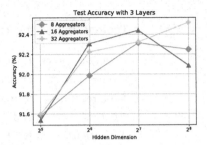

Fig. 7. Accuracy performance in respect to hidden dimension for two/three-layer models and different number of aggregators.

7   Conclusions

This paper details a novel Bipartite GNN architecture with a set transformer that uses the set attention mechanism to enhance the tracking with event features and ease the modeling of particle interactions in physics. Our model architecture benefits the pairwise interactions between tracks and allows a two-way scattering and gathering for effective information exchange and adaptive graph pooling. We empirically validate that BGN-ST outperforms all selected state-of-the-art methods. The paper adopts the physics-aware concept and introduces explicit physics properties such as transverse momentum. As a result, we improve the model accuracy and AUC score by about 15%.

References

1. Aaij, R., et al.: Prompt charm production in pp collisions at $\sqrt{s} = 7$ TeV. Nucl. Phys. B **871**(1), 1–20 (2013)
2. Aaij, R., et al.: Measurement of the b-quark production cross section in 7 and 13 TeV pp collisions. Phys. Rev. Lett. **118**(5), 052002 (2017)
3. Abreu, P., et al.: Classification of the hadronic decays of the Z^0 into b- and c-quark pairs using a neural network. Phys. Lett. B **295**(3–4), 383–395 (1992)
4. Adare, A., et al.: An upgrade proposal from the phenix collaboration. arXiv preprint arXiv:1501.06197 (2015)

5. Agostinelli, S., et al.: GEANT4: a simulation toolkit. Nucl. Instrum. Methods **A506**, 250 (2003)
6. Aguilar, M.R., et al.: PYTHIA8 underlying event tune for RHIC energies. Phys. Rev. D **105**(1), 016011 (2022)
7. Allison, J., Amako, K., Apostolakis, J., Araujo, H., Dubois, P., et al.: Geant4 developments and applications. IEEE Trans. Nucl. Sci. **53**, 270 (2006)
8. Ba, J.L., Kiros, J.R., Hinton, G.E.: Layer normalization. arXiv preprint arXiv:1607.06450 (2016)
9. Baldi, P., Bauer, K., Eng, C., Sadowski, P., Whiteson, D.: Jet substructure classification in high-energy physics with deep neural networks. Phys. Rev. D **93**(9), 094034 (2016)
10. Brown, T., et al.: Language models are few-shot learners. In: Advances in Neural Information Processing Systems, vol. 33, pp. 1877–1901 (2020)
11. Carleo, G., et al.: Machine learning and the physical sciences. Rev. Mod. Phys. **91**(4), 045002 (2019)
12. Carminati, F., et al.: Calorimetry with deep learning: particle classification, energy regression, and simulation for high-energy physics. In: Workshop on Deep Learning for Physical Sciences (DLPS 2017). NIPS (2017)
13. Choma, N., et al.: Graph neural networks for IceCube signal classification. In: 2018 17th IEEE International Conference on Machine Learning and Applications (ICMLA), pp. 386–391. IEEE (2018)
14. CMS Collaboration, et al.: CMS Phase 1 heavy flavour identification performance and developments. CMS Detector Performance Summary CMS-DP-2017-013 (2017)
15. De Oliveira, L., Nachman, B., Paganini, M.: Electromagnetic showers beyond shower shapes. Nucl. Instrum. Methods Phys. Res. Sect. A Accel. Spectrom. Detectors Assoc. Equip. **951**, 162879 (2020)
16. Denby, B.: Neural networks and cellular automata in experimental high energy physics. Comput. Phys. Commun. **49**(3), 429–448 (1988)
17. Devlin, J., Chang, M.W., Lee, K., Toutanova, K.: BERT: pre-training of deep bidirectional transformers for language understanding. arXiv preprint arXiv:1810.04805
18. Fey, M., Lenssen, J.E.: Fast graph representation learning with PyTorch geometric. In: ICLR Workshop on Representation Learning on Graphs and Manifolds (2019)
19. Guest, D., Cranmer, K., Whiteson, D.: Deep learning and its application to LHC physics. Annu. Rev. Nucl. Part. Sci. **68**, 161–181 (2018)
20. Guest, D., Collado, J., Baldi, P., Hsu, S.C., Urban, G., Whiteson, D.: Jet flavor classification in high-energy physics with deep neural networks. Phys. Rev. D **94**(11), 112002 (2016)
21. Jackson, P., et al.: Measurement of the total cross section from elastic scattering in pp collisions at $\sqrt{s} = 8$ TeV with the ATLAS detector. Phys. Lett. Sect. B Nucl. Elementary Part. High-Energy Phys. **761**, 158–178 (2016)
22. Kasieczka, G., et al.: The machine learning landscape of top taggers. SciPost Phys. **7**(1), 014 (2019)
23. Komiske, P.T., Metodiev, E.M., Schwartz, M.D.: Deep learning in color: towards automated quark/gluon jet discrimination. J. High Energy Phys. **2017**(1), 1–23 (2017). https://doi.org/10.1007/JHEP01(2017)110
24. Komiske, P.T., Metodiev, E.M., Thaler, J.: Energy flow networks: deep sets for particle jets. J. High Energy Phys. **2019**(1), 1–46 (2019). https://doi.org/10.1007/JHEP01(2019)121

25. Krizhevsky, A., Sutskever, I., Hinton, G.E.: ImageNet classification with deep convolutional neural networks. In: Advances in Neural Information Processing Systems, vol. 25 (2012)
26. LeCun, Y., Bengio, Y., Hinton, G.: Deep learning. Nature **521**(7553), 436–444 (2015)
27. Lee, J., Lee, Y., Kim, J., Kosiorek, A., Choi, S., Teh, Y.W.: Set transformer: a framework for attention-based permutation-invariant neural networks. In: International Conference on Machine Learning, pp. 3744–3753. PMLR (2019)
28. Louppe, G., Cho, K., Becot, C., Cranmer, K.: QCD-aware recursive neural networks for jet physics. J. High Energy Phys. **2019**(1), 1–23 (2019). https://doi.org/10.1007/JHEP01(2019)057
29. Mahesh, C., Dona, K., Miller, D.W., Chen, Y.: Towards an interpretable data-driven trigger system for high-throughput physics facilities. arXiv preprint arXiv:2104.06622 (2021)
30. Mesquita, D., Souza, A., Kaski, S.: Rethinking pooling in graph neural networks. In: Larochelle, H., Ranzato, M., Hadsell, R., Balcan, M., Lin, H. (eds.) Advances in Neural Information Processing Systems, vol. 33, pp. 2220–2231 (2020)
31. de Oliveira, L., Paganini, M., Nachman, B.: Learning particle physics by example: location-aware generative adversarial networks for physics synthesis. Comput. Softw. Big Sci. **1**(1), 1–24 (2017). https://doi.org/10.1007/s41781-017-0004-6
32. Skands, P., Carrazza, S., Rojo, J.: Tuning PYTHIA 8.1: the Monash 2013 tune. Eur. Phys. J. C **74** (2014). Article number: 3024. https://doi.org/10.1140/epjc/s10052-014-3024-y
33. Paganini, M., de Oliveira, L., Nachman, B.: CALOGAN: simulating 3D high energy particle showers in multilayer electromagnetic calorimeters with generative adversarial networks. Phys. Rev. D **97**(1), 014021 (2018)
34. Paszke, A., et al.: PyTorch: an imperative style, high-performance deep learning library. In: Wallach, H., Larochelle, H., Beygelzimer, A., d'Alché-Buc, F., Fox, E., Garnett, R. (eds.) Advances in Neural Information Processing Systems, vol. 32, pp. 8024–8035 (2019)
35. Peterson, C.: Track finding with neural networks. Nucl. Instrum. Methods Phys. Res., Sect. A **279**(3), 537–545 (1989)
36. Qasim, S.R., Kieseler, J., Iiyama, Y., Pierini, M.: Learning representations of irregular particle-detector geometry with distance-weighted graph networks. Eur. Phys. J. C **79**(7), 1–11 (2019). https://doi.org/10.1140/epjc/s10052-019-7113-9
37. Qu, H., Gouskos, L.: Jet tagging via particle clouds. Phys. Rev. D **101**(5), 056019 (2020)
38. Qu, H., Li, C., Qian, S.: Particle transformer for jet tagging. arXiv preprint arXiv:2202.03772 (2022)
39. Raffel, C., et al.: Exploring the limits of transfer learning with a unified text-to-text transformer. arXiv preprint arXiv:1910.10683 (2019)
40. Roe, B.P., Yang, H.J., Zhu, J., Liu, Y., Stancu, I., McGregor, G.: Boosted decision trees as an alternative to artificial neural networks for particle identification. Nucl. Instrum. Methods Phys. Res., Sect. A **543**(2–3), 577–584 (2005)
41. Sjöstrand, T., Mrenna, S., Skands, P.: A brief introduction to PYTHIA 8.1. Comput. Phys. Commun. **178**(11), 852–867 (2008)
42. Skands, P.: Tuning Monte Carlo generators: the Perugia tunes. Phys. Rev. D **82**(7), 074018 (2010)
43. Vaswani, A., et al.: Attention is all you need. In: Advances in Neural Information Processing Systems, vol. 30 (2017)

44. Xuan, T., Zhu, Y., Durao, F., Sun, Y.: End-to-end online sPHENIX trigger detection pipeline. In: Machine Learning and the Physical Sciences Workshop at the 35th Conference on Neural Information Processing Systems (2021)
45. Zhu, Y., Xuan, T., Borca-Tasciuc, G., Sun, Y.: A new sPHENIX heavy quark trigger algorithm based on graph neutral networks. In: Machine Learning and the Physical Sciences Workshop at the 35th Conference on Neural Information Processing Systems (2021)

Understanding Difficulty-Based Sample Weighting with a Universal Difficulty Measure

Xiaoling Zhou, Ou Wu$^{(\boxtimes)}$, Weiyao Zhu, and Ziyang Liang

Center for Applied Mathematics, Tianjin University, Tianjin, China
{xiaolingzhou,wuou,ziyangliang}@tju.edu.cn

Abstract. Sample weighting is widely used in deep learning. A large number of weighting methods essentially utilize the learning difficulty of training samples to calculate their weights. In this study, this scheme is called difficulty-based weighting. Two important issues arise when explaining this scheme. First, a unified difficulty measure that can be theoretically guaranteed for training samples does not exist. The learning difficulties of the samples are determined by multiple factors including noise level, imbalance degree, margin, and uncertainty. Nevertheless, existing measures only consider a single factor or in part, but not in their entirety. Second, a comprehensive theoretical explanation is lacking with respect to demonstrating why difficulty-based weighting schemes are effective in deep learning. In this study, we theoretically prove that the generalization error of a sample can be used as a universal difficulty measure. Furthermore, we provide formal theoretical justifications on the role of difficulty-based weighting for deep learning, consequently revealing its positive influences on both the optimization dynamics and generalization performance of deep models, which is instructive to existing weighting schemes.

Keywords: Learning difficulty · Generalization error · Sample weighting · Deep learning interpretability

1 Introduction

Treating each training sample unequally improves the learning performance. Two cues are typically considered in designing the weighting schemes of training samples [1]. The first cue is the application context of learning tasks. In applications such as medical diagnosis, samples with high gains/costs are assigned with high weights [2]. The second cue is the characteristics of the training data. For example, samples with low-confidence or noisy labels are assigned with low weights.

This study is supported by NSFC 62076178, TJF 19ZXAZNGX00050, and Zhijiang Fund 2019KB0AB03.

Supplementary Information The online version contains supplementary material available at https://doi.org/10.1007/978-3-031-26409-2_5.

Characteristic-aware weighting has attracted increasing attention owing to its effectiveness and universality [3–5].

Many existing characteristic-aware weighting methods are based on an intrinsic property of the training samples, i.e., their learning difficulty. The measures for the samples' learning difficulty can be roughly divided into five categories.

- Prediction-based measures. This category directly uses the loss [3,6,7] or the predicted probability of the ground truth [4,8] as the difficulty measures. This measure is simple yet effective and is widely used in various studies [3,4]. Their intention is that a large loss (a small probability) indicates a large learning difficulty.
- Gradient-based measures. This category applies the loss gradient in the measurement of the samples' learning difficulty [9,10]. Santiagoa et al. [9] uses the norm of the loss gradient as the difficulty measure. Their intuition is that the larger the norm of the gradient, the harder the sample.
- Category proportion-based measures. This category is mainly utilized in imbalanced learning [11], where the category proportion measures the samples' difficulty. People believe that the smaller the proportion of a category, the larger the learning difficulty of samples in this category [11,12].
- Margin-based measures. The term "margin" refers to the distance from the sample to the oracle classification boundary. The motivation is that the smaller the margin, the larger the difficulty of a sample [13].
- Uncertainty-based measures. This category uses the uncertainty of a sample to measure the difficulty. Aguilar et al. [14] identify hard samples based on epistemic uncertainty and leverage the Bayesian Neural Network [15] to infer it.

Varying difficulty measures have a huge impact on a difficulty-based weighting strategy. The underlying factors which influence samples' learning difficulty considered in the above measures include noise level [6,7], imbalance degree [11,12], margin [13], and uncertainty [14]. However, each measure only considers a single factor or in part, and comes from heuristic inspirations but not formal certifications, hindering the application scope of the measures. It is desirable to theoretically explore a universal measure capturing all of the above factors. Based on this measure, the role of difficulty-based sample weighting can be revealed more concretely. However, neither theoretical nor empirical investigations have been conducted to investigate a unified measure.

Moreover, despite the empirical success of various difficulty-based weighting methods, the process of how difficulty-based weighting positively influences the deep learning models remains unclear. Two recent studies have attempted to investigate the influence of weights in deep learning. Byrd and Lipton [16] empirically studied the training of over-parameterized networks with sample weights and found that these sample weights affect deep learning by influencing the implicit bias of gradient descent-a novel topic in deep learning interpretability, focusing on why over-parameterized models is biased toward solutions that generalize well. Existing studies on this topic [17–19][?] reveal that the direction of the parameters (for linear predictor) and the normalized margin (for nonlinear predictor) respectively converge to those of a max-margin solution.

Inspired by the finding of Byrd and Lipton [16], Xu et al. [20] dedicated to studying how the understandings for the implicit bias of gradient descent adjust to the weighted empirical risk minimization (ERM) setting. They concluded that assigning high weights to samples with small margins may accelerate optimization. In addition, they established a generalization bound for models that implement learning by using sample weights. However, they only discussed the measurement of difficulty by using one of the indicators (i.e., margin), resulting in that their conclusion is limited and inaccurate in some cases. Furthermore, their generalization bound cannot explicitly explain why hard samples are assigned with large weights in many studies. More analyses based on a universal difficulty measure are in urgent demand.

In this study, the manner of how the difficulty-based weighting affects the deep model training is deeply investigated. First, our analyses support that the generalization error of the training sample can be regarded as a universal difficulty measure for capturing all of the four factors described above. Second, based on this unified measure, we characterize the role of difficulty-based weighting on the implicit bias of gradient descent, especially for the convergence speed. Third, two new generalization bounds are constructed to demonstrate the explicit relationship between the sample weights and the generalization performance. The two bounds illuminate a new explanation for existing weighting strategies. Our study takes the first step of constructing a formal theory for difficulty-based sample weighting. In summary, our contributions are threefold.

- We theoretically prove the high relevance of the generalization error with four main factors influencing the samples' learning difficulty, further indicating that the generalization error can be used as a universal difficulty measure.
- We reveal how the difficulty-based sample weighting influences the optimization dynamics and the generalization performance for deep learning. Our results indicate that assigning high weights on hard samples can not only accelerate the convergence speed but also enhance the generalization performance.
- We bring to light the characteristics of a good set of weights from multiple perspectives to illuminate the deep understanding of numerous weighting strategies.

2   Preliminaries

2.1   Description of Symbols

Let $\mathcal{X}$ denote the input space and $\mathcal{Y}$ a set of classes. We assume that the training and test samples are drawn $i.i.d$ according to some distributions $\mathcal{D}^{tr}$ and $\mathcal{D}^{te}$ over $\mathcal{X} \times \mathcal{Y}$. The training set is denoted as $T = \{\boldsymbol{x}, y\} = \{(\boldsymbol{x}_i, y_i)\}_{i=1}^n$ that contains n training samples, where $\boldsymbol{x}_i$ denotes the i-th sample's feature, and y_i is the associated label. Let d_i and $w(d_i)$ be the learning difficulty and the difficulty-based weight of x_i. The learning difficulty can be approximated by several values, such as loss, uncertainty and generalization error which will be explained in Sect. 3.

The predictor is denoted by $f(\boldsymbol{\theta}, \boldsymbol{x})$ and $\mathcal{F} = \{f(\boldsymbol{\theta}, \cdot) \mid \boldsymbol{\theta} \in \boldsymbol{\Theta} \subset \mathbb{R}^d\}$. For the sake of notation, we focus on the binary setting $y_i \in \{-1, 1\}$ with $f(\boldsymbol{\theta}, \boldsymbol{x}) \in \mathbb{R}$. The sign of the model's output $f(\boldsymbol{\theta}, \boldsymbol{x}_i)$ is the predicted label. However, as to be clarified later, our results can be easily extended to the multi-class setting where $y_i \in \{1, 2, \cdots, C\}$. For multi-class setting, the softmax function is used to get the probability, and the logits are given by $\{f_{y_j}(\boldsymbol{\theta}, \boldsymbol{x})\}_{j=1}^C$. Given a non-negative loss ℓ and a classifier $f(\boldsymbol{\theta}, \cdot)$, the empirical risk can be expressed as $\mathcal{L}(\boldsymbol{\theta}, \boldsymbol{w}) = \frac{1}{n} \sum_{i=1}^n w(d_i) \cdot \ell(y_i f(\boldsymbol{\theta}, \boldsymbol{x}_i))$. We focus particularly on the exponential loss $\ell(u) = \exp(-u)$ and logistic loss $\ell(u) = \log(1 + \exp(-u))$. Let $\nabla l(u)$ be the loss gradient and $f(\boldsymbol{x}|T)$ is the trained model on T. The margin is denoted as $\gamma_i(T) = y_i f(\boldsymbol{\theta}, \boldsymbol{x}_i|T)$ for the binary setting, where it is equivalently denoted as $\gamma_i(T) = f_{y_i}(\boldsymbol{\theta}, \boldsymbol{x}_i|T) - \max_{i \neq j} f_{y_j}(\boldsymbol{\theta}, \boldsymbol{x}_i|T)$ for the multi-class setting.

2.2 Definition of the Generalization Error

Bias-variance tradeoff is a basic theory for the qualitative analysis of the generalization error [22]. This tradeoff is initially constructed via regression and mean square error, which is given by

$$
\begin{aligned}
Err &= \mathbb{E}_{\boldsymbol{x},y} \mathbb{E}_T[\||y - f(\boldsymbol{x}|T)\||_2^2] \\
&\approx \underbrace{\mathbb{E}_{\boldsymbol{x},y}[\||y - \overline{f}(\boldsymbol{x})\||_2^2]}_{Bias} + \underbrace{\mathbb{E}_{\boldsymbol{x},y} \mathbb{E}_T[\||f(\boldsymbol{x}|T) - \overline{f}(x)\||_2^2]}_{Variance},
\end{aligned}
\tag{1}
$$

where $\overline{f}(\boldsymbol{x}) = \mathbb{E}_T[f(\boldsymbol{x}|T)]$. Similarly, we define the generalization error of a single sample $\boldsymbol{x}_i$ as

$$
\mathrm{err}_i = \mathbb{E}_T[\ell(f(\boldsymbol{x}_i|T), y_i)] \approx B(\boldsymbol{x}_i) + V(\boldsymbol{x}_i),
\tag{2}
$$

where $B(\boldsymbol{x}_i)$ and $V(\boldsymbol{x}_i)$ are the bias and variance of $\boldsymbol{x}_i$.

2.3 Conditions and Definitions

Our theoretical analyses rely on the implicit bias of gradient descent. The gradient descent process is denoted as

$$
\boldsymbol{\theta}_{t+1}(\boldsymbol{w}) = \boldsymbol{\theta}_t(\boldsymbol{w}) - \eta_t \nabla \mathcal{L}(\boldsymbol{\theta}_t[\boldsymbol{w}(\boldsymbol{d}[t])]),
\tag{3}
$$

where η_t is the learning rate which can be a constant or step-independent, $\nabla \mathcal{L}(\boldsymbol{\theta}_t[\boldsymbol{w}(\boldsymbol{d}[t])])$ is the gradient of $\mathcal{L}$, and $\boldsymbol{w}(\boldsymbol{d}[t])$ is the difficulty-based weight of difficulty $\boldsymbol{d}$ at time t. The weight may be dynamic with respect to time t if difficulty measures, such as loss [3] and predicted probability [4], are used. To guarantee the convergence of the gradient descent, two conditions following the most recent study [20] are shown below.

- The loss ℓ has an exponential tail whose definition is shown in the supplementary file. Thus, $\lim_{u\to\infty} \ell(-u) = \lim_{u\to\infty} \nabla\ell(-u) = 0$.
- The predictor $f(\theta, \mathbf{x})$ is α-homogeneous such that $f(c \cdot \boldsymbol{\theta}, \boldsymbol{x}) = c^\alpha f(\cdot\boldsymbol{\theta}, \boldsymbol{x})$, $\forall c > 0$.

It is easy to verify that losses including the exponential loss, log loss, and cross-entropy loss satisfy the first condition. The second condition implies that the activation functions are homogeneous such as ReLU and LeakyReLU, and bias terms are disallowed. In addition, we need certain regularities from $f(\boldsymbol{\theta}, \boldsymbol{x})$ to ensure the existence of critical points and the convergence of gradient descent:

- For $\forall \boldsymbol{x} \in \mathcal{X}$, $f(\boldsymbol{\theta}, \boldsymbol{x})$ is β-smooth and l-Lipschitz on $\mathbb{R}^{\mathrm{d}}$.

The third condition is a common technical assumption whose practical implications are discussed in the supplementary file.

The generalization performance of deep learning models is measured by the generalization error of the test set $\hat{\mathcal{L}}(f)$ [21], defined as

$$\hat{\mathcal{L}}(f) = \mathbb{P}_{(\boldsymbol{x},y)\sim\mathcal{D}^{te}}[\gamma(f(\boldsymbol{x}, y)) \leq 0]. \tag{4}$$

2.4   Experiment Setup

Demonstrated experiments are performed to support our theoretical analyses. For the simulated data, the linear predictor is a regular regression model, and the nonlinear predictor is a two-layer MLP with five hidden units and ReLU as the activation function. Exponential loss and standard normal initialization are utilized. CIFAR10 [23] is experimented with, and ResNet32 [24] is adopted as the baseline model. For the imbalanced data, the imbalance setting follows Ref. [11]. For the noisy data, uniform and flip label noises are used and the noise setting follows Ref. [25]. The models are trained with a gradient descent by using 0.1 as the learning rate.

The model uncertainty is approximated by the predictive variance of five predictions. To approximate the generalization error, we adopt the five-fold cross-validation [26] to calculate the average learning error for each sample.

3   A Universal Difficulty Measure

As previously stated, four factors pointed out by existing studies, namely, noise, imbalance, margin, and uncertainty, greatly impact the learning difficulty of samples. Nevertheless, existing measures only consider one or part of them, and their conclusions are based on heuristic inspirations and empirical observations. In this section, we theoretically prove that the generalization error of samples is a universal difficulty measure reflecting all four factors. All proofs are presented in the supplementary file. Without increasing the ambiguity, the generalization error of the samples is termed as error for brevity.

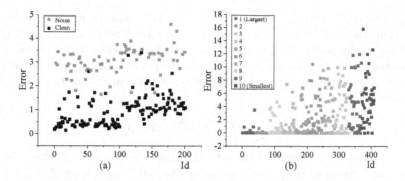

Fig. 1. (a) Generalization errors of clean and noisy samples on noisy data. The noise ratio is 10% (b) Generalization errors of samples in ten categories on imbalanced data. The imbalance ratio is 10:1. CIFAR10 and ResNet32 are used. Other values of noise ratio and imbalance ratio following Ref. [25] are also experimented with and the same conclusions can be obtained.

3.1   Noise Factor

Noise refers to data that is inaccurate in describing the scene. Numerous studies devoted to reducing the influence of noisy samples in the dataset on the deep learning models and these literature intuitively consider noisy samples as hard ones without formal certification [7, 27]. The two kinds of noise are feature noise [31] and label noise [27]. We offer two propositions to reveal the relationship between the generalization error and the noise factor. For feature noise, we offer the following proposition:

Proposition 1. *Let Δx_i be the perturbation of sample (x_i, y_i), which is extremely small in that $o(\Delta x_i)$ can be omitted. Let $\angle\varphi$ be the angle between the direction of Δx_i and the direction of $\mathbb{E}_T[f'(x_i|T)]$. If $\mathbb{E}_T[f'(x_i|T) \cdot \Delta x_i] < 0$ (i.e., $\angle\varphi > 90°$), then the error of the noisy sample is increased relative to the clean one. Alternatively, the direction of the perturbation Δx_i and that of $\mathbb{E}_T[f'(x_i|T)]$ are contradictory. Otherwise, if $\mathbb{E}_T[f'(x_i|T) \cdot \Delta x_i] > 0$, then $\angle\varphi < 90°$, and the error of the noisy sample is decreased.*

According to Proposition 1, feature noise can be divided into two categories, which increase or decrease the learning difficulty (generalization error) of the samples, respectively. In this paper, noise that increases the error is called the adversarial type, which is always used in the field of adversarial learning; otherwise, it is a promoted type, which refers to noise that decrease the learning difficulty of samples. Therefore, the variation of the error under feature noise is determined by the noise type. For example, as all feature noises are adversarial in adversarial learning [32], all of the samples' errors are increased with feature noise. For label noise, we offer the following proposition:

Proposition 2. *Let π be the label corruption rate (i.e., the probability of each label flipping to another one). Denote the probability of correct classification for the original samples as p. If $p > 0.5$, then the errors of the noisy samples are larger than those of the clean ones.*

This proposition implies that the errors of the samples with label noises are larger than those of the clean ones on the average. Specifically, if a sample is more likely to be predicted correctly, its generalization error is increased due to label noise. Let $\mathcal{L}^*$ be the global optimum of the generalization error of the clean dataset and y' be the corrupted label. When the noise in Proposition 2 is added, the empirical error $\mathcal{L}'$ is

$$\mathcal{L}' = (1 - \pi)\mathcal{L}^* + \pi\mathcal{L}\left(f\left(\boldsymbol{x}\right), y'\right), \tag{5}$$

where we have taken expectations over the noise. When $\pi \to 0$, the noise disappears, and the optimal generalization is attained. Proposition 2 is consistent with the empirical observation shown in Fig. 1(a), where the noisy samples have larger errors than the clean ones on the average.

3.2   Imbalance Factor

Besides noise, imbalance is another common deviation of real world datasets. The category distribution of the samples in the training set is non-uniform. Various methods solve this issue by assigning high weights on samples in tail categories which are considered to be hard ones [4,11]. Nevertheless, a theoretical justification about why these samples are harder lacks. The imbalance ratio is denoted by $c_r = \max\{c_1, c_2, \cdots, c_C\} : \min\{c_1, c_2, \cdots, c_C\}$. Then, we offer the following proposition.

Proposition 3. *If a predictor on an imbalanced dataset ($c_r > e : 1$) is an approximate Bayesian optimal classifier (as the exponential loss is an approximation for the zero-one loss), which is to minimize the total risk, then the average probability of the ground truth of the samples in the large category is greater than that of the samples in the small category.*

With Proposition 3, it is easy to obtain Proposition A.1 shown in the supplementary file that the average error of samples in the small category is larger than that of the samples in the large category, indicating there are more hard samples in the small category. This proposition is verified by the experiments, as shown in Fig. 1(b). The tail categories contain more samples with larger errors. To enhance the performance of the classification model, samples with larger errors should be assigned with higher weights, as most methods do [11]. Further experiments in Sect. 5 (Fig. 6) indicate that the classification performance of the small category can be improved by increasing its sample weights.

3.3   Margin Factor

The samples' margins measure the distances of the samples from the decision boundary. Some literature intuitively consider a small margin indicates a large learning difficulty and corresponds to a low confidence of the prediction [13,33]. However, a formal justification is lacking. We offer the following proposition.

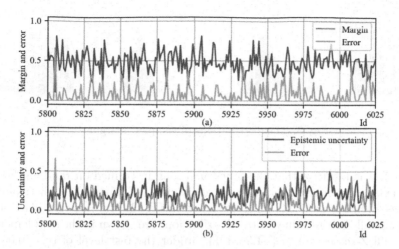

Fig. 2. (a) Correlation between generalization error and average margin. (b) Correlation between generalization error and epistemic uncertainty. CIFAR10 and ResNet32 are used in this experiment. All values are normalized.

Proposition 4. *Let μ_i be the true margin of $\boldsymbol{x}_i$ corresponding to the oracle decision boundary. The condition is that the functional margins of a sample trained on random datasets obey a Gaussian distribution. In other words, for sample $\boldsymbol{x}_i$, its functional margin γ_i obey a Gaussian distribution $\mathcal{N}(\mu_i, \sigma_i^2)$. For sample $\boldsymbol{x}_j$, $\gamma_j \sim \mathcal{N}(\mu_j, \sigma_j^2)$. When the margin variances of the two samples are same (i.e., $\sigma_i^2 = \sigma_j^2$), if $\mu_i \leq \mu_j$, then $\mathrm{err}_i \geq \mathrm{err}_j$. Similarly, when the true margins of the two samples are the same (i.e., $\mu_i = \mu_j$), if $\sigma_i^2 \geq \sigma_j^2$, then $\mathrm{err}_i \geq \mathrm{err}_j$.*

Proposition 5 indicates a fact that even a sample with a large true margin, as long as the margin variance is large, it may also have a high learning difficulty. Specifically, the true margin (i.e., the mean of the functional margin distribution) of a sample and error are negatively correlated when the margin variances of the samples are equal. By contrast, the margin variance and error are positively correlated when the true margins are equal. This illumination revises the current wisdom. The conclusion in which samples close to the oracle decision boundary are hard ones [20] is not completely correct. Indeed, the relation between the margin and error of sample $\boldsymbol{x}_i$ conforms with the following formula:

$$\mathrm{err}_i = \mathbb{E}_T[e^{-\gamma_i(T)}] = e^{-\mu_i + \frac{1}{2}\sigma_i^2}, \tag{6}$$

where $\mathrm{err_i}$, μ_i, and σ_i refer to the generalization error, the true margin, and the margin variance of sample x_i, respectively. For the two samples $\boldsymbol{x}_i$ and $\boldsymbol{x}_j$, if $\mu_i < \mu_j$ and $\sigma_i^2 < \sigma_j^2$, then we cannot judge whether err_i is greater than err_j. As shown in Fig. 2(a), the average margin and error are negatively correlated for most samples, but it is not absolute, which accords with the above analyses.

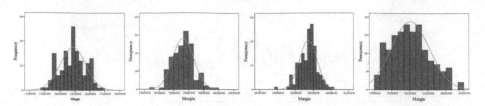

Fig. 3. The distributions of samples' margins.

Although it is intuitive that the functional margin trained on random datasets obeys a Gaussian distribution, we evaluate it via the Z-scores of the distributions' Kurtosis and Skewness [34] which is shown in Fig. 3. More margin distribution curves and all Z-score values of the distributions are shown in the supplementary file. As all Z-scores are in $[-1.96, 1.96]$, under the test level of $\alpha = 0.05$, the distribution of margin obeys the Gaussian distribution.

3.4   Uncertainty Factor

Uncertainties [37] in deep learning are classified into two types. The first type is aleatoric uncertainty (data uncertainty), which is caused by the noise in the observation data. Its correlation with the error has been discussed in Sect. 3.1. The second type is epistemic uncertainty (model uncertainty). It is used to indicate the consistency of multiple predictions. We give the analyses of the relationship between the generalization error and epistemic uncertainty.

Let T be a training set, and let $P(\boldsymbol{\theta}|T)$ be the distribution of the training models based on T. The predictive variance $Var(f(\boldsymbol{x}_i|\boldsymbol{\theta}_1), \cdots, f(\boldsymbol{x}_i|\boldsymbol{\theta}_K))$ plus a precision constant is a typical manner of estimating epistemic uncertainty [35, 36]. Take the mean square loss as an example[1], the epistemic uncertainty is

$$\widehat{\mathrm{Var}}\,[\boldsymbol{x}_i] := \tau^{-1} + \frac{1}{|K|}\sum\nolimits_k f(\boldsymbol{x}_i|\boldsymbol{\theta}_k)^{\mathsf{T}} f(\boldsymbol{x}_i|\boldsymbol{\theta}_k) - \mathbb{E}[f(\boldsymbol{x}_i|\boldsymbol{\theta}_k)]^{\mathsf{T}}\mathbb{E}[f(\boldsymbol{x}_i|\boldsymbol{\theta}_k)], \qquad (7)$$

where τ is a constant. The second term on the right side of Eq. (7) is the second raw moment of the predictive distribution and the third term is the square of the first moment. When $K \to \infty$ and the constant term is ignored, Eq. (7) becomes

$$\widehat{\mathrm{Var}}\,[\boldsymbol{x}_i] := \int_\theta \|f(\boldsymbol{x}_i|\boldsymbol{\theta}) - \overline{f}(\boldsymbol{x}_i)\|_2^2 dP(\boldsymbol{\theta}|T). \qquad (8)$$

If $P(\boldsymbol{\theta}|T)$ is approximated by the distribution of learned models on random training sets which conform to the Gaussian distribution $\mathcal{N}(T, \delta I)$, Eq. (8) is exactly the variance term of the error defined in Eq. (2) when the mean square loss is utilized.

As the bias term in the error can capture the aleatoric uncertainty and the variance term captures the epistemic uncertainty, the overall relationship

---

[1] For other losses, other methods can be used to calculate the predictive variance [26].

between uncertainty and error is positively correlated. Nevertheless, the relationship between epistemic uncertainty and error is not simply positively or negatively correlated. For some samples with heavy noises, their epistemic uncertainties will be small as their predictions remain erroneous. However, their errors are large due to their large bias. This phenomenon is consistent with the experimental results shown in Fig. 2(b). Epistemic uncertainty and error are positively correlated for some samples, and the two variables are negatively correlated for other samples.

3.5 Discussion About Generalization Error

The commonly used difficulty measures, such as loss [3] and gradient norm [9], are mainly related to the bias term. Shin et al. [27] emphasized that only using loss as the measurement cannot distinguish clean and noisy samples, especially for uniform label noise. There are also a few existing studies that use variance [28, 29]. For instance, Agarwal et al. [30] applied the variance of gradient norms as the difficulty measure. Indeed, both the variance and bias terms should not be underestimated when measuring the samples' learning difficulty. Our theoretical analyses support that generalization error including both the two terms can capture four main factors influencing the samples' learning difficulty. Thus, the error can be leveraged as a universal measure that is more reasonable than existing measures. Existing studies generally apply the K-fold cross-validation method [26] to calculate the generalization error. More efficient error calculation algorithms are supposed to be proposed which will be our future work.

4 Role of Difficulty-Based Weighting

This section aims to solve the second issue of explaining the difficulty-based weighting in deep learning. Based on the universal difficulty measure, the impacts of the difficulty-based weighting schemes on the optimization dynamics and the generalization performance in deep learning are investigated. Compared with the most recent conclusions [20] established only on the margin factor, our theoretical findings, which are based on our universal measure, are more applicable and precise.

4.1 Effects on Optimization Dynamics

Linear Predictor. We begin with the linear predictors allowing for a more refined analysis. Xu et al. [20] inferred an upper bound containing the term $D_{KL}(p\|w)$, where D_{KL} is the Kullback-Leibler divergence and p is the optimal dual coefficient vector. A smaller value of $D_{KL}(p\|w)$ means that the convergence may be accelerated. Therefore, to accelerate the convergence, they believe that the weights w should be consistent with the coefficients p. Alternatively, the samples with small functional margins will have large coefficients and thus should be assigned with large weights. However, the functional margin is not the true

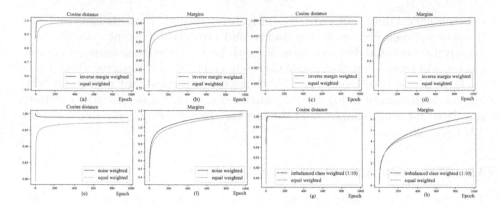

Fig. 4. "Cosine distance" represents the cosine of the angle between the decision boundary (at that epoch) and the max-margin solution. (a), (b) Cosine distance and average margin of equal weights and inverse margin weights using the linear predictor. (c), (d) Cosine distance and average margin of equal weights and inverse margin weights using the nonlinear predictor. (e), (f) Cosine distance and average margin of equal weights and increasing weights of noisy samples using the nonlinear predictor on the noisy data. (g), (h) Cosine distance and average margin of equal weights and increasing weights of samples in tail categories using the linear predictor on the imbalanced data. More results are placed in the supplementary file.

margin that corresponds to the oracle boundary. Therefore, their conclusion that samples close to the oracle classification boundary should be assigned with large weights [20] cannot be well-drawn according to their inference. We offer a more precise conclusion with the unified difficulty measure (i.e., generalization error). As before, we assume that the functional margins of a sample x_i obey a Gaussian distribution $\mathcal{N}(\mu_i, \sigma_i^2)$, where μ_i is the true margin and σ_i^2 is the margin variance of x_i. We offer the following proposition:

Proposition 5. *For two samples x_i and x_j, if $\mathrm{err}_i \geq \mathrm{err}_j$, then we have:*

(1) When the optimal dual coefficient p_i of x_i on a random training set T is a linear function of its functional margin γ_i on T, if $\mu_i \leq \mu_j$, then $\mathbb{E}_T[p_i] \geq \mathbb{E}_T[p_j]$ (i.e., $\mathbb{E}_T[w_i] \geq \mathbb{E}_T[w_j]$);

(2) When the optimal dual coefficient p_i of x_i on a random training set T is a natural exponential function of its functional margin γ_i on T, $\mathbb{E}_T[p_i] \geq \mathbb{E}_T[p_j]$ (i.e., $\mathbb{E}_T[w_i] \geq \mathbb{E}_T[w_j]$) always holds. Notably, even when $\mu_i > \mu_j$, $\mathbb{E}_T[p_i] > \mathbb{E}_T[p_j]$ may still hold.

The proof is presented in the supplementary file. $\mathbb{E}_T[p_i] > \mathbb{E}_T[p_j]$ implies that $w_i > w_j$ holds on the average. The conclusion that samples with small true margins should be assigned with large weights may not hold on some training sets when p_i is not a linear function of γ_i [17]. A sample with a small true margin may have a smaller weight than a sample with a large true margin yet a large error. Thus, a more general conclusion when p_i is not a linear function of γ_i is that increasing the weights of hard samples (samples with large generalization

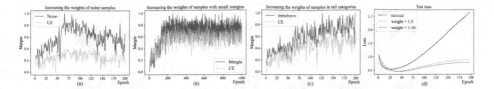

Fig. 5. (a)–(c) Normalized margin of increasing the weights of noisy samples/samples with small margins/samples in tail categories. CIFAR10 data is used. Uniform label noise is adopted. The noise ratio and imbalance ratio are 10% and 10:1. (d) Generalization error of the test set when the nonlinear model is trained with different weights on simulated imbalanced data with the imbalance ratio as 10:1. Other noise and imbalance settings are also experimented with and the same conclusions can be obtained.

errors) may accelerate the convergence, rather than just for samples with small margins. Other factors, including noise, imbalance, and uncertainty also affect samples' learning difficulty. Notably, the weights of the hard samples should not be excessively increased, as to be explained in the succeeding section. We reasonably increase the weights of the hard samples shown in Figs. 4 and A-3 in the supplementary file indicating that the optimization is accelerated.

We also prove that difficulty-based weights do not change the convergence direction to the max-margin solution shown in Theorem A.1 in the supplementary file. As shown in Fig. 3, the cosine distance and margin value are always increasing during the training procedure, indicating the direction of the asymptotic margin is the max-margin solution.

Nonlinear Predictor. Analyzing the gradient dynamics of the nonlinear predictors is insurmountable. The main conclusion obtained by Xu et al. [20] can also be established for difficulty-based weights only if the bound of weights is larger than zero. However, their theorem has only been proven for binary cases as the employed loss is inapplicable in multi-class cases. Here, we extend the theory to the multi-class setting with a regularization $\lambda\|\theta\|^r$ on the cross-entropy loss. Let $\theta_\lambda(w) \in \arg\min \mathcal{L}_\lambda(\theta, w)$. Formally, the dynamic regime for the nonlinear predictor can be described as follows:

Theorem 1. *Let* $w \in [b, B]^n$. *Denote the optimal normalized margin as*

$$\gamma^* = \max_{\|\theta(w)\| \leq 1} \min_i (f_{y_i}(\theta(w), x_i) - \max_{j \neq i}(f_{y_j}(\theta(w), x_i))) \qquad (9)$$

Let $\overline{\theta}_\lambda(w) = \theta_\lambda(w)/\|\theta_\lambda(w)\|$. *Then, it holds that (1) Denote the normalized margin as*

$$\gamma_\lambda(w) = \min_i (f_{y_i}(\overline{\theta}_\lambda(w), x_i) - \max_{j \neq i} f_{y_j}(\overline{\theta}_\lambda(w), x_i)) \qquad (10)$$

Then, $\gamma_\lambda(w) \to \gamma^*$, *as* $\lambda \to 0$.

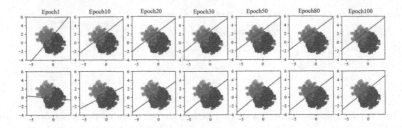

Fig. 6. Top: Equal weights of the two categories. Bottom: Samples in the small category are assigned with high weights, obtaining better performance for the small (red) category. The imbalance ratio is set to 10:1. The same conclusions can also be obtained for other imbalance ratios. (Color figure online)

(2) There exists a $\lambda := \lambda(r, a, \gamma^*, \boldsymbol{w})$. For $\alpha \leq 2$, let $\boldsymbol{\theta}'(\boldsymbol{w})$ denote a α-approximate minimizer of $\mathcal{L}_\lambda$. Thus, $\mathcal{L}_\lambda(\boldsymbol{\theta}'(\boldsymbol{w})) \leq \alpha L_\lambda(\boldsymbol{\theta}_\lambda(\boldsymbol{w}))$. Denote the normalized margin of $\boldsymbol{\theta}'(\boldsymbol{w})$ by $\gamma'(\boldsymbol{w})$. Then, $\gamma'(\boldsymbol{w}) \geq \frac{\gamma^*}{10\alpha^{a/r}}$.

The proof is presented in the supplementary file. When λ is sufficiently small, the difficulty-based weighting does not affect the asymptotic margin. According to Theorem 2, the weights do affect the convergence speed. A good property is that even though $L_\lambda(\boldsymbol{\theta}_\lambda(\boldsymbol{w}))$ has not yet converged but close enough to its optimum, the corresponding normalized margin has a reasonable lower bound. A good set of weights can help the deep learning model to achieve this property faster. However, the conditions in which a set of weights can accelerate the speed are not clearly illuminated. Notably, as shown in our experiments in Figs. 4 and A-3 in the supplementary file, assigning large weights for hard samples increases the convergence speed. The results on the multi-class cases (CIFAR10) indicate that assigning large weights on hard samples increases the margin, as shown in Figs. 5(a–c). However, some particular occasions of difficulty-based weights, such as SPL [3], do not satisfy the bounding condition because the lower bounds of these weights are zero instead of a positive real number. The theorem requires further revision to accommodate this situation.

4.2   Effects on Generalization Performance

Besides the role of difficulty-based weights on optimization dynamics, we are also concerned as to whether and how the difficulty-based weights affect the generalization performance. The generalization bound of Xu et al. [20] does not contain the sample weights, thus it cannot explicitly explain why hard samples are assigned with large weights. In addition, they assume that the source and target distributions are unequal, restricting the application of their conclusion. The two generalization bounds we propose offer good solutions to these issues. They illuminate how a weighting strategies can be designed.

Let P_s and P_t be the source (training) and target (testing) distributions, respectively, with the corresponding densities of $p_s(\cdot)$ and $p_t(\cdot)$. Assume that the two distributions have the same support. The training and test samples are drawn *i.i.d* according to distributions P_s and P_t, respectively. Learning with sample weights $\boldsymbol{w}(\boldsymbol{x})$ is equivalent to learning with a new training distribution $\widetilde{P}_s$. The density of the distribution of the weighted training set $\widetilde{P}_s$ is denoted as $\widetilde{p}_s(\boldsymbol{x})$ and $\widetilde{p}_s(\boldsymbol{x}) \sim \boldsymbol{w}(\boldsymbol{x})p_s(\boldsymbol{x})$. Pearson χ^2-divergence is used to measure the difference between $\widetilde{P}_s$ and P_t, i.e., $D_{\chi^2}(P_t\|\widetilde{P}_s) = \int[(d\widetilde{P}_s/dP_t)^2 - 1]d\widetilde{P}_s$. We consider depth-$q$ ($q \geq 2$) networks with the activation function ϕ. The binary setting is considered, in that the network computes a real value

$$f(\boldsymbol{x}) := \boldsymbol{W}_q\phi\left(\boldsymbol{W}_{q-1}\phi\left(\cdots\phi\left(\boldsymbol{W}_1\boldsymbol{x}\right)\cdots\right)\right), \tag{11}$$

where $\phi(\cdot)$ is the element-wise activation function (e.g., ReLU). The training set contains n samples. Denote the generalization error for a network f as $\hat{\mathcal{L}}(f)$. The generalization performance of f with weights can be described as follows.

Theorem 2. *Suppose ϕ is 1-Lipschitz and 1-positive-homogeneous. With a probability at least of $1 - \delta$, we have*

$$\hat{\mathcal{L}}(f) \leq \underbrace{\frac{1}{n}\sum_{i=1}^{n}\frac{p_t(\boldsymbol{x}_i)}{\widetilde{p}_s(\boldsymbol{x}_i)}\mathbb{1}(y_if(\boldsymbol{x}_i) < \gamma)}_{I} + \underbrace{\frac{L \cdot \sqrt{D_{\chi^2}\left(P_t\|\widetilde{P}_s\right) + 1}}{\gamma \cdot q^{(q-1)/2}\sqrt{n}}}_{(II)} + \underbrace{\epsilon(\gamma,n,\delta)}_{(III)},$$

$$\tag{12}$$

where $\epsilon(\gamma,n,\delta) = \sqrt{\frac{\log\log_2\frac{4L}{\gamma}}{n}} + \sqrt{\frac{\log(1/\delta)}{n}}$ and $L := \sup_{\boldsymbol{x}}\|\boldsymbol{x}\|$.

The proof is presented in the supplementary file. Compared with the findings of Xu et al. [20], the bound of the generalization error is directly related to the sample weights $\boldsymbol{w}(\boldsymbol{x})$ contained in $\widetilde{p}_s(\boldsymbol{x})$. In view of reducing the generalization error, a natural optimization strategy can be implemented as follows: 1) an optimal weight set $\boldsymbol{w}(\boldsymbol{x})$ (in $\widetilde{p}_s(x)$) is obtained according to decreasing the right side of Eq. (12) based on the current f; 2) f is then optimized under the new optimal weights $\boldsymbol{w}(x)$. In the first step, the reduction of generalization error can come from two aspects. One is to increase the weights of samples with small margins. The other is to make the test and training distributions close. Disappointingly, this strategy heavily relies on the current f which is unstable. Given a fixed training set, f depends on random variables (denoted as $\mathcal{V}$) such as hyperparameters and initialization. To obtain a more stable weighting strategy, we further propose the following proposition.

Proposition 6. *Suppose ϕ is 1-Lipschitz and 1-positive-homogeneous. With a probability of at least $1 - \delta$, we have*

$$\mathbb{E}_{\mathcal{V}}[\hat{\mathcal{L}}(f_{\mathcal{V}})] \leq \underbrace{\frac{1}{n}\sum_{i=1}^{n}\frac{p_t(\boldsymbol{x}_i)}{\widetilde{p}_s(\boldsymbol{x}_i)}\mathbb{E}_{\mathcal{V}}[\mathbb{1}(y_if_{\mathcal{V}}(\boldsymbol{x}_i) < \gamma)]}_{(I)} + \underbrace{\frac{L \cdot \sqrt{D_{\chi^2}\left(P_t\|\widetilde{P}_s\right) + 1}}{\gamma \cdot q^{(q-1)/2}\sqrt{n}}}_{(II)} + (III)$$

$$\tag{13}$$

Accordingly, increasing the $\widetilde{p}_s(\boldsymbol{x}_i)$ of the samples with large $\mathbb{E}_{\mathcal{V}}[\mathbb{1}(y_i f_{\mathcal{V}}(\boldsymbol{x}_i) < \gamma)]$ will reduce (I). In fact, samples with larger generalization errors will have larger values of $\mathbb{E}_{\mathcal{V}}[\mathbb{1}(y_i f_{\mathcal{V}}(\boldsymbol{x}_i) < \gamma)]$. The proof is placed in the supplementary file. Alternatively, increasing the weights of the hard samples will reduce (I). However, the weights of the hard samples cannot be increased arbitrarily as $D_{\chi^2}(P_t \| \widetilde{P}_s)$ may be large. Therefore, a tradeoff between (I) and (II) should be attained to obtain a good set of weights. Alternatively, a good set of weights should increase the weights of hard samples while ensuring that the distributions of the training set and the test set are close.

It is worth mentioning that our two above conclusions are still insightful when $P_t = P_s$ while the conclusion of Xu et al. [20] assumes $P_t \neq P_s$. Apparently, even when $P_t = P_s$, assigning weights according to the samples' difficulties is still beneficial as the tradeoff between (I) and (II) still takes effect.

5   Discussion

Our theoretical analyses in Sects. 3 and 4 provide answers to the two concerns described in Sect. 1.

First, the generalization error has been theoretically guaranteed as a generic difficulty measure. It is highly related to noise level, imbalance degree, margin, and uncertainty. Consequently, two directions are worth further investigating. The first direction pertains to investigating a more efficient and effective estimation method for the generalization error, enhancing its practicality. This will be our future work. As for the second direction, numerous existing and new weighting schemes can be improved or proposed using the generalization error as the difficulty measure. Our theoretical findings supplement or even correct the current understanding. For example, samples with large margins may also be hard-to-classify in some cases (e.g., with heterogeneous samples in their neighbors).

Second, the existing conclusions on convergence speed have been extended. For the linear predictors, the existing conclusion is extended by considering our difficulty measure, namely, the generalization error. For the nonlinear predictors, the conclusion is extended into the multi-class cases. Furthermore, the explicit relationship between the generalization gap and sample weights has been established. Our theorem indicates that assigning large weights on the hard samples may be more effective even when the source distribution P_s and target distribution P_t are equal.

Our theoretical findings of the generalization bounds provide better explanations to existing weighting schemes. For example, if heavy noise exists in the dataset, then the weights of the noisy samples should be decreased. As noisy samples are absent in the target distribution (i.e., $p_t(\boldsymbol{x}_i) = 0$), the weights of the noisy samples in a data set with heavy noise should be decreased to better match the source and target distributions. The experiments on the noisy data are shown in Fig. A-5 in which decreasing the weights of noisy samples obtain the best performance. In imbalanced learning, samples in small categories have higher errors on the average. Increasing the weights of the hard samples will not only accelerate the optimization but also improve the performance on the tail categories, as shown in Figs. 5(d) and 6. These high-level intuitions justify a number of difficulty-based weighting methods. Easy-first schemes, such as Super-loss [7] and Truncated loss [6], perform well on noisy data. Hard-first schemes, such as G-RW [12] and Focal Loss [4], are more suitable for imbalanced data.

6 Conclusion

This study theoretically investigates difficulty-based sample weighting. First, the generalization error is verified as a universal measure as a means of reflecting the four main factors influencing the learning difficulty of samples. Second, based on a universal difficulty measure, the role of the difficulty-based weighting strategy for deep learning is characterized in terms of convergence dynamics and the generalization bound. Theoretical findings are also presented. Increasing the weights of the hard samples may accelerate the optimization. A good set of weights should balance the tradeoff between the assigning of large weights on the hard samples (heavy training noises are absent) and keeping the test and the weighted training distributions close. These aspects enlighten the understanding and design of existing and future weighting schemes.

References

1. Zhou, X., Wu, O.: Which samples should be learned first: easy or hard? arXiv preprint arXiv:2110.05481 (2021)
2. Khan, S.-H., Hayat, M., Bennamoun, M., Sohel, F.-A., Togneri, R.: Cost-sensitive learning of deep feature representations from imbalanced data. IEEE Trans. Neural Netw. Learn. Syst. 29(8), 3573–3587 (2018)
3. Kuma, M.-P., Packer, B., Koller, D.: Self-paced learning for latent variable models. In: NeurIPS, pp. 1–9 (2010)
4. Lin, T.-Y., Goyal, P., Girshick, R., He, K., Dollar, P.: Focal loss for dense object detection. IEEE Trans. Pattern Anal. Mach. Intell. 42(2), 318–327 (2020)
5. Bengio, Y., Louradour, J., et al.: Curriculum learning. In: ICML, pp. 41–48 (2009)
6. Wang, W., Feng, F., He, X., Nie, L., Chua, T.-S.: Denoising implicit feedback for recommendation. In: WSDM, pp. 373–381 (2021)
7. Castells, T., Weinzaepfel, P., Revaud, J.: SuperLoss: a generic loss for robust curriculum learning. In: NeurIPS, pp. 1–12 (2020)
8. Emanuel, B.-B., et al.: Asymmetric loss for multi-label classification. arXiv preprint arXiv:2009.14119 (2020)
9. Santiago, C., Barata, C., Sasdelli, M., et al.: LOW: training deep neural networks by learning optimal sample weights. Pattern Recogn. 110(1), 107585 (2021)
10. Li, B., Liu, Y., Wang, X.: Gradient harmonized single-stage detector. In: AAAI, pp. 8577–8584 (2019)
11. Cui, Y., Jia, M., Lin, T.-Y., Song, Y., Belongie, S.: Class-balanced loss based on effective number of samples. In: CVPR, pp. 9260–9269 (2019)
12. Zhang, S., Li, Z., Yan, S., He, X., Sun, J.: Distribution alignment: a unified framework for long-tail visual recognition. In: CVPR, pp. 2361–2370 (2021)
13. Zhang, J., Zhu, J., Niu, G., Han, B., Sugiyama, M., Kankanhalli, M.: Geometry-aware instance-reweighted adversarial training. In: ICLR, pp. 1–29 (2021)

14. Aguilar, E., Nagarajan, B., Khatun, R., Bolaños, M., Radeva, P.: Uncertainty modeling and deep learning applied to food image analysis. In: Ye, X., et al. (eds.) BIOSTEC 2020. CCIS, vol. 1400, pp. 3–16. Springer, Cham (2021). https://doi.org/10.1007/978-3-030-72379-8_1

15. Xiao, Y., Wang, W.-Y.: Quantifying uncertainties in natural language processing tasks. In: AAAI, pp. 7322–7329 (2019)

16. Byrd, J., Lipton, Z.-C.: What is the effect of importance weighting in deep learning? In: ICML, pp. 1405–1419 (2019)

17. Soudry, D., Hoffer, E., Nacson, M.-S., Gunasekar, S., Srebro, N.: The implicit bias of gradient descent on separable data. J. Mach. Learn. Res. **19**(1), 1–14 (2018)

18. Chizat, L., Bach, F.: Implicit bias of gradient descent for wide two-layer neural networks trained with the logistic loss. arXiv preprint arXiv:2002.04486 (2020)

19. Lyu, K., Li, J.: Gradient descent maximizes the margin of homogeneous neural networks. arXiv preprint arXiv:1906.05890 (2019)

20. Xu, D., Ye, Y., Ruan, C.: Understanding the role of importance weighting for deep learning. In: ICLR, pp. 1–20 (2020)

21. Goodfellow, I., Bengio, Y., Courville, A.: Deep Learning (2016)

22. Heskes, T.: Bias/variance decompositions for likelihood-based estimators. Neural Comput. **10**(6), 1425–1433 (1998)

23. Alex, K., Hinton, G.: Learning multiple layers of features from tiny images. Technical report (2009)

24. He, K., Zhang, X., Ren, S., Sun, J.: Deep residual learning for image recognition. In: CVPR, pp. 770–778 (2016)

25. Shu, J., et al.: Meta-Weight-Net: learning an explicit mapping for sample weighting. In: NeurIPS, pp. 1–23 (2019)

26. Yang, Z., Yu, Y., You, C., Jacob, S., Yi, M.: Rethinking bias-variance trade-off for generalization of neural networks. In: ICML, pp. 10767–10777 (2020)

27. Shin, W., Ha, J.-W., Li, S., Cho, Y., et al.: Which strategies matter for noisy label classification? Insight into loss and uncertainty. arXiv preprint arXiv:2008.06218 (2020)

28. Chang, H.-S., Erik, L.-M., McCallum, A.: Active bias: training more accurate neural networks by emphasizing high variance samples. In: NeurIPS, pp. 1003–1013 (2017)

29. Swayamdipta, S., et al.: Dataset cartography: mapping and diagnosing datasets with training dynamics. arXiv preprint arXiv:2009.10795 (2020)

30. Agarwal, C., Hooker, S.: Estimating example difficulty using variance of gradients. arXiv preprint arXiv:2008.11600 (2020)

31. Wolterink, J.-M., Leiner, T., et al.: Generative adversarial networks for noise reduction in low-dose CT. IEEE Trans. Med. Imaging **36**(12), 2536–2545 (2017)

32. Lowd, D., Meek, C.: Adversarial learning. In: SIGKDD, pp. 641–647 (2005)

33. Elsayed, G.-F., Krishnan, D., Mobahi, H., Regan, K., Bengio, S.: Large margin deep networks for classification. In: NeurIPS, pp. 850–860 (2018)

34. Ghasemi, A., Zahediasl, S.: Normality tests for statistical analysis: a guide for non-statisticians. Int. J. Endocrinol. Metab. **10**(2), 486–489 (2012)

35. Gal, Y., Ghahramani, Z.: Dropout as a Bayesian approximation: representing model uncertainty in deep learning. In: ICML, pp. 1050–1059 (2016)

36. Abdar, M., et al.: A review of uncertainty quantification in deep learning: Techniques, applications and challenges. Inf. Fusion **76**(1), 243–297 (2021)

37. Kendall, A., Gal, Y.: What uncertainties do we need in Bayesian deep learning for computer vision? In: NeurIPS, pp. 5575–5585 (2017)

Avoiding Forgetting and Allowing Forward Transfer in Continual Learning via Sparse Networks

Ghada Sokar[1]([✉]), Decebal Constantin Mocanu[1,2], and Mykola Pechenizkiy[1]

[1] Eindhoven University of Technology, Eindhoven, The Netherlands
{g.a.z.n.sokar,m.pechenizkiy}@tue.nl
[2] University of Twente, Enschede, The Netherlands
d.c.mocanu@utwente.nl

Abstract. Using task-specific components within a neural network in continual learning (CL) is a compelling strategy to address the *stability-plasticity* dilemma in fixed-capacity models without access to past data. Current methods focus only on selecting a sub-network for a new task that reduces forgetting of past tasks. However, this selection could limit the forward transfer of *relevant* past knowledge that helps in future learning. Our study reveals that satisfying both objectives jointly is more challenging when a unified classifier is used for all classes of seen tasks–class-Incremental Learning (class-IL)–as it is prone to ambiguities between classes across tasks. Moreover, the challenge increases when the semantic similarity of classes across tasks increases. To address this challenge, we propose a new CL method, named AFAF (Code is available at: https://github.com/GhadaSokar/AFAF.), that aims to Avoid Forgetting and Allow Forward transfer in class-IL using fix-capacity models. AFAF allocates a sub-network that enables *selective* transfer of relevant knowledge to a new task while preserving past knowledge, *reusing* some of the previously allocated components to utilize the fixed-capacity, and addressing class-ambiguities when similarities exist. The experiments show the effectiveness of AFAF in providing models with multiple CL desirable properties, while outperforming state-of-the-art methods on various challenging benchmarks with different semantic similarities.

Keywords: Continual learning · Class-incremental learning · Stability plasticity dilemma · Sparse training

1 Introduction

Continual learning (CL) aims to build intelligent agents based on deep neural networks that can learn a sequence of tasks. The main challenge in this paradigm is the stability-plasticity dilemma [34]. Optimizing all model weights on a new task (*highest plasticity*) causes forgetting of past tasks [33]. While fixing

Supplementary Information The online version contains supplementary material available at https://doi.org/10.1007/978-3-031-26409-2_6.

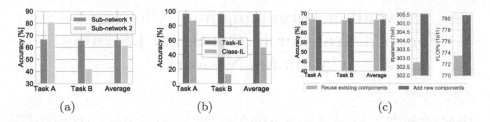

(a)   (b)   (c)

Fig. 1. (a) Two different sub-networks are evaluated for Task B in class-IL. One sub-network balances forgetting and forward transfer, while the other maintains the performance of Task A at the expense of Task B. (b) Performing the same model-altering scheme in task-IL and class-IL leads to different performance. (c) By reusing some of Task A's components during learning Task B, we can achieve similar performance in class-IL as adding new components while reducing the memory and computational costs, represented by model parameters (#params) and floating-point operations (FLOPs), respectively. Details are in Appendix B.

all weights (*highest stability*) hinders learning new tasks. Finding the balance between stability and plasticity is challenging. The challenge becomes more difficult when other CL requirements are considered, such as using fixed-capacity models without access to past data and limiting memory and computational costs [13].

Task-specific components strategy [12,30,48] offers some flexibility to address this dilemma by using different components (connections/neurons), i.e., sub-network within a model, for each task. The components of a new task are flexible to learn, while the components of past tasks are fixed. There are some challenges that need to be tackled in this strategy to balance multiple CL desiderata:

(a) **Selection of a new sub-network.** Current methods focus solely on forgetting and choose a sub-network for a new task that would maintain the performance of past tasks regardless of its effectiveness for learning the new task. This hinders the forward transfer of relevant past knowledge in future learning.

(b) **Managing the fixed-capacity and training efficiency.** Typically, new components are added for every new task in each layer, which may unnecessarily consume the available capacity and increase the computational costs. Using fixed-capacity models for CL requires utilizing the capacity efficiently.

(c) **Operating in the class-Incremental Learning setting (class-IL).** Unlike task-Incremental Learning (task-IL), where each task has a separate classifier under the assumption of the availability of task labels at inference, in class-IL, a *unified* classifier is used for all classes in seen tasks so far. The latter is more realistic, yet it brings additional challenges due to agnosticity to task labels: (1) **class ambiguities**, using past knowledge in learning new classes causes ambiguities between old and new classes. (2) **component-agnostic inference**, all model components are used at inference since the task label is not available to select the corresponding components. This increases interference between tasks and the performance dependency on the sub-network of each task.

In response to these challenges, we study the following question: *How to alter the model structure when a new task arrives to balance CL desiderata in class-IL? Specifically, which components should be added, updated, fixed, or reused?*
We summarize our findings below with illustrations shown in Fig. 1:

- The chosen sub-network of each task has a crucial role in the performance, affecting both forgetting and forward transfer (Fig. 1a).
- The optimal altering of a model differs in task-IL and class-IL. For example, all components from past similar tasks could be *reused* in learning a new task with minimal memory and computational costs in task-IL. However, this altering limits learning in class-IL due to class ambiguities (Fig. 1b).
- Reusability of past *relevant* components is applicable in class-IL under careful considerations for class ambiguity. It enhances memory and computational efficiency while maintaining performance (Fig. 1c).
- Neuron-level altering is crucial in class-IL (e.g., *fixing* some neurons for a task), while connection-level altering could be sufficient in task-IL since only one sub-network is selected at inference using the task label (Appendix C).
- The challenge in balancing CL desiderata increases in class-IL when similarity across tasks increases (Sect. 5.1).

Motivated by these findings, we propose a new CL method, named AFAF, based on sparse sub-networks within a fix-capacity model to address the above-mentioned challenges. Without access to past data, AFAF aims to Avoid Forgetting and Allow Forward transfer in class-IL. In particular, when a new task arrives, we identify the relevant knowledge from past tasks and allocate a new sub-network that enables selective forward transfer of this knowledge while maintaining past knowledge. Moreover, we reuse some of the allocated components from past tasks. To enable selective forward transfer and component re-usability jointly with forgetting avoidance in class-IL, AFAF considers the extra challenges of class-IL (class-ambiguities and agnostic component inference) in model altering. We propose two variants of standard CL benchmarks to study the challenging case where high similarity exists across tasks in class-IL. Experimental results show that AFAF outperforms baseline methods on various benchmarks while reducing memory and computation costs. In addition, our analyses reveal the challenges of class-IL over task-IL that necessitate different model altering.

2   Related Work

We divide CL methods into two categories: replay-free and replay-based.

Replay-Free. In this category, past data is inaccessible during future learning. It includes two strategies: **(1) *Task-specific components.*** *Specific* components are assigned to each task. Current methods either *extend* the initially allocated capacity of a model for new tasks [43,55] or use a *fixed-capacity* model and add a *sparse* sub-network for each task [29,30,32,48,52,54]. Typically, connections of past tasks are kept *fixed*, and the newly added ones are *updated* to learn the current task. The criteria used for *adding* new connections often focus solely on avoiding forgetting, limiting the selective transfer of relevant knowledge to learn future tasks. Moreover, most methods rely on task labels to pick the connections corresponding to the task at inference. SpaceNet [48] addressed component-agnostic inference by learning sparse representation during training and *fixing* a fraction of the most important neurons of each task to reduce interference. Sparse representations are learned by training each sparse sub-network using dynamic sparse training [14,36], where weights and the sparse topology are jointly optimized. Yet, selective transfer and class ambiguity are not addressed. **(2) *Regularization-based.*** A *fixed-capacity* model is used, and *all* weights are updated for each task. Forgetting is addressed by constraining changes in the important weights of past tasks [1,19,56] or via distillation loss [9,25].

Replay-Based. In this category, forgetting is addressed by *replaying*: (1) a subset of old samples [3,5,27,41,42], (2) pseudo-samples from a generative model [37,45,46], or (3) generative high-level features [50]. A buffer is used to store old samples or a generative model to generate them.

Attention to task similarities and forward transfer has recently increased. Most efforts are devoted to task-IL. In [40], the analysis showed that higher layers are more prone to forgetting, and intermediate semantic similarity across tasks leads to maximal forgetting. SAM [47] meta-trains a self-attention mechanism for selective transfer in dense networks. CAT [18] addresses the relation between task similarities and forward/backward transfer in task-IL, where task labels are used to find similar knowledge in dense models. In [23], an expectation-maximization method was proposed to select the shared or added *layers* to promote transfer in task-IL. Similarly, [51] uses a modular neural network architecture and searches for the optimal path for a new task by the composition of neural modules. A task-driven method is used to reduce the exponential search space. In [6], the lottery ticket hypothesis [11] is studied for CL.

3   Network Structure Altering

When a model faces a new task, task-specific components methods alter its structure via some actions to address the stability-plasticity dilemma. Next, we will present commonly used and our proposed actions.

3.1   Connection-Level Actions

Most state-of-the-art methods alter fixed-capacity models at the **connection level**. The connection-level actions include:

- "**Add**": New connections, parameterized by $\mathbf{W}^t$, are added for each new task t. Each task has a *sparse* sub-network resulting from either pruning dense connections [30] (Fig. 2b) or adding sparse ones from scratch [48] (Fig. 2c). The current practice is to **randomly** add connections in *each* layer using unimportant components of past tasks focusing solely on forgetting [30,48];

$\mathbf{W}^t = \{\mathbf{W}_l^t : 1 \leq l \leq L - 1\}$ where L is the number of layers. This will be addressed in Sect. 4 to enable **selective transfer** and **reusability**.

- "**Update**": $\mathbf{W}^t$ are updated during task t training (i.e., allows plasticity).
- "**Fix**": Once a task has been learned, $\mathbf{W}^t$ are frozen (i.e., controls stability).

Note that regularization methods *add* **dense** connections at step $t = 0$ (Fig. 2a). Each task *updates* all weights. Stability is controlled via regularization.

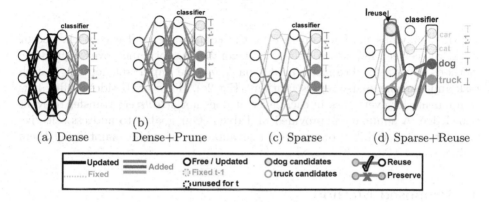

Fig. 2. Overview of a model when it faces a new task t using different methods. (a) Regularization methods use and update all connections with regularization for learning task t [1,19]. (b) Dense connections are added, and unimportant connections are pruned after training [30]. Connections of previous tasks are fixed. (c) Sparse connections are added from scratch [48]. Connections and important neurons of past tasks are fixed (dashed lines and filled circles). (d) Our proposed method, AFAF. Starting from layer l_{reuse}, new connections are added for each class using the candidate neurons for the class to enable selective transfer and free neurons to capture residual knowledge. New outgoing connections are allowed from fixed neurons at layer l but could only connect free neurons (unfilled circles) in layer $l + 1$ to preserve the old knowledge.

3.2   Neuron-Level Actions

Component-agnostic inference in class-IL makes reducing task interference at the connection level only more challenging since connections from different tasks share the same neurons (Fig. 2b). Reducing interference at the *representational* level is more efficient (Appendix C). Hence, we believe that the following neuron-level actions are needed:

- "**Fix**": After learning a task, its *important* neurons should be frozen to reduce the drift in its representation [48]. Other neurons are "free" to be updated.
- "**Reuse**": Reusing neurons that capture useful representation in learning future tasks. Fixed neurons could be reused by allowing **outgoing** connections only from these neurons. Details are provided in the next section.

Table 1. Actions used to alter the model components at the connection and neuron levels by different methods.

Method	Add weights	Fix old weights	Fix neurons	Reuse past components	Selective transfer
EWC [19], MAS [1]	Dense	×	×	×	×
PackNet [30]	Dense+Prune	√	×	×	×
SpaceNet [48]	Sparse	√	√	×	×
AFAF (ours)	Sparse	√	√	√	√

Most previous methods operate at the connection level except the recent work, SpaceNet [48], which shows its favorable performance over replay-free methods using a fixed-capacity model via *fixing* the important neurons of each task and learning sparse representations (Fig. 2c). However, besides adding new components for every task in each layer, it does not *selectively* transfer relevant knowledge. A summary is provided in Table 1. Our goal is to address selective transfer and reusability of previous relevant neurons while maintaining them stable.

4   Proposed Method

We consider the problem of learning T tasks sequentially using a *fix-capacity* model with a unified classifier. Each task t brings new C classes. The output layer is extended with new C neurons. Let $\mathcal{D}^t = \{D^{t_c}\}|_{c=1}^{C}$, where D^{t_c} is the data of class c in task t. Once a task has been trained, its data is *discarded*.

AFAF is a new task-specific component method that dynamically trains a sparse sub-network for each task from scratch. Our goal is to alter the model components to (i) reuse past components and (ii) selectively transfer relevant knowledge while (iii) reducing class ambiguity in class-IL. This relies on the selection of a new sub-network and its training. When a new task arrives, AFAF adds a sub-network such that past knowledge is transferred from relevant neurons while retaining it, and residual knowledge is learned. Section 4.1 introduces the selection mechanism of a sub-network and illustrates the added and reused components. Section 4.2 discusses the considerations for class ambiguity. In Sect. 4.3, we present task training and identify the fixed and updated components.

4.1   Selection of a New Sub-network

Notation. Let $l \in \{1, .., L\}$ represents a layer in a neural network model. A sparse connections W_l^t, with a sparsity ϵ_l, are allocated for task t in layer l between a subset of the neurons denoted by $\mathbf{h}_l^{alloc}$ and $\mathbf{h}_{l+1}^{alloc}$ in layers l and $l+1$, respectively. We use the term *neuron* to denote a node in multilayer perceptron networks or a feature map in convolution neural networks (CNNs). The selected neurons in each layer determine the initial sub-network for a new task.

Typically, when there is a semantic similarity between old and new classes, the existing learned components likely capture some useful knowledge for the current task. Hence, we propose to reuse some of these components. Namely, instead of adding new connections in each layer, we allocate new sparse connections starting from layer l_{reuse} (Fig. 2d). l_{reuse} is a hyper-parameter that controls the trade-off between adding new connections and reusing old components based on existing knowledge and the available capacity. Hence, the connections of a new task t are $\mathbf{W}^t = \{\mathbf{W}^t_l : l_{reuse} \leq l \leq L - 1\}$. Layers from $l = 1$ up to but excluding layer l_{reuse} remains unchanged. We show that reusing past components reduces memory and computational costs (Sect. 5.1), balances forward and backward transfer (Sect. 5.2), and utilizes the available fixed-capacity efficiently by allowing new dissimilar tasks to acquire more resources that are saved by reusability, i.e., higher density for sparse sub-networks (Appendix D).

Starting from l_{reuse}, we add new sparse connections $\mathbf{W}^t_l$ in each layer l. The selected neurons for allocation should allow: (i) selective transfer of relevant knowledge to promote forward transfer and (ii) preserving old representation to avoid forgetting. To this end, we select a set of *candidate* and *free* neurons in each layer, as we will discuss next. Details are provided in Algorithm 1.

Identify Candidate Neurons. To *selectively transfer* the relevant knowledge, for each layer $l \geq l_{reuse}$, we identify a set of candidate neurons $\mathcal{R}^c_l$ that has a high potential of being useful when "reused" in learning *class* c in a new task t. Classes with semantic similarities are most likely to share similar representations (Appendix F). Hence, we consider the average activation of a neuron as a metric to identify its potential for reusability. In particular, we feed the data of each class c in a new task t, $\mathcal{D}^{t_c}$, to the trained model at time step $t - 1$, $f^{t-1}(\mathbf{W}^{t-1})$, and calculate the average activation $\mathcal{A}^c_l$ in each layer l as follows:

$$\mathcal{A}^c_l(\mathbf{W}^{t-1}) = \frac{1}{|\mathcal{D}^{t_c}|} \sum_{i=1}^{|\mathcal{D}^{t_c}|} \mathbf{a}_l(x^{t_c}_i), \qquad (1)$$

where $\mathbf{a}_l$ is the vector of neurons activation at layer l when a sample $x^{t_c}_i \in \mathcal{D}^{t_c}$ is fed to the model $f^{t-1}(\mathbf{W}^{t-1})$, and $|\mathcal{D}^{t_c}|$ is the number of samples of class c. Once the activation is computed, neurons with the top-κ activation are selected as potential candidates $\mathcal{R}^c_l$ as follows:

$$\mathcal{R}^c_l = \{i | \mathcal{A}^c_{l_i} \in \text{top-} \kappa(\mathcal{A}^c_l)\}, \qquad (2)$$

where κ denotes the number of candidate neurons, and $\mathcal{A}^c_{l_i}$ is the average activation of neuron i in layer l. Neuron activity shows its effectiveness as an estimate of a neuron/connection importance in pruning neural networks for single task learning [7,15,28]. To assess our choice of this metric to identify the candidate neurons, we compare against two metrics: **(1) Random**, where the candidate neurons are randomly chosen from all neurons in a layer, and **(2) Lowest**, where the candidates are the neurons with the lowest activation (See Sect. 5.2).

Note that by exploiting the representation of candidate neurons $\mathcal{R}^c_{l_{reuse}}$ in layer l_{reuse}, we selectively *reuse* past components connected to these neurons in preceding layers ($l < l_{l_{reuse}}$). Hence, these components are stable but reusable.

Identify Free Neurons. To preserve past representation while allowing reusability of neurons, "fixed" neurons that might be selected as candidates should be stable. Hence, we allow *outgoing* connections from these neurons but not incoming connections. The outgoing connections from fixed neurons in layer l can be connected to "free" neurons in layer $l+1$. To this end, in each layer, we select a subset of the free neurons for *each task*, denoted as $\mathcal{S}_l^{Free}$. For these neurons, incoming and outgoing connections are allowed to be allocated and used to capture the specific representation of a new task.

Allocation. The neurons used to allocate new sparse weights $\mathbf{W}_l^c$ between layers l and $l+1$ for a class c are as follows:

Algorithm 1. AFAF Sub-network Allocation

1: **Require:** l_{reuse}, sparsity level ϵ_l, $|\mathbf{h}_l^{alloc}|$, κ, $\mathbf{h}_l^{Fix}$
2: **for each** class c in task t **do**   ▷ Get candidate neurons
3:     $\mathcal{A}_l^c \leftarrow$ calculate average activation of $\mathcal{D}^{t_c}$ ▷ Eq. 1
4:     $\mathcal{R}_l^c \leftarrow$ get candidates for class c $\forall l \geq l_{reuse}$ ▷ Eq. 2
5: **end for each**
6: $\mathcal{S}_l^{Free} \leftarrow$ randomly select subset of free neurons $\forall l \geq l_{reuse}$
7: $\mathcal{R}_{L-1}^c, \mathcal{R}_L^c \leftarrow \emptyset$ ▷ No candidate neurons used in last layer
8: $\mathcal{S}_L^{Free} \leftarrow \{c\}$ $\forall c \in t$ ▷ Output neurons for new classes
9: **for each** class c in task t **do**
10:     **for** $l \leftarrow l_{reuse}$ to $L-1$ **do**
11:         $\mathbf{h}_l^{alloc} \leftarrow \{\mathcal{R}_l^c \cup \mathcal{S}_l^{Free}\}$ ▷ Neurons for allocation
12:         $\mathbf{h}_{l+1}^{alloc} \leftarrow \{(\mathcal{R}_{l+1}^c \backslash \mathbf{h}_{l+1}^{Fix}) \cup \mathcal{S}_{l+1}^{Free}\}$ ▷ Neurons for allocation
13:         Allocate $\mathbf{W}_l^c$ with sparsity ϵ_l between $\mathbf{h}_l^{alloc}$ & $\mathbf{h}_{l+1}^{alloc}$
14:         $\mathbf{W}_l \leftarrow \mathbf{W}_l \cup \mathbf{W}_l^c$
15:     **end for each**
16: **end for each**

$$\mathbf{h}_l^{alloc} = \{\mathcal{R}_l^c \cup \mathcal{S}_l^{Free}\},$$
$$\mathbf{h}_{l+1}^{alloc} = \{(\mathcal{R}_{l+1}^c \setminus \mathbf{h}_{l+1}^{Fix}) \cup \mathcal{S}_{l+1}^{Free}\}, \tag{3}$$

where $\mathbf{h}_{l+1}^{Fix}$ is the set of fixed neurons at layer $l+1$.

4.2   Addressing Class Ambiguities

Unlike task-IL, reusing past relevant knowledge in future learning may result in class ambiguity in class-IL (Sect. 1). To this end, we propose three constraints for reusability that aim to (1) allow a new task to learn its specific representation and (2) increase the decision margin between classes (see Sect. 5.2 for analysis).

Learn Specific Representation. Learning specific representation reduces ambiguity between similar classes. Hence, we add two constraints. First, new connections should be added in at least one layer before the classification layer to capture the specific representation of the task (i.e., $l_{reuse} \in [2, L-2]$). Second, the output connections are allocated using free neurons only (i.e., no candidate neurons are used; $\mathbf{h}_{L-1}^{alloc} = \mathcal{S}_{L-1}^{Free}$) since candidate neurons in the highest-level layer are highly likely to capture the specific representation of past classes.

Increase the Decision Margin Between Classes. To learn more discriminative features and increase the decision margin between classes, we use orthogonal weights in the output layer [24]. To this end, once a task has been trained, we force all its neurons in the last layer, $\mathbf{h}_{L-1}^{alloc}$, to be fixed and not reusable.

4.3   Training

The new connections are trained with stochastic gradient descent. During training, the weights and important neurons of past tasks are kept fixed to protect past representation. We follow the approach in [48] to train a sparse topology (connections distribution) and identify the portion of the important neurons from $\mathbf{h}_l^{alloc}$ in each layer that will be fixed after training (Appendix G). In short, a sparse topology is optimized by a dynamic sparse training approach to produce sparse representation. To reuse important neurons of past tasks while protecting the representation, we block the gradient flow through all-important neurons of past tasks, even if they are reused as candidates for the current task. The gradient $\mathbf{g}_l$ through the neurons of layer l is:

$$\mathbf{g}_l = \mathbf{g}_l \otimes (1 - \mathbb{1}_{\mathbf{h}_l^{Fix}}), \tag{4}$$

where $\mathbb{1}$ is the indicator function. This *allows* not to forget past knowledge while reusing it for selective transfer. Since the important neurons of dissimilar classes are less likely to be involved in the sub-network selection, they are protected.

5   Experiments and Results

Baselines. Our study focuses on the replay-free setting using a fixed-capacity model. Therefore, we compare with several representative regularization methods that use *dense* fixed-capacity, **EWC** [19], **MAS** [1], and **LWF** [25]. In addition, we compare with task-specific components methods that use *sparse* sub-networks within a fixed-capacity model, **PackNet** [30] and **SpaceNet** [48].

Benchmarks. We performed our experiments on three sets of benchmarks: (1) standard split-CIFAR10, (2) sequences with high semantic similarity at the class level across tasks, and (3) sequence of mixed datasets where tasks come from different domains to study the stability-plasticity dilemma for sequences with concept drift and interfering tasks.

Standard Evaluation. We evaluate the standard **split-CIFAR10** benchmark with 5 tasks. Each task consists of 2 consecutive classes of CIFAR10 [21].

Similar Sequences. To assess replay-free methods under more challenging conditions, we design two new benchmarks with high semantic similarity across tasks. In the absence of past data, we test the unified classifier's ability to distinguish between similar classes when they are not presented together within the same task. **(1) sim-CIFAR10** is constructed from CIFAR-10 by shuffling the order of classes to increase the similarity across tasks (Appendix A; Table 5). It consists of 5 tasks. **(2) sim-CIFAR100** is constructed from CIFAR-100 [20]. Classes within the same superclass in CIFAR-100 have high semantic similarity. Hence, we construct a sequence of 8 tasks with two classes each and distribute the classes from the same superclass in different tasks (Appendix A; Table 6).

Mix Datasets. The considered datasets are CIFAR10 [20], MNIST [22], NotMNIST [4], and FashionMNIST [53]. We construct a sequence of 8 tasks with 5 classes each. The first four tasks are dissimilar, while the second four are similar to the first ones (Appendix A; Table 7).

Implementation Details. Motivated by the recent study on architectures for CL [35], we follow [18,44,51] to use an AlexNet-like architecture [21] that is trained from scratch using stochastic gradient descent. We start reusing relevant knowledge in future tasks after learning similar ones. Therefore, for Mix and sim-CIFAR100, we start reusing past components from task 5, while for split-CIFAR10 and sim-CIFAR10, we start from task 3. Earlier tasks add connections in each layer using the free neurons. For all benchmarks, we use l_{reuse} of 4.

Evaluation Metrics. To evaluate different CL requirements, we assess various metrics: (1) **Average accuracy (ACC)**, which measures the performance at the end of the learning experience, (2) **Backward Transfer (BWT)** [27], which measures the influence of learning a new task on previous tasks (large negative BWT means forgetting), (3) **Floating-point operations (FLOPs)**, which measure the required computational cost, and (4) **Model size (#params)**, which is the number of model parameters. More details are in Appendix A.

5.1   Results

Table 2 shows the performance on each benchmark. AFAF consistently outperforms regularization-based methods and other task-specific components methods on all benchmarks. The difference in performance between split-CIFAR10 and sim-CIFAR10, which have the same classes in a different order, reveals the challenge caused by having similar classes across tasks in class-IL. All studied methods have lower ACC and BWT on sim-CIFAR10 than split-CIFAR10. Yet, AFAF is the most robust method towards this challenge. When the similarity across tasks increases more, as in sim-CIFAR100, regularization methods and PackNet fail to achieve a good performance. We also observe that LWF outperforms other regularization-based methods in most cases except on the Mix datasets benchmark, where there is a big distribution shift across tasks from different domains. Most task-specific components methods outperform the regularization-based ones with much lower forgetting.

Analyzing task-specific components methods, we observe that altering the model at the connection level only by PackNet is not efficient in class-IL despite its high performance in task-IL (Appendix C). Besides the additional memory and computational overhead of pruning and retraining dense models, the performance is lower than other task-specific components methods. Altering at the connection and neuron levels, as in SpaceNet and AFAF, enables higher performance. The gap between these methods increases when the sequence has a larger number of tasks with high similarities (i.e., sim-CIFAR100 and Mix datasets).

AFAF consistently achieves higher ACC and BWT than SpaceNet on all benchmarks with various difficulty levels. Interestingly, the achieved performance is obtained by *reusing* relevant knowledge via selective transfer. AFAF exploits the similarity across tasks in learning future tasks while addressing class ambiguities. Moreover, the performance gain is accompanied by using a smaller memory and less computational cost than all the baselines.

5.2 Analysis

Effect of Selective Transfer. To measure the impact of selective transfer and the initially allocated sub-network on forward and backward transfer in AFAF, we compare our selection strategy for the candidate neurons to two other potential strategies discussed in Sect. 4.1: *Random* and *Lowest*. To reveal the role of selective transfer on performance, we also report the Learning Accuracy (LA) [42], which is the average accuracy for each task directly after it is learned (Appendix A). We calculate this metric starting from the first task that reuses past components in its learning onwards. Figure 3 shows the results on Mix datasets and sim-CIFAR10. We present the performance of the tasks that reuse past components (i.e., $t \geq 5$ and $t \geq 3$ for Mix and sim-CIFAR10, respectively) since the performance of other earlier tasks is the same for all baselines (i.e., the allocation is based on the free neurons). As shown in the figure, using the relevant neurons with the highest activation to allocate a new sub-network leads to higher LA on new tasks and lower negative BWT on past tasks than using random neurons. AFAF also has higher ACC than the Random baseline by 4.79% and 3.01% on Mix and sim-CIFAR10, respectively. On the other hand, the Lowest baseline limits learning new tasks. It has much lower ACC and LA than the other two baselines. Note that the high BWT of this Lowest baseline is a factor of its low LA. This analysis shows the effect of the *initial* sub-network on performance, although topological optimization occurs during training.

Class Ambiguities. We performed an ablation study to assess each of our proposed contributions in addressing class ambiguities in class-IL (Sect. 4.2). We performed this analysis on Mix datasets and sim-CIFAR10 benchmarks. To show that class ambiguity causes more challenges in class-IL, we also report the performance in task-IL. To compare only the effect of class ambiguities in both scenarios, we assume components-agnostic inference for task-IL. Yet, task labels are used to select the output neurons that belong to the task at hand.

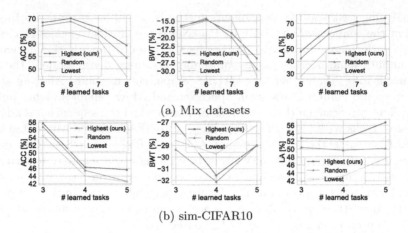

(a) Mix datasets

(b) sim-CIFAR10

Fig. 3. Performance of AFAF backbone with different strategies for selecting the candidate neurons used to add new connections.

Table 2. Evaluation results on four CL benchmarks in the replay-free class-IL setting with fixed-capacity models.

Method	Split-CIFAR10				Sim-CIFAR10			
	ACC (↑)	BWT (↑)	FLOPs (↓)	#params (↓)	ACC (↑)	BWT (↑)	FLOPs (↓)	#params (↓)
Dense model	–	–	1 × (14.97e14)	1 × (23459520)	–	–	1 × (14.97e14)	1 × (23459520)
EWC [19]	38.30 ± 0.81	−59.30 ± 2.03	1×	1×	28.90 ± 3.11	−66.10 ± 5.49	1×	1×
LWF [25]	48.10 ± 2.28	−42.33 ± 1.15	1×	1×	40.43 ± 1.22	−50.36 ± 1.98	1 ×	1×
MAS [1]	38.30 ± 1.06	−56.56 ± 3.30	1×	1×	28.93 ± 4.05	−61.86 ± 5.66	1×	1×
PackNet [30]	44.33 ± 0.85	−50.40 ± 1.45	3.081×	1×	32.63 ± 1.22	−61.96 ± 1.52	3.081×	1×
SpaceNet [48]	51.63 ± 1.28	−36.50 ± 1.53	0.154×	0.154×	42.86 ± 4.57	−30.69 ± 4.63	0.325×	0.325×
AFAF (ours)	**52.35 ± 2.35**	**−32.93 ± 3.19**	**0.148×**	**0.148×**	**45.23 ± 2.14**	**−29.35 ± 3.54**	**0.322×**	**0.322×**

Method	Sim-CIFAR100				Mix datasets			
	ACC (↑)	BWT (↑)	FLOPs (↓)	#params (↓)	ACC (↑)	BWT (↑)	FLOPs (↓)	#params (↓)
Dense Model	–	–	1×(23.96e13)	1 × (23471808)	–	–	1 × (5.600e15)	1 × (23520960)
EWC [19]	15.73 ± 1.89	−74.50 ± 5.10	1×	1×	54.63 ± 1.93	−31.33 ± 2.82	1×	1×
LWF [25]	14.40 ± 2.69	−52.86 ± 1.12	1×	1×	40.60 ± 3.25	−60.46 ± 3.54	1×	1×
MAS [1]	13.50 ± 0.66	−81.70 ± 1.02	1×	1×	56.86 ± 1.81	−26.83 ± 3.44	1×	1×
PackNet [30]	10.12 ± 2.55	−20.35 ± 3.83	5.241×	0.8×	16.61 ± 2.35	**−18.58 ± 1.38**	5.250×	0.8×
SpaceNet [48]	32.86 ± 2.73	−36.29 ± 2.78	0.089×	0.089×	56.25 ± 1.69	−30.02 ± 1.79	0.053×	0.053×
AFAF (ours)	**33.74 ± 2.18**	**−21.26 ± 2.21**	**0.088×**	**0.088×**	**59.02 ± 1.76**	−25.91 ± 1.51	**0.050×**	**0.050×**

Table 3. Effect of each contribution in addressing class ambiguities: orthogonal output weights, using free neurons only for allocating output weights, and constraining reusing all past components. ACC is reported in class-IL and task-IL.

	sim-CIFAR10		Mix datasets	
	Class-IL	Task-IL	Class-IL	Task-IL
AFAF (ours)	**45.23 ± 2.14**	**94.57 ± 0.05**	**59.02 ± 1.76**	**93.41 ± 0.26**
$w/o\ orth\ \mathbf{W}_L$	41.74 ± 2.05	93.20 ± 0.26	56.38 ± 3.34	93.39 ± 0.30
$w/\mathcal{R}_{L-1}$	40.30 ± 2.14	93.27 ± 0.17	56.83 ± 1.19	92.85 ± 0.28
$w/l_{reuse} = L-1$	22.27 ± 1.61	83.54 ± 1.64	40.98 ± 2.75	85.05 ± 1.01

Analysis 1: Effect of Orthogonal Output Weights. We evaluate a baseline that denotes AFAF without using orthogonal weights in the output layer "w/o orth W_L". We obtain this baseline by fixing part of the neurons in the last layer instead of fixing all neurons. We use the same fraction used for other fully connected layers (Appendix A). Table 3 shows the results. Having a large decision margin via orthogonal output weights increases the performance. We observe that the difference between AFAF and this baseline is larger in class-IL than task-IL, which indicates that task-IL is less affected by class ambiguities.

Analysis 2: Effect of Using Free Neurons Only in the Last Layer. To analyze this effect, we add another baseline that uses free S_{L-1}^{free} and candidate $\mathcal{R}_{L-1}$ neurons in allocating the output weights. We denote this baseline as "w/$\mathcal{R}_{L-1}$". As shown in Table 3, using free neurons only allows learning specific representation that decreases the ambiguities across tasks. The performance in class-IL is improved by 4.93% and 2.19% on sim-CIFAR10 and Mix datasets, respectively.

Analysis 3: Effect of Constraining Reusing all Past Components. We analyze the performance obtained by reusing all past components in learning similar tasks (i.e., $l_{reuse} = L-1$). We denoted this baseline as "w/$l_{reuse} = L-1$". As shown in Table 3, the performance has decreased dramatically. Despite that the degradation also occurs in task-IL, it has less effect. This shows the challenge of balancing performance, memory, and computational costs in class-IL.

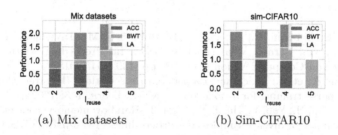

(a) Mix datasets   (b) Sim-CIFAR10

Fig. 4. Normalized performance of AFAF at different values of l_{reuse}.

Reusable Layers. We analyze the effect of reusing full layers on performance. We evaluate the performance at the possible values of $l_{reuse} \in [2, L-2]$ (Sect. 4.2). A min-max scaling is used to normalize the ACC, BWT, and LA; exact values are in Appendix E. Figure 4 shows the performance of Mix datasets and sim-CIFAR10. Adding new components in lower-level layers ($l_{reuse} \in \{2, 3\}$) enables new tasks to achieve high LA but increases forgetting (negative BWT), leading to lower ACC. Reusing the components in lower-level layers while adding new components to learn specific representations in the higher-level ones ($l_{reuse} = 4$) achieves a balance between ACC, BWT, and LA. While using a higher value for l_{reuse} limits the performance of new tasks due to class ambiguities, leading to the lowest LA and ACC.

We performed another study to illustrate the effect of reusing past components in utilizing the fixed-capacity. We show that reusability allows for using higher density for tasks that are dissimilar to the previously learned knowledge, which increases the performance. Details are provided in Appendix D.

6   Conclusion

Addressing the stability-plasticity dilemma while balancing CL desiderata is a challenging task. We showed that the challenge increases in the class-IL setting, especially when similar classes are not presented together within the same task. With our proposed task-specific components method, AFAF, we show that altering the model components based on exploiting past knowledge helps in achieving multiple desirable CL properties. Critically, the choice of the sub-network of a new task affects the forward and backward transfer. Hence, we proposed a selection mechanism that selectively transfers relevant knowledge while preserving it. Moreover, we showed that complete layers could be reused in learning similar tasks. Finally, we addressed the class ambiguity that arises in class-IL when similarities increase across tasks and showed that model altering at the connection and neuron levels is more efficient for component-agnostic inference.

References

1. Aljundi, R., Babiloni, F., Elhoseiny, M., Rohrbach, M., Tuytelaars, T.: Memory aware synapses: learning what (not) to forget. In: Proceedings of the European Conference on Computer Vision (ECCV), pp. 139–154 (2018)
2. Atashgahi, Z., et al.: Quick and robust feature selection: the strength of energy-efficient sparse training for autoencoders. Mach. Learn. 1–38 (2021)
3. Bang, J., Kim, H., Yoo, Y., Ha, J.W., Choi, J.: Rainbow memory: continual learning with a memory of diverse samples. In: Proceedings of the IEEE/CVF Conference on Computer Vision and Pattern Recognition, pp. 8218–8227 (2021)
4. Bulatov, Y.: Notmnist dataset. Technical report (2011). http://yaroslavvb.blogspot.it/2011/09/notmnist-dataset.html
5. Chaudhry, A., Ranzato, M., Rohrbach, M., Elhoseiny, M.: Efficient lifelong learning with a-gem. In: International Conference on Learning Representations (2018)
6. Chen, T., Zhang, Z., Liu, S., Chang, S., Wang, Z.: Long live the lottery: the existence of winning tickets in lifelong learning. In: International Conference on Learning Representations (2020)
7. Dekhovich, A., Tax, D.M., Sluiter, M.H., Bessa, M.A.: Neural network relief: a pruning algorithm based on neural activity. arXiv preprint arXiv:2109.10795 (2021)
8. Denil, M., Shakibi, B., Dinh, L., Ranzato, M., de Freitas, N.: Predicting parameters in deep learning. In: Proceedings of the 26th International Conference on Neural Information Processing Systems, vol. 2, pp. 2148–2156 (2013)
9. Dhar, P., Singh, R.V., Peng, K.C., Wu, Z., Chellappa, R.: Learning without memorizing. In: Proceedings of the IEEE/CVF Conference on Computer Vision and Pattern Recognition, pp. 5138–5146 (2019)

10. Evci, U., Gale, T., Menick, J., Castro, P.S., Elsen, E.: Rigging the lottery: making all tickets winners. In: International Conference on Machine Learning, pp. 2943–2952. PMLR (2020)
11. Frankle, J., Carbin, M.: The lottery ticket hypothesis: finding sparse, trainable neural networks. In: International Conference on Learning Representations (2018)
12. Golkar, S., Kagan, M., Cho, K.: Continual learning via neural pruning. arXiv preprint arXiv:1903.04476 (2019)
13. Hadsell, R., Rao, D., Rusu, A.A., Pascanu, R.: Embracing change: continual learning in deep neural networks. Trends Cogn. Sci. **24**, 1028–1040 (2020)
14. Hoefler, T., Alistarh, D., Ben-Nun, T., Dryden, N., Peste, A.: Sparsity in deep learning: pruning and growth for efficient inference and training in neural networks. J. Mach. Learn. Res. **22**(241), 1–124 (2021)
15. Hu, H., Peng, R., Tai, Y.W., Tang, C.K.: Network trimming: a data-driven neuron pruning approach towards efficient deep architectures. arXiv preprint arXiv:1607.03250 (2016)
16. Ioffe, S., Szegedy, C.: Batch normalization: accelerating deep network training by reducing internal covariate shift. In: International Conference on Machine Learning, pp. 448–456. PMLR (2015)
17. Jayakumar, S., Pascanu, R., Rae, J., Osindero, S., Elsen, E.: Top-KAST: top-k always sparse training. Adv. Neural. Inf. Process. Syst. **33**, 20744–20754 (2020)
18. Ke, Z., Liu, B., Huang, X.: Continual learning of a mixed sequence of similar and dissimilar tasks. Adv. Neural Inf. Process. Syst. **33**, 18493–18504 (2020)
19. Kirkpatrick, J., et al.: Overcoming catastrophic forgetting in neural networks. Proc. Natl. Acad. Sci. **114**(13), 3521–3526 (2017)
20. Krizhevsky, A., Hinton, G., et al.: Learning multiple layers of features from tiny images. Technical report. Citeseer (2009)
21. Krizhevsky, A., Sutskever, I., Hinton, G.E.: ImageNet classification with deep convolutional neural networks. Adv. Neural. Inf. Process. Syst. **25**, 1097–1105 (2012)
22. LeCun, Y.: The MNIST database of handwritten digits (1998). http://yann.lecun.com/exdb/mnist/
23. Lee, S., Behpour, S., Eaton, E.: Sharing less is more: lifelong learning in deep networks with selective layer transfer. In: International Conference on Machine Learning, pp. 6065–6075. PMLR (2021)
24. Li, X., et al.: OSLNet: deep small-sample classification with an orthogonal softmax layer. IEEE Trans. Image Process. **29**, 6482–6495 (2020)
25. Li, Z., Hoiem, D.: Learning without forgetting. IEEE Trans. Pattern Anal. Mach. Intell. **40**(12), 2935–2947 (2017)
26. Liu, S., et al.: Deep ensembling with no overhead for either training or testing: the all-round blessings of dynamic sparsity. arXiv preprint arXiv:2106.14568 (2021)
27. Lopez-Paz, D., Ranzato, M.: Gradient episodic memory for continual learning. In: Proceedings of the 31st International Conference on Neural Information Processing Systems, pp. 6470–6479 (2017)
28. Luo, J.H., Wu, J., Lin, W.: ThiNet: a filter level pruning method for deep neural network compression. In: Proceedings of the IEEE International Conference on Computer Vision, pp. 5058–5066 (2017)
29. Mallya, A., Davis, D., Lazebnik, S.: Piggyback: adapting a single network to multiple tasks by learning to mask weights. In: Proceedings of the European Conference on Computer Vision (ECCV), pp. 67–82 (2018)
30. Mallya, A., Lazebnik, S.: Packnet: Adding multiple tasks to a single network by iterative pruning. In: Proceedings of the IEEE Conference on Computer Vision and Pattern Recognition, pp. 7765–7773 (2018)

31. Masana, M., Liu, X., Twardowski, B., Menta, M., Bagdanov, A.D., van de Weijer, J.: Class-incremental learning: survey and performance evaluation. arXiv preprint arXiv:2010.15277 (2020)
32. Mazumder, P., Singh, P., Rai, P.: Few-shot lifelong learning. In: Proceedings of the AAAI Conference on Artificial Intelligence, vol. 35, pp. 2337–2345 (2021)
33. McCloskey, M., Cohen, N.J.: Catastrophic interference in connectionist networks: the sequential learning problem. Psychol. Learn. Motiv. **24**, 109–165 (1989)
34. Mermillod, M., Bugaiska, A., Bonin, P.: The stability-plasticity dilemma: Investigating the continuum from catastrophic forgetting to age-limited learning effects. Front. Psychol. **4**, 504 (2013)
35. Mirzadeh, S.I., et al.: Architecture matters in continual learning. arXiv preprint arXiv:2202.00275 (2022)
36. Mocanu, D.C., Mocanu, E., Stone, P., Nguyen, P.H., Gibescu, M., Liotta, A.: Scalable training of artificial neural networks with adaptive sparse connectivity inspired by network science. Nat. Commun. **9**(1), 1–12 (2018)
37. Mocanu, D.C., Vega, M.T., Eaton, E., Stone, P., Liotta, A.: Online contrastive divergence with generative replay: Experience replay without storing data. arXiv preprint arXiv:1610.05555 (2016)
38. Özdenizci, O., Legenstein, R.: Training adversarially robust sparse networks via Bayesian connectivity sampling. In: International Conference on Machine Learning, pp. 8314–8324. PMLR (2021)
39. Raihan, M.A., Aamodt, T.: Sparse weight activation training. In: Larochelle, H., Ranzato, M., Hadsell, R., Balcan, M.F., Lin, H. (eds.) Advances in Neural Information Processing Systems, vol. 33, pp. 15625–15638. Curran Associates, Inc. (2020)
40. Ramasesh, V.V., Dyer, E., Raghu, M.: Anatomy of catastrophic forgetting: hidden representations and task semantics. In: International Conference on Learning Representations (2020)
41. Rebuffi, S.A., Kolesnikov, A., Sperl, G., Lampert, C.H.: iCaRL: incremental classifier and representation learning. In: Proceedings of the IEEE Conference on Computer Vision and Pattern Recognition, pp. 2001–2010 (2017)
42. Riemer, M., Cases, I., Ajemian, R., Liu, M., Rish, I., Tu, Y., Tesauro, G.: Learning to learn without forgetting by maximizing transfer and minimizing interference. In: International Conference on Learning Representations (2018)
43. Rusu, A.A., et al.: Progressive neural networks. arXiv preprint arXiv:1606.04671 (2016)
44. Serra, J., Suris, D., Miron, M., Karatzoglou, A.: Overcoming catastrophic forgetting with hard attention to the task. In: International Conference on Machine Learning, pp. 4548–4557. PMLR (2018)
45. Shin, H., Lee, J.K., Kim, J., Kim, J.: Continual learning with deep generative replay. In: Advances in Neural Information Processing Systems, pp. 2990–2999 (2017)
46. Sokar, G., Mocanu, D.C., Pechenizkiy, M.: Learning invariant representation for continual learning. In: Meta-Learning for Computer Vision Workshop at the 35th AAAI Conference on Artificial Intelligence (AAAI-21) (2021)
47. Sokar, G., Mocanu, D.C., Pechenizkiy, M.: Self-attention meta-learner for continual learning. In: Proceedings of the 20th International Conference on Autonomous Agents and MultiAgent Systems, pp. 1658–1660 (2021)
48. Sokar, G., Mocanu, D.C., Pechenizkiy, M.: SpaceNet: make free space for continual learning. Neurocomputing **439**, 1–11 (2021)

49. Sokar, G., Mocanu, E., Mocanu, D.C., Pechenizkiy, M., Stone, P.: Dynamic sparse training for deep reinforcement learning. In: International Joint Conference on Artificial Intelligence (2022)
50. van de Ven, G.M., Siegelmann, H.T., Tolias, A.S.: Brain-inspired replay for continual learning with artificial neural networks. Nat. Commun. **11**(1), 1–14 (2020)
51. Veniat, T., Denoyer, L., Ranzato, M.: Efficient continual learning with modular networks and task-driven priors. In: International Conference on Learning Representations (2021)
52. Wortsman, M., et al.: Supermasks in superposition. Adv. Neural Inf. Process. Syst. **33**, 15173–15184 (2020)
53. Xiao, H., Rasul, K., Vollgraf, R.: Fashion-MNIST: a novel image dataset for benchmarking machine learning algorithms. arXiv preprint arXiv:1708.07747 (2017)
54. Yoon, J., Kim, S., Yang, E., Hwang, S.J.: Scalable and order-robust continual learning with additive parameter decomposition. In: International Conference on Learning Representations (2019)
55. Yoon, J., Yang, E., Lee, J., Hwang, S.J.: Lifelong learning with dynamically expandable networks. In: International Conference on Learning Representations (2018)
56. Zenke, F., Poole, B., Ganguli, S.: Continual learning through synaptic intelligence. In: International Conference on Machine Learning, pp. 3987–3995. PMLR (2017)
57. Zhu, H., Jin, Y.: Multi-objective evolutionary federated learning. IEEE Trans. Neural Netw. Learn. Syst. **31**(4), 1310–1322 (2019)

PrUE: Distilling Knowledge from Sparse Teacher Networks

Shaopu Wang[1,2], Xiaojun Chen[2(✉)], Mengzhen Kou[1,2], and Jinqiao Shi[3]

[1] School of Cyber Security, University of Chinese Academy of Sciences,
Beijing, China
[2] Institute of Information Engineering, Chinese Academy of Sciences, Beijing, China
{wangshaopu,chenxiaojun,koumengzhen}@iie.ac.cn
[3] Beijing University of Posts and Telecommunication, Beijing, China
shijinqiao@bupt.edu.cn

Abstract. Although deep neural networks have enjoyed remarkable success across a wide variety of tasks, their ever-increasing size also imposes significant overhead on deployment. To compress these models, knowledge distillation was proposed to transfer knowledge from a cumbersome (teacher) network into a lightweight (student) network. However, guidance from a teacher does not always improve the generalization of students, especially when the size gap between student and teacher is large. Previous works argued that it was due to the high certainty of the teacher, resulting in harder labels that were difficult to fit. To soften these labels, we present a pruning method termed Prediction Uncertainty Enlargement (PrUE) to simplify the teacher. Specifically, our method aims to decrease the teacher's certainty about data, thereby generating soft predictions for students. We empirically investigate the effectiveness of the proposed method with experiments on CIFAR-10/100, Tiny-ImageNet, and ImageNet. Results indicate that student networks trained with sparse teachers achieve better performance. Besides, our method allows researchers to distill knowledge from deeper networks to improve students further. Our code is made public at: https://github.com/wangshaopu/prue.

Keywords: Knowledge distillation · Network pruning · Deep learning

1 Introduction

Neural networks have gained remarkable practical success in many fields [3]. In practice, researchers usually introduce more layers and parameters to make the network deeper [37] and wider [15] for achieving better performance. However, these over-parameterized models also incur huge computational and storage overhead [5], which makes deploying them on edge devices impractical. Therefore, several methods have been proposed to shrink neural networks, e.g., network pruning [11,17], quantization [10], and knowledge distillation [13]. Among these approaches, knowledge distillation has been widely utilized in many fields [2,39].

© The Author(s), under exclusive license to Springer Nature Switzerland AG 2023
M.-R. Amini et al. (Eds.): ECML PKDD 2022, LNAI 13715, pp. 102–117, 2023.
https://doi.org/10.1007/978-3-031-26409-2_7

Table 1. The test accuracy in percentage of various teachers and ResNet-8 as the student.

	ResNet-8	ResNet-20 w/o LS	ResNet-20 w/LS	ResNet-32 w/o LS
Teacher Acc.	87.56($\pm$0.20)	91.72($\pm$0.21)	92.06($\pm$0.26)	92.99($\pm$0.12)
Student Acc.	–	88.05($\pm$0.18)	86.13($\pm$0.22)	87.60($\pm$0.08)

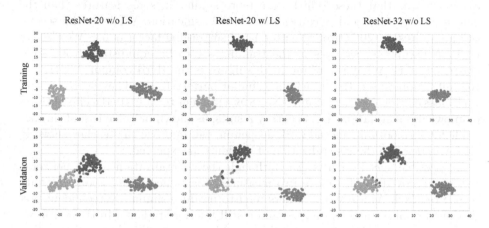

Fig. 1. Visualization of network predictions. We randomly select some training samples from the three classes of CIFAR-10 "airplane" (gray), "automobile" (blue), and "bird" (yellow), and then perform t-SNE dimensionality reduction [22] on network predictions. Note that the x-y axis has no real meaning here. (Color figure online)

Generally speaking, it utilizes a pre-trained teacher to produce supervision for students. In this way, a lightweight student network can achieve similar generalization as the teacher.

Although this paradigm of encouraging students to mimic teachers' behaviors has proven to be a promising way, some recent works [25,30] argued that knowledge distillation is not always effective. Specifically, it is found that well-behaved teachers failed to improve student generalization under certain circumstances. For instance, Müller et al. [25] discovered that teachers pre-trained with label smoothing (LS) [31], a commonly used technique to regularize models, will distill inferior students, even though the teacher's generalization has been improved. They attribute this phenomenon to the fact that LS tends to erase the relative information within a class. As a result, teachers generate harder labels that are difficult for students to fit. Meanwhile, Mirzadeh et al. [24] investigated another more common scenario. When there exists a large capacity gap between students and teachers, the former will perform worse. Coincidentally, their experiments lead to a similar conclusion that well-performed teachers fail to generate soft targets.

To investigate the relationship between network capacity and label smoothing, we train ResNet-20 and ResNet-32 on CIFAR-10 and report the results of visualizing their predictions for the classes "airplane", "automobile" and "bird" in Fig. 1. The first row represents examples from the training set, while the second row is from the validation set. As revealed in the first column, a ResNet-20 trained without label smoothing (w/o LS) produces predictions scattered in some broad clusters. We also notice that blue dots (automobile) and gray dots (airplanes) in the validation set tend to be mixed at the boundary. A possible explanation is that these vehicles are more similar in some features than the yellow dots (birds), and it causes some misclassification. While in the second column, a ResNet-20 is trained with a label smoothing factor of 0.2. We observe that LS encourages samples in the training set to be equidistant from other classes' centers. What is striking in this figure is the third column. We train a ResNet-32 and notice that it acts in a similar pattern to LS. They both compact each class cluster. Next, we use ResNet-8 as a student to validate the effectiveness of knowledge distillation. The accuracy results, as shown in Table 1, confirm that while label smoothing and network deepening can improve the teacher network, they will degrade the generalization of students as expected.

A possible speculation is that although the generalization of the networks can be improved by the above two measures, their uncertainty about the data is also reduced. As a result, teachers tend to produce similar overconfident predictions for all intra-class samples and distill inferior students. In this work, we propose to improve knowledge distillation by increase teachers' uncertainty. Fortunately, a statistical metric, which we term prediction uncertainty, has been proposed by [29] to quantify this phenomenon. Following this work, we propose a criterion to identify the effect of weights on uncertainty in the teacher network. Then we prune those less-contributing weights before distillation. Differing from traditional pruning algorithms that focus on generalization, our method aims to reduce the generalization error of student networks by softening teacher predictions. We name our method Prediction Uncertainty Enlargement (PrUE).

We evaluate our pruning method on CIFAR-10/100, Tiny-ImageNet, and ImageNet classification datasets with some modern neural networks. Specifically, we first verify that label smoothing and network deepening reduce generalization error with a sacrifice of prediction uncertainty. The following distillation experiments show a positive correlation between the student's accuracy and the teacher's prediction uncertainty. However, the teacher's accuracy does not play a crucial role in knowledge distillation. Generally, large networks struggle to distill stronger students despite their high accuracy. To bridge this gap, we apply PrUE to the aforementioned teacher networks and distill their knowledge to students. Results show that our method can increase the teacher's prediction uncertainty, resulting in better performance improvement for students than existing distillation methods. We also compare PrUE with several other pruning schemes and observe that sparse teacher networks distill good students, but PrUE usually presents better performance.

Contributions: our contributions in this paper are as follows.

- We empirically investigate the impact of label smoothing and network capacity on knowledge distillation. Interestingly, They both prevent teachers from generating soft labels and impair knowledge distillation, despite the improved accuracy of teachers themselves.

- We apply a statistical metric to quantify the softness of labels. Based on this, PrUE is proposed to increase the teacher's prediction uncertainty.
- We perform experiments on CIFAR-10/100, Tiny-ImageNet, and ImageNet with widely varying CNN networks. Results suggest that sparse teacher networks usually distill better students than dense ones. Besides, PrUE outperforms existing distillation and pruning schemes.

2 Related Work

Network Pruning. The motivation behind network pruning is that there is a mass of redundant parameters in the neural network [7]. Previous works have demonstrated that these parameters can be removed safely. Therefore, Lecun et al. [17] proposed removing parameters in an unstructured way by calculating the Hessian of the loss with respect to the weights. Furthermore, Han et al. [11] proposed a magnitude-based pruning method to remove all weights below s predefined threshold. Recently, Frankle et al. [8] proposed the "Lottery Ticker Hypothesis" that there exist sparse subnetworks that, when trained in isolation, can reach test accuracy comparable to the original network. Furthermore, Miao et al. [23] proposed a framework that can prune neural networks to any sparsity ratio with only one training.

Soft Labels. Theoretically, the widely used one-hot labels could lead to overfitting. Therefore, label smoothing was proposed to generate soft labels, thereby delivering regularization effects. On the other hand, there were usually some noisy labels in the dataset that mislead deep learning models, and a recent work [20] noted that label smoothing could help mitigate label noise. However, label smoothing could only add random noise and cannot reflect the relationship between labels. Another well-known paradigm for generating soft labels was knowledge distillation [13]. Differing label smoothing, knowledge distillation required a pretrained teacher to produce soft labels for each training example. Therefore, Yuan et al. [35] regarded it as a dynamic form of label smoothing. Although the original distillation scheme focused on transferring dark knowledge from large to small models, Zhang et al. [38] had found that these generated soft labels can be used for distributed machine learning. Therefore, some recent works [2,39] proposed distillation-based communication schemes to save bandwidth.

Pruning in Distillation. Both network pruning and knowledge distillation are widely used model compression methods. Therefore, some recent works proposed combining them together to achieve higher compression ratios. For instance, Xie et al. [33] used this paradigm to customize a compression scheme for the identification of Person re-identification (ReID). Chen et al. [4] proposed to use pruning and knowledge distillation to train a lightweight detection model, to achieve synthetic aperture radar ship real-time detection at a lower cost. Meanwhile, Aghli et al. [1] introduced a compression scheme of convolutional neural networks, mainly exploring how to combine pruning and knowledge distillation methods to reduce the scale of ResNet with the guarantee of accuracy. Neill et al. [26] proposed a pruning-based self-distillation scheme using distillation as the pruning criterion to maximize the similarity of network representations before and after pruning. Cui et al. [6] proposed a joint model compression method that combines structured pruning and dense knowledge distillation. However, these researches focused on simplifying student networks. In fact, they amplify the capacity gap between students and teachers.

3   Background

Producing soft labels has been shown to be an effective regularizer. In practice, encouraging networks to fit soft labels prevents overfitting. In this section, we introduce a statistical metric quantifying label softness.

3.1   Preliminaries

Notations. Given a K-class classification task, We denote by $\mathcal{D}$ the training dataset, consisting of m i.i.d tuples $\{(\boldsymbol{x}_1, \boldsymbol{y}_1), \ldots, (\boldsymbol{x}_m, \boldsymbol{y}_m)\}$ where $\boldsymbol{x}_i \in \mathbb{R}^{d \times 1}$ is the input data and $\boldsymbol{y}_i \in \{0, 1\}^K$ is the corresponding one-hot class label. Let $\boldsymbol{y}[i]$ be the i-th element in $\boldsymbol{y}$, and $\boldsymbol{y}[c]$ is 1 for the ground-truth class and 0 for others.

Knowledge Distillation. For a teacher network $f(\boldsymbol{w}_T)$ parameterized by $\boldsymbol{w}_T$, let $a(\boldsymbol{w}_T, \boldsymbol{x}_i)$ and $f(\boldsymbol{w}_T, \boldsymbol{x}_i)$ correspond to its logits and prediction for $\boldsymbol{x}_i$, respectively. In vanilla supervised learning, $f(\boldsymbol{w}_T)$ is usually trained on $\mathcal{D}$ with cross-entropy loss

$$\mathcal{L}_{CE} = -\sum_{i=1}^{m} \boldsymbol{y}_i \log f(\boldsymbol{w}_T, \boldsymbol{x}_i) \tag{1}$$

where $f(\boldsymbol{w}_T, \boldsymbol{x}_i) = softmax(a(\boldsymbol{w}_T, \boldsymbol{x}_i))$.

As for a student network $f(\boldsymbol{w}_S)$, its logits and prediction for $\boldsymbol{x}_i$ are denoted as $a(\boldsymbol{w}_S, \boldsymbol{x}_i)$ and $f(\boldsymbol{w}_S, \boldsymbol{x}_i)$. In knowledge distillation, $f(\boldsymbol{w}_S)$ is usually trained with a given temperature τ and KL-divergence loss

$$\mathcal{L}_{KD} = -\sum_{i=1}^{m} \tau^2 KL(a(\boldsymbol{w}_T, \boldsymbol{x}_i), a(\boldsymbol{w}_S, \boldsymbol{x}_i)) \tag{2}$$

When the hyperparameter τ is set to 1, we can regard the distillation process as training $f(\boldsymbol{w}_S)$ on a new dataset $\{(\boldsymbol{x}_1, f(\boldsymbol{w}_T, \boldsymbol{x}_1)), \ldots, (\boldsymbol{x}_m, f(\boldsymbol{w}_T, \boldsymbol{x}_m))\}$ with soft labels provided by a teacher. The key idea behind knowledge distillation is to encourage the student $f(\boldsymbol{w}_S)$ to mimic the behavior of the teacher $f(\boldsymbol{w}_T)$. In practice, researchers usually use correct labels to improve soft labels, especially when the generalization of teachers is poor. Therefore, the practical loss function for the student is modified as follows:

$$\mathcal{L}_{student} = \sum_{i=1}^{m}(1 - \lambda)\mathcal{L}_{CE} + \lambda\mathcal{L}_{KD} \qquad (3)$$

where λ is another hyperparameter that controls the trade-off between the two losses. We refer to this approach as *Logits(τ)* through the paper.

Label Smoothing. Similar to knowledge distillation, label smoothing aims to replace hard labels to penalize overfitting. Instead, it does not involve a teacher network. Specifically, label smoothing modifies one-hot hard label vector y with a mixture of weighted origin y and a uniform distribution:

$$y_c = \begin{cases} 1 - \alpha & \text{if } c = label, \\ \alpha/(K - 1) & \text{otherwise.} \end{cases} \qquad (4)$$

where $\alpha \in [0, 1]$ is the hyperparameter flattening the one-hot labels.

Label smoothing has been a widely used *trick* to improve network generalization. A prior work [29] observes that although the network trained with label smoothing suffers a higher cross-entropy loss on the validation set, its accuracy is better than that without label smoothing.

3.2 Prediction Uncertainty

To observe the effect of label smoothing on the penultimate layer representations, Müller et al. [25] proposed a visualization scheme based on squared Euclidean distance. Similarly, we use t-SNE in Sect. 1 to visualize the predictions. However, we cannot conduct numerical analysis on these intuitive presentations. To further measure the label softness quantitatively and address the *erasing* phenomenon caused by label smoothing, Shen et al. [29] propose a simple yet effective metric. The definition is as follows[1]:

$$\delta(w) = \frac{1}{K}\sum_{c=1}^{K}(\frac{1}{n_c}\sum_{i=1}^{n_c}\|f(w, x_i)[c] - \tilde{f}(w, x_i)[c]\|^2) \qquad (5)$$

where class c contains n_c samples. $\tilde{f}(x_i)[c]$ is the mean of in $f(x_i)$ class c. The key idea behind this metric is to use the variance of intra-class probabilities to measure the uncertainty of network predictions.

Now we discuss how prediction uncertainty influences knowledge distillation. Assume an ideal network classifies each input precisely, and it is absolutely certain of each prediction. Correspondingly, this network is commonly regarded as

---

[1] It was called *stability* in the origin paper. We modify it for the purpose of our work.

a perfect model that achieves excellent generalization and low loss on the validation set. However, it tends to produce one-hot labels that fail to inform student networks about the similarity between classes, i.e., dark knowledge. At this point, the certainty of the teacher network downgrades the knowledge distillation to vanilla training. Applying label smoothing to the distillation process could help to moderate the teacher's overconfidence. Unfortunately, this trick merely tells students that airplanes and birds have the same probability as automobiles. Therefore, we aim to make teachers feel uncertain between the automobile and the airplane, thus improving the generalization behavior of the student network.

We next work on simplifying the teacher network to enlarge its prediction uncertainty. Specifically, we utilize network pruning to close the capacity gap between teachers and students.

4   Prediction Uncertainty Enlargement

In deep model compression, network pruning aims to deliver the regularization effect to neural networks by simply removing parameters. Following the discussion above, we introduce auxiliary indicator variables $m \in \{0,1\}^l$ representing the pruning mask. Then the enlargement of prediction uncertainty is formulated as an optimization problem as:

$$\max_{m} \delta(m \odot w) = \max_{m} \frac{1}{K} \sum_{c=1}^{K} \left(\frac{1}{n_c} \sum_{i=1}^{n_c} \|f(m \odot w, x_i)[c] - \tilde{f}(m \odot w, x_i)[c]\|^2 \right),$$

$$s.t. \quad m \in \{0,1\}^l, \quad \|m\|_0 \le s,$$

$$(6)$$

where $\odot$ denotes the Hadamard product.

Solving such a combinatorial optimization problem requires computing its $\delta(m \odot w)$ for each candidate in the solution space, that is, it requires up to $l \times l$ forward passes over the training dataset. However, the number of network parameters has increased substantially recently. Since an arms race of training large models has begun, millions of calculations $\delta(m \odot w)$ are unacceptable.

Following [16,18], we next measure the impact of each weight on the network uncertainty and then prune less-contributing weights greedily. Since it is impractical to directly solve this optimization problem with respect to binary variables m, we first relax m into real variables $m \in [0,1]^l$. This change can be seen as a form of soft pruning, where the corresponding mask $m[j]$ is gradually reduced from 1 to 0. In this way, the optimization problem is differentiable with respect to m. We rewrite Optimization (6) as follows:

$$\max_{m} \delta(m \odot w) = \max_{m} \frac{1}{K} \sum_{c=1}^{K} \left(\frac{1}{n_c} \sum_{i=1}^{n_c} \|f(m \odot w, x_i)[c] - \tilde{f}(m \odot w, x_i)[c]\|^2 \right),$$

$$s.t. \quad m \in [0,1]^l, \quad \|m\|_0 \le s,$$

$$(7)$$

This modification allows us to perturb the mask instead of setting it to zero. For the weight $w[j]$, we add an infinitesimal perturbation ϵ to the mask $m[j]$ to obtain its influence on $\delta(m \odot w)$. Its magnitude of differential $\triangle\delta_j(m \odot w)$ indicates the dependence of $\delta(m \odot w)$ on $w[j]$. Next, we find the derivative of $\delta(m \odot w)$ with respect to $m[j]$ as follows:

$$\lim_{\epsilon\to 0} \frac{\delta(m \odot w) - \delta((1 - \epsilon e_j)m \odot w)}{\epsilon} = \lim_{\epsilon\to 0} \frac{\partial\delta(m \odot w)}{\partial m[j]} = g_j(w). \qquad (8)$$

where e_j is a one-hot vector $[0, ..., 0, 1, 0, ..., 0]$ with a 1 at position j.

Thus, we measure the importance of the weight $w[j]$ to the prediction uncertainty. To this end, we regard $|g_j(w)|$ as the proposed criterion. Given a desired sparsity s, we can achieve prediction uncertainty enlargement by pruning $s \times l$ weights that contribute less to the variance.

The key to our approach is to find the derivative of the uncertainty with respect to the pruning mask of each weight. However, restricted by the modern computing device, PrUE still faces some practical problems. Note that Optimization (7) calls $f(w)$ twice, which requires the automatic differentiation algorithm to perform two forward-backward pass through the computational graph. Modern deep learning frameworks like PyTorch usually free gradient tensors after the first backward pass to save memory. That is, our method consumes more resources due to retaining the computational graph.

On the other hand, our method requires computing the averaged intra-class probabilities for each class. In practice, researchers typically perform stochastic gradient descent by randomly selecting a mini-batch of training data, where the batchsize ranges from 128 to 1024. For a 10-class classification task like CIFAR-10, this batchsize is sufficient to estimate $\tilde{f}(x)[c]$, while not for ImageNet-1k containing 1000 classes. In fact, most classes in ImageNet-1k only appear once or twice in a batch, making accurate estimation of $\tilde{f}(x)[c]$ impractical.

One could take straightforward measures such as saving intermediate values of the graph or leveraging more devices, but this would result in additional overhead. Instead, we employ a simple yet effective trick to decompose the optimization into two steps. Specifically, we first compute $\tilde{f}(x)[c]$ for each class with the computational graph detached, then sort the dataset by labels, thus guaranteeing that only class c appears in each batch. Finally, $f(x)[c]$ can be estimated in the current batch. We empirically observed that this trick only slightly affects the results, but saves appreciable memory.

5   Experiments

In this section, we empirically investigate the effect of our proposed method on knowledge distillation. In addition, we compare PrUE with other distillation and pruning methods. The results show that our paradigm of distilling knowledge from sparse teacher networks tends to yield better students. Moreover, PrUE can exhibit better performance.

Table 2. Number of weights and training hyperparameters in our experiments.

Dataset	Network	# Weights	Epochs	Batchsize	Schedule
CIFAR-10	ResNet-8 [12]	78K	160	512	10× drop at 81, 122
	ResNet-20	272K			
	ResNet-32	466K			
	ResNet-56	855K			
	ResNet-110	1.7M			
CIFAR-100	ResNet-8	83.9K	160	512	10× drop at 81, 122
	ResNet-20	278K			
	ResNet-32	472K			
	ResNet-56	861K			
	ResNet-110	1.7M			
Tiny-ImageNet	ResNext-50 [34]	15.0M	200	256	5× drop at 60, 120, 160
	ShuffleNet V2 [21]	1.5M			
ImageNet	EffcientNet-B2 [32]	9.1M	90	256	10× drop at 30, 60
	MobileNet V3 Small [14]	2.5M			

Implementation Details. We conduct all experiments on 8 * NVIDIA Tesla A100 GPU. The sparsity level is defined to be $s = k/l \times 100(\%)$, where k is the number of zero weights, and l is the total number of network weights. All networks are trained with SGD with Nesterov momentum. We set the initial learning rate to 0.1, momentum to 0.9. Table 2 describes the number of parameters of all the networks and corresponding training hyperparameters. During distillation, we set λ to 1 for CIFAR-10 and 0.1 for the rest tasks.

5.1   The Effect of LS on Knowledge Distillation

We first investigate the compatibility of label smoothing and knowledge distillation on CIFAR-10 and CIFAR-100. Specifically, we train ResNet-20/32/56/100 with label smoothing turned on or turned off, then distill their knowledge into ResNet-8. Table 3 presents the accuracy of student networks supervised by various teachers. We also report the vanilla supervised training results of ResNet-8 for baseline comparison.

Table 3. The test accuracy of a fixed student with various teachers trained without (w/o) or with (w/) label smoothing. The vanilla supervised results of ResNet-8 is also reported.

	Vanilla		ResNet-20	ResNet-32	ResNet-56	ResNet-110
CIFAR-10	87.56	w/LS	86.62(±0.21)	85.56(±0.25)	85.61(±0.03)	85.88(±0.19)
		w/o LS	88.36(±0.12)	87.48(±0.22)	87.50(±0.10)	87.47(±0.17)
CIFAR-100	59.36	w/LS	58.75(±0.23)	58.73(±0.16)	59.14(±0.14)	58.52(±0.25)
		w/o LS	59.81(±0.19)	59.50(±0.24)	59.47(±0.17)	59.76(±0.09)

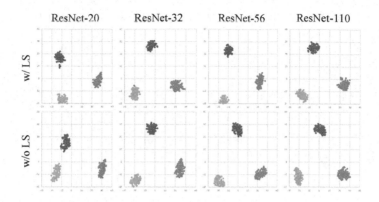

Fig. 2. Visualization of predictions of more network structures.

Although deep neural networks are well known for their generalization ability, they fail to bring proportional improvement for students. In particular, ResNet-20 tends to distill better students than other well-generalized teachers. Similarly, teachers trained with hard labels achieve better distillation results compared to those trained with label smoothing. To demonstrate this phenomenon, we provide visualizations of these teachers' predictions in Fig. 2. As we can see, network deepening and label smoothing compacts each cluster and thus impairs knowledge distillation in Table 3.

5.2   Comparison with Other Distillation Methods

Intuitively, improved teachers are overconfident in each sample, thus producing harder predictions containing low information. To enlarge teacher uncertainty without sacrificing generalization, we apply PrUE to prune them, and then fine-tune them to restore accuracy.

Table 4. The test accuracy (%) and uncertainty (1e−2) of teacher networks with varying sparsity.

Dataset	Sparsity	ResNet-20		ResNet-32		ResNet-56		ResNet-110	
		Acc.	Uncer.	Acc.	Uncer.	Acc.	Uncer.	Acc.	Uncer.
CIFAR-10	$s = 0$	91.72	6.40	93.17	3.34	93.40	2.04	93.38	1.55
	$s = 20\%$	92.82	6.12	93.54	3.19	93.75	2.12	94.14	1.31
	$s = 50\%$	91.97	8.19	93.08	4.23	93.77	2.39	93.75	1.74
	$s = 90\%$	87.98	22.87	90.63	17.49	91.64	13.00	92.13	7.28
CIFAR-100	$s = 0$	68.61	26.01	69.65	19.16	71.29	10.72	71.84	5.72
	$s = 20\%$	69.04	25.83	70.44	18.77	72.01	10.47	73.36	5.55
	$s = 50\%$	68.26	27.97	69.18	22.55	71.33	14.56	72.83	7.18
	$s = 90\%$	54.30	28.33	60.49	30.53	62.92	30.78	62.16	30.52

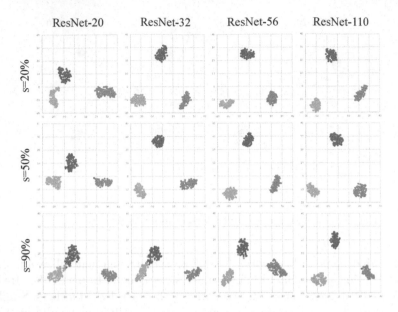

Fig. 3. Predictive visualization of networks with varying sparsity s. As the network deepens, the predictions get tighter. While the increasing sparsity spreads the predictions into broad clusters.

Figure 3 visualizes these sparse teacher networks. As the sparsity s increases, the teacher's predictions are scattered into wider clusters. We also observe that a higher sparsity is appropriate for deep networks such as ResNet-110. On the other hand, Table 4 provides quantitative results. It suggests that PrUE can effectively improve teachers' uncertainty with slight loss in performance.

Next, we distill knowledge from these sparse teachers to a ResNet-8. Meanwhile, we compare our method with other distillation methods. Table 5 and Table 6 depicts the results of students performance on CIFAR-10 and CIFAR-100, respectively. It is worth noting that λ is set to 0 on CIFAR-10, which means that our method can only obtain the teacher's prediction, while the others can receive the ground truth. Although this is an unfair comparison, PrUE still outperforms existing distillation methods notably. Another interesting observation is that teachers with high uncertainty distill better students, even when their accuracy is hurt by pruning. Therefore, we conclude that teacher uncertainty plays an important role in knowledge distillation, rather than accuracy.

5.3   Comparison with Other Pruning Methods

With promising results on distillation, we further compare PrUE with other pruning methods. In particular, we first train the teacher from scratch and apply several one-shot pruning algorithms (Magnitude [11,19], SNIP [18], Random [9], PrUE) to remove a portion of weights of the trained network, then fine-tune these

Table 5. The test accuracy of ResNet-8 on CIFAR-10 using different distillation methods. TA(20), TA(32) refers to using ResNet-20 and ResNet-32 as a teacher assistant, respectively.

	CIFAR-10			
	ResNet-20	ResNet-32	ResNet-56	ResNet-110
Logit ($\tau = 1$) [13]	88.36($\pm$0.16)	87.48($\pm$0.22)	87.50($\pm$0.29)	87.47($\pm$0.28)
Logit ($\tau = 4$)	88.72($\pm$0.26)	88.39($\pm$0.17)	88.66($\pm$0.21)	88.34($\pm$0.29)
FitNet [28]	87.00($\pm$0.24)	86.83($\pm$0.27)	86.68($\pm$0.08)	86.62($\pm$0.10)
AT [36]	86.64($\pm$0.14)	86.37($\pm$0.16)	86.71($\pm$0.09)	86.76($\pm$0.17)
PKT [27]	87.41($\pm$0.04)	87.30($\pm$0.24)	87.26($\pm$0.13)	87.08($\pm$0.19)
TA(20) [24]	–	87.55($\pm$0.26)	87.87($\pm$0.19)	87.66($\pm$0.16)
TA(32)	–	–	87.83($\pm$0.10)	87.37($\pm$0.28)
PrUE ($s = 20\%$)	88.89($\pm$0.11)	88.30($\pm$0.06)	88.49($\pm$0.19)	88.47($\pm$0.26)
PrUE ($s = 50\%$)	**89.17**($\pm$0.19)	**88.39**($\pm$0.07)	88.68($\pm$0.23)	89.22($\pm$0.15)
PrUE ($s = 90\%$)	87.01($\pm$0.20)	87.95($\pm$0.24)	**89.08**($\pm$0.24)	**89.27**($\pm$0.18)

pruned networks until convergence. We use ResNet-8 as a student to evaluate the distillation performance of these sparse teachers.

Table 6. The test accuracy of ResNet-8 on CIFAR-100 using different distillation methods.

	CIFAR-100			
	ResNet-20	ResNet-32	ResNet-56	ResNet-110
Logit ($\tau = 1$)	59.51($\pm$0.10)	59.25($\pm$0.23)	59.09($\pm$0.08)	59.56($\pm$0.26)
Logit ($\tau = 4$)	59.81($\pm$0.12)	59.50($\pm$0.05)	59.47($\pm$0.21)	59.76($\pm$0.12)
FitNet	58.92($\pm$0.20)	58.53($\pm$0.49)	58.59($\pm$0.07)	58.37($\pm$0.11)
AT	58.52($\pm$0.17)	58.74($\pm$0.09)	58.60($\pm$0.07)	57.87($\pm$0.24)
PKT	58.57($\pm$0.17)	58.74($\pm$0.05)	58.96($\pm$0.27)	58.81($\pm$0.06)
TA(20)	–	59.60($\pm$0.25)	59.45($\pm$0.09)	59.14($\pm$0.18)
TA(32)	–	–	59.68($\pm$0.12)	59.65($\pm$0.09)
PrUE ($s = 20\%$)	59.54($\pm$0.06)	59.66($\pm$0.12)	59.95($\pm$0.05)	59.56($\pm$0.14)
PrUE ($s = 50\%$)	**59.9**($\pm$0.30)	**59.71**($\pm$0.25)	**60.03**($\pm$0.37)	59.85($\pm$0.27)
PrUE ($s = 90\%$)	58.89($\pm$0.32)	59.44($\pm$0.19)	59.85($\pm$0.20)	**60.17**($\pm$0.08)

As illustrated in Fig. 4, our strategy of distilling knowledge from sparse networks can effectively improve the generalization behavior of student networks. Even if the weights in the network are randomly removed, students can still benefit from it. We also notice that PrUE could only exhibit similar performance to other pruning methods on shallower networks. Such as on 90% sparse ResNet-32, PrUE exhibits lower distillation performance (87.95%) than Magnitude (88.53%) and SNIP (88.14%). But as the network grows, our method achieves better results (up to 89.27%). This result suggests that while previous work has argued that the large capacity gap between teachers and students results in lower performance gains, our approach allows researchers to break the restriction and use deeper networks to obtain further improve student accuracy.

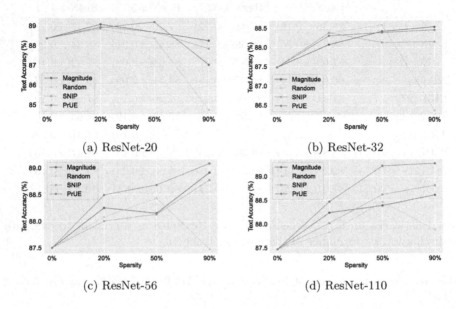

(a) ResNet-20   (b) ResNet-32

(c) ResNet-56   (d) ResNet-110

Fig. 4. Distillation accuracy of sparse teacher networks obtained using different pruning methods.

The Impact of Sparsity. We also find that inappropriate sparsity affects the distillation results of all pruning algorithms. For instance, ResNet-20 with 90% sparsity could face a 1–2% drop in distillation accuracy, although this result still outperforms traditional distillation methods in Table 5. While networks with more parameters like ResNet-110 can endure a higher sparsity ratio. Overall, if the size of teachers is much larger than that of students, we suggest a higher sparsity to bridge the capacity gap.

5.4   Distillation on Large-Scale Datasets

In this section, we consider practical applications on more challenging datasets. In practice, some large convolutional networks have been proposed to achieve better results on ImageNet tasks. On the other hand, researchers designed some lightweight networks to reduce overhead and accelerate inference. We aim to answer whether PrUE still works between these two different network structures.

We train ResNext-50 on Tiny-ImageNet as teacher network, while ShuffleNetV2 serves as the student. As for ImageNet, we distill knowledge from EfficientNet-B2 into MobileNetV3. Table 7 and Table 8 reports their own accuracy and distillation performance, respectively. Our method manages to improve

Table 7. The test accuracy (%) and uncertainty (1e−2) of sparse teacher networks on Tiny-ImageNet and ImageNet.

	$s = 0$		$s = 20\%$		$s = 50\%$		$s = 90\%$	
	Acc.	Uncer.	Acc.	Uncer.	Acc.	Uncer.	Acc.	Uncer.
ResNext-50	65.32	1.35	66.24	1.34	68.07	1.29	64.09	3.24
EfficientNet-B2	72.32	34.66	72.62	34.70	72.45	34.85	71.03	35.14

Table 8. The test accuracy of student network distilled by sparse teachers.

Teacher	Student	Vanilla	$s = 0$	$s = 20\%$	$s = 50\%$	$s = 90\%$
ResNext-50	ShuffleNet V2	62.09	63.28	63.45	64.09	64.65
EfficientNet-B2	MobileNet V3 Small	60.85	60.88	61.18	61.22	62.12

student generalization on real-world datasets. More interestingly, we observed on Tiny-ImageNet that the accuracy of the student network can sometimes exceed that of the teacher network. We believe this suggests that PrUE can be extended to a wider range of settings. Furthermore, we still lack understanding of knowledge distillation, and our proposed method could be a potential tool to shed light on it.

6   Conclusion

In this paper, we provided a data-dependent pruning method called PrUE to soften the network predictions, thereby improving its distillation performance. In particular, we proposed a computationally efficient criterion to estimate the effect of weights on uncertainty, and removed those less-contribution weights. We first showed a positive relationship between the uncertainty of the teacher network and its distillation effect through a visualization scheme. The following empirical experiments suggested that PrUE managed to increase the teacher uncertainty, thereby improving the distillation performance. Extensive experiments showed that our method notably outperformed traditional distillation methods. We also found that our strategy of distilling knowledge from sparse teacher networks could improve the generalization behavior of the student network, but the teacher pruned by PrUE tended to exhibit better performance.

Acknowledgements. This work is supported by The National Key Research and Development Program of China No. 2020YFE0200500 and National Natural Science Funds of China No. 61902394.

References

1. Aghli, N., Ribeiro, E.: Combining weight pruning and knowledge distillation for CNN compression. In: Proceedings of the IEEE/CVF Conference on Computer Vision and Pattern Recognition, pp. 3191–3198 (2021)

2. Anil, R., Pereyra, G., Passos, A., Ormandi, R., Dahl, G.E., Hinton, G.E.: Large scale distributed neural network training through online distillation. In: International Conference on Learning Representations (2018)
3. Brown, T.B., et al.: Language models are few-shot learners. arXiv preprint arXiv:14165 (2020)
4. Chen, S., Zhan, R., Wang, W., Zhang, J.: Learning slimming SAR ship object detector through network pruning and knowledge distillation. IEEE J. Sel. Top. Appl. Earth Obs. Remote Sens. **14**, 1267–1282 (2020)
5. Cheng, J., Wang, P., Li, G., Hu, Q., Lu, H.: Recent advances in efficient computation of deep convolutional neural networks. Front. Inf. Technol. Electron. Eng. **19**(1), 64–77 (2018). https://doi.org/10.1631/FITEE.1700789
6. Cui, B., Li, Y., Zhang, Z.: Joint structured pruning and dense knowledge distillation for efficient transformer model compression. Neurocomputing **458**, 56–69 (2021)
7. Denil, M., Shakibi, B., Dinh, L., Ranzato, M., De Freitas, N.: Predicting parameters in deep learning. In: Advances in Neural Information Processing Systems, vol. 26 (2013)
8. Frankle, J., Carbin, M.: The lottery ticket hypothesis: finding sparse, trainable neural networks. In: International Conference on Learning Representations (2019)
9. Frankle, J., Dziugaite, G.K., Roy, D., Carbin, M.: Pruning neural networks at initialization: why are we missing the mark? In: International Conference on Learning Representations (2021)
10. Gong, Y., Liu, L., Yang, M., Bourdev, L.: Compressing deep convolutional networks using vector quantization. In: Computer Vision and Pattern Recognition (2014). arXiv
11. Han, S., Pool, J., Tran, J., Dally, W.J.: Learning both weights and connections for efficient neural networks. In: Neural Information Processing Systems, pp. 1135–1143 (2015)
12. He, K., Zhang, X., Ren, S., Sun, J.: Deep residual learning for image recognition. In: Computer Vision and Pattern Recognition, pp. 770–778 (2016)
13. Hinton, G.E., Vinyals, O., Dean, J.: Distilling the knowledge in a neural network. In: Machine Learning (2015). arXiv
14. Howard, A., et al.: Searching for MobileNetV3. In: Proceedings of the IEEE/CVF International Conference on Computer Vision, pp. 1314–1324 (2019)
15. Ioffe, S., Szegedy, C.: Batch normalization: accelerating deep network training by reducing internal covariate shift. In: International Conference on Machine Learning, pp. 448–456. PMLR (2015)
16. Koh, P.W., Liang, P.: Understanding black-box predictions via influence functions. In: International Conference on Machine Learning, pp. 1885–1894. PMLR (2017)
17. Lecun, Y., Denker, J.S., Solla, S.A.: Optimal brain damage. In: Neural Information Processing Systems, vol. 2, pp. 598–605 (1989)
18. Lee, N., Ajanthan, T., Torr, P.H.S.: SNIP: single-shot network pruning based on connection sensitivity. In: International Conference on Learning Representations (2019)
19. Liu, N., et al.: Lottery ticket preserves weight correlation: is it desirable or not? In: International Conference on Machine Learning, pp. 7011–7020. PMLR (2021)
20. Lukasik, M., Bhojanapalli, S., Menon, A., Kumar, S.: Does label smoothing mitigate label noise? In: International Conference on Machine Learning, pp. 6448–6458. PMLR (2020)
21. Ma, N., Zhang, X., Zheng, H.-T., Sun, J.: ShuffleNet V2: practical guidelines for efficient CNN architecture design. In: Ferrari, V., Hebert, M., Sminchisescu, C.,

Weiss, Y. (eds.) Computer Vision – ECCV 2018. LNCS, vol. 11218, p. Shufflenet v2: Practical guidelines for efficient cnn architecture design-138. Springer, Cham (2018). https://doi.org/10.1007/978-3-030-01264-9_8

22. Van der Maaten, L., Hinton, G.: Visualizing data using t-SNE. J. Mach. Learn. Res. **9**(11), 2579–2605 (2008)

23. Miao, L., Luo, X., Chen, T., Chen, W., Liu, D., Wang, Z.: Learning pruning-friendly networks via frank-wolfe: one-shot, any-sparsity, and no retraining. In: International Conference on Learning Representations (2021)

24. Mirzadeh, S.I., Farajtabar, M., Li, A., Levine, N., Matsukawa, A., Ghasemzadeh, H.: Improved knowledge distillation via teacher assistant. In: Proceedings of the AAAI Conference on Artificial Intelligence, vol. 34, pp. 5191–5198 (2020)

25. Müller, R., Kornblith, S., Hinton, G.E.: When does label smoothing help? In: Advances in Neural Information Processing Systems, vol. 32 (2019)

26. Neill, J.O., Dutta, S., Assem, H.: Deep neural compression via concurrent pruning and self-distillation. arXiv preprint arXiv:2109.15014 (2021)

27. Passalis, N., Tefas, A.: Learning deep representations with probabilistic knowledge transfer. In: Ferrari, V., Hebert, M., Sminchisescu, C., Weiss, Y. (eds.) ECCV 2018. LNCS, vol. 11215, pp. 283–299. Springer, Cham (2018). https://doi.org/10.1007/978-3-030-01252-6_17

28. Romero, A., Ballas, N., Kahou, S.E., Chassang, A., Gatta, C., Bengio, Y.: FitNets: hints for thin deep nets. arXiv preprint arXiv:1412.6550 (2014)

29. Shen, Z., Liu, Z., Xu, D., Chen, Z., Cheng, K.T., Savvides, M.: Is label smoothing truly incompatible with knowledge distillation: an empirical study. In: International Conference on Learning Representations (2021)

30. Stanton, S., Izmailov, P., Kirichenko, P., Alemi, A.A., Wilson, A.G.: Does knowledge distillation really work? arXiv preprint arXiv:2106.05945 (2021)

31. Szegedy, C., Vanhoucke, V., Ioffe, S., Shlens, J., Wojna, Z.: Rethinking the inception architecture for computer vision. In: Proceedings of the IEEE Conference on Computer Vision and Pattern Recognition, pp. 2818–2826 (2016)

32. Tan, M., Le, Q.V.: EfficientNet: rethinking model scaling for convolutional neural networks. In: International Conference on Machine Learning, pp. 6105–6114 (2019)

33. Xie, H., Jiang, W., Luo, H., Yu, H.: Model compression via pruning and knowledge distillation for person re-identification. J. Ambient Intell. Human. Comput. **12**(2), 2149–2161 (2021). https://doi.org/10.1007/s12652-020-02312-4

34. Xie, S., Girshick, R., Dollár, P., Tu, Z., He, K.: Aggregated residual transformations for deep neural networks. In: Proceedings of the IEEE Conference on Computer Vision and Pattern Recognition, pp. 1492–1500 (2017)

35. Yuan, L., Tay, F.E., Li, G., Wang, T., Feng, J.: Revisit knowledge distillation: a teacher-free framework (2019)

36. Zagoruyko, S., Komodakis, N.: Paying more attention to attention: improving the performance of convolutional neural networks via attention transfer. arXiv preprint arXiv:1612.03928 (2016)

37. Zagoruyko, S., Komodakis, N.: Wide residual networks. arXiv preprint arXiv:1605.07146 (2016)

38. Zhang, Y., Xiang, T., Hospedales, T.M., Lu, H.: Deep mutual learning. In: Proceedings of the IEEE Conference on Computer Vision and Pattern Recognition, pp. 4320–4328 (2018)

39. Zhu, Z., Hong, J., Zhou, J.: Data-free knowledge distillation for heterogeneous federated learning. arXiv preprint arXiv:10056 (2021)

Robust and Adversarial Machine Learning

Fooling Partial Dependence via Data Poisoning

Hubert Baniecki$^{(\boxtimes)}$, Wojciech Kretowicz, and Przemyslaw Biecek

Warsaw University of Technology, Warsaw, Poland
{hubert.baniecki.stud,wojciech.kretowicz.stud,przemyslaw.biecek}@pw.edu.pl

Abstract. Many methods have been developed to understand complex predictive models and high expectations are placed on post-hoc model explainability. It turns out that such explanations are not robust nor trustworthy, and they can be fooled. This paper presents techniques for attacking Partial Dependence (plots, profiles, PDP), which are among the most popular methods of explaining any predictive model trained on tabular data. We showcase that PD can be manipulated in an adversarial manner, which is alarming, especially in financial or medical applications where auditability became a must-have trait supporting black-box machine learning. The fooling is performed via poisoning the data to bend and shift explanations in the desired direction using genetic and gradient algorithms. We believe this to be the first work using a genetic algorithm for manipulating explanations, which is transferable as it generalizes both ways: in a model-agnostic and an explanation-agnostic manner.

Keywords: Explainable AI · Adversarial ML · Interpretability

1 Introduction

Although supervised machine learning became state-of-the-art solutions to many predictive problems, there is an emerging discussion on the underspecification of such methods which exhibits differing model behaviour in training and practical setting [14]. This is especially crucial when proper accountability for the systems supporting decisions is required by the domain [35,38,42]. Living with black-boxes, several explainability methods were presented to help us understand models' behaviour [5,19,23,36,40], many are designed specifically for deep neural networks [6,34,44,49]. Explanations are widely used in practice through their (often estimation-based) implementations available to machine learning practitioners in various software contributions [4,8,11].

Nowadays, robustness and certainty become crucial when using explanations in the data science practice to understand black-box machine learning models; thus, facilitate rationale explanation, knowledge discovery and responsible decision-making [9,22]. Notably, several studies evaluate explanations

---

Supplementary Information The online version contains supplementary material available at https://doi.org/10.1007/978-3-031-26409-2_8.

M.-R. Amini et al. (Eds.): ECML PKDD 2022, LNAI 13715, pp. 121–136, 2023.
https://doi.org/10.1007/978-3-031-26409-2_8

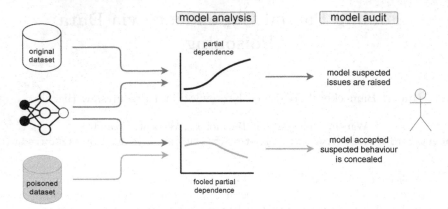

Fig. 1. Framework for fooling model explanations via data poisoning. The red color indicates the adversarial route, a potential security breach, which an attacker may use to manipulate the explanation. Researchers could use this method to provide a misleading rationale for a given phenomenon, while auditors may purposely conceal the suspected, e.g. biased or irresponsible, reasoning of a black-box model. (Color figure online)

[1, 2, 10, 27, 29, 51] showcasing their various flaws from which we perceive an existing robustness gap; in critical domains, one can call it a *security breach*. Apart from promoting wrong explanations, this phenomenon can be exploited to use adversarial attacks on model explanations to achieve manipulated results. In regulated areas, these types of attacks may be carried out to deceive an auditor.

Figure 1 illustrates a process in which the developer aims to conceal the undesired behaviour of the model by supplying a poisoned dataset for model audit. Not every explanation is equally good—just as models require proper performance validation, we need similar assessments of explainability methods. In this paper, we evaluate the robustness of Partial Dependence (PD) [19], moreover highlight the possibility of adversarial manipulation of PD (see Figs. 4 and 5 in the latter part of the paper). We summarize the contributions as follows:

(1) We introduce a novel concept of using a *genetic algorithm* for manipulating model explanations. This allows for a convenient generalization of the attacks in a model-agnostic and explanation-agnostic manner, which is not the case for most of the related work. Moreover, we use a gradient algorithm to perform fooling via data poisoning efficiently for neural networks.
(2) We explicitly target PD to highlight the potential of their adversarial manipulation, which was not done before. Our method provides a sanity check for the future use of PD by responsible machine learning practitioners. Evaluation of the constructed methods in extensive experiments shows that model complexity significantly affects the magnitude of the possible explanation manipulation.

2   Related Work

In the literature, there is a considerable amount of attacks on model explanations specific to deep neural networks [16, 21, 25, 30, 53]. At their core, they provide various algorithms for fooling neural network interpretability and explainability, mainly of image-based predictions. Such explanations are commonly presented through saliency maps [45], where each model input is given its attribution to the prediction [6,43,44,49]. Although explanations can be used to improve the adversarial robustness of machine learning models [37], we target explanations instead. When considering an explanation as a function of model and data, there is a possibility to change one of these variables to achieve a different result [54]. Heo et al. [25] and Dimanov et al. [15] propose fine-tuning a neural network to undermine its explainability capabilities and conceal model unfairness. The assumption is to alter the model's parameters without a drop in performance, which can be achieved with an objective function minimizing the distance between explanations and an arbitrarily set target. Note that [15] indirectly change partial dependence while changing the model. Aivodji et al. [3] propose creating a surrogate model aiming to approximate the unfair black-box model and explain its predictions in a fair manner, e.g. with relative variable dependence.

Alternate idea is to fool fairness and explainability via data change since its (background) distribution greatly affects interpretation results [28,30]. Solans et al. [48] and Fukuchi et al. [20] investigate concealing unfairness via data change by using gradient methods. Dombrowski et al. [16] propose an algorithm for saliency explanation manipulation using gradient-based data perturbations. In contrast, we introduce a genetic algorithm and focus on other machine learning predictive models trained on tabular data. Slack et al. [46] contributed adversarial attacks on post-hoc, model-agnostic explainability methods for *local-level* understanding; namely LIME [40] and SHAP [36]. The proposed framework provides a way to construct a biased classifier with safe explanations of the model's predictions.

Since we focus on *global-level* explanations, instead, the results will modify a view of overall model behaviour, not specific to a single data point or image. Lakkaraju and Bastani [32] conducted a thought-provoking study on misleading effects of manipulated Model Understanding through Subspace Explanations (MUSE) [33], which provide arguments for why such research becomes crucial to achieve responsibility in machine learning use. Further, the robustness of neural networks [13,50] and counterfactual explanations [47] have became important, as one wants to trust black-box models and extend their use to sensitive tasks. Our experiments further extend to global explanations the indication of Jia et al. [29] that there is a correlation between model complexity and explanation quality.

3   Partial Dependence

In this paper, we target one of the most popular explainability methods for tabular data, which at its core presents the expected value of the model's predictions as a function of a selected variable. Partial Dependence, formerly introduced as *plots* by Friedman [19], show the expected value fixed over the marginal joint distribution of other variables. These values can be relatively easily estimated and are widely incorporated into various tools for model explainability [7,8,11,24,39]. The theoretical explanation has its practical estimator used to compute the results, later visualized as a line plot showing the expected prediction for a given variable; also called *profiles* [12]. PD for model f and variable c in a random vector $\mathcal{X}$ is defined as $\mathcal{PD}_c(\mathcal{X}, z) := E_{\mathcal{X}_{-c}}\left[f(\mathcal{X}^{c|=z})\right]$, where $\mathcal{X}^{c|=z}$ stands for random vector $\mathcal{X}$, where c-th variable is replaced by value z. By $\mathcal{X}_{-c}$, we denote distribution of random vector $\mathcal{X}$ where c-th variable is set to a constant. We defined PD in point z as the expected value of model f given the c-th variable is set to z. The standard estimator of this value for data X is given by the following formula $\widehat{\mathcal{PD}}_c(X, z) := \frac{1}{N}\sum_{i=1}^{N} f\left(X_i^{c|=z}\right)$, where X_i is the i-th row of the matrix X and the previously mentioned symbols are used accordingly. To simplify the notation, we will use $\mathcal{PD}$, and omit z and c where context is clear.

4   Fooling Partial Dependence via Data Poisoning

Many explanations treat the dataset X as fixed; however, this is precisely a *single point of failure* on which we aim to conduct the attack. In what follows, we examine $\mathcal{PD}$ behaviour by looking at it as a function whose argument is an entire dataset. For example, if the dataset has N instances and P variables, then $\mathcal{PD}$ is treated as a function over $N \times P$ dimensions. Moreover, because of the complexity of black-box models, $\mathcal{PD}$ becomes an extremely high-dimensional space where variable interactions cause unpredictable behaviour. Explanations are computed using their estimators where a significant simplification may occur; thus, a slight shift of the dataset used to calculate $\mathcal{PD}$ may lead to unintended results (for example, see [26] and the references given there).

We aim to change the underlying dataset used to produce the explanation in a way to achieve the desired change in $\mathcal{PD}$. Figure 1 demonstrates the main threat of an adversarial attack on model explanation using data poisoning, which is *concealing the suspected behaviour of black-box models*. The critical assumption is that an adversary has a possibility to modify the dataset arbitrarily, e.g. in healthcare and financial audit, or research review. Even if this would not be the case, in modern machine learning, wherein practice dozens of variables are used to train complex models, such data shifts might be only a minor change that a person looking at a dataset or distribution will not be able to identify.

We approach fooling $\mathcal{PD}$ as an optimization problem for given criteria of attack efficiency, later called the attack loss. It originates from [16], where a similar loss function for manipulation of local-level model explanations for an image-based predictive task was introduced. This work introduces the attack loss that aims to change the output of a global-level explanation via data poisoning instead. The explanation's weaknesses concerning data distribution and causal inference are exploited using two ways of optimizing the loss:

– **Genetic-based**[1] algorithm that does not make any assumption about the model's structure – the black-box path from data inputs to the output prediction; thus, is model-agnostic. Further, we posit that for a vast number

---

[1] For convenience, we shorten the *algorithm based on the genetic algorithm* phrase to *genetic-based algorithm*.

of explanations, clearly post-hoc global-level, the algorithm does not make assumption about their structure either; thus, becomes *explanation-agnostic*.
- **Gradient-based** algorithm that is specifically designed for models with differentiable outputs, e.g. neural networks [15,16].

We discuss and evaluate two possible fooling strategies:

- **Targeted attack** changes the dataset to achieve the closest explanation result to the predefined desired function [16,25].
- **Robustness check** aims to achieve the most distant model explanation from the original one by changing the dataset, which refers to the sanity check [1].

For practical reasons, we define the distance between the two calculated $\mathcal{PD}$ vectors as $\|x - y\| := \frac{1}{I}\sum_{i=1}^{I}(x_i - y_i)^2$, yet other distance measures may be used to extend the method.

4.1   Attack Loss

The intuition behind the attacks is to find a modified dataset that minimizes the attack loss. A changed dataset denoted as $X \in \mathbb{R}^{N \times P}$ is an argument of that function; hence, an optimal X is a result of the attack. Let $Z \subset \mathbb{R}$ be the set of points used to calculate the explanation. Let $T : Z \to \mathbb{R}$ be the target explanation; we write just T to denote a vector over whole Z. Let $g_c^Z :$ $\mathbb{R}^{N \times P} \to \mathbb{R}^{|Z|}$ be the actual explanation calculated for points in Z; we write g_c for simplicity. Finally, let $X' \in \mathbb{R}^{N \times P}$ be the original (constant) dataset. We define the attack loss as $\mathcal{L}(X) := \mathcal{L}^{g,\,s}(X)$, where g is the explanation to be fooled, and an objective is minimized depending on the strategy of the attack, denoted as s. The aim is to minimize $\mathcal{L}$ with respect to the dataset X used to calculate the explanation. We never change values of the explained variable c in the dataset.

In the **targeted attack**, we aim to **minimize** the distance between the target model behaviour T and the result of model explanation calculated on the changed dataset. We denote this strategy by t and define $\mathcal{L}^{g,\,t}(X) = \|g_c(X) - T\|$. Since we focus on a specific model-agnostic explanation, we substitute $\mathcal{PD}$ in place of g to obtain $\mathcal{L}^{\mathcal{PD},\,t}(X) = \|\mathcal{PD}_c(X) - T\|$. This substitution can be generalized for various global-level model explanations, which rely on using a part of the dataset for computation.

In the **robustness check**, we aim to **maximize** the distance between the result of model explanation calculated on the original dataset $g_c(X')$, and the changed one; thus, minus sign is required. We denote this strategy by r and define $\mathcal{L}^{g,\,r}(X) = -\|g_c(X) - g_c(X')\|$. Accordingly, we substitute $\mathcal{PD}$ in place of g to obtain $\mathcal{L}^{\mathcal{PD},\,r}(X) = -\|\mathcal{PD}_c(X) - \mathcal{PD}_c(X')\|$. Note that $\mathcal{L}^{g,\,s}$ may vary depending on the explanation used, specifically for $\mathcal{PD}$ it is useful to centre the explanation before calculating the distances, which is the default behaviour in our implementation: $\mathcal{L}^{\overline{\mathcal{PD}},\,r}(X) = -\|\overline{\mathcal{PD}}_c(X) - \overline{\mathcal{PD}}_c(X')\|$, where $\overline{\mathcal{PD}}_c := \mathcal{PD}_c(X) - \frac{1}{|Z|}\sum_{z \in Z}\mathcal{PD}_c(X, z)$. We consider the second approach of comparing explanations using centred $\overline{\mathcal{PD}}$, as it forces changes in the shape of the explanation, instead of promoting to shift the profile vertically while the shape changes insignificantly.

4.2   Genetic-Based Algorithm

We introduce a novel strategy for fooling explanations based on the genetic algorithm as it is a simple yet powerful method for real parameter optimization [52]. We do not encode genes conventionally but deliberately use this term to distinguish from other types of evolutionary algorithms [18]. The method will be invariant to the model's definition and the considered explanations; thus, it becomes model-agnostic and explanation-agnostic. These traits are crucial when working with black-box machine learning as versatile solutions are convenient.

Fooling $\mathcal{PD}$ in both strategies include a similar genetic algorithm. The main idea is to define an individual as an instance of the dataset, iteratively perturb its values to achieve the desired explanation target, or perform the robustness check to observe the change. These individuals are initialized with a value of the original dataset X' to form a population. Subsequently, the initialization ends with mutating the individuals using a higher-than-default variance of perturbations. Then, in each iteration, they are randomly crossed, mutated, evaluated with the attack loss, and selected based on the loss values. **Crossover** swaps columns between individuals to produce new ones, which are then added to the population. The number of swapped columns can be randomized; also, the number of parents can be parameterized. **Mutation** adds Gaussian noise to the individuals using scaled standard deviations of the variables. It is possible to constrain the data change into the original range of variable values; also keep some variables unchanged. **Evaluation** calculates the loss for each individual, which requires to compute explanations for each dataset. **Selection** reduces the number of individuals using rank selection, and elitism to guarantee several best individuals to remain in the next population.

We considered the crossover through an exchange of rows between individuals, but it might drastically shift the datasets and move them apart. Additionally, a worthy mutation is to add or subtract whole numbers from the integer-encoded (categorical) variables. We further discuss the algorithm's details in the Supplementary material. The introduced attack is model-invariant because no derivatives are needed for optimization, which allows evaluating explanations of black-box models. While we found this method a sufficient generalization of our framework, there is a possibility to perform a more efficient optimization assuming the prior knowledge concerning the structure of model and explanation.

4.3   Gradient-Based Algorithm

Gradient-based methods are state-of-the-art optimization approaches, especially in the domain of deep neural networks [34]. This algorithm's main idea is to use gradient descent to optimize the attack loss, considering the differentiability of the model's output with respect to input data. Such assumption allows for a faster and more accurate convergence into a local minima using one of the stochastic optimizers; in our case, Adam [31]. Note that the differentiability assumption is with respect to input data, not with respect to the model's parameters. We shall derive the gradients $\nabla_{X_{-c}}\mathcal{L}^{g,\,s}$ for fooling explanations based on their estimators, not the theoretical definitions. This is because the input data is assumed to be a random variable in a theoretical definition of $\mathcal{PD}$, making it impossible to calculate a derivative over the input dataset. In practice, we do not derive our method directly from the definition as the estimator produces the explanation.

Although we specifically consider the usage of neural networks because of their strong relation to differentiation, the algorithm's theoretical derivation does not require this type of model. For brevity, we derive the theoretical definitions of gradients $\nabla_{X_{-c}}\mathcal{L}^{\mathcal{PD},\,t}$, $\nabla_{X_{-c}}\mathcal{L}^{\mathcal{PD},\,r}$, and $\nabla_{X_{-c}}\mathcal{L}^{\overline{\mathcal{PD}},\,r}$ in the Supplementary material. Overall, the gradient-based algorithm is similar to the genetic-based algorithm in that we aim to iteratively change the dataset used to calculate the explanation. Nevertheless, its main assumption is that the model provides an interface for the differentiation of output with respect to the input, which is not the case for black-box models.

5 Experiments

Setup. We conduct experiments on two predictive tasks to evaluate the algorithms and conclude with illustrative scenario examples, which refer to the framework shown in Fig. 1. The first dataset is a synthetic regression problem that refers to the Friedman's work [19] where inputs X are independent variables uniformly distributed on the interval $[0, 1]$, while the target y is created according to the formula: $y(X) = 10\sin(\pi \cdot X_1 \cdot X_2) + 20(X_3 - 0.5)^2 + 10X_4 + 5X_5$. Only 5 variables are actually used to compute y, while the remaining are independent of. We refer to this dataset as `friedman` and target explanations of the variable X_1. The second dataset is a real classification task from UCI [17], which has 5 continuous variables, 8 categorical variables, and an evenly-distributed binary target. We refer to this dataset as **heart** and target explanations of the variable **age**. Additionally, we set the discrete variables as constant during the performed fooling because we mainly rely on incremental change in the values of continuous variables, and categorical variables are out of the scope of this work.

Results. Figure 2 present the main result of the paper, which is that PD can be manipulated. We use the gradient-based algorithm to change the explanations of feedforward neural networks via data poisoning. The targeted attack aims to arbitrarily change the monotonicity of PD, which is evident in both predictive tasks. The robustness check finds the most distant explanation from the original one. We perform the fooling 30 times for each subplot, and the Y-axis denotes the model's architecture: layers×neurons. We observe that PD explanations are especially vulnerable in complex models.

Next, we aim to evaluate the PD of various state-of-the-art machine learning models and their complexity levels; we denote: Linear Model (LM), Random Forest (RF), Gradient Boosting Machine (GBM), Decision Tree (DT),

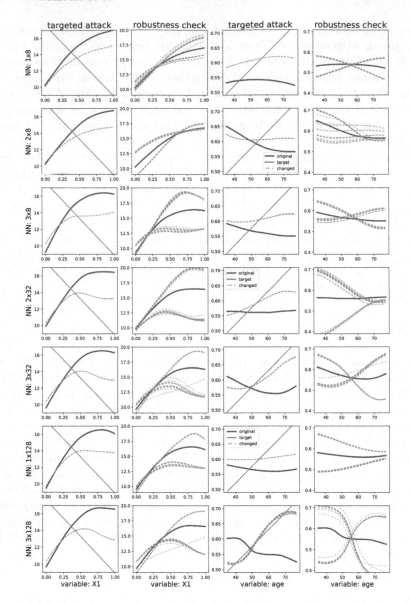

Fig. 2. Fooling Partial Dependence of neural network models (rows) fitted to the **friedman** and **heart** datasets (columns). We performed multiple randomly initiated gradient-based fooling algorithms on the explanations of variables X_1 and **age** respectively. The blue line denotes the original explanation, the red lines are the fooled explanations, and in the targeted attack, the grey line denotes the desired target. We observe that the explanations' vulnerability greatly increases with model complexity. Interestingly, the algorithm seems to converge to two contrary optima when no target is provided. (Color figure online)

K-Nearest Neighbours (KNN), feedforward Neural Network (NN). The model-agnostic nature of the genetic-based algorithm allows this comparison as it might be theoretically and/or practically impossible to differentiate the model's output with respect to the input; thus, differentiate the explanations and loss.

Table 1. Attack loss values of the robustness checks for Partial Dependence of various machine learning models (**top**), and complexity levels of tree-ensembles (**bottom**). Each value corresponds to the scaled distance between the original explanation and the changed one. We perform the fooling 6 times and report the mean±sd. We observe that the explanations' vulnerability increases with GBM complexity.

Task \ Model	LM	RF	GBM	DT	KNN	NN	SVM
friedman	$0_{\pm 0}$	$152_{\pm 76}$	$127_{\pm 71}$	$332_{\pm 172}$	$164_{\pm 61}$	$269_{\pm 189}$	$576_{\pm 580}$
heart	$2_{\pm 3}$	$20_{\pm 5}$	$77_{\pm 28}$	$798_{\pm 192}$	$133_{\pm 21}$	$501_{\pm 52}$	$451_{\pm 25}$

Task	Model \ Trees	10	20	40	80	160	320
friedman	GBM	$57_{\pm 12}$	$114_{\pm 20}$	$157_{\pm 37}$	$176_{\pm 20}$	$189_{\pm 8}$	$210_{\pm 9}$
	RF	$233_{\pm 22}$	$219_{\pm 25}$	$219_{\pm 9}$	$201_{\pm 23}$	$216_{\pm 13}$	$209_{\pm 15}$
heart	GBM	$1_{\pm 0}$	$3_{\pm 1}$	$29_{\pm 4}$	$70_{\pm 24}$	$152_{\pm 56}$	$321_{\pm 95}$
	RF	$62_{\pm 7}$	$55_{\pm 3}$	$29_{\pm 9}$	$21_{\pm 6}$	$14_{\pm 5}$	$13_{\pm 2}$

Table 1 presents the results of robustness checks for Partial Dependence of various machine learning models and complexity levels. Each value corresponds to the distance between the original explanation and the changed one; multiplied by 10^3 in `friedman` and 10^6 in `heart` for clarity. We perform the checks 6 times and report the mean ± standard deviation. Note that we cannot compare the values between tasks, as their magnitudes depend on the prediction range. We found the explanations of NN, SVM and deep DT the most vulnerable to the fooling methods (**top** Table). In contrast, RF seems to provide robust explanations; thus, we further investigate the relationship between the tree-models' complexity and the explanation stability (**bottom** Table) to conclude that an increasing complexity yields more vulnerable explanations, which is consistent with Fig. 2. We attribute the differences between the results for RF and GBM to the concept of bias-variance tradeoff. In some cases (**heart**, RF), explanations of too simple models become vulnerable too, since underfitted models may be as uncertain as overfitted ones.

Ablation Study. We further discuss the additional results that may be of interest to gain a broader context of this work. Figure 3 presents the distinction between the robustness check for centred Partial Dependence, which is the default algorithm, and the robustness check for *not centred* PD. We use the gradient-based algorithm to change the explanations of a 3 layers × 32 neurons ReLU neural network and perform the fooling 30 times for each subplot. We observe that centring the explanation in the attack loss definition is necessary to achieve the change in explanation shape. Alternatively, the explanation shifts upwards or downwards by essentially changing the mean of prediction. This observation was consistent across most of the models despite their complexity.

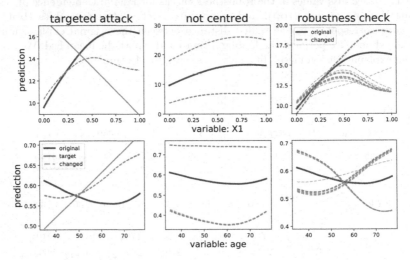

Fig. 3. Fooling Partial Dependence of a 3×32 neural network fitted to the `friedman` (**top** row) and `heart` (**bottom** row) datasets. We performed multiple randomly initiated gradient-based fooling algorithms on the explanations of variables X_1 and `age` respectively. We observe that centring PD is beneficial because it stops the manipulated explanation from shifting.

Table 2. Attack loss values of the robustness checks for PD of various ReLU neural networks. We add additional noise variables to the data before model fitting, e.g. `friedman+2` denotes the referenced dataset with 2 additional variables sampled from the normal distribution. We perform the fooling 30 times and report the mean $\pm$ sd. We observe that the explanations' vulnerability greatly increases with task complexity.

Task	NN						
	1×8	2×8	3×8	2×32	3×32	1×128	3×128
friedman	$25_{\pm 3}$	$33_{\pm 0}$	$75_{\pm 24}$	$100_{\pm 32}$	$98_{\pm 42}$	$54_{\pm 15}$	$97_{\pm 50}$
friedman+1	$31_{\pm 2}$	$40_{\pm 4}$	$50_{\pm 9}$	$106_{\pm 40}$	$115_{\pm 44}$	$57_{\pm 15}$	$114_{\pm 55}$
friedman+2	$34_{\pm 1}$	$40_{\pm 10}$	$50_{\pm 22}$	$106_{\pm 52}$	$115_{\pm 50}$	$50_{\pm 15}$	$137_{\pm 66}$
friedman+4	$46_{\pm 6}$	$33_{\pm 0}$	$83_{\pm 8}$	$145_{\pm 31}$	$163_{\pm 27}$	$40_{\pm 5}$	$140_{\pm 58}$
friedman+8	$71_{\pm 9}$	$47_{\pm 3}$	$89_{\pm 15}$	$204_{\pm 25}$	$176_{\pm 25}$	$39_{\pm 6}$	$156_{\pm 34}$
heart	$11_{\pm 0}$	$8_{\pm 1}$	$10_{\pm 0}$	$32_{\pm 3}$	$41_{\pm 5}$	$6_{\pm 1}$	$134_{\pm 14}$
heart+1	$10_{\pm 1}$	$17_{\pm 6}$	$17_{\pm 2}$	$44_{\pm 4}$	$57_{\pm 13}$	$6_{\pm 1}$	$128_{\pm 8}$
heart+2	$13_{\pm 1}$	$31_{\pm 13}$	$17_{\pm 5}$	$63_{\pm 4}$	$79_{\pm 10}$	$14_{\pm 2}$	$218_{\pm 82}$
heart+4	$13_{\pm 1}$	$21_{\pm 9}$	$30_{\pm 17}$	$113_{\pm 4}$	$139_{\pm 60}$	$29_{\pm 5}$	$232_{\pm 36}$
heart+8	$16_{\pm 0}$	$28_{\pm 18}$	$43_{\pm 20}$	$125_{\pm 49}$	$227_{\pm 28}$	$25_{\pm 8}$	$311_{\pm 283}$

Table 2 presents the impact of additional noise variables in data on the performed fooling. We observe that higher data dimensions favor vulnerable explanations (higher loss). The analogous results for targeted attack were consistent; however, showcased almost zero variance (partially observable in Figs. 2 and 3).

Adversarial Scenario. Following the framework shown in Fig. 1, we consider three stakeholders apparent in explainable machine learning: developer, auditor, and prediction recipients. Let us assume that the model predicting a heart attack should not take into account a patient's sex; although, it might be a valuable predictor. An auditor analyses the model using Partial Dependence; therefore, the developer supplies a poisoned dataset for this task. Figure 5 presents two possible outcomes of model audit: concealed and suspected, which are unequivocally bound to the explanation result and dataset. In first, the model is unchanged while the stated assumption of even dependence between sex is concealed (equal to about 0.5); thus, the prediction recipients become vulnerable. Additionally, we supply an alternative scenario where the developer wants to provide evidence of model unfairness to raise suspicion (dependence for class 0 equal to about 0.7).

Supportive Scenario. In this work, we consider an equation of three variables: data, model, and explanation; thus, we poison the data to fool the explanation while the model remains unchanged. Figures 4 and 5 showcase an exemplary data shift occurring in the dataset after the attack where changing only a few explanatory variables results in bending PD. We present a moderate change in data distribution to introduce a concept of analysing such relationships for explanatory purposes, e.g. the first result might suggest that resting blood pressure and maximum heart rate contribute to the explanation of age; the second result suggests how these variables contribute to the explanation of sex. We conclude that the data shift is worth exploring to analyse variable interactions in models.

6   Limitations and Future Work

We find these results both alarming and informative yet proceed to discuss the limitations of the study. First is the assumption that, in an adversarial scenario, the auditor has no access to the original (unknown) data, e.g. in research or healthcare audit. While the detectability of fooling is worth analyzing, our work focuses not only on an adversarial manipulation of PD, as we sincerely hope such data poisoning is not occurring in practice. Even more, we aim to underline the crucial context of data distribution in the interpretation of explanations and introduce a new way of evaluating PD; black-box explanations by generalizing the methods with genetic-based optimization.

Another limitation is the size of the used datasets. We have engaged with larger datasets during experiments but were turned off by a contradictory view that increasing the dataset size might be considered as exaggerating the results. PD clearly becomes more complex with increasing data dimensions; moreover,

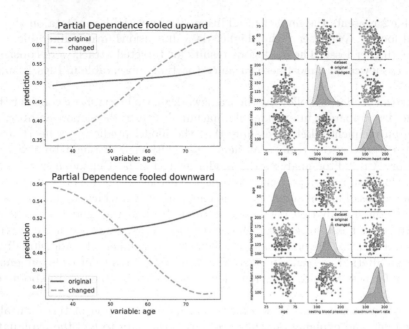

Fig. 4. Partial Dependence of `age` in the SVM model prediction of a heart attack (class 0). **Left:** Two manipulated explanations suggest an increasing or decreasing relationship between `age` and the predicted outcome depending on a desired outcome. **Right:** Distribution of the explained variable `age` and the two poisoned variables from the data, in which the remaining ten variables attributing to the explanation remain unchanged. The mean of the variables' Jensen-Shannon distance equals only 0.027 in the upward scenario and 0.021 in the downward scenario, which might seem like an insignificant change of the data distribution.

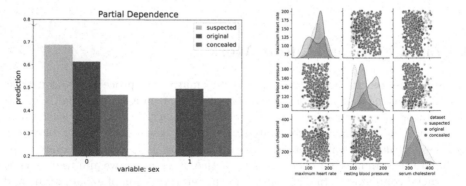

Fig. 5. Partial Dependence of `sex` in the SVM model prediction of a heart attack (class 0). **Left:** Two manipulated explanations present a suspected or concealed variable contribution into the predicted outcome. **Right:** Distribution of the three poisoned variables from the data, in which `sex` and the remaining nine variables attributing to the explanation remain unchanged. The mean of the variables' Jensen-Shannon distance equals only 0.023 in the suspected scenario and 0.026 in the concealed scenario.

higher-dimensional space should entail more possible ways of manipulation, which is evident in the ablation study. We note that in practice, the explanations might require 100–1000 observations for the estimation (e.g. kernel SHAP, PDP), hence the size of the datasets in this study. Finally, we omit datasets like Adult and COMPAS because they mainly consist of categorical variables.

Future Work. We foresee several directions for future work, e.g. evaluating the successor to PD – Accumulated Local Effects (ALE) [5]; although the practical estimation of ALE presents challenges. Second, the attack loss may be enhanced by regularization, e.g. penalty for substantial change in data or mean of model's prediction, to achieve more meaningful fooling with less evidence. We focus in this work on univariate PD, but targeting bivariate PD can also be examined. Overall, the landscape of global-level, post-hoc model explanations is a broad domain, and the potential of a security breach in other methods, e.g. SHAP, should be further examined. Enhancements to the model-agnostic and explanation-agnostic genetic algorithm are thereby welcomed.

Another future direction would be to enhance the stability of PD. Rieger and Hansen [41] present a defence strategy against the attack via data change [16] by aggregating various explanations, which produces robust results without changing the model.

7   Conclusion and Impact

We highlight that Partial Dependence can be maliciously altered, e.g. bent and shifted, with adversarial data perturbations. The introduced genetic-based algorithm allows evaluating explanations of any black-box model. Experimental results on various models and their sizes showcase the hidden debt of model complexity related to explainable machine learning. Explanations of low-variance models prove to be robust to the manipulation, while very complex models should not be explained with PD as they become vulnerable to change in reference data distribution. Robustness checks lead to varied modifications of the explanations depending on the setting, e.g. may propose two opposite PD, which is why it is advised to perform the checks multiple times.

This work investigates the vulnerability of global-level, post-hoc model explainability from the adversarial setting standpoint, which refers to the responsibility and security of the artificial intelligence use. Possible manipulation of PD leads to the conclusion that explanations used to explain black-box machine learning may be considered black-box themselves. These explainability methods are undeniably useful through implementations in various popular software. However, just as machine learning models cannot be developed without extensive testing and understanding of their behaviour, their explanations cannot be used without critical thinking. We recommend ensuring the reliability of the explanation results through the introduced methods, which can also be used to study models behaviour under the data shift. Code for this work is available at https://github.com/MI2DataLab/fooling-partial-dependence.

Acknowledgements. This work was financially supported by the NCN Opus grant 2017/27/B/ST6/01307 and NCN Sonata Bis-9 grant 2019/34/E/ST6/00052.

References

1. Adebayo, J., Gilmer, J., Muelly, M., Goodfellow, I., Hardt, M., Kim, B.: Sanity checks for saliency maps. In: NeurIPS (2018)
2. Adebayo, J., Muelly, M., Liccardi, I., Kim, B.: Debugging tests for model explanations. In: NeurIPS (2020)
3. Aivodji, U., Arai, H., Fortineau, O., Gambs, S., Hara, S., Tapp, A.: Fairwashing: the risk of rationalization. In: ICML (2019)
4. Alber, M., Lapuschkin, S., Seegerer, P., Hägele, M., Schütt, K.T., et al.: iNNvestigate neural networks! J. Mach. Learn. Res. **20**(93), 1–8 (2019)
5. Apley, D.W., Zhu, J.: Visualizing the effects of predictor variables in black box supervised learning models. J. Roy. Stat. Soc. Ser. B (Stat. Methodol.) **82**(4), 1059–1086 (2020)
6. Bach, S., Binder, A., Montavon, G., Klauschen, F., Müller, K.R., Samek, W.: On pixel-wise explanations for non-linear classifier decisions by layer-wise relevance propagation. PLoS ONE **10**(7), 1–46 (2015)
7. Baniecki, H., Biecek, P.: modelStudio: interactive studio with explanations for ML predictive models. J. Open Source Softw. **4**(43), 1798 (2019)
8. Baniecki, H., Kretowicz, W., Piatyszek, P., Wisniewski, J., Biecek, P.: dalex: responsible machine learning with interactive explainability and fairness in Python. J. Mach. Learn. Res. **22**(214), 1–7 (2021)
9. Barredo Arrieta, A., Díaz-Rodríguez, N., Del Ser, J., Bennetot, A., Tabik, S., et al.: Explainable Artificial Intelligence (XAI): concepts, taxonomies, opportunities and challenges toward responsible AI. Inf. Fusion **58**, 82–115 (2020)
10. Bhatt, U., Weller, A., Moura, J.M.F.: Evaluating and aggregating feature-based model explanations. In: IJCAI (2020)
11. Biecek, P.: DALEX: explainers for complex predictive models in R. J. Mach. Learn. Res. **19**(84), 1–5 (2018)
12. Biecek, P., Burzykowski, T.: Explanatory Model Analysis. Chapman and Hall/CRC (2021)
13. Boopathy, A., et al.: Proper network interpretability helps adversarial robustness in classification. In: ICML (2020)
14. D'Amour, A., Heller, K., Moldovan, D., Adlam, B., Alipanahi, B., et al.: Underspecification presents challenges for credibility in modern machine learning. arXiv preprint arXiv:2011.03395 (2020)
15. Dimanov, B., Bhatt, U., Jamnik, M., Weller, A.: You shouldn't trust me: learning models which conceal unfairness from multiple explanation methods. In: AAAI SafeAI (2020)
16. Dombrowski, A.K., Alber, M., Anders, C., Ackermann, M., Müller, K.R., Kessel, P.: Explanations can be manipulated and geometry is to blame. In: NeurIPS (2019)
17. Dua, D., Graff, C.: UCI Machine Learning Repository (2017). https://www.kaggle.com/ronitf/heart-disease-uci/version/1
18. Elbeltagi, E., Hegazy, T., Grierson, D.: Comparison among five evolutionary-based optimization algorithms. Adv. Eng. Inform. **19**(1), 43–53 (2005)
19. Friedman, J.H.: Greedy function approximation: a gradient boosting machine. Ann. Stat. **29**(5), 1189–1232 (2001)

20. Fukuchi, K., Hara, S., Maehara, T.: Faking fairness via stealthily biased sampling. In: AAAI (2020)
21. Ghorbani, A., Abid, A., Zou, J.: Interpretation of neural networks is fragile. In: AAAI (2019)
22. Gill, N., Hall, P., Montgomery, K., Schmidt, N.: A responsible machine learning workflow with focus on interpretable models, post-hoc explanation, and discrimination testing. Information 11(3), 137 (2020)
23. Goldstein, A., Kapelner, A., Bleich, J., Pitkin, E.: Peeking inside the black box: visualizing statistical learning with plots of individual conditional expectation. J. Comput. Graph. Stat. 24(1), 44–65 (2015)
24. Greenwell, B.M.: pdp: an R package for constructing partial dependence plots. R J. 9(1), 421–436 (2017)
25. Heo, J., Joo, S., Moon, T.: Fooling neural network interpretations via adversarial model manipulation. In: NeurIPS (2019)
26. Hooker, G.: Generalized functional ANOVA diagnostics for high-dimensional functions of dependent variables. J. Comput. Graph. Stat. 16(3), 709–732 (2007)
27. Hooker, S., Erhan, D., Kindermans, P.J., Kim, B.: A benchmark for interpretability methods in deep neural networks. In: NeurIPS (2019)
28. Janzing, D., Minorics, L., Blöbaum, P.: Feature relevance quantification in explainable AI: a causal problem. In: AISTATS (2020)
29. Jia, Y., Frank, E., Pfahringer, B., Bifet, A., Lim, N.: Studying and exploiting the relationship between model accuracy and explanation quality. In: Oliver, N., Pérez-Cruz, F., Kramer, S., Read, J., Lozano, J.A. (eds.) ECML PKDD 2021. LNCS (LNAI), vol. 12976, pp. 699–714. Springer, Cham (2021). https://doi.org/10.1007/978-3-030-86520-7_43
30. Kindermans, P.-J., et al.: The (un)reliability of saliency methods. In: Samek, W., Montavon, G., Vedaldi, A., Hansen, L.K., Müller, K.-R. (eds.) Explainable AI: Interpreting, Explaining and Visualizing Deep Learning. LNCS (LNAI), vol. 11700, pp. 267–280. Springer, Cham (2019). https://doi.org/10.1007/978-3-030-28954-6_14
31. Kingma, D.P., Ba, J.: Adam: a method for stochastic optimization. In: ICLR (2015)
32. Lakkaraju, H., Bastani, O.: "How do i fool you?": Manipulating user trust via misleading black box explanations. In: AIES (2020)
33. Lakkaraju, H., Kamar, E., Caruana, R., Leskovec, J.: Faithful and customizable explanations of black box models. In: AIES (2019)
34. LeCun, Y., Bengio, Y., Hinton, G.: Deep learning. Nature 521(7553), 436–444 (2015)
35. Lipton, Z.C.: The mythos of model interpretability. Queue 16(3), 31–57 (2018)
36. Lundberg, S.M., Lee, S.I.: A unified approach to interpreting model predictions. In: NeurIPS (2017)
37. Mangla, P., Singh, V., Balasubramanian, V.N.: On saliency maps and adversarial robustness. In: Hutter, F., Kersting, K., Lijffijt, J., Valera, I. (eds.) ECML PKDD 2020. LNCS (LNAI), vol. 12458, pp. 272–288. Springer, Cham (2021). https://doi.org/10.1007/978-3-030-67661-2_17
38. Miller, T.: Explanation in artificial intelligence: insights from the social sciences. Artif. Intell. 267, 1–38 (2019)
39. Molnar, C., Casalicchio, G., Bischl, B.: iml: an R package for interpretable machine learning. J. Open Source Softw. 3(26), 786 (2018)
40. Ribeiro, M.T., Singh, S., Guestrin, C.: "Why should i trust you?": Explaining the predictions of any classifier. In: KDD (2016)

41. Rieger, L., Hansen, L.K.: A simple defense against adversarial attacks on heatmap explanations. In: ICML WHI (2020)
42. Rudin, C.: Stop explaining black box machine learning models for high stakes decisions and use interpretable models instead. Nat. Mach. Intell. **1**, 206–215 (2019)
43. Selvaraju, R.R., Cogswell, M., Das, A., Vedantam, R., Parikh, D., Batra, D.: Grad-CAM: visual explanations from deep networks via gradient-based localization. Int. J. Comput. Vis. **128**(2), 336–359 (2020)
44. Shrikumar, A., Greenside, P., Kundaje, A.: Learning important features through propagating activation differences. In: ICML (2017)
45. Simonyan, K., Vedaldi, A., Zisserman, A.: Deep inside convolutional networks: visualising image classification models and saliency maps. In: ICLR (2014)
46. Slack, D., Hilgard, S., Jia, E., Singh, S., Lakkaraju, H.: Fooling LIME and SHAP: adversarial attacks on post hoc explanation methods. In: AIES (2020)
47. Slack, D., Hilgard, S., Lakkaraju, H., Singh, S.: Counterfactual explanations can be manipulated. In: NeurIPS (2021)
48. Solans, D., Biggio, B., Castillo, C.: Poisoning attacks on algorithmic fairness. In: Hutter, F., Kersting, K., Lijffijt, J., Valera, I. (eds.) ECML PKDD 2020. LNCS (LNAI), vol. 12457, pp. 162–177. Springer, Cham (2021). https://doi.org/10.1007/978-3-030-67658-2_10
49. Sundararajan, M., Taly, A., Yan, Q.: Axiomatic attribution for deep networks. In: ICML (2017)
50. Wang, Z., Wang, H., Ramkumar, S., Mardziel, P., Fredrikson, M., Datta, A.: Smoothed geometry for robust attribution. In: NeurIPS (2020)
51. Warnecke, A., Arp, D., Wressnegger, C., Rieck, K.: Evaluating explanation methods for deep learning in security. In: IEEE EuroS&P (2020)
52. Wright, A.H.: Genetic algorithms for real parameter optimization. Found. Genet. Algorithms **1**, 205–218 (1991)
53. Zhang, X., Wang, N., Shen, H., Ji, S., Luo, X., Wang, T.: Interpretable deep learning under fire. In: USENIX Security (2020)
54. Zhao, Q., Hastie, T.: Causal interpretations of black-box models. J. Bus. Econ. Stat. **39**(1), 272–281 (2019)

FROB: Few-Shot ROBust Model for Joint Classification and Out-of-Distribution Detection

Nikolaos Dionelis$^{(\boxtimes)}$ (ID), Sotirios A. Tsaftaris (ID), and Mehrdad Yaghoobi (ID)

School of Engineering, Electrical Engineering, Digital Communications,
The University of Edinburgh, Edinburgh, UK
nikolaos.dionelis@ed.ac.uk

Abstract. Classification and Out-of-Distribution (OoD) detection in the few-shot setting remain challenging aims, but are important for devising critical systems in security where samples are limited. OoD detection requires that classifiers are aware of when they do not know and avoid setting high confidence to OoD samples away from the training data distribution. To address such limitations, we propose the Few-shot ROBust (FROB) model with its key contributions being (a) the joint classification and few-shot OoD detection, (b) the sample generation on the boundary of the support of the normal class distribution, and (c) the incorporation of the learned distribution boundary as OoD data for contrastive negative training. FROB finds the boundary of the support of the normal class distribution, and uses it to improve the few-shot OoD detection performance. We propose a self-supervised learning methodology for sample generation on the normal class distribution confidence boundary based on generative and discriminative models, including classification. FROB implicitly generates adversarial samples, and forces samples from OoD, including our boundary, to be less confident by the classifier. By including the learned boundary, FROB reduces the threshold linked to the model's few-shot robustness in the number of few-shots, and maintains the OoD performance approximately constant, independent of the number of few-shots. The low- and few-shot robustness evaluation of FROB on different image datasets and on One-Class Classification (OCC) data shows that FROB achieves competitive performance and outperforms baselines in terms of robustness to the OoD few-shot population and variability.

Keywords: Out-of-Distribution detection · Few-shot anomaly detection

1 Introduction

In real-world settings, for AI-enabled systems to be operational, it is crucial to robustly perform joint classification and Out-of-Distribution (OoD) detection, and report an input as OoD rather than misclassifying it. The problem of detecting whether a sample is in-distribution, i.e. from the training distribution, or OoD is critical. This is crucial in safety and security as the consequences of failure to

© The Author(s), under exclusive license to Springer Nature Switzerland AG 2023
M.-R. Amini et al. (Eds.): ECML PKDD 2022, LNAI 13715, pp. 137–153, 2023.
https://doi.org/10.1007/978-3-031-26409-2_9

detect OoD objects can be severe and eventually fatal. However, deep neural networks produce overconfident predictions and do not distinguish in- and out-of-data-distribution. Adversarial examples, when small modifications of the input appear, can change the classifier decision. It is an important property of a classifier to address such limitations and provide robustness guarantees. In parallel, OoD detection is challenging as classifiers set high confidence to OoD samples away from the training data. In this paper, we propose the Few-shot ROBust (FROB) model to accurately perform simultaneous classification and OoD detection in the few-shot setting. To address rarity and the existence of limited samples in the few-shot setting [1, 2], we aim at reducing the number of few-shots of OoD data required, whilst maintaining accurate and robust performance.

Training with outlier sets of diverse data, available today in large quantities, can improve OoD detection [3–5]. General OoD datasets enable OoD generalization to detect unseen OoD with improved robustness and performance. Models trained with different outliers can detect OoD by learning cues for whether inputs lie within or out of the support of the normal class distribution. By exposing models to different OoD, the complement of the support of the normal class distribution is modelled. The detection of new types of OoD is enabled. OoD datasets improve the calibration of classifiers in the setting where a fraction of the data is OoD, addressing overconfidence issues when applied to OoD [3, 4].

The main benefits of FROB are that (a) joint classification and OoD detection is realistic, effective, and beneficial, (b) our proposed distribution boundary is a principled, effective, and beneficial approach to generate near OoD samples for negative training, and (c) contrastive training to include the learned negative data during training is effective and beneficial. Furthermore, the benefits of performing joint multi-class classification and OoD detection are that (i) this setting is more realistic and has wider applicability because in the real-world, models should be both operational and reliable and declare an input as OoD rather than misclassifying it, (ii) using discriminative classifier models leads to improved OoD detection performance, and (iii) in the few-shot setting, discriminative classifiers address the limited data problem with improved robustness. An additional benefit of performing simultaneous classification and OoD detection is that we take advantage of labelled data to achieve improved anomaly detection performance as they contain more information because of their labels and classes. Knowing the normal data better, as well as learning how the data are structured in clusters with class labels, helps us to detect OoD data better.

We address the rarity of near and relevant anomalies during training by performing sample generation on the boundary of the support of the underlying distribution of the data from the normal class. The benefit of this is improved robustness to the OoD few-shot population and variability. Task-specific OoD samples are hard to find in practice; in the real world, we also have budget limitations for (negative) sampling. FROB achieves significantly better robustness for few-shot OoD detection, while maintaining in-distribution accuracy. Aiming at solving the few-shot robustness problem with classification and OoD detection, the contribution of our FROB methodology is the development of an integrated robust framework for self-supervised few-shot negative data augmentation on the distribution confidence boundary, combined with few-shot OoD detection. FROB trains a generator to create low-confidence samples on the normal class boundary, and includes these learned samples in the training to improve the performance in the few-shot setting. The combination of the self-generated boundary and the imposition of low confidence at this learned boundary is a contribution of FROB, which improves robustness for few-shot OoD detection. The main benefits of our distribution boundary framework are that it is a principled approach based on distributions, it generates near-OoD samples that are well-sampled and evenly scattered, these near negative data are strong anomalies and adversarial anomalies [6,11], and these learned OoD data are the closest possible negative samples to the normal class. This latter characteristic of our algorithm leads to the tightest-possible OoD data description and characterization, and to self-generated negatives that are optimal in the sense that no unfilled space is allowed between the normal class data and the *learned* OoD samples. In this way, FROB uses the definition of anomaly and the delimitation of the support boundary of the normal class distribution, which are needed for improved robustness.

We achieve generalization to unseen OoD, with applicability to new unknown, in the wild, test sets that do not correlate to the training sets. FROB's evaluation in several settings, using cross-dataset and One-Class Classification (OCC) evaluations, shows that key methodological contributions such as generating samples on the normal class distribution boundary and few-shot adaptation, improve few-shot OoD detection. Our experiments show robustness to the number of OoD few-shots and to outlier variation, outperforming methods we compare with.

2   Proposed Methodology for Few-Shot OoD Detection

We propose FROB in Fig. 1 for joint classification and few-shot OoD detection combining discriminative and generative models. We aim for improved robustness and reliable confidence prediction, and force low confidence close and away from the data. Our key idea is to jointly learn a classifier but also a generative model that finds the boundary of the support of the in-distribution data. We use this generator to create adversarial samples on the boundary close to the in-distribution data. We combine these in a self-supervised learning manner, where the generated data act as a negative class. We propose a robustness loss to classify as less confident samples on, and out of, the learned boundary. FROB also uses few-shots of real OoD data naturally within the formulation we propose.

Loss Function. We denote the normal class data by $\mathbf{x}$ where $\mathbf{x}_i$ are the labeled data, with labels y_i between 1 and K. The few-shot OoD samples are $\mathbf{Z}_m$. The cost function of the classifier model, minimized during training, is

$$\arg\min_f - \frac{1}{N} \sum_{i=1}^{N} \log \frac{\exp(f_{y_i}(\mathbf{x}_i))}{\sum_{k=1}^{K} \exp(f_k(\mathbf{x}_i))} \tag{1}$$

$$- \lambda \frac{1}{M} \sum_{m=1}^{M} \log \left(1 - \max_{l=1,2,\ldots,K} \frac{\exp(f_l(\mathbf{Z}_m))}{\sum_{k=1}^{K} \exp(f_k(\mathbf{Z}_m))} \right),$$

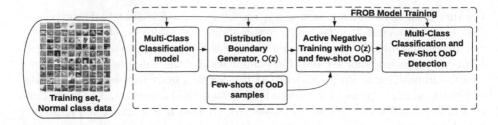

Fig. 1. FROB training with learned negative sampling, $O(\mathbf{z})$, and few-shot OoD.

where $f(\cdot)$ is the Convolutional Neural Network (CNN) discriminative model for multi-class classification with K classes. The proposed objective cost function has two loss terms and a hyperparameter. The two loss terms operate on different samples for positive and negative training, respectively. The first loss term is the cross-entropy between y_i and the predictions, softmax$(f(\mathbf{x}_i))$; the CNN is followed by the normalized exponential to obtain the probability over the classes. The second loss term enforces $f(\cdot)$ to more accurately detect outliers, in addition to performing multi-class classification. It is weighted by a hyperparameter, λ.

FROB then trains a generator to generate low-confidence samples on the normal class distribution boundary. Our algorithm includes these learned low-confidence samples in the training to improve the performance in the few-shot setting. We do not use a large general OoD dataset because general-purpose OoD datasets lead to an ad hoc selection of outliers that try to approximate data outside the support of the normal class distribution. Instead, we use negative data augmentation and self-supervised learning to model the boundary of the support of the normal class distribution. Our proposed FROB model generates outliers via a trained generator $O(\mathbf{z})$, which takes the form of a CNN. Here, O refers to OoD samples, and $\mathbf{z}$ are samples from a standard Gaussian distribution. The optimization of maximizing dispersion subject to being on the boundary is

$$\arg\ \min_O \frac{1}{N-1} \sum_{j=1,\ \mathbf{z}_j \neq \mathbf{z}}^{N} \frac{||\mathbf{z} - \mathbf{z}_j||_2}{||O(\mathbf{z}) - O(\mathbf{z}_j)||_2} + \nu \min_{j=1,2,\ldots,Q} ||O(\mathbf{z}) - \mathbf{x}_j||_2$$

$$+ \mu \max_{l=1,2,\ldots,K} \frac{\exp(f_l(O(\mathbf{z})) - f_l(\mathbf{x}))}{\sum_{k=1}^{K} \exp(f_k(O(\mathbf{z})) - f_k(\mathbf{x}))}, \tag{2}$$

where by using (2), we penalize the probability that $O(\mathbf{z})$ has higher confidence than the normal class. We make $O(\mathbf{z})$ have lower confidence than $\mathbf{x}$. FROB includes the learned low-confidence samples in the training by performing (1) with the self-generated boundary, $O(\mathbf{z})$, instead of $\mathbf{Z}$. Our self-supervised learning mechanism to calibrate prediction confidence in unforeseen scenarios is (2) followed by (1). We perform distribution boundary data augmentation in a learnable manner and set this distribution confidence boundary as negative data to improve few-shot OoD detection. This learned boundary includes strong and specifically adversarial anomalies close to the distribution support and near high probability normal class data. FROB sets samples just outside the data distribution boundary as OoD. We introduce relevant anomalies to more accurately and more robustly detect few shots of OoD [2]. We detect OoD samples by generating task-specific anomalous samples. We employ a nested optimization: an inner optimization to find $O(\mathbf{z})$ in (2), and an outer optimization based on cross-entropy with negative training in (1). For this nested optimization, if an optimum point is reached for the inner one, an optimum will also be reached for the outer.

FROB, for robust OoD detection, performs negative data augmentation on the support boundary of the normal class in a well-sampled manner. Specifically, FROB performs OoD sample description and characterization. By using (2), it does not allow for unfilled space between the normal class and the self-generated OoD. The second loss term of our loss function in (2) is designed to not permit unused and slack space between the learned negatives and the normal class data [6,11]. Our learned near-OoD samples have low point-to-set distance as measured by the second loss term of our proposed objective cost function shown in (2).

In the proposed self-supervised approach, the loss function for the parameter updation in the generator is (2). The first loss term is for scattering the generated samples. This measure reduces mode collapse and preserves distance proportionality in the latent and data spaces. The second loss term penalizes deviations from normality by using the distance from a point to a set. The third term in (2) guides to find the data distribution boundary by penalizing prediction confidence and pushing the generated samples OoD. In the second term, we denote the data by $(\mathbf{x}_j, y_j)_{j=1}^{Q}$, e.g. $\mathbf{x}_j$ is a vector of length 3072 for CIFAR-10.

By employing (2) followed by (1), FROB addresses the question of what OoD samples to introduce to our model for negative training in order to accurately and robustly detect few-shot data and achieve good few-shot generalization. FROB introduces self-supervised learning and learned negative data augmentation using the tightest-possible OoD data description algorithm of (2) followed by (1). Our distribution confidence boundary in (2) is robust to the problem of generators not capturing the entire data distribution and eventually learning only a Dirac distribution, which is known as mode collapse [6,7]. Using scattering, we achieve sample diversity by using the ratio of distances in the latent and data spaces. In addition, in (2), our FROB model also uses data space point-set distances.

FROB redesigns and streamlines the use of general OoD datasets to work for few-shot samples, even for zero-shots, using self-supervised learning to model the boundary of the support of the normal class distribution instead of using a large OoD set. Such general-purpose large OoD sets lead to an ad hoc selection of outliers trying to model the complement of the support of the normal class distribution. The boundary of the support of the normal class distribution, which FROB finds using (2), has and needs *less samples* than the entire complement of the support of the data distribution that big OoD sets try to approximate.

Inference. The Anomaly Score (AS) of FROB for *any* queried test sample, $\tilde{\mathbf{x}}$, in the data space, during inference and model deployment, is given by

$$AS(f, \tilde{\mathbf{x}}) = \max_{l=1,2,\ldots,K} \frac{\exp(f_l(\tilde{\mathbf{x}}))}{\sum_{k=1}^{K} \exp(f_k(\tilde{\mathbf{x}}))}, \tag{3}$$

where l is the decided class. If the AS of $\tilde{\mathbf{x}}$ has a value smaller than a predefined threshold, τ, i.e. AS $< \tau$, then $\tilde{\mathbf{x}}$ is OoD. Otherwise, $\tilde{\mathbf{x}}$ is in-distribution data.

3 Related Work on Classification and OoD Detection

General OoD Datasets. Training detectors using outliers from general OoD datasets can improve the OoD detection performance to detect unseen anomalies [5]. Using datasets disjoint from train and test data, models can learn representations for OoD detection. Confidence Enhancing Data Augmentation (CEDA), Adversarial Confidence Enhancing Training (ACET), and Guaranteed OoD Detection (GOOD) address the overconfidence of classifiers at OoD samples [3,4]. They enforce low confidence in a l_∞-norm ball around each OoD sample. CEDA employs point-wise robustness [13]. GOOD finds worst-case OoD detection guarantees. The models are trained on general OoD datasets that are, however, reduced by the normal class dataset. Disjoint distributions are used for positive and negative training, but the OoD samples are selected in an ad hoc manner. In contrast, FROB performs learned negative data augmentation on the normal class distribution confidence boundary to redesign few-shot OoD detection.

Human Prior. GOOD first defines the normal class, and then filters it out from the general-purpose OoD dataset. This filtering-out process of normality from the general OoD dataset is human-dependent. It is not practical and cannot be used in the real world as anomalies are not confined to a finite labelled closed set [15]. This modified dataset is set as anomalies. Next, GOOD learns the normal class, and sets low confidence to these OoD. This process is not automatic and data- and feature-dependent [10,11]. In contrast, FROB eliminates the need for feature extraction and human intervention which is the aim of deep learning, as they do not scale. FROB avoids application and dataset dependent processes.

Learned Negatives. The Confidence-Calibrated Classifier (CCC) uses a GAN to create samples out of, but also close to, the normal class distribution [9]. FROB substantially differs from CCC that finds a threshold and not the normal class distribution boundary. CCC uses a general OoD dataset, $U(\mathbf{y})$, where the labels follow a Uniform distribution, to compute this threshold. This can be limiting as the threshold depends on $U(\mathbf{y})$, which is an ad hoc selection of outliers that are located randomly somewhere in the data space. This leads to unfilled space between the OoD samples and the normal class which is suboptimal. In contrast, FROB finds the normal class distribution boundary and does not use $U(\mathbf{y})$ to find this boundary. Our distribution boundary is not a function of $U(\mathbf{y})$, as $U(\mathbf{y})$ is not necessary. For negative training, CCC defines a closeness metric (KL divergence), and penalizes it [11]. CCC suffers from mode collapse as it does not perform scattering for diversity. Confidence-aware classification is also performed in [9]. Self-Supervised outlier Detection (SSD) creates OoD samples in the Mahalanobis metric [8]. It is not a classifier, and it performs OoD detection with few-shot outliers. FROB achieves fast inference with (3), in contrast to [16] which is slow during inference. [16] does not address issues arising from detecting using nearest neighbours, while using a different composite loss for training.

4   Evaluation and Results

We evaluate FROB trained on different image datasets. For the evaluation of FROB, we examine the impact of different combinations of normal class datasets, OoD few-shots, and test datasets, in an alternating manner. We examine the generalization performance to few-shots of unseen OoD samples at the dataset level (out-of-dataset anomalies), which are different from the training sets.

Metrics. We report the Area Under the Receiver Operating Characteristic Curve (AUROC), Adversarial AUROC (AAUROC), and Guaranteed AUROC (GAUROC) [3,14]. To strengthen the robustness evaluation of FROB and to compare with benchmarks, in addition to AUROC, we also evaluate FROB with AAUROC and GAUROC. AAUROC and GAUROC are suitable for evaluating the robustness of OoD detection models focusing on the worst-case OoD detection performance using l_∞-norm perturbations for each of the OoD image samples. It uses the maximum confidence in the l_∞-norm ball around each OoD and finds a lower (upper respectively) bound on this maximum confidence. These worst-case confidences for the OoD samples are then used for the AUROC.

To examine the robustness to the number of few-shots, we decrease the number of OoD few-shots by dividing them by two, employing uniform sampling. We examine the influence of this on AUROC. Specifically, we examine the variation of AUROC, AAUROC, and GAUROC, which constitute the dependent variables, to changes of the independent variable, which is the provided number of few-shots of OoD samples. We examine the Breaking Point of our FROB algorithm and of benchmarks; we define this point as the number of few-shot data from which the OoD performance in AUROC decreases and then falls to 0.5.

Datasets. For the normal class, we use either CIFAR-10 or SVHN. For OoD few-shots, we use data from CIFAR-10, SVHN, CIFAR-100, and Low-Frequency Noise (LFN). To compare with baselines from the literature, for the general OoD datasets, we use SVHN, CIFAR-100, and the same general OoD dataset as in [3,5] but debiased, as in [18]. We evaluate our FROB model on the datasets CIFAR-100, SVHN, and CIFAR-10, as well as on LFN and Uniform noise.

Model Architecture. FROB uses a CNN discriminative model, as described in Sect. 2. We also train and use a generator that takes the form of a CNN. We implement FROB in PyTorch and use the optimizer Adam for training.

Baselines. We demonstrate that FROB is effective and outperforms baselines in the few-shot OoD detection setting. We compare FROB to the baselines GEOM, GOAD, DROCC, Hierarchical Transformation Discriminating Generator (HTD), Support Vector Data Description (SVDD), and Patch SVDD (PSVDD) in the few-shot setting, using OCC [1]. We also compare FROB to GOOD [3], CEDA, CCC, OE and ACET [4], and [5]. For many-samples OoD, [3,5] use a general OoD set, which is not representative of the few-shot OoD detection setting. General OoD sets result in a nonoptimal ad hoc selection of OoD, especially when operating on a fixed few-shot budget for sampling from the OoD class.

Table 1. OoD performance of FROB with the learned distribution boundary, $O(\mathbf{z})$, in AUROC using OCC and few-shots of 80 CIFAR-10 anomalies, and comparison to baselines, [1]. *FODS is FROB with the outlier OoD dataset SVHN.*

NORMAL	DROCC	GEOM	GOAD	HTD	SVDD	PSVDD	FROB	FODS
PLANE	0.790	0.699	0.521	0.748	0.609	0.340	**0.811**	**0.867**
CAR	0.432	0.853	0.592	**0.880**	0.601	0.638	0.862	0.861
BIRD	0.682	0.608	0.507	0.624	0.446	0.400	**0.721**	0.707
CAT	0.557	0.629	0.538	0.601	0.587	0.549	**0.748**	**0.787**
DEER	0.572	0.563	0.627	0.501	0.563	0.500	**0.742**	0.727
DOG	0.644	0.765	0.525	**0.784**	0.609	0.482	0.771	0.782
FROG	0.509	0.699	0.515	0.753	0.585	0.570	0.826	**0.884**
HORSE	0.476	0.799	0.521	**0.823**	0.609	0.567	0.792	0.815
SHIP	0.770	0.840	0.704	**0.874**	0.748	0.440	0.826	0.792
TRUCK	0.424	**0.834**	0.697	0.812	0.721	0.612	0.744	0.799
MEAN	0.585	0.735	0.562	0.756	0.608	0.510	**0.784**	**0.802**

Ablations. We evaluate FROB for few-shot OoD detection with ($\checkmark$) the learned distribution boundary, $O(\mathbf{z})$, i.e. FROB. For ablation, we also evaluate models that are trained without ($-$) $O(\mathbf{z})$ samples which we term FROBInit.

4.1   Evaluation of FROB

Evaluation of FROB Using OCC Compared to Baselines. We evaluate FROB using OCC for each CIFAR-10 class against several benchmarks in the few-shot setting of 80 samples [1]. FROB outperforms baselines in Table 1 which shows the mean performance of FROB when the normal class is a CIFAR-10 class. We compare our proposed FROB model to the baselines DROCC, GEOM, GOAD, HTD, SVDD, and PSVDD [1]. FROB with the self-learned $O(\mathbf{z})$ outperforms baselines for few-shot OoD detection in OCC when we have budget constraints and OoD sampling complexity limitations. We also evaluate our FROB model further retrained with the outlier OoD dataset SVHN, FODS, and show that using the OoD set is beneficial for few-shot OoD detection using OCC.

Robustness of FROB to the Number of Few-Shots. We evaluate FROB with few-shots of OoD samples from SVHN in decreasing number, setting the normal class as CIFAR-10. We experimentally demonstrate the effectiveness of FROB, and the results are shown in Table 2 and Fig. 2. We evaluate FROB on SVHN, as well as on CIFAR-100 and LFN, in Fig. 2 where the in-distribution data are from the CIFAR-10 dataset while the OoD are from SVHN, CIFAR-100, and LFN. Using FROB, the performance improves showing robustness even for a small number of OoD few-shots, pushing down the phase transition point in the number of few-shots in Fig. 2. When the few-shots are from the test set, i.e. SVHN in Fig. 2, FROB is effective and robust for few-shot OoD detection.

Table 2. OoD performance of FROB using the learned boundary, $O(\mathbf{z})$, and OoD few-shots, tested on SVHN. The normal class is CIFAR-10 (C10). The second column shows the training data of *OoD few-shots* and their number.

MODEL	OoD FEW-SHOTS	TEST SET	AUROC	AAUROC	GAUROC
FROB	SVHN: 1830	SVHN	0.997	0.997	0.990
FROB	SVHN: 915	SVHN	0.995	0.995	0.984
FROB	SVHN: 732	SVHN	0.995	0.995	0.981
FROB	SVHN: 457	SVHN	0.997	0.997	0.982
FROB	SVHN: 100	SVHN	**0.996**	**0.996**	**0.950**
FROB	SVHN: 80	SVHN	**0.995**	**0.995**	**0.928**

We experimentally demonstrate that first performing sample generation on the distribution boundary, $O(\mathbf{z})$, and then *including* these learned OoD samples in our training is beneficial. The improvement of FROB in AUROC is because of these well-sampled $O(\mathbf{z})$ samples. The component of FROB with the highest benefit is the self-generated distribution boundary, $O(\mathbf{z})$. Our proposed FROB model shows improved robustness to the number of OoD few-shots because with decreasing few-shots, the performance of FROB in AUROC is robust and approximately independent of the OoD few-shot number of samples in Fig. 2.

Performance on Unseen Datasets. We evaluate FROB on OoD samples from unseen, in the wild, datasets, i.e. on samples that are neither from the normal class nor from the OoD few-shots. We examine our proposed FROB model in the few-shot setting in Fig. 2 for normal CIFAR-10 with OoD few-shots from SVHN, and tested on the new CIFAR-100 and LFN. These are unseen as they are not the normal class or the OoD few-shots. The performance of FROB in this OoD few-shot setting in Fig. 2 is robust on CIFAR-100 and on LFN.

Next, exchanging the datasets, FROB with the normal class SVHN, and a variable number of CIFAR-10 OoD few-shots, is tested in Table 3 and in Fig. 3. In Table 3, compared to Table 2, FROB achieves comparable performance for normal class SVHN and few-shots of CIFAR-10, compared to for normal class CIFAR-10 and few-shots of SVHN, in all the AUC-type metrics. According to Fig. 3, when compared to Fig. 2, for the unseen test set CIFAR-100, FROB achieves better AUROC for normal SVHN compared to for normal CIFAR-10.

Effect of Domain and Normal Class. The performance of FROB in AUROC depends on the normal class. In Fig. 3, the OoD detection performance of FROB for small number of few-shots is higher for normal class SVHN than for normal CIFAR-10 in Fig. 2. FROB is robust and effective for normal SVHN on seen and unseen data. FROB is not sensitive to the number of few-shots for few-shot OoD detection, when we have OoD sample complexity constraints.

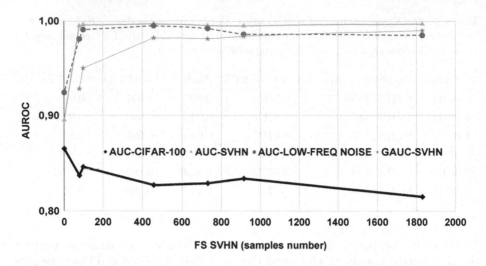

Fig. 2. FROB (normal C10) with SVHN OoD few-shots: AUROC and GAUROC.

Table 3. Evaluation of FROB for normal SVHN with the self-generated $O(\mathbf{z})$ and OoD few-shots, tested on CIFAR-10 (C10). According to the second and third columns, the OoD few-shots and the OoD test samples are from C10.

MODEL	OoD FEW-SHOTS	TEST SET	AUROC	AAUROC	GAUROC
FROB	C10: 600	C10	0.996	0.996	0.982
FROB	C10: 400	C10	0.994	0.994	0.964
FROB	C10: 200	C10	0.996	0.996	0.967
FROB	C10: 80	C10	0.991	0.991	0.951

OoD Detection Performance of FROB with OoD Few-Shots from the Test Set. In Tables 2 and 3 we experimentally demonstrate that FROB improves the AUROC and AAUROC when the few-shots and the test samples originate from the same dataset. We also show that FROB achieves high GAUROC.

OoD Detection Performance of FROB, OoD Few-Shots and Test are Different Sets. More empirical results in Figs. 2 and 3 show that FROB also improves the AUROC when the few-shots and OoD test samples originate from different sets, i.e. LFN and CIFAR-100. This shows *robustness* to the test set.

OoD Performance of FROB for OoD Few-Shots from the Test Set but Also Adding a General OoD Dataset. Table 4 shows the OoD detection performance of FROB for OoD few-shots from the test dataset, adding a general-purpose OoD dataset [3,5,18]. Compared to Table 2, FROB without the OoD dataset achieves *higher* AUC-metrics, and this is important. This happens

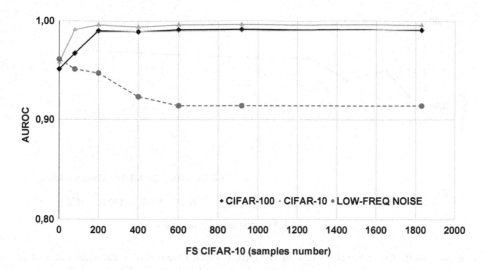

Fig. 3. Evaluation of FROB using the learned distribution boundary, $O(\mathbf{z})$, and few-shot OoD samples from CIFAR-10 in AUROC. The normal class is SVHN.

Table 4. OoD performance of FROB using the learned $O(\mathbf{z})$, OoD few-shots, and a general OoD dataset following the procedure in [3,18] resulting in 73257 samples, evaluated on SVHN. The normal class is CIFAR-10. *FS is Few-Shots.*

OoD FS	Outlier OoD	TEST	AUROC	AAUROC	GAUROC
SVHN: 1830	✓	SVHN	0.994	0.994	0.972
SVHN: 915	✓	SVHN	0.993	0.993	0.333
SVHN: 732	✓	SVHN	0.990	0.990	0.010
SVHN: 457	✓	SVHN	0.997	0.997	0.807
SVHN: 100	✓	SVHN	0.992	0.992	0.896
SVHN: 80	✓	SVHN	0.981	0.981	0.922
SVHN: 80	–	SVHN	0.995	0.995	0.928

because of including our proposed self-generated distribution boundary, $O(\mathbf{z})$, in our training. Adding a general-purpose OoD dataset leads to far-OoD samples which are not task-specific and might be irrelevant [17]. These far-OoD samples from the general benchmark OoD dataset are *far away* from the boundary of the support of the normal class distribution, have high point-to-set distance as measured by the second loss term of our loss function in (2), are unevenly scattered in the data space, and are non-uniformly dispersed. Notably, in Table 2, compared to Table 4, the AUROC of FROB for normal CIFAR-10 is 0.996 and 0.995 for 100 and 80 OoD few-shots from SVHN respectively, while the AUROC of FROB with a general OoD dataset is 0.992 and 0.981 respectively. This is an important

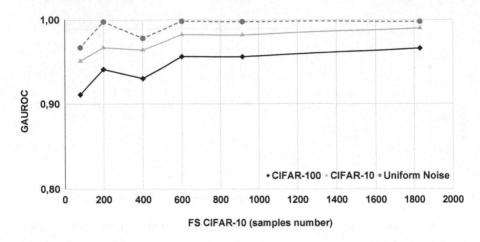

Fig. 4. FROB for normal SVHN in GAUROC with $O(\mathbf{z})$ and a variable number of OoD CIFAR-10 few-shots, tested on CIFAR-100, CIFAR-10, and Uniform Noise.

finding implying that a general OoD dataset is not needed, and that FROB with the self-generated $O(\mathbf{z})$ achieves state-of-the-art performance for few-shot OoD detection when the OoD few-shots originate from the test set. We hypothesise that the general OoD set is not required because $O(\mathbf{z})$ generates samples that are out of the data distribution that well cover the space between these samples and the normal (in-distribution) class. An external outlier OoD dataset likely provides samples that are *further out* and dispersed, and not task-specific.

We have thus shown in Table 4 that when FROB with the learned boundary, $O(\mathbf{z})$, is used during training, then the use of a general OoD dataset is not needed. Next, Fig. 4 shows the performance of FROB for normal SVHN and a variable number of OoD CIFAR-10 few-shots. In Figs. 4 and 3, compared to Fig. 2, we show that FROB achieves better performance for normal SVHN, compared to for normal CIFAR-10, in all AUC-type metrics, on the unseen CIFAR-100.

FROB Compared to Baselines. We compare our proposed FROB model to baselines for OoD detection. We focus on *all* the AUROC, AAUROC, and GAUROC, on the robustness of the models, and on the worst-case OoD detection performance using l_∞-norm perturbations for each of the OoD data samples (Table 5).

We examine the OoD detection performance of the baseline models CCC, CEDA, [5], ACET, and GOOD when using C10 as the normal class, a general OoD dataset [3,18], and the SVHN OoD dataset. We evaluate these baseline models on the SVHN set. FROB outperforms baselines, specifically when the three evaluation metrics AUROC, AAUROC, and GAUROC are considered.

Table 5. Performance of FROB with the self-generated $O(z)$, normal class C10, and general OoD set following the procedure in [3,18]. Comparison to baselines.

MODEL	$O(z)$	OoD DATASET	TEST	AUROC	AAUROC	GAUROC
FROB	✓	SVHN: 1830	SVHN	**0.997**	**0.997**	**0.990**
FROB	✓	[3,18], SVHN: 1830	SVHN	**0.994**	**0.994**	**0.972**
CCC	–	SVHN	SVHN	0.999	0.000	0.000
CEDA	–	[3,18]	SVHN	0.979	0.257	0.000
OE	–	[3,18]	SVHN	0.976	0.70	0.000
ACET	–	[3,18]	SVHN	0.966	0.880	0.000
GOOD	–	[3,18]	SVHN	0.757	0.589	0.569

4.2 Ablation Studies

Removing $O(z)$. We remove the learned distribution boundary, $O(z)$, in a model we term FROBInit. We compare with FROB using OoD few-shots from SVHN, using 1830 samples, in Table 6. The OoD detection performance of FROB in AUROC in Table 6 is 0.997 and that of FROBInit, which does not use the learned boundary, $O(z)$, is 0.847. FROB outperforms FROBInit in all AUC-based metrics, by approximately 18% in AUROC and AAUROC and 36% in GAUROC. These results demonstrate the effectiveness and efficacy of FROB.

FROB Generating the Boundary, $O(z)$, Leads to Robustness to the Number of OoD Few-Shots. Most existing methods from the literature are sensitive to the number of OoD few-shots. We demonstrate this sensitivity in Figs. 5 and 6, where we examine the performance of FROBInit which lacks the generator of boundary samples by varying the number of few-shot outliers. We also compare with FROB. Comparing Figs. 5 and 6 with Figs. 2 and 3, we see that ablating $O(z)$ leads to loss of robustness to a small number of few-shots.

In Figs. 5 and 6, the performance of FROBInit without the learned distribution boundary, $O(z)$, is not robust for few-shot OoD detection, i.e. for few-shots less than approximately 1800 samples. In Figs. 2 and 3, compared to Fig. 5, FROB achieves robust OoD detection performance as the number of OoD few-shots decreases. This indicates that $O(z)$ is effective and FROB is robust to the number of OoD few-shots, even to a small number of few-shot samples. We have shown that when the self-generated distribution boundary, $O(z)$, is not used, the OoD performance in AUROC *decreases* as the number of OoD few-shots decreases. The self-generated distribution boundary of FROB leads to a specific selection of anomalous samples that do not allow unfilled space in the data space, between the learned negatives and the normal class. FROB, because it generates samples on the distribution boundary, shows a more robust and improved OoD performance to the number of OoD few-shots when compared to FROBInit.

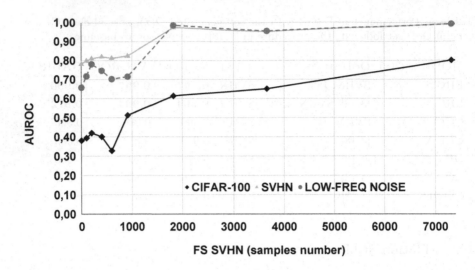

Fig. 5. OoD performance of FROBInit in AUROC, for the normal class CIFAR-10, without $O(\mathbf{z})$ and using OoD few-shots of variable number from SVHN.

Table 6. OoD performance of FROB with the learned distribution boundary, $O(\mathbf{z})$, and 1830 OoD samples from SVHN without and with a general OoD dataset following the procedure in [3,18]. The normal class is CIFAR-10.

MODEL	$O(\mathbf{z})$	OoD LOW-SHOTS	AUROC	AAUROC	GAUROC
FROB	✓	SVHN: 1830	**0.997**	**0.997**	**0.990**
FROBInit	–	SVHN: 1830	0.847	0.847	0.728

Further Evaluation of FROBInit and Its Breaking Point. To show the benefit of our proposed FROB model using our learned distribution boundary samples, $O(\mathbf{z})$, in (2), we now continue the evaluation of FROBInit in this ablation study analysis. We have demonstrated in Figs. 5 and 6 that the performance of FROBInit without the self-produced $O(\mathbf{z})$ data samples, when the normal class is the CIFAR-10 dataset, with a variable number of OoD few-shot samples from the SVHN dataset, when evaluated on different image datasets, decreases as the number of the few-shots of OoD data decreases.

The Break Point threshold at AUROC 0.5 is reached for approximately 800 few-shots for CIFAR-100. When the learned distribution boundary, $O(\mathbf{z})$, is not used, we do not achieve a robust performance for decreasing few-shots. The performance falls with the decreasing number of few-shots: *steep* decline for low-shots less than 1830 samples, tested on SVHN and on Low-Frequency Noise.

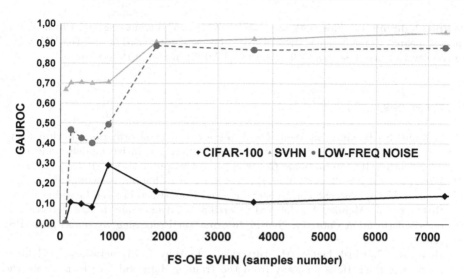

Fig. 6. Performance of FROBInit, without $O(\mathbf{z})$, in GAUROC for the normal class CIFAR-10, and with a variable number of OoD few-shots from SVHN.

5 Conclusion

We have proposed FROB which uses the self-generated support boundary of the normal class distribution to improve few-shot OoD detection. FROB tackles the few-shot problem using joint classification and OoD detection. Our contribution is the combination of the generated boundary in a self-supervised learning manner and the imposition of low confidence at this learned boundary leading to improved robust few-shot OoD detection performance. To improve robustness, FROB generates strong adversarial samples on the boundary, and enforces samples from OoD and on the boundary to be less confident. By including the self-produced boundary, we reduce the threshold linked to the model's few-shot robustness. FROB redesigns, restructures, and streamlines the use of general OoD datasets to work for few-shot samples. Our proposed FROB model performs classification and few-shot OoD detection with a high level of robustness in the real world, in the wild. FROB maintains the OoD performance approximately constant, independent of the few-shot number. The performance of FROB with the self-supervised boundary is robust and effective. Its performance is approximately stable as the OoD low- and few-shots decrease in number, while the performance of FROBInit, which is without $O(\mathbf{z})$, sharply falls as the few-shots decrease in number. The evaluation of FROB on several datasets, including the ones dissimilar to training and few-shot sets, shows that it is effective, achieves competitive state-of-the-art performance and generalization to unseen anomalies, with applicability to unknown, in the wild, test datasets, and outperforms baselines in the few-shot anomaly detection setting, in AUC-type metrics.

Acknowledgement. This work was supported by the Engineering and Physical Sciences Research Council of the UK (EPSRC) Grant EP/S000631/1 and the UK MOD University Defence Research Collaboration (UDRC) in Signal Processing.

References

1. Sheynin, S., Benaim, S., Wolf, L.: A hierarchical transformation-discriminating generative model for few shot anomaly detection generative models multi-class classification. In: International Conference on Computer Vision (ICCV) (2021)
2. Wang, K., Vicol, P., Triantafillou, E., Zemel, R.: Few-shot out-of-distribution detection. In: ICML Workshop on Uncertainty and Robustness in Deep Learning (2020)
3. Bitterwolf, J., Meinke, A., Hein, M.: Certifiably adversarially robust detection of out-of-distribution data. In: Neural Information Processing Systems (NeurIPS) (2020)
4. Hein, M., Andriushchenko, M., Bitterwolf, J.: Why ReLU networks yield high-confidence predictions far away from the training data and how to mitigate the problem. In: Conference on Computer Vision and Pattern Recognition (CVPR) (2019)
5. Hendrycks, D., Mazeika, M., Dietterich, T.: Deep anomaly detection with outlier exposure. In: International Conference on Learning Representations (ICLR) (2019)
6. Dionelis, N., Yaghoobi, M., Tsaftaris, S.: Boundary of distribution support generator (BDSG): sample generation on the boundary. In: IEEE International Conference on Image Processing (ICIP) (2020)
7. Dionelis, N., Yaghoobi, M., Tsaftaris, S.: Tail of distribution GAN (TailGAN): generative-adversarial- network-based boundary formation. In: IEEE Sensor Signal Processing for Defence (SSPD) (2020)
8. Sehwag, V., Chiang, M., Mittal, P.: SSD: a unified framework for self-supervised outlier detection. In: International Conference Learning Representations (ICLR) (2021)
9. Lee, K., Lee, K., Lee, H., Shin, J.: Training confidence-calibrated classifiers for detecting out-of-distribution samples. In: ICLR (2018)
10. Dionelis, N., Yaghoobi, M., Tsaftaris, S.: Few-shot adaptive detection of objects of concern using generative models with negative retraining. In: International Conference on Tools with Artificial Intelligence (ICTAI) (2021)
11. Dionelis, N., Yaghoobi, M., Tsaftaris, S.: OMASGAN: out-of-distribution minimum anomaly score GAN for sample generation on the boundary. arXiv:2110.15273 (2021)
12. Moon, J., Kim, J., Shin, Y., Hwang, S.: Confidence-aware learning for deep neural networks. In: International Conference on Machine Learning (ICML) (2020)
13. Bastani, O., Ioannou, Y., Lampropoulos, L., Vytiniotis, D., Nori, A., Criminisi, A.: Measuring neural net robustness with constraints. In: NeurIPS (2016)
14. Croce, F., Hein, M.: Reliable evaluation of adversarial robustness with an ensemble of diverse parameter-free attacks. In: ICML (2020)
15. Sensoy, M., Kaplan, L., Cerutti, F., Saleki, M.: Uncertainty-aware deep classifiers using generative models. In: AAAI Conference on Artificial Intelligence (2020)
16. Tack, J., Mo, S., Jeong, J., Shin, J.: CSI: novelty detection via contrastive learning on distributionally shifted instances. In: NeurIPS (2020)

17. Jeong, T., Kim, H.: OoD-MAML: meta-learning for few-shot out-of-distribution detection and classification. In: Proceedings of NeurIPS (2020)
18. Rafiee, N., Gholamipoorfard, R., Adaloglou, N., Jaxy, S., Ramakers, J., Kollmann, M.: Self-supervised anomaly detection by self-distillation and negative sampling. arXiv:2201.06378 (2022)

PRoA: A Probabilistic Robustness Assessment Against Functional Perturbations

Tianle Zhang⬛, Wenjie Ruan$^{(\boxtimes)}$⬛, and Jonathan E. Fieldsend⬛

University of Exeter, Exeter EX4 4PY, UK
{tz294,W.Ruan,J.E.Fieldsend}@exeter.ac.uk

Abstract. In safety-critical deep learning applications robustness measurement is a vital pre-deployment phase. However, existing robustness verification methods are not sufficiently practical for deploying machine learning systems in the real world. On the one hand, these methods attempt to claim that no perturbations can "fool" deep neural networks (DNNs), which may be too stringent in practice. On the other hand, existing works rigorously consider L_p bounded additive perturbations on the pixel space, although perturbations, such as colour shifting and geometric transformations, are more practically and frequently occurring in the real world. Thus, from the practical standpoint, we present a novel and general *probabilistic robustness assessment method* (PRoA) based on the adaptive concentration, and it can measure the robustness of deep learning models against functional perturbations. PRoA can provide statistical guarantees on the probabilistic robustness of a model, *i.e.*, the probability of failure encountered by the trained model after deployment. Our experiments demonstrate the effectiveness and flexibility of PRoA in terms of evaluating the probabilistic robustness against a broad range of functional perturbations, and PRoA can scale well to various large-scale deep neural networks compared to existing state-of-the-art baselines. For the purpose of reproducibility, we release our tool on GitHub: https://github.com/TrustAI/PRoA.

Keywords: Verification · Probabilistic robustness · Functional perturbations · Neural networks

1 Introduction

With the phenomenal success of Deep Neural Networks (DNNs), there is a growing and pressing need for reliable and trustworthy neural network components,

W. Ruan—Supported by Partnership Resource Fund (PRF) on Towards the Accountable and Explainable Learning-enabled Autonomous Robotic Systems from UK EPSRC project on Offshore Robotics for Certification of Assets (ORCA) [EP/R026173/1].
T. Zhang—Supported by Exeter-CSC scholarship [202108060090].

Supplementary Information The online version contains supplementary material available at https://doi.org/10.1007/978-3-031-26409-2_10.

particularly in safety-critical applications. Neural networks' inherent vulnerability to adversarial attacks has been receiving considerable attention from the research community [21]. Numerous empirical defence approaches, including adversarial training [14], have been developed recently in response to diverse adversarial attacks. Such defence strategies, however, are subsequently overwhelmed by elaborate and advanced adversarial attacks [10].

Therefore, in order to construct safe and trustworthy deep learning models with a certain confidence, a challenge has emerged: *how can we verify or certify our models under adversarial perturbations with guarantees?* Various earlier works have attempted to quantify the deterministic robustness of a given input x concerning a specific neural network; they seek to state that no adversarial examples exist within a neighbourhood of x [9]. However, such safety requirements are not always satisfied and applicable in practice. For instance, as ISO/IEC Guide 51 [6] suggests, *"safety risks and dangers are unavoidable; residual risks persist even after risk reduction measures have been implemented"*. Thus, in comparison to those ensuring deterministic robustness, it is a more practical assessment of robustness to properly confine the possibility of a failure event occurring. For example, no communication networks can guarantee that no message will be lost over a wireless communication route, and messages might be lost owing to collisions or noise contamination even with proper functioning network hardware. Occasional message loss is tolerated if the occurrence chance is within an acceptable level. However, it is still unexplored for such probabilistic robustness verification.

In the meantime, the majority of existing verification methods consider a narrow threat model with additive perturbations, *i.e.* adversarial examples are produced by adding slight tweaks (measured in L_p distance) to every single feature of normal inputs (e.g. counterexamples are generated by adding minor changes to every single pixel in an image classification task). While the additive threat model implies that the divergence between generated adversarial instances and original instances does not surpass a modest positive constant ϵ measured by L_p norm, other sorts of perturbations undetectable to humans are overlooked. For instance, cameras installed in self-driving cars may be vibrated on bumpy roads, leading to rotating or blurring photos. Resultant rotated and blurry photographs are likely to be misidentified by neural networks, even if they do not "hoodwink" human perception. Such risky and frequent scenarios motivate the robustness assessment against various general perturbations, e.g. geometric transformation like rotation and translation, and common corruptions.

In this paper, we propose a novel and scalable method called PRoA that can provide statistical guarantees on the probabilistic robustness of a large black-box neural network against functional threat models. Specifically, in this approach, we introduce functional perturbations, including random noise, image transformations and recolouring, which occur naturally and generally, and additive perturbation would be a specific instance in which perturbation functions add a modest adjustment to each feature of inputs. Instead of worst-case based verification, this method measures the probabilistic robustness, *i.e.* accurately bounds the tolerated risk of encountering counterexamples via adaptively randomly sampling perturbations. This robustness property is more appropriate in real-world circumstances. Furthermore, the proposed method makes no assumptions about the neural network, e.g. activation functions, layers, and neurons, etc. This grants our probabilistic robustness assessment method (PRoA) the scalability to evaluate state-of-the-art and large-scale DNNs. Our main contributions are threefold as follows:

- We propose a randomised algorithm-based framework for evaluating the probabilistic robustness of deep learning models using adaptive concentration inequality. This method is well-scalable and applicable to large and state-of-the-art black-box neural networks.
- The method is attack-agnostic and capable of providing a theoretical guarantee on the likelihood of encountering an adversarial example under parametric functional perturbations.
- Experimentally, we validate our certification method and demonstrate its practical applicability with different trained neural networks for various natural functional perturbations, e.g. geometric transformations, colour-shifted functions, and Gaussian blurring.

2   Related Work

Reachability Based Approaches. For a given input and a specified perturbation, reachability-based algorithms endeavour to determine the lower and upper bounds of the output. Thus, robustness can be evaluated by solving an output range analysis problem. Some reachability-based approaches employ layer-by-layer analysis to obtain the reachable range of outputs [13,18–20,22,26,27]. ExactReach [26] estimates a DNN's reachable set as a union of polytopes by setting the outputs of each layer with Relu activation to a union of polytopes. Yang et al. [27] present an exact reachability verification method utilising a facet-vertex incidence matrix. Additionally, another research approach is to employ global optimisation techniques to generate a reachable output interval. GeepGo [18] uses a global optimisation technique to find the upper and lower bounds of the outputs of Lipschitz-continuous networks. This algorithm is capable of operating on black-box DNNs. Reachability analysis can be used to address the challenge of safety verification; however, these methods often require that target networks be Lipschitz continuous over outputs, which limits their application.

Constraint Based Approaches. Constraint-based techniques generally transform a verification problem into a set of constraints, which can then be solved by a variety of programme solvers. In recent papers [1,8], Katz et al. [8] introduce an SMT-based technique called Reluplex for solving queries on DNNs with Relu activation by extending a simplex algorithm, while Amir et al. [1] propose another SMT-based method by splitting constraints into easier-to-solve linear constraints. For constraint-based techniques, all types of solvers can produce a deterministic answer with guarantees, *i.e.*, they can either satisfy or violate robustness conditions. However, these techniques suffer from a scalability issue and need to access the internal structure and parameters of the targeted DNN (in a white-box setting).

These deterministic verification approaches might be unduly pessimistic in realistic applications since they only account for the worst scenario. In contrast, PRoA focuses on the tail probability of the average case, which is more realistic in a wide range of real-world applications, and worst-case analysis can be a special case of tail risks when we take the tail probability (0%) of the most extreme performance into consideration.

Statistical Approaches. Unlike the above deterministic verification methods, statistics-based techniques aim to quantify the likelihood of finding a counterexample. For example, random sampling has lately emerged as an effective statistical strategy for providing certified adversarial robustness, e.g. randomised smoothing [3,28], cc-cert [16], and SRC [5], among others. Additionally, Webb et al. [24] propose an adaptive Monte Carlo approach, *i.e.* multi-level splitting, to estimate the probability of safety unsatisfiability, where failure occurs as an extremely rare occurrence in real-world circumstances. However, these statistics-based analyses focus on the pixel-level additive perturbations and always require assumptions upon target neural networks or distributions of input, which limits their applicability.

In contrast, we introduce a general adversarial threat model, *i.e.* functional perturbations, and PRoA aims to bound the failure chance with confidence under the functional threat model. Moreover, PRoA is able to provide rigorous robustness guarantees on black-box DNNs without any assumptions and scale to large-scale networks.

3 Preliminary

Classification Program. Given a training set with N distinct samples $S = \{(x_1, y_1), \ldots, (x_N, y_N)\}$ where $x_i \in \mathcal{X} = \mathbb{R}^n$ are i.i.d. samples with dimension n drawn from an unknown data distribution and $y_i \in \mathcal{R} = \{1, \ldots, K\}$ are corresponding labels. We consider a deterministic neural network $f : \mathcal{X} \to [0, 1]^K$ that maps any input to its associated output vector, and $f_k(\cdot) : \mathbb{R}^n \to [0, 1]$ is a deterministic function, representing the output confidence on label k. Our verification procedure solely requires blackbox assess to f, thus, it can obtain the corresponding output probability vector $f(x)$ for each input $x \in \mathcal{X}$.

Additive Perturbation. Given a neural network f and an input $x \in \mathcal{X}$, an adversarial example $\tilde{x}$ of x is crafted with a slight modification to the original input such that $\arg\max_{k \in \{1, \ldots, K\}} f_k(\tilde{x}) \neq \arg\max_{k \in \{1, \ldots, K\}} f_k(x)$; this means that the classifier assigns an incorrect label to $\tilde{x}$ but $\tilde{x}$ is perceptually indistinguishable from the original input x. Intuitively, slight perturbations $\delta \in \mathbb{R}^n$ can be added directly to x to yield adversarial examples $\tilde{x} = x + \delta$, in the meantime, a L_p norm bound is normally imposed on such additive perturbations, constraining $\tilde{x}$ to be fairly close to x. The relevant definition is as follows:

$$\tilde{x} = x + \delta \; and \; \|\delta\|_p \leq \epsilon \; s.t. \; \underset{k\in\{1,...,K\}}{\arg\max} \; f_k(\tilde{x}) \neq \underset{k\in\{1,...,K\}}{\arg\max} \; f_k(x).$$

Functional Perturbation. Unlike the additive perturbation, a normal input x is transformed using a perturbation function $\mathcal{F} : \mathcal{X} \to \mathcal{X}$ parameterised with $\theta \in \Theta$. That is to say, $x_{\mathcal{F}_\theta} = \mathcal{F}_\theta(x)$. It is worth noting that functional perturbation allows for a substantially larger pixel-based distance, which may be imperceptible to humans as well, since the perturbed version $x_{\mathcal{F}_\theta}$ consistently preserves semantic information underlying images, such as shape, boundary, and texture. Unfortunately, such perturbations may confuse the classifier $f(\cdot)$, which is capable of outputting the proper label to an undistorted image, i.e. $\underset{k\in\{1,...,K\}}{\arg\max} \; f_k(x) \neq \underset{k\in\{1,...,K\}}{\arg\max} \; f_k(x_{\mathcal{F}_\theta})$.

Prior literature on functional perturbations is surprisingly sparse. To our best knowledge, only one work involves a term *functional perturbations* [11], in which a functional threat model is proposed to produce adversarial examples by employing a single function to perturb all input features simultaneously. In contrast, we introduce a flexible and generalised functional threat model by removing the constraint of global uniform changes in images. Obviously, the additive threat model is a particular case of the functional threat model, when the perturbation function $\mathcal{F}_\theta$ manipulates pixels of an image by adding slight L_p bounded distortions.

Verification. The purpose of this paper is to verify the resilience of the classifier $f(\cdot)$ against perturbation functions $\mathcal{F}$ parameterised with $\theta \in \Theta$ while functional perturbations $\mathcal{F}_\theta$ would not change the oracle label from human perception if θ within parameter space Θ, or, more precisely, to provide guarantees that the classifier $f(\cdot)$ is probabilistically robust with regard to an input x when exposed to a particular functional perturbation $\mathcal{F}_\theta$. To this end, let k^* denote the ground truth class of the input sample. Assume that $\mathbb{S}_{\mathcal{F}}(x)$ is the space of all images $x_{\mathcal{F}_\theta}$ of x under perturbations induced by a perturbation function $\mathcal{F}_\theta$ and $\mathcal{P}$ is the probability measure on this space $\mathbb{S}_{\mathcal{F}}(x)$. This leads to the following robustness definitions:

Definition 1 (Deterministic robustness). *Let $\mathcal{F}_\theta$ be a specific perturbation function parametrized by θ, and Θ denotes a parameter space of a given perturbation function. Assume that $x_{\mathcal{F}_\theta} = \mathcal{F}_\theta(x)$ is the perturbed version of x given $\theta \in \Theta$, and $\mathbb{S}_{\mathcal{F}}(x)$ is the space of all images $x_{\mathcal{F}_\theta}$ of x under perturbation function $\mathcal{F}_\theta$. Given a K-class DNN f, an input x and a specific perturbation function $\mathcal{F}_\theta$ with $\theta \in \Theta$, we can say that f is deterministically robust w.r.t. the image x, i.e. x is correctly classified with probability one, if*

$$\underset{k\in\{1,...,K\}}{\arg\max} \; f_k(x_{\mathcal{F}}) = k^*, for \; all \; x_{\mathcal{F}} \sim \mathbb{S}_{\mathcal{F}}(x).$$

Definition 2 (Probabilistic Robustness). *Let $\mathcal{F}_\theta$ be a specific perturbation function parametrized by θ, and Θ denotes a parameter space of a given perturbation function. Assume that $x_{\mathcal{F}_\theta} = \mathcal{F}_\theta(x)$ is the perturbed version of x given $\theta \in \Theta$, and $\mathbb{S}_{\mathcal{F}}(x)$ is the space of all images $x_{\mathcal{F}_\theta}$ of x under perturbation function $\mathcal{F}_\theta$. Given a K-class DNN f, an input x, a specific perturbation function $\mathcal{F}_\theta$ with $\theta \in \Theta$, and a tolerated error rate τ, the K-class DNN f is said to be probabilistically robust with probability at least $1 - \delta$, if*

$$\mathbb{P}_{x_{\mathcal{F}} \sim \mathbb{S}_{\mathcal{F}}(x)} \left(p \left(\arg\max_{k \in \{1,\ldots,K\}} f_k\left(x_{\mathcal{F}}\right) \neq k^* \right) < \tau \right) \geq 1 - \delta. \tag{1}$$

Verifying deterministic robustness has been widely studied in the context of pixel-level additive perturbations and worst-case adversarial training; however, deterministic robustness is always too stringent to hold, and deterministic robustness and probabilistic robustness are "equivalent" to each other when we choose $\tau = 0$.

4   Verification of Probabilistic Robustness

We now present our proposed method, named PRoA, for verifying the probabilistic robustness of black-box classifiers against functional perturbations. A schematic overview of PRoA is illustrated in *Appendix A*.

4.1   Formulating Verification Problem

Our goal is to verify probabilistic robustness properties for a neural network classifier f, providing the classifier with probabilistic guarantees of its stability under functional perturbations. We formalise the robustness properties by examining substantial discrepancies of outputs w.r.t. input transformations [16]. Next, we describe how to formalise the robustness property using both original and perturbed images.

  We have a deterministic neural network $f : \mathbb{R}^n \to [0,1]^K$. Assume that a given input x and its perturbed image $x_{\mathcal{F}}$ are assigned by f with the output probability vectors $\mathbf{p} = f(x)$ and $\mathbf{p}_{\mathcal{F}} = f(x_{\mathcal{F}_\theta})$, respectively. Let $k^* = \arg\max \mathbf{p}$ and $\tilde{k} = \arg\max \mathbf{p}_{\mathcal{F}}$ denote the output labels assigned to original image x and perturbed version $x_{\mathcal{F}_\theta}$ and $d = \frac{p_1 - p_2}{2}$ be the half of the difference between two largest components of $\mathbf{p}$.

  Then, the certain perturbations would not change the label, *i.e.* $\tilde{c} = c$, if

$$\|\mathbf{p} - \mathbf{p}_{\mathcal{F}}\|_\infty < d. \tag{2}$$

where $\|\mathbf{p} - \mathbf{p}_{\mathcal{F}}\|_\infty = \max\left(|\mathbf{p}_1 - \mathbf{p}_{\mathcal{F}_1}|, \ldots, |\mathbf{p}_K - \mathbf{p}_{\mathcal{F}K}|\right)$.

  That means, if the maximum change caused by functional perturbations amongst all classes w.r.t. the output probability vectors, does not exceed half of the maximum difference d between the two largest components of $\mathbf{p}$, the classifier will retain the category to an input x. Thus, it is straightforward to provide the probabilistic guarantees that the class label assigned to an input x by a classifier f would not change under the transformation functions $\mathcal{F}_\theta$ by bounding the probability of the event $\|\mathbf{p} - \mathbf{p}_\mathcal{F}\|_\infty < d$ occurring.

Subsequently, we suggest applying adaptive concentration inequalities, which enable our algorithm iteratively to take more and more samples until the estimated probability of event occurrence is sufficiently accurate to be used to compute the probability satisfying Eq. (2). We establish some notation for the verification process that follows. For a random variable $Z \sim P_Z$ following any probability distribution P_Z, $\mu_Z = \mathbb{E}_{Z \sim P_Z}[Z]$ donates the expectation of Z. To fit the context of probabilistic robustness verification, we let

$$Z = \mathbb{1}\left[\|\mathbf{p} - \mathbf{p}_\mathcal{F}\|_\infty < d\right] \tag{3}$$

where $\mathbb{1}[x]$ is an indicator function that returns 1 if x is true and 0 otherwise. In this case, μ_Z represents certified stable probability of a data instance x under functional transformations $\mathcal{F}_\theta$ parameterised by θ, i.e.,

$$\mu_Z = \mathbb{E}_{Z \sim P_Z}[Z] = P_{Z \sim P_Z}[Z = 1]. \tag{4}$$

4.2   Adaptive Concentration Inequalities

Concentration inequalities [2], e.g. Chernoff inequality, Azuma's bound and Hoeffding's inequality, are fundamental statistical analytic techniques, widely applied to reliable decision-making with probabilistic guarantees. Hoeffding inequality is utilised to bound the probability of an event or the sum of bounded variables.

Let Z be a random variable with distribution P_Z, and $Z_1, Z_2, \ldots, Z_n$ are independent and identically distributed samples drawn from P_Z, then we can estimate μ_Z, which represents the expected value of Z using

$$\hat{\mu}_Z = \frac{1}{n} \sum_{i=1}^n Z_i. \tag{5}$$

Note that, regardless of the number of samples used, there must be some error ϵ between the estimated value $\hat{\mu}_Z$ and true expected value μ_Z. However, we can derive high-probability bounds on this error using Hoeffding inequality [4].

Definition 3 (Hoeffding Inequality [4]). *For any $\delta > 0$,*

$$P_{Z_1, \ldots, Z_n \sim P_Z}\left[|\hat{\mu}_Z - \mu_Z| \le \varepsilon\right] \ge 1 - \delta \tag{6}$$

holds for $\delta = 2e^{-2n\varepsilon^2}$, equivalently, $\varepsilon = \sqrt{\frac{1}{2n} \log \frac{2}{\delta}}$.

The number of samples n, on the other hand, must be independent of the underlying process and determined in advance, yet in most circumstances, we generally have no idea how many samples we will need to validate the robustness specification. Consequently, we would like the number of samples used during the verification procedure to be a random variable. We decide to incorporate *adaptive concentration inequality* into our algorithm, enabling our verification algorithm to take samples iteratively. Upon termination, n becomes a stopping time J, where J is a random variable, depending on the ongoing process. Then, the following adaptive Hoeffding inequality is utilised to guarantee the bound of the aforementioned probability since traditional concentration inequalities do not hold when the number of samples is stochastic.

Theorem 1 (Adaptive Hoeffding Inequality [30]**).** *Let Z_i be $1/2$-subgaussian random variables, and let $\hat{\mu}_Z^{(n)} = \frac{1}{n}\sum_{i=1}^n Z_i$, also let J be a random variable on $\mathbb{N} \cup \{\infty\}$ and let $\varepsilon(n) = \sqrt{\frac{a\log(\log_c n+1)+b}{n}}$ where $c > 1, a > c/2, b > 0$, and ζ is the Riemann-ζ function. Then, we have*

$$P\left[J < \infty \wedge \left(\left|\hat{\mu}_Z^{(J)}\right| \geq \varepsilon(J)\right)\right] \leq \delta_b \tag{7}$$

where $\delta_b = \zeta(2a/c)e^{-2b/c}$.

4.3 Verification Algorithm

In this section, we will describe how to verify the probabilistic robustness of a given classifier, deriving from adaptive Hoeffding inequality. To begin, we can derive a corollary from Theorem 1. Note that the values of a and c do not have a significant effect on the quality of the bound in practice [30] and we fix a and c with the recommended values in [30], 0.6 and 1.1, respectively.

Theorem 2. *Given a random variable Z as shown in Eq. (3) with unknown probability distribution P_Z, let $\{Z_i \sim P_Z\}_{i \in \mathbb{N}}$ be independent and identically distributed samples of Z. Let $\hat{\mu}_Z^{(n)} = \frac{1}{n}\sum_{i=1}^n Z_i$ be estimate of true value μ_Z, and let stopping time J be a random variable on $\mathbb{N} \cup \{\infty\}$ such that $P[J < \infty] = 1$. Then, for a given $\delta \in \mathbb{R}_+$,*

$$P\left[\left|\hat{\mu}_Z^{(J)} - \mu_Z\right| \leq \varepsilon(\delta, J)\right] \geq 1 - \delta \tag{8}$$

holds, where $\varepsilon(\delta, n) = \sqrt{\frac{0.6\cdot\log(\log_{1.1} n+1)+1.8^{-1}\cdot\log(24/\delta)}{n}}$.

We give a proof in *Appendix B*.

In the context of probabilistic robustness verification, we can certify the probabilistic robustness of a black-box neural network against functional threat models. Specifically, certified probability, μ_Z, is calculated by computing the proportion of the event $(Z < d)$ occurring through sampling the perturbed images surrounding an input x. For example, given a target neural network, we would like to verify whether there are at most τ (e.g. 1%) adversarial examples within a specific neighbouring area around an image x with greater than $1 - \delta$ (e.g. 99.9%) confidence. This means we would like to have more than 99.9%

confidence in asserting that the proportion of the adversarial examples is fewer than 1%.

Building upon this idea, the key of this statistical robustness verification is to prove the robustness specification of form $\mu_Z \geq 1 - \tau$ holds. If μ_Z is quite close to $1 - \tau$, then more additional samples are required to make ϵ to be small enough to ensure that $\hat{\mu}_Z$ is close to μ_Z. We use a hypothesis test parameterized by a given modest probability τ of accepted violation predefined by users.

○ $\mathcal{H}_0$: The probability of robustness satisfaction $\mu \geq 1 - \tau$. Thus, the classifier can be certified.
○ $\mathcal{H}_1$: The probability of robustness satisfaction $\mu < 1 - \tau$. Thus, the classifier should not be certified.

Alternatively, consider the hypothesis testing with two following conditions

$$\mathcal{H}_0 : \hat{\mu}_Z + \tau - \epsilon - 1 \geq 0$$
$$\mathcal{H}_1 : \hat{\mu}_Z + \tau + \epsilon - 1 < 0. \tag{9}$$

If H_0 holds, then together with $P\left[\left|\hat{\mu}_Z^{(J)} - \mu_Z\right| \leq \epsilon\right] \geq 1 - \delta$, we can assert that $\mu_Z \geq \hat{\mu}_Z - \epsilon \geq 1 - \tau$ with high confidence. Likewise, we can conclude that $\mu_Z \leq \hat{\mu}_Z + \epsilon < 1 - \tau$, if H_1 holds.

The full algorithm is summarized in Algorithm 1 in *Appendix A*.

5   Experiments

In order to evaluate the proposed method, an assessment is conducted involving various trained neural networks on public data sets CIFAR-10 and ImageNet.

Specifically, for neural networks certified on the CIFAR-10 dataset, we have trained three neural networks based on ResNet18 architecture: a naturally trained network (plain), an adversarial trained network augmented with adversarial examples generated by l_2 PGD attack (AT), and a perceptual adversarial trained network (PAT) against a perceptual attack [12]. In addition, four state-of-the-art neural networks, *i.e.* resnetv2_50, mobilenetv2_100, efficientnet_b0 and vit_base_patch16_244 are introduced for ImageNet dataset; all pre-trained models are available on a PyTorch library. For our models, selected details are described in Table 6 in *Appendix C.1*.

We provide the details about considered functional perturbations in the following subsection, and the results follow. *Nota bene*, we choose $\tau = 5\%$ for certifying the robustness of all models, as this is a widely accepted level in most practice. All the experiments are run on a desktop computer (i7-10700K CPU, GeForce RTX 3090 GPU).

5.1   Baseline Setting

To demonstrate the effectiveness and efficiency of PRoA[1], it is natural to compare the estimated probability of the event, *i.e.* a target model will not fail when

---
[1] Our code is released via https://github.com/TrustAI/PRoA.

Table 1. Comparison with related work in different aspects.

	SRC [5]	AMLS based [24]	Randomized smoothing [3,28]	DeepGo [18]	Reachability based [26]	Semantify-NN [15]	FVIM based [27]	CROWN [23,25,29]	PRoA
Deterministic robustness	✗	✗	✗	✓	✓	✓	✓	✓	✗
Probabilistic robustness	✓	✓	✓	✗	✗	✗	✗	✗	✓
Verifying robustness on functional perturbation	✗	✗	✗	✗	✗	✗	✗	✗	✓
Black-box model	✓	✓	✓	✓	✗	✓	✗	✗	✓

encountering functional perturbations, obtained by PRoA with the lower limit of the corresponding confidence interval, *i.e.* Agresti-Coull confidence interval (A-C CI), see *Appendix* C.2.

We list the relevant existing works in Table 1 and compare our method with these typical methods from five aspects. Specifically, DeepGo [18], Reachability based [26], Semantify-NN [15], FVIM based [27] and CROWN [23,25,29] only can evaluate deterministic robustness of neural networks. Although SRC [5], AMLS based [24], Randomized Smoothing [3,28] are able to certify probabilistic robustness, our work extensively consider models' probabilistic robustness under functional threat models.

To the best of our knowledge, there is no existing study in terms of certifying the probabilistic robustness of neural networks involving a functional threat model. Since [5] is the closest approach in spirit to our method amongst recent works, we use SRC [5] as our baseline algorithm. The proposal of SRC is to measure the probabilistic robustness of neural networks by finding the maximum perturbation radius using random sampling, and we extend it to be a baseline algorithm for computing the certified accuracy under functional perturbations.

5.2   Considered Functional Perturbations

PRoA is a general framework that is able to assess the robustness under any functional perturbations. In our experiments, we specifically study geometric transformation, colour-shifted function, and Gaussian blur in terms of verifying probabilistic robustness.

Gaussian Blur. Gaussian blurring is used to blur an image in order to reduce image noise and detail involving a Gaussian function

$$G_{\theta_g}(k) = \frac{1}{\sqrt{2\pi\theta_g}} \exp\left(-k^2/(2\theta_g)\right) \tag{10}$$

where θ_g is the squared kernel radius. For $x \in \mathcal{X}$, we define

$$\mathcal{F}_G(x) = x * G_{\theta_g} \tag{11}$$

as the corresponding function parameterised by θ_g where $*$ denotes the convolution operator.

Geometric Transformation. For geometric transformation, we consider three basic geometric transformations: rotation, translation and scaling. We implement the corresponding geometric functions in a unified manner using a spatial

Table 2. CIFAR-10 - Comparison of empirical robust accuracy (Grid and Rand.) and probabilistically certified accuracy (Cert. Acc) with respect to a specific model (ResNet18) with three training methods, shown in Table 6. Moreover, probabilistically certified accuracy is presented with three confidence levels $(1-\delta)$, low level of confidence $(\delta = 10^{-4})$, middle level of confidence $(\delta = 10^{-15})$ and high level of confidence $(\delta = 10^{-30})$, respectively.

Transformation	Parameters	Training type	Grid	Rand.	SRC Cert. Acc			PRoA Cert. Acc		
					$\delta = 10^{-30}$	$\delta = 10^{-15}$	$\delta = 10^{-4}$	$\delta = 10^{-30}$	$\delta = 10^{-15}$	$\delta = 10^{-4}$
Rotation	$\theta_r \in [-35°, 35°]$	plain	26.9%	76.8%	24.7%	24.8%	24.8%	30.3%	31.5%	32.0%
		PAT	16.7%	55.9%	8.1%	8.1%	8.1%	10.8%	12.4%	12.9%
		AT	16.5%	74.5%	11.2%	11.2%	11.2%	14.7%	15.2%	15.4%
Translation	$\theta_t \in [-30\%, 30\%]$	plain	62.8%	89.6%	64.9%	65.1%	66.6%	77.5%	78.8%	79.4%
		PAT	50.1%	77.7%	31.1%	31.7%	32.4%	47.5%	48.6%	49.5%
		AT	56.5%	79.3%	45.3%	45.7%	46.1%	58.9%	60.1%	61.7%
Scale	$\theta_s \in [-70\%, 130\%]$	plain	45.4 %	86.9%	48.7%	49.0%	49.7%	63.3%	65.2%	67.1%
		PAT	23.5%	73.1%	8.4%	8.7%	9.6%	20.9%	22.8%	24.7%
		AT	34.4%	74.4%	19.2%	19.4%	20.3%	32.4%	34.5%	35.9%
Hue	$\theta_h \in [-\frac{\pi}{2}, \frac{\pi}{2}]$	plain	76.9 %	89.9%	75.0%	75.0%	75.0%	79.5%	79.8%	79.6%
		PAT	63.0%	77.5%	53.6%	53.6%	53.6%	56.7%	57.6%	57.9%
		AT	57.6%	55.1%	**54.7%**	**54.7%**	54.7%	54.1%	54.5%	55.7%
Saturation	$\theta_s \in [-30\%, 30\%]$	plain	92.3 %	93.9%	95.3%	95.3%	95.3%	95.6%	95.6%	96.4%
		PAT	77.1%	80.8%	72.3%	72.3%	72.4%	75.6%	76.0%	77.3%
		AT	74.5%	76.4%	76.8%	77.0%	77.0%	79.4%	79.6%	80.3%
Brightness+Contrast	$\theta_b \in [-30\%, 30\%]$ $\theta_c \in [-30\%, 30\%]$	plain	72.6 %	92.7%	75.5%	75.7%	76.2%	83.8%	84.8%	84.1%
		PAT	36.1%	76.2%	20.5%	20.9%	21.6%	37.4%	38.2%	37.0%
		AT	31.5%	73.9%	17.8%	17.9%	18.4%	34.7%	38.1%	35.7%
Gaussian Blurring	$\theta_g \in [0, 9]$	plain	1.0%	18.1%	3.1%	3.1%	3.3%	3.6%	3.7%	3.4%
		PAT	2.9%	39.7%	11.0%	11.0%	11.0%	13.5%	13.7%	12.9%
		AT	3.7%	42.9%	18.7%	18.9%	**18.9%**	19.2%	19.3%	18.8%

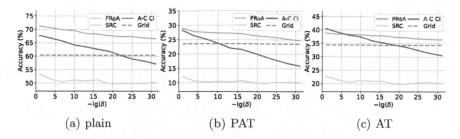

(a) plain   (b) PAT   (c) AT

Fig. 1. CIFAR-10 - An illustration of evaluating robustness of trained neural networks using PRoA, SRC and A-C CI with different confidence parameter δ against one specific perturbation, image scaling.

transformer block with a set of parameters of affine transformation, *i.e.* $\mathcal{T}(x, \theta)$, in [7], where

$$\theta = \begin{bmatrix} \theta_{11} & \theta_{12} & \theta_{13} \\ \theta_{21} & \theta_{22} & \theta_{23} \end{bmatrix} \tag{12}$$

is an affine matrix determined by θ_r, θ_t as well as θ_s.

Colour-Shifted Function. Regarding colour shifting, we change the colour of images based on HSB (Hue, Saturation and Brightness) space instead of RGB space since HSB give us a more intuitive and semantic sense for understanding the perceptual effect of the colour transformation. We also consider a combination attack using brightness and contrast.

All mathematical expressions of these functional perturbations as well as their parameter ranges are presented in *Appendix C.3*.

5.3   Quantitative Results of Experiments on CIFAR-10

To evaluate our method, we calculate the probabilistically certified accuracy of 1,000 images randomly from a test set for various functional perturbations by PRoA and SRC, in dependence on the user-defined confidence level. Furthermore, empirical robust accuracy is computed against random and grid search adversaries as well.

To begin, we validate the effectiveness of our method over three ResNet18 models (plain, AT and PAT) trained with different training protocols against all considered functional perturbations on the CIFAR-10 dataset as mentioned previously. As a result of the experiments, we present considered perturbation functions, accompanying parameters, and quantitative results in terms of probabilistically certified accuracy (Cert. Acc), empirical accuracy (Rand.) and empirical robust accuracy (Grid) in Table 2. Clearly, the results of the proposed method align well with the validation results obtained by exhaustive search and random perturbation, and PRoA is able to achieve higher certified accuracy than SRC in almost all scenarios. Thus, the effectiveness of PRoA can be demonstrated.

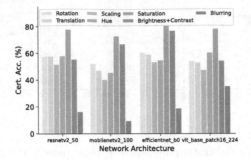

Fig. 2. ImageNet - Comparison of probabilistic certified accuracy with confidence level $1 - 10^{-4}$, computed over the 500 randomly selected ImageNet images, amongst the models described in Table 6.

Table 3. CIFAR-10 - Confusion matrix comparing SRC and PRoA [plain model under brightness+contrast perturbation, $\delta = 10^{-10}$].

		PRoA		
		Certified	Uncertified	Termination
SRC	Certified	329	32	12
	Uncertified	36	9	0
	Infeasible	69	11	2

An illustration of model verification using A-C CI, SRC and PRoA with various confidence levels ($90\% \sim 1 - 10^{-30}$) against the picture scaling function on CIFAR-10 is depicted in Fig. 1. For instance, according to Fig. 1(a), we have 90% confidence ($\delta = 10^{-1}$) that this considered trained model will correctly identify roughly 71% of images in CIFAR-10 after a no more than 30% image scaling with a greater than 95% chance ($\tau = 5\%$). In contrast, we have $1 - 10^{-30}$ ($\delta = 10^{-30}$) confidence that the proportion of images with a misclassification probability below our accepted level 5% would be 67%. Clearly, accuracy certified by PRoA reduces along with the growth of confidence, but it is not significantly changed for SRC and progressively diminishes for A-C CI. In addition, as compared with baselines, the proposed method achieves remarkable higher certified robust accuracy and a narrower gap to empirical robust accuracy along with our confidence increasing, see Fig. 1. Moreover, Grid is an approximated accuracy to the extreme case with zero tolerance ($\tau = 0$) to perturbations. However, certified accuracy with a 5% tolerance level obtained by SRC and A-C CI always tends to be below the Grid without tolerance as the confidence level increases, which causes underestimation of the probabilistic robustness.

We apply SRC and PRoA for verifying the robustness of 500 images, which are randomly chosen from the test set on CIFAR-10. The corresponding confusion matrix is shown in Table 3, which takes into account the cases in which the SRC outputs an "infeasible" status when it fails to obtain a deterministic certification result, and PRoA reaches sample limitation (set to 10,000) as a termination condition. Unsurprisingly, our method can take a certification decision in most cases when SRC returns an "infeasible", even though 14 images obtain a "termination" status due to adaptive sampling reaching sample limitation.

5.4 Comparing Probabilistic Robustness Across Models on ImageNet

We also use our method to analyse four large state-of-the-art neural networks against perturbation functions as mentioned earlier with 500 images randomly picked from the ImageNet test set. Figure 2 demonstrates the robustness comparison of different models when subjected to diverse functional perturbations. All validation results of different models are shown as percentages in Fig. 2. For the 'rotation' scenario, the certified accuracy of resnetv2_50 produced by PRoA is 57.8%, which means we have 99.99% confidence in the claim that on average, in resnetv2_50, 57.8% of images will produce an adversarial example with a chance of more than 5% in the 'rotation' scenario, e.g. camera rotation.

Table 4. ImageNet - Comparison Agresti-Coull, SRC and PRoA $[\delta = 10^{-10}]$.

Model	Perturbation	Certified (%)			Avg. runtime (sec.)		
		Agresti-Coull	SRC	PRoA	Agresti-Coull	SRC	PRoA
Mobilenetv2_100	Rotation	38	40	**43**	**5.08**	5.10	8.35
	Translation	41	34	**47**	**5.14**	5.20	8.96
	Scaling	38	30	**44**	5.32	**5.07**	8.64
	Hue	40	**48**	**48**	5.64	**5.16**	5.19
	Saturation	65	71	**72**	5.60	**5.16**	7.26
	Brightness+Contrast	47	54	**62**	5.58	**5.17**	7.03
	Gaussian Blurring	3	6	**8**	6.38	5.72	**3.89**
efficientnet_b0	Rotation	46	47	**49**	**5.08**	6.25	6.28
	Translation	49	44	**57**	**5.14**	6.24	7.77
	Scaling	46	44	**51**	**5.32**	6.25	9.83
	Hue	48	55	**57**	**5.64**	6.53	8.69
	Saturation	73	79	**81**	**5.60**	6.53	9.83
	Brightness+Contrast	55	56	**65**	**5.59**	6.53	12.37
	Gaussian Blurring	10	14	**17**	6.38	7.11	**5.61**
Resnetv2_50	Rotation	51	46	**54**	12.77	**9.68**	15.76
	Translation	58	44	**57**	12.80	**9.56**	18.88
	Scaling	51	38	**54**	12.80	**9.54**	17.89
	Hue	61	61	**63**	13.10	**9.81**	13.04
	Saturation	77	83	**86**	13.15	**9.82**	16.87
	Brightness+Contrast	39	32	**40**	14.02	**9.81**	20.51
	Gaussian Blurring	15	14	**17**	14.16	10.43	**6.66**
vit_base_patch16_224	Rotation	39	34	**41**	34.68	**33.04**	49.62
	Translation	47	32	**49**	34.43	**33.06**	59.18
	Scaling	40	33	**43**	34.32	**33.00**	63.21
	Hue	**63**	53	54	34.61	**33.31**	45.70
	Saturation	70	71	**73**	37.54	**33.31**	41.37
	Brightness+Contrast	32	24	**34**	36.83	**34.19**	70.69
	Gaussian Blurring	**32**	28	30	35.99	34.88	**33.96**

Table 5. ImageNet - PRoA $[\delta = 10^{-10}]$.

Model	Perturbation	Avg. runtime (sec.$\pm std$)	Avg. sample num.	Certified (%)
resnetv2_50	Rotation	15.76 ± 17.74	5820	54
	Brightness+Contrast	20.51 ± 33.05	7930	40
	Blurring	6.66 ± 9.64	2420	17
mobilenetv2_100	Rotation	8.35 ± 13.40	7860	43
	Brightness+Contrast	7.03 ± 8.23	6650	62
	Blurring	3.88±10.55	3180	6
efficientnet_b0	Rotation	6.28 ± 7.28	4970	49
	Brightness+Contrast	12.37 ± 13.81	9370	65
	Blurring	5.61 ± 2.50	3790	17
vit_base_patch16_224	Rotation	49.62 ± 80.11	7260	41
	Brightness+Contrast	70.69 ± 106.84	9950	34
	Blurring	35.96 ± 69.26	5020	30

We also compare our algorithm to the Agresti-Coull confidence interval and SRC with a moderate confidence level, *i.e.* $\delta = 10^{-10}$, as shown in Table 4. On the one hand, our method provides the highest certified accuracy for practically all scenarios and models; on the other hand, the average runtime of our method is comparatively longer than baselines, due to the error bounds of the estimate, which are not tight enough to make decisions and necessitate more samples. Interestingly, our algorithm takes the shortest time to certify images under a sophisticated functional perturbation, the Gaussian blurring, whereas the computation time of A-C CI and SRC increases. This is because, instead of a predetermined and decided a priori number of samples, our method terminates at any runtime J depending on the ongoing process once it is capable of delivering a result, avoiding superfluous samples.

Finally, the average number of samples and the average runtime for a single image are reported in Table 5. As one can notice, our method can be easily scaled to various SOTA network architectures, and the computation time and required samples increase reasonably with network size and complexity of perturbation function.

6   Conclusion

This paper aims to certify the probabilistic robustness of a target neural network to a functional threat model with an adaptive process inspired by the Adaptive Concentration Inequalities. With PRoA, we can certify that a trained neural network is robust if the estimated probability of the failure is within a tolerance level. PRoA is dependent on the ongoing hypothesis test, avoiding a-prior sample size. The tool is scalable, efficient and generic to black-box classifiers, and it also comes with provable guarantees. In this paper, the hypothesis testing and adaptive sampling procedure are sequential and bring difficulty for parallelization, so one interesting future direction lies in how to further boost PRoA's efficiency, e.g., by enabling parallelization on GPUs. Another interesting future work is to bridge the gap between worst-case certification and chance-case certification.

References

1. Amir, G., Wu, H., Barrett, C., Katz, G.: An SMT-based approach for verifying binarized neural networks. In: TACAS 2021. LNCS, vol. 12652, pp. 203–222. Springer, Cham (2021). https://doi.org/10.1007/978-3-030-72013-1_11
2. Boucheron, S., Lugosi, G., Massart, P.: Concentration Inequalities: A Nonasymptotic Theory of Independence, 1st edn. Oxford University Press, Oxford (2013)
3. Cohen, J., Rosenfeld, E., Kolter, Z.: Certified adversarial robustness via randomized smoothing. In: ICML, pp. 1310–1320. PMLR, California (2019)
4. Hoeffding, W.: Probability inequalities for sums of bounded random variables. In: Fisher, N.I., Sen, P.K. (eds.) The Collected Works of Wassily Hoeffding, pp. 409–426. Springer, New York (1994)
5. Huang, C., Hu, Z., Huang, X., Pei, K.: Statistical certification of acceptable robustness for neural networks. In: Farkaš, I., Masulli, P., Otte, S., Wermter, S. (eds.) ICANN 2021. LNCS, vol. 12891, pp. 79–90. Springer, Cham (2021). https://doi.org/10.1007/978-3-030-86362-3_7
6. ISO, I.: Iso/iec guide 51: Safety aspects-guidelines for their inclusion in standards. Geneva, Switzerland (2014)
7. Jaderberg, M., Simonyan, K., Zisserman, A., Kavukcuoglu, K.: Spatial transformer networks. In: NeurIPS, pp. 2017–2025. The MIT Press, Quebec (2015)
8. Katz, G., Barrett, C., Dill, D.L., Julian, K., Kochenderfer, M.J.: Reluplex: a calculus for reasoning about deep neural networks. Formal Methods Syst. Des., 1–30 (2021). https://doi.org/10.1007/s10703-021-00363-7
9. Katz, G., Barrett, C., Dill, D.L., Julian, K., Kochenderfer, M.J.: Reluplex: an efficient SMT solver for verifying deep neural networks. In: Majumdar, R., Kunčak, V. (eds.) CAV 2017. LNCS, vol. 10426, pp. 97–117. Springer, Cham (2017). https://doi.org/10.1007/978-3-319-63387-9_5
10. Kurakin, A., et al.: Adversarial attacks and defences competition. In: Escalera, S., Weimer, M. (eds.) The NIPS '17 Competition: Building Intelligent Systems. TSSCML, pp. 195–231. Springer, Cham (2018). https://doi.org/10.1007/978-3-319-94042-7_11
11. Laidlaw, C., Feizi, S.: Functional adversarial attacks. In: NeurIPS, pp. 10408–10418. The MIT Press, Vancouver (2019)
12. Laidlaw, C., Singla, S., Feizi, S.: Perceptual adversarial robustness: Defense against unseen threat models. In: ICLR. Austria (2021)
13. Li, J., Liu, J., Yang, P., Chen, L., Huang, X., Zhang, L.: Analyzing deep neural networks with symbolic propagation: towards higher precision and faster verification. In: Chang, B.-Y.E. (ed.) SAS 2019. LNCS, vol. 11822, pp. 296–319. Springer, Cham (2019). https://doi.org/10.1007/978-3-030-32304-2_15
14. Madry, A., Makelov, A., Schmidt, L., Tsipras, D., Vladu, A.: Towards deep learning models resistant to adversarial attacks. In: ICLR. Vancouver (2018)
15. Mohapatra, J., Weng, T.W., Chen, P.Y., Liu, S., Daniel, L.: Towards verifying robustness of neural networks against a family of semantic perturbations. In: CVPR, pp. 244–252. IEEE, Seattle (2020)

16. Pautov, M., Tursynbek, N., Munkhoeva, M., Muravev, N., Petiushko, A., Oseledets, I.: Cc-cert: A probabilistic approach to certify general robustness of neural networks. arXiv preprint. arXiv:2109.10696 (2021)
17. Rivasplata, O.: Subgaussian random variables: An expository note (2012)
18. Ruan, W., Huang, X., Kwiatkowska, M.: Reachability analysis of deep neural networks with provable guarantees. In: IJCAI, pp. 2651–2659. IJCAI.org, Stockholm (2018)
19. Singh, G., Ganvir, R., Püschel, M., Vechev, M.T.: Beyond the single neuron convex barrier for neural network certification. In: NeurIPS, pp. 15072–15083. The MIT Press, Vancouver (2019)
20. Singh, G., Gehr, T., Püschel, M., Vechev, M.T.: An abstract domain for certifying neural networks. Proc. ACM Program. Lang. 3(POPL), 1–30 (2019)
21. Szegedy, C., et al.: Intriguing properties of neural networks. In: ICLR. Banff (2014)
22. Tran, H.-D., et al.: NNV: the neural network verification tool for deep neural networks and learning-enabled cyber-physical systems. In: Lahiri, S.K., Wang, C. (eds.) CAV 2020. LNCS, vol. 12224, pp. 3–17. Springer, Cham (2020). https://doi.org/10.1007/978-3-030-53288-8_1
23. Wang, S., et al.: Beta-crown: Efficient bound propagation with per-neuron split constraints for neural network robustness verification. In: NeurIPS. The MIT Press, Virtual (2021)
24. Webb, S., Rainforth, T., Teh, Y.W., Kumar, M.P.: A statistical approach to assessing neural network robustness. In: ICLR. New Orleans (2019)
25. Weng, L., et al.: Towards fast computation of certified robustness for ReLU networks. In: ICML, pp. 5276–5285. PMLR, Stockholm (2018)
26. Xiang, W., Tran, H.D., Johnson, T.T.: Output reachable set estimation and verification for multilayer neural networks. IEEE Trans. Neural Netw. Learn. Syst. 29(11), 5777–5783 (2018)
27. Yang, X., Johnson, T.T., Tran, H.D., Yamaguchi, T., Hoxha, B., Prokhorov, D.V.: Reachability analysis of deep ReLU neural networks using facet-vertex incidence. In: HSCC, pp. 18:1–18:7. ACM, Nashville (2021)
28. Zhang, D., Ye, M., Gong, C., Zhu, Z., Liu, Q.: Black-box certification with randomized smoothing: a functional optimization based framework. In: NeurIPS. The MIT Press, Virtual (2020)
29. Zhang, H., Weng, T.W., Chen, P.Y., Hsieh, C.J., Daniel, L.: Efficient neural network robustness certification with general activation functions. In: NeurIPS. The MIT Press, Montréal (2018)
30. Zhao, S., Zhou, E., Sabharwal, A., Ermon, S.: Adaptive concentration inequalities for sequential decision problems. In: NeurIPS, pp. 1343–1351. The MIT Press, Barcelona (2016)

Hypothesis Testing for Class-Conditional Label Noise

Rafael Poyiadzi$^{(\boxtimes)}$, Weisong Yang, Niall Twomey, and Raul Santos-Rodriguez

University of Bristol, Bristol, UK
{rp13102,ws.yang,niall.twomey,enrsr}@bristol.ac.uk

Abstract. In this paper we aim to provide machine learning practitioners with tools to answer the question: *have the labels in a dataset been corrupted?* In order to simplify the problem, we assume the practitioner already has preconceptions on possible distortions that may have affected the labels, which allow us to pose the task as the design of hypothesis tests. As a first approach, we focus on scenarios where a given dataset of instance-label pairs has been corrupted with *class-conditional label noise*, as opposed to *uniform label noise*, with the former biasing learning, while the latter – under mild conditions – does not. While previous works explore the direct estimation of the noise rates, this is known to be hard in practice and does not offer a real understanding of how trustworthy the estimates are. These methods typically require *anchor points* – examples whose true posterior is either 0 or 1. Differently, in this paper we assume we have access to a set of anchor points whose true posterior is approximately 1/2. The proposed hypothesis tests are built upon the asymptotic properties of Maximum Likelihood Estimators for Logistic Regression models. We establish the main properties of the tests, including a theoretical and empirical analysis of the dependence of the power on the test on the training sample size, the number of anchor points, the difference of the noise rates and the use of relaxed anchors.

1 Introduction

When a machine learning practitioner is presented with a new dataset, a first question is that of data quality [24] as this will affect any subsequent machine learning tasks. This has led to tools to address transparency and accountability of data [27,28]. However, in supervised learning, an equally important concern is the quality of labels. For instance, in standard data collections, data curators usually rely on annotators from online platforms, where individual annotators cannot be unconditionally trusted as they have been shown to perform inconsistently [25]. Labels are also expected to not be ideal in situations where the data is harvested directly from the web [31,32]. In general this is a consequence of annotations not being carried out by domain experts [13].

The existing literature primarily focuses on directly estimating the distortion(s) present in the labels and mainly during the learning process (see Sect. 4). In this paper we argue that, in most cases, that is too hard a problem and might

M.-R. Amini et al. (Eds.): ECML PKDD 2022, LNAI 13715, pp. 171–186, 2023.
https://doi.org/10.1007/978-3-031-26409-2_11

lead to suboptimal outcomes. Instead, we suggest modifying this approach in two ways. First, we leverage the practitioner's prior knowledge on the type possible distortions affecting the labels and use their preconceptions to design hypothesis testing procedures that would allow us (under certain assumptions we state later) to provide a measure of evidence for the presence of the distortion. This is of course a much simple task than addressing the estimation of any possible distortion. As an example, in this paper, we focus on class-conditional noise, as opposed to uniform noise (as we discuss later, class-conditional noise biases the learning procedure, while uniform noise under mild conditions does not). Secondly, with this information at hand, and given that the tests are performed right after data collection and annotation and before learning takes place, the practitioner can then make more informed decisions. If the quality of the labels is deemed poor, then the practitioner could resort to: (1) a modified data labelling procedure (e.g., active learning in the presence of noise [29]), (2) seek methods to make the training robust (e.g., algorithms for learning from noisy labels [30]), or (3) drop the dataset altogether.

Let us introduce the binary classification setting, where the goal is to train a classifier $g : \mathcal{X} \rightarrow \{-1, +1\}$, from a labelled dataset $\mathcal{D}_n^{train} = \{(\boldsymbol{x}_i, y_i)\}_{i=1}^n \in (\mathbb{R}^d \times \{-1, 1\})$, with the objective of achieving a low miss-classification error: $\mathbb{P}_{X,Y}(g(X) \neq Y)$. While it is generally assumed that the training dataset is drawn from the distribution for which we wish to minimise the error for $\mathcal{D}_n^{train} \sim p(X, Y)$, as mentioned above, this is often not the case. Instead, the task requires us to train a classifier on a corrupted version of the dataset $\tilde{\mathcal{D}}_n^{train} \sim p(X, \tilde{Y})$ whilst still hoping to achieve a low error rate on the clean distribution.

In this work we focus on a particular type of corruption, *instance-independent label noise*, where labels are flipped with a certain rate, that can either be uniform across the entire data-generating distribution or conditioned on the true class of the data point. A motivating example of class-conditional noise is given in [12] in the form of medical case-control studies, where different tests may be used for subject and control. An essential ingredient in our procedure is the input from the user in the form of a set of *anchor points*. Differently from previous works, we assume anchor points for which the true posterior distribution $\mathbb{P}(Y = 1 \mid X = x)$ is (approximately) ½. For an instance $\boldsymbol{x}$ this requirement means that an expert would not be able to provide *any* help to identify the correct class label. While this will be shown to be convenient for theoretical purposes, finding such anchor points might be rather difficult to accomplish in practice, so we show how to relax this notion to a more realistic $\eta(x) \approx 1/2$.

The tests rely on the asymptotic properties of the *Maximum Likelihood Estimate* (MLE) solution for Logistic Regression models, and the relationship between the true and noisy posteriors. On the theoretical side, we show that when the asymptotic properties of MLE hold and the user provides a single anchor point, we can devise hypothesis tests to assess the presence of class-conditional label corruption in the dataset. We then further extend these ideas to allow for richer sets of anchor points and illustrate how these lead to gains in the *power* of the test. In Sect. 2 we cover the necessary background on MLE, noisy labels and define the necessary tools. In Sect. 3 we illustrate how to carry a z-test using anchor points on the presence of class-conditional noise. In Sect. 4 we discuss related work and in Sect. 5 we present experimental findings.

2   Background

We are provided with a dataset $(\boldsymbol{X}, \boldsymbol{y}) = \{(\boldsymbol{x}_i, y_i)\}_{i=1}^n \in (\mathbb{R}^d \times \{-1, 1\})$, and our task is to assess whether the labels have been corrupted with class-conditional flipping noise. We use y to denote the true label, and $\tilde{y}$ to denote the noisy label. We assume the feature vectors $(\boldsymbol{x})$ have been augmented with *ones* such that we have $\boldsymbol{x} \to (1, \boldsymbol{x})$. We assume the following model:

$$y_i \sim \text{Bernoulli}(\eta_i),$$

$$\eta_i = \sigma(\theta_0^\top \boldsymbol{x}_i) = \frac{1}{1 + \exp\left(-\theta_0^\top \boldsymbol{x}_i\right)}.$$

Following the MLE procedure we have:

$$\hat{\theta}_n = \underset{\theta \in \Theta}{\text{argmax}}\ \ell_n(\theta \mid D_n) = \underset{\theta \in \Theta}{\text{argmax}} \prod_{i=1}^n \ell_i(\theta \mid \boldsymbol{x}_i,\ y_i)$$

where:

$$\ell(\theta \mid \boldsymbol{x}_i,\ y_i) = \frac{y_i + 1}{2} \cdot \log \eta_i + \frac{1 - y_i}{2} \cdot \log(1 - \eta_i)$$

In this setting, the following can be shown (See for example Chap. 4 of [15]):

$$\sqrt{n}\left(\hat{\theta}_n - \theta_0\right) \xrightarrow{D} \mathcal{N}\left(0,\ I_n(\theta_0)^{-1}\right) \tag{1}$$

where I_{θ_0} denotes the Fisher-Information Matrix:

$$I_n(\theta_0) = \mathbb{E}_\theta\left(-\frac{\partial^2 \ell_n(\theta; Y \mid x)}{\partial\theta\partial\theta^\top}\right) = \mathbb{E}_\theta\left(-H_n(\theta; Y \mid x)\right)$$

where the expectation is with respect to the conditional distribution, and H_n is the Hessian matrix.

We will consider two types of flipping noise and in both cases the noise rates are independent of the instance: $\mathbb{P}(\tilde{Y} = -i \mid Y = i, X = x) = \mathbb{P}(\tilde{Y} = -i \mid Y = i)$ for $i \in \{-1, 1\}$.

Definition 1. Bounded Uniform Noise (UN)
In this setting the per-class noise rates are identical: $\mathbb{P}(\tilde{Y} = 1 \mid Y = -1) = \mathbb{P}(\tilde{Y} = -1 \mid Y = 1) = \tau$ *and bounded:* $\tau < 0.50$. *We will denote this setting with* $UN(\tau)$, *and a dataset* $\mathcal{D} = (\boldsymbol{X}, \boldsymbol{y})$ *inflicted by* $UN(\tau)$ *by:* $\mathcal{D}_\tau$.

Definition 2. Bounded Class-Conditional Noise (CCN)
In this setting the per-class noise rates are different, $\alpha \neq \beta$ *and bounded* $\alpha + \beta < 1$ *with:* $\mathbb{P}(\tilde{Y} = -1 \mid Y = 1) = \alpha$ *and* $\mathbb{P}(\tilde{Y} = 1 \mid Y = -1) = \beta$. *We will denote this setting with* $CCN(\alpha, \beta)$, *and a dataset* $\mathcal{D} = (\boldsymbol{X}, \boldsymbol{y})$ *inflicted by* $CCN(\alpha, \beta)$ *by:* $\mathcal{D}_{\alpha, \beta}$.

An object of central interest in classification settings is the posterior predictive distribution: $\eta(\boldsymbol{x}) = \mathbb{P}(Y = 1 \mid X = \boldsymbol{x})$. Its noisy counterpart, $\tilde{\eta}(\boldsymbol{x}) = \mathbb{P}(\tilde{Y} = 1 \mid X = \boldsymbol{x})$, under the two settings, $UN(\tau)$ and $CCN(\alpha, \beta)$, can be expressed as: (See Appendix 8.1 for full derivation)

$$\tilde{\eta}(\boldsymbol{x}) = \begin{cases} (1 - \alpha - \beta) \cdot \eta(\boldsymbol{x}) + \beta & \text{if (CCN)} \\ (1 - 2\tau) \cdot \eta(\boldsymbol{x}) + \tau & \text{if (UN)} \end{cases} \tag{2}$$

We consider loss functions that have the margin property: $\ell(y, f(x)) = \psi(yf(x))$, where $f : \mathbb{R}^d \to \mathbb{R}$ is a scorer, and $g(\boldsymbol{x}) = sign(f(\boldsymbol{x}))$ is the predictor. Let $f^* = \arg\min_{f \in \mathcal{F}} \mathbb{E}_{X,Y} \psi(Yf(X))$ and $\tilde{f}^* = \arg\min_{f \in \mathcal{F}} \mathbb{E}_{X,\tilde{Y}} \psi(\tilde{Y}f(X))$ denote the minimisers under the clean and noisy distributions, under model-class $\mathcal{F}$.

Definition 3. Uniform Noise robustness [14]
Empirical risk minimization under loss function ℓ is said to be noise-tolerant if $\mathbb{P}_{X,Y}(g^*(X) = Y) = \mathbb{P}_{X,Y}(\tilde{g}^*(X) = Y)$.

Theorem 1. Sufficient conditions for robustness to uniform noise
Under uniform noise $\tau < 0.50$, and a margin loss function, $\ell(y, f(x)) = \psi(yf(x))$ satisfying: $\psi(f(x)) + \psi(-f(x)) = K$ for a positive constant K, we have that $\tilde{g}^*(x) = sign(\tilde{f}^*(x))$ obtained from: $\tilde{f}^* = \arg\min_{f \in \mathcal{F}} \mathbb{E}_{X,\tilde{Y}} \psi(\tilde{Y}f(X))$ is robust to uniform noise.

For the proof see Appendix 8.2. Several loss functions satisfy this, such as: the *square*, *unhinged* (linear), *logistic*, and more. We now introduce our definition of anchor points[1].

Definition 4. (Anchor Points) *An instance $\boldsymbol{x}$ is called an anchor point if we are provided with its true posterior $\eta(\boldsymbol{x})$. Let $\mathcal{A}_s^k$ denote a collection of k anchor points, with $\eta(\boldsymbol{x}) = s \; \forall \boldsymbol{x} \in \mathcal{A}_s^k$. Furthermore, let us also define $\mathcal{A}_{s,\delta}^k$, to imply that $\eta(\boldsymbol{x}_i) = s + \epsilon_i$, for $\epsilon_i \sim \mathbb{U}([-\delta, \; \delta])$, with $0 \le \delta \ll 1$ (respecting $0 \le \eta(\boldsymbol{x}) \le 1$). Also let $\mathcal{A}_{s,\delta} = \mathcal{A}_{s,\delta}^1$.*

$$\mathcal{A}_1^k \quad \to \quad \eta(\boldsymbol{x}) = 1 \quad \to \quad \tilde{\eta}(\boldsymbol{x}) = 1 - \alpha$$

$$\mathcal{A}_{1/2}^k \quad \to \quad \eta(\boldsymbol{x}) = 1/2 \quad \to \quad \tilde{\eta}(\boldsymbol{x}) = \frac{1 - \alpha + \beta}{2}$$

$$\mathcal{A}_0^k \quad \to \quad \eta(\boldsymbol{x}) = 0 \quad \to \quad \tilde{\eta}(\boldsymbol{x}) = \beta$$

The cases we will be referring to are shown to the right. The first and last, $\mathcal{A}_1^k$ and $\mathcal{A}_0^k$, have been used in the past in different scenarios. In this work we will make use of the second case, $\mathcal{A}_{1/2}^k$.

---

[1] Different notions -related to our definition- of anchor points have been used before in the literature under different names. We review their uses and assumptions in Sect. 4.

3   Hypothesis Tests Based on Anchor Points

In this section we introduce our framework for devising hypothesis tests to examine the presence of class-conditional label noise in a given dataset (with uniform noise, as the alternative), assuming we are provided with an anchor point(s). Our procedure is based on a two-sided *z-test* (see for example Chap. 8 of [33]) with a simple null hypothesis, and a composite alternative hypothesis (Eq. 5). We first define the distribution under the null hypothesis (Eq. 6), and under the alternative hypothesis (Eq. 7), when provided with one strict anchor point ($\eta(x) = 1/2$). In this setting, for a fixed *level of significance* (Type I error) (Eq. 8), we first derive a region for retaining the null hypothesis (Eq. 9), and then we analyse the *power* (Prop. 1) of the test (where we have that Type II Error = 1 - *power*). We then extend the approach to examine scenarios that include: (1) having multiple strict anchors ($\eta(x_i) = 1/2$, $\forall i \in [k]$, $k > 1$), (2) having multiple relaxed anchors ($\eta(x_i) \approx 1/2$, $\forall i \in [k]$, $k > 1$), and (3) having no anchors.

With the application of the *delta method* (See for example Chap. 3 of [15]) on Eq. 1, we can get an asymptotic distribution for the predictive posterior:

$$\sqrt{n}(\hat{\eta}(\boldsymbol{x}) - \eta(\boldsymbol{x})) \xrightarrow{D} \mathcal{N}\left(0, \; (\eta(\boldsymbol{x})(1 - \eta(\boldsymbol{x})))^2 \cdot \boldsymbol{x}^\top \boldsymbol{I}_{\theta_0}^{-1} \boldsymbol{x}\right) \tag{3}$$

This would not work in the case of $\eta(\boldsymbol{x}) \in \{0, 1\}$, so instead we work with $1/2$. Which, together with the approximation of the Fisher-Information matrix with the empirical Hessian, we get:

$$\hat{\eta}(\boldsymbol{x}) \xrightarrow{D} \mathcal{N}\left(\frac{1}{2}, \; \frac{1}{16} \cdot \boldsymbol{x}^\top \hat{H}_n \boldsymbol{x}\right) \tag{4}$$

where $\hat{H}_n = (X^\top D X)^{-1}$, where D is a diagonal matrix, with $D_{ii} = \hat{\eta}_i(1 - \hat{\eta}_i)$, where $\hat{\eta}_i = \sigma(\boldsymbol{x}_i^\top \hat{\theta})$.

For the settings: $(\mathcal{D}, \mathcal{A}_{1/2}^k)$ and $(\mathcal{D}_\tau, \mathcal{A}_{1/2}^k)$, for an $\boldsymbol{x} \in \mathcal{A}_{1/2}^k$ we get: $\tilde{\eta}(\boldsymbol{x}) = \frac{1}{2}$. While for $(\mathcal{D}_{\alpha,\beta}, \mathcal{A}_{1/2}^k)$ we get: $\tilde{\eta}(\boldsymbol{x}) = \frac{1-\alpha+\beta}{2}$. Note that under $(\mathcal{D}_\tau, \mathcal{A}_{1/2}^k)$, we also have $\left(\tilde{\eta}(\boldsymbol{x})(1 - \tilde{\eta}(\boldsymbol{x}))\right)^2 = \frac{1}{16}$ similarly to $(\mathcal{D}, \mathcal{A}_{1/2}^k)$.

3.1   A Hypothesis Test for Class-Conditional Label Noise

We now define our null hypothesis ($\mathcal{H}_0$) and (implicit) alternative hypothesis ($\mathcal{H}_1$) as follows:

$$\mathcal{H}_0 : \alpha = \beta \quad \& \quad \mathcal{H}_1 : \alpha \neq \beta \tag{5}$$

Under the null and the alternative hypotheses, we have the following distributions for the estimated posterior of the anchor:

$$\mathcal{H}_0 : \hat{\eta}(\boldsymbol{x}) \sim \mathcal{N}\left(\frac{1}{2}, \ \frac{1}{16} \cdot \boldsymbol{x}^\top \hat{H} \boldsymbol{x}\right)$$

$$= \mathcal{N}\left(\frac{1}{2}, \ v(\boldsymbol{x})\right) \tag{6}$$

$$\mathcal{H}_1 : \hat{\eta}(\boldsymbol{x}) \sim \mathcal{N}\left(\frac{1+\alpha-\beta}{2}, \ \tilde{v}(\boldsymbol{x})\right) \tag{7}$$

where

$$\tilde{v}(\boldsymbol{x}) = \frac{\left((1-\alpha+\beta)(\beta-\alpha)\right)^2}{16} \cdot \boldsymbol{x}^\top \hat{H} \boldsymbol{x}$$

Level of Significance and Power of the Test. The *level of significance* (also known as Type I Error) is defined as follows:

$$a = \mathbb{P}(\text{reject } \mathcal{H}_0 \mid \mathcal{H}_0 \text{ is True}) \tag{8}$$

Rearranging Eq. 6 we get: $\frac{\hat{\eta}(\boldsymbol{x})-1/2}{\sqrt{v(\boldsymbol{x})}} \sim \mathcal{N}(0, \ 1)$, under the null. Which for a chosen level of *significance* (a) allows us to define a region of retaining the null $\mathcal{H}_0$. We let $z_{a/2}$ and $z_{1-a/2}$ denote the lower and upper critical values for retaining the null at a level of significance of a.

Retain $\mathcal{H}_0$ if:

$$z_{a/2} \cdot \sqrt{v(x)} + 1/2 \ \leq \ \hat{\eta}(x) \ \leq \ z_{1-a/2} \cdot \sqrt{v(x)} + 1/2 \tag{9}$$

Using the region of retaining the null hypothesis, we can now derive the *power* of the test.

Proposition 1. Power of the test *(See Appendix 8.3 for the full derivation.) Under the distributions for the estimated posterior under the null and alternative hypotheses in Eqs. 6 & 7, based on the definition of the hypotheses in Eq. 5, the test has power:* $\mathbb{P}(\text{reject } \mathcal{H}_0 \mid \mathcal{H}_0 \text{ is False}) = 1 - b_1$, *where:*

$$b_1 = \Phi\left(\frac{z \cdot \sqrt{v(x)} + \frac{\beta-\alpha}{2}}{\sqrt{\tilde{v}(x)}}\right) - \Phi\left(\frac{-z \cdot \sqrt{v(x)} + \frac{\beta-\alpha}{2}}{\sqrt{\tilde{v}(x)}}\right) \tag{10}$$

3.2 Multiple Anchor Points

In this section we discuss how the properties of the test change in the setting where multiple anchors points are provided.

Let $\hat{\eta}_i$ correspond to the ith instance in $\mathcal{A}_{1/2}^k$. Then for $\bar{\eta} = \frac{1}{k}\sum_{i=1}^{k} \hat{\eta}_i$ we have:

$$\bar{\eta} \sim \mathcal{N}\left(\frac{1}{2}, \ \frac{1}{16} \cdot \bar{\boldsymbol{x}}^\top H \bar{\boldsymbol{x}}\right)$$

where $\bar{\boldsymbol{x}} = \frac{1}{k}\sum_{i=1}^{k} \boldsymbol{x}_i$ with $\boldsymbol{x}_i \in \mathcal{A}_{1/2}^k \ \forall i$. For the full derivation see Appendix 8.4.

Anchors Chosen at Random. We have that $x \in \mathcal{A}^k_{1/2} \to x^\top \beta_0 = 0$, so that for an orthonormal basis U, $x = Ur$. Without loss of generalisation we let $U_{:,0} = \frac{\beta_0}{\|\beta_0\|_2}$, and therefore $\eta(x) = 1/2 \to r_0 = 0$. In words: $\forall x \in \mathcal{A}^k_{1/2}$ we have that x's component in the direction of β_0 is 0.

Now we make the assumption that x's are random with $r_j \sim \mathbb{U}([-c, \ c])$. Therefore, $\mathbb{E}r_j = 0$, and $\mathbb{V}r_j = \frac{c^2}{3}$. In the following we use the subscript S in the operator $\mathbb{E}_S$ to denote the randomness in choosing the set $\mathcal{A}$. In words: we assume that the set $\mathcal{A}^k_{1/2}$ is chosen uniformly at random from the set of all anchor points.

Combining these we get:

$$\mathbb{E}_S v(x) = \mathbb{E}_S x^\top H x = \mathbb{E}_S r^\top U H U^\top r$$
$$= \frac{dc^2}{3} \cdot tr(UHU^\top) = \frac{dc^2 q}{3}$$

where $q = tr(H)$. While for k anchor points chosen independently at random, we get:

$$\mathbb{E}_S v(\bar{x}) = \mathbb{E}_S \left[\frac{1}{k^2} \sum_{i,j}^{k} x_i^\top H x_j \right]$$

$$= \mathbb{E}_S \left[\frac{1}{k^2} \sum_{i,j}^{k} r_i^\top U H U^\top r_j \right]$$

$$= \frac{dc^2}{3k} \cdot tr(UHU^\top) = \frac{dc^2 q}{3k}$$

Following the same derivation as above we get:

$$b_k = \Phi \left(\frac{z \cdot \sqrt{v(\bar{x})} + \frac{\beta - \alpha}{2}}{\sqrt{\tilde{v}(\bar{x})}} \right) - \Phi \left(\frac{-z \cdot \sqrt{v(\bar{x})} + \frac{\beta - \alpha}{2}}{\sqrt{\tilde{v}(\bar{x})}} \right)$$

If we let $v = \mathbb{E}_S v(x)$ (similarly $\tilde{v} = \mathbb{E}_S \tilde{v}(x)$), then we have seen that $\mathbb{E}_S v(\bar{x}) = \frac{v}{k}$ (Reminder: expectations are with respect to the randomness in picking the anchor points). Then we have:

$$\frac{b_k}{b_1} = \frac{\Phi \left(\frac{z\sqrt{v} + h\sqrt{k}}{\sqrt{\tilde{v}}} \right) - \Phi \left(\frac{-z\sqrt{v} + h\sqrt{k}}{\sqrt{\tilde{v}}} \right)}{\Phi \left(\frac{z\sqrt{v} + h}{\sqrt{\tilde{v}}} \right) - \Phi \left(\frac{-z\sqrt{v} + h}{\sqrt{\tilde{v}}} \right)} \leq 1 \qquad (11)$$

with $h = \frac{\beta - \alpha}{2}$.

3.3   Multiple Relaxed Anchors-Points

In this section we see how the properties of the test change in the setting where the anchors do not have a perfect $\eta(\boldsymbol{x}) = 1/2$. We now consider the case of $\mathcal{A}^k_{1/2,\delta}$. Let $\boldsymbol{x}$ be such that $\eta(\boldsymbol{x}) = \frac{1}{2} + \epsilon$, where $\epsilon \sim \mathbb{U}([-\delta, \delta])$ with $0 < \delta \ll 1$. (Note: by definition $\delta \leq 1/2$.)

For one instance we have the following: $\mathbb{E}_{\hat{\theta}}\hat{\eta} = 1/2 + \epsilon,$ and $\mathbb{E}_S\mathbb{E}_{\hat{\theta}}\hat{\eta} = 1/2$

For the variance component we have: $(\hat{\eta}(1 - \hat{\eta}))^2 = \left(\left(\frac{1}{2} + \epsilon\right)\left(\frac{1}{2} - \epsilon\right)\right)^2 \approx \frac{1}{16} - \frac{\epsilon^2}{2}$, ignoring terms of order higher than ϵ^2, under the assumption that $\delta \ll 1$.

Under the *law of total variance* we have:

$$\mathbb{V}(\eta) = \mathbb{E}\left(\mathbb{V}\left(\eta \mid \epsilon\right)\right) + \mathbb{V}\left(\mathbb{E}\left(\eta \mid \epsilon\right)\right)$$

$$= \mathbb{E}\left(\left(\frac{1}{16} - \frac{\epsilon^2}{2}\right) \cdot \boldsymbol{x}^\top H \boldsymbol{x}\right) + \mathbb{V}\left(\frac{1}{k}\sum_{i=1}^{k}\hat{\eta}_i\right)$$

$$= \left(\frac{1}{16} - \frac{\delta^2}{6}\right) \cdot \boldsymbol{x}^\top H \boldsymbol{x} + \mathbb{V}\left(\frac{1}{2} + \frac{1}{k}\sum_{i=1}^{k}\epsilon_i\right)$$

$$= \left(\frac{1}{16} - \frac{\delta^2}{6}\right) \cdot \boldsymbol{x}^\top H \boldsymbol{x} + \frac{\delta^2}{3k} \tag{12}$$

For the full derivation see Appendix 8.6. Finally, bringing everything together and ignoring δ^2 terms we get:

$$\bar{\eta} \sim \mathcal{N}\left(\frac{1}{2}, \left(\frac{1}{16} - \frac{\delta^2}{6}\right) \cdot \bar{\boldsymbol{x}}^\top H \bar{\boldsymbol{x}}\right)$$

$$\approx \mathcal{N}\left(\frac{1}{2}, \frac{1}{16} \cdot \bar{\boldsymbol{x}}^\top H \bar{\boldsymbol{x}}\right)$$

3.4   What if We have No Anchor Points?

We have shown that we can relax the hard constraint on the anchor points to be exactly $\eta = 1/2$, to $\eta \approx 1/2$. It is natural then to ask if we need anchor points at all. If instead we were to sample points at random, then we would have the following: $\mathbb{E}_{p(X)}\eta(X) = \pi$. The importance of needing for set of anchor points, either $\mathcal{A}^k_{1/2}$ or $\mathcal{A}^k_{1/2,\delta}$, is that, the anchor points would be centered around a known value $1/2$, as opposed to having no anchor points and sampling at random, where the anchor points would end up being centered around π. Knowledge of the class priors could allow for a different type of hypothesis tests to asses the presence of label noise. We do not continue this discussion in the main document as it relies on different type of information, but provide pointers in the Appendix 7.8.

3.5   Practical Considerations and Limitations

Beyond Logistic Regression. Our approach relies on the asymptotic properties of MLE estimators, and specifically of Logistic Regression. More complex models can be constructed in a similar fashion through polynomial feature expansion. However the extension of these tests to richer model-classes, such as Gaussian Processes, remains open.

Multi-class Classification. Multi-class classification setting can be reduced to *one-vs-all, all-vs-all,* or more general error-correcting output codes setups as described in [23], which rely on multiple runs of binary classification. In these settings then we could apply the proposed framework. The challenge would then be how to interpret $\eta = 1/2$.

Finding Anchor Points. While it might not be straightforward for the user to provide instances whose true posterior is $\eta(x) = 1/2$, we do show how this can be relaxed, by allowing $\eta(x) \approx 1/2$. We then show how multiple anchor points can be stacked, improving the properties of the test.

Model Misspecification. Our work relies on properties of the MLE and its asymptotic distribution (Eq. 1). These assume the model is *exactly* correct. Similarly, under the null in the scenario of $\alpha = \beta > 0$, we are at risk of model misspecification. This is not a new problem for Maximum Likelihood estimators, and one remedy is the so-called *Huber Sandwich Estimator* [34] which replaces the Fisher Information Matrix, with a more robust alternative.

Instance-Dependent Noise (IDN). In IDN the probability of label flipping depends on the features. It can be seen as a generalisation over UN (which is unbiased under mild conditions (See Theorem 1) and CCN (where learning is in general biased). Our theoretical framework for CCN serves as a starting point to devise tests of IDN.

4   Related Work

Previous works have focused on the importance of (automatic) data preparation and data quality assessment [24,36–38]. These data quality measures refer to aspects such as the presence of noise in data, missing values, outliers, imbalanced classes, inconsistency, redundancy, timeliness and more [36,38]. Within this context, in this work we focus on label noise and, in particular, assessing the presence of class-conditional label noise, as opposed to uniform label noise. Related approaches include the identification specific corrupted instances, or distilled examples, and the direct estimation of the noise rates. These are discussed below.

Noisy Examples. As presented in [11,26], the aim is to identify the *specific* examples that have been inflicted with noise. This is a non-trivial task unless certain assumptions can be made about the per-class distributions, and their shape. For example, if we can assume that the supports of the two classes do not overlap (i.e. $\eta(x)(1 - \eta(x)) \in \{0, 1\}\ \forall x$), then we can identify mislabelled instances using per-class densities. If this is not the case, then it would be difficult to differentiate between a mislabelled instance and an instance for which $\eta(x)(1 - \eta(x)) \in (0,1)$. A different assumption could be uni-modality, which would again provide a prescription for identifying mislabelled instances through density estimation tools.

Distilled Examples. The authors in [16] go in the opposite direction by trying to identify instances that *have not been corrupted* → the *distilled examples.* As a first step the authors assume knowledge of an upper-bound[2] (Theorem 2 of [16]) which allows them to define sufficient conditions for identifying whether an instance is *clean.* As a second step they aim at estimating the (local) noise rate based on the neighbourhood of an instance (Theorem 3 of [16]).

Anchor Points and Perfect Samples. Finally, we can aim to directly estimate noise rates (or general distortions) while training [17,39]. A common approach is to proceed by correcting the loss to be minimised, by introducing the notion of a *mixing matrix* $M \in [0,1]^{c \times c}$, where $M_{i,j} = \mathbb{P}(\tilde{y} = e^j \mid y = e^i)$ [8]. Using these formulations, we are in a position where, if we have access to M, we can correct the training procedure to obtain an unbiased estimator. However, M is rarely known and difficult to estimate. Works on estimating M rely on having access to *perfect samples* and can be traced back to [3], and the idea was later adapted and generalised in [4,5,17] to the multi-class setting. Interestingly, in [1] authors do not explicitly define these perfect samples, but rather assume they do exist in a large enough (validation) dataset X' – obtaining good experimental results. Similarly, [18] also work by not explicitly requiring anchor points, but rather assuming their existence.

5   Experiments

In order to illustrate the properties of the tests, for the experiments we consider a synthetic dataset where the per-class distributions are Gaussians, with means $[1, 1]^\top$ and $[-1, -1]^\top$, with identity as scale. For this setup we know that anchor points should lie on the line $y = -x$, and draw them uniformly at random $x \in [-4, 4]$. We analyse the following parameters of interest:

1. $N \in [500, 1000, 2000, 5000]$: the training sample size.
2. $(\alpha - \beta) \in [-0.05, 0.10, 0.20]$: the difference between the per-class noise rates.
3. $k \in [1, 2, 4, 8, 16, 32]$: the number of anchor points.
4. $\delta \in [0, 0.05, 0.10]$: how relaxed the anchor points are: $\eta(x) \in [0.50 - \delta, 0.50 + \delta]$.

For all combinations of N and $(\alpha - \beta)$ we perform 500 runs. In each run, we generate a clean version of the data $\mathcal{D}$, and then proceed by corrupting it to

---

[2] The paper aims at tackling instance-dependent noise.

obtain a separate version: $\mathcal{D}_{\alpha,\beta}$. For both datasets, we fit a Logistic Regression model. We sample both the anchor points and relaxed anchor points. Finally, we then compute the z-scores, and subsequently the corresponding p-values[3].

The box-plots should be read as follows: $Q1$, $Q2$ & $Q3$ separate the data into 4 equal parts. The inner box starts (at the bottom) at $Q1$ and ends (at the top) at $Q3$, with the horizontal line inside denoting the median ($Q2$). The whiskers extend to show $Q1 - 1.5 \cdot IQR$, and $Q3 + 1.5 \cdot IQR$. IQR denotes the *Interquartile Range* and $IQR = Q3 - Q1$.

In Figs. 1, 2 and 3 we have the following: moving to the right we increase the relaxation of anchor points, and moving downwards we increase the training sample-size. On the subplot level, on the x-axis we vary the number of anchor points, and on the y-axis we have the p-values. In all subplots we indicate with a red dashed line the mark of 0.10, and with a blue one the mark of 0.05, which would serve as rejection thresholds for the null hypothesis.

The experiments are illustrative of the claims made earlier in the paper. Below we discuss the findings in the experiments and what they mean with regards to Type I and Type II errors. We discuss these points in two parts; we first discuss the effect on sample size (N), difference in noise rates ($|\alpha - \beta|$) and number of anchor points (k).

Size of Training Set (N). As the size of training set (N) increases, the power increases. This can be seen Figs. 1, 2 & 3. By moving down the first column, and fixing a value for k, where N increases, we see the range of the purple box-plots decreasing, and essentially a larger volume of tests falling under the cut-off levels of significance (red and blue dashed lines). This is expected given that the variance of the MLE $\hat{\theta}_{MLE}$ vanishes as N increases, as is seen in Eq. 1 and discussion underneath it.

Difference in noise rates ($|\alpha - \beta|$) As $|\alpha - \beta|$ increases, the power increases. This can be seen in Figs. 1, 2 & 3, by fixing a particular subplot in the first column (for example, top-left one), and a value for k, we see again that the volume moves down. As presented in Eq. 10, as $\beta - \alpha$ increases, the power also increases.

Number of Anchor Points (k). The same applies to the number of anchor points – as the number of anchor points (k) increases, the power of the test increases. This can be seen in all three figures by focusing in any subplot in the first column, and considering the purple box-plots moving to the right. In Eq. 11 we see effect of k on the power.

---

[3] What we have so far presented is aligned with the Neyman-Pearson theory of hypothesis testing. We have shown how to utilise anchor points to obtain the p-value – a continuous measure of evidence against the null hypothesis- and then leverage the implicit *alternative hypothesis* of class-conditional noise and a *significance level* to analyse the *power* of the test. In this case, the p-value is the basis of formal decision-making process of rejecting, or failing to reject, the null hypothesis. Differently, in Fisher's theory of significance testing, the p-value is the end-product [35]. Both the p-value and the output of the test can be used as part of a broader decision process that considers other important factors.

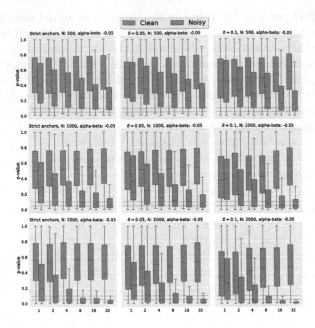

Fig. 1. Fixed $|\beta - \alpha| = 0.05$. Red dotted line at 0.10, and blue at 0.05. (Color figure online)

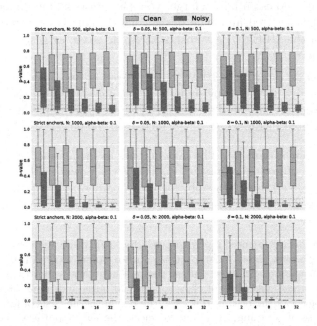

Fig. 2. Fixed $|\beta - \alpha| = 0.10$. Red dotted line at 0.10, and blue at 0.05. (Color figure online)

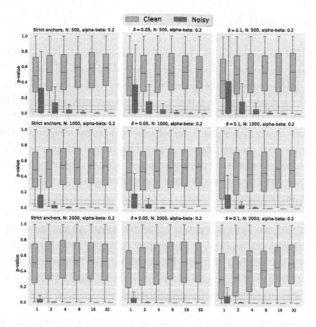

Fig. 3. Fixed $|\beta - \alpha| = 0.20$. Red dotted line at 0.10, and blue at 0.05. (Color figure online)

In all three discussions above we focused on the first column of each of the figures – which shows results from experiments on strict anchors. What we also observe in this case (the first column of all figures) is that the p-values follow the uniform distribution under the null (as expected, given the null hypothesis is true) – shown by the green box-plots. Therefore the portion of Type I Errors $= a$ (the level of significance Eq. 8). When we relax the requirements for strict anchors to allow for values close to $1/2$, we introduce a bias in the lower and upper bounds in Eq. 9 of $+\epsilon$. While $\mathbb{E}\epsilon = 0$ this shift on the boundaries of the retention region will increase Type I Error. On the other hand, in Eq. 12 we see how this bias decreases as you increase the number of anchor points. Both of these phenomena are also shown experimentally by looking at the latter two columns of the figures.

Anchor Point Relaxation (δ). Lastly, we examine the effect of relaxing the strictness of the anchors (δ), $\eta(x) \in [0.50 - \delta, \ 0.50 + \delta]$ on the properties of the test. As just discussed we see that as we increase the number of anchor points Type I Error decreases (volume of green box-plots under each of the cut-off points). We also observe that, as compared to only allowing strict anchors, the power is not affected significantly – with the effect decreasing as the number of anchor points increases. Furthermore, in the latter two columns we also observe the phenomena mentioned in the discussion concerning the first column only.

6   Conclusion and Future Work

In this work we introduce the first statistical hypothesis test for class-conditional label noise. Our approach requires the specification of anchor points, i.e. instances whose labels are highly uncertain under the true posterior probability distribution, and we show that the test's significance and power is preserved over several relaxations on the requirements for these anchor points. Our experimental analysis, which confirms the soundness of our test, explores many configurations of practical interest for practitioners using this test. Of particular importance for practitioners, since anchor specification is under their control, is the high correspondence shown theoretically and experimentally between the number of anchors and test significance.

Future work will cover both theoretical and experimental components. On the theoretical front, we are interested in understanding the test's value under a richer set of classification models, and further relaxing requirements on true posterior uncertainty for anchor points. Experimentally, we are particularly interested in applying the tests to diagnostically challenging healthcare problems and utilising clinical experts for anchor specification.

Acknowledgements. This work was funded by the UKRI Turing AI Fellowship EP/V024817/1.

References

1. Patrini, G., et al.: Making deep neural networks robust to label noise: a loss correction approach. In: Proceedings of the IEEE Conference on Computer Vision and Pattern Recognition (2017)
2. Wang, D., Cui, P., Zhu, W.: Structural deep network embedding. In: Proceedings of the 22nd ACM SIGKDD International Conference on Knowledge Discovery and Data Mining (2016)
3. Blanchard, G., et al.: Classification with asymmetric label noise: consistency and maximal denoising. Electron. J. Stat. **10**(2), 2780–2824 (2016)
4. Menon, A., et al.: Learning from corrupted binary labels via class-probability estimation. In: International Conference on Machine Learning. PMLR (2015)
5. Liu, T., Tao, D.: Classification with noisy labels by importance reweighting. IEEE Trans. Pattern Anal. Mach. Intell. **38**(3), 447–461 (2015)
6. Patrini, G.: Weakly supervised learning via statistical sufficiency (2016)
7. Van Rooyen, B.: Machine learning via transitions (2015)
8. Cid-Sueiro, J.: Proper losses for learning from partial labels. Adv. Neural Inf. Process. Syst. **25** (2012)
9. Cid-Sueiro, J., García-García, D., Santos-Rodríguez, R.: Consistency of losses for learning from weak labels. In: Calders, T., Esposito, F., Hüllermeier, E., Meo, R. (eds.) ECML PKDD 2014. LNCS (LNAI), vol. 8724, pp. 197–210. Springer, Heidelberg (2014). https://doi.org/10.1007/978-3-662-44848-9_13
10. Perelló-Nieto, M., Santos-Rodríguez, R., Cid-Sueiro, J.: Adapting supervised classification algorithms to arbitrary weak label scenarios. In: Adams, N., Tucker, A., Weston, D. (eds.) IDA 2017. LNCS, vol. 10584, pp. 247–259. Springer, Cham (2017). https://doi.org/10.1007/978-3-319-68765-0_21

11. Northcutt, C., Jiang, L., Chuang, I.: Confident learning: estimating uncertainty in dataset labels. J. Artif. Intell. Res. **70**, 1373–1411 (2021)
12. Frénay, B., Verleysen, M.: Classification in the presence of label noise: a survey. IEEE Tran. Neural Netw. Learn. Syst. **25**(5), 845–869 (2013)
13. Poyiadzi, R., et al.: The weak supervision landscape. In: 2022 IEEE International Conference on Pervasive Computing and Communications Workshops and other Affiliated Events (PerCom Workshops). IEEE (2022)
14. Ghosh, A., Manwani, N., Sastry, P.S.: Making risk minimization tolerant to label noise. Neurocomputing **160**, 93–107 (2015)
15. Van der Vaart, A.W.: Asymptotic Statistics, vol. 3. Cambridge University Press, Cambridge (2000)
16. Cheng, J., et al.: Learning with bounded instance and label-dependent label noise. In: International Conference on Machine Learning. PMLR (2020)
17. Perello-Nieto, M., et al.: Recycling weak labels for multiclass classification. Neurocomputing **400**, 206–215 (2020)
18. Xia, X., et al.: Are anchor points really indispensable in label-noise learning? Adv. Neural Inf. Process. Syst. **32**, 1–12 (2019)
19. Bedrick, E.J., Christensen, R., Johnson, W.: A new perspective on priors for generalized linear models. J. Am. Stat. Assoc. **91**(436), 1450–1460 (1996)
20. Greenland, S.: Putting background information about relative risks into conjugate prior distributions. Biometrics **57**(3), 663–670 (2001)
21. Gelman, A., et al.: A weakly informative default prior distribution for logistic and other regression models. Ann. Appl. Stat. **2**(4), 1360–1383 (2008)
22. Garthwaite, P.H., Kadane, J.B., O'Hagan, A.: Statistical methods for eliciting probability distributions. J. Am. Stat. Assoc. **100**(470), 680–701 (2005)
23. Dietterich, T.G., Bakiri, G.: Solving multiclass learning problems via error-correcting output codes. J. Artif. Intell. Res. **2**, 263–286 (1994)
24. Lawrence, N.D.: Data readiness levels. arXiv preprint arXiv:1705.02245 (2017)
25. Jindal, I., Nokleby, M., Chen, X.: Learning deep networks from noisy labels with dropout regularization. In: 2016 IEEE 16th International Conference on Data Mining (ICDM). IEEE (2016)
26. Northcutt, C.G., Wu, T., Chuang, I.L.: Learning with confident examples: rank pruning for robust classification with noisy labels. arXiv preprint arXiv:1705.01936 (2017)
27. Gebru, T., et al.: Datasheets for datasets. Commun. ACM **64**(12), 86–92 (2021)
28. Sokol, K., Santos-Rodriguez, R., Flach, P.: FAT forensics: a python toolbox for algorithmic fairness, accountability and transparency. arXiv preprint arXiv:1909.05167 (2019)
29. Zhao, L., Sukthankar, G., Sukthankar, R.: Incremental relabeling for active learning with noisy crowdsourced annotations. In: 2011 IEEE Third International Conference on Privacy, Security, Risk and Trust and 2011 IEEE Third International Conference on Social Computing. IEEE (2011)
30. Bacaicoa-Barber, D., Perello-Nieto, M., Santos-Rodríguez, R., Cid-Sueiro, J.: On the selection of loss functions under known weak label models. In: Farkaš, I., Masulli, P., Otte, S., Wermter, S. (eds.) ICANN 2021. LNCS, vol. 12892, pp. 332–343. Springer, Cham (2021). https://doi.org/10.1007/978-3-030-86340-1_27
31. Fergus, R., et al.: Learning object categories from google's image search. In: Tenth IEEE International Conference on Computer Vision (ICCV 2005), vol. 1. IEEE (2005)
32. Schroff, F., Criminisi, A., Zisserman, A.: Harvesting image databases from the web. IEEE Trans. Pattern Anal. Mach. Intell. **33**(4), 754–766 (2010)

33. Casella, G., Berger, R.L.: Statistical inference. Cengage Learning (2021)
34. Freedman, D.A.: On the so-called "Huber sandwich estimator" and "robust standard errors". Am. Stat. **60**(4), 299–302 (2006)
35. Perezgonzalez, J.D.: Fisher, Neyman-Pearson or NHST? a tutorial for teaching data testing. Front. Psychol. **6**, 223 (2015)
36. Gupta, N., et al.: Data quality toolkit: automatic assessment of data quality and remediation for machine learning datasets. arXiv preprint arXiv:2108.05935 (2021)
37. Afzal, S., et al.: Data readiness report. In: 2021 IEEE International Conference on Smart Data Services (SMDS). IEEE (2021)
38. Corrales, D.C., Ledezma, A., Corrales, J.C.: From theory to practice: a data quality framework for classification tasks. Symmetry **10**(7), 248 (2018)
39. Chu, Z., Ma, J., Wang, H.: Learning from crowds by modeling common confusions. In: AAAI (2021)

On the Prediction Instability of Graph Neural Networks

Max Klabunde[✉][iD] and Florian Lemmerich[iD]

Faculty of Computer Science and Mathematics, University of Passau,
Passau, Germany
{max.klabunde,florian.lemmerich}@uni-passau.de

Abstract. Instability of trained models, i.e., the dependence of individual node predictions on random factors, can affect reproducibility, reliability, and trust in machine learning systems. In this paper, we systematically assess the prediction instability of node classification with state-of-the-art Graph Neural Networks (GNNs). With our experiments, we establish that multiple instantiations of popular GNN models trained on the same data with the same model hyperparameters result in almost identical aggregated performance, but display substantial disagreement in the predictions for individual nodes. We find that up to 30% of the incorrectly classified nodes differ across algorithm runs. We identify correlations between hyperparameters, node properties, and the size of the training set with the stability of predictions. In general, maximizing model performance implicitly also reduces model instability.

Keywords: Prediction churn · Reproducibility · Graph neural networks

1 Introduction

Intuitively, if we fit any machine learning model with the same hyperparameters and the same data twice, we would expect to end up with the same fitted model twice. However, recent research has found that due to random factors, such as random initializations or undetermined orderings of parallel operations on GPUs, different training runs can lead to significantly different predictions for a significant part of the (test) instances, see for example [2,17,21]. This *prediction instability* (also called *prediction differences* or *prediction churn*, see Fig. 1) is undesirable for several reasons, including reproducibility, system reliability, and potential impact on user experience. For example, if a service is offered based on some classification of users with regularly retrained models, prediction instability can lead to fluctuating recommendations although there was no change of the users. Furthermore, if only one part of a machine learning system

Supplementary Information The online version contains supplementary material available at https://doi.org/10.1007/978-3-031-26409-2_12.

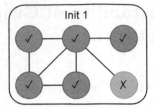 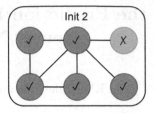

Fig. 1. An example of prediction instability. Green nodes (✓) denote correct predictions, red ones (✗) false predictions. The two sparsely connected nodes on the right are predicted differently depending on the initialisation. Although the performance is identical between runs, one third of the predictions are different. (Color figure online)

is retrained, the subsequent parts may not be able to adapt to the difference in predictions and overall system performance deteriorates unpredictably despite improvement of the retrained model [9]. Finally, the reproducibility of individual predictions is important in critical domains such as finance or medicine, in which recommendations reliant on (for example) random initializations might not be acceptable.

Due to this importance, there has been a recent surge of work studying the prediction instability of machine learning models [2,6,9,12,15,17,21]. However, research on the instability of models in graphs/network settings, such as node classification, has received little attention so far. As an exception, two recent studies [14,19] assessed the stability of unsupervised node embedding methods, mainly from a geometrical perspective. An evaluation of state-of-the-art supervised node classification algorithms based on Graph Neural Networks (GNNs) has—to the best of the authors' knowledge—not yet been performed.

This paper aims to fill this research gap by presenting an extensive and systematic experimental evaluation of the prediction instability of graph neural networks. In addition, we set out to understand how design, data, and training setup affect prediction stability.[1] In more detail, we summarize the major contributions of our paper as follows:

1. We demonstrate that the popular Graph Convolutional Networks [7] and Graph Attention Networks [18] exhibit significant prediction instability (Sect. 3.1). As a key result, we establish that while the aggregated accuracy of the algorithms is mostly stable, up to a third of incorrectly classified nodes differ between training runs of a model.
2. We empirically study the influence of node properties (Sect. 3.2), model hyperparameters and the training setup on prediction instability (Sect. 3.3). We find that nodes that are central in the network are less likely to be unstably predicted. High width, L2 regularization, low dropout rate, and low depth all show a tendency to help decrease prediction instability.

---

[1] Code and supplementary material are available at https://github.com/mklabunde/gnn-prediction-instability.

3. By introspecting individual deep GNN models with centered kernel alignment [8], we discover a trend that deeper layers (closer to the output) are less stable (Sect. 3.4).

In general, we find that the objectives of maximizing performance and minimizing prediction instability almost always align. Our results have direct implications for practitioners who seek to minimize prediction instability, such as in high-stakes decision-recommendation scenarios.

2   Preliminaries and Experimental Setup

This section introduces our problem setting and describes the models, datasets, and instability measures used in our study.

Multiclass Node Classification. We focus on the multiclass classification problem on graphs $G = (V, E)$, where every node $v \in V$ has some features $x \in \mathcal{X}$ and a one-hot encoded label $y \in \{0, 1\}^C$ with C the number of different classes, and E is the set of edges. We only consider the transductive case, in which all edges and nodes including their features are known during training, but only a subset of node labels is available.

Graph Neural Networks. Graph Neural Networks [13] operate by propagating information over the graph edges. For a specific node, the propagated information from neighboring nodes is aggregated and combined with its own representation to update its representation. For node classification, the representations of the last layer can be used to predict the node labels.

In particular, we study in this paper two of the currently most popular state-of-the-art models for node classification: Graph Convolutional Networks (GCN) [7] and Graph Attention Networks (GAT) [18]. GAT and GCN differ, as the aggregation mechanism of GCN uses a static normalization based on the degree of nodes, whereas GAT aggregation employs a trained multi-head attention mechanism. We select the hyperparameters as stated in their respective papers. The hidden dimension is 64 for GCN, GAT uses 8 attention heads with 8 dimensions each. The models have two layers, with the second layer producing the classification output. For the final GAT layer, the outputs of the different heads are summed up, in contrast to concatenation in the earlier layer. In addition to the convolutional layers, we use dropout on the input and activations of the first layer with $p = 0.6$. We apply the same dropout rate to the attention weights in the GAT layer. GCN uses ReLU activation, and GAT uses ELU. For more details on the investigated models, we refer to the original publications.

Training Procedure. We train the models for a maximum of 500 epochs with an early stopping period of 40 epochs on the validation loss using the Adam optimizer with a learning rate of 0.01. In all cases, we use full batch training. For each dataset, we train 50 models with different initializations. We keep all other known sources of randomness constant including low-level operations.

Datasets. We focus on the node classification task and use the following publicly available standard datasets: CiteSeer and Pubmed [20], Coauthor CS and Physics [16], Amazon Photo and Computers [16], and WikiCS [11]. Since the datasets from [16] do not have public train/validation/test splits, we create splits for them by randomly taking 20 nodes from each class for training and using 500 nodes as validation data. The rest are used for testing. All datasets are treated as undirected graphs. While this could affect performance negatively, e.g., when a class depends on the ratio of incoming to outgoing edges, we follow the common approach in literature [7,16] and note that disregarding directions lead to improved performance in preliminary experiments.

Measuring Prediction Instability. To quantify prediction instability, we now define several measures that capture differences in the model output. These measures have two main differences: They either use the predicted labels as input, i.e., the argmax of the model output, or the softmax-normalized output. While the first approach directly follows intuition and has the advantage of being interpretable, it has the downside of using the discontinuous argmax, which means that even slight differences in model output can lead to different outcomes.

We follow the definition of Madani et al. [10] and define the (expected) prediction disagreement as follows:

$$d = \mathbb{E}_{x,f_1,f_2} \mathbb{1}\{\arg\max f_1(x) \neq \arg\max f_2(x)\} \tag{1}$$

where $f_i \in F$ are instantiations of a model family F and $\mathbb{1}\{\cdot\}$ is the indicator function. The disagreement is easily calculated in practice by training a number of models and then averaging the pairwise disagreement of their predictions of individual nodes. This measure is also known as churn [1,2,6,12] and jitter [9].

The theoretically possible value of the disagreement of a pair of model instantiations is bounded by their performance. For example, when two models, f_1 and f_2, perform with 95% accuracy, then the minimal disagreement equals zero, which occurs when the predictions are identical. Maximal disagreement occurs when 90% of predictions are identical, and f_1 is correct on the 5% of data where f_2 is incorrect. In general, it holds [2]:

$$|Err_{f_1} - Err_{f_2}| \leq d_{f_1,f_2} \leq \min(1, Err_{f_1} + Err_{f_2}), \tag{2}$$

where Err is the error rate of a model and d_{f_1,f_2} is the empirical disagreement between f_1 and f_2. While two models with high error rates do not necessarily have large disagreement, we later show that disagreement and error rate are in fact highly correlated, a finding which has not received much attention so far.

Disagreement is an intuitive and straightforward measure of disagreement. However, to better understand the relationship between disagreement and error rate, we define the min-max normalized disagreement, which gives the disagreement relative to its minimal and maximal possible values:

$$d_{norm} = \mathbb{E}_{f_1,f_2} \left[\frac{d_{f_1,f_2} - \min d_{f_1,f_2}}{\max d_{f_1,f_2} - \min d_{f_1,f_2}} \right]. \tag{3}$$

A natural extension of the aforementioned measures is to condition the computation on specific subgroups of predictions, e.g., the correct or incorrect predictions. As Milani Fard et al. [12], we define those as true disagreement d_{True} and false disagreement d_{False}, respectively:

$$d_{True} = \mathbb{E}_{(x,y),f_1,f_2} \mathbb{1}\{\arg\max f_1(x) \neq \arg\max f_2(x) | \arg\max f_1(x) = y\}. \quad (4)$$

d_{False} is computed analogously. False disagreement is less reliant on model performance as it is always possible that incorrectly predicted nodes are predicted differently in another run if the second model has equal or lower performance. False disagreement and normalized disagreement are especially important to disentangle model performance and stability since model hyperparameters, such as width, may affect performance and disagreement jointly.

All of the above measures work on the hard predictions of the models. To measure the difference between the output distributions, we use the mean absolute error, where C is the number of classes: $d_{MAE} = \mathbb{E}_{x,f_1,f_2} \left[\frac{1}{C} \|f_1(x) - f_2(x)\|_1 \right]$.

3   Results

This section presents our main results on the instability of GNNs. We showcase overall results before introducing detailed analyses on the effect of node properties and model design. Finally, we describe results of inspecting the instability of GNNs layer-by-layer. We always report results on examples not seen during training. For all experiments, we only show a subset of the results due to space constraints. The complete results can be viewed in the supplementary material (See footnote 1).

3.1   Overall Prediction Instability of GNNs

We now demonstrate the prediction instability of GAT and GCN on several well-known datasets.

We show the results in Table 1. The prediction disagreement d is between three and four percent, with the exceptions of CiteSeer and Computers, where the disagreement is almost ten percent. Models with higher accuracy tend to have lower disagreement, but the datasets must be taken into account. For example, the classification accuracies are higher on the Computers dataset compared to Pubmed, but disagreement is higher as well. We find that d_{norm} is lower than 25%, which means that disagreement is relatively close to the minimum that variation in model performance allows. Interestingly, nodes that are falsely predicted by one model have a high probability of being predicted differently by another model. False disagreement is almost always at least one magnitude larger than true disagreement, which is partially explained by the high performance of the models. Finally, the mean absolute error reveals that the predicted probabilities of classes are not much different between models. Overall, GNNs clearly demonstrate prediction instability to a significant degree.

In the following, we focus in our discussion on the prediction disagreement measure d. Typically, identical tendencies can be observed for the other measures.

Table 1. Prediction disagreement (in %) and its standard deviation.

Dataset	Model	Accuracy	d	d_{norm}	d_{True}	d_{False}	MAE
CiteSeer	GAT	69.0 ± 1.0	10.5 ± 1.7	15.4 ± 2.5	5.2 ± 1.4	22.3 ± 3.8	3.4 ± 0.6
	GCN	69.2 ± 0.7	7.1 ± 1.0	10.3 ± 1.6	3.5 ± 0.9	15.1 ± 2.4	2.9 ± 0.3
Pubmed	GAT	75.7 ± 0.6	3.7 ± 1.4	6.4 ± 2.7	2.4 ± 1.0	8.0 ± 3.3	2.3 ± 0.7
	GCN	76.8 ± 0.5	2.4 ± 0.7	4.1 ± 1.4	1.5 ± 0.6	5.6 ± 2.2	2.5 ± 1.0
CS	GAT	90.7 ± 0.5	3.7 ± 0.5	17.3 ± 2.0	1.7 ± 0.4	22.0 ± 3.6	0.7 ± 0.1
	GCN	90.7 ± 0.5	3.3 ± 0.6	15.4 ± 2.7	1.6 ± 0.5	19.9 ± 4.1	0.7 ± 0.2
Physics	GAT	92.0 ± 0.7	3.8 ± 0.8	19.7 ± 4.2	1.8 ± 0.6	25.7 ± 6.4	2.0 ± 0.4
	GCN	92.7 ± 0.3	1.6 ± 0.4	8.6 ± 2.7	0.8 ± 0.3	12.2 ± 4.3	1.2 ± 0.4
Computers	GAT	81.0 ± 1.5	9.5 ± 2.2	21.6 ± 5.6	4.8 ± 1.8	29.6 ± 7.3	2.3 ± 0.5
	GCN	81.2 ± 0.9	9.9 ± 1.9	24.2 ± 4.9	4.8 ± 1.3	31.9 ± 6.0	2.3 ± 0.4
Photo	GAT	90.3 ± 0.8	4.4 ± 1.1	18.9 ± 4.9	2.0 ± 0.8	26.0 ± 6.9	1.5 ± 0.3
	GCN	90.8 ± 0.5	3.7 ± 0.8	17.5 ± 3.7	1.6 ± 0.5	24.1 ± 5.5	1.4 ± 0.3
WikiCS	GAT	79.6 ± 0.3	3.8 ± 0.5	8.6 ± 1.3	1.7 ± 0.3	11.7 ± 1.8	0.9 ± 0.1
	GCN	79.4 ± 0.2	3.3 ± 0.4	7.6 ± 1.0	1.6 ± 0.3	10.1 ± 1.4	0.7 ± 0.1

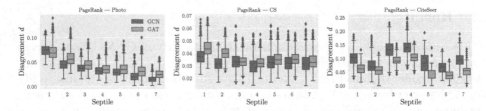

Fig. 2. Prediction disagreement for PageRank septiles on selected datasets. Low septile index corresponds to low Pagerank. On most datasets, we see that low PageRank nodes are less stably predicted than high PageRank nodes. However, on CiteSeer the trend is unclear.

3.2   The Effect of Node Properties

Next, we examine the unveiled prediction instability in more detail and set them in relation to data properties. We consider four node properties: i) PageRank, ii) clustering coefficient, iii) k-core, and iv) class label. The first three are related to the graph structure, whereas the class label is related to underlying node features. PageRank measures the centrality of a node, the clustering coefficient the connectivity of a node neighborhood, and the k-core of a node is the maximal k, for which the node is part of a maximal subgraph containing only nodes with a degree of at least k. Thus, the k-core gives an indication of both connectivity and centrality. See the supplementary material for formal definitions.

For the structural properties, we divide the nodes into seven equal-sized parts (septiles) with respect to the analyzed property. Then, we record prediction instability and model performance for each of these subgroups of the data. We use the same models as in the previous section.

We give a representative overview of prediction disagreement in relation to the values of structural properties in Fig. 2. Central nodes show lower prediction

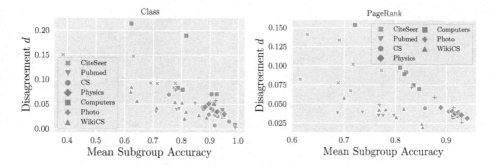

Fig. 3. Relation between subgroup accuracy and prediction disagreement for GAT with respect to the node classes (left) and PageRank septiles (right). The higher the accuracy, the lower the disagreement. The correlation between accuracy and disagreement suggests that central nodes are stably predicted due to high model performance on that subgroup of the data. Results for the other properties and GCN are similar.

instability. Low clustering negatively impacts stability, but there is no consistent relation over all datasets for other septiles. A high core number reduces the risk of prediction instability. For all properties, the results depend on the dataset to some extent. We attribute many of the differences in prediction instability to differences in model performance in different subgroups (Fig. 3). Now, we describe the results in more detail.

Structural Properties. Nodes with higher PageRank have lower prediction disagreement on all datasets except Pubmed. However, the magnitude of the difference varies depending on the dataset. Interestingly, the disagreement of falsely predicted nodes is roughly constant in many cases. This also holds for normalized disagreement, which shows that differences between subgroups can be explained to a large degree by differences in accuracy. The MAE of the output distributions almost always decreases with higher PageRank.

There is no consistent relationship between the clustering coefficient and the prediction instability in our results. The only common trend is that low clustering is correlated with high prediction disagreement. Higher clustering coincides with lower prediction disagreement on WikiCS and CS, Physics and Computers display an U-like relationship. On Photo, there is no clear trend. For Pubmed and CiteSeer, more than 65% of the nodes have a clustering coefficient of zero, so a comparison of equal bins is not possible. Similar to PageRank, false and normalized disagreement are almost constant in many cases, which again highlights that the prediction disagreement is closely related to the model performance.

On all datasets, the group with the highest k-core has the lowest prediction disagreement, and the group with the lowest k-core has the highest disagreement. In between, we do not observe a clear trend. Furthermore, the variance of the disagreement decreases with increasing k-core. Again, false disagreement and normalized disagreement do not show large differences.

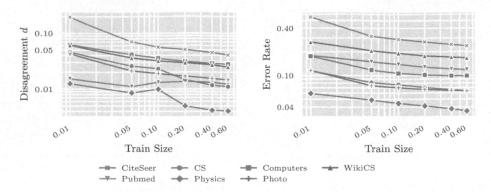

Fig. 4. Effect of training data on disagreement (left) and error rate (right) for GCN.

Class Label. There are large differences between classes with respect to both the average disagreement and its variance. Differences shrink for false disagreement and when normalizing for accuracy. MAE behaves similar to disagreement.

Classes with few examples are not less stable than large classes. As shown in Fig. 3, the average accuracy explains much of the differences in disagreement. Interestingly, the variance of the accuracy does not impact the prediction disagreement, although it affects the lower bound of prediction disagreement, which depends on the performance difference of two models.

3.3   The Effect of Model Design and Training Setup

In the previous section, we find evidence that prediction stability is related to model performance. Model design and training setup reasonably influence model performance (and thus may impact prediction stability), but how exactly they correlate with prediction stability and if they influence stability beyond the performance is unclear. We test the influence of individual hyperparameters on prediction stability by following the training protocol of Sect. 2, but changing one hyperparameter per experiment, if not specified otherwise. Since including standard deviations decreases readability of the visualizations, we focus on the mean value here and refer to the supplementary material for more details.

Training Data. To analyze the effect of training data availability on prediction stability, we vary the number of node labels available for training between 1 and 60% of all nodes and use a fixed-size validation set of 15% of the data. We sample the nodes of each class proportionally to their total class size, so that nodes of all classes are present in the test set. Further, to avoid dependency on specific data splits, we repeat the experiment with 10 different data splits per graph. In total, we train 42000 models for this experiment.

Results are shown on a log-log scale in Fig. 4. We find that both the disagreement and the error rate decrease significantly with increasing available training data. This underlines the correlation between model performance and

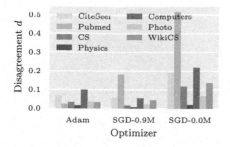
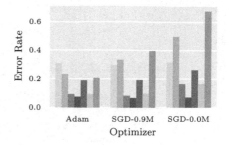

Fig. 5. Prediction disagreement (left) and error rate (right) for training GCN with Adam, SGD with momentum of 0.9, and SGD without momentum.

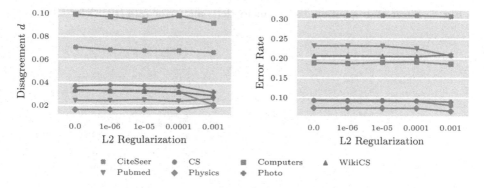

Fig. 6. Effect of L2 reg. on disagreement (left) and error rate (right) for GCN.

disagreement. The same trends can also be observed for the other measures of disagreement. Only Pubmed shows significantly different behaviour; that is, the disagreement does not decrease. GAT results are highly similar and also show a smooth decrease in disagreement and error rate.

Optimizer. We train the models with different optimizers: Adam, Stochastic Gradient Descent (SGD), and SGD with momentum (SGD-M), which we set to 0.9. We show the results for GCN in Fig. 5. For prediction disagreement, SGD performs much worse than SGD-M and Adam. SGD-M performs on par with or better than Adam, with the exception of Pubmed. Overall, disagreement and error rate are correlated, but SGD-M optimization leads to lower disagreement. Furthermore, SGD-M decreases the average MAE between the output distributions of the models more than Adam. For GAT, Adam and SGD-M perform similarly with a slight edge to Adam. SGD performs much worse with respect to both error rate and disagreement. Based on the discrepancy between GCN and GAT results, there does not appear to be a simple rule to select one of the tested optimizers to generally minimize prediction disagreement.

L2 Regularization. We show the results of GCN for varying L2 regularization in Fig. 6. Disagreement decreases only slightly with moderate regularization. We

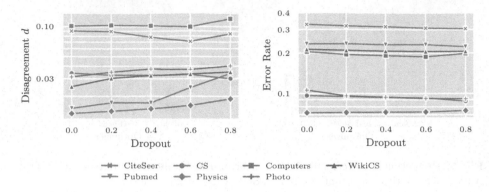

Fig. 7. Effect of dropout on disagreement (left) and error rate (right) for GCN.

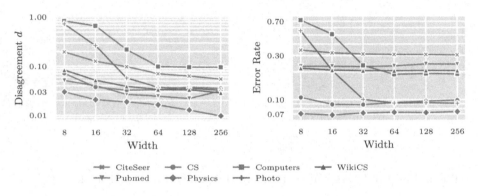

Fig. 8. Effect of width on disagreement (left) and error rate (right) for GCN.

observe the largest changes, both for the disagreement and error rate, with the maximal value of L2 regularization. Strong regularization reduces disagreement (all measures) on 5 out of 7 datasets compared to without regularization, though changes are small. Interestingly, disagreement decreases even when the error rate stays roughly constant.

Dropout. We show the results for varying dropout rates in Fig. 7. Large dropout rates increase disagreement for 6 of the 7 datasets for GCN. In contrast to previous observations, change in disagreement does not follow the error rate. Instead, the more dropout, the more prediction disagreement in most cases. Although the effect is small in absolute terms, dropout influences prediction stability negatively. Even so, a finely tuned dropout rate can improve disagreement in some cases while also decreasing the error rate.

Width. We vary the width of the models on a logarithmic scale between 8 and 256. GAT always has eight attention heads, which means that the dimension per head varies between 1 and 32. Figure 8 shows the absolute disagreement d and the error rate in relation to the width for GCN. Wider models have less

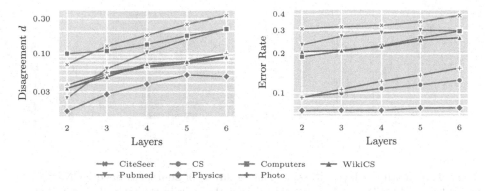

Fig. 9. Effect of depth on disagreement (left) and error rate (right) for GCN.

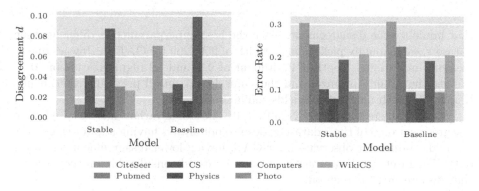

Fig. 10. Comparison of baseline GCN model and a stable variant, which implements all hyperparameters as suggested by the previous experiments.

prediction disagreement, which holds even for models for which the error rate does not decrease. This relation is mirrored in all stability measures.

Depth. We change the number of layers from 2 to 6. Between every layer, there are dropout and activation functions, while otherwise following the previously used training procedure. Figure 9 shows the results for GCN. Prediction disagreement increases with depth of the model. Similarly, the error rate increases, which can be explained by a lack of training techniques for deep GNNs, e.g., residual connections or normalization. Nevertheless, even when the model performance does not decrease much, prediction stability decreases, e.g., on the Physics dataset, the absolute disagreement increases almost four-fold. We make the same observation for GAT, which suggests that the depth of a model negatively affects its prediction stability.

Combining Optimal Hyperparameters. To test whether the observations so far can inform model selection, we now train "stable variants" of GAT and GCN and compare them with the baseline models, as described in Sect. 2. We select hyperparameters of the models according to the previous experiments, i.e., those

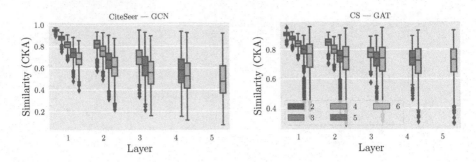

Fig. 11. Similarity of layers to corresponding layers of another trained GCN on Cite-Seer (left) and GAT on CS (right). The deeper the layer in the trained model, the less similar they are.

that minimize the disagreement. Since the best hyperparameters differ between datasets, we manually pick them as width of 256 (200 for GAT on Physics due to memory limitations), depth of 2, dropout of 0.2, and L2 regularization of 10^{-4}. We use the Adam optimizer and the same data as in all previous experiments. Using SGD with momentum yields similar results.

We show the results for GCN in Fig. 10. The stable variant has less prediction disagreement on 6 of the 7 datasets, despite not always having a lower error rate. We make a different observation for GAT, having lower disagreement on only 2 of the 7 datasets as the average error rate increased considerably on the datasets that disagreement is high on.

3.4   Layer-Wise Model Introspection

In the last part of our analysis, we aim to obtain a better understanding about where in the deep neural architecture instability primarily arises. For that purpose, we investigate the (in-)stability of internal representations with centered kernel alignment (CKA) [8] to measure the similarity of representations from corresponding layers in different models, see below for a slightly more extensive description. That is, we compare layer 1 of model A with layer 1 of model B, layer 2 of A with layer 2 of B, etc. We focus on the similarity of models with varying depths and train the models according to Sect. 2, but again vary the number of layers and add dropout layers between them.

Centered Kernel Alignment. CKA is a state-of-the-art method for measuring the similarity of neural network representations. Roughly speaking, CKA compares two matrices of pairwise similarities by vectorizing them and calculating the dot product. We use the linear variant of CKA, i.e., the pairwise similarities are calculated with the dot product, since it is efficient and other variants, such as using the RBF kernel for similarity computation, do not show better performance consistently. For details, we refer to the original publication [8] and the supplementary material.

Results. Figure 11 shows some exemplary results. In these plots, each color of boxplots refers to one model with a specific number of layers while the different groups of boxplots from left to right refer to the position of the layer within the model architecture. In general, the more layers the model has, the lower the similarity of the layers. Moreover, the deeper the layer (closer to the output), the lower the similarity. Similarly, the variance of the similarity increases with depth. Some outliers exist, for which changes in similarity between layers are small, or the similarity increases with depth (GCN WikiCS and Computers, 6 layer GAT on Pubmed). Overall, however, deep GNNs have more variance in their internal structure and representations.

Although the first layers may suffer from vanishing gradients, they are much more self-similar than the deeper layers, which should receive much larger updates. On the one hand, these large updates could make deep layers less similar as they may need to adapt to varying outputs of the earlier layers. On the other hand, the first layers are extremely similar, although they start from a random initialization. As a consequence, these layers provide very similar representations to deep layers, questioning why the deep layers are dissimilar. We leave a more detailed analysis for future work and hope that this observation sparks further research into the learned representations of GNNs.

4   Discussion

We discuss limitations and implications of our work, as well as future work.

Limitations. The models we study do not use popular techniques for deep GNNs, such as normalization or residual connections. Furthermore, we avoid mini-batching and distributed training. Although relatively shallow GNNs work well on many tasks, recent work introduces new benchmarks that benefit greatly from more complex models [3,5]. Therefore, interesting future work would be to explore how these techniques, combined with larger models and larger graphs, affect prediction stability.

We find statistical relationships between model hyperparameters and prediction stability. However, it is not transparent how different aspects, such as model performance, model hyperparameters, and prediction stability, causally influence each other. Attribution of changes to specific variables is difficult; hence, we only propose heuristics on how to select hyperparameters that minimize prediction instability. However, as our experiments show, training a model with hyperparameters jointly selected according to these rules does decrease prediction instability if care is taken with respect to performance. Causal attribution and consequent robust rules for model selection with respect to prediction stability is another avenue for future work.

Dataset Dependency. Repeatedly, models behave differently on the Pubmed dataset compared to the others. This suggests that the dataset plays a crucial role in determining prediction stability. We did not identify a single property of the dataset that sensibly explains the effect, which highlights the opportunity of examining prediction stability from the data perspective.

Implementation as a Source of Instability. We investigated the sources of instability by doing additional experiments with fixed random seeds and models trained on GPU vs CPU. We do not restrict the implementation to deterministic low-level algorithms, see the supplementary material for full results. We find that with a fixed random seed on the exact same data, GCN behaves completely deterministic while GAT in some cases still exhibits considerable instability. Surprisingly, even when GAT is trained on a CPU with fixed initialization and training, minor instability remains. Since the instability does not show all the time, we speculate that small implementations issues could be hidden initially and only influence stability later, e.g., after data updates. Overall, GPU instabilities are much smaller compared to differences introduced by changing initialization. This is good news from a pure reproducibility perspective, but we consider the instabilities established and analyzed in this paper still as crucial in many practical scenarios since they emphasize the sensitivity of predictions on implementation details.

Influence on Model Selection. Our results have direct implications on model selection. If we have to decide between multiple models that perform equivalently, and we are interested in minimizing prediction instability, then we can select the model with higher width and L2 regularisation, and lower dropout rate and depth. While this may not be a straightforward decision, as tradeoffs between different variables have to be made, the proposed rule can be a rough guide.

5   Related Work

Our work is related to previous research that we outline in the following section.

Stability of Node Embeddings. Wang et al. [19] and Schumacher et al. [14] study the influence of randomness on unsupervised node embeddings. These embeddings, mainly computed via random walk-based models or matrix factorization, capture some notion of proximity of nodes, which should then be reflected in the geometry of the embedding space. They both measure large variability in the geometry of the embedding spaces, e.g., in the nearest neighbors of embeddings. Schumacher et al. find that the aggregated performance of downstream models does not change much, but individual predictions vary. Wang et al. further demonstrate that less stable nodes are less likely to be predicted correctly. In contrast to their work, we focus on supervised GNNs and prediction instability instead of geometrical instability.

Impact of Tooling. Zhuang et al. [21] find that prediction instability arises along the entire stack of software, algorithm design, and hardware. Modifying model training to be perfectly reproducible incurs highly variable costs, in some cases more than tripling the computation time. Moreover, they observe that subgroups of the data are affected to different extents from random factors in training. Introducing batch normalization reduces performance variability but increases prediction disagreements. They focus mainly on large CNNs, whereas we study comparatively much smaller GNNs.

Increasing Prediction Stability. Several recent works successfully introduce techniques to increase prediction stability. For example, by regularizing labels [1, 12], distillation [2,6], ensembling techniques [15,17], or data augmentation [17]. It is noted that these techniques sometimes also increase model performance, but a general relationship between prediction stability and model performance is not highlighted. For the CNNs in the work by Summers and Dineen [17], even single bit changes lead to significantly different models.

Model Influence. Liu et al. [9] study how data updates affect prediction stability in the domain of language processing. Moreover, they compare whether model architecture, model complexity, or usage of pretrained word embeddings improve stability. They identify a trade-off between prediction stability and model performance. In our experiments, the trade-off is small or nonexistent, validating experiments of prediction stability in different domains.

GNN Robustness. The stability of GNNs can be viewed from a different perspective: adversarial unnoticable perturbations of the graph data can significantly reduce model performance and thus prediction stability [22]. Zügner and Günnemann [23] propose a method to certify robustness of nodes against such attacks. Further, stochastic perturbations in the graph structure can lead to instability of predictions. Gao et al. [4] show that increased width and depth increase possible changes of outputs of GNNs. In this paper, we assume that only the initialization changes, i.e., there are no changes to the graph.

6   Conclusion

In this paper, we systematically assessed the instability of Graph Neural Network predictions with respect to multiple aspects: random initialization, model architecture, data, and training setup. We found that up to 30% of the falsely predicted nodes are different between training runs that use the same data and hyperparameters but change the initialization. Nodes on the periphery of a graph are less likely to be stably predicted. Furthermore, models with higher width, higher L2 regularization, lower depth, and a lower dropout rate are more stable in their predictions. Instability of deep GNNs is reflected in their internal representations. Finally, maximizing model performance almost always implicitly minimizes prediction instability.

Future work may study prediction instability of GNNs from the perspective of larger, more complex models or data properties. Furthermore, finding clear causal relationships may be beneficial to select models that are more stable with respect to their predictions. Lastly, it would be interesting to see whether existing techniques aiming to reduce model instability for other types of models perform well for GNNs.

References

1. Bahri, D., Jiang, H.: Locally adaptive label smoothing improves predictive churn. In: ICML (2021)

2. Bhojanapalli, S., et al.: On the reproducibility of neural network predictions. arXiv preprint arXiv:2102.03349 (2021)
3. Dwivedi, V.P., Joshi, C.K., Laurent, T., Bengio, Y., Bresson, X.: Benchmarking graph neural networks. arXiv preprint arXiv:2003.00982 (2020)
4. Gao, Z., Isufi, E., Ribeiro, A.: Stability of graph convolutional neural networks to stochastic perturbations. Signal Process. **188**, 108216 (2021)
5. Hu, W., et al.: Open graph benchmark: Datasets for machine learning on graphs. arXiv preprint arXiv:2005.00687 (2020)
6. Jiang, H., Narasimhan, H., Bahri, D., Cotter, A., Rostamizadeh, A.: Churn reduction via distillation. In: ICLR (2022)
7. Kipf, T.N., Welling, M.: Semi-supervised classification with graph convolutional networks. In: ICLR (2017)
8. Kornblith, S., Norouzi, M., Lee, H., Hinton, G.: Similarity of neural network representations revisited. In: ICML (2019)
9. Liu, H., S., A.P.V., Patwardhan, S., Grasch, P., Agarwal, S.: Model stability with continuous data updates. arXiv preprint arXiv:2201.05692 (2022)
10. Madani, O., Pennock, D., Flake, G.: Co-validation: using model disagreement on unlabeled data to validate classification algorithms. In: NeurIPS (2004)
11. Mernyei, P., Cangea, C.: Wiki-CS: a Wikipedia-based benchmark for graph neural networks. arXiv preprint arXiv:2007.02901 (2020)
12. Milani Fard, M., Cormier, Q., Canini, K., Gupta, M.: Launch and iterate: reducing prediction churn. In: NeurIPS (2016)
13. Scarselli, F., Gori, M., Tsoi, A.C., Hagenbuchner, M., Monfardini, G.: The graph neural network model. IEEE Trans. Neural Netw. **20**(1), 61–80 (2009)
14. Schumacher, T., Wolf, H., Ritzert, M., Lemmerich, F., Grohe, M., Strohmaier, M.: The effects of randomness on the stability of node embeddings. In: Workshop on Graph Embedding and Mining, co-located with ECML PKDD (2021)
15. Shamir, G.I., Coviello, L.: Anti-distillation: improving reproducibility of deep networks. arXiv preprint arXiv:2010.09923 (2020)
16. Shchur, O., Mumme, M., Bojchevski, A., Günnemann, S.: Pitfalls of graph neural network evaluation. arXiv preprint arXiv:1811.05868 (2018)
17. Summers, C., Dinneen, M.J.: Nondeterminism and instability in neural network optimization. In: ICML (2021)
18. Veličković, P., Cucurull, G., Casanova, A., Romero, A., Liò, P., Bengio, Y.: Graph attention networks. In: ICLR (2018)
19. Wang, C., Rao, W., Guo, W., Wang, P., Liu, J., Guan, X.: Towards understanding the instability of network embedding. IEEE Trans. Knowl. Data Eng. **34**(2), 927–941 (2022)
20. Yang, Z., Cohen, W., Salakhudinov, R.: Revisiting semi-supervised learning with graph embeddings. In: ICML (2016)
21. Zhuang, D., Zhang, X., Song, S.L., Hooker, S.: Randomness in neural network training: characterizing the impact of tooling. arXiv preprint arXiv:2106.11872 (2021)
22. Zügner, D., Akbarnejad, A., Günnemann, S.: Adversarial attacks on neural networks for graph data. In: KDD (2018)
23. Zügner, D., Günnemann, S.: Certifiable robustness and robust training for graph convolutional networks. In: KDD (2019)

Adversarially Robust Decision Tree Relabeling

Daniël Vos[(✉)] [iD] and Sicco Verwer [iD]

Delft University of Technology, Delft, The Netherlands
{d.a.vos,s.e.verwer}@tudelft.nl

Abstract. Decision trees are popular models for their interpretation properties and their success in ensemble models for structured data. However, common decision tree learning algorithms produce models that suffer from adversarial examples. Recent work on robust decision tree learning mitigates this issue by taking adversarial perturbations into account during training. While these methods generate robust shallow trees, their relative quality reduces when training deeper trees due the methods being greedy. In this work we propose robust relabeling, a post-learning procedure that optimally changes the prediction labels of decision tree leaves to maximize adversarial robustness. We show this can be achieved in polynomial time in terms of the number of samples and leaves. Our results on 10 datasets show a significant improvement in adversarial accuracy both for single decision trees and tree ensembles. Decision trees and random forests trained with a state-of-the-art robust learning algorithm also benefited from robust relabeling.

Keywords: Decision trees · Pruning · Adversarial examples

1 Introduction

With the increasing interest in trustworthy machine learning, decision trees have become important models [17]. Due to their simple structure humans can interpret the behavior of size-limited decision trees. Additionally, decision trees are popular for use within ensemble models where random forests [3] and particularly gradient boosting ensembles [6,7,9,15] achieve top performance on prediction tasks with tabular data. However, decision trees are optimized without considering robustness which results in models that misclassify many data points after adding tiny perturbations [14,26], i.e. adversarial examples. Therefore we are interested in training tree-based models that correctly predict data points not only at their original coordinates but also in a radius around these coordinates.

Supplementary Information The online version contains supplementary material available at https://doi.org/10.1007/978-3-031-26409-2_13.

Recent work has proposed decision tree learning algorithms that take adversarial perturbations into account during training to improve adversarial robustness [4,5,22]. These methods significantly improved robustness for shallow decision trees, but lacked performance for deeper trees due to their greedy nature. Optimal methods for robust decision tree learning [12,23] have also been proposed but they use combinatorial optimization solvers which makes them scale poorly in terms of both tree depth and data size. Adversarial pruning [24,25] is a method that pre-processes datasets by removing a minimal number of samples to make the dataset well-separated. While this method helps ignore samples that will only worsen robustness when predicted correctly, the learning algorithm is unchanged so the resulting models still suffer from adversarial examples. It is important to be able to train deeper robust trees as shallow trees can significantly underfit the data. Particularly in random forests where we aim to ensemble unbiased models [2] we need to be able to train very deep trees.

To improve the performance of robust decision trees we propose Robust Relabeling[1]. This post-learning procedure optimally changes the prediction labels of the decision tree leaves to maximize accuracy against adversarial examples. We assume that the user specifies an arbitrary region around each sample that represents the set of all possible perturbations of the sample. Then, we only consider a sample to be correctly predicted under adversarial attacks if there is no way for an attacker to perturb the sample such that the prediction is different from the label. We prove that in binary classification the optimal robust relabeling is induced by the minimum vertex cover of a bipartite graph. This property allows us to compute the relabeling in polynomial time in terms of the number of samples and leaves.

We compare the classification performance of decision trees and tree ensemble models on 10 datasets from the UCI Machine Learning Repository [8] and find that robust relabeling improves the average adversarial accuracy for all models. We also evaluate the performance when relabeling robust decision trees trained with a state-of-the-art method GROOT [22]. The resulting models improve adversarial accuracy compared to the default GROOT models by up to 20%. Additionally, we study the effects of standard Cost Complexity Pruning against robust relabeling. Both methods reduce the size of the learned decision trees and can improve both regular accuracy and adversarial accuracy compared to unpruned models. While, Cost Complexity Pruning performs better on regular accuracy, robust relabeling results in better adversarial accuracy.

2   Background Information

2.1   Decision Tree Learning

Decision trees are simple models that execute a series of logical tests to arrive at a prediction value. In our work, we focus on decision trees where a node k performs an operation of the type 'feature value a_k is less than or equal to some

---

[1] https://github.com/tudelft-cda-lab/robust-relabeling.

value b_k'. When we follow the path of such decision nodes to a leaf node t, we find the prediction value c_t.

The most popular methods to learn such decision trees are greedy algorithms that recursively create decision nodes to improve predictive accuracy. For instance, CART [1] starts by creating a root node and tests all possible combinations of feature a_k and value b_k to use for a split. These splits are all scored with the Gini impurity and the best one is selected. The samples are then sorted into a left side and right side of this split and the algorithm continues recursively on both sides until no improvements can be made or a user-defined stopping criterion is reached. While methods like CART have been hugely popular, their splitting criteria (e.g. Gini impurity or information gain) do not account for adversarial attacks. Therefore recent work has focused on modifying such algorithms in a way that the learned trees are robust to perturbations.

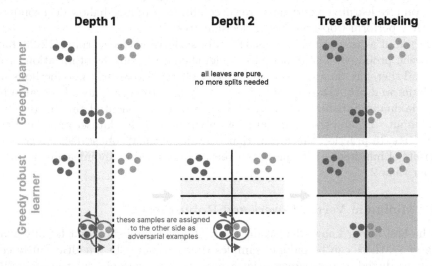

Fig. 1. Training decision trees using a (regular) greedy learner and robust greedy learner. The robust learner greedily perturbed samples close to the split which caused it to assign sub-optimal predictions to its leaves. This effect increases with the tree depth. By relabeling these leaves we can improve robustness.

2.2   Robust Decision Tree Learning

In the field of robust decision trees we usually assume that an adversary modifies our data points at test time in order to cause misclassifications. Then, we aim to train a decision tree that is maximally robust to such modifications. The type of modifications that we allow the adversary to make strongly influences the learned trees. In this work we consider an adversary that can make arbitrary changes to each test data point i within a radius ϵ of the original point. In line with previous works [5,22] we measure this distance with the l_∞ norm. Therefore the set of all possible perturbations applied to data point i is $S(i) = \{x + \delta \mid ||\delta||_\infty \leq \epsilon\}$. For a decision tree $\mathcal{T}$ it is especially important to know what leaves $\mathcal{T}_L$ sample i can reach after applying perturbations, we refer to this set as $\mathcal{T}_L^{S(i)}$.

To improve the adversarial robustness of decision trees different methods have been proposed [4,5,22] that take adversarial perturbations into account during training. These methods are based on the same greedy algorithm that is used to train regular decision trees, but they use a different function to score the quality of splits. Then, when a locally optimal split is found, they apply perturbations to samples that are close to the split in the worst possible arrangement. This means that some number of samples that were originally on the left side of the split will be sent to the right and vice versa. Although this generally improves robustness, the fact that these samples are perturbed greedily can be detrimental to the quality of the learned tree after creating successive splits. For example in Fig. 1 the fact that samples were greedily perturbed caused the learned decision tree to create to leaves with bad predictions. Due to their greedy nature, these robust decision tree learning algorithms are successful in training shallow decision trees but they perform worse on deeper decision trees.

In recent work, optimal methods for robust decision tree learning [12,23] have also been proposed. These methods model the entire robust optimization problem and therefore do not suffer from greedy effects. However, these methods use combinatorial optimization solvers such as Mixed-Integer Linear Programming and Maximum Satisfiability. These solvers run in exponential time in terms of their input size, and the inputs grow with the size of the dataset and the depth of the tree. In practice, this means that training optimal robust decision trees on datasets of hundreds of samples is currently computationally infeasible for trees deeper than 2.

2.3   Minimum Vertex Covers and Robustness

To the best of our knowledge, Wang et al. [24] first published that for any given dataset D, there can be pairs of samples that can never be simultaneously correctly predicted against adversarial examples. For example, when considering perturbations within some radius ϵ, two samples with different labels that are within distance 2ϵ cannot both be correctly predicted when accounting for these perturbations. Given this fact, one can create a graph G with each vertex representing a sample and connect all such pairs. When we compute the minimum vertex cover of this graph, we find the minimum number of data points C to remove from D such that $D \setminus C$ can be correctly predicted. Although $D \setminus C$ *can* be correctly predicted, non-robust learning algorithms can and will still learn models that suffer from adversarial examples, e.g., because decision planes are placed too close to the remaining data points.

Wang et al. [24] used this minimum vertex cover idea by removing C from the training data in order to learn a robust nearest neighbor classifier. Adversarial pruning [25] uses a similar method to train nearest neighbor models, decision trees and tree ensembles from $D \setminus C$. The authors of ROCT [23] also used the minimum vertex cover but to compute an upper bound on adversarial accuracy which improved the time needed to train optimal robust decision trees.

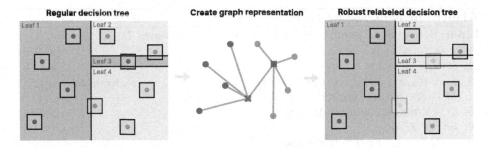

Fig. 2. Example of the robust relabeling procedure applied to a decision tree that suffers from adversarial examples. We first create a bipartite graph that connects samples with different labels that can reach the same leaf with perturbations. After removing the samples corresponding to the graph's minimum vertex cover we can relabel the decision tree to correctly predict the remaining samples. The resulting labeling is maximally robust to adversarial perturbations.

2.4   Relabeling and Pruning Decision Trees

Improving the quality of decision trees with respect to some metric by changing their leaf predictions is not a new idea. Many pruning algorithms have been proposed that remove parts of the decision tree to improve generalization. For example, Cost Complexity Pruning [1] is a widely used method that merges leaves when this improves the trade-off between the size of the tree and its predictive performance. Similarly, ideas to relabel decision trees have been used to improve performance for objectives such as fairness [13] and monotonicity [20]. Such metrics are not aligned with the objective that is optimized during training. To the best of our knowledge, we are the first to propose using leaf relabeling to improve adversarial robustness. Since relabeling methods never add new leaves, they can be seen as pruning methods since they reduce the size of the trees (after merging leaves that have the same label) (Fig. 2).

3   Robust Relabeling

Since regular decision trees and ensembles suffer from adversarial examples we are interested in post-processing the learned models to improve their robustness. In this work, we propose 'robust relabeling' (Algorithm 1) a method that keeps the decision tree structure intact but changes the predictions in the leaves to maximize adversarial accuracy. Robust relabeling is closely related to earlier works that determine minimum vertex covers to improve robustness [23–25]. In these works, the authors leverage the fact that samples with overlapping perturbation ranges and different labels can never be simultaneously classified correctly under optimal adversarial perturbations. We notice that in decision trees, two samples cannot be simultaneously classified correctly under optimal adversarial perturbations when they both reach the same leaf. Using this property we can

---

Algorithm 1. Robust relabeling decision tree

Input: dataset X (n samples, m features), labels y, tree leaves $\mathcal{T}_L$

1: $L \leftarrow \{i \mid y_i = 0\}$ $\triangleright \; \mathcal{O}(n)$
2: $R \leftarrow \{i \mid y_i = 1\}$ $\triangleright \; \mathcal{O}(n)$
3: $E \leftarrow \{(u,v) \mid u \in L, v \in R, \mathcal{T}_L^{S(u)} \cap \mathcal{T}_L^{S(v)} \neq \emptyset\}$ $\triangleright \; \mathcal{O}(nm|\mathcal{T}_L| + n^2|\mathcal{T}_L|)$
4: $M \leftarrow$ MAXIMUM_MATCHING(L, R, E) $\triangleright \; \mathcal{O}(n^{2.5})$
5: $C \leftarrow$ KÖNIG'S_THEOREM(M, L, R, E) $\triangleright \; \mathcal{O}(n)$
6: **for** $t \in \mathcal{T}_L$ **do** $\triangleright \; \mathcal{O}(n|\mathcal{T}_L|)$
7:      **if** $\{i \in L \mid t \in \mathcal{T}_L^{S(i)}\} \neq \emptyset$ **then**
8:          $c_t \leftarrow 0$
9:      **else**
10:        $c_t \leftarrow 1$
11:      **end if**
12: **end for**

---

find the smallest set of samples to remove from the dataset such that all remaining samples *can* be classified correctly under perturbations. These samples then induce a labeling of the decision tree that correctly classifies the largest possible set of samples against adversarial perturbations.

To robustly relabel a decision tree $\mathcal{T}$ we create a bipartite graph $G = (L, R, E)$ where L represents the set of samples with label $y_i = 0$ and R the set of samples with label $y_j = 1$. The set of edges E is defined by connecting all pairs of samples (i, j) that have different labels $y_i \neq y_j$ and overlapping perturbation ranges $\mathcal{T}_L^{S(i)} \cap \mathcal{T}_L^{S(j)} \neq \emptyset$. Here $S(i)$ is the set of all possible perturbations applied to data point i and $\mathcal{T}_L^{S(i)}$ is the set of leaves that are reached by $S(i)$. We find the minimum vertex cover C of G and remove the samples represented by C from the dataset. We can then relabel the decision tree to classify all remaining samples correctly even under optimal adversarial attacks.

In this paper, we consider only the case where $S(i)$ describes an l_∞ radius around each data point as this is common in research on robust decision trees. However, our proof does not make use of this fact and robust relabeling can easily be extended to other attack models such as different l-norms or arbitrary sets of perturbations.

Theorem 1. *The optimal adversarially robust relabeling for decision tree leaves $\mathcal{T}_L$ is determined by the minimum vertex cover of the bipartite graph where samples i, j with different labels $y_i \neq y_j$ are represented by vertices that share an edge when their perturbations can reach any same leaf ($\mathcal{T}_L^{S(i)} \cap \mathcal{T}_L^{S(j)} \neq \emptyset$).*

Proof. For a sample i to be correctly classified by a decision tree $\mathcal{T}$, all leaves $\mathcal{T}_L^{S(i)}$ reachable by adversarial perturbations applied to X_i need to predict the correct label, i.e. $\forall t \in \mathcal{T}_L^{S(i)}, c_t = y_i$ (otherwise an adversarial example exists). Given two samples i, j with different labels $y_i \neq y_j$ and overlapping sets of reachable leaves by perturbations $T_L^{S(i)} \cap T_L^{S(j)} \neq \emptyset$ these samples cannot be correctly robustly predicted as there exists a leaf t that is in both sets $T_L^{S(i)}$ and $T_L^{S(j)}$ and that misclassifies one of the samples. Create the bipartite graph $G = (L, R, E)$ where $L = \{i \mid y_i = 0\}$, $R = \{i \mid y_i = 1\}$ and $E = \{(u,v) \mid u \in L, v \in R, T_L^{S(u)} \cap T_L^{S(v)} \neq \emptyset\}$. By removing the minimum vertex cover C from G, we are left with the largest graph $G' = (L \setminus C, R \setminus C, \emptyset)$ for which no edges remain. Since none of the remaining vertices (representing samples) share an edge we are able to set $\forall t \in T_L^{S(i)}, \forall i \in (L' \cup R') : c_t = y_i$, so all remaining samples get correctly robustly classified. Since C is of minimum cardinality the induced relabeling maximizes the adversarial accuracy. $\qquad \square$

Where a naive relabeling algorithm would take exponential time to enumerate all $2^{|T_L|}$ labelings, the above relabeling procedure runs in polynomial time in terms of the dataset size ($n \times m$ matrix) and the number of leaves $|T_L|$. When building the graph G the runtime is dominated by computing the edges which takes $\mathcal{O}(nm|T_L| + n^2|T_L|)$ time. This is because we first build a mapping for each sample i to their reachable leaves $T_L^S(i)$ in $\mathcal{O}(nm|T_L|)$ time, then compute samples that reach any same leaf in $\mathcal{O}(n^2|T_L|)$ time. Given the bipartite graph G we use the Hopcroft-Karp algorithm [11] to compute a maximum matching in $\mathcal{O}(n^{2.5})$ time and convert this in linear time into a minimum vertex cover using König's theorem. Combining all steps, robust relabeling runs in worst-case $\mathcal{O}(nm|T_L| + n^2|T_L| + n^{2.5})$ time.

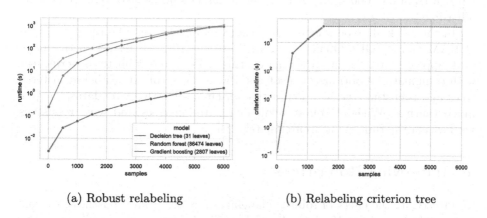

(a) Robust relabeling   (b) Relabeling criterion tree

Fig. 3. Runtime of robust relabeling and relabeling criterion trees on samples of the Wine dataset. While decision tree relabeling runs in within seconds, relabeling ensembles takes more time due to the number of trees and increase in tree size. The runtime for relabeling criterion trees quickly increases with the number of samples.

3.1   Robust Relabeling as Splitting Criterion

While the robust relabeling procedure described before provides an intuitive use case as a post-processing step for decision tree learners, we can also use the procedure to select splits during learning. In greedy decision tree learning the learner finds a locally optimal split, partitions the samples into a left and right set (including perturbed samples) and continues this process recursively. While this approach finds an optimal split for the top decision node, the detrimental effect of choosing splits greedily increases with the depth of the tree. We will try to reduce the impact of greedily perturbing samples by using the robust relabeling procedure. To do this we can consider all samples each time we score a split and use the cardinality of the maximum matching M as a splitting criterion. By choosing splits that minimize this criterion we are then directly optimizing the adversarial accuracy of the decision tree. The pseudo-code for this algorithm is given in the appendix. We will refer to this method as relabeling criterion trees.

3.2   Runtime Comparison

We compare the runtimes of robust relabeling and relabeling criterion trees on different sample sizes of the Wine dataset in Fig. 3. Robust relabeling decision trees runs in a matter of seconds since the number of trees is small. In tree ensembles where there are 100 trees to relabel and many more leaves the runtime quickly increases. We find that relabeling criterion trees take more than an hour to train on 2000 samples, training on larger sample sizes quickly becomes infeasible.

In this work all experiments ran without parallelism on a laptop with 16 GB of RAM and a 2 GHz Quad-Core Intel Core i5 CPU. All results in this paper took approximately a day to compute, this is including robustness verification with combinatorial optimization solvers. Particularly robust relabeling criterion trees and robustness verification of tree ensembles for the wine dataset require much runtime. Without robust relabeling criterion trees and wine robustness verification the runtime is approximately 2 h.

4   Improving Robustness

To investigate the effect of robust relabeling on adversarial robustness, we compare performance on 10 datasets with a fixed perturbation radius for each dataset. We used datasets from the UCI Machine Learning Repository [8] retrieved through OpenML [19]. All datasets, their properties and perturbation radii ϵ are listed in Table 1. We pre-process each dataset by scaling the features to the range $[0, 1]$. This way, we can interpret ϵ as representing a fraction of each feature's range. We compare robust relabeling to regular decision trees and ensembles trained with Scikit-learn [16], robust decision trees trained with GROOT [22] and adversarial pruning [25]. All adversarial accuracy scores were

Table 1. Summary of datasets used. Features are scaled to $[0, 1]$ so the l_∞ perturbation radius ϵ represents a fraction of each feature's range.

Dataset	ϵ	Samples	Features	Majority class
Banknote-authentication	.05	1,372	4	.56
Breast-cancer-diagnostic	.05	569	30	.63
Breast-cancer	.1	683	9	.65
Connectionist-bench-sonar	.05	208	60	.53
Ionosphere	.05	351	34	.64
Parkinsons	.05	195	22	.75
Pima-Indians-diabetes	.01	768	8	.65
Qsar-biodegradation	.05	1,055	41	.66
Spectf-heart	.005	349	44	.73
Wine	.025	6,497	11	.63

computed with optimal adversarial attacks using the GROOT toolbox[2]. For single trees, computing optimal adversarial attacks is done by enumerating all the leaves and for tree ensembles by solving the Mixed-Integer Linear Programming formulation by Kantchelian et al. [14] using GUROBI 9.1 [10].

4.1 Decision Trees

Decision trees have the desirable property that they are interpretable when constrained to be small enough. What exactly is the number of leaves that allow a decision tree to be interpretable is not well defined. In this work, we decide to train single trees up to a depth of 5 which enforces a maximum number of leaves of $2^5 = 32$. In Table 2 we compare the performance of regular, robust GROOT [22] trees and relabeling criterion trees defined in Sect. 3.1. We score the regular and GROOT trees before and after relabeling but we skip this step for relabeling criterion trees as this does not affect the learned tree.

Robust relabeling improves the performance of regular and GROOT trees significantly on most datasets and never reduces the mean adversarial accuracy. Relabeling criterion trees and relabeled GROOT trees performed similarly on average but relabeled GROOT trees run orders of magnitude faster.

4.2 Decision Tree Ensembles

For tasks that do not require model interpretability, it is a popular choice to ensemble multiple decision trees to create stronger models. We experiment with the robust relabeling of random forests, GROOT random forests and gradient boosting ensembles, all limited to 100 decision trees. For the gradient boosting ensembles we limit the trees to a depth of 5 to prevent overfitting. This

---

[2] https://github.com/tudelft-cda-lab/GROOT.

Table 2. Mean **adversarial accuracy** scores of decision trees of depth 5 on 5-fold cross validation. GROOT trees with robust relabeling and relabeling criterion trees score best against adversarial attacks. However, GROOT with relabeling runs orders of magnitude faster.

Dataset	Tree	Tree relabeled	GROOT	GROOT relabeled	Relabeling criterion
Banknote	.734 ± .077	.823 ± .035	.794 ± .049	**.824** ± .038	.811 ± .049
Breast-cancer	.874 ± .013	.903 ± .025	.912 ± .035	.922 ± .012	**.925** ± .013
Breast-cancer-d	.617 ± .158	.810 ± .026	.835 ± .013	.847 ± .014	**.851** ± .038
Sonar	.482 ± .140	.573 ± .073	.601 ± .048	**.606** ± .048	.582 ± .070
Ionosphere	.689 ± .071	.792 ± .045	.892 ± .030	.889 ± .028	**.895** ± .028
Parkinsons	.513 ± .139	.759 ± .126	.749 ± .071	.790 ± .058	**.795** ± .075
Diabetes	.708 ± .009	**.712** ± .025	.677 ± .053	**.712** ± .025	.710 ± .032
Qsar-bio	.292 ± .060	.661 ± .004	.704 ± .029	**.736** ± .050	.686 ± .014
Spectf-heart	**.840** ± .041	**.840** ± .041	.831 ± .044	.831 ± .044	.768 ± .016
Wine	.526 ± .027	.610 ± .047	**.618** ± .043	**.618** ± .052	Timeout

is not required for random forests where one purposefully ensembles low bias, high variance models [2], i.e., unconstrained decision trees. We did not compare to random forests trained with the robust relabeling criterion as this was computationally infeasible. The adversarial accuracy scores before and after robust relabeling are presented in Table 3. On the Wine dataset we only used 100 test samples to limit the runtime.

Robust relabeling increases the mean adversarial accuracy over 5-fold cross-validation on all datasets and models. On average, the GROOT random forests with robust relabeling performed best. Clearly, the combination of robust splits and robust labeling is better than regular splits and robust labeling. Additionally we find that relabeled GROOT forests (Table 3) outperform relabeled GROOT trees (Table 2) on many datasets. This is in contrast with the original GROOT paper [22]. In that paper, large values were used for ϵ that did not allow for the models to achieve significant improvements over predicting the majority class.

4.3   Adversarial Pruning

Adversarial pruning [25] is a technique that implicitly prunes models by removing samples from the training dataset that are not well separated ($D \setminus C$). This intuitively makes models more robust as the models then explicitly ignore samples that will make the models more susceptible to adversarial attacks. Using decision tree learning algorithms as-is on this dataset ($D \setminus C$), without taking robustness into account, still provides models that suffer from adversarial examples. In Table 4 we compare the adversarial robustness of models trained with adversarial pruning and robust relabeling. We notice that on many datasets, adversarial pruning only removes a small number of samples which results in models that are similar to the fragile models produced by regular decision tree learning algorithms.

Table 3. Mean **adversarial accuracy** scores of decision tree ensembles on 5-fold cross validation. GROOT trees with robust relabeling score best against adversarial attacks, relabeled regular trees perform on average similarly to robust GROOT trees that did not use relabeling.

Dataset	Boosting	Forest	GROOT forest	Boosting relabeled	Forest relabeled	GROOT forest rel
Banknote	.786 ± .072	.846 ± .032	.851 ± .037	.822 ± .052	.849 ± .039	**.862 ± .039**
Breast-cancer	.873 ± .027	.908 ± .020	.946 ± .017	.937 ± .020	.930 ± .011	**.952 ± .017**
Breast-cancer-d	.606 ± .070	.745 ± .035	.805 ± .025	.821 ± .019	.842 ± .022	**.847 ± .021**
Sonar	.438 ± .089	.389 ± .051	.510 ± .069	**.616 ± .076**	.582 ± .034	.577 ± .048
Ionosphere	.635 ± .163	.812 ± .037	.903 ± .018	.815 ± .010	.872 ± .017	**.912 ± .021**
Parkinsons	.492 ± .177	.508 ± .170	.728 ± .092	.759 ± .126	.749 ± .021	**.826 ± .066**
Diabetes	.596 ± .043	.668 ± .049	.703 ± .052	.697 ± .032	.729 ± .036	**.730 ± .047**
Qsar-bio	.078 ± .025	.173 ± .015	.648 ± .046	.663 ± .002	.663 ± .002	**.781 ± .026**
Spectf-heart	.863 ± .034	.891 ± .028	.888 ± .023	.877 ± .037	**.897 ± .025**	.894 ± .029
Wine	.202 ± .044	.184 ± .050	.384 ± .051	.494 ± .081	.482 ± .090	**.606 ± .038**

4.4 Accuracy Robustness Trade-Off

Since we optimize robustness by enforcing samples to be correctly classified in a region around each sample, there can be a cost in regular accuracy. In Table 5 we compare the accuracy of regular models with and without robust relabeling. We find that indeed robust relabeling reduces regular accuracy in approximately two out of three cases that we tested. However, there are also instances where accuracy actually improves, such as in the case of gradient boosting on the breast cancer datasets. We expect that robustification has a helpful regularization effect in these situations.

5 Regularizing Decision Trees

Robust relabeling regularizes decision trees and tree ensembles by changing the leaf labels to maximize adversarial robustness. To understand the regularization effect we first visualize models before and after robust relabeling on toy datasets. Additionally, we contrast the regularization effect of robust relabeling with Cost Complexity Pruning. We show that while both methods can improve test accuracy, robust relabeling favors robustness while Cost Complexity Pruning favors regular accuracy.

5.1 Toy Datasets

To understand the effects of robust relabeling we generate three two-dimensional datasets and visualize the decision regions before and after relabeling. In Fig. 4 we train decision trees, random forests and gradient boosting with 5% label noise and adversarial attacks within an l_∞ radius of $\epsilon = 0.05$. All features are scaled to the range $[0, 1]$ therefore ϵ represents 5% of each feature's range. The boosted and single decision trees were limited to a depth of 5.

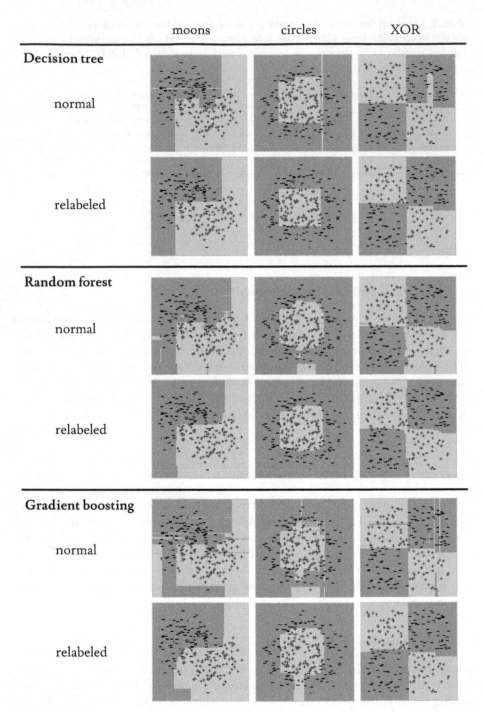

Fig. 4. Decision regions of tree models before and after robust relabeling. Robust relabeling effectively removes fragile regions resulting in visually simpler models.

Table 4. Comparison of **adversarial accuracy** scores for adversarial pruning [25] and robust relabeling (ours). Adversarial pruning does not take into account that the learner can select non-robust splits where relabeling effectively removes such splits thus producing more robust models.

Dataset	Decision tree		Gradient boosting		Random forest	
	Pruning	Relabeling	Pruning	Relabeling	Pruning	Relabeling
Banknote	.718 ± .060	**.823** ± .035	.809 ± .053	**.822** ± .052	**.855** ± .032	.849 ± .039
Breast-cancer	.867 ± .016	**.903** ± .025	.868 ± .031	**.937** ± .020	.906 ± .016	**.930** ± .011
Breast-cancer-d	.617 ± .158	**.810** ± .026	.619 ± .081	**.821** ± .019	.749 ± .035	**.842** ± .022
Sonar	.482 ± .140	**.573** ± .073	.438 ± .089	**.616** ± .076	.389 ± .051	**.582** ± .034
Ionosphere	.689 ± .071	**.792** ± .045	.635 ± .163	**.815** ± .010	.812 ± .037	**.872** ± .017
Parkinsons	.513 ± .139	**.759** ± .126	.492 ± .177	**.759** ± .126	.508 ± .170	**.749** ± .021
Diabetes	.708 ± .009	**.712** ± .025	.596 ± .043	**.697** ± .032	.668 ± .049	**.729** ± .036
Qsar-bio.	.262 ± .052	**.661** ± .004	.149 ± .024	**.663** ± .002	.183 ± .010	**.663** ± .002
Spectf-heart	**.840** ± .041	**.840** ± .041	.863 ± .034	**.877** ± .037	.891 ± .028	**.897** ± .025
Wine	.562 ± .030	**.610** ± .047	.212 ± .094	**.494** ± .081	.240 ± .047	**.482** ± .090

In all types of models we see that there are small regions with a wrong prediction in areas where the model predicts the correct label. For instance, in the normal decision tree trained on 'moons', we see a slim orange leaf in the region that is otherwise predicted as blue. This severely reduces the robustness of the model against adversarial attacks since nearby samples can be perturbed into those regions. Robust relabeling effectively removes these leaves from the models to improve adversarial robustness. Although the effect of regularization of decision trees is hard to quantify, we intuitively see that the relabeled models are more consistent in their predictions. We expect this property to also improve the explainability of the models by methods such as counterfactual explanations [21].

5.2   Comparison with Cost Complexity Pruning

Cost Complexity Pruning is a method that reduces the size of decision trees to improve their generalization capabilities. This method iteratively merges leaves that have a lower increase in predictive performance than some user-defined threshold α. In Fig. 5 we compare the effects of Cost Complexity Pruning and robust relabeling on the Diabetes dataset. Here, we trained decision trees without size constraints and varied the hyperparameters α and ϵ then measured accuracy before and after adversarial attacks.

While the effectiveness of cost complexity pruning and robust relabeling varies between datasets we find that generally both methods can increase test accuracy compared to the baseline model. However, cost complexity pruning achieved better accuracy scores on average while robust relabeling achieved better adversarial accuracy scores. Clearly, there is a difference between regularization for generalization and adversarial robustness. Such a trade-off between accuracy and robustness has been widely described in the literature [18,27].

Table 5. Comparison of **regular accuracy** scores before and after applying robust relabeling. Since robustness is generally at odds with accuracy robust relabeling loses out on accuracy in about 2 out of 3 cases. However, in some cases robustness actually improves accuracy as a type of regularization.

Dataset	Decision tree		Gradient boosting		Random forest	
	Before	After	Before	After	Before	After
Banknote	**.967** ± .022	.948 ± .036	**.991** ± .007	.941 ± .038	**.994** ± .005	.956 ± .020
Breast-cancer	**.969** ± .014	.958 ± .019	.962 ± .008	**.965** ± .009	.968 ± .008	**.969** ± .003
Breast-cancer-d	**.930** ± .034	.912 ± .032	.917 ± .036	**.931** ± .024	**.954** + .017	.947 ± .028
Sonar	**.740** ± .058	.716 ± .054	.731 ± .044	**.740** ± .046	**.803** ± .031	.755 ± .054
Ionosphere	**.883** ± .034	.857 ± .047	**.906** ± .041	.863 ± .022	.926 ± .026	**.932** ± .033
Parkinsons	**.872** ± .065	.851 ± .071	**.867** ± .071	.851 ± .071	**.913** ± .082	.759 ± .014
Diabetes	.737 ± .026	**.738** ± .034	**.742** ± .026	.719 ± .028	**.769** ± .039	.768 ± .040
Qsar-bio	**.819** ± .042	.661 ± .004	**.872** ± .023	.663 ± .002	**.868** ± .023	.663 ± .002
Spectf-heart	**.840** ± .041	**.840** ± .041	.871 ± .028	**.880** ± .033	**.897** ± .025	**.897** ± .025
Wine	**.696** ± .048	.686 ± .047	**.774** ± .024	.494 ± .081	**.786** ± .019	.498 ± .080

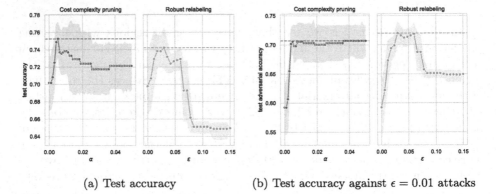

(a) Test accuracy   (b) Test accuracy against $\epsilon = 0.01$ attacks

Fig. 5. Test scores on the Diabetes dataset when varying the hyperparameters of cost complexity pruning and robust relabeling. Both improve upon unpruned trees ($\alpha = \epsilon = 0$) but cost complexity pruning performs better at regular accuracy while robust relabeling enhances adversarial accuracy.

6   Conclusions

In this work, we studied relabeling as a method to improve the adversarial robustness of decision trees and their ensembles. As training optimal robust decision trees is expensive and training heuristic robust trees inexact, we propose a polynomial-time post-learning algorithm to overcome these problems: robust relabeling. Our results show that robust relabeling significantly improves the robustness of regular and robust tree models. Robustly relabeling models trained with the state-of-the-art robust tree heuristic GROOT further improved the performance. While we can also use the robust relabeling method during the tree learning procedure this took up to hours of runtime and produced decision trees that were approximately as robust as relabeled GROOT trees.

We expect robust relabeling in combination with methods such as GROOT to become important for training models that get deployed in adversarial contexts such as fraud or malware detection. The result that finding an optimal robust labeling can be done in polynomial time can help to further improve methods for optimal robust decision tree learning. In future work, we will explore the regularity effects of robust models for instance for improved counterfactual explanations.

References

1. Breiman, L., Friedman, J., Stone, C., Olshen, R.: Classification and Regression Trees. Taylor & Francis, Milton Park (1984)
2. Breiman, L.: Bagging predictors. Mach. Learn. **24**(2), 123–140 (1996)
3. Breiman, L.: Random forests. Mach. Learn. **45**(1), 5–32 (2001)
4. Calzavara, S., Lucchese, C., Tolomei, G., Abebe, S.A., Orlando, S.: Treant: training evasion-aware decision trees. Data Min. Knowl. Disc. **34**(5), 1390–1420 (2020)
5. Chen, H., Zhang, H., Boning, D., Hsieh, C.J.: Robust decision trees against adversarial examples. In: International Conference on Machine Learning, pp. 1122–1131. PMLR (2019)
6. Chen, T., Guestrin, C.: XGBoost: a scalable tree boosting system. In: Proceedings of the 22nd ACM SIGKDD International Conference on Knowledge Discovery and Data Mining, pp. 785–794 (2016)
7. Dorogush, A.V., Ershov, V., Gulin, A.: CatBoost: gradient boosting with categorical features support. arXiv preprint arXiv:1810.11363 (2018)
8. Dua, D., Graff, C.: UCI machine learning repository (2017). http://archive.ics.uci.edu/ml
9. Friedman, J.H.: Greedy function approximation: a gradient boosting machine. Ann. Stat. 1189–1232 (2001)
10. Gurobi Optimization, LLC: Gurobi Optimizer Reference Manual (2022). https://www.gurobi.com
11. Hopcroft, J.E., Karp, R.M.: An n^5/2 algorithm for maximum matchings in bipartite graphs. SIAM J. Comput. **2**(4), 225–231 (1973)
12. Justin, N., Aghaei, S., Gomez, A., Vayanos, P.: Optimal robust classification trees. In: The AAAI-22 Workshop on Adversarial Machine Learning and Beyond (2021)
13. Kamiran, F., Calders, T., Pechenizkiy, M.: Discrimination aware decision tree learning. In: 2010 IEEE International Conference on Data Mining, pp. 869–874. IEEE (2010)
14. Kantchelian, A., Tygar, J.D., Joseph, A.: Evasion and hardening of tree ensemble classifiers. In: International Conference on Machine Learning, pp. 2387–2396. PMLR (2016)
15. Ke, G., et al.: LightGBM: a highly efficient gradient boosting decision tree. In: Advances in Neural Information Processing Systems, vol. 30 (2017)
16. Pedregosa, F., et al.: Scikit-learn: machine learning in python. J. Mach. Learn. Res. **12**, 2825–2830 (2011)
17. Rudin, C.: Stop explaining black box machine learning models for high stakes decisions and use interpretable models instead. Nat. Mach. Intell. **1**(5), 206–215 (2019)

18. Tsipras, D., Santurkar, S., Engstrom, L., Turner, A., Madry, A.: Robustness may be at odds with accuracy. arXiv preprint arXiv:1805.12152 (2018)
19. Vanschoren, J., van Rijn, J.N., Bischl, B., Torgo, L.: OpenML: networked science in machine learning. SIGKDD Explor. **15**(2), 49–60 (2013). https://doi.org/10.1145/2641190.2641198. http://doi.acm.org/10.1145/2641190.2641198
20. Velikova, M., Daniels, H.: Decision trees for monotone price models. CMS **1**(3), 231–244 (2004)
21. Verma, S., Dickerson, J., Hines, K.: Counterfactual explanations for machine learning: a review. arXiv preprint arXiv:2010.10596 (2020)
22. Vos, D., Verwer, S.: Efficient training of robust decision trees against adversarial examples. In: International Conference on Machine Learning, pp. 10586–10595. PMLR (2021)
23. Vos, D., Verwer, S.: Robust optimal classification trees against adversarial examples. arXiv preprint arXiv:2109.03857 (2021)
24. Wang, Y., Jha, S., Chaudhuri, K.: Analyzing the robustness of nearest neighbors to adversarial examples. In: International Conference on Machine Learning, pp. 5133–5142. PMLR (2018)
25. Yang, Y.Y., Rashtchian, C., Wang, Y., Chaudhuri, K.: Robustness for non-parametric classification: a generic attack and defense. In: International Conference on Artificial Intelligence and Statistics, pp. 941–951. PMLR (2020)
26. Zhang, C., Zhang, H., Hsieh, C.J.: An efficient adversarial attack for tree ensembles. Adv. Neural. Inf. Process. Syst. **33**, 16165–16176 (2020)
27. Zhang, H., Yu, Y., Jiao, J., Xing, E., El Ghaoui, L., Jordan, M.: Theoretically principled trade-off between robustness and accuracy. In: International Conference on Machine Learning, pp. 7472–7482. PMLR (2019)

Calibrating Distance Metrics Under Uncertainty

Wenye Li[1,2](✉) [ID] and Fangchen Yu[1] [ID]

[1] The Chinese University of Hong Kong, Shenzhen, China
wyli@cuhk.edu.cn, fangchenyu@link.cuhk.edu.cn
[2] Shenzhen Research Institute of Big Data, Shenzhen, China

Abstract. Estimating distance metrics for given data samples is essential in machine learning algorithms with various applications. Accurately determining the metric becomes impossible if there are observation noises or missing values. In this work, we proposed an approach to calibrating distance metrics. Compared with standard practices that primarily reside on data imputation, our proposal makes fewer assumptions about the data. It provides a solid theoretical guarantee in improving the quality of the estimate. We developed a simple, efficient, yet effective computing procedure that scales up to realize the calibration process. The experimental results from a series of empirical evaluations justified the benefits of the proposed approach and demonstrated its high potential in practical applications.

Keywords: Missing data · Metric calibration · Alternating projection

1 Introduction

In data processing, a distance metric, or a distance matrix, is used to measure the pairwise dis-similarity relationship between data samples. It is crucial and lays a foundation in many supervised and unsupervised learning models, such as the K-means clustering algorithm, the nearest neighbor classifier, support vector machines [9,18,30,35].

Calculating pairwise distance is straightforward if the data samples are clean and fully observed. Unfortunately, with observation noises or missing values, which are natural and common in practice, obtaining a high-quality distance metric becomes a challenging task, and nontrivial challenges arise to learning algorithms based on the distance estimation between data samples.

Significant research attention has been devoted to handling the difficulty brought by missing values. Various imputation techniques were designed as a routine treatment, which has greatly influenced the progress in various disciplines [11,25]. These techniques complete the data by replacing the missing values with substituted ones based on various assumptions. Based on the imputed data, the pairwise distances can be calculated accordingly.

M.-R. Amini et al. (Eds.): ECML PKDD 2022, LNAI 13715, pp. 219–234, 2023.
https://doi.org/10.1007/978-3-031-26409-2_14

Despite the popularity received by data imputation approaches, nontrivial challenges still exist. When the assumptions made by the imputation techniques are violated, there is no guarantee at all on the quality of the imputed values, needless to say, the impact on subsequent data analysis tasks. Furthermore, with a large portion of missing values, the imputation can be highly demanding or even prohibitive in computation, which becomes another serious concern.

As a remedy, we carried out a series of work in two directions. Firstly and as the main contribution of this paper, our work proposed a metric calibration model that avoids data imputation in Sect. 3.2. It starts from an approximate metric estimated from incomplete samples or prior knowledge and then calibrates the metric iteratively. The calibrated metric is guaranteed to be better than the initial metric in terms of a shorter Frobenius distance to the accurate unknown metric, except in rare cases, the two metrics are identical. Secondly, our work applied Dykstra's projection algorithm to realize the calibration process and designed a cyclic projection algorithm as a more scalable alternative.

Compared with the popular imputation methods in handling missing data, the calibration approaches seemed to rely less on the assumption of the correlation among data features or the data's intrinsically low dimension/rank. As a result of the less dependency, the approaches reported more robust and reliable results in empirical evaluations. The improvement from the calibration approaches is especially significant when the missing ratio is high, or the noisy level is high, which exhibited their high potential in handling missing and noisy data in practical applications.

The paper is organized as follows. Section 2 introduces the background. Section 3 presents our model and algorithms. Section 4 reports the experimental results, followed by the conclusion in Sect. 5.

2   Background

2.1   Missing Data and Imputation

Missing observations are everywhere and pose nontrivial challenges to numerous data analysis applications in science and engineering. Developing techniques to process incompletely observed data becomes one of the most critical tasks in statistical sciences [11,25].

A common approach to dealing with missing observations is through data imputation. A missing value may be replaced by a zero value, the feature's mean, median, or the most frequent value among the nearest neighbor samples or all observed samples.

A more rigorous treatment is based on the expectation-maximization (EM) algorithm [6]. The approach assumes the existence of specific latent structures and variables. By alternatively estimating the model parameters and the missing values with the fitted model parameters, the approach generates a maximum likelihood or a maximum a posterior estimate for each missing observation.

Another imputation approach, the low-rank matrix completion approach developed more recently, makes assumptions on the rank of the data matrix to be completed. Efficient algorithms were designed to achieve exact reconstruction with high probability and reported quite successful results, such as in recommender systems [3,19]. In recent work, based on the assumption that two random batches from the same dataset share the same distribution, a measure of optimal transport distances is applied as an optimization objective for missing data imputation, which achieves excellent performances on some practical tasks [27].

Despite the success and the popularity that has been achieved, an inherent challenge exists. All imputation approaches, either explicitly or implicitly, have assumed the low dimensionality or the low-rank structure of the data. However, when the assumption does not hold, all these approaches will lose the performance guarantee on the imputation quality.

2.2   Metric Calibration

Instead of imputations, a matrix calibration approach can be applied to improve a metric obtained from incomplete or noisy data. As an example, let us consider the *metric nearness* model [2]. Denote the set of all $n \times n$ matrices by $\mathcal{M}_n$, which is a closed, convex polyhedral cone. Assume we are given n incomplete samples and an estimate of their distance matrix $D^0 = \left\{d_{ij}^0\right\}_{i,j=1}^n \in \mathcal{M}_n$. The estimate is inaccurate and might violate the triangle inequality property that the true metric possesses. As a remedy, we consider the following model:

$$\min_{D \in \mathcal{M}_n} \left\| D - D^0 \right\|_F^2 \tag{1}$$

s.t.,

$$d_{ij} \geq 0, \ d_{ii} = 0, \ d_{ij} = d_{ji}, \ \text{and} \ d_{ij} \leq d_{ik} + d_{kj},$$

for all $1 \leq i, j, k \leq n$.

The model above seeks a new matrix $D = \left\{d_{ij}\right\}_{i,j-1}^n$ that *best approximates* the input matrix D^0 in Frobenius norm, from a feasible region of matrices that meet the desired constraints. After calibration, the result will restore the property that the true distance metric should possess.

The calibration approach has an implicit but key benefit [23]. Suppose the feasible region of the distance matrix of interest is appropriately defined. In that case, although the factual matrix is never known to us, the new calibrated matrix can be guaranteed to be nearer to the ground truth than the initial estimate D^0, except in rare cases that they are identical.

The metric nearness model defined in Eq. (1) can be formulated as a quadratic program and solved by modern convex optimization packages [1]. Besides, an elegant *triangle fixing* algorithm [2] was developed, which exploited the inherent structure of the triangle inequalities and improved running efficiency. Besides, we can also consider a stochastic sampling of constraints or Lagrangian formulations to seek an algorithmic solution [31]. Despite the partial success that has been achieved along this line, however, the intrinsic complexity from $O\left(n^3\right)$ inequality constraints to Eq. (1) makes the model hard to scale up, which significantly limits the application of the model.

3 Model

3.1 A Kernel's Trick

Our work resides on a mild assumption that the data samples in the study are isometrically embeddable in a real Hilbert space, or equivalently, the samples can be represented as real vectors. Recall the definition of isometrical embedding.

Definition 1. *Consider a separable metric space $\mathcal{X}$ with a distance function ρ, having the properties that $\rho(x, x') = \rho(x', x) \geq 0$ and $\rho(x, x) = 0$ for all points x and x' in $\mathcal{X}$. $(\mathcal{X}, \rho)$ is said to be isometrically embeddable in a real Hilbert space $\mathcal{H}$ (or embeddable, for short) if there exists a map $\phi : \mathcal{X} \mapsto \mathcal{H}$ such that*

$$\|\phi(x) - \phi(x')\| = \rho(x, x')$$

for all points x and x' in $\mathcal{X}$.

A classical result on isometrical embedding [29,34] states that:

Theorem 2. *Assume $(\mathcal{X}, \rho)$ is embeddable. Then, for each $\gamma > 0$ and $0 < \alpha < 1$,*

$$\sum_{i,j=1}^{n} \exp\left(-\gamma \rho^{2\alpha}(x_i, x_j)\right) \xi_i \xi_j \geq 0$$

holds for every choice of points $x_1, \cdots, x_n$ in $\mathcal{X}$ and real $\xi_1, \cdots, \xi_n$.

For any finite subset $\{x_1, \cdots, x_n\} \subseteq \mathcal{X}$ $(n \geq 2)$, denote by $D^* = \left\{d_{ij}^*\right\}_{i,j=1}^{n}$ with $d_{ij}^* = \rho(x_i, x_j)$ for each i, j and $\exp\left(-\gamma D^*\right) = \left\{\exp\left(-\gamma d_{ij}^*\right)\right\}_{i,j=1}^{n}$. By choosing $\alpha = \frac{1}{2}$, we have, if $(\mathcal{X}, \rho)$ is embeddable, the matrix $\exp\left(-\gamma D^*\right)$ is positive semi-definite and we also say that the matrix D^* is embeddable.

In the machine learning area, the positive definite function $\exp\left(-\gamma \|x - x'\|\right)$ is known as the Laplacian kernel, popularly used in the context of kernel-based algorithms [30]. In our application, the function connects an embeddable metric D^* and a positive semi-definiteness matrix $\exp\left(-\gamma D^*\right)$.

3.2 Direct Calibration

For given samples $\{x_1, \cdots, x_n\}$ in $\mathcal{X}$, let $D^0 = \left\{d_{ij}^0\right\}_{i,j=1}^{n}$ be an input distance matrix between the samples. Assume that, due to observation noise or missing values, the metric D^0 is not accurate. From the relationship between isometrical embedding and positive semi-definiteness in Sect. 3.1, we naturally investigate the following model to calibrate the matrix D^0 to a better estimate:

$$\min_{D \in \mathcal{M}_n} \left\|D - D^0\right\|_F^2, \tag{2}$$

s.t.

$$\exp(-\gamma D) \succeq 0, \quad d_{ii} = 0 \ (1 \leq i \leq n), \text{ and } d_{ij} \geq 0 \ (1 \leq i \neq j \leq n),$$

where $\succeq 0$ denotes the positive semi-definiteness constraint on a matrix.

Solving the optimization problem in Eq. (2) is not straightforward. Here we develop an efficient approximation. Let $\mu = \max\{\rho(x_i, x_j), 1 \leq i, j \leq n\}$ be a normalizing factor, and $\gamma = \frac{\epsilon}{\mu}$ where ϵ is a small positive number[1]. Denote $E^0 = \{e_{ij}^0\}_{i,j=1}^n = \exp(-\gamma D^0)$, and we reach a known problem in literature [23]:

$$\min_{E \in \mathcal{M}_n} \left\| E - E^0 \right\|_F^2 \tag{3}$$

s.t.

$$E \succeq 0, \quad e_{ii} = 1 \;\; (1 \leq i \leq n), \text{ and } e_{ij} \in [1 - \epsilon, 1] \;\; (1 \leq i \neq j \leq n).$$

Let $\mathcal{R} = \{X \in \mathcal{M}_n | X \succeq 0, x_{ii} = 1, x_{ij} \in [1 - \epsilon, 1] \text{ for all } i, j\}$ be a closed convex subset of $\mathcal{M}_n$. The optimal solution to Eq. (3) is the projection of E^0 onto $\mathcal{R}$, denoted by $E_{\mathcal{R}}^0$. Let $E^* = \exp(-\gamma D^*)$, where D^* is the true but unknown metric. Obviously $E^* \in \mathcal{R}$, and

$$\left\| E^* - E_{\mathcal{R}}^0 \right\|_F^2 \leq \left\| E^* - E_{\mathcal{R}}^0 \right\|_F^2 - 2 \left\langle E^* - E_{\mathcal{R}}^0, E^0 - E_{\mathcal{R}}^0 \right\rangle$$
$$\leq \left\| \left(E^* - E_{\mathcal{R}}^0 \right) - \left(E^0 - E_{\mathcal{R}}^0 \right) \right\|_F^2. \tag{4}$$

The first "$\leq$" holds due to Kolmogrov's criterion [7], which states that the projection of E^0 onto $\mathcal{R}$ is unique and characterized by:

$$E_{\mathcal{R}}^0 \in \mathcal{R} \text{ and } \left\langle E - E_{\mathcal{R}}^0, E^0 - E_{\mathcal{R}}^0 \right\rangle \leq 0, \text{ for all } E \in \mathcal{R}.$$

The equality holds if and only if $E_{\mathcal{R}}^0 = E^0$, i.e., $E^0 \in \mathcal{R}$.

Equation (4) gives $\left\| E^* - E_{\mathcal{R}}^0 \right\|_F^2 \leq \left\| E^* - E^0 \right\|_F^2$, which shows that $E_{\mathcal{R}}^0$ is an improved estimate towards the unknown E^*. Next, let $D_{\mathcal{R}}^0$ be obtained from $E_{\mathcal{R}}^0 = \exp(-\gamma D_{\mathcal{R}}^0)$. From Taylor-series expansion:

$$e^z = 1 + z + O\left(z^2\right) \approx 1 + z, \text{ for } |z| \ll 1,$$

we have:

$$\left\| E^* - E_{\mathcal{R}}^0 \right\|_F^2 = \frac{\epsilon^2}{\mu^2} \left\| D^* - D_{\mathcal{R}}^0 \right\|_F^2 + O(\epsilon^4), \tag{5}$$

and

$$\left\| E^* - E^0 \right\|_F^2 = \frac{\epsilon^2}{\mu^2} \left\| D^* - D^0 \right\|_F^2 + O(\epsilon^4). \tag{6}$$

If $E^0 \notin \mathcal{R}$, we have $\left\| E^* - E_{\mathcal{R}}^0 \right\|_F^2 < \left\| E^* - E^0 \right\|_F^2$. For a sufficiently small ϵ, we have $\left\| D^* - D_{\mathcal{R}}^0 \right\|_F^2 < \left\| D^* - D^0 \right\|_F^2$ based on Eqs. (5) and (6). If $E^0 \in \mathcal{R}$, we have $E_{\mathcal{R}}^0 = E^0$, which implies $D_{\mathcal{R}}^0 = D^0$. Considering both cases, we have:

---

[1] We set $\mu = \max\{d_{ij}^0\}$ and $\epsilon = 0.02$ in the study.

$$\left\| D^* - D_{\mathcal{R}}^0 \right\|_F^2 \leq \left\| D^* - D^0 \right\|_F^2. \tag{7}$$

The equality holds if and only if $\exp(-\lambda D^0) \succeq 0$. The result shows that, except in the special case $D_{\mathcal{R}}^0 = D^0$ that happens when $E^0 \in \mathcal{R}$, the calibrated $D_{\mathcal{R}}^0$ is a better estimate than the input D^0 in terms of a smaller Frobenius distance to the true but unknown metric D^*.

3.3 Dykstra's Algorithm

Solving Eq. (3) to find the projection of E^0 onto set $\mathcal{R}$ is well-studied in the optimization community. Several algorithms are available with quite good performances [28]. Similarly to the work of [16,23], we resort to a simple and flexible procedure based on Dykstra's alternating projection algorithm [10], also called *direct calibration* in the sequel.

Equip the closed convex set $\mathcal{M}_n$ with an inner product that induces the Frobenius norm:

$$\langle X, Y \rangle = trace\left(X^T Y \right), \text{ for } X, Y \in \mathcal{M}_n.$$

Define two nonempty, closed and convex subsets of $\mathcal{M}_n$:

$$\mathcal{S} = \{X \in \mathcal{M}_n | X \succeq 0\}, \text{ and } \mathcal{T} = \{X \in \mathcal{M}_n | x_{ii} = 1, x_{ij} \in [1 - \epsilon, 1] \text{ for all } i, j\}.$$

Obviously $\mathcal{R} = \mathcal{S} \cap \mathcal{T}$. Directly projecting E^0 onto $\mathcal{R}$ is expensive, while projecting it onto $\mathcal{S}$ and $\mathcal{T}$ respectively is easier. Denote by $P_{\mathcal{S}}$ the projection onto $\mathcal{S}$, and $P_{\mathcal{T}}$ the projection onto $\mathcal{T}$. For $P_{\mathcal{S}}$ and $P_{\mathcal{T}}$, we have the following two results.

Fact 1. *Let $X \in \mathcal{M}_n$ and $U\Sigma V^T$ be its singular value decomposition with $\Sigma = diag\left(\lambda_1, \cdots, \lambda_n\right)$. The projection of X onto $\mathcal{S}$ is given by: $X_{\mathcal{S}} = P_{\mathcal{S}}\left(X\right) = U\Sigma' V^T$ where $\Sigma' = diag\left(\lambda_1', \cdots, \lambda_n'\right)$ and each $\lambda_i' = \max\{\lambda_i, 0\}$.*

Fact 2. *The projection of $X \in \mathcal{M}_n$ onto $\mathcal{T}$ is given by: $X_{\mathcal{T}} = P_{\mathcal{T}}\left(X\right) = \left\{ \left(x_{\mathcal{T}}\right)_{ij} \right\}_{i,j=1}^n$ where $\left(x_{\mathcal{T}}\right)_{ij} = med\{1 - \epsilon, x_{ij}, 1\}$, i.e., the median of the three numbers.*

Dykstra's projection algorithm can be applied to find the minimizer to Eq. (3). Starting from E^0, it generates a sequence of iterates $\left\{ E_{\mathcal{S}}^t, E_{\mathcal{T}}^t \right\}$ and increments $\left\{ I_{\mathcal{S}}^t, I_{\mathcal{T}}^t \right\}$, for $t = 1, 2, \cdots$, by:

$$E_{\mathcal{S}}^t = P_{\mathcal{S}}\left(E_{\mathcal{T}}^{t-1} - I_{\mathcal{S}}^{t-1} \right) \tag{8}$$

$$I_{\mathcal{S}}^t = E_{\mathcal{S}}^t - \left(E_{\mathcal{T}}^{t-1} - I_{\mathcal{S}}^{t-1} \right) \tag{9}$$

$$E_{\mathcal{T}}^t = P_{\mathcal{T}}\left(E_{\mathcal{S}}^t - I_{\mathcal{T}}^{t-1} \right) \tag{10}$$

$$I_{\mathcal{T}}^t = E_{\mathcal{T}}^t - \left(E_{\mathcal{S}}^t - I_{\mathcal{T}}^{t-1} \right) \tag{11}$$

where $E_T^0 = E^0$, $I_S^0 = 0$, $I_T^0 = 0$ and 0 is an all-zero matrix of proper size. The sequences $\left\{ E_S^t \right\}$ and $\left\{ E_T^t \right\}$ converge to the optimal solution E_R^0 as $t \to \infty$.

3.4 Cyclic Calibration

Based on the proposed calibration model and the Dykstra's alternating projection algorithm presented in Sects. 3.2 and 3.3 respectively, a more scalable calibration algorithm, called *cyclic calibration* [24] in the sequel, can be designed based on the following result.

Fact 3. *Let $\mathcal{R}$ be a closed convex subset of $\mathcal{M}_n$ and $E^* \in \mathcal{R}$. Let $\mathcal{C}$ be a closed convex superset of $\mathcal{R}$ and $\mathcal{C} \subseteq \mathcal{M}_n$. For any $E^0 \in \mathcal{M}_n$, we have $\left\| E^* - E_\mathcal{C}^0 \right\|_F^2 \leq \left\| E^* - E^0 \right\|_F^2$. The equality holds if and only if $E_\mathcal{C}^0 = E^0$, i.e., $E^0 \in \mathcal{C}$.*

This result can be obtained similarly to Eq. (4). It states that the projection of E^0 onto $\mathcal{C}$ provides an improved estimate towards E^*. Based on the observation, we can design a domain decomposition algorithm that avoids factorizing the full $n \times n$ matrix. Let $\mathcal{C}_1, \cdots, \mathcal{C}_r$ be r closed convex sets that satisfy $\mathcal{R} \subseteq \bigcap_{k=1}^r \mathcal{C}_k$ and $\bigcup_{k=1}^r \mathcal{C}_k \subseteq \mathcal{M}_n$. Starting from $E^0 \in \mathcal{M}_n$, again we apply Dykstra's projection which generates the iterates $\{E_k^t\}$ and the increments $\{I_k^t\}$ cyclically by:

$$E_0^t = E_r^{t-1} \tag{12}$$
$$E_k^t = P_{\mathcal{C}_k} \left(E_{k-1}^t - I_k^{t-1} \right) \tag{13}$$
$$I_k^t = E_k^t - \left(E_{k-1}^t - I_k^{t-1} \right) \tag{14}$$

where $k = 1, \cdots, r$ and $t = 1, 2, \cdots$. The initial values are given by $E_r^0 = E^0$ and $I_k^0 = 0$ $(1 \leq k \leq r)$. The sequences of $\{E_k^t\}$ converges to the projection of E^0 onto $\bigcap_{k=1}^r \mathcal{C}_k$ [10].

Theorem 3. *Let $\mathcal{C}_1, \cdots, \mathcal{C}_r$ be closed and convex subsets of $\mathcal{M}_n$ such that $\mathcal{C} = \bigcap_{k=1}^r \mathcal{C}_k$ is not empty. For any $E^0 \in \mathcal{M}_n$ and any $k = 1, \cdots, r$, the sequence $\{E_k^t\}$ converges strongly to $E_\mathcal{C}^0 = P_\mathcal{C}(E^0)$, i.e., $\left\| E_k^t - E_\mathcal{C}^0 \right\|_F^2 \to 0$ as $t \to \infty$.*

To realize the cyclic calibration approach, we define the r supersets $\mathcal{C}_1, \cdots, \mathcal{C}_r$ of $\mathcal{R}$ as follows. Denote r nonempty index sets by $\mathcal{I}_1, \cdots, \mathcal{I}_r$, which satisfies $\bigcup_{k=1}^r \mathcal{I}_k = \{1, \cdots, n\}$. For any matrix $A \in \mathcal{M}_n$, denote by A_k the principal submatrix formed by selecting the same rows and columns of A indicated by $\mathcal{I}_k$. Then for each $\mathcal{I}_k$ $(1 \leq k \leq r)$, define

$$\mathcal{S}_k = \{A \in \mathcal{M}_n | A_k \succeq 0\} \text{ , and, } \mathcal{C}_k = \mathcal{S}_k \cap \mathcal{T}.$$

Recall that a matrix is positive semi-definite if and only if all its principal submatrices are positive semi-definite [14,17], and we know that $\mathcal{R} \subseteq \mathcal{C} = \bigcap_{k=1}^r \mathcal{C}_k$. So by projecting E^0 onto each $\mathcal{C}_k$ successively with Dykstra's procedure, we will obtain the projection onto $\mathcal{C}$, which provides an improved estimate towards the unknown E^*, with the following steps:

1. For given r, randomly generate index sets $\mathcal{I}_1, \ldots, \mathcal{I}_r$;
2. Calibrate the matrix by projecting it onto $\mathcal{C}_1, \cdots, \mathcal{C}_r$ cyclically;
3. Repeat steps 1 and 2 until convergence.

Cyclic calibration can be regarded as an extension of the direct calibration presented in Sect. 3.3. When $r = 1$, the cyclic algorithm reduces exactly to the direct algorithm.

Let $D_{\mathcal{C}}^0$ be obtained from $E_{\mathcal{C}}^0$. Similarly to the result in Eq. (7), we have:

$$\left\| D^* - D_{\mathcal{C}}^0 \right\|_F^2 \leq \left\| D^* - D^0 \right\|_F^2, \tag{15}$$

which shows that $D_{\mathcal{C}}^0$ improves D^0 and gets nearer to the unknown D^*.

3.5   Complexity Analysis

To project an $n \times n$ matrix directly onto the convex set $\mathcal{S}$ via SVD, the complexity is $O\left(n^3\right)$ per iteration [5,15]. With cyclic calibration, we set the cardinality of I_k to $O\left(n/r\right)$ and project an input matrix onto $\mathcal{S}_k$ $(1 \leq k \leq r)$ successively. We need to decompose r principal submatrices in each iteration. The complexity is $O\left(n^3/r^3\right)$ to decompose one submatrix, and $O\left(n^3/r^2\right)$ for r decompositions, which significantly improves the complexity of the direct approach.

For the number of iterations to converge, theoretically, the convergence rate of Dykstra's alternating projection for polyhedral sets is known to be linear [10,12]. Empirically the direct approach converged in around 20 iterations, and the cyclic approach converged in around 40 iterations on a problem with $n = 10,000$ and $r = 10$ in our evaluation.

For memory requirement, if the whole distance matrix is stored in memory, both calibration approaches have a storage complexity of $O\left(n^2\right)$. For the cyclic approach, it is also possible to reduce the storage complexity to $O\left(n^2/r^2\right)$ by only keeping the working principal submatrix in memory, at the cost of swapping-in and swapping-out operations on other matrix elements from time to time.

4   Evaluation

4.1   Settings

We carried out empirical studies to evaluate the proposed model and calibration algorithms, specifically with the objectives of investigating their effectiveness in:

- reducing the noise of distance matrices;
- computing distance metrics from incomplete data;
- performances in classification applications;
- running speed and scalability.

We used five benchmark datasets that are publicly available. These datasets cover a reasonably wide range of application domains, including:

Table 1. Relative squared deviations on calibration of noisy distance metrics. Each item has two values, corresponding to $\zeta = 0.1$ and 0.5 respectively: the smaller RSD value, the better performance. Direct calibration reported the best calibration quality on almost all experiments.

Dataset	TRIFIX	DIRECT	CYCLIC
MNIST	.976/.362	**.137/.034**	.164/.039
CIFAR10	.904/.369	**.146/.031**	.181/.037
PROTEIN	.992/.358	**.178**/.031	.222/**.025**
RCV1	.999/.356	**.184/.023**	.233/.023
SENSEIT	.778/.351	**.103/.032**	.123/.045

- MNIST: images of handwritten digits with 28×28 pixels each [21];
- CIFAR10: ten classes of color images with 32×32 pixels each [20];
- PROTEIN: 357-dimensional sparse binary bio-samples in three classes [4];
- RCV1: $47,236$-dimensional sparse newswires from Reuters in two classes [22];
- SENSEIT: 100-dimensional samples from a vehicle net in three classes [8].

We implemented the calibration approaches in the MATLAB platform. For the cyclic calibration approach, the number of partitions was set to $r = 10$ unless otherwise specified. All results were recorded on a server with 28 CPU cores and 192GB memory enabled for computation.

4.2 Noise Reduction on Distance Metrics

One specific application scenario of the proposed approaches is noise reduction in given distance metrics. In each run of the experiment, we randomly chose $1,000$ samples from the MNIST dataset and computed their pairwise Euclidean distance matrix (D^*) as the ground truth metric. Next we added certain amounts of white noise to D^* and obtain a noisy metric D^0 with each $d_{ij}^0 = \max\{0, d_{ij}^* + \zeta\mu v\}$, where μ is the mean of all elements in D^*, ζ was set to $0.1/0.5$ respectively and $v \sim N(0,1)$ is a standard Gaussian random variable.

We applied the direct calibration approach (denoted by DIRECT) and the cyclic calibration approach (denoted by CYCLIC) on D^0 and obtained two calibrated matrices $(D_{\mathcal{R}}^0)$. The relative squared deviation (RSD) from D^*, calculated as $\frac{\left\|D_{\mathcal{R}}^0 - D^*\right\|_F^2}{\left\|D^0 - D^*\right\|_F^2}$, was recorded to measure the performance of each calibration method.

We repeated the experiment for ten runs and reported the mean of the results in Table 1. Compared with the noisy matrix D^0, the direct calibration reduced more than 86% of squared deviation when $\zeta = 0.1$ and more than 96% when $\zeta = 0.5$, and the cyclic calibration reported comparable improvements.

Table 2. Relative squared deviations on calibration of approximate metrics from incomplete samples. Each item has two values, corresponding to $p = 0.1$ and 0.5 respectively: the smaller RSD value, the better performance. Direct calibration reported the best performances on almost all experiments.

DATASET	IMPUTATION			CALIBRATION		
	MEAN	kNN	SVT	TRIFIX	DIRECT	CYCLIC
MNIST	2.80/8.20	7.96/18.5	2.51/8.57	1.00/.998	**.998/.767**	1.12/.813
CIFAR10	765./243	119./292	25.1/61.5	1.00/1.00	**.991/.979**	1.37/.997
PROTEIN	53.0/17.7	1.77/3.40	1.79/3.12	.994/.924	**.975**/.506	1.05/**.464**
RCV1	1.48/3.00	1.52/3.07	1.44/2.89	.999/.931	**.806/.429**	.975/.430
SENSEIT	82.8/27.9	.908/1.13	.890/.861	.922/.502	**.874/.489**	.917/.543

We also recorded the performance of triangle fixing (TRIFIX) algorithm[2] (cf. Sect. 2.2), which calibrates the noisy metric to restore the triangle inequalities. The triangle fixing algorithm reduced around 2% and 64% squared deviations, respectively. As a comparison, our proposed approaches reported significantly superior calibration results. In addition to the MNIST dataset, we carried out the same experiment on the other datasets and found very similar results.

4.3   Distance Metrics from Incomplete Data

The second experiment was on estimating the distance metric from incomplete observations. In each of the ten runs, we randomly chose a subset of $1,000$ samples from the MNIST dataset and computed the pairwise distance matrix D^* as the ground truth.

Then, we randomly marked different portions ($p = 0.1/0.5$ respectively) of features as missing for each sample. For any two incomplete samples x_i and x_j in the dataset, denote $x_i(x_j)$ a new vector formed by keeping those features of x_i that are observed in both x_i and x_j. Based on the common features, an approximate distance for the two incomplete samples was given by:

$$d_{ij}^0 = \|x_i(x_j) - x_j(x_i)\| \sqrt{\frac{q}{q_{ij}}}$$

where $q = 784$ is the dimension of the MNIST samples and q_{ij} is the number of features observed in both samples.

Let a distance matrix $D^0 = \{d_{ij}^0\}_{i,j=1}^n$. The matrix is often not embeddable, which leaves potential room for further calibration. Accordingly, we calibrated D^0 to a new estimate $D_{\mathcal{R}}^0$ by our proposed approaches, and computed their RSD values from the ground truth D^* as described in Sect. 4.2.

The two proposed approaches were compared with the triangle fixing algorithm (cf. Sect. 2.2) on the quality of the calibration. In addition, the results

---

[2] Implementation downloaded from http://optml.mit.edu/software.html.

from several imputation methods, which were popularly used in practice, were also included as a baseline. These imputation methods include:

- MEAN: Replacing missings by the observed mean of the feature;
- kNN: Replacing missings by weighted mean of $k = 5$ nearest samples [33];
- SVT: Low-rank matrix completion with singular value thresholding[3] [3].

We applied these imputation methods to replace the missing features with substituted values, calculated the distance matrix based on the imputed data, and recorded the corresponding RSD values. In the experiment, we also tested two implementations [13,26] of the classical expectation-maximization algorithm and the recent optimal transport algorithm [31] to impute the data. Unfortunately, different from their known excellent performances on low-dimensional data, both algorithms failed to execute on most of these high-dimensional data samples with a large portion of missing values. So their results were not available.

The results are given in Table 2. Compared with the un-calibrated D^0, we can see that the calibration approaches brought significant drops in squared deviations from the true D^*. Direct calibration reported the best results in RSD values on most of the datasets and the settings. When $p = 0.5$, it reduced around 23% to 57% squared deviations on most datasets. The only exception is on the CIFAR10 dataset, where the reduction of squared deviations is not that significant. However, the improvement from calibration approaches over the imputation methods is still significant.

At the same time, we can see that the imputation approaches had no guarantee of the quality of RSD values. The imputed data's distance matrix may be far from the ground truth. For example, naïvely filling the mean to the missing values on the CIFAR10 dataset produced a distance matrix that was more than seven hundred times away from the ground truth than that of D^0. Comparatively, the calibration approaches consistently reduced the squared deviation from the ground truth by calibrating the input matrix as expected.

4.4   Classification on Incomplete Samples

Having justified the capability of removing metric noises by the proposed approaches, we would like to investigate whether the calibrated results benefit real applications. Specifically, we applied the calibrated metrics in nearest neighbor classification tasks. Given a training set of samples with class labels, we tried to predict the labels of the samples in the testing set. For each testing sample, its label was predicted by the label of the nearest neighbor in the training set. Then the predicted label was compared against the accurate label to measure the classification performance.

We carried out one-versus-all cross-validation on the classification task. Each sample was used, in turn, as the testing sample, while all other samples formed the training set. We averaged all testing samples' classification errors and

---

[3] Implementation downloaded from https://candes.su.domains/software/.

Table 3. Ten-fold mean classification errors on incomplete samples by nearest neighbor classifier. Each item has two values, corresponding to $p = 0.1$ and 0.5 respectively: the smaller MCE values, the better performance. Direct calibration reported the best performances on almost all experiments.

	D^0	IMPUTATION			CALIBRATION		
		MEAN	kNN	SVT	TRIFIX	DIRECT	CYCLIC
MN	.127/.203	.145/.503	.132/.272	.150/.450	.127/.200	**.126/.192**	.127/**.191**
CI	.767/.765	.786/.878	.781/.842	.780/.838	.767/.765	**.751/.757**	.771/.758
PR	.581/.615	.604/.634	.638/.698	.595/.620	.581/.624	**.573/.610**	.574/.617
RC	.339/.442	.324/.432	.324/.435	.334/.437	.338/.463	**.321/.419**	.327/.427
SE	.299/.377	.394/.503	.299/.389	.292/.402	.292/.362	**.288/.357**	.301/.366

Table 4. Ten-fold mean classification errors on incomplete samples by hard-margin SVM with Gaussian kernel and default parameters. One-versus-all strategy was applied for classifying more than two classes. Each item has two values, corresponding to $p = 0.1$ and 0.5 respectively: the smaller MCE values, the better performance. Direct calibration reported the best performances on almost all experiments.

	D^0	IMPUTATION			CALIBRATION		
		MEAN	kNN	SVT	TRIFIX	DIRECT	CYCLIC
MN	.100/.141	.105/.158	.098/.894	.105/.159	.100/.142	**.096/.136**	.097/.138
CI	.661/.664	.671/.694	.785/.903	.665/.691	.661/.664	**.656/.658**	.658/.670
PR	.516/.606	.399/.485	.401/.686	.400/.486	.516/.610	.452/.550	**.397/.484**
RC	.247/.370	.127/.287	.497/.523	.138/.258	.136/.384	**.124/.237**	.135/.248
SE	.373/.478	.239/.293	.249/.741	.240/.291	.369/.473	.264/.388	**.231/.289**

recorded the mean of average classification errors (MCE) over ten runs. Similar to the RSD results in Table 2, the calibration approaches reported improved MCE results in Table 3. With different missing ratios $p = 0.1$ and $p = 0.5$, the calibration approaches consistently reduce the classification errors over the approximate metric D^0 on all datasets, among which the direct calibration approach performed the best. Comparatively, the metrics from imputation-based approaches sometimes performed even worse than D^0.

We further experimented with the support vector machines (SVM) algorithm [30]. SVM seeks a linear boundary with the maximum margin to separate two classes of samples in the feature space. To apply SVM, a positive semi-definite kernel matrix needs to be provided as the input to the algorithm. In the evaluation, we used the popular Gaussian kernel to construct the kernel matrix, $\exp(-\alpha D^2)$, where D^2 is the element-wise square of the metric obtained from each algorithm and α is the default kernel parameter set by the LibSVM package [4]. In case the kernel matrix constructed is not positive semi-definite (namely,

D^0 and TRIFIX), a small positive number will be added to the diagonal elements to shift the matrix to be positive semi-definite. The one-versus-all strategy was applied for classifying more than two classes.

The MCE results are shown in Table 4. The proposed calibration methods reported similarly improved accuracies over the un-calibrated metric and the imputation approaches. The most significant improvement over the performance of D^0 was on the RCV1 dataset, from 0.247 to 0.124 and from 0.370 to 0.237 respectively. Consistent improvements were observed on the other datasets. Similarly, the calibration approaches reported superior results over the imputation approaches on most experiments. In the evaluation, we found that when the missing ratio is high, the performances of the imputation approaches become relatively unstable. For example, when $p = 0.5$, the misclassification error with kNN imputation significantly increased to 0.894 on the MNIST dataset and 0.903 on the CIFAR10 dataset, like a random guess. Comparatively, the calibration approaches' performances are much more reliable.

When comparing the proposed calibration approaches with the triangle fixing algorithm, we can find a similar trend of improvement in classification errors, although not as significant as the improvement over the imputation approaches. The improved classification accuracies are consistent with the results reported in Sects. 4.2 and 4.3, which again justifies the benefits from the better-calibrated metrics to the unknown ground truth metric.

4.5   Scalability

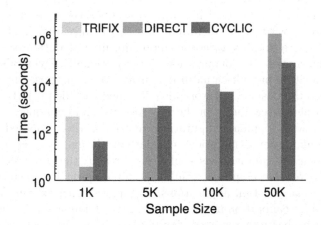

Fig. 1. Running time on the MNIST dataset with different sample sizes. For $n = 1K/5K/10K$, $r = 10$; for $n = 50K$, $r = 50$. $|I_k| \approx \frac{2n}{r}$ ($1 \le k \le r$). The triangle fixing algorithm failed to execute other than $n = 1K$. Cyclic calibration exhibited evidently improved scalability when the sample size is large.

The last experiment was to evaluate the scalability. Similarly to the setting described in Sect. 4.2, white noise was added to the factual distance matrix, and then the calibration approaches were carried out. Figure 1 shows the running time in seconds of our two proposed approaches and the triangle fixing algorithm on the MNIST dataset with different numbers of training samples from $n = 1,000$ to $n = 50,000$.

With $n = 1,000$ training samples, the direct calibration approach took around three seconds, about a hundred times faster than the triangle fixing algorithm with the default parameter setting. When $n = 2,000$ (not shown in the figure) or larger, the triangle fixing algorithm failed to execute on our platform due to prohibitive memory requirement caused by the $O(n^3)$ triangle inequalities, so the results were not available here.

When comparing the direct and the cyclic calibration approaches, we can see that the cyclic approach did not report advantage with a small number of samples. However, when the number of samples got sufficiently large, e.g., $n = 10,000$, the cyclic approach began to exhibit its superiority. When $n = 50,000$, the cyclic approach was around twenty times faster than the direct approach to converge, being consistent with the complexity analysis in Sect. 3.5 and confirming a more scalable solution.

5   Conclusion

Estimating distance metrics between samples is a fundamental problem in data processing with various applications. To deal with the challenge, we suggested calibrating an approximate metric, which avoids the difficulty in imputation and returns an improved estimate with a solid guarantee. By connecting isometrical embedding and positive semi-definiteness of a distance matrix, the proposed approach provides a simple yet rigorous model for missing data processing, which forms the main contribution of our work. Computationally, Dykstra's alternating projection algorithm provides a natural solution to our proposed model and can be applied directly. Besides, our work also designed a cyclic projection algorithm that provided better scalability in the way of divide and conquer.

Compared with popular imputation methods, the proposed calibration approaches make fewer assumptions on the correlations among data features and the intrinsic data dimensions/ranks. As a result, the proposed approaches reported more reliable empirical results in our empirical evaluations of noise reduction and classification applications. Compared with existing models that can be applied for calibration purposes, such as the triangle fixing algorithm, the proposed approaches also reported significantly improved speed and accuracy. Although preliminary, all the results clearly justified the proposed approaches' benefits and demonstrated their high potential in practical tasks.

Despite the achieved results, more work along this line deserves to be investigated. The improved performance of our work relies on the assumption that the data samples can be isometrically embeddable in a Hilbert/Euclidean space. However, this assumption may not hold for general metrics. For example, the

Robinson-Foulds distance metric [32] defined on trees satisfies the triangle inequalities but is typically not embeddable. Can we extend the proposed approach to calibrate such metrics? It deserves our investigation. Another potential topic, although the cyclic calibration approach exhibited better scalability, it still seems demanding when handling big data, and the scalability issue deserves further consideration.

Acknowledgments. We thank the reviewers for the helpful comments. The work is supported by Guangdong Basic and Applied Basic Research Foundation (2021A1515011825), Shenzhen Science and Technology Program (CUHK-SZWDZC0004), and Shenzhen Research Institute of Big Data.

References

1. Boyd, S., Vandenberghe, L.: Convex Optimization. Cambridge University Press, New York (2004)
2. Brickell, J., Dhillon, I., Sra, S., Tropp, J.: The metric nearness problem. SIAM J. Matrix Anal. Appl. **30**(1), 375–396 (2008)
3. Cai, J.F., Candès, E., Shen, Z.: A singular value thresholding algorithm for matrix completion. SIAM J. Optim. **20**(4), 1956–1982 (2010)
4. Chang, C.C., Lin, C.J.: LIBSVM: a library for support vector machines. ACM Trans. Intell. Syst. Technol. **2**(3), 1–27 (2011)
5. Cline, A., Dhillon, I.: Computation of the singular value decomposition. In: Handbook of Linear Algebra, pp. 45–1. Chapman and Hall/CRC (2006)
6. Dempster, A., Laird, N., Rubin, D.: Maximum likelihood from incomplete data via the EM algorithm. J. Roy. Stat. Soc. B **39**, 1–38 (1977)
7. Deutsch, F.: Best Approximation in Inner Product Spaces. Springer, New York (2001)
8. Duarte, M., Hu, Y.: Vehicle classification in distributed sensor networks. J. Parallel Distrib. Comput. **64**(7), 826–838 (2004)
9. Duda, R., Hart, P.: Pattern Classification. Wiley, Hoboken (2000)
10. Dykstra, R.: An algorithm for restricted least squares regression. J. Am. Stat. Assoc. **78**(384), 837–842 (1983)
11. Enders, C.: Applied Missing Data Analysis. Guilford Press (2010)
12. Escalante, R., Raydan, M.: Alternating Projection Methods. SIAM, Philadelphia (2011)
13. Ghahramani, Z., Jordan, M.: Supervised learning from incomplete data via an EM approach. Adv. Neural. Inf. Process. Syst. **6**, 120–127 (1994)
14. Gilbert, G.: Positive definite matrices and Sylvester's criterion. Am. Math. Mon. **98**(1), 44–46 (1991)
15. Golub, G., Van Loan, C.: Matrix Computations. Johns Hopkins University Press, Baltimore (1996)
16. Higham, N.: Computing the nearest correlation matrix - a problem from finance. IMA J. Numer. Anal. **22**, 329–343 (2002)
17. Horn, R., Johnson, C.: Matrix Analysis. Cambridge University Press, Cambridge (2012)
18. Jain, A., Murty, M., Flynn, P.: Data clustering: a review. ACM Comput. Surv. **31**(3), 264–323 (1999)

19. Jannach, D., Resnick, P., Tuzhilin, A., Zanker, M.: Recommender systems—beyond matrix completion. Commun. ACM **59**(11), 94–102 (2016)
20. Krizhevsky, A., Hinton, G., et al.: Learning multiple layers of features from tiny images (2009)
21. LeCun, Y., Bottou, L., Bengio, Y., Haffner, P.: Gradient-based learning applied to document recognition. Proc. IEEE **86**(11), 2278–2324 (1998)
22. Lewis, D., Yang, Y., Rose, T., Li, F.: RCV1: a new benchmark collection for text categorization research. J. Mach. Learn. Res. **5**(Apr), 361–397 (2004)
23. Li, W.: Estimating Jaccard index with missing observations: a matrix calibration approach. Adv. Neural. Inf. Process. Syst. **28**, 2620–2628 (2015)
24. Li, W.: Scalable calibration of affinity matrices from incomplete observations. In: Asian Conference on Machine Learning, pp. 753–768 (2020)
25. Little, R., Rubin, D.: Statistical Analysis with Missing Data, vol. 793. Wiley, Hoboken (2019)
26. Murphy, K.: Machine Learning: a Probabilistic Perspective. MIT Press, Cambridge (2012)
27. Muzellec, B., Josse, J., Boyer, C., Cuturi, M.: Missing data imputation using optimal transport. In: International Conference on Machine Learning, pp. 7130–7140. PMLR (2020)
28. Qi, H., Sun, D.: An augmented Lagrangian dual approach for the H-weighted nearest correlation matrix problem. IMA J. Numer. Anal. **31**(2), 491–511 (2011)
29. Schoenberg, I.: Metric spaces and positive definite functions. Trans. Am. Math. Soc. **44**(3), 522–536 (1938)
30. Schölkopf, B., Smola, A., Bach, F., et al.: Learning with Kernels: Support Vector Machines, Regularization, Optimization, and Beyond. MIT Press, Cambridge (2002)
31. Sonthalia, R., Gilbert, A.C.: Project and forget: solving large-scale metric constrained problems. arXiv preprint arXiv:2005.03853 (2020)
32. Stockham, C., Wang, L.S., Warnow, T.: Statistically based postprocessing of phylogenetic analysis by clustering. Bioinformatics **18**(suppl_1), S285–S293 (2002)
33. Troyanskaya, O., Cantor, M., Sherlock, G., Brown, P., Hastie, R., et al.: Missing value estimation methods for DNA microarrays. Bioinformatics **17**(6), 520–525 (2001)
34. Wells, J., Williams, L.: Embeddings and Extensions in Analysis, vol. 84. Springer, Heidelberg (1975). https://doi.org/10.1007/978-3-642-66037-5
35. Xing, E., Jordan, M., Russell, S., Ng, A.: Distance metric learning with application to clustering with side-information. Adv. Neural. Inf. Process. Syst. **15**, 521–528 (2002)

Defending Observation Attacks in Deep Reinforcement Learning via Detection and Denoising

Zikang Xiong[1][(✉)], Joe Eappen[1], He Zhu[2], and Suresh Jagannathan[1]

[1] Purdue University, West Lafayette, IN 47906, USA
{xiong84,jeappen}@purdue.edu, suresh@cs.purdue.edu
[2] Rutgers University, New Brunswick, NJ 08854, USA
hz375@cs.rutgers.edu

Abstract. Neural network policies trained using Deep Reinforcement Learning (DRL) are well-known to be susceptible to adversarial attacks. In this paper, we consider attacks manifesting as perturbations in the observation space managed by the external environment. These attacks have been shown to downgrade policy performance significantly. We focus our attention on well-trained deterministic and stochastic neural network policies in the context of continuous control benchmarks subject to four well-studied observation space adversarial attacks. To defend against these attacks, we propose a novel defense strategy using a detect-and-denoise schema. Unlike previous adversarial training approaches that sample data in adversarial scenarios, our solution does not require sampling data in an environment under attack, thereby greatly reducing risk during training. Detailed experimental results show that our technique is comparable with state-of-the-art adversarial training approaches.

1 Introduction

Deep Reinforcement Learning (DRL) has achieved promising results in many challenging continuous control tasks. However, DRL controllers have proven vulnerable to adversarial attacks that trigger performance deterioration or even unsafe behaviors. For example, the operation of an unmanned aerial navigation system may be degraded or even maliciously affected if the training of its control policy does not carefully account for observation noises introduced by sensor errors, weather, topography, obstacles, etc. Consequently, building robust DRL policies remains an important ongoing challenge in architecting learning-enabled applications.

There have been several different formulations of DRL robustness that have been considered previously. [13,18] consider DRL robustness against perturbations of physical environment parameters. More generally, [7] has formalized DRL

Supplementary Information The online version contains supplementary material available at https://doi.org/10.1007/978-3-031-26409-2_15.

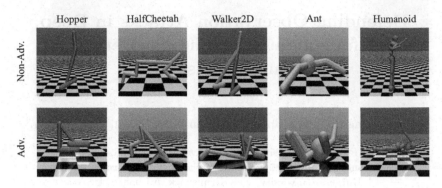

Fig. 1. Robots we evaluated in non-adversarial and adversarial scenarios. Robots fall down and gain less rewards when they are under attack.

robustness against uncertain state transitions, and [21] has studied DRL robustness against action attacks. Similar to [27], our work considers DRL robustness against *observation attacks*. Prior work has demonstrated a range of strong attacks in the observation space of a DRL policy [9,11,16,20,26,27], all of which can significantly reduce a learning-enabled system's performance or cause it to make unsafe decisions. Because observations can be easily perturbed, robustness to these kinds of adversarial attacks is an important consideration that must be taken into account as part of a DRL learning framework. There have been a number of efforts that seek to improve DRL robustness in response to these concerns. These include enhancing DRL robustness by adding a regularizer to optimize goals [1,27] and defending against adversarial attacks via switching policies [5,24]. There have also been numerous proposals to improve robustness using adversarial training methods. These often require sampling observations under *online* attacks (e.g., during simulation) [9,16,26]. However, while these approaches provide more robust policies, it has been shown that such approaches can negatively impact policy performance in non-adversarial scenarios. Moreover, a large number of unsafe behaviors may be exhibited during online attacks, potentially damaging the system controlled by the learning agent if adversarial training takes place in a physical rather than simulated environment.

To address the aforementioned challenges, we propose a new algorithm that strengthens the robustness of a DRL policy *without* sampling data under adversarial scenarios, avoiding the drawbacks that ensue from encountering safety violations during an online training process. Our method is depicted in Fig. 2. Given a DRL policy π, our defense algorithm *retains* π's parameters and trains a *detector* and *denoiser* with offline data augmentation. The detector and denoiser address problems on when and how to defend against an attack, resp. When defending π in a possibly adversarial environment, the detector identifies anomalous observations generated by the adversary, and the denoiser processes these observations to reverse the effect of the attacks. With assistance from the detector and the denoiser, the algorithm overcomes adversarial attacks in the policy's observation space while retaining performance in terms of the achieved total reward.

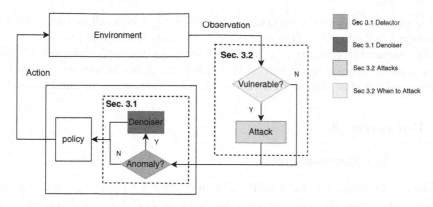

Fig. 2. Framework

Both the detector and denoiser are modeled with Gated Recurrent Unit Variational Auto-Encoders (GRU-VAE). This design choice is inspired by recent work [12,15,23,25] that has demonstrated the power of such anomaly detectors and denoisers. After anomalies enforced by attacks are detected, we need to reverse the effect of the attacks with a denoiser. However, training such a denoiser requires the observations under attack as input, but sampling such adversarial observations online is unappealing. To avoid unsafe sampling, our algorithm instead conducts adversarial attacks using offline data augmentation on a dataset of observations collected by the policy in a non-adversarial environment.

Our approach provides several important benefits compared with previous online adversarial training approaches. First, because we do not retrain victim policies, our approach naturally *retains* a policy's performance in non-adversarial scenarios. Second, unlike adversarial training methods that need to sample data under online adversarial attacks, we only require sampled observations with a pretrained policy in a normal environment not subject to attacks. Third, the stochastic components in our detect-and-denoise pipeline (i.e., the prior distribution in the variational autoencoders) provide a natural barrier to defeat adversarial attacks [10,14]. We have evaluated our approach on a range of challenging MuJoCo [22] continuous control tasks for both deterministic TD3 policies [3] and stochastic PPO policies [19]. Our experimental results show that compared with the state-of-the-art online adversarial training approaches [26], our algorithm does not compromise policy performance in perturbation-free environments and achieves comparable policy performance in environments subject to adversarial attacks.

To summarize, our contributions are as follows:

- We integrate autoencoder-style anomaly detection and denoising into a defense mechanism for DRL policy robustness and show that the defense mechanism is effective under environments with strong known attacks as well as their variants and does not compromise policy performance in normal environments.

- We propose an adversarial training approach that uses offline data augmentation to avoid risky online adversarial observation sampling.
- We extensively evaluate our defense mechanism for both deterministic and stochastic policies using four well-studied categories of strong observation space adversarial attacks to demonstrate the effectiveness of our approach.

2   Background

2.1   Markov Decision Process

A Markov Decision Process (MDP) is widely used for modeling reinforcement learning problems. It is described as a tuple $(S, A, T, R, \gamma, O, \phi)$. S and A represent the state and action space, resp. $T(s, a) : S \times A \rightarrow \mathbb{P}(S)$ is the transition probability distribution. Given current state s and the action a, the Markov probability transition function $T(s, a)$ returns the probability of a new state s'. $R(s, a, s') : S \times A \times R \rightarrow \mathbb{R}$ is the reward function that measures the performance of a given transition (s, a, s'). Let the cumulative discounted reward be $\mathcal{R}$, and the reward at time t be $R(s_t, a_t, s_{t+1})$. Then, $\mathcal{R} = \sum_{t=0}^{T} \gamma^t R(s_t, a_t, s_{t+1})$, where $\gamma \in [0, 1)$ is the discounted factor and T is the maximum time horizon. The last element in the MDP tuple is an observation function $\phi : S \rightarrow O$ which transforms states in the state space S to the observation space O. The task of solving an MDP is tantamount to finding an optimal policy $\pi : O \rightarrow A$ that maximizes the discounted cumulative reward $\mathcal{R}$.

2.2   Observation Attack

Given a pretrained policy π, the observation attack $\mathcal{A}_\mathcal{B}$ injects noise to the observation to downgrade the cumulative reward $\mathcal{R}$. $\mathcal{B}$ quantifies this noise term. Typically, $\mathcal{B}$ is an ℓ_n-norm region around the ground-truth observation. Given an observation o_t, $\mathcal{B}(o_t) = \{\hat{o}_t \mid ||\hat{o}_t - o_t||_n < \varepsilon\}$, where ε is the radius of the ℓ_n norm region. Additionally, attacks can choose when to inject noise. Since it is crucial to downgrade performance using as few attacks as possible, it is typical to define a vulnerability indicator $\mathbb{1}_{vul} : O \rightarrow \{\text{True}, \text{False}\}$. Given an observation o_t, if $\mathbb{1}_{vul}(o_t)$ is False, the policy π receives the perturbation-free observation o_t as input; otherwise, the input will be an adversarial observation $\hat{o}_t = \mathcal{A}_\mathcal{B}(o_t)$.

2.3   Defense via Detection and Denoising

The MDP tuple becomes $(S, A, T, R, \gamma, O, \phi, \mathcal{A}_\mathcal{B}, \mathbb{1}_{vul})$ after incorporating an adversary. One way to defend against adversarial attacks is to retrain a policy for the new MDP. However, such an approach ignores the fact that we already have a trained policy that performs well in non-adversarial scenarios. Additionally, the solution of such an MDP may not yield an optimal policy [27]. In contrast, our approach considers removing the effects introduced by $\mathcal{A}_\mathcal{B}, \mathbb{1}_{vul}$ by casting the adversarial MDP problem back into a standard MDP.

To do so, we exploit the trained policy and avoid the possibility of failing to find an optimal policy, even in non-adversarial scenarios. Notably, our approach eliminates the effect introduced by the vulnerability indicator $\mathbb{1}_{vul}$ and observation attack $\mathcal{A}_{\mathcal{B}}$ by using an anomaly detector and a denoiser, resp. Given a sequence of observations $h_t = \{o_0, ..., o_t\}$, the detector is tasked with predicting whether an attack happens in the latest observation o_t. Conversely, the denoiser predicts the ground-truth observation of o_t with h_t. If the detector finds an anomaly, the denoiser's prediction is used to replace the current observation o_t with the ground-truth observation. Our defense only intervenes when the detector reports an anomaly, which preserves the performance of pretrained policies when no adversary appears. Training a VAE denoiser typically requires both the groundtruth inputs (i.e., the actual observations) and the perturbed inputs (i.e., the adversarial observations). However, sampling the adversarial observations under online adversarial attacks can be risky. Thus, we prefer sampling adversarial observations offline.

2.4 Online and Offline Sampling

The difference between online and offline sampling manifests in whether we need to sample data via executing an action in an environment. Adversarial attacks can downgrade performance by triggering unsafe behaviors (e.g., flipping an ant robot, letting a humanoid robot fall), and hence online sampling adversarial observations can be risky. In contrast, offline sampling does not collect data via executing actions in an environment and thus does not suffer from potential safety violations when performing the sampling online. Here, adversarial observations are sampled offline by running adversarial attacks on a normal observation dataset (i.e., observations generated in non-adversarial scenarios).

3   Approach

The overall framework of our approach is shown in Fig. 2. Our defense technique is presented in Sect. 3.1. It consists of two components: a detector and a denoiser. First, the anomaly detector checks whether the current environment observation is an anomaly due to an adversarial attack. When an anomaly is detected, the denoiser reverses the attack by denoising the perturbed observation. We evaluate our defense strategy over four attacks described in Sect. 3.2. Similar to [9,11], our framework allows an adversary, when given an observation, to decide whether the observation is vulnerable to an attack.

3.1   Defense

Adversarial training has broad applications to improve the robustness of machine learning models by augmenting the training dataset with samples generated by adversarial attacks. In the context of deep reinforcement learning, previous approaches [9,13,16,26] conduct policy searches in environments subject to such attacks, leading to robust policies under observations generated from adversarial distributions. As mentioned earlier, our method is differentiated from these approaches by using a detect-and-denoise schema learnt from offline data augmentation while keeping pretrained policies.

Prior work has shown that the LSTM-Autoencoder structure outperforms other methods in various anomaly settings [12, 23, 25] including anomaly detection in real-world robotic tasks [15]. Inspired by the success of this design choice, we choose to implement both the detector and denoiser as Gated Recurrent Unit Variational Auto-Encoders (GRU-VAE).

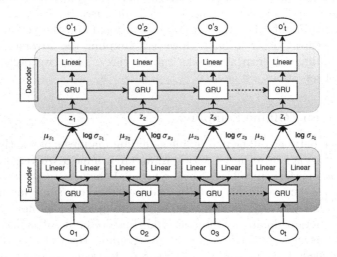

Fig. 3. GRU-VAE. The input of the encoder is a sequence of observations. These observations pass a GRU layer and two different linear layers to generate the mean μ_t and log variance $log\ \sigma_t$ of a Gaussian distribution. The latent variable z_t is sampled from this Gaussian distribution, and is passed to the decoder. The decoder decodes z_t with a GRU and a linear layer. The decoder is a deterministic model. For the detector, the output of the decoder is trained to be the same as the input observation sequence. For the denoiser, the output is trained to remove perturbations injected by adversaries.

Detector. The structure of our detector is depicted in Fig. 3. The detector learns what normal observation sequences should be. We train it with an observation dataset $\mathcal{D}_{normal}$ sampled online with a pretrained policy in non-adversarial environments. The objective function is the standard variational autoencoder lower bound [2],

$$L_{\text{det}} = \mathbb{E}_{q_{\theta_q}(z_t|o_t, h_t^o)} \left[\log p_{\theta_p}(o_t \mid z_t, h_t^z) \right] - D_{KL}\left(q_{\theta_q}(z_t \mid o_t, h_t^o) \| pr(z_t) \right)$$

where θ_q is the parameters of encoder q_{θ_q} and θ_p is the parameters of decoder p_{θ_p}; $o_t \in \mathcal{D}_{normal}$ is the observation at time t; h_t^o and h_t^z are the hidden states for the encoder and decoder, resp.; and z_t is sampled from the distribution parameterized by q_{θ_q}. Decoding the latent variable z_t reconstructs the input observation o_t. $\mathbb{E}_{q_{\theta_q}(z_t|o_t,h_t^o)}\left[\log p_{\theta_p}(o_t \mid z_t, h_t^z)\right]$ is known as the reconstruction objective, the maximization of which increases the likelihood of reconstructing the observations sampled by the pre-trained policy. $D_{KL}\left(q_{\theta_q}(z_t \mid o_t, h_t^o)\|pr(z_t)\right)$ is the KL-divergence between the distribution $q_{\theta_q}(z_t \mid o_t, h_t^o)$ generated by the encoder and the prior distribution $pr(z_t)$, which serves as the KL regularizer that makes these two distributions similar. Following [15], we set $pr(z_t)$ as a Gaussian distribution whose covariance is the identity matrix I, but leave the mean of $pr(z_t)$ to be μ_{z_t} instead of 0. The learnable μ_{z_t} allows the mean of the prior distribution to be conditioned on input observations. This modified GRU-VAE is different from a general GRU-VAE model which assumes the prior distribution is a fixed normal distribution. It depends on the decoder to provide prior distributions, which is crucial for a detector as shown in [15].

The detector reports an anomaly observation when a decoded observation is significantly different from the encoded observation, measured by the ℓ_∞-norm between the input observation o_t and the output observation o_t'. The detector reports the anomaly if the ℓ_∞-norm between o_t and o_t' is greater than a threshold $C_{anomaly}$ using by the anomaly detection indicator function:

$$\mathbb{1}_{anomaly}(o_t, o_t') := \ell_\infty(o_t, o_t') > C_{anomaly}$$

Denoiser. The denoiser learns to map anomaly observations found by the detector to the ground-truth normal observations. The objective function of the denoiser is:

$$L_{\text{den}} = \mathbb{E}_{q'_{\theta_{q'}}(z_t|\bar{o}_t, h^{\bar{o}}t)}\left[\log p'_{\theta_{p'}}(o_t \mid z_t, h_t^z)\right] - D_{KL}\left(q'_{\theta_{q'}}(z_t \mid \bar{o}_t, h_t^{\bar{o}})\|pr(z_t)\right)$$

Compared to the detector's objective function L_{det}, the input to the encoder $q'_{\theta_{q'}}$ is replaced by an observation $\bar{o}_t \in \mathcal{D}_{adv} \cup \mathcal{D}_{normal}$ and hidden state $h_t^{\bar{o}}$. The encoder of the denoiser maps $\bar{o}_t$ and hidden state h_t^o to a latent variable z_t that is used by the decoder $p'_{\theta_{p'}}$ to generate the ground-truth observation o_t. We also leave the mean of $pr(z_t)$ to be μ_{z_t} as in the detector. Training the denoiser requires the observation $\bar{o}_t$, which we have to sample by conducting adversarial attacks.

Sampling adversarial observations online is generally viewed to be a costly requirement because it must handle potentially unsafe behaviors that might manifest; these behaviors could damage physical agents (e.g., robots) during training (e.g., by causing a robot to fall down). In contrast, we generate these adversarial observations offline. First, since a well-trained policy already exists, we can sample a normal observation dataset $\mathcal{D}_{normal}$ online. Then, we directly apply adversarial attacks to this dataset. Given an adversary $\mathcal{A}_\mathcal{B}$, we build an adversarial dataset $\mathcal{D}_{adv} = \{\mathcal{A}_\mathcal{B}(o) \mid o \in \mathcal{D}_{normal}\}$. In Sect. 3.2, we will demonstrate why two types of adversarial attacks can generate the $\mathcal{D}_{adv}$ without interacting with an environment under adversarial attacks.

Robustness Regularizer. A robustness regularizer [27] can also be integrated into our defense schema. The intuition behind the robustness regularizer is that if we can minimize the difference between the action distribution under normal observations and the action distribution under attacks, the robustness of our network can be improved. A robustness regularizer measures this difference.

Assuming a denoiser den and pretrained policy π, the action $a = \pi(\text{den}(o))$. We treat π and den as one network π_{den}. Given an attack $\hat{o} = \mathcal{A}_{\mathcal{B}}(o)$ and the policy covariance matrix Σ, the robustness regularizer for stochastic PPO is

$$R_{\text{ppo}} = (\pi_{\text{den}}(\mathcal{A}_{\mathcal{B}}(o)) - \pi_{\text{den}}(o)) \cdot \Sigma^{-1} \cdot (\pi_{\text{den}}(\mathcal{A}_{\mathcal{B}}(o)) - \pi_{\text{den}}(o))$$

and the robustness regularizer for deterministic TD3 is:

$$R_{\text{td3}} = \|\pi_{\text{den}}(\mathcal{A}_{\mathcal{B}}(o)) - \pi_{\text{den}}(o)\|_2.$$

Following [26], the attack $\mathcal{A}_{\mathcal{B}}$ considered here is the opposite attack that will be introduced in Sect. 3.2. The opposite attack depends on the policy network. When computing the robustness regularizer, we attack π_{den} instead of π. The theoretical foundation for minimizing the difference between action distributions is provided by Theorem 5 in [27]. It shows the total variance between the normal action distribution and the action distribution generated by observation $\hat{o}$ under attack can bound the value function (i.e., performance) difference. However, unlike [26] that trains a *policy* with a robustness regularizer, we achieve this by training the parameters of the *denoiser* den, and retain the parameters of the pretrained policy π.

The regularizers can be added with the denoiser's objective function directly. Then, according to the policy type, we optimize $L_{\text{den}} + R_{\text{ppo}}$ or $L_{\text{den}} + R_{\text{td3}}$ to update the denoiser's parameters. Optimizing the denoiser's objective function and robustness regularizer focus on different goals. A small value of L_{den} means the output of the denoiser is close to the groundtruth observations, while a small R_{ppo} or R_{td3} means the action distributions in adversarial and non-adversarial scenarios are similar.

3.2   Observation Attacks

Attacks. We evaluate our defense on four well-studied categories of observation attacks. In this section, we briefly introduce these attacks and explain why the opposite attack and Q-function attack can be used to generate offline adversarial datasets without sampling under adversarial scenarios.

Opposite Attack. The opposite attack appears in [6,11,16,27]. By perturbing observations, this attack either minimizes the likelihood of the action with the highest probability [6,16,27] or maximizes the likelihood of the least-similar action [11] in discrete action domains. We choose to minimize the likelihood of the preferred action. The attacked observation $\hat{o}_t$ is computed as:

$$\hat{o}_t = \underset{\hat{o}_t \in \mathcal{B}(o_t)}{\arg\max} \, l_{op}(o_t, \hat{o}_t), \tag{1}$$

where $\mathcal{B}(o_t)$ signifies all the allowed perturbed observations around o_t. For stochastic policies, $l_{op}(o_t, \hat{o}_t) = (\pi(o_t) - \pi(\hat{o}_t))\Sigma^{-1}(\pi(o_t) - \pi(\hat{o}_t))$, where $\pi(o_t)$ and $\pi(\hat{o}_t)$ are the mean of the predicted Gaussian distribution, and Σ is the policy covariance matrix. For a deterministic TD3 algorithm, the difference is defined as the Euclidean distance between the predicted actions, $l_{op}(o_t, \hat{o}_t) = ||\pi(o_t) - \pi(\hat{o}_t)||_2$. This attack only depends on the policy π. Given a normal dataset $\mathcal{D}_{normal}$, we can apply this attack on every observation in $\mathcal{D}_{normal}$ to generate the adversarial dataset $\mathcal{D}_{adv} = \{\mathcal{A}_{\mathcal{B}}(o) \mid o \in \mathcal{D}_{normal}\}$ without any interaction with the environment. Since generating $\mathcal{D}_{normal}$ and applying the opposite attack does not sample under adversarial scenarios. Thus, generating $\mathcal{D}_{adv}$ does not require sampling under adversarial scenarios.

Q-function Attack. [9,16,27] compute observation perturbations with the Q-function $Q(o_t, a_t)$. This attack only depends on the Q function $Q(o_t, a_t)$. The Q function sometimes comes with trained policies (e.g., TD3). When the Q function is not accompanied by trained policies (e.g., PPO), the Q-function can be learnt under non-adversarial scenarios [26]. We want to find a $\hat{o}_t$ such that it minimizes the Q under budget $\mathcal{B}$. Thus, the attacked observation $\hat{o}_t$ is computed as

$$\hat{o}_t = \underset{\hat{o}_t \in \mathcal{B}(o_t)}{\operatorname{argmin}} Q(o_t, \pi(\hat{o}_t)) \tag{2}$$

We can generate the adversarial dataset $\mathcal{D}_{adv}$ with $\mathcal{D}_{normal}$ and $Q(o_t, a_t)$. Notice that getting $\mathcal{D}_{normal}$ and $Q(o_t, a_t)$ does not require interacting with environments under attacks. Therefore, the Q-function attack can also generate the $\mathcal{D}_{adv} = \{\mathcal{A}_{\mathcal{B}}(o) \mid o \in \mathcal{D}_{normal}\}$ without sampling under adversarial scenarios.

Optimal Attack. The optimal attack learns an adversarial policy π_{adv} adding perturbation Δ_{o_t} to the observation o_t. For example, [26] demonstrated this strong attack over MuJoCo benchmarks. [4] learns such an adversarial policy in two-player environments. The action outputted by the adversarial policy is Δ_{o_t}, and the input of π_{adv} is o_t. The perturbed observation is

$$\hat{o}_t = \operatorname{proj}_{\mathcal{B}}(o_t + \Delta_{o_t}) \tag{3}$$

where $\operatorname{proj}_{\mathcal{B}}$ is a projection function that constrains the perturbed observation $\hat{o}_t$ to satisfy the attack budget $\mathcal{B}$. The adversarial policy is trained to *minimize* the cumulative discounted reward $\mathcal{R}$. Importantly, training this adversarial policy π_{adv} requires adversarial sampling online. Thus, we did not adopt it to generate our adversarial dataset $\mathcal{D}_{adv}$.

Enchanting Attack. This type of attack first appeared in [11]. It integrates a planner into the attack loop. The planner generates a sequence of adversarial actions, and the adversary crafts perturbations to mislead neural network policies to output adversarial action sequences. At time step t, an adversarial motion planner generates a sequence of adversarial actions $[a_{t,0}, a_{t,1}, ..., a_{t,T-t}]$ guiding the agent to perform poorly. Since we attack the observation space and cannot change the action directly, we need to perturb observations to mislead the policy to predict the planner's adversarial actions. Given the policy network π, the perturbed observation is

$$\hat{o}_t = \underset{\hat{o}_t \in \mathcal{B}}{\arg\min} \, ||\pi(\hat{o}_t) - a_{t,0}||_2 \tag{4}$$

The $a_{t,0}$ is the target adversarial action. In our attack, we call the planner at every step and use the first action as an adversarial action, which avoids the errors caused by the deviation between the actual trajectory and planned trajectory, and thus strengthens the enchanting attack. For the continuous control problem, we use a Cross-Entropy Motion (CEM) planner [8] for adversarial planning. Generating or applying an adversarial planner typically requires online adversarial sampling. Therefore, we did not generate the $\mathcal{D}_{adv}$ with the enchanting attack.

To summarize, we evaluate our defense over four types of attacks. However, we only generate the adversarial dataset $\mathcal{D}_{adv}$ with the opposite attack and Q-function attack because they do not require risky online adversarial sampling. Section 4 shows that the denoiser trained with the adversarial dataset generated from these two attacks alone performs surprisingly well even when used in defense against all the four attacks we consider.

When to Attack. Since we want to minimize the reward with as few perturbations as possible, it is crucial to attack when the agent is vulnerable. We use the value function approximation as the indicator of vulnerability. When the value function predicts a certain observation has a small future value, such an observation is likely to cause a lower cumulative reward. A lower cumulative reward shows either the vulnerability of this observation itself (e.g., a running robot is about to fall) or the vulnerability of the corresponding policy (i.e., the policy would perform poorly given this observation). Thus, we can use the value function approximation to choose the time to trigger our attack. Given an observation o_t and the value function V, by choosing a threshold C_{vul}, we only trigger the attack when $V(o_t) < C_{vul}$. The vulnerability indicator is

$$\mathbb{1}_{vul}(o_t) := V(o_t) < C_{vul}$$

We use the value function learned during training for the PPO policy. Because $V(o_t) = \int_{a_t \sim \pi(o_t)} Q(o_t, a_t)$ and $a_t = \pi(o_t)$ for a deterministic policy, $V(o_t) = Q(o_t, \pi(o_t))$. Hence, we can compute the value function of TD3 with the learned Q function and policy. We tune C_{vul} to achieve the strongest attack while minimizing the number of perturbations triggered.

4   Experiments

We evaluate our approach on five continuous control tasks with respect to a stochastic PPO policy and a deterministic TD3 policy. The PPO policies were trained by ourselves, and the TD3 policies use pretrained models from [17]. Our experiments answer the following questions.

Q1. Does our defense improve robustness against adversarial attacks?
Q2. How does our defense impact performance in non-adversarial scenarios?
Q3. How does our approach compare with state-of-the-art online adversarial training approach?
Q4. How is the performance of our detectors and denoisers in terms of accuracy?
Q5. How does our defense perform under adaptive attacks?

4.1 Rewards Under Attack w/wo Defense (Q1)

In this section, we show how our defense improves robustness. We report attack and defense results on the pretrained policies in Table 1. The "Benchmark" and "Algo" columns are the continuous control tasks and the reinforcement learning policies, resp. The "Dimension" column contains the dimensionality information of state and action space. The ε column shows the ε of attack budget $\mathcal{B}$. The ε of "Hopper", "HalfCheetah", and "Ant" are the same as the attack budget provided in [26]; we increased ε in "Walker2d" to 0.1. The "Humanoid" with the highest observation and action dimension is not evaluated in [26]. We choose $\varepsilon = 0.15$ for "Humanoid".

Table 1. Benchmark information and rewards under attack w/wo defense

Benchmark	Algo	ε	Dimension		Attack/Defense			
			State	Action	Opposite	Q-function	Optimal	Enchanting
Hopper	TD3	0.075	11	3	390/2219	960/3328	267/2814	1629/3287
	PPO				271/2615	700/3569	247/3068	217/2751
Walker2d	TD3	0.1	17	6	751/4005	478/4329	187/4772	762/4538
	PPO				241/1785	3510/4737	−38/1393	1582/1741
HalfCheetah	TD3	0.15	17	6	1770/8946	1603/8471	1017/8174	1802/8838
	PPO				1072/6115	1665/4218	833/3765	274/4477
Ant	TD3	0.15	111	8	603/3516	−46/2137	−893/2809	522/4729
	PPO				−351/5404	−157/1042	558/4574	196/5497
Humanoid	TD3	0.15	376	17	431/4849	454/4042	585/5130	420/5125
	PPO				531/3161	406/3508	415/3630	396/1695

We provide attack and defense results in the "Attack/Defense" column. The four sub-columns in this column are the attacks we described in Sect. 3.2. The numbers before the slash are the cumulative rewards gained under attack. In this table, we assume the adversary is not aware of our defense's existence. The experiment results show that these strong attacks can significantly decrease the benchmarks' rewards, and our defense significantly improved rewards for all attacks.

4.2   Non-adversarial Scenarios (Q2) and Comparison (Q3)

We evaluate rewards in non-adversarial scenarios and compare them with ATLA [26], a state-of-the-art online adversarial training approach, in this section. Adversarial attacks do not always happen. Therefore, maintaining strong performance in normal cases is essential. The "Non-adversarial" column summarizes the reward gained by policies without any adversarial attack injected. Rewards are computed as the average reward over 100 rollouts. The "Pre." column shows the cumulative reward of pretrained policies, while the "ATLA" column is the reward gained by the ATLA policy in [26]. The "Ours" column is the reward gained by the policies under our defense. The numbers in parentheses are the percentages of rewards preserved when compared with the pretrained policies, which are computed with reward in "Ours" divided by reward in "Pre.". Observe that the introduction of the detector preserves the performance of pretrained policies. Because our defense only intervenes when it detects anomalies, it has a mild impact on the pretrained policies in non-adversarial cases. In contrast, ATLA policies do not perform as well as our defended policies when no adversary appears on all the benchmarks (Table 2).

Table 2. Rewards in non-adversarial scenarios and comparison

Benchmark	Algo	Non-adversarial			Avg./Min (Best attack)	
		Pre.	ATLA	Ours	ATLA	Ours
Hopper	TD3	3607	3220	3506(0.97)	2192/1761(opt)	2912/2219(ops)
	PPO	3206		3201(1.00)		3001/2615(ops)
Walker2d	TD3	4719	3819	4712(1.00)	1988/1430(opt)	4411/4005(ops)
	PPO	4007		3980(0.99)		2414/1393(opt)
HalfCheetah	TD3	9790	6294	8935(0.91)	5104/4617(enc)	8607/8174(opt)
	PPO	8069		7634(0.95)		4644/3765(opt)
Ant	TD3	5805	5313	5804(1.00)	4310/3765(q)	3298/2137(q)
	PPO	5698		5538(0.97)		4129/1042(q)
Humanoid	TD3	5531	4108	5438(0.98)	3311/2719(q)	4786/4042(q)
	PPO	4568		4429(0.97)		2999/1695(enc)

The column "Avg./Min.(Best Attack)" show statistics comparing ATLA and our defense under the four attacks. The numbers before the slash are the average reward gained under attacks, and the numbers after the slash are the lowest rewards among all the attacks. The abbreviations in parentheses are the best attack that achieves the lowest reward, where "ops" means the opposite attack, "q" means the Q-function attack, "opt" means the optimal attack, and "enc" means the enchanting attack. The results show that our defense trained with data sampled under non-adversarial scenarios provides comparable results with the riskier online adversarial training approach. Observe that 6 out of 10 benchmarks have a higher reward than ATLA for the average rewards over attacks. For the worst rewards over attacks, our defense has a higher reward than the ATLA on 5 of 10 benchmarks. The result is surprising considering that we do not sample any adversarial observations online.

4.3   Detector and Denoiser (Q4)

The detector's performance is crucial for our defense since it prevents unnecessary interventions. We report the detectors' accuracy in non-adversarial scenarios and their F1 scores and false-negative rates under attack. The accuracy measures detectors' performance when no attack appears, and the F1 score measures how well the detectors perform when policies are under attack. Meanwhile, the false-negative rate tells us the percentage of adversarial attacks that are not detected. We present these results in the "Detector" column of Table 3.

Table 3. Detector and denoiser performance

Benchmark	Algo	Detector									Denoiser			
		Acc.	F1 score				False negative rate				Mean absolute error			
		Normal	Ops	Q	Opt	Enc	Ops	Q	Opt	Enc	Ops	Q	Opt	Enc
Hopper	TD3	0.99	0.82	0.98	0.88	0.99	0.00	0.01	0.07	0.00	0.030	0.023	0.032	0.024
	PPO	0.99	0.97	0.94	0.94	0.95	0.00	0.00	0.00	0.00	0.026	0.018	0.034	0.038
Walker2d	TD3	0.99	0.99	0.99	0.99	0.99	0.01	0.01	0.02	0.01	0.030	0.032	0.042	0.045
	PPO	0.95	0.97	0.93	0.95	0.99	0.05	0.01	0.00	0.01	0.041	0.030	0.033	0.037
HalfCheetah	TD3	0.95	0.99	0.98	0.98	0.96	0.00	0.00	0.00	0.00	0.049	0.048	0.050	0.043
	PPO	0.99	0.99	0.98	0.96	0.96	0.00	0.01	0.02	0.01	0.057	0.041	0.046	0.048
Ant	TD3	0.99	0.99	0.99	0.99	0.99	0.00	0.00	0.00	0.00	0.022	0.022	0.022	0.023
	PPO	0.99	0.99	0.99	0.99	0.99	0.00	0.00	0.00	0.00	0.023	0.024	0.027	0.026
Humanoid	TD3	0.99	0.96	0.97	1.00	0.96	0.08	0.04	0.00	0.07	0.048	0.047	0.043	0.046
	PPO	0.99	0.99	0.99	0.99	0.99	0.00	0.00	0.00	0.00	0.055	0.045	0.048	0.050

The detector is expected only to report negative in non-adversarial scenarios. Since there is no adversarial observation (i.e., positive sample) in non-adversarial scenarios, we measure the detector's quality with accuracy instead of the F1 score. The 3rd column in Table 3 reports the accuracy of all the detectors in non-adversarial scenarios. The worst accuracy is 0.95. The high accuracy explains why our defense retains the performance of pretrained policies in non-adversarial cases. We measure the quality of detectors under attack with the F1 scores and false-negative rate. When attacking Hopper's PPO policy with the opposite attack and optimal attack, the F1 scores are 0.82 and 0.88, respectively. However, their false negative rates are 0.00 and 0.01, respectively. The low false-negative rates show that our detectors ensure the denoiser would be triggered under attack. Moreover, the data shows that the relatively low F1 score was caused by false positives, which means the defense will be cautious and use denoised observations more often. The left data has an F1 score higher than 0.94 and a false negative rate lower than 0.04, which supports our claim that the detector works well when the policies are under attack.

The Mean Absolute Errors (MAEs) between the outputs of the denoiser and groundtruth observations are reported in the "Denoiser" column in Table 3. Although we only train the denoiser with the augmented data generated with the opposite and Q-function attack, the MAE of the optimal attack and enchanting attack is close to the MAE of the opposite attack and Q-function attack. This explains why our defense also works well on the opposite and enchanting attacks, as shown in Table 1.

4.4   Adaptive Attack (Q5)

We further evaluated the robustness of our defense under adaptive attacks. The defense in Sect. 4.1 is evaluated when the attacks are not aware of the existence of our defense. However, once the adversaries realize that we have upgraded our defense, they can jointly attack our defense and pretrained policies. When the adversary can access both the detector and denoiser, it can mislead the detector to ignore anomalies with adversarial observations. We briefly introduce the key idea of adaptive attacks here. A more formal description of our adaptive attack design is provided in Appendix B.

The adversary needs to attack our defense and the pretrained policy jointly. Firstly, we consider how to attack the denoiser. Under our defense, the action a_t is computed with a sequential model $a_t = \pi(\text{den}(o_t))$; we thus replace the pretrained policy $\pi(o_t)$ with $\pi(\text{den}(o_t))$ and attack this sequential model. Secondly, adaptive attacks also need to fool the detector. Because the anomaly is defined with respect to being greater than a threshold, a malicious observation should decrease the ℓ_∞-norm in $\mathbb{1}_{anomaly}$. This objective can be defined with a loss term $l_{det}(o_t) = ||\text{det}(o_t) - o_t||_\infty$. For the opposite attack, q-function attack, and enchanting attack, in addition to using $\pi(\text{den}(o_t))$ to replace $\pi(o_t)$, we optimize $l_{det}(o_t)$ jointly with Eq. (1), Eq. (2), and Eq. (4) respectively. For the optimal attack, we train the adversarial policy with the involvement of our defense.

We use the defense rewards in Table 1 (numbers after the slash) as baselines and report the percentages by which reward changes under adaptive attacks in Table 4. The benchmark column contains the task names and the policy types. We have introduced the attack name abbreviations in Sect. 4.1, and the rewards changes under these attacks are reported from column 2 to column 5. The "Min" and "Max" columns are the minimal and maximal changes comparable with the baseline rewards. In the worst case, the adaptive attack causes the performance on Ant-TD3 to decrease 28% under the opposite attack. We can observe that some rewards increase under the adaptive attack. This is because jointly attacking the detector can be challenging for the adversary. Since the detector is also a GRU-VAE, the first problem the adversary needs to address is the stochasticity introduced by the detector and denoiser themselves. Moreover, the adversary needs to fool the policy and detector simultaneously, which increases the difficulty of attacking our defense.

Table 4. Adaptive attack (% change in reward)

Benchmark	Algo	Ops	Q	Opt	Enc	Min	Max
Hopper	TD3	0.63	0.06	0.17	0.04	0.04	0.63
	PPO	−0.16	−0.18	−0.12	−0.11	−0.18	−0.11
Walker2d	TD3	0.19	0.10	0.00	0.05	0.00	0.19
	PPO	0.12	−0.18	−0.16	0.17	−0.18	0.17
HalfCheetah	TD3	−0.14	0.06	0.08	0.03	−0.14	0.08
	PPO	−0.23	0.10	0.17	0.20	−0.23	0.20
Ant	TD3	−0.28	−0.17	−0.14	−0.18	−0.28	−0.14
	PPO	−0.24	0.40	0.11	−0.22	−0.24	0.40
Humanoid	TD3	0.09	0.28	−0.03	−0.06	−0.06	0.28
	PPO	0.28	0.23	0.01	0.06	0.01	0.28

5   Conclusion

This paper proposes a detect-and-denoise defense against the observation attacks on deep reinforcement learning. Our defense samples the adversarial observations offline and thus avoids the risky online sampling under adversarial attacks. In the absence of an adversary, our defense does not compromise performance. We evaluated our approach over four strong attacks with five continuous control tasks under both stochastic and deterministic policies. Experiment results show that our approach is comparable to previous online adversarial training approaches, provides reasonable performance under adaptive attacks, and does not sacrifice performance in normal (non-adversarial) settings.

Acknowledgment. This work was supported in part by C-BRIC, one of six centers in JUMP, a Semiconductor Research Corporation (SRC) program sponsored by DARPA. He Zhu thanks the support from NSF Award #CCF-2007799.

References

1. Achiam, J., Held, D., Tamar, A., Abbeel, P.: Constrained policy optimization. In: International Conference on Machine Learning, pp. 22–31. PMLR (2017)
2. Doersch, C.: Tutorial on variational autoencoders. arXiv preprint arXiv:1606.05908 (2016)
3. Fujimoto, S., Hoof, H., Meger, D.: Addressing function approximation error in actor-critic methods. In: ICML (2018)
4. Gleave, A., Dennis, M., Wild, C., Kant, N., Levine, S., Russell, S.: Adversarial policies: attacking deep reinforcement learning. arXiv:1905.10615 (2019)
5. Havens, A.J., Jiang, Z., Sarkar, S.: Online robust policy learning in the presence of unknown adversaries. arXiv preprint arXiv:1807.06064 (2018)
6. Huang, S., Papernot, N., Goodfellow, I., Duan, Y., Abbeel, P.: Adversarial attacks on neural network policies. arXiv preprint arXiv:1702.02284 (2017)

7. Iyengar, G.N.: Robust dynamic programming. Math. Oper. Res. **30**(2), 257–280 (2005)
8. Kobilarov, M.: Cross-entropy motion planning. Int. J. Robot. Res. **31**(7), 855–871 (2012)
9. Kos, J., Song, D.: Delving into adversarial attacks on deep policies. arXiv preprint arXiv:1705.06452 (2017)
10. Li, B., Chen, C., Wang, W., Carin, L.: Certified adversarial robustness with additive noise. arXiv preprint arXiv:1809.03113 (2018)
11. Lin, Y.C., Hong, Z.W., Liao, Y.H., Shih, M.L., Liu, M.Y., Sun, M.: Tactics of adversarial attack on deep reinforcement learning agents. arXiv preprint arXiv:1703.06748 (2017)
12. Malhotra, P., Ramakrishnan, A., Anand, G., Vig, L., Agarwal, P., Shroff, G.: LSTM-based encoder-decoder for multi-sensor anomaly detection. arXiv preprint arXiv:1607.00148 (2016)
13. Mandlekar, A., Zhu, Y., Garg, A., Fei-Fei, L., Savarese, S.: Adversarially robust policy learning: active construction of physically-plausible perturbations. In: IEEE/RSJ IROS, pp. 3932–3939. IEEE (2017)
14. Panda, P., Roy, K.: Implicit generative modeling of random noise during training for adversarial robustness. arXiv preprint arXiv:1807.02188 (2018)
15. Park, D., Hoshi, Y., Kemp, C.C.: A multimodal anomaly detector for robot-assisted feeding using an LSTM-based variational autoencoder. IEEE Robot. Autom. Lett. **3**(3), 1544–1551 (2018)
16. Pattanaik, A., Tang, Z., Liu, S., Bommannan, G., Chowdhary, G.: Robust deep reinforcement learning with adversarial attacks. arXiv preprint arXiv:1712.03632 (2017)
17. Raffin, A.: RL baselines3 Zoo (2020). https://github.com/DLR-RM/rl-baselines3-zoo
18. Rajeswaran, A., Ghotra, S., Ravindran, B., Levine, S.: EPOpt: learning robust neural network policies using model ensembles. arXiv preprint arXiv:1610.01283 (2016)
19. Schulman, J., Wolski, F., Dhariwal, P., Radford, A., Klimov, O.: Proximal policy optimization algorithms. arXiv preprint arXiv:1707.06347 (2017)
20. Sun, J., et al.: Stealthy and efficient adversarial attacks against deep reinforcement learning. In: AAAI (2020)
21. Tessler, C., Efroni, Y., Mannor, S.: Action robust reinforcement learning and applications in continuous control. In: ICML, pp. 6215–6224. PMLR (2019)
22. Todorov, E., Erez, T., Tassa, Y.: MuJoCo: a physics engine for model-based control. In: 2012 IEEE/RSJ International Conference on Intelligent Robots and Systems (2012)
23. Vincent, P., Larochelle, H., Bengio, Y., Manzagol, P.A.: Extracting and composing robust features with denoising autoencoders. In: ICML (2008)
24. Xiong, Z., Eappen, J., Zhu, H., Jagannathan, S.: Robustness to adversarial attacks in learning-enabled controllers. arXiv preprint arXiv:2006.06861 (2020)
25. Zhang, C., et al.: A deep neural network for unsupervised anomaly detection and diagnosis in multivariate time series data. In: AAAI (2019)
26. Zhang, H., Chen, H., Boning, D., Hsieh, C.J.: Robust reinforcement learning on state observations with learned optimal adversary. arXiv:2101.08452 (2021)
27. Zhang, H., et al.: Robust deep reinforcement learning against adversarial perturbations on state observations. arXiv:2003.08938 (2020)

Resisting Graph Adversarial Attack via Cooperative Homophilous Augmentation

Zhihao Zhu[1], Chenwang Wu[1], Min Zhou[2], Hao Liao[3], Defu Lian[1(✉)], and Enhong Chen[1]

[1] University of Science and Technology of China, Hefei, China
{zzh98,wcw1996}@mail.ustc.edu.cn, {liandefu,cheneh}@ustc.edu.cn
[2] Huawei Technologies Co. Ltd., Shenzhen, China
[3] Shenzhen University, Shenzhen, China
haoliao@szu.edu.cn

Abstract. Recent studies show that Graph Neural Networks (GNNs) are vulnerable and easily fooled by small perturbations, which has raised considerable concerns for adapting GNNs in various safety-critical applications. In this work, we focus on the emerging but critical attack, namely, Graph Injection Attack (GIA), in which the adversary poisons the graph by injecting fake nodes instead of modifying existing structures or node attributes. Inspired by findings that the adversarial attacks are related to the increased heterophily on perturbed graphs (the adversary tends to connect dissimilar nodes), we propose a general defense framework CHAGNN against GIA through cooperative homophilous augmentation of graph data and model. Specifically, the model generates pseudo-labels for unlabeled nodes in each round of training to reduce heterophilous edges of nodes with distinct labels. The cleaner graph is fed back to the model, producing more informative pseudo-labels. In such an iterative manner, model robustness is then promisingly enhanced. We present the theoretical analysis of the effect of homophilous augmentation and provide the guarantee of the proposal's validity. Experimental results empirically demonstrate the effectiveness of CHAGNN in comparison with recent state-of-the-art defense methods on diverse real-world datasets.

Keywords: Graph neural network · Adversarial attack · Defense

1 Introduction

In recent years, graph neural networks (GNNs) have been successfully applied in social networks [1], knowledge graphs [2] and recommender systems [3] due to it's good performance in analyzing graph data. In spite of the popularity and success of GNNs, they have shown to be vulnerable to adversarial attacks [4–6]. Classification accuracy of GNNs on the target node might be significantly

Supplementary Information The online version contains supplementary material available at https://doi.org/10.1007/978-3-031-26409-2_16.

degraded by imperceptible perturbations, posing certain practical difficulties. For instance, an attacker can disguise his credit rating in credit prediction by establishing links with others. Financial surveillance enables attackers to conceal the holding relationship in order to carry out a hostile takeover. Due to the widespread use of graph data, it is critical to design a robust GNN model capable of defending against adversarial attacks.

Existing graph adversarial attacks mainly focus on two aspects [7]. The first one is the graph modification attack(GMA), which poisons graphs by modifying the edges and features of original nodes. The other attack method, graph injection attack (GIA), significantly lowers the performance of graph embedding algorithms on graphs by introducing fake nodes and associated characteristics. The latter type of attack appears to be more promising. For instance, it is unquestionably easier for attackers to establish fake users than to manipulate authentic data in recommender systems. Promoting attackers' influence in social media via registering fake accounts is less likely to be detected than modifying system data. Given the flexibility and concealment of GIA, it is crucial to develop defense strategies. However, the fact is that there are currently fewer methods to defend GIA in comparison to GMA. In this paper, we propose a defense method against GIA.

Defense methods are mainly categorized into two groups [8]. One approach is to begin with models and then improve their robustness, for example, through adversarial training [9]. The other seeks to recover the poisoned graph's original data. Obviously, the first sort of approach is firmly connected to the model, which implies that it may not work with new models. By contrast, approaches based on data modification are disconnected from concrete models, which is the subject of this paper.

Recent studies [10,17] have shown that adversarial attacks are related to the increased heterophily on the perturbed graph, which has inspired works about data cleaning. Existing works [17] rely heavily on the similarity of features, e.g., utilizing Jaccard similarity or Cosine similarity, to eliminate potentially dirty data. However, they ignored critical local subgraphs in the graph. More specifically, existing methods only measure heterophilous anomalies based on descriptive features while ignoring meaningful interactions between nodes, which leads to biased judgments of heterophily, normal data cleaning, and model performance decline. Additionally, while eliminating heterophily makes empirical sense, it lacks theoretical support.

In this work, we propose a general defense framework, CHAGNN, to resist adversarial attacks by cooperative homophilous augmentation. To begin, in order to fully use graph information, we propose using GCN labels rather than descriptive attributes to determine heterophily. This is because GCNs' robust representation capability takes into account both the feature and adjacency of nodes. However, it is undeniable that the model must guarantee good performance to provide credible pseudo-labels. Thus, we further propose a cooperative homophily enhancement of both the model and the graph. To be more precise, during each round of training, the model assigns prediction labels to the graph data in order to find and clean heterophilous regions, while the cleaned data is supplied back to the model for training in order to get more informative samples. In this self-enhancing manner, the model's robustness and performance are steadily improved. Notably, we theoretically demonstrate the effectiveness of homophilous augmentation in resisting adversarial attacks, which has not been demonstrated in prior research [10,17]. Homophilous augmentation can considerably reduce the risk of graph injection attacks and increase the performance of the model. The experimental results indicate that after only a few rounds of cleaning, the model outperforms alternative protection approaches.

The contributions of this paper are summarized as follows:

- We propose a homophily-augmented model to resist graph injection attacks. The model and data increase the graph homophily in a cooperative manner, thereby improving model robustness.
- We theoretically prove that the benefit of heterophilous edge removal process is greater than the penalty of misoperation, which guarantees the effectiveness of our method.
- Our experiments consistently demonstrate that our method significantly outperforms over baselines against various GIA methods across different datasets.

2   Preliminaries

2.1   Graph Convolutional Network

Let $G = (V, E)$ be a graph, where V is the set of N nodes, and E is the set of edges. These edges can be formalized as a sparse adjacency matrix $\boldsymbol{A} \in \mathbb{R}^{N \times N}$, and the features of nodes can be represented as a matrix $\boldsymbol{X} \in \mathbb{R}^{N \times D}$, where D is the feature dimension. Besides, in the semi-supervised node classification task, nodes can be divided into labeled nodes V_L and unlabeled nodes V_U.

In the node classification task we focus on, the model f_θ is trained based on $G = (\boldsymbol{A}, \boldsymbol{X})$ and the labeled nodes V_L to predict all unlabeled nodes V_U as correctly as possible. θ is the model's parameter. The model's objective function can be defined as:

$$\max_\theta \sum_{v_i \in V_U} I(argmax(f_\theta(G)_i) = y_i), \tag{1}$$

where $f_\theta(G)_i \in [0, 1]^C$, C is the number of categories of nodes.

GCN [25], one of the most widely used models in GNNs, aggregates the structural information and attribute information of the graph in the message passing process. Due to GCN's excellent learning ability and considerable time complexity, it has been applied in various real-world tasks, e.g., traffic prediction and recommender systems. Therefore, it is important to study improve the robustness of GCN against adversarial attacks. Given G and V_L as input, a two-layer GCN with $\theta = (\boldsymbol{W_1}, \boldsymbol{W_2})$ implements $f_\theta(G)$ as

$$f_\theta(G) = softmax(\hat{\boldsymbol{A}}\sigma(\hat{\boldsymbol{A}}\boldsymbol{X}\boldsymbol{W_1})\boldsymbol{W_2}), \tag{2}$$

where $\hat{\boldsymbol{A}} = \tilde{\boldsymbol{D}}^{-1/2}(\boldsymbol{A} + \boldsymbol{I})\tilde{\boldsymbol{D}}^{-1/2}$ and $\tilde{\boldsymbol{D}}$ is the diagonal matrix of $\boldsymbol{A} + \boldsymbol{I}$. σ represents an activation function, e.g., ReLU.

2.2 Graph Adversarial Attack

The attacker's goal is to reduce the node classification accuracy of the model on the target nodes T as much as possible. A poisoning attack on a graph can be formally defined as

$$\min_{G'} \max_{\theta} \sum_{v_i \in T} I(argmax(f_\theta(G')_i) = y_i), \qquad (3)$$

$$s.t. \quad G' = (A', X'), \|A' - A\| + \|X' - X\| \leq \Delta,$$

where A' and X' are modified adjacency and feature matrix, and the predefined Δ is used to ensure that the perturbation on the graph is small enough.

The attacker only makes changes in the original graph without introducing new nodes, which is called graph modification attack (GMA). Inversely, the attack that does not destroy the original graph but injects new nodes on graphs is defined as the graph injection attack (GIA). Modifying existing nodes is often impractical, e.g., manipulating other users in a social network. However, creating new accounts in social media is feasible and difficult to be detected. Due to the practicality and concealment of GIA, we focus more on it. The difference between GMA and GIA is shown in Fig. 1.

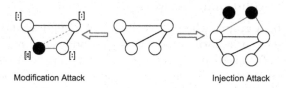

Modification Attack   Injection Attack

Fig. 1. GMA vs GIA

Now, we give GIA's formal definition. An attacker is limited to inject N_I nodes with well-crafted features into the graph. If the injected nodes are represented by V_I, then the injected adjacency matrix and feature matrix can be formalized as follows:

$$A' = \begin{bmatrix} A & A_{OI} \\ A_{OI}^T & A_I \end{bmatrix}, A \in \mathrm{R}^{N \times N}, \ A_{OI} \in \mathrm{R}^{N \times N_I}, \ A_I \in \mathrm{R}^{N_I \times N_I}, \qquad (4)$$

$$X' = \begin{bmatrix} X \\ X_I \end{bmatrix}, X \in \mathrm{R}^{N \times D}, \ X_I \in \mathbb{R}^{N_I \times D}, \qquad (5)$$

where A_{OI} is the connections between original nodes and injected nodes.

Following the settings of KDD-CUP 2020, V_U and V_I are mixed. The defender does not know which unlabeled nodes belong to V_U or V_I. Given G' and V_L as input, the defender's goal is to maximize the classification accuracy of the model on V_U.

$$\max_{\theta} \sum_{v_i \in V_U} I(argmax(f_\theta(G')_i) = y_i). \qquad (6)$$

3   The Proposed Framework

Based on the practicability and harm of graph injection attacks, we present an efficient method to resist it in this section. In Sect. 3.1, we relate heterophily to adversarial attacks and defense, and reveal the motivation for our method. Section 3.2 proposes a defensive framework by homophilous augmentation while leveraging the cooperation of the graph and the model to boost robustness. Moreover, in Sect. 3.3, we theoretically demonstrate the effectiveness of the proposed method.

3.1   Heterophily and Attack

Before formally tracing the source of the attack, we give the definition of heterophily and homophily in the graph. If the labels of nodes at both ends of a path are the same, we call it a homophilous path. Conversely, a heterophilous path indicates that the labels of nodes at both ends of it are not the same. Following [20,21], we use the homophily ratio h to quantify the degree of homophily, which is defined as the fraction of homophilous edges among all the edges in a graph:

$$h = |\{(u,v) \in E | y_u = y_v\}| / |E| \tag{7}$$

Assume that the nodes in a graph are randomly connected, then for a balanced class, the expectation for h is $\frac{1}{C}$. If the homophily ratio h satisfies $h >> \frac{1}{C}$, we call the graph a homophilous graph. On the other hand, it is a heterophilous graph if $h << \frac{1}{C}$. In this paper, we focus on the homophilous graph due to it's ubiquity.

Many research [17,18] shows that extremely destructive attacks tend to increase the heterophily of the homophilous graph. It seems plausible since neighbor relationships in graph networks provide critical insights for GNN predictions. The attacker cannot destroy these relationships, but can only weaken the connection by connecting heterophilous edges. This empirical finding also inspired subsequent research work based on data cleaning. We do not have a god-view to know the label of each node, so GCNJaccard [17] measures heterophily based on the similarity of features, such as using Jaccard similarity or cosine similarity. Removing heterophilous paths thereby increases the homophily of the graph. However, they only measure heterophilous anomalies based on descriptive features, ignoring the more critical local subgraphs in the graph. The similarity is stronger if a node and its neighbors share similar hobbies, but it cannot be measured based on descriptive features. Therefore, the unreasonable homophily measures may lead to biased judgments of heterophily and reduce model performance. In addition, these studies are only reasonable assumptions based on experience, and how to guarantee their validity theoretically is challenging.

Algorithm 1: Eliminating heterophilous Edges

Input: Poisoned graph G', modified nodes V_M, labeled nodes V_L, pseudo-labels $\hat{Y}$, soft-labels $f_\theta(G')$, elimination rate q

Output: Cleaned Graph $\hat{G}'$

1   $H_e = \emptyset$
2   **for** $u \in V_M$ **do**
3      $\hat{y}_u \leftarrow$ pseudo-label of node u; $N_u \leftarrow$ u's neighbors
4   **for** $v \in N_u$ **do**
5   **if** $(v \in V_L \text{ and } y_v \neq \hat{y}_u)$ **or** $(v \notin V_L \text{ and } \hat{y}_v \neq \hat{y}_u)$ **then**
6           $H_e \leftarrow H_e \cup \{(u,v)\}$

7   **for** $(u,v) \in H_e$ **do**
8      $f_\theta(G')_u \leftarrow$ soft label of node u; $f_\theta(G')_v \leftarrow$ soft label of node v
9   The degree of heterophily of (u,v) is $\bar{h}_{u,v} = JS(f_\theta(G')_u, f_\theta(G')_v)$

10 Pick and eliminate $q \cdot |H_e|$ heterophilous edges according to the sampling probability vector $\boldsymbol{p}$, which is calculated by Eq.(9)
11 Output the cleaned graph $\hat{G}'$

3.2 Cooperative Homophilous Augmentation

Considering the deficiencies of existing methods discussed in last subsection, we propose a synergistic homophily augmentation strategy to resist attacks. As mentioned before, using the similarity of features without graph's structure information to represent heterophily is biased. Thus, we propose to increase the graph's homophily by pseudo-labels that contain the information of both features and structure.

In GIA scenarios, fake connections must be the edges of unlabeled nodes. Therefore, our method focuses on this region of the graph. Due to the gap between pseudo-labels and labels, using pseudo-labels to discriminate and remove heterophilous edges may lead to mistakenly eliminating homophilous edges. Besides, pseudo-labels can not quantify the strength of heterophily of edges. For example, suppose the predictions of nodes u, v, and w are $[0.99, 0.01]$, $[0.49, 0.51]$ and $[0.01, 0.99]$ respectively. Pseudo-labels will treat edge (u,v) and (u,w) as identical, which is unreasonable. Compared to pseudo-labels, a node's soft label can more specifically reflect the probability that the node belongs to each category. Therefore, we use the JS divergence of the soft labels of nodes at both ends of the edge to measure the degree of heterophily of the edge (u,v). Heterophilous edges with a high degree of heterophily are more likely to be removed.

$$\bar{h}_{u,v} = JS\left(f_\theta(G')_u, f_\theta(G')_v\right)$$
$$= \frac{1}{2} \sum_{i=1}^{C} f_\theta(G')_u^i \log \frac{2f_\theta(G')_u^i}{f_\theta(G')_u^i + f_\theta(G')_v^i} + \frac{1}{2} \sum_{i=1}^{C} f_\theta(G')_v^i \log \frac{2f_\theta(G')_v^i}{f_\theta(G')_u^i + f_\theta(G')_v^i},$$

(8)

where $f_\theta(G')_u$ is the soft label of node u, $f_\theta(G')_u^i$ denotes the probability that node u belongs to class i.

The value range of JS divergence is $[0, 1]$. The value of the JS divergence is closer to 0 as the two probability distributions are more similar. It means that the smaller the value, the more likely the edge is a homophilous edge. We normalize the vector $\bar{h}$, which stores the degree of heterophily of all the heterogeneous edges.

$$p_{i,j} = exp(\bar{h}_{i,j})/ \sum_{(u,v)\in H_e} exp(\bar{h}_{u,v}), \qquad (9)$$

where H_e is the heterophilous edges set, $p_{i,j}$ is the probability that (i,j) is sampled to be removed. Then we pick out some heterophilous edges according to the sampling probability vector p. Edges with a higher degree of heterophily are more likely to be picked out for removal. The process of eliminating heterophilous edges is described in Algorithm 1.

However, the result of Algorithm 1 strongly depends on the authenticity of the pseudo-label. To achieve better performance, we propose to enhance the homophily of graph via cooperatively cleaning graph and improving model performance. Specifically, the model provides pseudo-labels to clean the data, while the purified graph guides the model by providing more reliable pseudo-labels. The model and data thus cooperatively increase classification accuracy.

Next we give the implementation details of CHAGNN. In GIA scenarios, the poisoned regions are consumingly related to the unlabeled nodes, including the unlabeled nodes V_U in the original graph and injected nodes V_I. The nodes selected to modify their edges are called modified nodes (V_M). In order to accurately remove maliciously injected edges, we simply define V_M as $V_U \cup V_I$. At the beginning of our algorithm, we first use poisoned graph to conduct pre-training process on the model. Then we obtain all nodes' pseudo-labels and soft labels. The pseudo-labels and soft labels are input to Algorithm 1 to generate a purified graph $\hat{G}$. After that, model parameters will be fine-tuned on $\hat{G}$. This process will dramatically improve classification performance in a few rounds. The details of our method are shown in Algorithm 2. The algorithm flowchart of CHAGNN is shown in Fig. 2.

We found that AdaEdge [29] also used pseudo-labels in their algorithm, but our proposal is quite different from it. Unlike AdaEdge, which directly removes the heterophilous edges based on pseudo labels, we introduce JS divergence to quantify the degree of heterophily of heterophilous edges before the elimination process, which greatly reduces the possibility of misoperation. Besides, the subjects of the research are different. AdaEdge focuses on solving the over-smoothing problem, and they perform a cleanup operation on the entire graph. Instead, we consider the scenario of graph injection attacks, applying heterophilous edge removal to potentially injected edges in the graph and providing the corresponding theoretical guarantee. The experimental comparison results of the two algorithms are shown in Sect. 4.2 and 4.3.

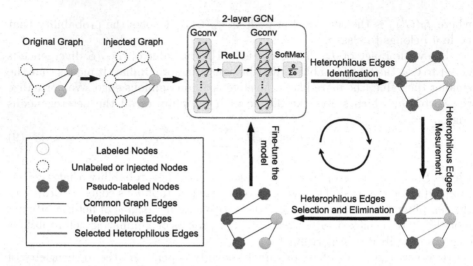

Fig. 2. Algorithm flowchart of CHAGNN

Algorithm 2: CHAGNN

Input: Poisoned graph G', labeled nodes V_L, modified nodes V_M, elimination
rate q, max iterations max_iter
Output: Prediction on test set
1 Pretrain model parameters θ on G'
2 **for** $i=1,...,max\_iter$ **do**
3   Obtain the pseudo-labels P and soft-labels $f_\theta(G')$ of all nodes
4     $G' \leftarrow Algorithm1(G', V_M, V_L, P, f_\theta(G'), q)$
5   Fine-tune θ on G'
6 Output the prediction on test set

3.3 Theoretical Guarantee

In general, removing heterophilous edges benefits model, while homophilous
edges deletion brings model penalties. We want to mitigate the damage of the
graph data by heterophilous edges as much as possible. However, identifying
and eliminating heterophilous edges via nodes' pseudo-labels may mistakenly
delete homophilous edges. In this section, we prove that given an arbitrary model
accuracy, the expected benefits of the proposed strategy outweigh the expected
penalties. Specifically, we firstly use the variation of loss to represent the impact
of eliminating heterophilous(homophilous) edges. Then we analyze the probability of deleting homophilous edge by mistake with model's accuracy. Combining
these two parts, we guarantee the reliability of CHAGNN theoretically. We give
the proof of theorems in appendix.

To simplify the proof, we employ the SGC model which removes the activation function compared with GCN model. Given $G = (A, X)$ and Y_L as input, a two-layer SGC with $\theta = W$ implements $f_\theta(G)$ as

$$f_\theta(G) = softmax(\hat{A}^2 X W). \tag{10}$$

Following [10]'s setting, we assume that G is a d-regular graph which means that each node of G has d connections with other nodes. For each node of G, proportion h of their neighbors belong to the same class, while proportion $\frac{1-h}{C-1}$ of them belong to any other class uniformly. The features of node v are defined as $x_v = p \cdot onehot(y_v) + \frac{1-p}{C}$, where y_v means the node's label.

We use the change of CM loss of the model to analyze the influence of injecting nodes to the graph. The CM loss of node v is defined as:

$$Z = \hat{A}^2 X W, loss_v = Z_{vy_v} - \max_{j \neq y_v} Z_{vj}. \tag{11}$$

Define the CM loss of node v on clean graph as L_0. After we generate nodes to inject homophilous edges to the graph, the CM loss changes to L_1. Correspondingly, the CM loss is called L_2 after we generate nodes to inject heterophilous edges to the graph. Assume that the proportion of node v's edges which is connected to class y_v before poisoned is h_0(including self-loop of node v), and the proportion of other classes is h_1. After we inject nodes to the graph, the proportion of node v's edges which is connected to class y_v is r_0, and the proportion of other classes is r_1. For convenience, we separate the proportions of injected edges from r_0 or r_1. The proportion of injected edges is denoted as r_2.

Theorem 1. *Consider target attack and direct attack which means that the inject nodes are directly connected to the target node v. Then we have:*

$$\frac{L_1 - L_0}{L_0 - L_2} = \frac{(r_0 - r_1 \mid r_2) - (h_0 - h_1)}{(h_0 - h_1) - (r_0 - r_1 - r_2)} \tag{12}$$

Remark 1. heterophilous edges elimination is actually the reverse process of attack. According to Theorem 1, we can estimate the ratio between the penalty of deleting a homophilous edge and the benefit of deleting a heterophilous edge.

Based on the relation between the penalty and benefit stated in Theorem 1, we analyze the expected benefit and expected penalty of the edge deletion operation at the specified model accuracy. For simplicity, we focus on judging whether a node belongs to a specific class in Theorem 2, which is a binary classification problem. Referring to [32], it is easily extensible and applicable to multi-class scenarios.

Assume that the prediction accuracy of model on unlabeled nodes is p. The prediction accuracy of different nodes is independent. Suppose we judge that there is a heterogeneous edge between nodes u and v according to pseudo-labels and then we delete e_{uv}. The probability that e_{uv} is actually a homogeneous edge is p_1 while the probability that e_{uv} is actually a heterogeneous edge is p_2. The pseudo-labels of u and v are $\hat{y}_u$ and $\hat{y}_v$, $\hat{y}_u \neq \hat{y}_v$. The labels of u and v are y_u and y_v. Then we have:

Theorem 2. *The ratio of expected penalty to expected benefit for eliminating an edge in CHAGNN is related to the prediction accuracy p.*

$$\frac{e_1}{e_2} < 2p(1-p) < 1 \tag{13}$$

Remark 2. Theorem 2 shows that the expected benefit is always greater than the expected penalty in our algorithm. For a binary classification problem, an effective classifier should have an accuracy greater than 50%. This means that we can reduce the ratio in Theorem 2 by continuously improving the accuracy of an effective classifier.

4 Experiment

In this section, we compare the proposed CHAGNN with state-of-the-art defense strategies. The experiment primarily validates our algorithm's excellent performance by answering the following research questions:

- RQ1. How well does CHAGNN perform compared to other state-of-the-art defense methods under different graph injection attacks?
- RQ2. How well does CHAGNN perform with different injected nodes ratio under the state-of-the-art GIA methods?
- RQ3. How much does the deletion rate affect the performance of CHAGNN?

4.1 Experimental Setup

Dataset. We evaluate the proposed algorithms with four widely used citation network datasets, including Cora-ml, Cora [22,23], Citeseer [24], and Pubmed. The statistics of datasets are summarized in Table 1. Following [12], we only consider the largest connected component (LCC) of each graph data.

To evaluate the effectiveness of our method, we compared it with the state-of-the-art defense models. The compared algorithms and attack methods are introduced in the next two subsections.

Compared Algorithms

- GCN [25]: We compare our algorithm with other methods with GCN, one of the most widely used models in GNNs.
- GCNSVD [16]: GCNSVD is a preprocessing method to resist adversarial attacks. It use a low-rank approximation of the graph to train GCN.

Table 1. Statistics of benchmark datasets

	N_{LCC}	E_{LCC}	Classes	Features
Cora-ml	2810	7981	7	2879
Cora	2485	5069	7	1433
Citeseer	2110	3668	6	3703
Pubmed	19717	44338	3	500

- GCN-Jaccard [17]: Another preprocessing method to resist adversarial attacks. They identity and eliminate heterophilous edges with nodes' features.
- GNNGUARD [19]: GNNGUARD added the attention mechanism to defend against adversarial attacks. It learns how to best assign higher weights to edges connecting similar nodes while pruning edges between unrelated nodes.
- ORH [10]: ORH mitigates the damage to the graph structure on account of the addition of heterophilouss edges by increasing the node's weight.
- VPN [31]: VPN replaces the graph convolutional operator A with the weighted sum of adjacency matrices with different powers.
- AdaEdge [29]: AdaEdge uses pseudo-labels to remove heterophilous edges to solve model's over-smoothing problem. Unlike our method, AdaEgde does not consider actual attack scenarios. Moreover, the judged heterophilous edges are directly removed without screening, which can easily lead to the mistaken deletion of homophilous edges.

Attack Methods

- TDGIA [7]: TDGIA first introduces the topological defective edge selection strategy to choose the original nodes for connecting with the injected ones. It then designs the smooth feature optimization objective to generate the features for the injected nodes.
- FGA [26]: A framework to generate adversarial networks based on the gradient information in GCN.
- MGA [27]: This paper proposes a Momentum Gradient Attack (MGA) against the GCN model, which can achieve more aggressive attacks with fewer rewiring links than FGA.

FGA and MGA are not directly applicable in GIA scenario. We modify them to work for GIA setting. They are performed on the graph poisoned by a heuristic injection.

Parameter Settings. For each dataset, we randomly split the nodes into labeled nodes for training procedure (10%), labeled nodes for validation (10%), and unlabeled nodes as test set to evaluate the model (80%). The hyperparameters of all the models are tuned based on the loss and accuracy on validation set. We report the average performance of 5 runs for each experiment.

Table 2. Node classification performance (Accuracy $\pm$ Std) under non-targeted attack

Attack	Defense	Cora-ml	Cora	Citeseer	Pubmed
TDGIA	No Attack	84.64 ± 0.34	81.26 ± 1.20	71.46 ± 0.30	85.00 ± 0.09
	Attack	67.80 ± 0.44	71.52 ± 0.33	60.18 ± 1.19	73.18 ± 0.25
	GCNSVD	68.44 ± 0.28	73.56 ± 0.26	63.24 ± 0.55	78.06 ± 0.28
	GCNJaccard	65.10 ± 1.35	70.04 ± 0.48	61.60 ± 0.70	76.48 ± 0.23
	GNNGUARD	65.26 ± 1.35	72.30 ± 0.23	64.38 ± 0.61	70.46 ± 0.66
	ORH	56.30 ± 1.13	65.16 ± 0.55	55.92 ± 1.34	70.58 ± 0.46
	VPN	69.03 ± 1.11	75.76 ± 0.39	66.90 ± 0.83	78.83 ± 0.17
	AdaEdge	76.34 ± 1.43	76.86 ± 0.48	67.62 ± 0.44	78.62 ± 0.20
	CHAGNN	**79.52 ± 0.32**	**77.84 ± 0.29**	**69.22 ± 0.59**	**79.66 ± 0.37**
FGA	No Attack	84.64 ± 0.34	81.26 ± 1.20	71.46 ± 0.3	85.00 ± 0.09
	Attack	82.08 ± 0.33	79.38 ± 0.58	70.72 ± 0.56	80.62 ± 0.32
	GCNSVD	78.84 ± 0.16	76.66 ± 0.22	67.66 ± 0.3	80.96 ± 0.05
	GCNJaccard	79.66 ± 0.91	79.38 ± 0.32	**70.92 ± 0.27**	81.52 ± 0.17
	GNNGUARD	74.98 ± 0.41	74.42 ± 0.55	69.26 ± 0.84	80.38 ± 0.07
	ORH	73.34 ± 1.61	74.22 ± 0.84	66.70 ± 2.74	72.20 ± 0.25
	VPN	78.53 ± 0.71	74.32 ± 0.64	69.76 ± 0.62	81.48 ± 0.13
	AdaEdge	83.00 ± 0.41	78.66 ± 0.66	70.08 ± 1.12	81.40 ± 0.14
	CHAGNN	**83.06 ± 0.56**	**79.64 ± 0.6**	70.36 ± 0.83	**81.56 ± 0.14**
MGA	No Attack	84.64 ± 0.34	81.26 ± 1.20	71.46 ± 0.30	85.00 ± 0.09
	Attack	81.98 ± 0.70	76.06 ± 0.54	69.68 ± 0.45	80.74 ± 0.10
	GCNSVD	80.38 ± 0.30	74.42 ± 0.47	67.82 ± 0.17	80.96 ± 0.05
	GCNJaccard	80.10 ± 0.82	75.42 ± 0.43	69.94 ± 0.38	81.30 ± 0.44
	GNNGUARD	73.56 ± 0.37	72.18 ± 0.55	67.28 ± 0.77	80.62 ± 0.69
	ORH	72.46 ± 1.66	69.24 ± 0.95	66.68 ± 1.20	74.66 ± 0.48
	VPN	78.93 ± 0.84	74.68 ± 0.36	69.84 ± 0.64	81.20 ± 0.05
	AdaEdge	83.26 ± 0.67	77.64 ± 0.68	70.04 ± 0.27	81.12 ± 0.12
	CHAGNN	**83.66 ± 0.62**	**78.02 ± 0.23**	**70.14 ± 0.34**	**81.38 ± 0.15**

To avoid excessive cleaning of the graph, we fixed the elimination rate in each iteration at 10% and the maximum number of iterations at 5.

4.2 Defense Performance Against Non-targeted Adversarial Attacks

We compare the performance of different methods at 10% injected nodes rate on four datasets. The results are shown in Table 2. We highlight the best performance in bold. From the table, we have the following observations and discussions.

- CHAGNN significantly outperforms all compared algorithms on most settings, indicating that the validity of targeted design towards GIA and cooperative homophilous augmentation.
- The performance of FGA and MGA is not significant compared to TDGIA. It makes sense because TDGIA was designed specifically for GIA scenarios, whereas FGA and MGA were originally designed for GMA. When resisting weak attacks, such as FGA and MGA, the defense performance of several compared models is poor or even worse than the vanilla GCN. We think it is due to the fact that the graph considered by the defense algorithm is severely damaged. However, when dealing with less poisoned or clean graphs, the performance of most defense algorithms may decrease. For instance, GCNSVD uses low-rank representation of the graph, leading to the loss of information carried in the original graph structure. The performance of GCNSVD on the original graph will be worse than the vanilla GCN. We can also see this phenomenon in [18]'s experiment.
- The performance of ORH and VPN fluctuates greatly. We think it is because the performance of both algorithms depends on the choice of hyperparameters. Specifically, the performance of ORH depends on the weight of the node's own information and neighbors' information in the message passing process. The performance of VPN depends on the weights of different powered graphs.

4.3 Defense Performance Under Different Injected Nodes Ratio

We compare the performance of different algorithms under different injected nodes ratio. We choose TDGIA, the attack method with the best results in our experiment, to evaluate the performance of defense methods under different injected nodes ratios. The results are reported in Fig. 3. Observations and discussions are listed as follows.

- Our method is effective against more powerful attacks. Even with a high injected nodes ratio, our approach can significantly improve model's performance. Vanilla GCN shows poor performance under 20% injected nodes ratio. Our method can improve it by 27%, 12%, 13% and 10% on the four datasets respectively.
- TDGIA shows better performance as the injected nodes ratio increases. Under different injected nodes ratios, our method outperforms others in most cases, exhibiting excellent defensive performance. It illustrates that heterophilous edges elimination can indeed enhance the robustness of the model against adversarial attacks.
- The performance of AdaEdge is better than GCNJaccard, which illustrates the effectiveness of using nodes' pseudo-labels to discriminate heterophily is more effective than using nodes' features. The performance of AdaEdge is second only to CHAGNN on multiple datasets. It shows that the process of screening the discriminated heterophilous edges can effectively reduce the possibility of homophilous edges being mistakenly removed, which brings stronger defense performance in CHAGNN.

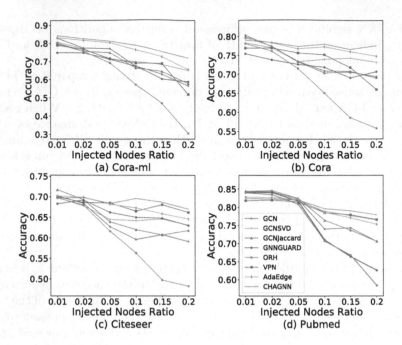

Fig. 3. Node classification performance under different injected nodes ratio on Cora-ml, Cora, Citeseer and Pubmed

4.4 Parameter Sensitivity on Eliminating Rate

In this part, we conduct sensitivity analysis with respect to the eliminating rate. We only report the results for the Cora-ml dataset at 20% and 2% injected nodes rates, since the results for other datasets share similar trends. The performance of node classification with different eliminating rates under TDGIA is shown in Fig. 4. We fixed the maximum number of iterations at 10. The following are some observations.

- The classification performance improves overall as the number of iterations increases. Our method has a certain defensive effect on most eliminating rates.
- It is not true that the higher the eliminating rate, the better our method performs. An excessive eliminating rate on a graph with few injected nodes can cause our method to perform poorly. This is because in a graph with few injected nodes, it is very easy to remove homophilous edges by mistake. In the future, we will devise some efficient methods to find the most appropriate eliminating rate.

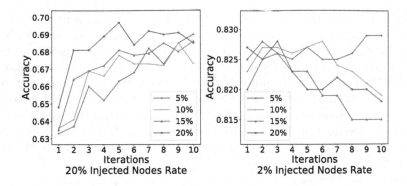

Fig. 4. Node classification performance under different eliminating rates

5 Related Work

5.1 Adversarial Attacks on GNNs

Nettack [11] stated that adding unnoticeable perturbations to the graph can fool GCN into incorrectly predicting. They generated perturbations to lead GCN to misclassify the target node while preserving the features' co-occurrences and the graph's degree distribution. Metattack [12] is proposed to reduce the overall performance of the model based on meta-learning. Most attacks are based on modifying nodes in the original graph. A more realistic scenario, graph injection attack (GIA), is studied in [13,14], which injects new vicious nodes instead of modifying the original graph. A greedy algorithm [15] is proposed to generate edges of malicious nodes and their corresponding features aiming to minimize the classification accuracy on the target nodes. NIPA [13] modeled the critical steps of graph injection attack based on reinforcement learning strategy. TDGIA [7] presented an analysis on the topological vulnerability of GNNs under GIA setting and proposed the topological defective graph injection attack (TDGIA) for effective injection attacks.

5.2 Defenses on GNNs

GCNSVD [16] found that Nettack has a greater impact on the high-rank part of the network. Then they proposed to use a low-rank approximation of the graph to train GCN by Singular Value Decomposition (SVD). GCNJaccard [17] stated that the attacks tend to connect the target node to nodes with different features. They removed the edges connecting the nodes that share few similarities to the target node by jaccard similarity. Pro-GNN [18] explored both properties mentioned before and designed a general framework to jointly learn a structural graph and a robust graph neural network model guided by these properties. GNNGUARD [19] detected and quantified the relationship between the graph structure and node features and then exploited that relationship to mitigate negative effects of the adversarial attacks. In addition to the defense methods against graph adversarial attacks, some methods based on data augmentation can also mitigate the influence of the model on graph adversarial attacks. VPN [31] designed the robust GCN via graph powering. They proposed a new convolution operator that is provably robust in the spectral domain. They incorporated it in the GCN architecture to improve model's expressivity and interpretability. AdaEdge [29] optimizes the graph topology based on the model predictions for relieving the over-smoothing issue. They simply remove the heterophilous edges without considering the effect of mistakenly removing the homophilous edges in this process. And the method does not consider the scenario of graph adversarial attacks. GAUG [30] used GAE to help improve GCN's robustness. The model's effectiveness at defending graph adversarial attacks depends on GAE's performance. However, all the defense methods mentioned are designed for GMA. As there are currently few methods to defend GIA, this paper defines this problem, which may provide critical insights for future research.

6   Conclusion

A more realistic scenario, graph injection attack (GIA), demonstrated effective attack performance on GNNs. However, there were few specific defense methods against GIA, a scenario that is easier for attackers to implement. In this paper, we formalized the anti-GIA defense scenario and designed the corresponding algorithm. Our experiments showed that our method significantly outperforms state-of-the-art baselines and improves the overall robustness under various GIA methods. Theoretically, the proposed strategy could work in various graph adversarial attacks. However, in the more practical GIA scenario, we can strictly guarantee the effectiveness from empirical and theoretical aspects. In the future, we plan to apply this strategy to more attack scenarios.

References

1. Zhong, T., Wang, T., Wang, J., et al.: Multiple-aspect attentional graph neural networks for online social network user localization. IEEE Access **8**, 95223–95234 (2020)
2. Liu, X., Tan, H., Chen, Q., et al.: RAGAT: relation aware graph attention network for knowledge graph completion. IEEE Access **9**, 20840–20849 (2021)
3. Wu, S., Tang, Y., Zhu, Y., et al.: Session-based recommendation with graph neural networks. In: Proceedings of the AAAI 2019, vol. 33, no. 01, pp. 346–353 (2019)
4. Li, J., Zhang, H., Han, Z., et al.: Adversarial attack on community detection by hiding individuals. In: Proceedings of the Web Conference 2020, pp. 917–927 (2020)
5. Zhang, M., Hu, L., Shi, C., et al.: Adversarial label-flipping attack and defense for graph neural networks. In: Proceedings of ICDM 2020, pp. 791–800. IEEE (2020)
6. Ma, J., Ding, S., Mei, Q.: Towards more practical adversarial attacks on graph neural networks. Adv. Neural. Inf. Process. Syst. **33**, 4756–4766 (2020)
7. Zou, X., Zheng, Q., Dong, Y., et al.: TDGIA: effective injection attacks on graph neural networks. In: Proceedings of KDD 2021, pp. 2461–2471 (2021)

8. Akhtar, N., Mian, A.: Threat of adversarial attacks on deep learning in computer vision: a survey. IEEE Access **6**, 14410–14430 (2018)
9. Feng, F., He, X., Tang, J., et al.: Graph adversarial training: dynamically regularizing based on graph structure. IEEE Trans. Knowl. Data Eng. **33**(6), 2493–2504 (2019)
10. Zhu, J., Jin, J., Loveland, D., et al.: On the Relationship between Heterophily and Robustness of Graph Neural Networks. arXiv preprint arXiv:2106.07767 (2021)
11. Zügner, D., Akbarnejad, A., Günnemann, S.: Adversarial attacks on neural networks for graph data. In: Proceedings of KDD 2018, pp. 2847–2856 (2018)
12. Zügner, D., Günnemann, S.: Adversarial attacks on graph neural networks via meta learning. arXiv preprint arXiv:1902.08412 (2019)
13. Sun, Y., Wang, S., Tang, X., et al.: Adversarial attacks on graph neural networks via node injections: a hierarchical reinforcement learning approach. In: Proceedings of the Web Conference 2020, pp. 673–683 (2020)
14. Wang, J., Luo, M., Suya, F., et al.: Scalable attack on graph data by injecting vicious nodes. Data Min. Knowl. Disc. **34**(5), 1363–1389 (2020)
15. Wang, X., Cheng, M., Eaton, J., et al.: Attack graph convolutional networks by adding fake nodes. arXiv preprint arXiv:1810.10751 (2018)
16. Entezari, N., Al-Sayouri, S.A., Darvishzadeh, A., et al.: All you need is low (rank) defending against adversarial attacks on graphs. In: Proceedings of WSDM 2020, pp. 169–177 (2020)
17. Wu, H., Wang, C., Tyshetskiy, Y., et al.: Adversarial examples on graph data: deep insights into attack and defense. arXiv preprint arXiv:1903.01610 (2019)
18. Jin, W., Ma, Y., Liu, X., et al.: Graph structure learning for robust graph neural networks. In: Proceedings of KDD 2020, pp. 66–74 (2020)
19. Zhang, X., Zitnik, M.: GNNGuard: defending graph neural networks against adversarial attacks. Adv. Neural. Inf. Process. Syst. **33**, 9263–9275 (2020)
20. Lim, D., Li, X., Hohne, F., et al.: New benchmarks for learning on non-homophilous graphs. arXiv preprint arXiv:2104.01404 (2021)
21. Zhu, J., Yan, Y., Zhao, L., et al.: Beyond homophily in graph neural networks: current limitations and effective designs. In: Advances in NeurIPS 2020, vol. 33, pp. 7793–7804 (2020)
22. Bojchevski, A., Günnemann, S.: Deep gaussian embedding of graphs: unsupervised inductive learning via ranking. arXiv preprint arXiv:1707.03815 (2017)
23. McCallum, A.K., Nigam, K., Rennie, J., et al.: Automating the construction of internet portals with machine learning. Inf. Retrieval **3**(2), 127–163 (2000)
24. Giles, C.L., Bollacker, K.D., Lawrence, S.: CiteSeer: an automatic citation indexing system. In: Proceedings of the Third ACM Conference on Digital Libraries, pp. 89–98 (1998)
25. Kipf, T.N., Welling, M.: Semi-supervised classification with graph convolutional networks. arXiv preprint arXiv:1609.02907 (2016)
26. Chen, J., Wu, Y., Xu, X., et al.: Fast gradient attack on network embedding. arXiv preprint arXiv:1809.02797 (2018)
27. Chen, J., Chen, Y., Zheng, H., et al.: MGA: momentum gradient attack on network. IEEE Trans. Comput. Soc. Syst. **8**(1), 99–109 (2020)
28. Li, Y., Jin, W., Xu, H., et al.: Deeprobust: a pytorch library for adversarial attacks and defenses. arXiv preprint arXiv:2005.06149 (2020)
29. Chen, D., Lin, Y., Li, W., et al.: Measuring and relieving the over-smoothing problem for graph neural networks from the topological view. In: Proceedings of the AAAI 2020, vol. 34, no. 04, pp. 3438–3445 (2020)

30. Zhao, T., Liu, Y., Neves, L., et al.: Data augmentation for graph neural networks. arXiv preprint arXiv:2006.06830 (2020)
31. Jin, M., Chang, H., Zhu, W., et al.: Power up! robust graph convolutional network via graph powering. In: 35th AAAI Conference on Artificial Intelligence (2021)
32. Mohri, M., Rostamizadeh, A., Talwalkar, A.: Foundations of Machine Learning. MIT Press, Cambridge (2018)

Securing Cyber-Physical Systems: Physics-Enhanced Adversarial Learning for Autonomous Platoons

Guoxin Sun[1(✉)], Tansu Alpcan[1], Benjamin I. P. Rubinstein[1], and Seyit Camtepe[2]

[1] University of Melbourne, Melbourne, Australia
guoxins@student.unimelb.edu.au, {tansu.alpcan,brubinstein}@unimelb.edu.au
[2] CSIRO Data61, Sydney, Australia
seyit.camtepe@data61.csiro.au

Abstract. The rapid development of cyber-physical systems in high-stakes safety-critical areas requires innovations in protecting them against malicious adversaries. Data-driven attack detection mechanisms based on deep learning (DL) have emerged as powerful tools to fulfil this need. However, it is well-known that adversarial attacks deceive DL models with specifically crafted perturbations added to clean data samples. This work combines cyber-physical system characteristics with DL to develop a hybrid attack detection system. Using knowledge from both physical dynamics and data, we defend against both cyber-physical attacks and adversarial attacks. This approach paves the way to use classical theories from the application domain to mitigate the deficiency of DL, complementing existing adversarial defence methods such as adversarial training. We implement our defence system for an autonomous vehicle platoon test-bed in a sophisticated simulator, where our approach doubles the detection F1 score and increases the minimum inter-vehicle distances compared to existing baselines. Hence, we greatly improve the safety and security of the target system against adversarially-masked cyber-physical attacks.

Keywords: Cyber-physical attacks · Adversarial machine learning · Autonomous platoons

1 Introduction

Cyber-physical systems, where sensor networks and embedded computing are intertwined with the physical environment, are fast becoming a key driving force of today's economy. Such systems observe and interact with the changes

We gratefully acknowledge support from the DSTG Next Generation Technology Fund and CSIRO Data61 CRP on 'Adversarial Machine Learning for Cyber' and CSIRO Data61 PhD scholarship.

Supplementary Information The online version contains supplementary material available at https://doi.org/10.1007/978-3-031-26409-2_17.

M.-R. Amini et al. (Eds.): ECML PKDD 2022, LNAI 13715, pp. 269–285, 2023.
https://doi.org/10.1007/978-3-031-26409-2_17

in surrounding environments to achieve high levels of reliability and context-aware autonomy. As a paradigm and example of such a cyber-physical system, *autonomous vehicle platoons* attract attention with potentially improved driving experience and energy efficiency, reduced pollution as well as increased traffic throughput. This concept involves a string of vehicles travelling as a single unit from an origin to a destination. Each platoon member obtains other vehicles' dynamics and manoeuvre-related information through existing and emerging vehicle-to-vehicle communication networks and embedded sensors in order to adapt its own behaviour to maintain a narrow inter-vehicle distance and relative velocity.

The high levels of connectivity and open communication implementations highlight vehicle platoons as appealing targets for cyber-physical attacks causing degradation of their dependability or even catastrophic incidents. The potential impact of security vulnerabilities has motivated the development of attack detection methods. Due to the rapid development of deep learning (DL), researchers have shown an increased interest in applying data-driven techniques, especially in the form of deep neural networks, to study and classify the complex patterns of system behaviour. Although DL-based attack detection demonstrates excellent defence performance against conventional cyber-physical attacks, they are also known to be vulnerable to adversarial attacks, in which specifically crafted perturbations are added on top of clean data with the aim of evading detection.

Motivation and Problem. If a DL-based attack detection system fails to detect cyber-physical attacks masked with adversarial perturbations, then the system is exposed to a much wider range of safety risks, since any conventional attack can be masked to evade detection this way. Traditionally, physical systems have been designed and analysed with classical modelling techniques, which constitute the foundation of control theory. Although data-driven approaches are becoming dominant in many areas, those classical tools still have an important role to play in cyber-physical systems such as vehicle platoons.

In this context, recent work [9] generates adversarial attacks against an anomaly detector for a water treatment problem considering the effects of a 'rule-checker'. Yet, the rules are derived mainly from observations instead of physical laws from first principles. While [1] combine a data-driven algorithm - generalized Extreme Studentized Deviate (ESD) - with the physical laws of kinematics to perform real-time anomaly detection, they have not considered the effects of adversarial attacks in their work. Similarly, model-based approaches alone are not a 'silver bullet' to the cyber-physical security problem either. A well-educated attacker could derive and leverage the underlying system model to increase the level of stealthiness [6].

Novelty and Contributions. This paper presents a novel combination of engineering modelling techniques with DL and proposes a hybrid attack detection system using knowledge from both physical dynamics and data to defend against both cyber-physical attacks and adversarial attacks. This approach paves the way to use classical theories from the application domain to make up for the deficiencies of DL and vice versa. Our approach is also applied in combination with existing adversarial defence techniques such as adversarial training to further improve its robustness. The contributions of this paper include:

(1) We provide a novel physics-enhanced data-driven attack detection system for cyber-physical systems that leverages knowledge from both data and physics.
(2) We illustrate that classical physics-modelling techniques can help to mitigate the deficiency of deep learning-based approaches, which extends the applicability of many state-of-the-art DL-based approaches for cyber-physical systems.
(3) As a demonstration, we successfully improve the security and dependability of vehicle platoons. Our defence system provides excellent detection performance against an informed white-box attacker.
(4) Our results are demonstrated both analytically and visually using sophisticated, system-level simulations. It outperforms standard baseline attack detection methods and proves the potential to be applied with existing adversarial defence techniques for better performance.

Related Work. Sumra et al. [18] provide a comprehensive survey of the cyber-physical attacks on major security goals, i.e., confidentiality, integrity and availability. For example, data integrity attacks corrupt the legitimacy of transmitted information, which allows malicious or Sybil vehicles to gain the privilege of the road or to cause traffic congestion and even serious collisions [3]. Malicious attackers may conduct eavesdropping attacks to steal and misuse confidential information [21].

In terms of data-driven learning-based attack detection approaches, [11] apply both feed-forward deep neural networks and convolutional neural networks to identify a malicious attacker who tries to cause collisions by altering the controller gains. [22] propose an ensemble model consisting of 4 tree-based algorithms to detect attacks against the Controller Area Network (CAN) bus. To better utilise the embedded temporal information within the time-series data from such systems, several attempts [2] have been made to solve the attack detection problem by examining the deviations of system behaviour and model predictions with machine learning models.

In the past, adversarial attacks have been extensively studied mainly in domains such as image and audio and far less attention has been paid to cyber-physical domains especially systems involved with time-series data. Existing research on the subject has also been mostly restricted to a few pre-generated datasets. For instance, [10] investigate the effects of adversarial attacks against time-series classifiers based on the UCR archive with data generated a posteriori of various types (e.g., motion, sensor etc.). In our work, we investigate the targeted cyber-physical system in various types of fringe and dangerous situations where the data and its corresponding adversarial examples are generated in an online fashion.

The vulnerabilities of DL models have motivated the development of adversarial defences. Adversarial training is a simple yet effective defence approach, which is to include adversarial examples directly as part of the training dataset [8]. Although the improved model is aware of adversarial examples in advance thereby more robust, the defender needs knowledge of the adversarial attacks and efforts to generate those examples a priori. Other defence methods including data distortion [23], defence distillation [14] have been proved to have their own advantages and limitations. Recently, physical knowledge has been exploited to enhance the training procedure or overall performance of neural networks in the targeted domain. Physical models of the underlying system become part of the loss function to bound the space of admissible solutions to the neural network parameters [5]. Nevertheless, few researchers have been able to draw any systematic study on incorporating physical knowledge for adversarial defence.

2   Problem Definition

A typical cyber-physical system (e.g., autonomous vehicles, smart grids, etc.) acquires necessary real-time information via onboard sensors or wireless communication with other parties. Malicious adversaries often target these communication networks and onboard sensors to destabilize or break down such safety-critical systems via cyber-physical attacks. If the system contains machine learning components, a range of adversarial attacks can be utilized by the attacker to perform so-called adversarially-perturbed cyber-physical attacks. The attacker's ultimate goal is to maximize physical damage while remaining stealthy.

This work presents a hybrid defence method that utilizes knowledge both from data and physics to address such security challenges. The data-driven component of our approach learns the complex physical dynamics of a real system purely from data when existing modelling techniques fail to model accurately and reliably. The physics component with a simple system model helps when learning-based methods suffer from adversarial perturbations. Specifically, the underlying system structure can be modelled by physical first principles with differential equations in the form of $\dot{x} = g(x)$, where x contains the states of the system. For example, the motion of autonomous vehicle platoons can be modelled by *the kinematic model* whereas power system dynamics can be modelled by *the swing equation* [20]. As a general defence framework, the physics part of our proposed defence framework could be substituted accordingly based on the underlying cyber-physical system. The deep-learning model could also be replaced by feed-forward neural networks, convolutional neural networks etc. We would utilize the kinematic model for vehicle platoons as a case study in the rest of the work. To generalize from autonomous vehicles to the smart grids, for example, one can replace the kinematic model used by our physics component with the swing equation.

3   Attacker Model

As a paradigm of cyber-physical attacks, false data injection corrupts the content of wirelessly transmitted messages or sensor observations to cause performance degradation or catastrophic failure of safety-critical systems. In the present work, we consider two attack approaches as presented in Sect. 3.1 and Sect. 3.2.

3.1   Conventional Cyber-Physical Attacks

Vanilla False Data Injection Attack. We extend the message falsification attacks [2, 19] from only affecting communication messages to attacking both communication and sensor observations [4] in a subtle way and name it vanilla false data injection (v-FDI). In particular, the adversary progressively increases the attack intensity to achieve its malicious objectives (e.g., causing collisions) while evading detection. Take acceleration modification in the vehicle platoon case as an example, the modified acceleration value is similar to the original one at the beginning of the attack. As attack effects progressively build up, it might become too late for the defence system to react since the attack may have already led to limited response time or even collision.

Model-Aware False Data Injection Attack. Model-aware false data injection (m-FDI) can be seen as an evolved version of its vanilla counterpart. Instead of injecting arbitrary modifications, the adversary utilizes the knowledge of the underlying system model to conduct malicious modifications concurrently on a range of observations. Following the acceleration modification example, as the acceleration modification progressively increases, the attacker computes the resulting velocity and position quantities based on the system model and injects velocity and position modifications accordingly. In this way, the modified data is consistent with the underlying system model (i.e., the kinematic model) thereby increasing its stealthiness level and attack strength. We will show in later sections how this type of attack can bypass model-based detection methods but not ours.

3.2   Adversarially-Masked Cyber-Physical Attacks

While attacking the cyber-physical systems with conventional cyber-physical attacks, intelligent adversaries may also create carefully-crafted adversarial perturbations to deceive DL-based attack detection systems. In contrast to conventional adversarial attacks against classifiers, we investigate similar attack methods but applied against regression models in cyber-physical domains. Inspired by the linear behaviour of modern machine learning models, the basic iterative method (BIM) [13] uses the first-order information of the loss function and generates adversarial examples iteratively. It is adopted in our work because of its improved attack performance with an even reduced perturbation level compared to other gradient-based attack methods such as the fast gradient sign method.

3.3   Attacker Capabilities

For simplicity, we consider only the dynamic information of a single vehicle (e.g., the preceding vehicle) will be modified by the attacker, which includes the transmitted acceleration messages via wireless communication as well as velocity and position information measured by a rangefinder (e.g., radar). Different levels of a priori knowledge (e.g., white-box knowledge of the controller, the DL-based attack detection system, the underlying system model as well as full access to the onboard memory) are assumed to conduct different types of malicious attacks, which are summarized in Table 1. Attack abbreviation followed by (adv. masked) denotes that adversarial perturbations are added to deceive a machine learning model which in our case is a DL-based anomaly detector.

Table 1. Knowledge required by the attacker to conduct different attacks.

Attack types	Access to				
	Sensors	Communication	DL model	System model	Memory
v-FDI	✓	✓	✗	✗	✗
m-FDI	✓	✓	✗	✓	✗
v-FDI (adv. masked)	✓	✓	✓	✗	✓
m-FDI (adv. masked)	✓	✓	✓	✓	✓

4   Physics-Enhanced Defense Approach

The proposed defence system consists of a data-driven component powered by deep neural networks to detect conventional cyber-physical attacks (e.g., false data injections) and a physics-inspired component to assist in reporting adversarial perturbations to compensate for the deficiency of deep learning models. We apply this general approach to the specific case of autonomous vehicle platoons as an illustrative example. Figure 1a shows the overall structure of our proposed defence system along with the pseudo-code of this new Double-Insured Anomaly Detection (DAD) method presented in Algorithm 1.

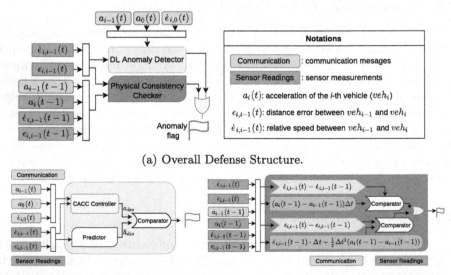

(a) Overall Defense Structure.

(b) Structure of the Deep Learning-based Anomaly Detector.

(c) Structure of the Physical Consistency Checker.

Fig. 1. The defence system applied to the vehicle platoon case study.

Algorithm 1. Double-Insured Anomaly Detection (DAD)

Input: Communication messages S and sensor readings R
Output: Anomaly flag

1: Initialization()
2: **while** Destination is not reached **do**
3:     Vehicle receives S and measures R
4:     $hist \leftarrow$ Load one-step history data
5:     $flag1 \leftarrow AnomalyDetector(R, hist)$
6:     $flag2 \leftarrow PhysicalConsistencyChecker(S, R, hist)$
7:     **if** $flag1$ or $flag2$ is TRUE **then**
8:         Anomaly flag $\leftarrow$ Anomaly
9:     **else**
10:        Anomaly flag $\leftarrow$ Normal
11:    **end if**
12: **end while**

4.1 Case Study: Autonomous Vehicle Platoons

Platoon Control Policy. We consider a vehicle platoon consisting of N vehicles travelling on a straight highway segment and let veh_i denote the i-th vehicle within the platoon, where $i \in [0, N-1]$. Each vehicle member adopts the predecessor-leader following information flow topology. Specifically, veh_i obtains dynamic information including location, speed, acceleration, etc., from both the platoon leader veh_0 and its immediate preceding vehicle veh_{i-1}. Based on this topology, the vehicle's longitudinal motion is governed by the cooperative adaptive cruise control (CACC) policy, which computes the desired acceleration for veh_i by:

$$\ddot{x}_i = a_{des} = \alpha_1 a_{i-1} + \alpha_2 a_0 + \alpha_3 \dot{\epsilon}_{i,i-1} + \alpha_4 \dot{\epsilon}_{i,0} + \alpha_5 \epsilon_{i,i-1}, \tag{1}$$

$$\epsilon_{i,i-1} = x_i - x_{i-1} + L, \quad \dot{\epsilon}_{i,i-1} = v_i - v_{i-1},$$

where α's are controller gains taken from [16]. a_{i-1} and a_0 are the accelerations of the preceding vehicle and the leader respectively. The distance error $\epsilon_{i,i-1}$ is calculated based on a desired gap distance L and the obtained inter-vehicle distance between veh_i and veh_{i-1}. Similarly, their corresponding relative speed is represented as $\dot{\epsilon}_{i,i-1}$ with v_i denoting the speed of veh_i.

Kinematic Model. The longitudinal motion of each vehicle can be modelled as uniformly accelerated motion along a line by Eq. (2). This motion arises when an object is subjected to a constant acceleration. The acceleration value determines the gradient of the velocity-time function with an initial velocity labelled as $v_i(0)$. Similarly, the steady changing velocity determines the gradient of the position-time function with an arbitrary initial position $x_i(0)$.

$$v_i(t) = v_i(0) + a_i t, \quad x_i(t) = x_i(0) + v_i(0)t + \frac{1}{2}a_i t^2, \tag{2}$$

where the acceleration and velocity variables are defined in Eq. (1). This kinematic model can be used to approximate the local behaviour of general longitudinal motions by modeling the object's motion within two consecutive sampling steps $t - 1$ and t as uniformly accelerated motion. In general, its approximation strength increases with increased sampling frequency.

4.2   Data-Driven Anomaly Detector

At each time instance t, veh_i obtains the relative speed and distance with respect to its predecessor via a rangefinder (e.g., a radar sensor), which are the inputs to our DL attack detector. As a bonus, although veh_i also receives communication messages from other vehicles, only sensor measurements are used as detector inputs because they are more difficult to modify in practice and inherently immutable to communication-related attacks resulting in high detection success rate when only communication channels are compromised. The overall structure is shown in Fig. 1b, which consists of two parts:

1. **Predictor**, trained with data from normal manoeuvre behaviour based on a sliding window, outputs the expected desired acceleration value $\tilde{a}_{des}$ at the current time instance. In our work, we use a multivariate time-series regression model and a sliding window to fully extract the temporal information within the data.

2. **Comparator** computes the difference between the inputs, i.e., the controller output a_{des} and the predicted value $\tilde{a}_{des}$. We use a sliding window to compute the mean absolute error $\bar{e}$ in order to reduce the false alarm rate. Consider an error window of size M, $\bar{e}$ at time t is computed as

$$\bar{e}(t) = \frac{1}{M} \sum_{m=t-M+1}^{i=t} \|a_{des}(m) - \tilde{a}_{des}(m)\|. \tag{3}$$

An anomaly is flagged when $\bar{e}$ is greater than a threshold pre-determined in a benign driving environment.

4.3   Physical Consistency Checker

Corrupted controller inputs may not obey the underlying physical processes of the platoon system. Based on the kinematic model, the physics-based component of the proposed defence method - the physical consistency checker - consists of two components: the distance checker and speed checker.

Distance Checker. The change of inter-vehicle distance $(\Delta\epsilon_{i,i-1})$ within consecutive sampling instances can be directly calculated based on transmitted location information from the preceding vehicle veh_{i-1} and its own location readings. The same quantity $(\Delta\tilde{\epsilon}_{i,i-1})$ can also be computed based on locally measured speed and acceleration information according to the kinematic model.

$$\Delta\epsilon_{i,i-1}(t) = \epsilon_{i,i-1}(t) - \epsilon_{i,i-1}(t-1),$$

$$\Delta\tilde{\epsilon}_{i,i-1}(t) = \dot{\epsilon}_{i,i-1} \cdot \Delta t + \frac{1}{2}\Delta t^2 \left(a_i(t-1) - a_{i-1}(t-1)\right)$$

Speed Checker. Similarly, the change of relative speed can be computed directly by the subtraction of speed measurements or by the kinematic model utilizing acceleration information.

$$\Delta\dot{\epsilon}_{i,i-1}(t) = \dot{\epsilon}_{i,i-1}(t) - \dot{\epsilon}_i(t-1),$$

$$\Delta\tilde{\dot{\epsilon}}_{i,i-1}(t) = \left(a_i(t-1) - a_{i-1}(t-1)\right)\Delta t.$$

Both the direct calculation and physical model-based calculation produce similar results when there are no adversarial attacks against the anomaly detector or the proposed defence system in general. However, they deviate in an adversarial environment. If the deviation is greater than a pre-defined threshold, it triggers our physical consistency checker to report anomalies. Note that, these thresholds are domain-specific. In our evaluation, they are determined to balance out the false positive and false negative rates in a benign driving environment, which contains various types of highway driving scenarios. The overall structure is shown in Fig. 1c along with the pseudo-code presented in Algorithm 2.

Algorithm 2. Physical Consistency Checker (PCC)

Input: Communication messages S, sensor readings R and one-step history *hist*
Output: TRUE or FALSE

1: $\Delta\dot{\epsilon} \leftarrow \dot{\epsilon}_{i,i-1}(t) - \dot{\epsilon}_{i,i-1}(t-1)$ {Speed check}
2: $\Delta\tilde{\dot{\epsilon}} \leftarrow (a_i(t-1) - a_{i-1}(t-1))\Delta t$
3: *Anomaly flag* 1 $\leftarrow$ *Comparator*($\Delta\dot{\epsilon}, \Delta\tilde{\dot{\epsilon}}$)
4: $\Delta\epsilon \leftarrow \epsilon_{i,i-1}(t) - \epsilon_{i,i-1}(t-1)$ {Distance check}
5: $\Delta\tilde{\epsilon} \leftarrow \dot{\epsilon}_{i,i-1}(t-1) \cdot \Delta t + \frac{1}{2}\Delta t^2(a_i(t-1) - a_{i-1}(t-1))$
6: *Anomaly flag* 2 $\leftarrow$ *Comparator*($\Delta\epsilon, \Delta\tilde{\epsilon}$)
7: **if** *Anomaly flag* 1 or *Anomaly flag* 2 is TRUE **then**
8:      *Anomaly flag* $\leftarrow$ TRUE
9: **else**
10:     *Anomaly flag* $\leftarrow$ FALSE
11: **end if**
12: **return** *Anomaly flag*

---

5 Experimental Results

5.1 Simulation Setup

To provide a comprehensive evaluation of our proposed detection method, we use *Webots* as our simulation platform, which provides a broad range of calibrated vehicle models, sensor modules as well as static objects and materials to realize different simulation scenarios with high physical accuracy. It is a cost-efficient approach to generating adequate training data and constructing different cyber-physical attacks. Our data sets and implementations are available on Github at https://garrisonsun.github.io/Securing-Cyber-Physical-Systems/.

Platoon and Traffic Simulation. We simulate a vehicle platoon of 4 *BMW X5* vehicles driving along a highway segment. Multiple sensors are embedded in each vehicle to measure, transmit and receive critical driving information. For example, veh_i uses a radar sensor to measure the inter-vehicle distance and relative speed with respect to its predecessor veh_{i-1}. Radar noise is calibrated according to the datasheet of a real-world radar (*Delphi ESR 2.5 pulse Doppler cruise control radar*). Other control inputs (e.g., leader's dynamics used in Eq. 1) are obtained via wireless communication. In addition, each vehicle reads its own speed, acceleration, etc. directly from the speedometer and accelerometer respectively.

We generate a large number of vehicles in real-time in *Webots* interfacing with Simulation of Urban MObility (SUMO) [15] in order to construct a more realistic driving environment. Traffic flows involve four types of vehicles (i.e., motorcycles, light-weight vehicles, trucks, and trailers) with various driving characteristics (cooperative or competitive) and intentions to merge, which generates many random situations.

5.2   Double-Insured Anomaly Detection (DAD)

Data-Driven Anomaly Detector. For this work, we train an LSTM network from normal data as our predictor due to its outstanding performance for time-series prediction. It is a many-to-one prediction model, which consists of a normalization layer, and two stacked LSTM layers with 200 and 100 hidden units respectively. Each LSTM layer is followed by a dropout layer (rate = 0.3) to avoid overfitting. The last dropout layer is connected with two fully connected layers with 50 and 1 hidden units, respectively. The model takes a sequence of historical sensor measurements (controlled by the sliding window size) and outputs a prediction of the desired acceleration $\tilde{a}_{des}$ for the next time instance. An anomaly is reported if this predicted value significantly deviates from the controller output.

Physical Consistency Checker. Since the physical consistency checker assumes vehicle motion within consecutive sampling time instances as uniformly accelerated motion, high-frequency sensor noise could degrade its anomaly detection performance when noise level and sampling frequency are both high. Therefore, we apply a digital Butterworth low-pass filter to remove the noise and reveal the underlying trend of the residuals between direct and model calculations. Note that, its performance is expected to be improved with low-noise sensors specifically designed for vehicle platoon applications.

5.3 Evaluation Setup

We employ Keras 2.4.3 and Python 3.8.10 to implement DAD and all the baselines on Ubuntu 20.04 operating system with a commodity i7-10510U CPU. The models are trained using an Adam optimizer with a learning rate of 0.001 for up to 500 epochs with early stopping (patience = 10). Mean squared error is chosen as the loss function. We found a window size of 20 (input size 20×2) results in the best prediction performance. For the comparator, a window size of 40 and a detection threshold of 2 can effectively smooth out the prediction residuals and reduce the false alarm rate without degrading prediction performance.

Metrics. We prioritize F1 score [24] for this evaluation because a high F1 score indicates a combination of high precision and recall. Missing an attack is often more costly for such safety-critical systems, potentially causing catastrophic damages. Therefore, detection recall (Rec) is also included as our secondary comparison metric.

Attacks. Conventional cyber-physical attacks along with their adversarially-masked versions are examined in our evaluation:

Vanilla false data injection (v-FDI), as described in Sect. 3.1, and progressively modify received/measured dynamics information from the proceeding vehicle. For v-FDI, the modifications can be posed on a single variable such as on acceleration (v-FDI-Acce.) or in combination (e.g., v-FDI-Acce.Speed that alters both acceleration and speed).

Model-aware false data injection (m-FDI), is seen as an evolved version of v-FDI. We consider the acceleration is modified with the maximum allowable modification as 5 m/s^2 (since higher accelerations are unrealistic in practice) and both the speed and location magnitudes are also modified based on the underlying system model to improve stealthiness.

Adversarial attack, the BIM attack approach [13] in particular, is used to mask these cyber-physical attacks in order to deceive the deployed attack detector (e.g., adversarially masked m-FDI is denoted as m-FDI adv. masked). The max-norm ball ϵ is chosen to be a small value as 0.4. In this way, the attack data is only slightly modified thereby retaining the original attack effects of the cyber-physical attack. Note that, the ϵ value is prefixed in this evaluation and a grid search may be required to find the optimal ϵ for different detectors under different attacks.

Baselines Attack Detectors. To demonstrate the effectiveness and robustness of our proposed defence system, we compare the detection results with pure data-driven and model-based detection approaches and study the impact of each component of our proposed method. In addition, we also consider adversarial training, based on the BIM attack, as an adversarial defence baseline in our evaluation with a robustified LSTM reconstruction-based attack detector (D1: LSTM*). The state-of-the-art data-driven defence baselines include an LSTM reconstruction-based attack detector (D1: LSTM) as in [7] and a CNN

reconstruction-based attack detector (D2: CNN) similar to [12]. Some literature [17] also recognizes the effectiveness of autoencoders in performing classification or anomaly detection tasks based on reconstruction errors. For completeness, we also implement a convolutional autoencoder-based detector (D3: AE).
Besides, the physics component of our proposed method - physical consistency
checker (PCC) - is also used for comparison as a standalone attack detector.

5.4   Attack Detection Results

We demonstrate that our proposed attack detection system (DAD) provides
improved attack detection performance against both conventional cyber-physical
attacks and their adversarially-masked counterparts. In total, eight conventional
cyber-physical attacks are examined including 7 variants of vanilla false data
injection and 1 model-aware false data injection. Each type of attack has been
performed five times. The complete detection *F1 score* is summarized in Fig. 2
along with error bars at the top. The detection results for the model-aware false
data injection (m-FDI) are summarized in Table 2.

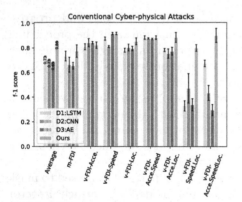 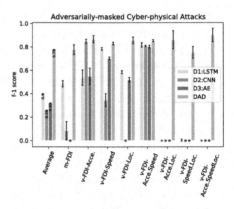

(a) Conventional cyber-physical attacks.   (b) Adv. masked cyber-physical attacks.

Fig. 2. Detection F1 scores comparing our method with alternatives.

Conventional Cyber-Physical Attacks. The detection F1 scores for conventional cyber-physical attacks are summarized in Fig. 2a. In general, data-driven
methods such as LSTM, CNN and AE reconstruction-based attack detectors perform well against these attacks. Depending on the exact attack type, one may
slightly outperform the other. The physical consistency checker (PCC), as seen in
the left half of Table 2, misses most of the attack instances when acting alone to
detect m-FDI resulting in a recall of 0.18 and an F1 score of 0.29. This highlights
the necessity of data-driven approaches to capture complex system behaviour. In
comparison, our proposed method takes the advantage of both data-driven and
physics-inspired methods achieving an F1 score of 0.86 on average (Fig. 2a) over
all considered conventional attacks and outperforms other popular data-driven
approaches.

Adversarially-Masked Cyber-Physical Attacks. The physics-inspired component of our method, the physical consistency checker, starts to shine when the cyber-physical attacks are masked with adversarial perturbations. The detection performance, as shown in Fig. 2b and right half of Table 2, is greatly reduced for all data-driven attack detectors, which exposes the cyber-physical system (i.e., the vehicle platoon) to a wide range of safety risks. Although some attacks require larger perturbations to fully deceive the detector, the average F1 scores are reduced to 0.40, 0.26 and 0.32 from 0.73, 0.71, 0.69 respectively for the LSTM, CNN, and AE based detectors. Because the generated adversarial perturbations are inconsistent with the physics model, our proposed method is able to detect the adversarially-perturbed cyber-physical attacks with an average F1 score of 0.78, which doubles the detection F1 score compared to existing baselines.

Adversarial Training Variants. As seen at the bottom of Table 2, our proposed method can be applied along with existing adversarial defence approaches (e.g., adversarial training). Combining the robustified model with physics knowledge would result in a better detection system DAD*, increasing detection recall and F1 score from 0.77 and 0.78 to 0.84 and 0.81, respectively, against adversarial-perturbed attacks. Although adversarial training might slightly sacrifice detection performance against classical cyber-physical attacks, it is demonstrated that our defence framework has the potential to provide better performance with an improved DL model and/or with other advanced adversarial defence methods against much stronger adversaries.

Table 2. Attack detection results against m-FDI with different detection methods. * denotes adversarial training.

Attack	m-FDI		m-FDI (adv. masked)		
Defense	Rec	F1	Defense	Rec	F1
D1: LSTM	0.70	0.73	D1: LSTM	0.39	0.49
D2: CNN	0.57	0.66	D2: CNN	0.05	0.08
D3: AE	0.56	0.66	D3: AE	0.00	0.00
PCC	0.18	0.29	PCC	0.63	0.75
Ours: DAD	**0.77**	**0.77**	**Ours: DAD**	**0.77**	**0.78**
D1: LSTM*	0.70	0.73	D1: LSTM*	0.48	0.56
Ours: DAD*	**0.75**	**0.76**	**Ours: DAD***	**0.84**	**0.81**

5.5 Simulation Demonstration for the M-FDI Attack

In this subsection, we use the inter-vehicle distance as a measuring metric to examine the dangerous level of the compromised vehicle under attack. The entire simulation process can be roughly divided into four stages as shown in Fig. 3a and Fig. 3c. It starts from the *preparation stage*, where each vehicle starts with zero velocity and accelerates from an arbitrary position with a random inter-vehicle distance. Once the vehicle platoon is established, all platoon members quickly enter the *transient stage* and gradually reach the desired inter-vehicle distance (2 m in this case). This distance will be maintained throughout the simulation with only minor fluctuations when traffic condition changes with the power of the CACC platoon controller. The steady stage ends when an attack is initiated and we start to observe the resulting inter-vehicle distances for different defence methods. In this demonstration, we assume the vehicle would request a manual manoeuvre (not affected by data false injection attacks) as a simple mitigation strategy when the defence system reports an attack.

- As indicated in Fig. 3a and Fig. 3b, our method as well as other data-driven baselines can maintain a relatively safe inter-vehicle distance under conventional cyber-physical attacks (i.e., m-FDI). However, our method results in the best detection performance against m-FDI with nearly unnoticeable fluctuation throughout the entire attack period. In comparison, the model-based detection method PCC leads to a minimum distance of 0.25 m, which greatly increases safety risks, especially in the highway driving scenario, and highlights the importance of data-driven approaches for cyber-physical systems.
- Figure 3c and Fig. 3d indicate that DL-based detectors suffer the most under adversarially-masked cyber-physical attacks with CNN and AE-based detec-

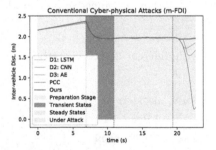

(a) Inter-vehicle distance comparison.

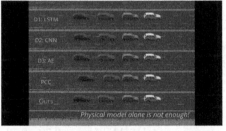

(b) Simulation screenshot of the vehicle platoon.

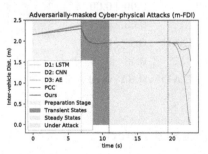

(c) Inter-vehicle distance comparison.

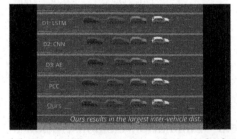

(d) Simulation screenshot of the vehicle platoon.

Fig. 3. Comparison between different defense methods under attacks: (a)&(b)-Conventional cyber-physical attacks (m-FDI), (c)&(d)-Adversarially-masked cyber-physical attacks (m-FDI (adv. masked)).

tors leading to catastrophic collisions. Our proposed method again significantly improves system safety and security against such a powerful adversary with white-box knowledge of both the DL and physics models. It is also important to point out that the model-based detector PCC alone cannot detect such attacks accurately during the entire attack period.

6 Conclusions

In this paper, we have presented a novel physics-enhanced attack detection system for autonomous vehicle platoons as a critical cyber-physical system. Our approach and algorithms greatly improve platoon security and dependability against both classical cyber-physical and adversarial attacks. Our methods inherit the advantages of existing data-driven attack detection systems based on recent advances in deep learning as well as utilize physics modelling techniques to improve robustness against adversarial attacks to cyber-physical systems. We consider a powerful white-box attacker and demonstrate that our approach outperforms conventional detection methods with a sophisticated simulator, which highlights its potential to perform even better when dealing with real-world attackers who normally only have limited information about the system. Future research will evaluate the extension of this resiliency architecture to other cyber-physical systems (e.g., smart grids) with various data-driven defence approaches. The scope of adversarial attacks in this work is limited to existing approaches developed mainly in the vision domain. Therefore, a further study could incorporate the physics model with the adversarial example generation process to create a stronger adversarial attack method and investigate its attack effects on the cyber-physical system and evasion strength against the proposed defence method.

References

1. Alotibi, F., Abdelhakim, M.: Anomaly detection for cooperative adaptive cruise control in autonomous vehicles using statistical learning and kinematic model. IEEE Trans. Intell. Transp. Syst. **22**(6), 3468–3478 (2020)
2. Boddupalli, S., Rao, A.S., Ray, S.: Resilient cooperative adaptive cruise control for autonomous vehicles using machine learning. IEEE Trans. Intell. Transp. Syst. **23**(9), 15655–15672 (2022)
3. Boeira, F., Barcellos, M.P., de Freitas, E.P., Vinel, A., Asplund, M.: Effects of colluding sybil nodes in message falsification attacks for vehicular platooning. In: 2017 IEEE Vehicular Networking Conference (VNC), pp. 53–60. IEEE (2017)
4. Cao, Y., et al.: Adversarial sensor attack on lidar-based perception in autonomous driving. In: Proceedings of the 2019 ACM SIGSAC Conference on Computer and Communications Security, pp. 2267–2281 (2019)
5. Daw, A., Karpatne, A., Watkins, W., Read, J., Kumar, V.: Physics-guided neural networks (PGNN): an application in lake temperature modeling. arXiv preprint arXiv:1710.11431 (2017)

6. Garcia, L., Brasser, F., Cintuglu, M.H., Sadeghi, A.R., Mohammed, O.A., Zonouz, S.A.: Hey, my malware knows physics! attacking PLCS with physical model aware rootkit. In: NDSS (2017)
7. Goh, J., Adepu, S., Tan, M., Lee, Z.S.: Anomaly detection in cyber physical systems using recurrent neural networks. In: 2017 IEEE 18th International Symposium on High Assurance Systems Engineering (HASE), pp. 140–145. IEEE (2017)
8. Goodfellow, I.J., Shlens, J., Szegedy, C.: Explaining and harnessing adversarial examples. arXiv preprint arXiv:1412.6572 (2014)
9. Jia, Y., Wang, J., Poskitt, C.M., Chattopadhyay, S., Sun, J., Chen, Y.: Adversarial attacks and mitigation for anomaly detectors of cyber-physical systems. Int. J. Crit. Infrastruct. Prot. **34**, 100452 (2021)
10. Karim, F., Majumdar, S., Darabi, H.: Adversarial attacks on time series. IEEE Trans. Pattern Anal. Mach. Intell. **43**(10), 3309–3320 (2020)
11. Khanapuri, E., Chintalapati, T., Sharma, R., Gerdes, R.: Learning-based adversarial agent detection and identification in cyber physical systems applied to autonomous vehicular platoon. In: 2019 IEEE/ACM 5th International Workshop on Software Engineering for Smart Cyber-Physical Systems (SEsCPS), pp. 39–45. IEEE (2019)
12. Kravchik, M., Shabtai, A.: Detecting cyber attacks in industrial control systems using convolutional neural networks. In: Proceedings of the 2018 Workshop on Cyber-Physical Systems Security and PrivaCy, pp. 72–83 (2018)
13. Kurakin, A., Goodfellow, I., Bengio, S., et al.: Adversarial examples in the physical world (2016)
14. Li, J., Liu, Y., Chen, T., Xiao, Z., Li, Z., Wang, J.: Adversarial attacks and defenses on cyber-physical systems: a survey. IEEE Internet Things J. **7**(6), 5103–5115 (2020)
15. Lopez, P.A., et al.: Microscopic traffic simulation using sumo. In: The 21st IEEE International Conference on Intelligent Transportation Systems. IEEE (2018). https://elib.dlr.de/124092/
16. Segata, M., Joerer, S., Bloessl, B., Sommer, C., Dressler, F., Cigno, R.L.: Plexe: a platooning extension for veins. In: 2014 IEEE Vehicular Networking Conference (VNC), pp. 53–60. IEEE (2014)
17. Seyfioğlu, M.S., Özbayoğlu, A.M., Gürbüz, S.Z.: Deep convolutional autoencoder for radar-based classification of similar aided and unaided human activities. IEEE Trans. Aerosp. Electron. Syst. **54**(4), 1709–1723 (2018)
18. Sumra, I.A., Hasbullah, H.B., AbManan, J.B.: Attacks on security goals (confidentiality, integrity, availability) in VANET: a survey. In: Laouiti, A., Qayyum, A., Mohamad Saad, M.N. (eds.) Vehicular Ad-hoc Networks for Smart Cities. AISC, vol. 306, pp. 51–61. Springer, Singapore (2015). https://doi.org/10.1007/978-981-287-158-9_5
19. Sun, G., Alpcan, T., Rubinstein, B.I.P., Camtepe, S.: Strategic mitigation against wireless attacks on autonomous platoons. In: Joint European Conference on Machine Learning and Knowledge Discovery in Databases. ECML-PKDD (2021)
20. Tielens, P., Van Hertem, D.: The relevance of inertia in power systems. Renew. Sustain. Energy Rev. **55**, 999–1009 (2016)
21. Wiedersheim, B., Ma, Z., Kargl, F., Papadimitratos, P.: Privacy in inter-vehicular networks: why simple pseudonym change is not enough. In: 2010 Seventh International Conference on Wireless On-demand Network Systems and Services (WONS), pp. 176–183. IEEE (2010)

22. Yang, L., Moubayed, A., Hamieh, I., Shami, A.: Tree-based intelligent intrusion detection system in internet of vehicles. In: 2019 IEEE Global Communications Conference (GLOBECOM), pp. 1–6. IEEE (2019)

23. Yuan, X., et al.: Commandersong: a systematic approach for practical adversarial voice recognition. In: 27th USENIX Security Symposium (USENIX Security 2018), pp. 49–64 (2018)

24. Zhang, C., et al.: A deep neural network for unsupervised anomaly detection and diagnosis in multivariate time series data. In: Proceedings of the AAAI Conference on Artificial Intelligence, vol. 33, pp. 1409–1416 (2019)

MEAD: A Multi-Armed Approach for Evaluation of Adversarial Examples Detectors

Federica Granese[1][(✉)], Marine Picot[2,3], Marco Romanelli[2], Francesco Messina[5], and Pablo Piantanida[4]

[1] Lix, Inria, Institute Polytechnique de Paris, Sapienza University of Rome, Rome, Italy
federica.granese@inria.fr
[2] Laboratoire des signaux et systèmes (L2S), Université Paris-Saclay, CNRS, CentraleSupélec, Gif-sur-Yvette, France
{marine.picot,marco.romanelli}@centralesupelec.fr
[3] McGill University, Montreal, Canada
[4] International Laboratory on Learning Systems (ILLS), McGill - ETS - MILA - CNRS - Université Paris-Saclay, CentraleSupélec, Saint-Aubin, Canada
pablo.piantanida@centralesupelec.fr
[5] Universidad de Buenos Aires, Buenos, Argentina
fmessina@fi.uba.ar

Abstract. Detection of adversarial examples has been a hot topic in the last years due to its importance for safely deploying machine learning algorithms in critical applications. However, the detection methods are generally validated by assuming a single implicitly known attack strategy, which does not necessarily account for real-life threats. Indeed, this can lead to an overoptimistic assessment of the detectors' performance and may induce some bias in the comparison between competing detection schemes. We propose a novel multi-armed framework, called MEAD, for evaluating detectors based on several attack strategies to overcome this limitation. Among them, we make use of three new objectives to generate attacks. The proposed performance metric is based on the worst-case scenario: detection is successful if and only if all different attacks are correctly recognized. Empirically, we show the effectiveness of our approach. Moreover, the poor performance obtained for state-of-the-art detectors opens a new exciting line of research.

Keywords: Adversarial examples · Detection · Security

F. Granese and M. Picot—These authors contributed equally to this work.

Supplementary Information The online version contains supplementary material available at https://doi.org/10.1007/978-3-031-26409-2_18.

1   Introduction

Despite recent advances in the application of machine learning, the vulnerability of deep learning models to maliciously crafted examples [34] is still an open problem of great interest for safety-critical applications [2,4,10,39]. Over time, a large body of literature has been produced on the topic of defense methods against adversarial examples. On the one hand, interest in detecting adversarial examples given a pre-trained model is gaining momentum [17,24,27]. On the other hand, several techniques have been proposed to train models with improved robustness to future attacks [26,31,38]. Interestingly, Croce et al. have recently pointed out that, due to the large number of proposed methods, the problem of crafting an objective approach to evaluate the quality of methods to train robust models is not trivial. To this end, they have presented *RobustBench* [9], a standardized benchmark to assess adversarial robustness. To the best of our knowledge, we claim that an equivalent benchmark does not exist in the case of methods to detect adversarial examples given a pre-trained model. Therefore, in this work, we provide a general framework to evaluate the performance of adversarial detection methods. Our idea stems from the following key observation. Generally, the performance of current state-of-the-art (SOTA) adversarial examples detection methods is evaluated assuming a unique and thus implicitly known attack strategy, which does not necessarily correspond to real-life threats. We further argue that this type of evaluation has two main flaws: the performance of detection methods may be overestimated, and the comparison between detection schemes may be biased. We propose a two-fold solution to overcome the aforementioned limitations, leading to a less biased evaluation of different approaches. This is accomplished by evaluating the detection methods on simultaneous attacks on the target classification model using different adversarial strategies, considering the most popular attack techniques in the literature, and incorporating three new attack objectives to extend the generality of the proposed framework. Indeed, we argue that additional attack objectives result in new types of adversarial examples that cannot be constructed otherwise. In particular, we translate such an evaluation scheme in MEAD.

MEAD is a novel evaluation framework that uses a simple but still effective "multi-armed" attack to remove the implicit assumption that detectors know the attacker's strategy. More specifically, for each natural sample, we consider the detection to be successful *if and only if* the detector is able to identify all the different attacks perpetrated by perturbing the testing sample at hand. We deploy the proposed framework to evaluate the performance of SOTA adversarial examples detection methods over multiple benchmarks of visual datasets. Overall, the collected results are consistent throughout the experiments. The main takeaway is that considering a multi-armed evaluation criterion exposes the weakness of SOTA detection methods, yielding, in some cases, relatively poor performances. The proposed framework, although not exhaustive, sheds light on the fact that evaluations so far presented in the literature are highly biased and unrealistic. Indeed, the same detector achieves very different performances when it is informed about the current attack as opposed to when it is not. Not surprisingly, supervised and unsupervised methods achieve comparable performances with the multi-armed framework, meaning that training the detectors knowing a specific attack used at testing time does not generalize to other attacks enough. Indeed the goal of MEAD is not to show that new attacks can always fool robust classifiers but to show that the detectors that may work well when evaluated with a unique attack strategy end up being defeated by new attacks.

1.1   Summary of Contributions

We propose MEAD, a novel multi-armed evaluation framework for adversarial examples detectors involving several attackers to ensure that the detector is not overfitted to a particular attack strategy. The proposed metric is based on the following criterion. Each adversarial sample is correctly detected if and only if all the possible attacks on it are successfully detected. We show that this approach is less biased and yields a more effective metric than the one obtained by assuming only a single attack at evaluation time (see Sect. 4).

We make use of three new objective functions which, to the best of our knowledge, have never been used for the purpose of generating adversarial examples at testing time. These are *KL divergence*, *Gini Impurity* and *Fisher-Rao distance*. Moreover, we argue that each of them contributes to jointly creating competitive attacks that cannot be created by a single function (see Sect. 3.2).

We perform an extensive numerical evaluation of SOTA and uncover their limitations, suggesting new research perspectives in this research line (see Sect. 5).

The remaining paper is organized as follows. First, in Sect. 2, we present a detailed overview of the recent related works. In Sect. 3, we describe the adversarial problem along with the new objectives we introduce within the proposed evaluation framework, MEAD, which is further explained in Sect. 4. We extensively experimentally validate MEAD in Sect. 5. Finally, in Sect. 6, we provide the summary together with concluding remarks.

2   Related Works

State-of-the-art methods to detect adversarial examples can be separated in two main groups [2]: supervised and unsupervised methods. In the supervised setting, detectors can make use of the knowledge of the attacker's procedure. The *network invariant model approach* extracts natural and adversarial features from the activation values of the network's layers [7,23,28], while the *statistical approach* extract features using statistical tools (e.g. maximum mean discrepancy [16], PCA [21], kernel density estimation [13], local intrinsic dimensionality [25], model uncertainty [13] or natural scene statistics [17]) to separate in-training and out-of-training data distribution/manifolds. To overcome the intrinsic limitation of the necessity to have prior knowledge of attacks, unsupervised detection methods consider only clean data at training time. The features extraction can rely on different techniques (e.g., *feature squeezing* [22,36], *denoiser approach* [27], *network invariant* [24], *auxiliary model* [1,32,40]). Moreover, detection methods of adversarial examples can also act on the underlying classifier by considering *a novel training procedure* (e.g., reverse cross-entropy [30]; the rejection option [1,32]) and a thresholding test strategy towards robust detection of adversarial examples. Finally, detection methods can also be impacted by the learning task of the underlying network (e.g., for human recognition tasks [35]).

2.1 Considered Detection Methods

Supervised Methods. Supervised methods can make use of the knowledge of how adversarial examples are crafted. They often use statistical properties of either the input samples or the output of hidden layers. *NSS* [17] extract the *Natural Scene Statistics* of the natural and adversarial examples, while *LID* [25] extract the *local intrinsic dimensionality* features of the output of hidden layers for natural, noisy and adversarial inputs. *KD-BU* [13] estimates the *kernel density* of the last hidden layer in the feature space, then estimates the *bayesian uncertainty* of the input sample, following the intuition that the adversarial examples lie off the data manifold. Once those features are extracted, all methods train a detector to discriminate between natural and adversarial samples.

Unsupervised Methods. Unsupervised method can only rely on features of the natural samples. *FS* [36] is an unsupervised method that uses *feature squeezing* to compare the model's predictions. Following the idea of estimating the distance between the test examples and the boundary of the manifold of normal examples, *MagNet* [27] comprises detectors based on *reconstruction error* and detectors based on *probability divergence*.

2.2 Considered Attack Mechanisms

The attack mechanisms can be divided into two categories: whitebox attacks, where the adversary has complete knowledge about the targeted classifier (its architecture and weights), and blackbox attacks where the adversary has no access to the internals of the target classifier.

Whitebox Attacks. One of the first introduced attack mechanisms is what we call the Fast Gradient Sign Method (**FGSM**) [14]. It relies on computing the direction gradient of a given objective function with respect to (w.r.t.) the input of the targeted classifier and modifying the original sample following it. This method has been improved multiple times. Basic Iterative Method (**BIM**) [19] and Projected Gradient Descent (**PGD**) [26] are two iteration extensions of **FGSM**. They were introduced at the same time, and the main difference between the two is that **BIM** initializes the algorithm to the original sample while **PGD** initializes it to the original sample plus a random noise. Despite that **PGD** was introduced under the L_∞-norm constraint, it can be extended to any L_p-norm constraint. Deepfool (**DF**) [29] was later introduced. It is an iterative method based on a local linearization of the targeted classifier and the resolution of this simplified problem. Finally, the Carlini&Wagner method (**CW**) aims at finding the smaller noise to solve the adversarial problem. To do so, they present a relaxation based on the minimization of specific objectives that can be chosen depending on the attacker's goal.

Blackbox Attacks. Blackbox attacks can only rely on queries to attack specific models. Square Attack (**SA**) [3] is an iterative method that randomly searches for a perturbation that will increase the attacker's objective at each step, Hop Skip Jump (**HOP**) [8] tries to estimate the gradient direction to perturb, and Spatial Transformation Attack (**STA**) [12] applies small translations and rotations to the original sample to fool the targeted classifier.

3   Adversarial Examples and Novel Objectives

Let $\mathcal{X} \subseteq \mathbb{R}^d$ be the input space and let $\mathcal{Y} = \{1, \dots, C\}$ be the label space related to some task of interest. We denote by P_{XY} the unknown data distribution over $\mathcal{X} \times \mathcal{Y}$. Throughout the paper we refer to the classifier $q_{\widehat{Y}|X}(y|\mathbf{x}; \theta)$ to be the parametric soft-probability model, where $\theta \in \Theta$ are the parameters, $y \in \mathcal{Y}$ the label and $f_\theta : \mathcal{X} \to \mathcal{Y}$ s.t. $f_\theta(\mathbf{x}) = \arg\max_{y \in \mathcal{Y}} q_{\widehat{Y}|X}(y|\mathbf{x}; \theta)$ to be its induced hard decision. Finally, we denote by $\mathbf{x}' \in \mathbb{R}^d$ an adversarial example, by $\ell(\mathbf{x}, \mathbf{x}'; \theta)$ the objective function used by the attacker to generate that sample, and $a_\ell(\cdot; \varepsilon, p)$ the attack mechanism according to a objective function ℓ, with ε the maximal perturbation allow and p the L_p-norm constraint.

3.1   Generating Adversarial Examples

Adversarial examples are slightly modified inputs that can fool a target classifier. Concretely, Szegedy *et al.* [33] define the adversarial generation problem as:

$$\mathbf{x}' = \underset{\mathbf{x}' \in \mathbb{R}^d \,:\, \|\mathbf{x}'-\mathbf{x}\|_p < \varepsilon}{\arg\min} \|\mathbf{x}' - \mathbf{x}\| \quad \text{s.t.} \quad f_\theta(\mathbf{x}') \neq y, \tag{1}$$

where y is the true label (supervision) associated to the sample $\mathbf{x}$. Since this problem is difficult to tackle, it is commonly relaxed as follows [6][1]:

$$a_\ell(\mathbf{x}; \varepsilon, p) \equiv \mathbf{x}'_\ell = \underset{\mathbf{x}'_\ell \in \mathbb{R}^d \,:\, \|\mathbf{x}'_\ell-\mathbf{x}\|_p < \varepsilon}{\arg\max} \ell(\mathbf{x}, \mathbf{x}'_\ell; \theta). \tag{2}$$

It is worth to emphasize that the choice of the objective $\ell(\mathbf{x}, \mathbf{x}'_\ell; \theta)$ plays a crucial role in generating powerful adversarial examples $\mathbf{x}'_\ell$. The objective function ℓ traditionally used is the Adversarial Cross-Entropy (ACE) [26]:

$$\ell_{\mathrm{ACE}}(\mathbf{x}, \mathbf{x}'_\ell; \theta) = \mathbb{E}_{Y|\mathbf{x}}\big[-\log q_{\widehat{Y}|X}(Y|\mathbf{x}'_\ell; \theta)\big], \tag{3}$$

It is possible to use any objective function ℓ to craft adversarial samples. We present the three losses that we use to generate adversarial examples in the following. While these losses have already been considered in detection/robustness cases, to the best of our knowledge, they have never been used to craft attacks to test the performances of detection methods.

---

[1] Throughout the paper, when the values of ε and p are clear from the context, we denote the attack mechanism as $a_\ell(\cdot)$.

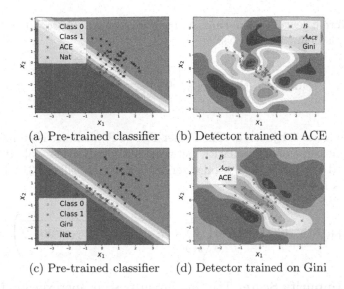

(a) Pre-trained classifier   (b) Detector trained on ACE

(c) Pre-trained classifier   (d) Detector trained on Gini

Fig. 1. Decision boundary for the binary classifier 1a–1c: the decision region for class 1 is green, the decision region of class 0 is pink. The natural testing samples belonging to class 0 are reported in blue, the corresponding adversarial examples crafted using ACE (1a) and Gini Impurity (1c) in red. Decision boundary of the detectors 1b–1d: $\mathcal{B}$, the decision region of the natural examples; $\mathcal{A}_\ell$, reported in red shades, the decision region of the adversarial examples when the detector is trained on data points crafted via $\ell \in \{\text{ACE, Gini}\}$ as objective. The darker shades stand for higher confidence. The red points represent the adversarial examples created with the opposite loss (respectively $\ell \in \{\text{Gini, ACE}\}$).

3.2 Three New Objective Functions

The Kullback-Leibler Divergence. The Kullback-Leibler (KL) divergence between the natural and the adversarial probability distributions has been widely used in different learning problems, as building training losses for robust models [37]. KL is defined as follows:

$$\ell_{KL}(\mathbf{x}, \mathbf{x}'_\ell; \theta) = \mathbb{E}_{\widehat{Y}|\mathbf{x};\theta}\left[\log\left(\frac{q_{\widehat{Y}|X}(\widehat{Y}|\mathbf{x}; \theta)}{q_{\widehat{Y}|X}(\widehat{Y}|\mathbf{x}'_\ell; \theta)}\right)\right]. \tag{4}$$

The Fisher-Rao Objective. The Fisher-Rao (FR) distance is an information-geometric measure of dissimilarity between soft-predictions [5]. It has been recently used to craft a new regularizer for robust classifiers [31]. FR can be computed as follows:

$$\ell_{FR}(\mathbf{x}, \mathbf{x}'_\ell; \theta) = 2\arccos\left(\sum_{y \in \mathcal{Y}} \sqrt{q_{\widehat{Y}|X}(y|\mathbf{x}; \theta)q_{\widehat{Y}|X}(y|\mathbf{x}'_\ell; \theta)}\right). \tag{5}$$

The Gini Impurity Score. The Gini Impurity score approximates the probability of incorrectly classifying the input $\mathbf{x}$ if it was randomly labeled according to the model's output distribution $q_{\widehat{Y}|X}(y|\mathbf{x}'_\ell; \theta)$. It was recently used in [15] to determine whether a sample is correctly or incorrectly classified.

$$\ell_{Gini}(\cdot, \mathbf{x}'_\ell; \theta) = 1 - \sqrt{\sum_{y \in \mathcal{Y}} q^2_{\widehat{Y}|X}(y|\mathbf{x}'_\ell; \theta)}. \tag{6}$$

3.3   A Case Study: ACE vs. Gini Impurity

In Fig. 1 we provide insights on why we need to evaluate the detectors on attacks crafted through different objectives. We create a synthetic dataset that consists of 300 data points drawn from $\mathcal{N}_0 = \mathcal{N}(\mu_0, \sigma^2\mathbf{I})$ and 300 data points drawn from $\mathcal{N}_1 = \mathcal{N}(\mu_1, \sigma^2\mathbf{I})$, where $\mu_0 = [1\ 1]$, $\mu_1 = [-1\ -1]$ and $\sigma = 1$. To each data point $\mathbf{x}$ is assigned true label 0 or 1 depending on whether $\mathbf{x} \sim \mathcal{N}_0$ or $\mathbf{x} \sim \mathcal{N}_1$, respectively. The data points have been split between the training set (70%) and the testing set (30%). We finally train a simple binary classifier with one single hidden layer and a learning rate of 0.01 for 20 epochs. We attack the classifier by generating adversarial examples with PGD under the L_∞-norm constraints with $\varepsilon = 1.2$ for the ACE attacks and $\varepsilon = 5$ for the Gini Impurity attacks to have a classification accuracy (classifier performance) of 50% on the corrupted data points. In Fig. 1a–1c we plot the decision boundary of the binary classifier together with the adversarial and natural examples belonging to class 0. As can be seen, ACE creates points that lie in the opposite decision region with respect to the original points (Fig. 1a). Conversely, Gini Impurity tends to create new data points in the region of maximal uncertainty of the classifier (Fig. 1c). Consider the scenario where we train a simple Radial Basis Function (RBF) kernel SVM on a subset of the testing set of the natural points together with the attacked examples, generated with the ACE or the Gini Impurity score

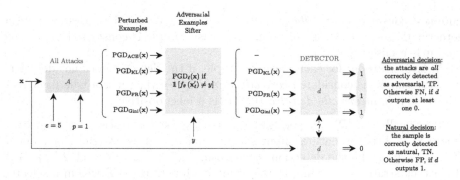

Fig. 2. MEAD: $\mathbf{x}$ is the natural example, $\varepsilon = 5$ is the perturbation magnitude, L_1 is the norm. From the set of all the possible existing attacks $\mathcal{A}$ we consider the ones using PGD. The sifter discards all the perturbed samples that do not fool the classifier f_θ. d is the detector.

depending on the case (Fig. 1b–1d). We then test the detector on the data points originated with the opposite loss, Fig. 1b and Fig. 1d respectively. The decision region of the detector for natural examples is in blue, and the one for the adversarial examples is in red. The intensity of the color corresponds to the level of certainty of the detector. The accuracy of the detector on natural and adversarial data points decreases from 71% to 62% when changing to the opposite loss in Fig. 1b, and from 87% to 63% in Fig. 1d. Hence, testing on samples crafted using a different loss in Eq. (2) means changing the attack and, consequently, evaluating detectors without taking into consideration this possibility leads to a more biased and unrealistic estimation of their performance. When the detector is trained on the adversarial examples created with both the losses, the accuracy is 79.8% when testing on Gini and 66.3% when testing on ACE, which is a better trade-off in adversarial detection performances.

The aforementioned losses will be included in the following section to design MEAD, our *multi-armed evaluation framework*, a new method to evaluate the performance of adversarial detection with low bias.

4 Evaluation with a Multi-armed Attacker

The proposed evaluation framework, MEAD, consists in testing an adversarial examples detection method on a large collection of attacks grouped w.r.t. the L_p-norm and the maximal perturbation ε they consider. Each given natural input example is perturbed according to the collection of attacks. Note that, for every attack, a perturbed example is considered adversarial *if and only if* it fools the classifier. Otherwise, it is discarded and will not influence the evaluation. We then feed all the natural and successful adversarial examples to the detector and gather all the predictions. Finally, based on the detection decisions, we evaluate the detector according to a worst-case scenario:

i) Adversarial decision: for each natural example, we gather all the successful adversarial examples. If the detector detects *all* of them, then the perturbed sample is considered *correctly detected* (i.e., it is a true positive). However, if the detector misses at least one of them, the noisy sample is considered *undetected* (i.e., it is a false negative).

ii) Natural decision: for each natural sample, if the detector does not detect it, then the sample is considered *correctly non-detected* (i.e., it is a true negative); otherwise it is *incorrectly detected* (i.e., it is a false positive).

Specifically, let $\mathcal{D}_m = \{(\mathbf{x}_i, y_i)\}_{i=1}^m \sim P_{XY}$ be the testing set of size m, where $\mathbf{x}_i \in \mathcal{X}$ is the natural input sample and $y_i \in \mathcal{Y}$ is its true label. Let $d: \mathcal{X} \times \mathbb{R} \to \{0, 1\}$ be the detection mechanism and $a_\ell : \mathcal{X} \times \mathbb{R} \times \{1, 2, \infty\} \to \mathcal{X}$ the attack strategy according to the objective function $\ell \in \mathcal{L}$ within a selected collection of objectives $\mathcal{L}$ as described in Sect. 3. For every considered L_p-norm, $p \in \{1, 2, \infty\}$, maximal perturbation $\varepsilon \in \mathbb{R}$, and every threshold $\gamma \in \mathbb{R}^2$:

$$TP_{\varepsilon,p}(\gamma) = \left\{ (\mathbf{x}, y) \in \mathcal{D}_m : \forall \ell \in \mathcal{L} \left\{ f_\theta\big(a_\ell(\mathbf{x})\big) \neq y \right\} \wedge \left\{ d\big(a_\ell(\mathbf{x}), \gamma\big) = 1 \right\} \right\} \quad (7)$$

$$FN_{\varepsilon,p}(\gamma) = \left\{ (\mathbf{x}, y) \in \mathcal{D}_m : \exists \ell \in \mathcal{L} \left\{ f_\theta\big(a_\ell(\mathbf{x})\big) \neq y \right\} \wedge \left\{ d\big(a_\ell(\mathbf{x}), \gamma\big) = 0 \right\} \right\} \quad (8)$$

$$TN_{\varepsilon,p}(\gamma) = \{ (\mathbf{x}, y) \in \mathcal{D}_m : d(\mathbf{x}, \gamma) = 0 \} \quad (9)$$

$$FP_{\varepsilon,p}(\gamma) = \{ (\mathbf{x}, y) \in \mathcal{D}_m : d(\mathbf{x}, \gamma) = 1 \}. \quad (10)$$

In Fig. 2 we provide a graphical interpretation of MEAD when the perturbation magnitude and the norm are fixed.

5   Experiments

In this section, we assess the effectiveness of the proposed evaluation framework, MEAD. The code is available at https://github.com/meadsubmission/MEAD.

5.1   Experimental Setting

Evaluation Metrics. For each L_p-norm and each considered ε, we apply our multi-armed detection scheme. We gather the global result considering all the attacks and all the objectives. Moreover, we also report the results per objective. The performance is measured in terms of the AUROC ↑ [11] and in terms of FPR ↓95%. The first metric is the *Area Under the Receiver Operating Characteristic curve* and represents the ability of the detector to discriminate between adversarial and natural examples (higher is better). The second metric represents the percentage of natural examples detected as adversarial when 95% of the adversarial examples are detected, i.e., FPR at 95% TPR (lower is better).

---

[2] With an abuse of notation, $\forall \ell \in \mathcal{L}$ stands for all the considered attack mechanisms for specific values of ε, p within a collection of objectives $\mathcal{L}$.

Table 1. Overall performances on CIFAR10 of all the detectors per objective and in MEAD. The worst results among all the settings is in **bold**; the ones in the single-armed setting is underlined. No norm denotes the group of attacks that do not depend on the norm constraint.

CIFAR10	MEAD		ACE		KL		Gini		FR	
NSS	AUROC↑%	FPR↓95%	AUROC↑%	FPR↓95%	AUROC↑%	FPR↓95%	AUROC↑%	FPR↓95%	AUROC↑%	FPR↓95%
L_1 Average	62.9	81.6	67.4	75.7	67.1	76.0	67.8	78.2	67.6	75.6
L_2 Average	64.0	82.0	68.7	71.0	68.5	70.9	65.1	82.0	68.6	71.1
L_∞ Average	71.9	62.0	76.9	40.1	77.2	39.5	73.7	59.6	74.1	57.2
No norm	**88.5**	**38.8**	**88.5**	**38.8**	**88.5**	**38.8**	**88.5**	**38.8**	**88.5**	**38.8**
KD-BU	AUROC↑%	FPR↓95%	AUROC↑%	FPR↓95%	AUROC↑%	FPR↓95%	AUROC↑%	FPR↓95%	AUROC↑%	FPR↓95%
L_1 Average	50.9	95.7	70.0	88.6	70.0	88.4	74.3	92.3	69.8	88.4
L_2 Average	59.0	94.1	71.6	71.9	71.7	71.6	70.6	92.8	71.7	71.8
L_∞ Average	36.8	96.9	64.8	92.1	68.1	91.3	53.7	95.6	67.8	91.7
No norm	**65.4**	**94.2**	**65.4**	**94.2**	**65.4**	**94.2**	**65.4**	**94.2**	**65.4**	**94.2**
LID	AUROC↑%	FPR↓95%	AUROC↑%	FPR↓95%	AUROC↑%	FPR↓95%	AUROC↑%	FPR↓95%	AUROC↑%	FPR↓95%
L_1 Average	50.8	95.4	69.6	82.1	69.4	82.9	88.9	49.9	69.1	83.7
L_2 Average	63.5	83.1	73.7	70.1	73.4	70.7	82.5	61.3	73.2	71.3
L_∞ Average	53.8	90.8	75.7	56.8	79.9	57.6	71.3	79.7	82.0	51.4
No norm	**88.0**	**58.1**	**88.0**	**58.1**	**88.0**	**58.1**	**88.0**	**58.1**	**88.0**	**58.1**
FS	AUROC↑%	FPR↓95%	AUROC↑%	FPR↓95%	AUROC↑%	FPR↓95%	AUROC↑%	FPR↓95%	AUROC↑%	FPR↓95%
L_1 Average	75.4	64.8	92.8	25.1	92.9	24.9	73.5	67.6	92.9	24.6
L_2 Average	74.9	65.8	87.4	31.2	87.6	36.9	73.7	67.2	87.4	37.5
L_∞ Average	52.7	81.1	73.0	60.1	77.5	55.7	58.2	78.8	75.7	58.5
No norm	**62.7**	**82.5**	**62.7**	**82.5**	**62.7**	**82.5**	**62.7**	**82.5**	**62.7**	**82.5**
MagNet	AUROC↑%	FPR↓95%	AUROC↑%	FPR↓95%	AUROC↑%	FPR↓95%	AUROC↑%	FPR↓95%	AUROC↑%	FPR↓95%
L_1 Average	49.6	93.7	49.8	93.5	49.7	93.3	50.1	93.2	**49.1**	**93.8**
L_2 Average	50.9	93.1	52.3	89.6	52.3	89.4	50.5	93.3	51.8	91.4
L_∞ Average	78.0	46.1	79.2	44.6	80.2	44.1	79.2	44.6	80.0	44.6
No norm	79.9	45.7	79.9	45.7	79.9	45.7	79.9	45.7	79.9	45.7

Table 2. Overall performances on MNIST of all the detectors per objective and in MEAD. The worst results among all the settings is in **bold**; the ones in the single-armed setting is underlined. No norm denotes the group of attacks that do not depend on the norm constraint.

MNIST	MEAD		ACE		KL		Gini		FR	
NSS	AUROC↑%	FPR↓95%%	AUROC↑%	FPR↓95%%	AUROC↑%	FPR↓95%%	AUROC↑%	FPR↓95%%	AUROC↑%	FPR↓95%%
L_1 Average	**96.8**	**9.4**	97.0	8.2	97.1	8.6	97.4	7.0	97.1	8.1
L_2 Average	**90.3**	**26.5**	90.7	25.8	90.8	25.4	91.4	23.7	90.6	**26.5**
L_∞ Average	88.7	23.5	89.5	23.5	89.5	**23.6**	90.0	**23.6**	89.8	23.5
No norm	87.1	57.8	87.1	57.8	87.1	57.8	87.1	57.8	87.1	57.8
KD-BU	AUROC↑%	FPR↓95%%	AUROC↑%	FPR↓95%%	AUROC↑%	FPR↓95%%	AUROC↑%	FPR↓95%%	AUROC↑%	FPR↓95%%
L_1 Average	45.6	95.7	59.9	93.0	59.3	93.1	61.4	92.7	58.9	93.3
L_2 Average	50.3	94.8	59.9	93.0	59.7	93.1	59.3	93.2	59.8	93.0
L_∞ Average	34.1	96.7	42.8	96.0	44.7	95.8	48.6	95.3	44.9	95.8
No norm	76.0	88.2	76.0	88.2	76.0	88.2	76.0	88.2	76.0	88.2
LID	AUROC↑%	FPR↓95%%	AUROC↑%	FPR↓95%%	AUROC↑%	FPR↓95%%	AUROC↑%	FPR↓95%%	AUROC↑%	FPR↓95%%
L_1 Average	79.9	54.9	83.7	48.2	84.0	50.0	90.4	52.1	84.1	50.2
L_2 Average	85.6	46.2	87.4	44.1	87.0	45.1	87.6	44.4	86.1	45.4
L_∞ Average	77.9	55.1	83.3	46.3	83.6	47.8	88.7	38.8	83.0	49.5
No norm	98.1	8.2	98.1	8.2	98.1	8.2	98.1	8.2	98.1	8.2
FS	AUROC↑%	FPR↓95%%	AUROC↑%	FPR↓95%%	AUROC↑%	FPR↓95%%	AUROC↑%	FPR↓95%%	AUROC↑%	FPR↓95%%
L_1 Average	79.8	66.8	83.4	57.6	83.5	57.1	83.2	53.0	83.4	57.4
L_2 Average	73.5	69.0	75.6	65.0	75.5	65.4	74.5	67.0	74.7	65.7
L_∞ Average	76.4	63.5	80.8	54.6	80.2	54.6	79.0	58.7	80.4	58.2
No norm	61.5	85.9	61.5	85.9	61.5	85.9	61.5	85.9	61.5	85.9
MagNet	AUROC↑%	FPR↓95%%	AUROC↑%	FPR↓95%%	AUROC↑%	FPR↓95%%	AUROC↑%	FPR↓95%%	AUROC↑%	FPR↓95%%
L_1 Average	98.1	5.7	98.2	5.4	98.3	5.6	98.3	5.2	98.1	5.6
L_2 Average	90.0	28.7	90.6	27.6	90.8	27.8	90.6	29.1	89.7	28.1
L_∞ Average	98.5	10.3	98.5	10.3	98.4	10.6	98.5	10.5	98.5	10.4
No norm	86.9	74.3	86.9	74.3	86.9	74.3	86.9	74.3	86.9	74.3

Datasets and Classifiers. We run the experiments on MNIST [20] and CIFAR10 [18]. The underlying classifiers are a simple CNN for MNIST, consisting of two blocks of two convolutional layers, a max-pooling layer, one fully-connected layer, one dropout layer, two fully-connected layers, and ResNet-18 for CIFAR10. The training procedures involve 100 epochs with Stochastic Gradient Descent (SGD) optimizer using a learning rate of 0.01 for the simple CNN and 0.1 for ResNet-18; a momentum of 0.9 and a weight decay of 10^{-5} for ResNet-18. Once trained, these networks are fixed and never modified again.

Grouping Attacks. We test the methods on the attacks presented in Sect. 2.2, and we present them based on the norm constraint used to construct the attacks. Under the L_1-norm fall PGD with ε in $\{5, 10, 15, 20, 25, 30, 40\}$. Under the L_2-norm fall PGD with ε in $\{0.125, 0.25, 0.3125, 0.5, 1, 1.5, 2\}$, CW with $\varepsilon = 0.01$, HOP with $\varepsilon = 0.1$, and DF which has no constraint on ε. Under the L_∞-norm fall FGSM, BIM and PGD with ε in $\{0.0315, 0.0625, 0.125, 0.25, 0.3125, 0.5\}$, CW with $\varepsilon = 0.3125$, and SA with $\varepsilon = 0.3125$ for MNIST and $\varepsilon = 0.125$ for CIFAR10. Finally, ST is not constrained by a norm or a maximum perturbation, as it is limited in maximum rotation (30 for CIFAR10 and 60 for MNIST) and translation (8 for CIFAR10 and 10 for MNIST).

Detection Methods. We tested detection methods introduced in Sect. 2.1. In the supervised case, we train the detectors using adversarial examples created by perturbing the samples in the original training sets with PGD under L_∞-norm and $\varepsilon = 0.03125$. In the unsupervised case, the detectors only need natural samples in the training sets. They are tested on all the previously mentioned attacks, generated on the testing sets.

5.2   Experimental Results

In this section, we refer to *single-armed setting* when we consider the setup where the adversarial examples are generated w.r.t. one of the objectives in Sect. 3. We provide the average of the performances of all the detection methods on CIFAR10 in Table 1 and on MNIST in Table 2. Due to space constraints, the detailed tables for each detection method (i.e., *NSS*, *LID*, *KD-BU*, *MagNet*, and *FS*) and for each dataset (i.e., CIFAR10 and MNIST) are reported in Appendix A.

MEAD and the Single-Armed Setting. Table 1 shows a decrease in the performance of all the detectors when going from the single-armed setting to MEAD. *NSS* is the more robust among the supervised methods when passing from the single-armed setting to the proposed setting. Indeed, (cf Table 1), in terms of AUROC↑, it registers a decrease of up 4.9% points under the L_1-norm constraint, 4.7 under the L_2-norm constraint, and 5.3 under the L_∞-norm constraint. This can be explained by the fact that the network in *NSS* is trained on the natural scene statistics extracted from the trained samples differently from the other detectors. In particular, these statistical properties are altered by the presence of adversarial perturbations and hence are found to be a good candidate to determine if a sample is adversarial or not. By looking closely at the results for *NSS* in Table 5, it comes out that it performs better when evaluated on L_∞ norm constraint. Indeed, in this case, the adversarial examples at testing time are similar to those used at training time. Not surprisingly, the performance decreases when evaluated on other kinds of attacks. Notice that, in the single-armed setting, all the supervised methods turn out to be much more inefficient than when presented in the original papers. Indeed, as already explained in Sect. 5.1, we train the detectors using adversarial examples created by perturbing the samples in the original training sets with PGD under L_∞-norm and $\varepsilon = 0.03125$, and then we test them on a variety of attacks. Hence, we do not train a different detector for each kind of attack seen at testing time. On the other side, the unsupervised detector *MagNet* appears to be more robust than *FS* when changing from the single-armed setting to MEAD. Indeed, in terms of AUROC↑, it loses at most 2.2% points (L_∞ norm case). On average, *FS* is the unsupervised detector that achieves the best performance on CIFAR10, while *MagNet* is the one to achieve the best performance on MNIST.

Remark: Some single-armed setting results turn out to be worse than the corresponding results in MEAD (cf Table 5–9 and Table 11–15 in Appendix A). We provide here an explanation of this phenomenon. Given a natural input sample $\mathbf{x}$, let $\mathbf{x}_\ell$ denotes the perturbed version of $\mathbf{x}$ according to some fixed norm p, fixed perturbation magnitude ε and objective function ℓ between ACE, KL, Gini and FR. Suppose $f_\theta(\mathbf{x}_{\text{ACE}}) = y$, where y is the ground true label of $\mathbf{x}$, this means that $\mathbf{x}_{\text{ACE}}$ is a perturbed version of the natural example but not adversarial. Assume instead $f_\theta(\mathbf{x}_{\text{KL}}) \neq y$, $f_\theta(\mathbf{x}_{\text{Gini}}) \neq y$ and $f_\theta(\mathbf{x}_{\text{FR}}) \neq y$. If at testing time the detector is able to recognize all of them as being positive (i.e., adversarial), then under MEAD, $\mathbf{x}_{\text{KL}}, \mathbf{x}_{\text{Gini}}, \mathbf{x}_{\text{FR}}$ would be considered a *true positive*. This example, counting as a true positive under MEAD, would instead be discarded under the single-armed setting of ACE, as $\mathbf{x}_{\text{ACE}}$ is neither a clean example nor an adversarial one. Then, the larger amount of true positives in MEAD can potentially lead to an increase in the global AUROC↑.

Effectiveness of the Proposed Objective Functions. In Table 4 and Table 10, relegated to the Appendix due to space constraints, we report the averaged number of successful adversarial examples under the multi-armed setting as well as the details per single-armed settings on CIFAR10 and MNIST, respectively. The attacks are most successful when the value of the constraint ε for every L_p-norm increases. Generating adversarial examples using the ACE for each attack scheme creates more harmful (adversarial) examples for the classifier than using any other objective. However, using either the Gini Impurity score, the Fisher-Rao objective, or the Kullback-Leibler divergence seems to create examples that are either equally or more difficult to be detected by the detection methods. For this purpose, we provide two examples. First, by looking at the results in Table 7, we can deduce that *LID* finds it difficult to recognize

Table 3. Performances of each detection method under the MEAD framework on CIFAR10 and MNIST averaged over the norm-based constraint. The best results among all the methods is in **bold**; the ones per type of detection method (i.e. Supervised and Unsupervised) are <u>underlined</u>.

	Supervised methods						Unsupervised methods			
	NSS		KD-BU		LID		FS		MagNet	
	AUROC↑%	FPR↓$_{95\%}$%	AUROC↑%	FPR↓$_{95\%}$%	AUROC↑%	FPR↓$_{95\%}$%	AUROC↑%	FPR↓$_{95\%}$%	AUROC↑%	FPR↓$_{95\%}$%
MNIST	<u>90.7</u>	<u>29.3</u>	51.5	93.9	85.4	41.1	72.8	71.3	**93.4**	29.8
CIFAR10	<u>71.8</u>	<u>66.1</u>	53.0	95.2	64.0	81.8	<u>66.4</u>	73.6	64.6	<u>69.7</u>

the attacks based on KL and FR objective functions but not the ones created through Gini. For example, with PGD1 and $\varepsilon = 40$, we register a decrease in AUROC↑ of 9.5% points when going from the single-armed setting of Gini to the one of FR. Similarly, the decrease is 8.3% points in the case of KL. This behavior is even more remarkable when we look at the results in terms of FPR↓$_{95\%}$: the gap between the best FPR↓$_{95\%}$ values (obtained via Gini) and the worst (via FR) is 30.7% points. On the other side, the situation is reversed if we look at the results in Table 8 as FS turns out to be highly inefficient at recognizing adversarial examples generated via the Gini Impurity score. By considering the results associated to the highest value of ε for each norm, namely $\varepsilon = 40$ for L_1-norm; $\varepsilon = 2$ for L_2-norm; $\varepsilon = 0.5$ for L_∞-norm, the gap between best FPR↓$_{95\%}$ values (obtained via KL divergence) and the worst (via Gini Impurity score), varies from a minimum of 41.7 (L_∞-norm) to a maximum of 64.4 (L_2-norm) percentage points. This example, in agreement with Sect. 3.3, testify on real data that testing the detectors without taking into consideration the possibility of creating attacks through different objective functions leads to a biased and unrealistic estimation of their performances.

Comparison Between Supervised and Unsupervised Detectors. The unsupervised methods find it challenging to recognize attacks crafted using the Gini Impurity score. Indeed, according to Sect. 3.3, that objective function creates attacks on the decision boundary of the pre-trained classifier. Consequently, the unsupervised detectors can easily associate such input samples with the cluster of naturals. Supervised methods detect Adversarial Cross-Entropy loss-based attacks more and, therefore, more volatile when it comes to other types of loss-based attacks. Overall, by looking at the results in Table 3 on both the datasets, most of the supervised and unsupervised methods achieve comparable performances with the multi-armed framework, meaning that the current use of the knowledge about the specific attack is not general enough. The exception to this is NSS, which, as already explained, seems to be the most general detector.

On the effects of the norm and ε. The detection methods recognize attacks with a large perturbation more easily than other attacks (cf Table 5–9 and Table 11–15). L_∞-norm attacks are less easily detectable than any other L_p-norm attack. Indeed, multiple attacks are tested simultaneously for a single ε under the L_∞ norm constraint. For example, in CIFAR10 with $\varepsilon = 0.3125$ and L_∞,

PGD, FGSM, BIM, and CW are tested together, whereas, with any other norm constraint, only one typology of attack is examined. Indeed the more attack we consider for a given ε, the more likely at least one attack will remain undetected. Globally, under the L_∞-norm constraint, Gini Impurity score-based attacks are the least detected attacks. However, each method has different behaviors under L_1 and L_2. *NSS* is more sensitive to Kullback-Leibler divergence-based attacks while *MagNet* is more volatile to the Fisher-Rao distance-based attacks. As already pointed out, *FS* achieves inferior performance when evaluated against attacks crafted through the Gini Impurity objective, while the sensitivity of *LID* and *KD-BU* to a specific objective depends on the L_p-norm constraint.

6   Summary and Concluding Remarks

We introduced MEAD a new framework to evaluate detection methods of adversarial examples. Contrary to what is generally assumed, the proposed setup ensures that the detector does not know the attacks at the testing time and is evaluated based on simultaneous attack strategies. Our experiments showed that the SOTA detectors for adversarial examples (both supervised and unsupervised) mostly fail when evaluated in MEAD with a remarkable deterioration in performance compared to single-armed settings. We enrich the proposed evaluation framework by involving three new objective functions to generate adversarial examples that create adversarial examples which can simultaneously fool the classifier while not being successfully identified by the investigated detectors. The poor performance of the current SOTA adversarial examples detectors should be seen as a challenge when developing novel methods. However, our evaluation framework assumes that the attackers do not know the detection method. As future work we plan to enrich the framework to a complete whitebox scenario.

Acknowledgements. The work of Federica Granese was supported by the European Research Council (ERC) project HYPATIA under the European Union's Horizon 2020 research and innovation program. Grant agreement №835294. This work has been supported by the project PSPC AIDA: 2019-PSPC-09 funded by BPI-France. This work was performed using HPC resources from GENCI-IDRIS (Grant 2022-[AD011012352R1]) and thanks to the Saclay-IA computing platform.

References

1. Aldahdooh, A., Hamidouche, W., Déforges, O.: Revisiting model's uncertainty and confidences for adversarial example detection. Appl. Intell. **53**, 509–531 (2021)
2. Aldahdooh, A., Hamidouche, W., Fezza, S.A., Déforges, O.: Adversarial example detection for DNN models: a review. CoRR abs/2105.00203 (2021)
3. Andriushchenko, M., Croce, F., Flammarion, N., Hein, M.: Square attack: a query-efficient black-box adversarial attack via random search. In: Vedaldi, A., Bischof, H., Brox, T., Frahm, J.-M. (eds.) ECCV 2020. LNCS, vol. 12368, pp. 484–501. Springer, Cham (2020). https://doi.org/10.1007/978-3-030-58592-1_29

4. Athalye, A., Carlini, N., Wagner, D.A.: Obfuscated gradients give a false sense of security: circumventing defenses to adversarial examples. In: Proceedings of the 35th International Conference on Machine Learning, ICML 2018, Stockholmsmässan, Stockholm, Sweden, 10–15 July 2018. Proceedings of Machine Learning Research, vol. 80, pp. 274–283. PMLR (2018)
5. Atkinson, C., Mitchell, A.F.S.: Rao's distance measure. Sankhyā: the Indian J. Stat. Series A (1961–2002), **43**(3), 345–365 (1981)
6. Carlini, N., Wagner, D.: Towards evaluating the robustness of neural networks. In: 2017 IEEE Symposium on Security and Privacy (SP), pp. 39–57. IEEE (2017)
7. Carrara, F., Becarelli, R., Caldelli, R., Falchi, F., Amato, G.: Adversarial examples detection in features distance spaces. In: Leal-Taixé, L., Roth, S. (eds.) ECCV 2018. LNCS, vol. 11130, pp. 313–327. Springer, Cham (2019). https://doi.org/10.1007/978-3-030-11012-3_26
8. Chen, J., Jordan, M.I., Wainwright, M.J.: Hopskipjumpattack: a query-efficient decision-based attack. In: 2020 IEEE Symposium on Security and Privacy (SP), pp. 1277–1294. IEEE (2020)
9. Croce, F., et al.: Robustbench: a standardized adversarial robustness benchmark. CoRR abs/2010.09670 (2020)
10. Croce, F., Hein, M.: Minimally distorted adversarial examples with a fast adaptive boundary attack. In: International Conference on Machine Learning, pp. 2196–2205. PMLR (2020)
11. Davis, J., Goadrich, M.: The relationship between precision-recall and roc curves. In: Proceedings of the 23rd International Conference on Machine Learning, pp. 233–240 (2006)
12. Engstrom, L., Tran, B., Tsipras, D., Schmidt, L., Madry, A.: Exploring the landscape of spatial robustness. In: International Conference on Machine Learning, pp. 1802–1811. PMLR (2019)
13. Feinman, R., Curtin, R.R., Shintre, S., Gardner, A.B.: Detecting adversarial samples from artifacts. CoRR abs/1703.00410 (2017)
14. Goodfellow, I.J., Shlens, J., Szegedy, C.: Explaining and harnessing adversarial examples. In: International Conference on Learning Representations (2015)
15. Granese, F., Romanelli, M., Gorla, D., Palamidessi, C., Piantanida, P.: DOCTOR: a simple method for detecting misclassification errors. In: Advances in Neural Information Processing Systems 34: Annual Conference on Neural Information Processing Systems 2021, NeurIPS 2021, 6–14 December 2021, virtual, pp. 5669–5681 (2021)
16. Grosse, K., Manoharan, P., Papernot, N., Backes, M., McDaniel, P.D.: On the (statistical) detection of adversarial examples. CoRR abs/1702.06280 (2017)
17. Kherchouche, A., Fezza, S.A., Hamidouche, W., Déforges, O.: Natural scene statistics for detecting adversarial examples in deep neural networks. In: 22nd IEEE International Workshop on Multimedia Signal Processing, MMSP 2020, Tampere, Finland, 21–24 September 2020, pp. 1–6. IEEE (2020)
18. Krizhevsky, A.: Learning multiple layers of features from tiny images. Technical report (2009)
19. Kurakin, A., Goodfellow, I., Bengio, S., et al.: Adversarial examples in the physical world (2016)
20. LeCun, Y., Cortes, C., Burges, C.: MNIST handwritten digit database (2010)
21. Li, X., Li, F.: Adversarial examples detection in deep networks with convolutional filter statistics. In: IEEE International Conference on Computer Vision, ICCV 2017, Venice, Italy, 22–29 October 2017, pp. 5775–5783. IEEE Computer Society (2017)

22. Liang, B., Li, H., Su, M., Li, X., Shi, W., Wang, X.: Detecting adversarial image examples in deep neural networks with adaptive noise reduction. IEEE Trans. Dependable Secur. Comput. **18**(1), 72–85 (2021)
23. Lu, J., Issaranon, T., Forsyth, D.A.: SafetyNet: detecting and rejecting adversarial examples robustly. In: IEEE International Conference on Computer Vision, ICCV 2017, Venice, Italy, 22–29 October 2017, pp. 446–454. IEEE Computer Society (2017)
24. Ma, S., Liu, Y., Tao, G., Lee, W., Zhang, X.: NIC: detecting adversarial samples with neural network invariant checking. In: 26th Annual Network and Distributed System Security Symposium, NDSS 2019, San Diego, California, USA, 24–27 February 2019. The Internet Society (2019)
25. Ma, X., et al.: Characterizing adversarial subspaces using local intrinsic dimensionality. In: 6th International Conference on Learning Representations, ICLR 2018, Vancouver, BC, Canada, April 30–May 3 2018, Conference Track Proceedings. OpenReview.net (2018)
26. Madry, A., Makelov, A., Schmidt, L., Tsipras, D., Vladu, A.: Towards deep learning models resistant to adversarial attacks. In: International Conference on Learning Representations (2018)
27. Meng, D., Chen, H.: MagNet: a two-pronged defense against adversarial examples. In: Proceedings of the 2017 ACM SIGSAC Conference on Computer and Communications Security, CCS 2017, Dallas, TX, USA, October 30–November 03 2017, pp. 135–147. ACM (2017)
28. Metzen, J.H., Genewein, T., Fischer, V., Bischoff, B.: On detecting adversarial perturbations. In: 5th International Conference on Learning Representations, ICLR 2017, Toulon, France, 24–26 April 2017, Conference Track Proceedings. OpenReview.net (2017)
29. Moosavi-Dezfooli, S.M., Fawzi, A., Frossard, P.: DeepFool: a simple and accurate method to fool deep neural networks. In: Proceedings of the IEEE Conference on Computer Vision and Pattern Recognition, pp. 2574–2582 (2016)
30. Pang, T., Du, C., Dong, Y., Zhu, J.: Towards robust detection of adversarial examples. In: Advances in Neural Information Processing Systems 31: Annual Conference on Neural Information Processing Systems 2018, NeurIPS 2018, 3–8 December 2018, Montréal, Canada, pp. 4584–4594 (2018)
31. Picot, M., Messina, F., Boudiaf, M., Labeau, F., Ben Ayed, I., Piantanida, P.: Adversarial robustness via fisher-Rao regularization. IEEE Trans. Pattern Anal. Mach. Intell. **45**, 1–1 (2022)
32. Sotgiu, A., Demontis, A., Melis, M., Biggio, B., Fumera, G., Feng, X., Roli, F.: Deep neural rejection against adversarial examples. EURASIP J. Inf. Secur. **2020**(1), 1–10 (2020). https://doi.org/10.1186/s13635-020-00105-y
33. Szegedy, C., et al.: Intriguing properties of neural networks. In: International Conference on Learning Representations (2014)
34. Szegedy, C., et al.: Intriguing properties of neural networks. In: 2nd International Conference on Learning Representations, ICLR 2014, Banff, AB, Canada, 14–16 April 2014, Conference Track Proceedings (2014)
35. Tao, G., Ma, S., Liu, Y., Zhang, X.: Attacks meet interpretability: Attribute-steered detection of adversarial samples. In: Advances in Neural Information Processing Systems 31: Annual Conference on Neural Information Processing Systems 2018, NeurIPS 2018, 3–8 December 2018, Montréal, Canada, pp. 7728–7739 (2018)
36. Xu, W., Evans, D., Qi, Y.: Feature squeezing: detecting adversarial examples in deep neural networks. In: 25th Annual Network and Distributed System Security

Symposium, NDSS 2018, San Diego, California, USA, 18–21 February 2018. The Internet Society (2018)

37. Zhang, H., Yu, Y., Jiao, J., Xing, E.P., Ghaoui, L.E., Jordan, M.I.: Theoretically principled trade-off between robustness and accuracy. In: International Conference on Machine Learning, pp. 1–11 (2019)

38. Zheng, S., Song, Y., Leung, T., Goodfellow, I.J.: Improving the robustness of deep neural networks via stability training. In: 2016 IEEE Conference on Computer Vision and Pattern Recognition, CVPR 2016, Las Vegas, NV, USA, 27–30 June 2016, pp. 4480–4488. IEEE Computer Society (2016)

39. Zheng, T., Chen, C., Ren, K.: Distributionally adversarial attack. In: The Thirty-Third AAAI Conference on Artificial Intelligence, AAAI 2019, The Thirty-First Innovative Applications of Artificial Intelligence Conference, IAAI 2019, The Ninth AAAI Symposium on Educational Advances in Artificial Intelligence, EAAI 2019, Honolulu, Hawaii, USA, January 27–February 1 2019, pp. 2253–2260. AAAI Press (2019)

40. Zheng, Z., Hong, P.: Robust detection of adversarial attacks by modeling the intrinsic properties of deep neural networks. In: Advances in Neural Information Processing Systems 31: Annual Conference on Neural Information Processing Systems 2018, NeurIPS 2018, 3–8 December 2018, Montréal, Canada, pp. 7924–7933 (2018)

Adversarial Mask: Real-World Universal Adversarial Attack on Face Recognition Models

Alon Zolfi[1]([✉])[iD], Shai Avidan[2], Yuval Elovici[1][iD], and Asaf Shabtai[1][iD]

[1] Ben-Gurion University of the Negev, Be'er Sheva, Israel
zolfi@post.bgu.ac.il, {elovici,shabtaia}@bgu.ac.il
[2] Tel Aviv University, Tel Aviv, Israel
avidan@tauex.tau.ac.il

Abstract. Deep learning-based facial recognition (FR) models have demonstrated state-of-the-art performance in the past few years, even when wearing protective medical face masks became commonplace during the COVID-19 pandemic. Given the outstanding performance of these models, the machine learning research community has shown increasing interest in challenging their robustness. Initially, researchers presented adversarial attacks in the digital domain, and later the attacks were transferred to the physical domain. However, in many cases, attacks in the physical domain are conspicuous, and thus may raise suspicion in real-world environments (e.g., airports). In this paper, we propose *Adversarial Mask*, a physical universal adversarial perturbation (UAP) against state-of-the-art FR models that is applied on face masks in the form of a carefully crafted pattern. In our experiments, we examined the transferability of our adversarial mask to a wide range of FR model architectures and datasets. In addition, we validated our adversarial mask's effectiveness in real-world experiments (CCTV use case) by printing the adversarial pattern on a fabric face mask. In these experiments, the FR system was only able to identify 3.34% of the participants wearing the mask (compared to a minimum of 83.34% with other evaluated masks). A demo of our experiments can be found at: https://youtu.be/_TXkDO5z11w.

Keywords: Adversarial attack · Face recognition · Face mask

1 Introduction

For the past two years, the coronavirus has impacted every aspect of our lives, and its impact will continue for the foreseeable future. Since its emergence, various suggestions have been made to reduce its spread. While the effectiveness of some actions is questionable, there is no doubt that face masks are a key factor in preventing the spread of the virus in crowded and enclosed spaces. The widespread adoption of face masks and the ever-increasing use of deep learning-based facial recognition (FR) models in everyday systems can be leveraged to perpetrate targeted adversarial attacks that will enable attackers to evade such models and compromise their robustness, without raising an alarm.

© The Author(s), under exclusive license to Springer Nature Switzerland AG 2023
M.-R. Amini et al. (Eds.): ECML PKDD 2022, LNAI 13715, pp. 304–320, 2023.
https://doi.org/10.1007/978-3-031-26409-2_19

Fig. 1. Illustrating the effect of an adversarial pattern printed on a fabric mask (right), which results in the failure of the FR system to detect the person wearing it, compared to the FR system's ability to detect the same individual without a mask, as well as with a standard disposable mask.

Adversarial attacks in the computer vision domain have gained a lot of interest in recent years, and various ways of fooling image classifiers [9,22] and object detectors [21,23,32] have been proposed. Attacks against FR systems have also been shown to be effective. For example, research has demonstrated that face synthesis in the digital domain can be used to fool FR models [28]. In the physical domain, some of the proposed methods involved wearing adversarial eyeglasses [18], projecting lights on human faces [20], wearing a hat containing an adversarial sticker [14], and using adversarial makeup [10]. However, the proposed attacks are conspicuous and do not allow the attacker to blend in naturally in real-world scenarios, potentially triggering defense systems.

In this work, we propose a *universal* adversarial attack that can be used to physically evade FR systems; in this case, an adversarial pattern is printed on a fabric face mask, as shown in Fig. 1. To create the adversarial pattern, we use a gradient-based optimization process that aims to cause *all* identities wearing the mask to be misclassified by the FR model. We first demonstrate the attack's ability to fool state-of-the-art models (e.g., ArcFace [7]) in the digital domain by applying the face mask to every facial image in the dataset (dynamically) using 3D face reconstruction. Then, we print the adversarial pattern on an actual fabric face mask and test it under real-world conditions. The results in the digital domain show that our adversarial mask performs better than all evaluated masks and is transferable to other models. In the physical domain, we show that 96.66% of the participants wearing our mask evaded the detection by the FR system.

The contributions of our research can be summarized as follows:

- We are the first to present a **physical universal** adversarial attack that fools FR models, i.e., we craft a single perturbation that causes the FR model to falsely classify all potential attackers as unknown identities, even under diverse conditions (angles, scales, etc.) in a real-world environment (fully-automated CCTV scenario).
- In the digital domain, we study the transferability of our attack across different model architectures and datasets.
- We present a fully differentiable novel digital masking method that can accurately place any kind of mask on any face, regardless of the position of the head. This method can be used for other computer-vision tasks (e.g., training masked-face detection models).

- We craft an inconspicuous pattern that "continues" the contour of the face, allowing a potential attacker to easily blend in with a crowd without raising an alarm, given the variety and widespread use of face masks during the COVID-19 pandemic.
- We propose various countermeasures that can be used during the FR model training and inference phases.

2   Background and Related Work

2.1   Adversarial Attacks

Digital Attacks. Initially, attacks in the digital domain aimed at fooling classification models were introduced [9,22]. While those earlier attacks are based on methods that generate a perturbation for a single image, Moosavi-Dezfooli *et al.* [17] proposed universal adversarial perturbations (UAPs), which enable any image that is blended with the UAP to fool a DNN. Digital attacks on models that perform more complex computer vision tasks (e.g., face recognition and object detection) have also emerged. Yang *et al.* [28] designed a digital patch which is placed on a person's forehead to deceive face detectors. Recent studies targeting FR models suggested various techniques. Deb *et al.* [6] proposed automated adversarial face synthesis, using a generative adversarial network (GAN) to create minimal perturbations. Agarwal *et al.* [1] and Amada *et al.* [2] proposed UAPs that can deceive FR models for multiple identities simultaneously. However, these attacks only call attention to the potential threat inherent to such models but cannot be transferred to the physical world.

Physical Attacks. Physical attacks differ from digital attacks in the way real-world constraints are considered throughout the process of generating the perturbation. Consequently, these constraints allow the perturbations to transfer more easily to the physical world. In recent years, physical attacks on object detectors have gained attention. Chen *et al.* [5] printed stop signs containing adversarial patterns that evaded detection by the object detector, and Sitawarin *et al.* [21] deceived autonomous car systems by crafting toxic traffic signs that look similar to the original traffic signs. Methods against person detectors have also been proposed. Thys *et al.* [23] suggested attaching a small adversarial cardboard plate to a person's body to evade detection. Continuing this line of research, other studies involved printing adversarial patterns on t-shirts, which resulted in a more realistic article of clothing that blends into the environment more naturally [26,27]. A slightly different approach, in which the perturbation affects the sensor's perception of the object by applying a translucent patch on the camera's lens, was also introduced [32].

Numerous studies have demonstrated different ways of fooling FR systems. For example, Shen *et al.* [20] introduced the visible light-based attack, where lights are projected on human faces. Other studies showed that carefully applied makeup patterns can negatively affect the performance of FR systems [10,30].

Accessories were also shown to be effective; for example, Sharif *et al.* [18] suggested wearing adversarial eyeglass frames that were crafted using gradient-based methods. Later, GAN methods were used to generate an enhanced version of the adversarial eyeglass frames [19]. Recently, Komkov *et al.* [14] printed an adversarial paper sticker and placed it on a hat to fool the state-of-the-art *ArcFace* [7] FR model. However, when implemented on a person, these methods may call attention to the person by causing them to stand out in a crowd given their unnatural appearance. In contrast, we propose a method in which the perturbation is placed on a face mask, a safety measure widely used in the COVID-19 era; in addition, unlike prior work in which the proposed attacks craft tailor-made perturbations (target a single image or person), our universal attack can be applied more widely without the need for an expert to train a tailor-made one. Furthermore, we demonstrate the effectiveness of our method in a real-world use case involving a CCTV system, an aspect not addressed by previous studies.

2.2 Face Recognition

Models. FR models can be categorized by two main attributes, the model's backbone and the novel loss function, both of which are involved in the training phase. The main architecture used as the backbone in these models is the ResNet [12] architecture, which varies in terms of the number of layers it contains, also referred to as the backbone *depth*. On top of the backbone, an additional layer (or more) is added, usually containing a novel loss function that is used to train the backbone weights [7,16,24]. Later, when the FR model is used for inference, only the backbone layers are used to generate the embedding vector.

Systems. The end-to-end procedure of a fully automated FR system consists of several main steps: (a) Record - a camera records the environment and then produces a series of frames (a video stream); (b) Detect - each frame is analyzed by a face detector to extract cropped faces; (c) Align - the cropped faces are aligned according to the FR model's alignment method; (d) Embed - the aligned facial images serve as input to an FR model f that maps a facial image I_{face} to a vector $f(I_{face})$, also referred to as an *embedding* vector; (e) Verify - the embedding vector is compared to a list of precalculated embedding vectors (also referred to as ground-truth embedding vectors) using a similarity measure (e.g., cosine similarity). The identity with the highest similarity score is marked as a potential candidate and eventually confirmed if its similarity score surpasses a predefined verification threshold (which depends on the system's use case).

3 Method

The objective of our research is to generate an adversarial pattern that can be printed on a face mask and cause FR systems to classify a registered identity as an unknown identity. Further, we aim to create an adversarial pattern that is: (a) universal - it must be effective on any identity from multiple views and angles, and at multiple scales, (b) practical - the pattern should remain adversarial when printed on a fabric mask in the real world, and (c) transferable - it must be effective on different models (backbone depths and loss functions).

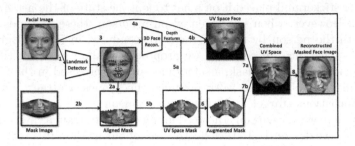

Fig. 2. Overview of our mask projection method pipeline.

3.1 Mask Projection

In order to digitally train our adversarial mask, we first need to simulate the mask overlay on a person's face in the real world. Therefore, we use 3D face reconstruction to digitally apply a mask on a facial image. Feng *et al.* [8] introduced an end-to-end approach called *UV position map* that records the 3D coordinates of a complete facial point cloud using a 2D image. This map records the position information of a 3D face and provides dense correspondence to the semantic meaning of each point in the UV space, allowing us to achieve near-real approximation of the mask on the face, which is essential to the creation of a successful adversarial mask in the real world.

More formally, we consider our mask $M_{\mathrm{adv}} \in \mathbb{R}^{w \times h \times 3}$ and a rendering function $\mathcal{R}_{\theta}$. The rendering function (partially inspired from [25]) takes a mask M_{adv} and a facial image x_{face}, and applies the mask on the face, resulting in a masked face image $\mathcal{R}_{\theta}(M_{\mathrm{adv}}, x_{\mathrm{face}})$. As shown in Fig. 2, the pipeline of the mask's projection on the facial image is as follows:

1. Detect the landmark points of the face - given a landmark detector, we extract the landmark points of the face.
2. Map the mask pixels to the facial image - the landmark points of the face extracted in the previous step of the pipeline are used to map the mask pixels to the corresponding location on the facial images.
3. Extract depth features of the face - the facial image is passed to the 3D face reconstruction model to obtain depth features.
4. Transfer 2D facial image to the UV space - the depth features are used to remap the facial image to the UV space.
5. Transfer 2D mask image to the UV space - the depth features are used to remap the mask image to the UV space.

6. Augment mask - to improve the robustness of our adversarial mask, random geometric transformations and color-based augmentations (parameterized by θ) are applied: (i) geometric transformations - random translation and rotation are added to simulate possible distortions in the mask's placement on the face in the real world, and (ii) color-based augmentations - random contrast, brightness, and noise are added to simulate changes in the appearance of the mask that might result from various factors (e.g., lighting, noise or blurring caused when the camera captures the image).
7. Combine and reconstruct - the UV representations of the facial image and the mask are combined, and the combined image is reconstructed back to the regular 2D space, resulting in a masked face image.

Usually, adversarial attacks that employ textile-like objects (e.g., wearable t-shirt [26,27]) use thin plate splines (TPSs) [4] to simulate fabric distortions. In contrast to these studies, although we aim to craft a textile-based mask, in our case, the mask form on the face remains steady and is not subject to significant distortions. In addition, our 3D approach allows us to simulate smaller distortions (e.g., caused by the nose shape) without actively using TPSs.

Above all, it is important to note that the entire process presented is completely differentiable and allows us to backpropagate and update the mask pixels.

3.2 Patch Optimization

To optimize our mask's pixels, we propose an iterative optimization process. In each iteration, we select a random batch of facial images of multiple identities and digitally project the mask on each facial image. We then feed the masked face images to the FR model and obtain the embedding representations. Since our goal is to cause an attacker to be unknown to FR models, we aim to create a patch M_{adv} that will decrease the similarity between the output embedding and the ground-truth embedding e_{gt} (precalculated) for each identity.

More formally, an FR model $f : \mathcal{X}^{w \times h \times 3} \rightarrow \mathbb{R}^N$ receives a facial image $x \in \mathcal{X}$ (in our case, a masked face image $\mathcal{R}_\theta(M_{adv}, x)$) as input and outputs the embedding representation $f(\mathcal{R}_\theta(M_{adv}, x))$. Therefore, we minimize the cosine similarity between the embedding vectors and use the following loss function:

$$\ell_{sim}(M_{adv}) = \mathbb{E}_{\theta,x}[\cos(f(\mathcal{R}_\theta(M_{adv}, x)), e_{gt})] \qquad (1)$$

Since our method is not system-dependent (i.e., does not use a fixed verification threshold determined by a specific use case), we aim to decrease the similarity to the fullest extent possible, in order to perform the most successful attack.

To improve the mask's transferability to other models, we train our patch using an ensemble of FR models, denoted as J. We replace 1 with the following:

$$\ell_{sim}(M_{adv}) = \mathbb{E}_{\theta,x} \frac{1}{|J|} \sum_j \cos(f^{(j)}(\mathcal{R}_\theta(M_{adv}, x)), e_{gt}^{(j)}), \qquad (2)$$

where $f^{(j)}$ denotes the jth model and $e_{gt}^{(j)}$ denotes the embedding representation calculated using the jth model.

We also include the *total variation (TV)* [18] factor to ensure that the optimizer favors smooth color transitions between neighboring pixels and is calculated on the mask pixels as follows:

$$\ell_{TV} = \sum_{i,k} \sqrt{(p_{i,k} - p_{i+1,k})^2 + (p_{i,k} - p_{i,k+1})^2} \tag{3}$$

When neighboring pixels are not similar, the penalty of this component is greater.

To be more precise, since the output of ℓ_{sim} is in the range of $[-1, 1]$ and the output of ℓ_{TV} is in the range of $[0, 1]$, we transform ℓ_{sim} so it is in the same range ($[0, 1]$); thus, we replace 2 with the following:

$$\ell_{sim}(M_{adv}) = \mathbb{E}_{\theta,x} \frac{1}{|J|} \sum_j \frac{\cos(f^{(j)}(\mathcal{R}_\theta(M_{adv}, x)), e_{gt}^{(j)}) + 1}{2} \tag{4}$$

Finally, the optimization problem we solve is as follows:

$$\min_{M_{adv}} [\ell_{sim}(M_{adv}) + \lambda * \ell_{TV}(M_{adv})], \tag{5}$$

where λ is set at a low value.

4 Evaluation

In our evaluation, we first run experiments in the digital domain by applying the mask to facial images, using the rendering function R_θ (as explained in Sect. 3). Then, we evaluate the performance of our adversarial pattern in the physical domain (i.e., real world) by printing it on a fabric mask.

Models. We use three different types of loss functions that were originally used to train the models, which are considered state-of-the-art: ArcFace [7], CosFace [24], and MagFace [16]. Specifically, we use pretrained models which were trained using the ArcFace and CosFace loss functions [3], with four different ResNet depths (18, 34, 50, and 100) each, and a pretrained ResNet100 backbone originally trained with the MagFace [16] loss function, for a total of nine different models. We examine multiple training variations, using one or more (i.e., ensemble) models to train the adversarial mask and then test it in a white-box setting to evaluate the performance. We also evaluate the transferability of our mask to other unknown models (i.e., black-box setting).

Datasets. Throughout this paper, we use three commonly used datasets in the face recognition domain: CASIA-WebFace [29], CelebA [15], and MS-Celeb [11].

For the training phase, we randomly choose 100 different identities (50 men and 50 women) from the CASIA-WebFace dataset. We extract five random facial images for each identity, for a total of 500 facial images.

For the evaluation phase, we use 200 identities from each dataset (an equal number of men and women from each dataset), evaluating both the performance on the same distribution (different identities from the CASIA-WebFace dataset, ~20K images) and the transferability to other datasets (CelebA and MS-Celeb, ~6K and ~24K images, respectively).

Metrics. In our experiments, we quantify the performance of our attack as the ability to decrease the similarity score - specifically the cosine similarity (an approach originally presented in [14]). The cosine similarity calculation is a step required prior to making a binary decision based on a predefined threshold. This evaluation approach does not require a system-dependent predefined threshold and demonstrates our attack's effectiveness. In the physical domain, we also quantify the effectiveness of our attack using two additional metrics, each of which relates to a different stage of an end-to-end FR system:

- *Recognition rate* (RR) $= |F_{rec}| / |F_{det}|$, where $|F_{rec}|$ denotes the total number of frames in which the identity was correctly recognized (the cosine similarity between the ground-truth embedding and the output embedding surpasses the verification threshold), and $|F_{det}|$ denotes the total number of frames in which a face was detected and analyzed by the FR system.
- *Persistence detection* - since the goal of our adversarial mask is to ensure that an attacker is not identified by the system, we propose a metric that indicates whether the goal was met. An attacker is considered as identified if, within a window of $N_{\text{sliding window}}$ frames, the attacker was recognized in $N_{\text{recognized}}$ frames (where $N_{\text{recognized}} \leq N_{\text{sliding window}}$).

Implementation Details. The models we work with in this research only take size $3 \times 112 \times 112$ facial images as input. Therefore, We set the size of our patch to be $3 \times 60 \times 112$ to avoid significant downsampling when dynamically rendering the mask to the facial image, and we set the initial color of the mask to white. The pixels are updated using the Adam optimizer [13], where the initial learning rate is set at 10^{-2}. The weight factor of the TV component in the loss function λ is manually set at 0.1. The source code is available online.[1]

Types of Face Masks Evaluated. Since we are the first to present a physical universal perturbation, we compare the effectiveness of our mask with several control masks: (a) Clean - the original facial image without a mask, (b) Adv - our optimized adversarial mask, (c) Random - a mask with randomly colored pixels, and (d) Blue - a standard disposable blue mask (simple black and white masks were also tested and yielded the same results). In addition, due to our trained mask's resemblance to a human face, the lower face area of a female and male are used as control masks and will be referred to as *Female Face* and *Male Face*, respectively. The masks compared in our evaluation are shown in Fig. 3.

Evaluation Setup. Since the state-of-the-art models discussed above were not specifically designed to address the issue of masked faces, we first examine the model's (ResNet100@ArcFace) performance on a number of simple face masks. For this evaluation, we use 100 identities from the CASIA-WebFace dataset, where five images of each identity are used to calculate the ground-truth embedding, and the remaining images are applied with different types of masks.

---

[1] https://github.com/AlonZolfi/AdversarialMask.

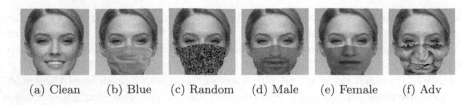

(a) Clean   (b) Blue   (c) Random   (d) Male   (e) Female   (f) Adv

Fig. 3. Examples of facial images w/o mask (a), and when various masks are digitally applied to them (b)–(f). (Color figure online)

Table 1. Cosine similarity comparison between two ground-truth embedding generation methods on the Resnet100@ArcFace. Bold indicates better performance.

	Mask type	No mask	Blue	Black	White
Cosine similarity	w/o Mask*	**.732**	.399	.407	.428
	w/Mask**	.682	**.547**	**.549**	**.561**

*Embedding vectors created using original facial images.
**A masked version of the original images is added to the embedding calculation.

To the best of our knowledge, the scientific community has not reached a consensus on the way in which masked face images should be dealt with by FR models. Therefore, we use two approaches for generating the ground-truth embedding: (a) the current approach for unmasked face models - averaging the embedding vectors of the original images only, and (b) an extension of the first approach - in addition to the original images, we create a masked face version for each image (the specific mask is randomly chosen from blue, black, and white masks) and average the embedding vectors of the two versions of the images. We then calculate the cosine similarity between the masked face images' embedding vectors and the two versions of ground-truth embedding vectors generation.

In Table 1 we can see that although the first approach (w/o Mask) performs better on unmasked images, its performance on masked images is unsatisfactory. On the other hand, the cosine similarity for the second approach (w/Mask) only slightly decreases the cosine similarity on unmasked images (~0.05 decrease) and performs significantly better on masked images (~0.1–0.15 increase). Thus, throughout this section the results we present are obtained using the second approach (the ground-truth embedding vectors used for the training procedure are generated using first approach). It is important to note that by choosing the second approach, we increase the difficulty of deceiving these models, since the ground-truth embedding vectors encapsulate the use of a face mask.

4.1   Digital Attacks

We conduct digital experiments to quantify our adversarial mask's effectiveness using the rendering function R_θ (see Sect. 3), which allows us to dynamically apply masks to the facial images in the test set.

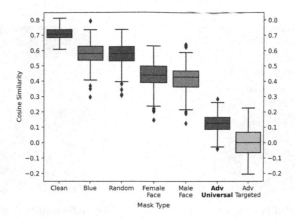

Fig. 4. Distribution of the cosine similarity score across different masks. 'Adv Universal' represents our optimized universal mask, and 'Adv Targeted' represents a tailormade mask for each identity.

Effectiveness of the Adversarial Mask in a White-Box Setting. We examine the effectiveness of our attack in a *white-box* setting in which our mask is optimized and tested on the ResNet100@ArcFace. As shown in Fig. 4, our adversarial mask has a significant impact compared to the no mask case, in which the average cosine similarity decreased from ∼0.7 to ∼0.1. As the case of no mask images represents the upper bound of the cosine similarity, we also perform a targeted attack in which a mask is tailored to each person, to determine the lower bound. The targeted mask results are averaged across all identities in the test set. We can see that the universal mask performs almost as good as a tailor-made mask (∼0.1 difference). The tailor-made masks represent an attack that is more difficult to detect, since the adversarial pattern varies among different identities. In addition, while the female and male face control masks are also able to decrease the cosine similarity to a lower level (∼0.45), our mask outperforms them both for almost all tested identities.

Transferability Across Backbone Depth. We also examine whether our mask can deceive FR models it was not trained on. Since the majority of the models use the ResNet architecture, we evaluate the performance across different depths of the ResNet@ArcFace. The results are presented in Fig. 5a. In the figure, we can see that the use of our adversarial mask can cause the cosine similarity to decrease regardless of the model used for training. It can also be seen that our attack generalizes better to unknown models whose architecture depth is closer to that of the trained model. For example, an adversarial mask trained on a model with 100 layers performs better on the models with 34 and 50 layers (decreasing the cosine similarity to 0.182 and 0.168, respectively) than on the 18-layer model (0.282). In addition, we see that the mask trained on an ensemble of all models does not outperform a mask trained on a single model in a white-box setting, however the ensemble's effectiveness is seen over all models combined.

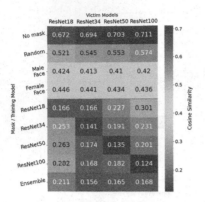
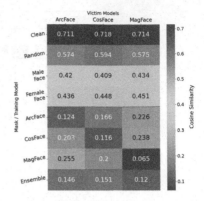

(a) **Transferability** across various ResNet backbone **depths** originally trained using the ArcFace loss function.

(b) **Transferability** across various ResNet100 backbones originally trained with different **loss functions**.

Fig. 5. Transferability experiments measured in terms of cosine similarity. Rows are divided into three groups: control masks, masks trained using a single model, mask trained using all of the models.

Transferability Across Different Loss Functions. We further demonstrate the adversarial mask's transferability across different model loss functions. We use the ResNet100 backbone in which the weights were trained using one of the following loss functions: ArcFace, CosFace, and MagFace. In Fig. 5b, we observe that our method is loss-agnostic, as the decrease in the cosine similarity is seen on for all tested models. However, a mask that was trained using the MagFace model does not generalize as well as the masks trained with other models, where the cosine similarity decreased to 0.065 in the white-box setting but only decreased to 0.255 and 0.2 on the ArcFace and CosFace models, respectively. It is interesting to examine the mask trained by each model (presented in Fig. 6). Whereas there is a resemblance in the contour of the optimized masks, the mask trained using the ResNet100@MagFace backbone (Fig. 6c) learns completely different colors than the other two, in some way providing a possible explanation for its decreased ability to generalize to the ArcFace and CosFace models.

Transferability Across Datasets. We also find our mask to be effective across different datasets. In another experiment, we train our mask using images from one of the examined datasets (presented earlier in this section) and study its effectiveness on the other datasets (i.e., the ground-truth embedding vectors are generated using another dataset's images). We train all of the masks using the ResNet100@ArcFace. The results show that the impact of using a specific dataset is insignificant, since our mask generalizes over all datasets. For example, when

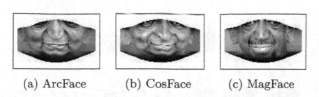

(a) ArcFace   (b) CosFace   (c) MagFace

Fig. 6. Illustrations of our adversarial masks trained on different ResNet100 backbones, which vary in terms of the original loss function they were trained on.

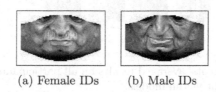

(a) Female IDs   (b) Male IDs

Fig. 7. The adversarial masks trained on the ResNet100@ArcFace using single gender identities.

training the mask on the CASIA-WebFace dataset and testing it on the CelebA and MS-Celeb datasets, we respectively obtained an average cosine similarity of 0.128 and 0.114, similar to the white-box setting results (mask trained and tested on images from the CASIA-WebFace dataset, Fig. 4).

Effect of Gender. Another aspect we studied is the effect of a specific gender on the trained mask. The experiments include optimization of the adversarial mask using only female or male identities, and the final masks are presented in Fig. 7a and Fig. 7b, respectively. The results show that even when training the mask on facial images of a single gender, the cosine similarity decreases to the same level as the mask trained on both genders ($\sim$0.1). In addition, masks trained by a single gender were able to transfer very well to the other gender (male $\rightarrow$ female = 0.097, and female $\rightarrow$ male = 0.145).

Generally, the contour of the trained masks (including the mask trained on both genders, Fig. 6a) is quite interesting. Despite the fact that only facial images of female identities were used to train the mask (Fig. 7a), the optimized mask has an high resemblance to a male face. More generally, the resemblance of all the trained masks to a male face might indicate there is an underlying bias hidden in these models.

4.2 Physical Attacks

Finally, to evaluate the effectiveness of our attack in the real world, we print our digital pattern on two surfaces: on regular paper cut in the shape of a face mask and on a white fabric mask, as shown in Fig. 9. In addition, we create a testbed that operates an end-to-end fully automated FR system (explained in Sect. 2), simulating a CCTV use case.

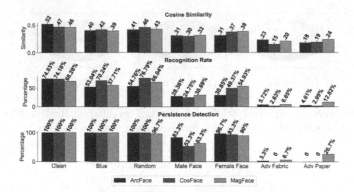

Fig. 8. Physical experiments' averaged results on all participants across different evaluated masks and different victim models.

Setup. The system contains: (a) a *Dahua IPC-HDBW1431E* network camera which records a long corridor, (b) an MTCNN [31] detection model for face detection, preprocessing, and alignment, and (c) an attacked model - we perform a white-box attack in which the model used for training the adversarial mask is also the model under attack, a ResNet100@ArcFace. In addition, we perform an "offline" analysis in a black-box setting, in which the facial images are cropped from the original frames and compared to ground-truth embedding vectors generated using other models.

To calculate the specific verification threshold (set at 0.38), we use a subset of 1,000 identities from the CASIA-WebFace dataset and perform the following procedure. Various face masks are applied (digitally) to each identity's original facial images. Then, we calculate the cosine similarity between the identity's embedding vector and each masked face image. Since we employ a semi-critical security use case (CCTV), we chose the threshold that led to a false acceptance rate (FAR) of 1%. Furthermore, to minimize false positive alarms, we used a persistence threshold of $N_{\text{recognized}} = 7$ frames and a sliding window of $N_{\text{sliding window}} = 10$ frames to designate a candidate identity as a valid one.

We recruited a group of 15 male and 15 female participants (after approval was granted by the university's ethics committee). Each participant was asked to walk along the corridor seven times, once with each mask evaluated (clean, blue, random, male face, and female face), similar to the digital experiments, and two more times with our adversarial masks printed on paper and fabric. The ground-truth embedding of each participant was calculated using two facial images, where a standard face mask was applied (digitally) to each image, for a total of four facial images.

Results. The results of our experiments are shown in Fig. 8 where we can see that our adversarial masks (paper and fabric) performed significantly better than the other masks evaluated on every metric, with a high correlation to the cosine similarity results obtained in the digital domain.

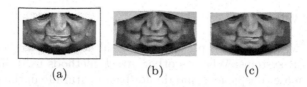

(a)   (b)   (c)

Fig. 9. An illustration of: (a) the digital adversarial mask trained on the ResNet100@ArcFace; (b) the digital pattern printed on fabric mask; and (c) the digital pattern printed on paper.

In terms of the RR, the performance of the FR model for the different masks can be divided into four groups (listed in decreasing order): (a) the unmasked version (74.83%), (b) blue and random masks (53.04% and 54.76%, respectively), (c) male and female masks (30.85% and 28.36%, respectively), and (d) our fabric and paper adversarial masks (5.72% and 4.61%, respectively).

In a realistic case of CCTV use in which an attacker tries to evade the detection of the system, our adversarial fabric mask was able to conceal the identity of 29 out of 30 participants (which represents a persistence detection value of 3.34%), as opposed to the control masks which were able to conceal 5 out of 30 participants at most (persistence detection value of 83.34%).

We also examine the effectiveness of our masks on models they were not trained on. The results presented in Fig. 8 show that our masks have similar adversarial effect on FR models in a black-box setting as in a white-box setting.

Another aspect we examined in our physical evaluation is the ability to print the adversarial pattern on a real surface. Figures 9b and 9c present the digital adversarial pattern (Fig. 9a) printed on the different surfaces. Due to the limited ability of a printer to accurately output the original colors onto the fabric, we can see that there is a slight difference in the performance of the masks. Nonetheless, both of our adversarial masks outperformed the other masks evaluated.

5 Countermeasures

We propose two ways in which our digital masking method can be used to defend against adversarial masks: (a) adversarial training – adversarial (universal and tailor-made) masked face images could be provided to the model during training to improve its robustness; and (b) mask substitution – during the inference phase, every masked face image could be preprocessed so that the worn mask is replaced digitally with a standard one (e.g., blue mask Fig. 3b), where the models had satisfactory performance, as shown in Sect. 4, eliminating the potential threat of an adversarial face mask. An implementation of the mask substitution method on facial images of 100 identities ($\sim$10K images) from the CASIA-WebFace dataset increased the RR from 0.4% (the adversarial mask is applied to the facial images) to 65.5% (the blue mask is applied to the adversarial images). In a physical experiment, in which the blue mask was digitally placed on facial images extracted from the videos frames (videos of participants wearing the adversarial mask), the RR increased from 5.72% to 57.3%.

6 Conclusion

In this paper, we presented a physical universal attack in the form of a face mask against FR systems. Whereas other attack methods used different accessories that are more conspicuous and do not blend naturally in the environment, our mask will not raise any suspicion due to the widespread use of face masks during the COVID-19 pandemic. We demonstrated the effectiveness of our mask in the digital domain, both under white-box and black-box settings. In the physical domain, we showed how our mask is able to prevent the detection of multiple participants in a CCTV use case system. Moreover, we proposed possible countermeasures to deal with such attacks. To sum up, in this research, we highlight the potential risk FR models face from an adversary simply wearing a carefully crafted adversarial face mask in the COVID-19 era.

References

1. Agarwal, A., Singh, R., Vatsa, M., Ratha, N.: Are image-agnostic universal adversarial perturbations for face recognition difficult to detect? In: 2018 IEEE 9th International Conference on Biometrics Theory, Applications and Systems (BTAS), pp. 1–7. IEEE (2018)
2. Amada, T., Liew, S.P., Kakizaki, K., Araki, T.: Universal adversarial spoofing attacks against face recognition. In: 2021 IEEE International Joint Conference on Biometrics (IJCB), pp. 1–7. IEEE (2021)
3. An, X., et al.: Partial FC: training 10 million identities on a single machine. arxiv:2010.05222 (2020)
4. Bookstein, F.L.: Principal warps: thin-plate splines and the decomposition of deformations. IEEE Trans. Pattern Anal. Mach. Intell. **11**(6), 567–585 (1989)
5. Chen, S.T., Cornelius, C., Martin, J., Chau, D.H.P.: Shapeshifter: robust physical adversarial attack on faster R-CNN object detector. In: Berlingerio, M., Bonchi, F., Gärtner, T., Hurley, N., Ifrim, G. (eds.) ECML PKDD 2018. LNCS, pp. 52–68. Springer, Heidelberg (2018). https://doi.org/10.1007/978-3-030-10925-7_4
6. Deb, D., Zhang, J., Jain, A.K.: AdvFaces: adversarial face synthesis. In: 2020 IEEE International Joint Conference on Biometrics (IJCB), pp. 1–10. IEEE (2020)
7. Deng, J., Guo, J., Xue, N., Zafeiriou, S.: ArcFace: additive angular margin loss for deep face recognition. In: Proceedings of the IEEE/CVF Conference on Computer Vision and Pattern Recognition, pp. 4690–4699 (2019)
8. Feng, Y., Wu, F., Shao, X., Wang, Y., Zhou, X.: Joint 3D face reconstruction and dense alignment with position map regression network. In: Proceedings of the European Conference on Computer Vision (ECCV), pp. 534–551 (2018)
9. Goodfellow, I.J., Shlens, J., Szegedy, C.: Explaining and harnessing adversarial examples. arXiv preprint arXiv:1412.6572 (2014)
10. Guetta, N., Shabtai, A., Singh, I., Momiyama, S., Elovici, Y.: Dodging attack using carefully crafted natural makeup. arXiv preprint arXiv:2109.06467 (2021)
11. Guo, Y., Zhang, L., Hu, Y., He, X., Gao, J.: MS-Celeb-1M: a dataset and benchmark for large-scale face recognition. In: Leibe, B., Matas, J., Sebe, N., Welling, M. (eds.) ECCV 2016. LNCS, vol. 9907, pp. 87–102. Springer, Cham (2016). https://doi.org/10.1007/978-3-319-46487-9_6

12. He, K., Zhang, X., Ren, S., Sun, J.: Deep residual learning for image recognition. In: Proceedings of the IEEE Conference on Computer Vision and Pattern Recognition, pp. 770–778 (2016)
13. Kingma, D.P., Ba, J.: Adam: a method for stochastic optimization. arXiv preprint arXiv:1412.6980 (2014)
14. Komkov, S., Petiushko, A.: AdvHat: real-world adversarial attack on ArcFace face ID system. In: 2020 25th International Conference on Pattern Recognition (ICPR), pp. 819–826. IEEE (2021)
15. Liu, Z., Luo, P., Wang, X., Tang, X.: Large-scale CelebFaces attributes (CelebA) dataset. Retrieved August 15(2018), 11 (2018)
16. Meng, Q., Zhao, S., Huang, Z., Zhou, F.: MagFace: a universal representation for face recognition and quality assessment. In: Proceedings of the IEEE/CVF Conference on Computer Vision and Pattern Recognition, pp. 14225–14234 (2021)
17. Moosavi-Dezfooli, S.M., Fawzi, A., Fawzi, O., Frossard, P.: Universal adversarial perturbations. In: Proceedings of the IEEE Conference on Computer Vision and Pattern Recognition, pp. 1765–1773 (2017)
18. Sharif, M., Bhagavatula, S., Bauer, L., Reiter, M.K.: Accessorize to a crime: real and stealthy attacks on state-of-the-art face recognition. In: Proceedings of the 2016 ACM SIGSAC Conference on Computer and Communications Security, pp. 1528–1540 (2016)
19. Sharif, M., Bhagavatula, S., Bauer, L., Reiter, M.K.: A general framework for adversarial examples with objectives. ACM Trans. Priv. Secur. (TOPS) 22(3), 1–30 (2019)
20. Shen, M., Liao, Z., Zhu, L., Xu, K., Du, X.: VLA: a practical visible light-based attack on face recognition systems in physical world. Proc. ACM Interact. Mob. Wearable Ubiquit. Technol. 3(3), 1–19 (2019)
21. Sitawarin, C., Bhagoji, A.N., Mosenia, A., Chiang, M., Mittal, P.: DARTS: deceiving autonomous cars with toxic signs. arXiv preprint arXiv:1802.06430 (2018)
22. Szegedy, C., et al.: Intriguing properties of neural networks. arXiv preprint arXiv:1312.6199 (2013)
23. Thys, S., Van Ranst, W., Goedemé, T.: Fooling automated surveillance cameras: adversarial patches to attack person detection. In: Proceedings of the IEEE Conference on Computer Vision and Pattern Recognition Workshops (2019)
24. Wang, H., et al.: CosFace: large margin cosine loss for deep face recognition. In: Proceedings of the IEEE Conference on Computer Vision and Pattern Recognition, pp. 5265–5274 (2018)
25. Wang, J., Liu, Y., Hu, Y., Shi, H., Mei, T.: FaceX-Zoo: a Pytorch toolbox for face recognition. In: Proceedings of the 29th ACM International Conference on Multimedia, pp. 3779–3782 (2021)
26. Wu, Z., Lim, S.-N., Davis, L.S., Goldstein, T.: Making an invisibility cloak: real world adversarial attacks on object detectors. In: Vedaldi, A., Bischof, H., Brox, T., Frahm, J.-M. (eds.) ECCV 2020. LNCS, vol. 12349, pp. 1–17. Springer, Cham (2020). https://doi.org/10.1007/978-3-030-58548-8_1
27. Xu, K., et al.: Adversarial T-shirt! evading person detectors in a physical world. In: Vedaldi, A., Bischof, H., Brox, T., Frahm, J.-M. (eds.) ECCV 2020. LNCS, vol. 12350, pp. 665–681. Springer, Cham (2020). https://doi.org/10.1007/978-3-030-58558-7_39
28. Yang, X., Wei, F., Zhang, H., Zhu, J.: Design and interpretation of universal adversarial patches in face detection. In: Vedaldi, A., Bischof, H., Brox, T., Frahm, J.-M. (eds.) ECCV 2020. LNCS, vol. 12362, pp. 174–191. Springer, Cham (2020). https://doi.org/10.1007/978-3-030-58520-4_11

29. Yi, D., Lei, Z., Liao, S., Li, S.Z.: Learning face representation from scratch. arXiv preprint arXiv:1411.7923 (2014)
30. Yin, B., et al.: Adv-makeup: a new imperceptible and transferable attack on face recognition. arXiv preprint arXiv:2105.03162 (2021)
31. Zhang, K., Zhang, Z., Li, Z., Qiao, Y.: Joint face detection and alignment using multitask cascaded convolutional networks. IEEE Signal Process. Lett. **23**(10), 1499–1503 (2016)
32. Zolfi, A., Kravchik, M., Elovici, Y., Shabtai, A.: The translucent patch: a physical and universal attack on object detectors. In: Proceedings of the IEEE/CVF Conference on Computer Vision and Pattern Recognition, pp. 15232–15241 (2021)

Generative Models

TrafficFlowGAN: Physics-Informed Flow Based Generative Adversarial Network for Uncertainty Quantification

Zhaobin Mo[1], Yongjie Fu[1], Daran Xu[2], and Xuan Di[1,3(✉)]

[1] Department of Civil Engineering and Engineering Mechanics, Columbia University, New York, USA
{zm2302,yf2578,xd2187}@columbia.edu
[2] Department of Statistics, Columbia University, New York, USA
dx2207@columbia.edu
[3] Data Science Institute, Columbia University, New York, USA

Abstract. This paper proposes the TrafficFlowGAN, a physics-informed flow based generative adversarial network (GAN), for uncertainty quantification (UQ) of dynamical systems. TrafficFlowGAN adopts a normalizing flow model as the generator to explicitly estimate the data likelihood. This flow model is trained to maximize the data likelihood and to generate synthetic data that can fool a convolutional discriminator. We further regularize this training process using prior physics information, so-called physics-informed deep learning (PIDL). To the best of our knowledge, we are the first to propose an integration of normalizing flow, GAN and PIDL for the UQ problems. We take the traffic state estimation (TSE), which aims to estimate the traffic variables (e.g. traffic density and velocity) using partially observed data, as an example to demonstrate the performance of our proposed model. We conduct numerical experiments where the proposed model is applied to learn the solutions of stochastic differential equations. The results demonstrate the robustness and accuracy of the proposed model, together with the ability to learn a machine learning surrogate model. We also test it on a real-world dataset, the Next Generation SIMulation (NGSIM), to show that the proposed TrafficFlowGAN can outperform the baselines, including the pure flow model, the physics-informed flow model, and the flow based GAN model. Source code and data are available at https://github.com/ZhaobinMo/TrafficFlowGAN.

Keywords: Uncertainty Quantification (UQ) · Normalizing flow · Generative Adversarial Networks (GAN) · Physics-informed Deep Learning (PIDL)

Supplementary Information The online version contains supplementary material available at https://doi.org/10.1007/978-3-031-26409-2_20.

1   Introduction

Uncertainty quantification (UQ) is the process of characterizing the uncertainty of system dynamics, accounting for two main sources of uncertainty [4]. The *aleatoric uncertainty* (or data uncertainty) refers to uncertainty arising from external factors, such as measurement noise, and random initial or boundary conditions. The *epistemic uncertainty* arises from the inadequate knowledge of the underlying model, such as inherent stochasticity in human behavior. With these random factors, UQ of dynamical systems is crucial to avoid potential system oscillation or cascading errors.

Two classes of methods are developed to characterize the aforementioned sources of uncertainty. The **physics-based** method assumes that the observations are generated from underlying physics imposed by Gaussian noise; thus filtering methods or Bayesian inference can be applied to propagate uncertainty. However, the physics-based method suffers from limitations such as non-Gaussian likelihoods and high-dimensional posterior distributions [22]. In contrast, the **data-driven** method, such as generative adversarial networks (GAN) [7], tries to characterize any distribution of data directly without making any assumption of noise. Recently, there is a growing trend in integrating physics-based models into the data-driven framework, namely, **physics-informed deep learning** (PIDL) [15]. PIDL-based UQ methods can characterize generic data distribution while ensuring physics consistency.

Among all PIDL models for UQ, the physics-informed GAN is the most widely used, which has been applied to solve stochastic differential equations [5,21,22] and quantify uncertainty in various domains [12–14,17–19]. Although GAN generates high-quality samples [9] through adversarial training, it has stability and convergence issues. Moreover, as GAN cannot calculate the model likelihood, it may miss important modes of the data distribution, namely, *mode collapse* [20]. In contrast, normalizing flow [6] calculates the exact data likelihood and is trained using maximum likelihood estimation (MLE), which is an effective way to avoid mode collapse. However, applying PIDL to the normalizing flow is still at its nascent stage and we only find one relevant work [10].

Leveraging the pros of both the MLE and adversarial training, Flow-GAN, a combination of normalizing flow and GAN, is first introduced in [9], which can achieve both high data likelihood and good sample qualities. Flow-GAN has been applied to manifold learning [3] and image-to-image translation [8]. Little research has been documented that applies Flow-GAN to UQ problems.

In this paper, we propose TrafficFlowGAN that leverages likelihood training, adversarial training, and PIDL for the UQ problems. To the best of our knowledge, we are the first to integrate these three methods for the UQ problems. Main contributions of this paper include:

- We propose a hybrid generative model, TrafficFlowGAN, combining normalizing flow and GAN to achieve both high likelihoods and good sample qualities and to avoid mode collapse.
- We incorporate physics information into the TrafficFlowGAN model for estimation accuracy and data efficiency, and use neural network surrogate models to learn the inter-relations between the physics variables at the same time.

- We apply the TrafficFlowGAN model to learn solutions of second-order stochastic partial differential equations (PDEs), and demonstrate the performance of TrafficFlowGAN by applying it to a traffic state estimation (TSE) problem with real-world data.

The rest of this paper is organized as follows: Sect. 2 introduces the background and related work. Section 3 introduces the structure of TrafficFlowGAN for the UQ problems. Section 4 demonstrates how TrafficFlowGAN learns solutions of a PDE and the relations between physics variables using a neural network surrogate model. Section 5 demonstrates how TrafficFlowGAN characterizes uncertainty from the real-world data in the TSE problem, where two traffic models, i.e. the Aw-Rascle-Zhang (ARZ) [2] and the Lighthill-Whitham-Richards (LWR) [11] models, are used as the physics components. Section 6 concludes our work and projects future directions in this promising arena.

2   Background and Related Work

2.1   Normalizing Flow

The flow model aims to learn an invertible function $z = f_\theta(u) : \mathbb{R}^D \mapsto \mathbb{R}^D$, where data u is sampled from a distribution $p_{\text{data}}(u)$ and $z \sim p_z(z)$ is a random noise of the same dimension as the data. The data likelihood $p_\theta(u)$ can be explicitly expressed by using the *change of variable formula*:

$$p_\theta(u) = p_z(z) \left| \det\left(\frac{\partial f_\theta^{-1}(z)}{\partial z} \right) \right|^{-1}. \tag{1}$$

To compute $p_\theta(u)$, it is nontrivial to choose a latent variable z that has an easy form and to design the invertible function f_θ so that the Jacobian determinant can be easily computed. A common selection of the latent variable z is the standard Gaussian, i.e. $p_z(z) \sim \mathcal{N}(0, I_D)$. To compute the Jacobian determinant, RealNVP [6] designs the invertible function f_θ as an *affine coupling transformation* following the equations below:

$$f_\theta := \begin{cases} z_{1:d} = u_{1:d} \\ z_{d+1:D} = u_{d+1:D} \odot e^{k_\theta(u_{1:d})} + b_\theta(u_{1:d}) \end{cases}, \tag{2}$$

where u and z are split into two partitions at the dth elements. The *scale function* k_θ and the *translation function* b_θ are neural networks to be learned, which constitute the affine transformation of the partition $u_{1:d}$. $\odot$ is the Hadamard product or element-wise product. By this design of invertible function, the Jacobian determinant in Eq. 1 can be computed by

$$\left| \det\left(\frac{\partial f_\theta^{-1}(z)}{\partial z} \right) \right|^{-1} = e^{\sum_j [k_\theta(z_{1:d})]_j}, \tag{3}$$

where j is the index of the element of $k_\theta(z_{1:d})$. The inverse function f_θ^{-1} can also be obtained by

$$f_\theta^{-1} := \begin{cases} u_{1:d} = z_{1:d} \\ u_{d+1:D} = (z_{d+1:D} - b_\theta(z_{1:d}))/e^{k_\theta(z_{1:d})} \end{cases}. \tag{4}$$

To better accommodate the complex data distribution, f_θ is further modeled as a sequence of affine coupling transformations: $f_\theta = f_L \circ \ldots \circ f_1$, where L is the total number of transformations. Let f_l be the lth invertible mapping and $h^{(l)}$ be the lth latent variable that satisfies $h^{(l)} = f_l(h^{(l-1)})$, where $h^{(0)} = u$ and $h^{(L)} = z$. Then the log-likelihood of u can be computed by:

$$\log p_\theta(u) = \log p_z(z) + \sum_{l=0}^{L-1} \log \left| \det \left(\frac{\partial f_l^{-1}(h^{(l)})}{\partial h^{(l)}} \right) \right|^{-1}. \tag{5}$$

The computation of the log-likelihood of z is straightforward as z is assumed to follow a standard Gaussian distribution, and each Jacobian determinant can be calculated following Eq. 3. Thus, the exact data likelihood is tractable and the flow model can be trained by the MLE.

2.2 Generative Adversarial Network (GAN)

GAN aims to train a generator G_θ to learn the mapping from a random noise z to the corresponding state variables u, i.e. $G_\theta : z \to u$. The objective of the generator G_θ is to fool an adversarially trained discriminator D_ϕ. Different GAN variants use different metrics to evaluate the divergence between the prediction distribution and the data distribution, such as the Kullback-Leibler (KL) divergence, the Jensen-Shannon divergence, and the Wasserstein distance [1]. Among these metrics, the Wasserstein distance has received growing popularity for its stability, which optimizes the following objective:

$$\min_\theta \max_{\phi \in \mathcal{F}} \mathbb{E}_{p_{\text{data}}(u)} \left[D_\phi(u) \right] - \mathbb{E}_{p_z(z)} \left[D_\phi(G_\theta(z)) \right], \tag{6}$$

where θ and ϕ are the parameters of the generator and the discriminator, respectively. $\mathcal{F}$ is defined such that D_ϕ is 1-Lipschitz.

3   Framework of TrafficFlowGAN

3.1   Problem Statement

Define the spatial and temporal domains as $\mathcal{X}$ and $\mathcal{T}$, respectively. $(x,t) \in \mathcal{X} \times \mathcal{T}$ is the spatio-temporal coordinate ("coordinate" for short). It is assumed that the state variable u can only be observed by limited number of sensors placed at fixed locations and at a specific frequency. Thus, we further define the *observed (labeled) region* $O \subseteq \mathcal{X} \times \mathcal{T}$ as the spatio-temporal region where the state variable u is observed, and thereby the *unobserved (unlabeled) region*

$C = \mathcal{X} \times \mathcal{T} \setminus O$. We represent the continuous domain in a discrete manner using grid points. Thus, the observed region O and the unobserved region C can be represented as collections of discrete coordinates: $O = \{(x_o^{(i)}, t_o^{(i)})\}_{i=1}^{N_o}$ and $C = \{(x_c^{(j)}, t_c^{(j)})\}_{j=1}^{N_c}$, where i and j are the indices of observed and unobserved coordinates, respectively; N_o, N_c are the numbers of observed and unobserved coordinates, respectively.

The state variable u is a random variable for each coordinate, i.e. $u \sim p_{data}(u|x,t)$. Our goal is to train a generator such that its prediction distribution distribution $p_\theta(\hat{u}|x,t)$ matches the data distribution $p_{data}(u|x,t)$. Below we will introduce how to achieve this goal with our proposed TrafficFlowGAN.

3.2   Overview of TrafficFlowGAN Structure

An overview of TrafficFlowGAN is illustrated in Fig. 1, which consists of three main components, namely, a conditional flow f_θ, a physics-based computational graph, and a convolutional discriminator D_ϕ. The data is illustrated as a heatmap in the spatio-temporal domain. We assume the data is measured by sensors at fixed locations. Due to limited range each sensor can cover, the observation region consists of separate horizontal "strips." The observed and unobserved coordinates are fed into the conditional flow model to generate predictions $\hat{u}_o$ and $\hat{u}_c$, respectively. Those predictions bifurcate into two branches. In the upper branch, $\hat{u}_c$ are fed into a physics-based computational graph, which encodes physics laws, to calculate the physics loss function. This process of calculating physics loss from the unobserved coordinates is illustrated by grey arrows. In the lower branch, the prediction states $\hat{u}_o$ and the observed states u_o are then reshaped to constitute the prediction matrix $\hat{M}$ and the observation matrix M, respectively. These two matrices are then fed into the convolutional discriminator. The process of calculating the adversarial loss from observed coordinates and states is illustrated by blue arrows.

We will detail each component sequentially and explain how we integrate those components in the following subsections.

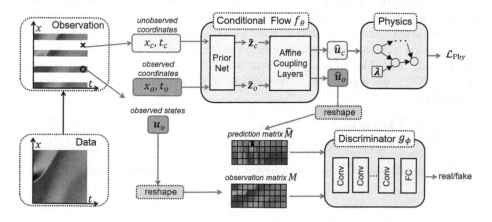

Fig. 1. Structure of the TrafficFlowGAN.

3.3 Convolutional Neural Network as the Discriminator

Existing physics-informed GAN models construct the discriminator as a *fully connected network*. By explicitly adding the spatio-temporal coordinates into the input space, the discriminator is expected to make decisions based on the *spatio-temporal pattern*, which will be represented better by the convolutional neural network (CNN). In this work, we propose to use CNN as the discriminator.

The discriminator D_ϕ consists of a sequence of convolutional layers (Conv) followed by a fully connected layer (FC). This FC layer outputs a 1×1 scalar indicating if the input matrix is from observations or predictions. The pooling layer is not used in this structure, as there is no requirement for compression in our task.

The reshape of the observation $\boldsymbol{u}$ is straightforward. As we represent the spatio-temporal domain in a discrete manner, $\boldsymbol{u}(x, t)$ for each coordinate can be viewed as a "pixel", and the dimension of $\boldsymbol{u}$ is its number of channels. This is the same for reshaping the prediction $\hat{\boldsymbol{u}}$ to get the prediction matrix $\hat{M}$. Note that due to randomness in data and predictions, we can sample multiple observation matrices $\{M^{(i)}\}_{i=1}^{N_\omega}$ and prediction matrices $\{\hat{M}^{(i)}\}_{i=1}^{N_\omega}$, where N_ω is the total number of sampling.

The discriminator D_ϕ can be updated by minimizing the Wasserstein loss:

$$\mathcal{L}_D(\phi) = -\frac{1}{N_\omega} \sum_{i=1}^{N_\omega} D_\phi(M^{(i)}) - D_\phi(\hat{M}^{(i)}). \tag{7}$$

3.4 Conditional Flow as the Generator

We construct a conditional flow as our generator, as illustrated in Fig. 2. Assume $\boldsymbol{u}$ has two elements, i.e. u_1 and u_2. Different from the tradition normalizing flow, we add a *prior network* (p-net) to transform the standard Gaussian prior $\boldsymbol{z} = (z_1, z_2)$ to $\tilde{\boldsymbol{z}} = (\tilde{z}_1, \tilde{z}_2)$ with shifting and scaling, considering that the magnitude of uncertainty at different (x, t) coordinates can be different. The prior network takes as input the coordinate (x, t) and outputs the prior mean $\boldsymbol{\mu} = (\mu_1, \mu_2)$ and prior standard deviation $\boldsymbol{\sigma} = (\sigma_1, \sigma_2)$; thus $\tilde{\boldsymbol{z}} \sim \mathcal{N}(\boldsymbol{\mu}, \boldsymbol{\sigma})$. The prior network is followed by affine coupling layers. Each affine coupling layer consists of a scale function (k-net) and a translation function (b-net), as introduced in Sect. 2.1.

Based on [23] and our experiment, the exponential operation in Eq. 2 is numerically unstable, which may result in gradient explosion. Instead of using the RealNVP, we replace the exponential operation in Eq. 2 with a Sigmoid operation:

$$f_\theta := \begin{cases} \tilde{\boldsymbol{z}}_{1:d} = \boldsymbol{u}_{1:d} \\ \tilde{\boldsymbol{z}}_{d+1:D} = \boldsymbol{u}_{d+1:D} \odot \mathrm{Sigmoid}(k_\theta(\boldsymbol{u}_{1:d}; x, t)) + b_\theta(\boldsymbol{u}_{1:d}; x, t) \end{cases}, \tag{8}$$

and the calculation of Jacobian determinant in Eq. 5 is thus changed to

$$\left| \det\left(\frac{\partial f_\theta^{-1}(\tilde{z})}{\partial \tilde{z}} \right) \right|^{-1} = \sum_j [k_\theta(\tilde{z}_{1:d}; x, t)]_j, \tag{9}$$

where j is the element index of $k_\theta(\tilde{z}_{1:d}; x, t)$. We define a likelihood loss function for the generator as:

$$
\begin{aligned}
\mathcal{L}_{\text{NLL}}(\theta) &= -\sum_{i=1}^{N_\omega} \sum_{(x_o, t_o) \in O} \log p_\theta(\boldsymbol{u}|x_o, t_o, \omega^{(i)}) \\
&= -\sum_{i=1}^{N_\omega} \sum_{(x_o, t_o) \in O} \log p_{\tilde{z}}(\tilde{z}|x_o, t_o) + \sum_{l=0}^{L-1} \log \left| \det \left(\frac{\partial f_l^{-1}(\boldsymbol{h}^{(l)})}{\partial \boldsymbol{h}^{(l)}} \right) \right|^{-1},
\end{aligned}
\tag{10}
$$

which is the summation of negative log-likelihood (NLL) over all observed coordinates and random events.

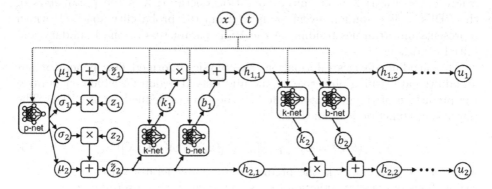

Fig. 2. Structure of the conditional flow with a prior network.

Apart from the likelihood loss $\mathcal{L}_{\text{NLL}}$, the flow generator f_θ can also be trained with the discriminator D_ϕ through adversarial training. The adversarial loss for the generator is depicted as:

$$
\mathcal{L}_{\text{Adv}}(\theta) = -\frac{1}{N_\omega} \sum_{i=1}^{N_\omega} D_\phi(\hat{M}^{(i)}),
\tag{11}
$$

which uses the Wasserstein objective defined in Eq. 6; $\hat{M}$ is the prediction matrix and N_ω is the total number of sampling.

Using the adversarial loss $\mathcal{L}_{\text{Adv}}$ alone is prone to mode collapse. We demonstrate below how the likelihood loss $\mathcal{L}_{\text{NLL}}$ can mitigate the mode collapse by one example. Suppose the data is generated from a mixture model of two Gaussian distributions $\mathcal{N}(-1, 1)$ and $\mathcal{N}(1, 1)$. By adversarial training, the generator may end up only generating one mode, say $\mathcal{N}(-1, 1)$. In this case, the discriminator cannot distinguish between the samples and the ground truth. If the MLE is used, the likelihood of the missing mode is very low, resulting in a high overall NLL. Thus, the likelihood loss $\mathcal{L}_{\text{NLL}}$ can guide the generator to leave the current local optimum.

3.5   Physics Regularization

The conditional flow model is further regularized by the physics-informed computational graph, which encodes physics prior knowledge like partial differential equations (PDE).

Suppose data follows laws that can be depicted as stochastic PDEs below:

$$
\begin{aligned}
\boldsymbol{u}_t(x,t;\omega) + \mathcal{N}_x[\boldsymbol{u}(x,t;\omega); \lambda(\omega)] &= 0, & (x,t) \in \mathcal{X} \times \mathcal{T}, \omega \in \Omega, \\
\mathcal{B}[\boldsymbol{u}(x,t;\omega)] &= 0, & (x,t) \in \partial\mathcal{X} \times \mathcal{T}, \\
\mathcal{I}[\boldsymbol{u}(x,0;\omega)] &= 0, & x \in \mathcal{X},
\end{aligned}
\tag{12}
$$

where, $\boldsymbol{u}_t$ is its partial derivative of $\boldsymbol{u}$ with regard to t; $\partial\mathcal{X}$ is the boundary of the space domain $\mathcal{X}$; $\mathcal{N}_x$ is a non-linear differential operator; $\mathcal{B}$ is a boundary condition operator; $\mathcal{I}$ is an initial condition operator; λ is the parameters of the PDEs. ω is a random event sampled from the probability space Ω, which represents uncertainties residing in the PDE parameters or the boundary and initial conditions.

By encoding physics information, the physics-informed flow generator has an additional learning objective on the unobserved region C, which encourages the prediction of the generator to follow the physics defined by the PDE. The physics loss function is defined as:

$$
\mathcal{L}_{\text{Phy}}(\theta, \lambda) = \mathbb{E}_{q(x_c, t_c)} \left| \mathbb{E}_{p_z(z)} \left[(\hat{\boldsymbol{u}}_c)_t + \mathcal{N}_x[\hat{\boldsymbol{u}}_c; \lambda] \right] \right|^2,
\tag{13}
$$

where $\hat{\boldsymbol{u}}_c = f_\theta(x_c, t_c, z)$ is the prediction of the generator on the unobserved region. This physics loss function serves as a regularization term for the generator. If the flow generator f_θ is well trained, the physics loss needs to be as close to zero as possible.

3.6   Training of TrafficFlowGAN

The loss function of the flow model is a weighted sum of the likelihood loss, adversarial loss, and physics loss:

$$
\mathcal{L}_f(\theta) = \alpha\mathcal{L}_{\text{NLL}}(\theta) + \beta\mathcal{L}_{\text{Adv}}(\theta) + \gamma\mathcal{L}_{\text{Phy}}(\theta, \lambda),
\tag{14}
$$

where $\alpha, \beta, \gamma \in (0,1]$ are hyperparameters that determine the contribution of each loss component. With the generator loss $\mathcal{L}_f(\theta)$, the discriminator loss $\mathcal{L}_D(\phi)$ defined in Eq. 7, and physics loss $\mathcal{L}_{\text{Phy}}(\theta, \lambda)$ defined in Eq. 13, we are ready to introduce the training algorithm as shown in Algorithm 1.

4   Numerical Experiment: Learning Solutions of a Known Second-Order PDE

In this experiment, we apply TrafficFlowGAN to learn solutions of a known PDE and also to learn the relations of the PDE's parameters.

Algorithm 1. TrafficFlowGAN Training Algorithm.

Initialization:
Initialized physics parameters λ^0; Initialized networks parameters θ^0, ϕ^0; Training iterations $Iter$; Batch size m; Learning rate lr; Weights of loss functions α, β, and γ.

Input: The observation data $\{(x_o^{(i)}, t_o^{(i)}, u_o^{(i)})\}_{i=1}^{N_o}$ and unobserved coordinates $\{x_c^{(j)}, t_c^{(j)}\}_{j=1}^{N_c}$.

1: **for** $k \in \{0, ..., Iter\}$ **do**
2:    Sample batches $\{(x_o^{(i)}, t_o^{(i)}, u_o^{(i)})\}_{i=1}^{m}$ and $\{x_c^{(j)}, t_c^{(j)}\}_{j=1}^{m}$ from the observation data and unobserved coordinates, respectively
      `// update the discriminator`
3:    Calculate $\mathcal{L}_D$ by Eq. 7
4:    $\phi^{k+1} \leftarrow \phi^k - lr \cdot \text{Adam}(\phi^k, \nabla_\phi \mathcal{L}_D)$
      `// update the generator`
5:    Calculate $\mathcal{L}_{\text{NLL}}$ by Eq. 10, $\mathcal{L}_{\text{Adv}}$ by Eq. 11, and $\mathcal{L}_{\text{Phy}}$ by Eq. 13
6:    Calculate $\mathcal{L}_f$ by Eq. 14
7:    $\theta^{k+1} \leftarrow \theta^k - lr \cdot \text{Adam}(\theta^k, \nabla_\theta \mathcal{L}_f)$
      `// update the physics`
8:    $\lambda^{k+1} \leftarrow \lambda^k - lr \cdot \text{Adam}(\lambda^k, \nabla_\lambda \mathcal{L}_{\text{Phy}})$
9: **end for**

4.1 Numerical Data

The numerical data is generated from the ARZ model [2], which is a second-order PDE that is used to describe the traffic dynamics. It is depicted as

$$\begin{cases} \rho_t + (\rho u)_x = 0, \\ (u + h(\rho))_t + u(u + h(\rho))_x - (U_{eq}(\rho) - u)/\tau, \end{cases} \tag{15}$$

where,

$$h(\rho) = U_{eq}(0) - U_{eq}(\rho) \tag{16}$$

is the hesitation function and

$$U_{eq}(\rho) = u_{max}(1 - \rho/\rho_{max}) \tag{17}$$

is the equilibrium traffic velocity; traffic density ρ and traffic velocity u are the state variables, i.e. $\mathbf{u} = (\rho, u)$; τ is the relaxation parameter; ρ_{max} and u_{max} are the maximum traffic density and the maximum traffic velocity, respectively. In this experiment, we study a "ring road" in $t \in [0,3]$ and $x \in [0,1]$ with a boundary condition $\mathbf{u}(0,t) = \mathbf{u}(1,t)$. We set the parameters as $\rho_{max} = 1.13$, $u_{max} = 1.02$, and $\tau = 0.02$. We set the initial conditions of ρ and u as bell-shaped functions shown in Fig. 3(a). The x-axis is the space domain, and the y-axis is the initial value of ρ (blue line) and u (red line). We solve Eq. 15 using the Lax-Friedrichs scheme on a spatio-temporal grid of sizes 240×960, and the solutions

$\rho(x, t)$ and $u(x, t)$ are shown in Fig. 3(b) and Fig. 3(c), respectively. The dashed black lines indicate 4 locations where data is observed (the top and bottom dashed black lines indicate the same position as it is a ring road). We then add a white noise $\epsilon \sim \mathcal{N}(0, 0.02)$ to the solution to represent the uncertainty.

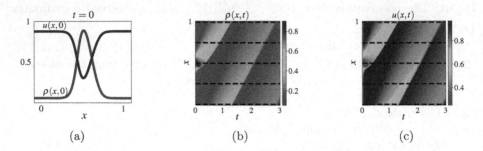

(a)   (b)   (c)

Fig. 3. Numerical data generator from the ARZ model. (a) is the bell-shaped initial ρ and u over $x \in [0, 1]$; (b) and (c) are numerical solutions for ρ and u, respectively. (Color figure online)

4.2 Physics-Based Computational Graph

Figure 4(a) illustrates the physics-based computational graph assosciated with the ARZ. We assume that the exact form of U_{eq} is unknown, and we use a *surrogate network* (s-net) to learn an approximate $\hat{U}_{eq}$. The corresponding physics loss for each line of Eq. 15 is as below:

$$\begin{cases} \mathcal{L}_{ARZ}^{(1)} = |\mathbb{E}_{\mathbf{z}} [\hat{\rho}_t + (\hat{\rho}\hat{u})_x]|^2 \\ L_{ARZ}^{(2)} = \left| \mathbb{E}_{\mathbf{z}} \left[(\hat{u} + h(\hat{\rho}))_t + \hat{u}(\hat{u} + h(\hat{\rho}))_x - (\hat{U}_{eq}(\hat{\rho}) - \hat{u})/\tau \right] \right|^2. \end{cases} \tag{18}$$

In implementation, derivatives with regard to x and t can be easily calculated by the Pytorch module `torch.autograd`.

Additionally, we add a *shape constraint* to regularize the s-net to be monotonically decreasing for the ARZ physics loss, given the domain knowledge that the equilibrium speed U_{eq} decreases as the density ρ increases. This shape constraint is depicted as follows:

$$\mathcal{L}_{reg} = \int_a^b \max \left(0, \frac{\partial \hat{U}_{eq}(\rho)}{\partial \rho} \right) d\rho, \tag{19}$$

where hyperparameters a and b determine the interval where the shape constraint takes effect, e.g. $a = 0$ and $b = 1.13$ in this ring road experiment. Summarizing the loss terms defined in Eq. 18 and Eq. 19, the final physics loss can thus be written as:

$$\mathcal{L}_{\text{Phy}} = \eta \mathcal{L}_{\text{ARZ}}^{(1)} + (1 - \eta)\mathcal{L}_{\text{ARZ}}^{(2)} + \xi \mathcal{L}_{\text{reg}}, \tag{20}$$

where $\eta \in (0, 1]$ and $\xi \in (0, \infty)$ are hyperparameters that control the weights.

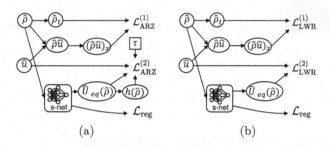

(a)   (b)

Fig. 4. Physics with surrogate models. (a) is the physics for ARZ and (b) is the physics for LWR, which will be introduced in Sect. 5.2.

4.3   Experiment Setting

Experiments are conducted on a Google cloud workstation with 8 Intel Xeon E5-2686 v4 processors and an NVIDIA V100 Tensor Core GPU with 16 GB memory in Ubuntu 18.04.3. The learning rate for the Adam optimizer is 0.0005, and other configurations are kept as default. The configuration of the discriminator is different for different loop detectors, which are detailed in the supplementary materials.

4.4   Results

The results of the TrafficFlowGAN are shown in Fig. 5. Figure 5(a) is the prediction of the traffic density. It demonstrates that the TrafficFlowGAN can reconstruct the traffic density with observations from 4 sensors. Two prediction snapshots at $t = 0.078$ and $t = 1.0$ shown in Fig. 5(b) and Fig. 5(c), respectively. The blue line stands for the mean of the ground truth; the dashed red line represents the mean of the prediction, and the yellow band is the prediction interval. Figure 5(d) illustrates the relation between traffic density and traffic velocity learned by the s-net, i.e. $\hat{U}_{eq}(\hat{\rho})$. The solid blue line is the ground-truth relation $U_{eq}(\rho)$ that is defined in Eq. 17. The dashed black line and the dashed red line are the $\hat{U}_{eq}(\hat{\rho})$ at the 1st and the 15000th epochs, respectively. We can see that s-net manages to recover the underlying traffic density-velocity relation. The reason for the relatively poor performance for $\rho > 0.9$ is that the numerical data does not contain ρ that is bigger than 0.9, as indicated by the colorbar of Fig. 3(b).

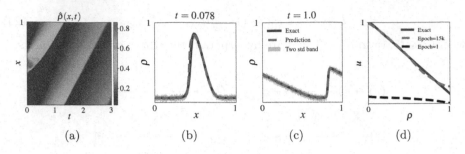

Fig. 5. Results of the TrafficFlowGAN on the numerical data generated by ARZ.

5   Case Study: Traffic State Estimation

Traffic state estimation (TSE) is an important traffic engineering problem that aims to infer traffic state variables represented by traffic density and velocity along a road segment from partial observations. In a nutshell, the goal of TSE is to learn a mapping from a spatio-temporal domain to traffic states, i.e. $f : (x, t) \rightarrow (\rho, u)$, using partial observations from fixed sensors (e.g. loop detectors).

5.1   Dataset

The Next Generation SIMulation (NGSIM)[1] is a real-world dataset that collects vehicle trajectory every 0.1 s. We focus on a 15-min data fragments collected on highway US 101. The traffic density and velocity are shown in Fig. 6.

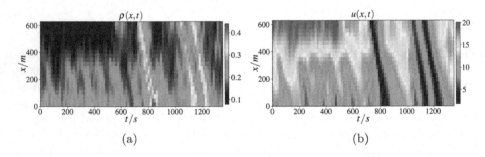

Fig. 6. NGSIM dataset, where (a) is the traffic density and (b) is the traffic velocity.

---

[1] www.fhwa.dot.gov/publications/research/operations/07030/index.cfm.

5.2   Physics-Based Computational Graph

As the underlying physics for the real-world scenario is unknown, in addition to the ARZ, we also adopt the LWR model as our physics. LWR is depicted as below:

$$\begin{cases} \rho_t + (\rho u)_x = 0, \\ u = U_{eq}(\rho) \triangleq u_{max}(1 - \rho/\rho_{max}), \end{cases} \tag{21}$$

which shares the same physics parameters, i.e. ρ_{max} and u_{max}, as the ARZ. The physics computational graph associated with LWR is illustrated in Fig. 4(b). The corresponding physics losses are as below:

$$\begin{cases} \mathcal{L}_{LWR}^{(1)} = |\mathbb{E}_z [\hat{\rho}_t + (\hat{\rho}\hat{u})_x]|^2 \\ L_{LWR}^{(2)} = \left|\mathbb{E}_z \left[(\hat{U}_{eq}(\hat{\rho}) - \hat{u})\right]\right|^2 \end{cases} . \tag{22}$$

5.3   Baselines and Metrics

We adopt the following baselines for comparison: the pure flow model, the physics-informed flow with ARZ as the physics (PhysFlow-ARZ), the physics-informed flow with LWR as the physics (PhysFlow-LWR), FLowGAN, and the ARZ-based extended Kalman filter (EKF) [16]. EKF applies a non-linear version of the Kalman filter and is widely used in nonlinear systems like the TSE.

We use the $\mathbb{L}^2$ relative percentage error (RE) to measure the difference between the mean of the prediction and that of the ground truth. The reason for choosing this metric is to mitigate the influence of the scale of the ground truth. In addition, the reverse Kullback-Leibler (KL) divergence is used to measure the difference between the prediction distribution and the sample distribution.

5.4   Results

Figure 7 shows the REs (left two) and KL divergences (right two) of traffic density ρ and velocity u of TrafficFlowGAN and the baselines. The x-axis is the number of loop detectors. Different scatter types and colors are used to distinguish with different models. From this figure, we can see that TrafficFlowGAN-ARZ outperforms others across nearly all numbers of loop detectors for REs, and TrafficFlowGAN-LWR achieves the best performance for KL divergences. We also record the training time of each model when the number of loops is 10. The Flow-based models, including Flow and PhysFlow, cost 0.11 s per epoch. This means that the extra computational time from calculating the physics loss is negligible. The training time of the FlowGAN-based models, including TrafficFlowGAN and FlowGAN, is 0.31 s per epoch.

Figure 8 show the predictions of traffic density (top row) and traffic velocity (bottom row). Figure 8(a) and Fig. 8(d) present the heatmaps of traffic density and traffic velocity in spatio-temporal space. Those two predictions are close to the ground truth shown in Fig. 6. The other 4 subfigures show the snapshots of the prediction intervals of the traffic density and velocity.

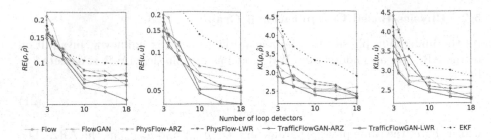

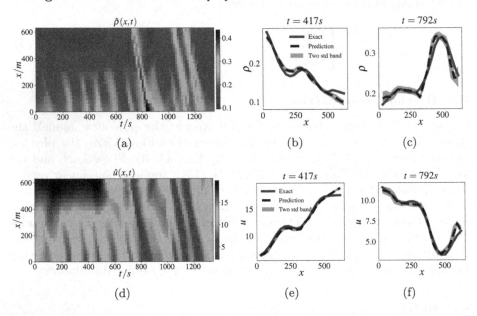

Fig. 7. The REs and KL of our proposed TraficFlowGAN and the baselines.

Fig. 8. Predictions of the traffic density (top row) and the traffic velocity (bottom row) of the TrafficFlowGAN.

Figure 9 presents the comparison between the ground-truth traffic density distribution and that predicted by the TrafficFlowGAN model, each subfigure for a randomly sampled spatio-temporal coordinates. Most parts of the predicted and ground-truth distributions overlap with each other, which demonstrates that our proposed model can estimate the real-world traffic states uncertainties well.

Figure 10 shows the learned traffic density-velocity relation by TrafficFlow-GAN. When the number of loop detectors is 4, TrafficFlowGAN can already capture this relation well. Increasing the number of loop detectors helps TrafficFlowGAN learn a subtler pattern.

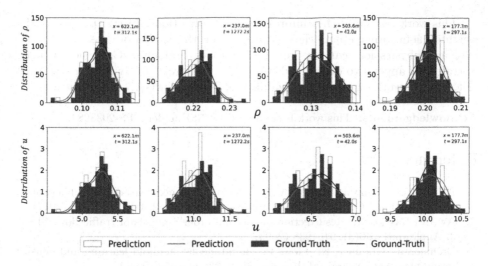

Fig. 9. Prediction distributions of TrafficFlowGAN for the NGSIM dataset.

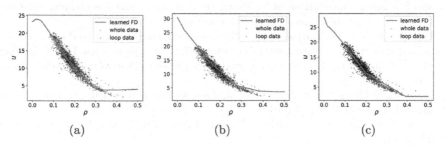

Fig. 10. Traffic density-velocity relations learned by the surrogate model (s-net) for different loop detector numbers. (a) loop detector number is 4. (b) loop detector number is 6. (c) loop detector number is 10.

6   Conclusions

This paper proposes TrafficFlowGAN to quantify the uncertainty in dynamical systems. TrafficFlowGAN leverages MLE, adversarial training, and PIDL to generate high-quality samples with exact data likelihood and efficient data usage. To verify that TrafficFlowGAN can learn the solutions of a second-order PDE, we conduct a numerical experiment where data is generated from a known ARZ equation. Numerical results show that TrafficFlowGAN manages to reconstruct the PDE solutions and recover the underlying relation between state variables. We further apply TrafficFlowGAN to a TSE problem using real-world data to demonstrate its performance. Results show that TrafficFlowGAN can better capture the real-world uncertainty than baselines, including the pure flow, the physics-informed flow, and the flow based GAN. We also show that TrafficFlowGAN can learn the real-world traffic density-velocity relation simultaneously.

This work can be further improved in two directions. First, apart from the weighted sum, other approaches to integrating the likelihood loss, adversarial loss, and physics loss can be proposed. Second, TrafficFlowGAN needs to be re-trained if applied to other roads or to the same road but within a new time slot. We will work on the generalizability of TrafficFlowGAN in the future.

Acknowledgements. This work is sponsored by NSF under CPS-2038984.

References

1. Arjovsky, M., Chintala, S., Bottou, L.: Wasserstein generative adversarial networks. In: International Conference on Machine Learning, pp. 214–223. PMLR (2017)
2. Aw, A., Rascle, M.: Resurrection of "second order" models of traffic flow. SIAM J. Appl. Math. **60**(3), 916–938 (2000)
3. Brehmer, J., Cranmer, K.: Flows for simultaneous manifold learning and density estimation. Adv. Neural Inf. Process. Syst. **33**, 442–453 (2020)
4. Council, N.R., et al.: Assessing the Reliability of Complex Models: Mathematical and Statistical Foundations of Verification, Validation, and Uncertainty Quantification. National Academies Press, Washington, DC (2012)
5. Daw, A., Maruf, M., Karpatne, A.: Pid-gan: a gan framework based on a physics-informed discriminator for uncertainty quantification with physics. In: Proceedings of the 27th ACM SIGKDD Conference on Knowledge Discovery & Data Mining, pp. 237–247 (2021)
6. Dinh, L., Sohl-Dickstein, J., Bengio, S.: Density estimation using real nvp. arXiv preprint arXiv:1605.08803 (2016)
7. Goodfellow, I., et al.: Generative adversarial nets. Adv. Neural Inf. Process. Syst. **27** (2014)
8. Grover, A., Chute, C., Shu, R., Cao, Z., Ermon, S.: Alignflow: cycle consistent learning from multiple domains via normalizing flows. In: Proceedings of the AAAI Conference on Artificial Intelligence, vol. 34, pp. 4028–4035 (2020)
9. Grover, A., Dhar, M., Ermon, S.: Flow-gan: combining maximum likelihood and adversarial learning in generative models. In: Proceedings of the AAAI Conference on Artificial Intelligence, vol. 32 (2018)
10. Guo, L., Wu, H., Zhou, T.: Normalizing field flows: solving forward and inverse stochastic differential equations using physics-informed flow models. arXiv preprint arXiv:2108.12956 (2021)
11. Lighthill, M.J., Whitham, G.B.: On kinematic waves II: a theory of traffic flow on long crowded roads. Proc. Roy. Soc. Lond. Ser. A. Math. Phys. Sci. **229**(1178), 317–345 (1955)
12. Mo, Z., Di, X.: Uncertainty quantification of car-following behaviors: physics-informed generative adversarial networks. In: The 28th ACM SIGKDD in Conjunction with the 11th International Workshop on Urban Computing (UrbComp2022) (2022)
13. Mo, Z., Fu, Y., Di, X.: Quantifying uncertainty in traffic state estimation using generative adversarial networks. In: 2022 IEEE 25th International Conference on Intelligent Transportation Systems (ITSC), pp. 2769–2774. IEEE (2022)
14. Mo, Z., Shi, R., Di, X.: A physics-informed deep learning paradigm for car-following models. Transp. Res. Part C: Emerg. Technol. **130**, 103240 (2021)

15. Raissi, M.: Deep hidden physics models: deep learning of nonlinear partial differential equations. J. Mach. Learn. Res. **19**(1), 932–955 (2018)
16. Seo, T., Bayen, A.M.: Traffic state estimation method with efficient data fusion based on the aw-rascle-zhang model. In: 2017 IEEE 20th International Conference on Intelligent Transportation Systems (ITSC), pp. 1–6. IEEE (2017)
17. Shi, R., Mo, Z., Di, X.: Physics-informed deep learning for traffic state estimation: a hybrid paradigm informed by second-order traffic models. In: Proceedings of the AAAI Conference on Artificial Intelligence, vol. 35, pp. 540–547 (2021)
18. Shi, R., Mo, Z., Huang, K., Di, X., Du, Q.: A physics-informed deep learning paradigm for traffic state and fundamental diagram estimation. IEEE Trans. Intell. Transp. Syst. **23**, 11688–11698 (2021)
19. Siddani, B., Balachandar, S., Moore, W.C., Yang, Y., Fang, R.: Machine learning for physics-informed generation of dispersed multiphase flow using generative adversarial networks. Theor. Comput. Fluid Dyn. **35**(6), 807–830 (2021). https://doi.org/10.1007/s00162-021-00593-9
20. Theis, L., Oord, A.V.D., Bethge, M.: A note on the evaluation of generative models. arXiv preprint arXiv:1511.01844 (2015)
21. Yang, L., Zhang, D., Karniadakis, G.E.: Physics-informed generative adversarial networks for stochastic differential equations. SIAM J. Sci. Comput. **42**(1), A292–A317 (2020)
22. Yang, Y., Perdikaris, P.: Adversarial uncertainty quantification in physics-informed neural networks. J. Comput. Phys. **394**, 136–152 (2019)
23. Zang, C., Wang, F.: Moflow: an invertible flow model for generating molecular graphs. In: Proceedings of the 26th ACM SIGKDD International Conference on Knowledge Discovery & Data Mining (2020)

STGEN: Deep Continuous-Time Spatiotemporal Graph Generation

Chen Ling[1], Hengning Cao[2], and Liang Zhao[1(✉)]

[1] Emory University, Atlanta, GA, USA
{chen.ling,liang.zhao}@emory.edu
[2] Cornell University, Ithaca, NY, USA
hc2225@cornell.edu

Abstract. Spatiotemporal graph generation has realistic social significance since it unscrambles the underlying distribution of spatiotemporal graphs from another perspective and fuels substantial spatiotemporal data mining tasks. Generative models for temporal and spatial networks respectively cannot be easily generalized to spatiotemporal graph generation due to their incapability of capturing: 1) mutually influenced graph and spatiotemporal distribution, 2) spatiotemporal-validity constraints, and 3) characteristics of multi-modal spatiotemporal properties. To this end, we propose a generic and end-to-end framework for spatiotemporal graph generation (STGEN) that jointly captures the graph, temporal, and spatial distributions of spatiotemporal graphs. Particularly, STGEN learns the multi-modal distribution of spatiotemporal graphs via learning the distribution of spatiotemporal walks based on a new heterogeneous probabilistic sequential model. Auxiliary activation layers are proposed to retain the spatiotemporal validity of the generated graphs. In addition, a new boosted strategy for the ensemble of discriminators is proposed to distinguish the generated and real spatiotemporal walks from multi-dimensions and capture the combinatorial patterns among them. Finally, extensive experiments are conducted on both synthetic/real-world spatiotemporal graphs and demonstrated the efficacy of the proposed model.

Keywords: Deep graph generation · Spatiotemporal graph · Deep generative model

1 Introduction

Many complex systems can be modeled as graphs, which characterize the objects (i.e., nodes) and their interactions (i.e., edges) [31]. In many graph systems, the nodes and edges need to be embedded in space and evolve over time. The former is denoted as spatial network [4] while the latter is named temporal graphs [14], both of which are well-explored domains by network science models such as spatial small worlds model [29], optimal network [1], and epidemic temporal

---

Supplementary Information The online version contains supplementary material available at https://doi.org/10.1007/978-3-031-26409-2_21.

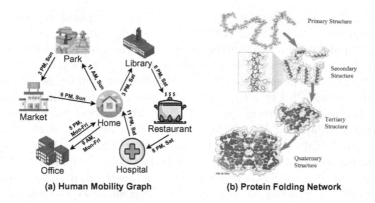

(a) Human Mobility Graph   (b) Protein Folding Network

Fig. 1. The example of real world STGs. a) the human mobility graph describes one week's living trajectory between different locations of an adult with timestamps on directed edges. b) The protein folding process [21] with amino acids in different folding phases, which is a spatial graph evolves with time.

network [20]. These conventional methods propose to utilize prescribed structural assumptions (e.g., temporal exponentiality, network shortest distance) to characterize spatiotemporal graphs (STGs) through synthesizing them. However, traditional methods are limited in modeling and interpreting STGs since the intrinsically complex spatiotemporal patterns are hard to be modeled only by prior knowledge. Such prior knowledge is not always available especially considering the limited information of human beings on many real-world complex networks such as brain network dynamics [23], the folding of protein structure [21], and catastrophic failures in power grids [27]. Therefore, it is desired to have a model with high expressiveness in learning the dynamics directly from data without detailed handcrafted rules.

Recently, there has been a surge of research efforts on deep generative models in the task of graph generation. For example, enormous works [7,25,32,34,35] have achieved promising performance in generating realistic static or temporal graphs or separately considering the spatial properties. On the other hand, there is also a fast-growing research body on discriminative learning for STG data and their applications, such as traffic prediction [33], emotion perception [5], and object recognition [22]. However, joint generative consideration of spatial, temporal, and graph aspects is still under-explored.

To fill this gap, we focus on the generic problem of STG generation, which cannot be easily handled by combining existing works because *1) Difficulties in jointly learning both graph and spatiotemporal distribution of STGs.* As shown in Fig. 1(a), the human mobility behavior follows the distribution characterized jointly by the spatial, temporal, and graph patterns. More important, these three patterns are strongly correlated. For example, the time "9 AM on workday" may correlate to the edge "going to work" and the location "traffic from home to downtown". Existing static and dynamic graph generative models cannot be combined to model it, which will discard the synergies among all the patterns simultaneously [8,10]. *2) Difficulties in ensuring spatiotemporal validity in the generated STGs*. STGs need to follow spatial and temporal constraints. The former means that the locations of the nodes and edges are confined in a specific geometric topology, while the latter means that the nodes and edges need to respect their temporal order. For example, in Fig. 1(a), for a path in the human mobility graph, the timestamp value of the first edge must be smaller than that for the second. In Fig. 1(b), a pair of amino acids with direct connections must be close to each other in space. Spatial and temporal constraints directly determine the validity of STGs. And because they are hard constraints, they cannot be intrinsically merged into the distribution of STGs, which are typically continuous according to the common statistical models. Therefore, establishing a model that can generate STGs while maintaining their validity is imperative yet challenging. *3) Difficulties in identifying the dependencies and independencies of spatial, temporal, and graph patterns.* A STG is composed of multi-modal components: graph information, temporal information, and spatial information. Some of the information are correlated while some are independent, which forms combinatorial among them into STG patterns, such as spatial and temporal graph patterns. Taking Fig. 1(a) and (b) as examples: in human mobility STG, spatial, temporal, and graph patterns are strongly correlated. In the folding process of protein (spatial graph of amino acids), the correlation between spatial and network patterns is even stronger than that between temporal and some graph properties (e.g., edge connections).

To address all the challenges, we propose a novel continuous-time STG generation framework, called *STGEN*, which coherently models both graph topology and spatiotemporal dynamics of observed STGs through a new generative adversarial model. Specifically, we propose to decompose STGs into spatiotemporal walks by developing a novel spatiotemporal walk generator to jointly capture the graph and spatiotemporal distribution. Novel STGs can be assembled through conditionally concatenating generated spatiotemporal walks with the guarantee of spatiotemporal validity. On top of that, we design a new discriminator which is an ensemble of multi-modal sub discriminators with different combinations of spatial, temporal, and graph patterns. We summarize contributions as follows:

- **The development of a new generative framework for STG generation.** We formally define the problem of STG generation and propose STGEN to tackle its unique challenges arising from real applications. It generates STGs with ensuring graph, temporal, and spatial validity.
- **The design of a novel spatiotemporal walk generator.** We develop a novel spatiotemporal walk generator with spatiotemporal information decoders to capture the underlying dynamics of observed STGs. Auxiliary activation layers are leveraged to ensure the validity of the generated STG.
- **The proposal of the ensemble of multi-modal sub discriminators for stronger STG adversarial training.** Multiple sub discriminators are designed and synergized by adaptively boosting extension in order to coherently examine the different combinations of spatial, temporal, and graph patterns. To extend WGAN by our ensemble discriminator, we propose a well-modularized learning objective and optimization algorithm.

- **The conduct of extensive experiments to validate the effectiveness of the proposed model.** Extensive experiments and case analysis on both synthetic and real-world datasets demonstrate the capability of STGEN in generating the most realistic STGs compared to other baselines in terms of both temporal and spatial similarities.

2 Related Works

Spatial and Temporal Graphs. The graph is a mathematical subject that describes the interactions (edges) between a set of objects (nodes). Such a graph structure can also be employed as the infrastructure of real-world dynamic systems. However, for many dynamic systems, one might have more information than just about who interacts. Transportation and mobility networks [3], internet of things [18], and social networks [24], are all examples where time and space information are significant and where graph topology alone does not contain all the information. Nowadays, both spatial and temporal graphs have become indispensable extensions of static graphs and achieved success in various dynamic system modeling tasks [28,30,33]. We refer readers to recent surveys [4,14] for more details of both spatial and temporal graphs.

Deep Graph Generation. A number of deep generative models for static graphs have emerged in the past few years [12]. Specifically, GraphVAE [25] try to generate new graphs' adjacency matrix in a variational auto-encoder way, while [32] treats graph generation as a node/edge sequence generation process based on LSTM. NetGAN and its variant [7,19] generates new graphs through modeling the random walk distribution in observed graphs. Other than the static graph generation, the other line of works [11,34,35] have achieved success in generating temporal or spatial graphs. Both lines of works leverage temporal walks and spatial attribute to model the temporal and spatial properties of the observed graphs, respectively. However, none of the above works can be directly adapted to generate STGs since they neither can effectively decode the joint distribution of topology and spatiotemporal properties in STGs nor ensure the spatiotemporal validity (i.e., temporal ordering, semantics of spatial coordinates, and physical constraints of spatial properties) of the generated graph [4,14]. Recently, STGD-VAE [9] is proposed to model deep generative processes of composing discrete spatiotemporal networks, which is a specific type of general spatiotemporal graphs that reduces time information into ordinal values. Hence it does not directly generate continuous-time spatiotemporal graphs, spatiotemporal validity constraints, and various spatiotemporal distributions. In this paper, we propose a generic framework to jointly model the distribution of multi-modal properties of observed STGs and generate novel ones with the spatiotemporal validity guarantee. To the best of our knowledge, STGEN is the first-of-its-kind deep generative model designed for continuous STGs with validity constraints.

Spatiotemporal Deep Learning. With the prevalence of deep learning techniques, various models [16,33] have been proposed to model spatiotemporal data

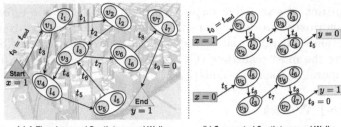

(a) A Time-inversed Spatiotemporal Walk   (b) Segmented Spatiotemporal Walks

Fig. 2. Figure 2(a) illustrates a spatiotemporal walk, and (b) indicates the decomposition process of a spatiotemporal walk to multiple smaller-sized segmented spatiotemporal walk. For the sake of simplicity, we omit the turning angle ϕ_i and velocity ξ_i attached on each edge e_i in above figures.

through decoding its underlying distribution in a discriminative way without applying any hand-crafted rules. These approaches generally achieve promising results in plenty of predictive tasks through analyzing specific patterns (e.g., spatial proximity [16] and temporal correlations [5,33]) of the spatiotemporal data. However, interpreting spatiotemporal data from the generation perspective has received less attention since it is a more challenging task and requires one to comprehensively capture the underlying dynamics among multi-modal spatiotemporal properties and their intricate and entangled dependencies. A few tries of utilizing deep generative models have been observed in spatiotemporal data generation. Additionally, [26] converts spatiotemporal data (e.g., trajectory) to images and applies GAN for the generation. Another work SVAE [15] utilizes VAE modules to learn variables from Gaussian distribution and generate novel human mobility accordingly. However, existing works still cannot explicitly consider the spatiotemporal validity during the generation. STGEN is the first-of-its-kind generative model that generates spatiotemporal data in the form of graphs with spatiotemporal validity.

3   Problem Setting

A continuous-time spatiotemporal graph (STG) is a directed graph $G = \{e_1, ..., e_i, ...\}$, where each edge $e_i = \left(u_i, v_i, t_i, l_u^{(i)}, l_v^{(i)}\right)$. $u_i, v_i \in V$ are two end nodes of e_i, $t_i \in [0, t_{\text{end}}]$ is the timestamp on the edge, and $[0, t_{\text{end}}]$ is the time span of the STG with $t_{\text{end}} \in \mathbb{R}^+$. Each node v_i in a STG is associated with a spatial attribute $l_v^{(i)}$. The spatial attribute $l_v^{(i)} = (l_{\text{lat}}^{v_i}, l_{\text{lon}}^{v_i})$ can be interpreted as a Global Positioning System (GPS) location with specific latitude and longitude on earth.

Definition 1 (*Spatiotemporal Walk*) *A spatiotemporal walk* $\mathbf{s} = \{t_0, e_1, ..., e_{L_\mathbf{s}}, t_{L_\mathbf{s}}\}$ *is defined as a sequence of spatiotemporal edges and a pair of initial time budget* t_0 *and end time budget* $t_{L_\mathbf{s}}$, *where* $L_\mathbf{s}$ *is the length of the spatiotemporal walk* $\mathbf{s}$ *and* $\forall e_i \in G$. *Specifically, a spatiotemporal edge is defined* as: $e_i = \left(u_i, v_i, t_i, l_u^{(i)}, l_v^{(i)}\right) \in E$, where each $t_i \in [0, t_{end}], t_i < t_{i-1}$ is called the "time budget" for e_i indicating the total timespan consumed in e_i. Intuitively, the initial time budget $t_0 = t_{end}$ and end time budget $t_{L_s} = 0$. An example of the spatiotemporal walk is illustrated in Fig. 2(a).

A continuous-time STG can be denoted as the union of all the spatiotemporal walks $G = \bigcup_{s \sim P_r(s)} s$, where the $P_r(s)$ is the distribution of all walks in graph G. It is straightforward that we can leverage fixed-length sequential models to learn the distribution of the spatiotemporal walks in order to capture the overall STG distribution. However, the nature of continuous-time STGs decides one cannot arbitrarily sample random walks as was done in static graphs [7] but needs to follow certain spatiotemporal orders. More specifically, the length of a spatiotemporal walk in STG is regulated by the starting and ending point based on the spatiotemporal information on edges. The length of the spatiotemporal walk varies that may easily reach a million-scale (especially when the time granularity is small), which cannot be learned effectively and efficiently by sequence learning methods. Therefore, the definition of spatiotemporal walk needs to be enriched into the following extension.

Definition 2 (*Segmented Spatiotemporal Walk*) *A segmented spatiotemporal walk is defined as a sequence* $\bar{s} = \{(x, y, t_0), e_1, e_2, ..., e_L\}$ *with its profile informatiom* (x, y, t_0), *which is segmented from an originated spatiotemporal walk. Specifically, the* $L \leq L_s$ *is the length of each segmented spatiotemporal walk. The profile information* (x, y, t_0) *includes* $x \in \{0, 1\}$ *and* $y \in \{0, 1\}$, *which denote whether* s *is the respective starting or ending segmentation* $(x = 1, y = 0$ *or* $x = 0, y = 1)$ *or neither of them* $(x, y = 0)$ *in its originated spatiotemporal walk. The whole process of decomposing a spatiotemporal walk to segmented spatiotemporal walks is elucidated in Fig. 2(b).*

With all the aforementioned notions, we can formalize the problem of STG generation as follows:

*Problem 1 (**Spatiotemporal Graph Generation**). The problem of the STG generation is to learn an overall distribution* $P_r(G)$ *from the observed STGs, where each* G *is denoted as the union of all the spatiotemporal walks. Novel STGs* $\hat{G}$ *can be sampled from the distribution such that* $\hat{G} \sim P_r(\cdot)$.

There are several challenges in solving the novel STG generation problem: *1)* It is difficult to capture the joint distribution of the multi-modal properties in $\bar{s}$ since its properties are characterized with both categorical distributions (e.g., x, y, and (u_i, v_i)) and continuous-value distribution (e.g., t_i and (u_i, v_i)). *2)* Correctly decoding all the information from the learned distribution poses another challenge. The validity of each graph component $((u_i, v_i)$, t_i, and the calculated spatial locations $(l_u^{(i)}, l_v^{(i)})$) requires extra attention since the generated STGs need to have realistic semantic meaning (i.e., $t_{L_s} \leq t_i \leq t_0$ and $(l_u^{(i)}, l_v^{(i)})$ is valid in a prescribed spatial region). *3)* It is also difficult to characterize the dependency among multi-modal properties in different STGs due to the interplay among spatial, temporal, and graph information.

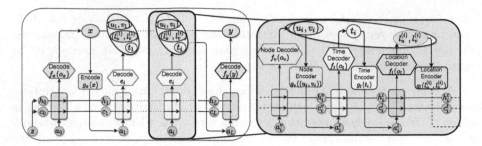

Fig. 3. Overview of spatiotemporal walk generator

4   Generative Model for Spatiotemporal Graph

In this section, we first introduce the backbone of our framework - STGEN for generating continuous-time STG. Then, we elaborate on each component of the generative framework, namely the spatiotemporal walk generator $\mathcal{G}$ and the boosted spatiotemporal walk discriminator $\mathcal{D}$.

4.1   Overall Architecture

STGEN learns the distribution of STG through a generative adversarial architecture, which consists of two parts: a recurrent-structure-based spatiotemporal walk generator $\mathcal{G}$ (as shown in Fig. 3), and a boosted spatiotemporal walk discriminator $\mathcal{D}$ (as shown in Fig. 6). The training of both generator $\mathcal{G}$ and discriminator $\mathcal{D}$ are conducted under the framework of Wasserstein GAN (WGAN) [2] to maximize the discrepancy $W(P_r, P_\theta)$ between the real STG distribution P_r and the generated STG distribution P_θ such that:

$$W(P_r, P_\theta) = \max \left[\mathbb{E}_{\mathbf{s}_r \sim P_r}[\mathcal{D}(\mathbf{s}_r)] - \mathbb{E}_{\mathbf{s}_\theta \sim P_\theta}[\mathcal{D}(\mathbf{s}_\theta)] \right]$$
$$s.t. \; \mathbf{s}_\theta = \mathcal{G}(z) \sim P_\theta, \mathcal{T}(\mathbf{s}_\theta) \in \mathbb{T}, \mathcal{K}(\mathbf{s}_\theta) \in \mathbb{K} \tag{1}$$

where $\mathbf{s}_r$ and $\mathbf{s}_\theta$ are real and fake spatiotemporal walks sampled from P_r and P_θ, respectively. z is a latent noise sampled from the standard normal distribution. Specifically, the spatiotemporal walk generator $\mathcal{G}$ trains a fixed-length LSTM whose output $\mathbf{s}_\theta = \mathcal{G}(z)$ is a *segmented* spatiotemporal walk. The time budgets as well as the spatial attributes of these generated spatiotemporal walks $\mathbf{s}_\theta$ are regulated by a temporal activation layer $\mathcal{T}$ and a spatial activation layer $\mathcal{K}$, such that $\mathcal{T}(\mathbf{s}_\theta)$ and $\mathcal{K}(\mathbf{s}_\theta)$ are valid in respect of both temporal constraint $\mathbb{T}$ and spatial constraint $\mathbb{K}$. We introduce both the spatiotemporal walk generator and the corresponding STG assembler in Sect. 4.2. Moreover, due to the multi-modal nature of the spatiotemporal walk (i.e., node sequence, time budget on the edge, and the spatial attribute), a boosted discriminator is proposed for each combination of the spatiotemporal walk components to characterize dependencies among all components. We give further details of the boosted discriminator and theoretical analysis in Sect. 4.3.

Model Complexity. Existing static and dynamic graph generative models usually generate graphs through generating adjacency matrices or snapshots, and they require at least $O(N^2 \cdot M_s)$ time complexity, where N is the number of nodes, and M_s is the number of snapshots. For large-scale or dynamic graphs with a large timespan but small granularity, these models [9,25,32] would suffer from the scalability and information loss. However, since a continuous-time STG is composed of spatiotemporal walks and does not involve with an adjacency matrix, it makes the overall complexity of STGEN to be $O(L_\mathbf{s} \cdot M_e)$ since STGEN only needs to generate segmented spatiotemporal walks and assemble them as a whole, where $L_\mathbf{s}$ is the maximal length of all the spatiotemporal walks and M_e is the number of spatiotemporal walks needed to compose an STG.

4.2   Spatiotemporal Walk Generator with Validity Constraints

Segmented spatiotemporal walk is a heterogeneous sequence where some elements are categorical signals (starting point x and ending point y), continuous-value scalars $(t_i, l_u^{(i)}, l_v^{(i)})$, and edges (u_i, v_i). To effectively characterize all the modalities in segmented spatiotemporal walks, we propose a novel heterogeneous recurrent-structured generator with various encoding/decoding functions for various modalities. In addition, we propose new activation functions to enforce spatiotemporal validity constraints on the generated spatiotemporal walk patterns. Finally, an STG assembler is proposed to compose final STGs by conditionally generating spatiotemporal walks given other spatiotemporal walks.

Segmented Spatiotemporal Walk Generation. The generator $\mathcal{G}$ defines an implicit probabilistic model for generating segmented spatiotemporal walks $\bar{\mathbf{s}}$ that are similar to the real spatiotemporal walks $\mathbf{s}$ sampled from P_r, and its overall architecture is summarized in Fig. 3. Specifically, the generator is modeled by a fixed-length LSTM model. Each LSTM unit keeps a hidden state h_i and cell state c_i as memory state, takes a_i as input, and returns o_i as output. The generator outputs the generated time budget t_i, node pair (u_i, v_i), and the spatial attributes $(l_u^{(i)}, l_v^{(i)})$ through decoding o_i in different decoding functions in a sequential order. The decoding functions can be divided into two categories: *discrete-value* and *continuous-value* decoding functions. *Discrete-value* decoding functions include a node decoding function: $f_v(o_{v_i})$ that outputs the node v_i, a starting point decoding function $f_x(o_x)$ that outputs the starting point x, and a ending point decoding function $f_y(o_y)$ that outputs the ending point y. *Continuous-value* decoding functions include a time decoding function $f_t(o_t)$ that outputs a residual time budget t_i on the generated edge, and the time budget t_i is regulated by the temporal activation layer $\mathcal{T}$ in order to meet the temporal constraint $\mathbb{T}$. In addition, STGEN also contains a location decoding function $f_l(o_l)$ that outputs spatial attributes $l_u^{(i)}$ and $l_v^{(i)}$ for both end nodes u_i and v_i on the generated edge. Likewise, the generated spatial attributes are also regulated by a spatial activation layer $\mathcal{K}$ to meet the spatial constraint $\mathbb{K}$. Other than the decoding functions, the generator $\mathcal{G}$ also uses different encoding functions to encode each generated components back to the next LSTM unit input a_{i+1}.

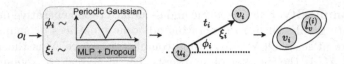

Fig. 4. Example of the sampling procedure of location decoder $f_l(\cdot)$.

We employ Multi-layer Perceptron (MLP) structures for all encoding functions. For discrete-value decoding functions, a Gumbel-softmax trick [17] is applied to make the sampling procedure differentiable. We describe the continuous-value decoding functions in more details in the following part. Due to space limit, we summarize the overall generative process in the supplementary material.

With Spatiotemporal Validity Constrained Multi-modal Decoding Functions. The time budget t_i as well as the spatial attributes $(l_u^{(i)}, l_v^{(i)})$ (e.g., GPS coordinates) of a spatiotemporal walk in many real world situations are typically irregular but following different underlying distributions, which makes both components cannot be trivially decoded by deterministic functions. Therefore, t_i and $(l_u^{(i)}, l_v^{(i)})$ are assumed to be sampled from a latent distribution. The sampling procedure is handled by both time decoder $f_t(\cdot)$ and location decoder $f_l(\cdot)$. Particularly, $f_t(\cdot)$ is an end-to-end sampling function that convert the latent representation o_i to parameters of an prescribed distribution (e.g., μ and σ in Gaussian distribution). Then, $t_i \sim f_t(o_t)$ could be sampled directly. In order to fulfill the temporal constraint, we apply a activation layer $\mathcal{T}$ to ensure the temporal validity of the generated segmented spatiotemporal walks. Specifically, we propose to impose a *Min-max Bounding* in the activation layer $\mathcal{T}$:

$$\mathcal{T}(t_i) = \begin{cases} t_i = t_i - \min(\{t_i\}), & \text{if } \min(\{t_i\}) \leq \epsilon \\ t_i = t_i / \max(\{t_i\}), & \text{if } \max(\{t_i\}) > 1 \\ t_i = t_i, & \text{otherwise} \end{cases}$$

where $\min(\{t_i\})$ and $\max(\{t_i\})$ are the min and max time budget in the generated mini-batch $\{t_i\}$, respectively. ϵ is a threshold with small value (i.e., $1e-3$) to prevent zero value for t_i. On the other side, spatial locations in many real-world situations have higher variance and could not to be typically described by known distributions. Instead of directly sampling exact locations, in this work, we sample the relative turning angle ϕ_i and the speed ξ in the spatiotemporal edge e_i for computing the node location $l_v^{(i)}$ and $l_u^{(i)}$. Specifically, we leverage MLP and dropout to mimic the sampling operation to obtain a continuous-valued speed ξ_i on spatiotemporal edge e_i from the LSTM unit's latent input o_l. Moreover, we model the distribution of turning angle ϕ_i to fit the Periodic Gaussian distribution [6]. Based on the sampled time budget t_i, speed ξ_i, and turning angle ϕ_i, the relative distance can be directly computed. By assigning the initial location for the first node in the generated spatiotemporal walk, the locations for the subsequent node can also be determined. We visualize the whole procedure in Fig. 4.

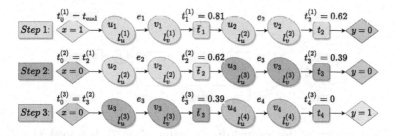

Fig. 5. Example of the spatiotemporal walk assembly. For step 1 (first line), the generator initially generates a segmented spatiotemporal walk with $x = 1$ and $y = 0, t_0^{(1)} = t_{\text{end}}$ containing two spatiotemporal edges (orange ovals) indicating it is the start of a spatiotemporal walk. At Step 2 and 3, the generator generates one additional edge e_3 (purple oval) and e_4 (blue oval) conditioned on the inputs of last edge, respectively. The final time budget $t_4^{(3)} = 0$ indicating the end of the spatiotemporal walk ($x = 0$ and $y = 1$) has reached.

To further regulate the generated node locations to have semantic meanings, we propose a *spatial activation layer* $\mathcal{K}$ with *Geographical Bounding*:

$$\mathcal{K}(l_v^{(i)}) = \begin{cases} l_v^{(i)} = l_{\xi_j}, & \text{if } \tau(l_v^{(i)}) \notin \bigcup_{j=1}^{H} \tau(\psi_j) \\ l_v^{(i)} = l_v^{(i)}, & \text{otherwise} \end{cases}$$

where $\{\psi_j\}$ is a set of prescribed geographical areas with size H. Particularly, for each generated geo-location $l_v^{(i)}$ generated in the set $\{l_v^{(i)}\}$, we project both $l_v^{(i)}$ and $\{\psi_j\}$ to the same space with a geographical projection function $\tau(\cdot)$ (e.g., Universal Transverse Mercator projection). If the projected $\tau(l_v^{(i)})$ belongs to any $\tau(\psi_j)$, this generated geo-location is valid and has realistic semantic meaning. Otherwise, this geo-location will be replaced by the closest point l_{ψ_j} on the prescribed area ψ_j that has the minimal distance to the original $l_v^{(i)}$.

Spatiotemporal Walk Assembler. The next step is to compose spatiotemporal walks from these segmented spatiotemporal walks generated from $\mathcal{G}$. In order to force the consistency of the underlying spatiotemporal diffusion pattern when we concatenate two segmented spatiotemporal walk, we may not chronologically concatenate walks purely based on their residual time budget t_i. Instead, we start by generating an initial segmented spatiotemporal walk $\bar{s}_1 = (e_1^{(1)}, ..., e_L^{(1)})$ with $x = 1, y = 0$, and $t_0 = t_{\text{end}}$ that contains L spatiotemporal edges. The last spatiotemporal edge $e_L^{(1)}$ of $\bar{s}_1$ is taken as the input to generate the next segmented spatiotemporal walk $\bar{s}_2 = (e_L^{(1)}, e_2^{(2)}, ..., e_L^{(2)})$ starting from $e_L^{(2)}$. In this case, $\bar{s}_2$ can be appended to $\bar{s}_1$ with the guarantee of following the underlying diffusion pattern. We incrementally appending additional segmented spatiotemporal walks until we run out of the time budget (i.e., $t_i = 0$) to form one final spatiotemporal walk (with the ending flag $x = 0, y = 1$). The overall process of assembling a spatiotemporal walk is illustrated in Fig. 5.

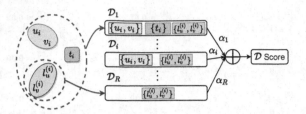

Fig. 6. The overview of the boosted discriminator. Each $\mathcal{D}_i$ is a weak classifier that takes a certain combination of spatiotemporal walk components.

4.3   Spatiotemporal Walk Discriminator

Discriminator $\mathcal{D}$ is employed to distinguish real and generated spatiotemporal walks, which is difficult due to their multi-modal nature. We aim to impose a stronger discriminator that can guide the generator characterizing both dependency and independency among the spatial, temporal, and graph modalities in STGs. Beyond a single discriminator that merely jointly considers spatial, temporal, and graph aspects of the walk, we propose to consider all the combinatorial of these aspects, such as spatial-temporal discriminator, temporal graph discriminator, spatial graph discriminator, etc. Our method is achieved by leveraging boosting strategy, which is well-recognized to enable the ensemble of models to outperform each individual model. Using a unified discriminator would also bring up the well-known training instability and potential mode collapse since the GAN may fall into recognizing only one of the generated modalities as the real sample while neglecting other modalities. Thus, we consider a multi-discriminator structure to better approximate Eq. (1) and constantly provide a harsher critique to the generator by considering the combinatorial of all modalities. A detailed architecture of the boosted discriminator is shown in Fig. 6.

Particularly, we adopt an adaptive boosted structure of the discriminator for discriminating each combination $\mathbf{s}^{(i)}$ of the spatiotemporal walk components, and there are a total of R combinations, such as spatiotemporal component (t_i and $(l_u^{(i)}, l_v^{(i)})$) and joint STG element ((u_i, v_i), $(l_u^{(i)}, l_v^{(i)})$, and t_i), etc. The boosted discriminator $\mathcal{D}$ takes the voting result over R sub discriminators $\mathcal{D}_i$ so that each sub discriminator $\mathcal{D}_i$ is performed as a weak-classifier. Such a boosted discriminator structure forces $\mathcal{G}$ to generate high fidelity samples that must hold up under the scrutiny of all R discriminators. The major voting strategy in our adaptive boosting can be induced into the WGAN objective and lead to a well-modularized objective function in Theorem 1.

Table 1. Dataset Description

	Syn_100	Syn_500	Syn_2000	Taxi	Check-in	Citation
Node	100	500	2,000	66	70	628
Temporal Edge	5,606	5,750	5,750	28,532	17,045	914
Temporal Samples	60	110	120	30	30	30

Theorem 1. *The aforementioned adaptive boosting strategy extends the objective function* $W(P_r, P_\theta)$ *in Eq. (1) of STGEN into the following:*

$$W(P_r, P_\theta) = \max \sum_{i=1}^{R} \alpha_i \Big[\mathbb{E}_{\mathbf{s}_r^{(i)} \sim P_r}[\mathcal{D}_i(\mathbf{s}_r^{(i)})] - \mathbb{E}_{\mathbf{s}_\theta^{(i)} \sim P_\theta}[\mathcal{D}_i(\mathbf{s}_\theta^{(i)})] \Big],$$

$$s.t., \ \mathbf{s}_\theta = \mathcal{G}(z) \sim P_\theta, \mathcal{G}(z) \in \{\mathbb{T}, \mathbb{K}\}, \sum_{i=1}^{R} \alpha_i = \frac{R}{4}.$$

With the objective of maximizing the above objective function, we can minimize the loss function of the generator $\mathcal{L}_G = -\sum_{i=1}^{R} \alpha_i \cdot \mathbb{E}_{\mathbf{s}_\theta^{(i)} \sim P_\theta}[\mathcal{D}_i(\mathbf{s}_\theta^{(i)})]$, where as well as the discriminator's overall loss function $\mathcal{L}_\mathcal{D} = \sum_{i=1}^{R} \alpha_i \cdot \mathcal{L}_{\mathcal{D}_i}$, where each sub discriminator's loss function is defined as: $\mathcal{L}_{\mathcal{D}_i} = \big[\mathbb{E}_{\mathbf{s}_\theta^{(i)} \sim P_\theta}[\mathcal{D}_i(\mathbf{s}_\theta^{(i)})] - \mathbb{E}_{\mathbf{s}_r^{(i)} \sim P_r}[\mathcal{D}_i(\mathbf{s}_r^{(i)})]\big]$. The proof of Theorem 1 can be found in Appendix. We illustrate the overall training framework in the supplementary material.

5   Experiment

In this section, we demonstrate the performance of our proposed STGEN framework across various synthetic and real world STGs. Basic experiment settings are illustrated here. Additional experiments (e.g., sensitivity analysis) are provided in the supplementary material. Code and data are made available[1].

Data. We performed experiments on three synthetic and three real-world STGs with different graph sizes and characteristics, where basic statistics are shown in Table 1. All graphs contain continuous timestamps as temporal information and geo-coordinates (latitude and longitude) as spatial information. Due to the space limit, details of all graphs can be found in the supplementary material.

Comparison Methods. Since there is no existing methods handling the STG generation problem, STGEN is compared with two categories of methods: *deep graph generation methods*: a) GraphRNN [32], b) NetGAN [7], c) TagGen [35], d) TG-GAN [34], and e) STGD-VAE [9]; and *spatial attribute generation methods*: a) LSTM [13], b) SVAE and its variant SVAE-γ [15], and c) IGMM-GAN [26].

---

[1] github.com/lingchen0331/STGEN.

Table 2. Performance comparison between real and generated graph samples in maximum mean discrepancy (MMD) (the lower the better), models with the best performance are marked with black.

Dataset	Method	AND	AGS	AGN	ACN	GD	Method	MCD	VCP
Syn_100	GraphRNN	$4.2\,e-3$	$7.12\,e-3$	0.175	$7.2\,e-2$	$7.12\,e-2$	LSTM	$1.09\,e-1$	51.37%
	NetGAN	$3.64\,e-5$	$1.89\,e-3$	0.0613	$2.77\,e-2$	$5.17\,e-3$	SVAE-y	$4.17\,e-1$	61.24%
	TagGen	$4.14\,e-4$	$2.72\,e-3$	0.0911	$6.18\,e-3$	$4.11\,e-2$	SVAE	$2.41\,e-1$	50.19%
	TG-GAN	$3.16\,e-5$	$5.31\,e-3$	**0.0221**	$3.06\,e-3$	$4.17\,e-3$	IGMM-GAN	$9.12\,e-3$	69.13%
	STGD-VAE	$2.53\,e-5$	$3.57\,e-3$	0.0367	$5.15\,e-3$	$7.92\,e-3$	–	–	–
	STGEN	**$2.31\,e-5$**	$1.96\,e-3$	0.0315	**$2.14\,e-3$**	**$3.89\,e-3$**	STGEN	**$3.52\,e-3$**	100%
Syn_500	GraphRNN	$7.64\,e-4$	$3.73\,e-5$	0.0525	$3.72\,e-2$	$4.13\,e-2$	LSTM	$4.56\,e-1$	57.15%
	NetGAN	$2.19\,e-4$	$6.12\,e-4$	0.023	$7.71\,e-3$	$7.18\,e-3$	SVAE-y	$7.12\,e-2$	69.21%
	TagGen	$3.24\,e-5$	$1.23\,e-2$	0.3912	$1.71\,e-2$	**$5.86\,e-3$**	SVAE	$6.12\,e-2$	59.19%
	TG-GAN	$1.44\,e-4$	$8.5\,e-3$	**0.0023**	$1.24\,e-2$	$5.12\,e-2$	IGMM-GAN	$9.12\,e-2$	49.14%
	STGD-VAE	$7.98\,e-5$	$2.67\,e-4$	0.0024	$3.97\,e-3$	$8.43\,e-3$	–	–	–
	STGEN	**$2.29\,e-5$**	$4.71\,e-3$	0.009	**$2.19\,e-3$**	$7.41\,e-3$	STGEN	**$9.18\,e-3$**	100%
Syn_2000	GraphRNN	$2.79\,e-4$	$7.4\,e-2$	0.042	$2.88\,e-2$	$2.15\,e-2$	LSTM	$4.22\,e-1$	51.33%
	NetGAN	$3.67\,e-5$	$2.69\,e-2$	0.0472	$6.92\,e-3$	$6.22\,e-3$	SVAE-y	$2.33\,e-1$	48.12%
	TagGen	$2.67\,e-4$	$1.98\,e-2$	0.031	$5.57\,e-3$	$5.66\,e-3$	SVAE	$7.28\,e-2$	50.67%
	TG-GAN	$2.66\,e-5$	**$4.19\,e-4$**	0.012	$4.95\,e-3$	**$2.97\,e-3$**	IGMM-GAN	$5.43\,e-2$	33.27%
	STGD-VAE	$5.48\,e-4$	$2.67\,e-2$	0.037	$9.42\,e-3$	$6.43\,e-3$	–	–	–
	STGEN	**$2.56\,e-5$**	$4.34\,e-3$	**0.009**	**$4.49\,e-3$**	$9.76\,e-3$	STGEN	**$1.07\,e-2$**	100%
Taxi	GraphRNN	$4.73\,e-3$	$4.63\,e-3$	0.0226	$3.2\,e-3$	$2.57\,e-2$	LSTM	1.1852	17.23%
	NetGAN	$8.29\,e-1$	$5.25\,e-6$	0.0189	$5.63\,e-4$	$7.87\,e-3$	SVAE-y	$2.37\,e-2$	21.44%
	TagGen	$3.92\,e-2$	$7.24\,e-4$	0.0221	$3.58\,e-4$	$3.91\,e-3$	SVAE	$3.11\,e-2$	20.46%
	TG-GAN	$6.69\,e-4$	$8.87\,e-6$	**0.0132**	$1.06\,e-5$	$2.67\,e-3$	IGMM-GAN	$1.67\,e-1$	9.54%
	STGD-VAE	**$9.61\,e-6$**	$3.61\,e-4$	0.0233	$7.26\,e-4$	$4.74\,e-3$	–	–	–
	STGEN	$9.85\,e-5$	**$3.09\,e-6$**	0.0165	$1.17\,e-5$	**$2.55\,e-3$**	STGEN	**$3.39\,e-3$**	100%
Check-in	GraphRNN	$3.5\,e-3$	$2.89\,e-2$	0.0312	$4.78\,e-4$	$5.32\,e-2$	LSTM	0.0963	3.77%
	NetGAN	$4.37\,e-3$	$2.54\,e-4$	0.063	$4.38\,e-4$	**$2.38\,e-3$**	SVAE-y	$2.67\,e-2$	23.11%
	TagGen	$1.27\,e-2$	$1.77\,e-2$	0.0292	$3.58\,e-4$	$3.91\,e-3$	SVAE	$1.79\,e-2$	19.24%
	TG-GAN	$7.69\,e-4$	**$3.76\,e-6$**	0.0139	$1.79\,e-5$	$3.99\,e-3$	IGMM-GAN	$2.37\,e-1$	15.23%
	STGD-VAE	$1.33\,e-4$	$6.71\,e-3$	0.0283	$3.12\,e-4$	$9.95\,e-2$	–	–	–
	STGEN	**$9.17\,e-5$**	$5.74\,e-6$	**0.0126**	**$1.39\,e-5$**	$3.57\,e-3$	STGEN	**$1.58\,e-4$**	100%
Citation	GraphRNN	$3.24\,e-1$	$3.12\,e-2$	0.0465	$1.98\,e-3$	$4.21\,e-2$	LSTM	1.6857	45.18%
	NetGAN	$5.2\,e-1$	$3.77\,e-3$	0.0577	$3.13\,e-3$	$2.18\,e-2$	SVAE-y	$7.41\,e-1$	8.33%
	TagGen	$3.34\,e-2$	$3.24\,e-3$	0.0561	$7.71\,e-5$	$3.67\,e-3$	SVAE	$5.91\,e-1$	10.84%
	TG-GAN	$1.24\,e-2$	$1.78\,e-5$	0.0587	$1.13\,e-4$	**$1.38\,e-4$**	IGMM-GAN	$1.47\,e-1$	3.21%
	STGD-VAE	$3.75\,e-1$	$3.64\,e-3$	0.0493	$5.67\,e-4$	$2.15\,e-2$	–	–	–
	STGEN	**$3.04\,e-3$**	**$1.23\,e-5$**	**0.0398**	**$3.66\,e-5$**	$2.77\,e-3$	STGEN	**$3.77\,e-2$**	100%

Details of all baselines can be found in the supplementary material. Note that GraphRNN, NetGAN, TagGen, and STGD-VAE are discrete graph generative models, and they cannot generate continuous-time STGs. We instead modify them to generate multiple discrete-snapshot graphs and convert them into a continuous-time temporal graph. In addition, STGD-VAE cannot generate realistic spatial attributes (i.e., GPS locations) so that we only compare STGD-VAE in generating graph properties. **Evaluation Metrics.** A set of metrics, as elucidated in Table 2, are used to measure the similarity between the generated and real STGs in terms of both temporal and spatial graph attributes. For temporal attributes measurement, we adopt AND: Average Node Degree, AGS: Average

Group Size, AGN: Average Group Number, ACN: Average Coordination Number, and GD: Group Duration. For spatial attributes measurement, we leverage MCD: Mean Coordinates Distribution and VCP: Valid Coordinates Percentage.

5.1 Quantitative Performance

The overall performance comparisons are shown in Table 2. The proposed STGEN generally outperforms other methods in terms of both temporal and spatial attributes generation with only a few exceptions. Specifically, in terms of the similarity in temporal graph properties, STGEN performs better than static graph generation methods (i.e., GraphRNN and NetGAN) by, on average, two orders of magnitudes in several temporal graph properties (e.g., AND, AGS, and ACN). The main reason is that static graph generation methods generate dynamic graphs via generating a series of snap-shots, which may cause severe information loss [14]. STGEN also consistently achieves competitive results with dynamic graph generation methods (i.e., TagGen and TG-GAN) among all datasets since STGEN can effectively capture the underlying multi-modal distribution of STGs. Compared with the only discrete STG generation method - STGD-VAE, STGEN still exhibits an overall better performance in generating continuous STGs. In terms of the spatial graph properties generation, STGEN exceeds other approaches with an evident margin (two orders of magnitudes on average) in MCD while achieving a 100% validity rate of the generated spatial properties among all datasets. With the applied spatiotemporal constraints, the coordinates generated by STGEN can always be regulated in a valid semantic region and guarantee a 100% validity rate, while other methods can only achieve at most 70% validity rate. In other words, a large portion of the coordinates generated by baseline methods do not have valid semantic meanings.

5.2 Case Study

Spatiotemporal walks in STGs are typically associated with various purposes, such as "wandering in attractions" and "picking-up people from the airport". In other words, each node in STGs has a semantic meaning that can be projected to a certain geographic area. We thus conduct a case study to evaluate the quality of the generated spatial information by projecting each of the generated GPS coordinates to a real-world map. Taking the Manhattan taxi trip STG as an example, we project a batch of generated coordinates for each method to the Manhattan map, and the results are shown in Fig. 7. Compared with the true coordinates (Fig. 7e), our proposed method STGEN (Fig. 7d) generates coordinates that all lie within the valid region by successfully characterizing the spatiotemporal properties. However, other deep methods like LSTM generate a large portion of coordinates scattered all over the New York City area since LSTM cannot effectively decode the spatial information from the multi-modal spatiotemporal distribution of STG. As can be seen from Fig. 7a to 7c, coordinates generated by comparison methods are disorganized (coordinates are not in

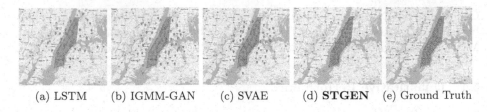

(a) LSTM   (b) IGMM-GAN   (c) SVAE   (d) **STGEN**   (e) Ground Truth

Fig. 7. The comparison of generated coordinates by each baseline.

Table 3. Average runtime comparison (in minutes).

	GraphRNN	NetGAN	TagGen	TG-GAN	STGD-VAE	STGEN
Syn_100	23.3561	0.8126	1.1682	0.9533	1.9781	0.9783
Syn_500	71.5622	1.9346	2.2653	1.4861	3.9923	1.6458
Syn_2000	337.7614	10.3302	8.5324	6.0182	10.8779	6.9633

the Manhattan area) and lack real semantic meaning (coordinates are projected on the sea) since they cannot consider any constraints during the model learning.

5.3 Model Scalability

The runtime of graph generative methods is often composed of model training and graph assembling. Therefore, we record the average training time per epoch plus the graph assembling time, and the results are shown in Table 3. All the runtimes are shown with respect to the growth of graph size for all synthetic STGs. As can be seen from the table, TagGen, TG-GAN, and STGEN have linear growth regarding the graph size because these dynamic graph generation methods decode dynamic graphs into dynamic walks instead of utilizing snapshots, which makes both training and graph assembling processes in these methods are not sensitive to the overall graph size. Although NetGAN also utilizes random walk to learn the static graph distribution, but its random walk sampling process limits its capability in generating large dynamic graphs with many snapshots. Finally, the runtime growth of GraphRNN is exponential because of its quadratic complexity in modeling the whole graph as a sequence.

6   Conclusion

In this paper, we propose a novel generative framework for continuous-time STG generation, which can effectively model the underlying dynamics of STGs while maintaining the spatiotemporal validity. Our framework captures both graph and spatiotemporal distribution through utilizing a novel heterogeneous recurrent-structured generator to learn the distribution of sampled spatiotemporal walks. A novel boosted discriminator is proposed to characterize correlations between all modalities in STGs. Extensive experiments are conducted on generating both synthetic and real world STGs. Experimental results and analysis demonstrate the advantages of STGEN over existing deep graph generative models in terms of generating the most similar and realistic STGs.

Acknowledgement. This work was supported by the National Science Foundation (NSF) Grant No. 1755850, No. 1841520, No. 2007716, No. 2007976, No. 1942594, No. 1907805, a Jeffress Memorial Trust Award, Amazon Research Award, NVIDIA GPU Grant, and Design Knowledge Company (subcontract number: 10827.002.120.04).

References

1. Ahuja, R.K., Magnanti, T.L., Orlin, J.B.: Network flows (1988)
2. Arjovsky, M., Chintala, S., Bottou, L.: Wasserstein generative adversarial networks. In: ICML (2017)
3. Banavar, J.R., Maritan, A., Rinaldo, A.: Size and form in efficient transportation networks. Nature (1999)
4. Barthélemy, M.: Spatial networks. Phys. Rep. 499(1–3) (2011)
5. Bhattacharya, U., Mittal, T., Chandra, R., Randhavane, T., Bera, A., Manocha, D.: Step: spatial temporal graph convolutional networks for emotion perception from gaits. In: AAAI (2020)
6. Bishop, C.M., Nasrabadi, N.M.: Pattern recognition and machine learning (2006)
7. Bojchevski, A., Shchur, O., Zügner, D., Günnemann, S.: Netgan: generating graphs via random walks. In: ICML (2018)
8. Cui, Z., et al.: DyGCN: dynamic graph embedding with graph convolutional network. arXiv (2021)
9. Du, Y., Guo, X., Cao, H., Ye, Y., Zhao, L.: Disentangled spatiotemporal graph generative models. In: AAAI (2022)
10. Goyal, P., Kamra, N., He, X., Liu, Y.: DynGEM: deep embedding method for dynamic graphs. arXiv preprint arXiv:1805.11273 (2018)
11. Guo, X., Du, Y., Zhao, L.: Deep generative models for spatial networks. In: KDD, pp. 505–515 (2021)
12. Guo, X., Zhao, L.: A systematic survey on deep generative models for graph generation. arXiv (2020)
13. Hochreiter, S., Schmidhuber, J.: Long short-term memory. Neural Comput. 9(8), 1735–1780 (1997)
14. Holme, P.: Modern temporal network theory: a colloquium. Eur. Phys. J. B 88, 234 (2015). https://doi.org/10.1140/epjb/e2015-60657-4
15. Huang, D., et al.: A variational autoencoder based generative model of urban human mobility. In: MIPR (2019)
16. Jain, A., Zamir, A.R., Savarese, S., Saxena, A.: Structural-RNN: deep learning on spatio-temporal graphs. In: CVPR (2016)
17. Jang, E., Gu, S., Poole, B.: Categorical reparameterization with gumbel-softmax. arXiv (2016)
18. Li, S., Xu, L.D., Zhao, S.: The internet of things: a survey. Inf. Syst. Front. 17(2), 243–259 (2014). https://doi.org/10.1007/s10796-014-9492-7
19. Ling, C., Yang, C., Zhao, L.: Deep generation of heterogeneous networks. In: ICDM, pp. 379–388 (2021)

20. Masuda, N., Holme, P.: Predicting and controlling infectious disease epidemics using temporal networks. F1000prime reports 5 (2013)
21. Melo-Vega, A., et al.: Protein folding problem in the case of peptides solved by hybrid simulated annealing algorithms. In: Fuzzy Logic Augmentation of Neural and Optimization Algorithms: Theoretical Aspects and Real Applications (2018)
22. Sabirin, H., Kim, M.: Moving object detection and tracking using a spatio-temporal graph in h. 264/avc bitstreams for video surveillance. TOM (2012)
23. Schröter, M.S., et al.: Spatiotemporal reconfiguration of large-scale brain functional networks during propofol-induced loss of consciousness. J. Neurosci. (2012)
24. Scott, J.: Social network analysis. Sociology 22 (1988)
25. Simonovsky, M., Komodakis, N.: Graphvae: towards generation of small graphs using variational autoencoders. In: ICANN (2018)
26. Smolyak, D., Gray, K., Badirli, S., Mohler, G.: Coupled IGMM-GANs with applications to anomaly detection in human mobility data. ACM TSAS (2020)
27. Solé, R.V., Rosas-Casals, M., Corominas-Murtra, B., Valverde, S.: Robustness of the European power grids under intentional attack. Phys. Rev. E (2008)
28. Wang, S., Guo, X., Zhao, L.: Deep generative model for periodic graphs. arXiv preprint arXiv:2201.11932 (2022)
29. Watts, D.J., Strogatz, S.H.: Collective dynamics of 'small-world'networks. Nature (1998)
30. Wiewel, S., Becher, M., Thuerey, N.: Latent space physics: towards learning the temporal evolution of fluid flow. In: Computer graphics forum (2019)
31. Wu, L., Cui, P., Pei, J., Zhao, L., Song, L.: Graph neural networks. In: Graph Neural Networks: Foundations, Frontiers, and Applications (2022)
32. You, J., Ying, R., Ren, X., Hamilton, W., Leskovec, J.: GraphRNN: generating realistic graphs with deep auto-regressive models. In: ICML, pp. 5708–5717 (2018)
33. Yu, B., Yin, H., Zhu, Z.: Spatio-temporal graph convolutional networks: a deep learning framework for traffic forecasting. In: IJCAI (2018)
34. Zhang, L., Zhao, L., Qin, S., Pfoser, D., Ling, C.: TG-GAN: continuous-time temporal graph generation with deep generative models. In: WebConf (2020)
35. Zhou, D., Zheng, L., Han, J., He, J.: A data-driven graph generative model for temporal interaction networks. In: KDD (2020)

Direct Evolutionary Optimization of Variational Autoencoders with Binary Latents

Jakob Drefs[1]([✉]), Enrico Guiraud[2]([✉]), Filippos Panagiotou[1], and Jörg Lücke[1][ID]

[1] Machine Learning Lab, University of Oldenburg, 26129 Oldenburg, Germany
{jakob.drefs,filippos.panagiotou,joerg.luecke}@uol.de
[2] CERN, 1211 Geneva, Switzerland
enrico.guiraud@cern.ch

Abstract. Many types of data are generated at least partly by discrete causes. Deep generative models such as variational autoencoders (VAEs) with binary latents consequently became of interest. Because of discrete latents, standard VAE training is not possible, and the goal of previous approaches has therefore been to amend (i.e, typically anneal) discrete priors to allow for a training analogously to conventional VAEs. Here, we divert more strongly from conventional VAE optimization: We ask if the discrete nature of the latents can be fully maintained by applying a direct, discrete optimization for the encoding model. In doing so, we sidestep standard VAE mechanisms such as sampling approximation, reparameterization and amortization. Direct optimization of VAEs is enabled by a combination of evolutionary algorithms and truncated posteriors as variational distributions. Such a combination has recently been suggested, and we here for the first time investigate how it can be applied to a deep model. Concretely, we (A) tie the variational method into gradient ascent for network weights, and (B) show how the decoder is used for the optimization of variational parameters. Using image data, we observed the approach to result in much sparser codes compared to conventionally trained binary VAEs. Considering the for sparse codes prototypical application to image patches, we observed very competitive performance in tasks such as 'zero-shot' denoising and inpainting. The dense codes emerging from conventional VAE optimization, on the other hand, seem preferable on other data, e.g., collections of images of whole single objects (CIFAR etc.), but less preferable for image patches. More generally, the realization of a very different type of optimization for binary VAEs allows for investigating advantages and disadvantages of the training method itself. And we here observed a strong influence of the method on the learned encoding with significant impact on VAE performance for different tasks.

Keywords: Variational autoencoder · Evolutionary optimization · Sparse encoding · Variational optimization · Binary latents

---

J. Drefs and E. Guiraud—Joint first authorship.

Supplementary Information The online version contains supplementary material available at https://doi.org/10.1007/978-3-031-26409-2_22.

M.-R. Amini et al. (Eds.): ECML PKDD 2022, LNAI 13715, pp. 357–372, 2023.
https://doi.org/10.1007/978-3-031-26409-2_22

1 Introduction and Related Work

Objects or edges in images are either present or absent, which suggests the use of discrete latents for their representation. There are also typically only few objects per image (of all possible objects) or only few edges in any given image patch (of all possible edges), which suggests a sparse code (e.g., [20,43,56,61]). In order to model such and similar data, we study a novel, direct optimization approach for variational autoencoders (VAEs), which can learn discrete and potentially sparse encodings. VAEs [32,51] in their many different variations, have successfully been applied to a large number of tasks including semi-supervised learning (e.g., [40]), anomaly detection (e.g., [33]) or sentence and music interpolation [5,52] to name just a few. The success of VAEs, in these tasks, rests on a series of methods that enable the derivation of scalable training algorithms to optimize VAE parameters. These methods were originally developed for Gaussian priors [32,51]. To account for VAEs with discrete latents, novel methodology had to be introduced (we elaborate below and later in Sec. S1).

The training objective of VAEs is derived from a likelihood objective, i.e., we seek model parameters Θ of a VAE that maximize the data log-likelihood, $L(\Theta) = \sum_n \log\left(p_\Theta(\vec{x}^{(n)})\right)$, where we denote by $\vec{x}^{(1:N)}$ a set of N observed data points, and where $p_\Theta(\vec{x})$ denotes the modeled data distribution. Like conventional autoencoders (e.g., [1]), VAEs use a deep neural network (DNN) to generate (or decode) observables $\vec{x} \in \mathbb{R}^D$, from a latent code $\vec{z}$. Unlike conventional autoencoders, however, the generation of data $\vec{x}$ is not deterministic but it takes the form of a probabilistic generative model. For VAEs with binary latents, we here consider a generative model with Bernoulli prior:

$$p_\Theta(\vec{z}) = \prod_h \left(\pi_h^{z_h}(1-\pi_h)^{(1-z_h)}\right), \quad p_\Theta(\vec{x}\,|\,\vec{z}) = \mathcal{N}\left(\vec{x}; \vec{\mu}(\vec{z}; W), \sigma^2 \mathbb{I}\right), \qquad (1)$$

with $\vec{z} \in \{0,1\}^H$ being a binary code, $\vec{\pi} \in [0,1]^H$ being parameters of the prior on $\vec{z}$, and the non-linear function $\vec{\mu}(\vec{z}; W)$ being a DNN (that sets the mean of a Gaussian distribution). $p_\Theta(\vec{x}\,|\,\vec{z})$ is commonly referred to as *decoder*. The set of model parameters is $\Theta = \{\vec{\pi}, W, \sigma^2\}$, where W incorporates DNN weights and biases. Here, we assume homoscedasticity of the Gaussian distribution, but note that there is no obstacle to generalizing the model by inserting a DNN non-linearity that outputs a covariance matrix. Similarly, the algorithm could easily be generalized to different noise distributions should the task at hand call for it. Here, however, we will focus on the elementary VAEs given by Eq. (1).

For conventional and discrete VAEs, essentially all optimization approaches seek to approximately maximize the log-likelihood using the following series of methods (we elaborate in Sec. S1):

(A) Instead of the log-likelihood, a variational lower-bound (a.k.a. ELBO) is optimized.
(B) VAE posteriors are approximated by an *encoding model*, i.e., by a specific distribution (usually Gaussian) parameterized by one or more DNNs.
(C) The variational parameters of the encoder are optimized using gradient ascent on the lower bound, where the gradient is evaluated based on sam-

pling and the reparameterization trick (which allows for sufficiently low-variance and yet efficiently computable estimates).

(D) Using samples from the encoder, the parameters of the decoder are optimized using gradient ascent on the variational lower bound.

Optimization procedures for VAEs with discrete latents follow the same steps (Points A to D). However, discrete or binary latents pose substantial further obstacles for learning, mainly due to the fact that backpropagation through discrete variables is generally not possible or biased [2,53]. Widely used stochastic gradient estimators for discrete random variables typically either exploit the REINFORCE [65] estimator in combination with variance control techniques [11,13,34,38] or reparameterization of continuous relaxations of discrete distributions [29,41]; reparameterization is also combined with REINFORCE [22] or generalized to non-reparameterizable distributions [8]. Also a recent approach by Berliner et al. [3] is related to REINFORCE but uses natural evolution strategies (not to be confused with evolutionary optimization we apply here) to derive low-variance estimates for gradients (also see Related Work and Sec. S1). While accomplishing, in different senses, the goal of maintaining standard VAE training as developed for continuous latents (i.e., learning procedures and/or learning objectives that allow for gradient-based optimization of the encoder and decoder DNNs), gradient estimation methods usually apply significant amounts of methodology *additional* to the learning methods conventionally applied for VAE optimization (see Fig. S2). These additional methods, their accompanying design decisions and used hyper-parameters increase the complexity of the system. Furthermore, the additional methods usually impact the learned representations. For instance, softening of discrete distributions, e.g., by using 'Gumbel-softmax' [29] or 'tanh' approximations [18] seems to favor dense codes. While dense codes (as also used by conventional VAEs and generative adversarial networks [21]) can result in competitive performance for a subset of the above discussed tasks, other recent contributions point out advantages of sparse codes, e.g., in terms of disentanglement [63] or robustness [45,60].

In order to avoid adding methods for discrete latents to those already in place for standard VAEs, it may be reasonable to investigate more direct optimization procedures that do not require, e.g., a softening of discrete distributions or other mechanisms. Such a direct approach is challenging, however, because once DNNs are used to define the encoding model (as commonly done), we require methodologies for discrete latents to estimate gradients for the encoder (as done via sampling and reparameterization; see Points C and D). A direct optimization procedure, as we investigate here, consequently has to change VAE training substantially. For the data model of Eq. (1), we will maintain the variational setting (Point A) and a decoding model with DNNs as non-linearity. However, we will not use an encoding model parameterized by DNNs (Point B). Instead, the variational bound will be increased w.r.t. an implicitly defined encoding model which allows for an efficient discrete optimization. The procedure does not require gradients to be computed for the encoder such that discrete latents are addressed without the use of reparameterization trick and sampling approximations.

Related Work. In order to maintain the general VAE framework for encoder optimization in the case of discrete latents, different groups have suggested different possible solutions (for discussion of numerical evaluations of related approaches, see Sec. S1.3): Rolfe [53], for instance, extends VAEs with discrete latents by auxiliary continuous latents such that gradients can still be computed. Work on the concrete distribution [41] or Gumbel-softmax distribution [29] proposes newly defined continuous distributions that contain discrete distributions as limit cases. Lorberbom et al. [39] merge the Gumbel-Max reparameterization with the use of direct loss minimization for gradient estimation, enabling efficient training on structured latent spaces (also compare [48,49] for further improved Gumbel-softmax versions). Furthermore, work, e.g., by van den Oord et al. [44] combines VAEs with a vector quantization (VQ) stage in the latent layer. Latents become discrete through quantization but gradients for learning are adapted from latent values before they are processed by the VQ stage. Similarly, Tomczak & Welling [62] use, what they call, (learnable) pseudo-inputs which determine a mixture distribution as prior, and the ELBO then contains an additional regularization for consistency between prior and average posterior. Tonolini et al. [63] extend this work and introduce an additional DNN classifier which selects pseudo-inputs and whose weights are learned instead of the pseudo-inputs themselves. Tonolini et al. also argue for the benefits not only of discrete latents but of a sparse encoding in the latent layer in general. Fajtl et al. [18] base their approach on a deterministic autoencoder and use a tanh-approximation of binary latents and projections to spheres in order to treat binary values. Targeting not only the optimization of discrete latent VAEs but also more general approaches such as probabilistic programming or general stochastic automatic differentiation, Bingham et al. [4] and van Krieken et al. [35] apply gradient estimators for discrete random variables which optimize surrogate losses [54] derived based on the score function [19] or other methods [35].

2   Direct Variational Optimization

Let us consider the variational lower bound of the likelihood. If we denote by $q_\Phi^{(n)}(\vec{z})$ the variational distributions with parameters $\Phi = (\Phi^{(1)}, \ldots, \Phi^{(N)})$, then the lower bound is given by:

$$\mathcal{F}(\Phi, \Theta) = \sum_n \mathbb{E}_{q_\Phi^{(n)}} \big[\log \big(p_\Theta(\vec{x}^{(n)} \mid \vec{z}) \, p_\Theta(\vec{z}) \big) \big] - \sum_n \mathbb{E}_{q_\Phi^{(n)}} \big[\log \big(q_\Phi^{(n)}(\vec{z}) \big) \big], \quad (2)$$

where we sum over all data points $\vec{x}^{(1:N)}$, and where $\mathbb{E}_{q_\Phi^{(n)}} \big[h(\vec{z}) \big]$ denotes the expectation value of a function $h(\vec{z})$ w.r.t. $q_\Phi^{(n)}(\vec{z})$. The general challenge for the maximization of $\mathcal{F}(\Phi, \Theta)$ is the optimization of the encoding model $q_\Phi^{(n)}$. VAEs with discrete latents, as an additional challenge, have to address the question how gradients w.r.t. discrete latents can be computed. Seeking to avoid the problem of gradients w.r.t. discrete variables, we do *not* use a DNN for the encoding model. Consequently, we need to define an *alternative* encoding model

$q_{\Phi}^{(n)}$, which has to remain sufficiently efficient. Considering prior work on generative models with discrete latents, variational distributions based on truncated posteriors offer themselves as such an alternative. Truncated posteriors have previously been considered to be functionally competitive (e.g., [27,56,58]). Most relevant for our purposes are very efficient and fully variational approaches that allow mixture models [17,26] and shallow generative approaches [14] to be very efficiently scaled to large model sizes. In all these previous applications, optimization of truncated variational distributions used standard expectation maximization based on closed-form or pseudo-closed form M-steps available for the shallow decoder models considered. In the context of VAEs with discrete latents, the important question arising is if or how efficient optimization with truncated variational distributions can be performed for deep generative models.

Optimization of the Encoding Model. Encoder optimization is usually based on a reformulation of the variational bound of Eq. (2) given by:

$$\mathcal{F}(\Phi, \Theta) = \sum_n \mathbb{E}_{q_{\Phi}^{(n)}} \left[\log \left(p_{\Theta}(\vec{x}^{(n)} \mid \vec{z}) \right) \right] - \sum_n D_{\mathrm{KL}} \left[q_{\Phi}^{(n)}(\vec{z}); p_{\Theta}(\vec{z}) \right]. \quad (3)$$

For discrete latent VAEs, the variational distributions in Eq. (3) are commonly replaced by an amortized encoding model $q_{\Phi}(\vec{z})$ with a DNN-based parameterization. When expectations w.r.t. $q_{\Phi}(\vec{z})$ are approximated (as usual) via sampling, the encoder optimization requires gradient estimation methods for discrete random variables (cf. Related Work and Sec. S1). At this point truncated posteriors represent alternative variational distributions which avoid gradients w.r.t. discrete latents. Given a data point $\vec{x}^{(n)}$, a truncated posterior is the posterior itself truncated to a subset $\Phi^{(n)}$ of the latent space, i.e., for $\vec{z} \in \Phi^{(n)}$ applies:

$$q_{\Phi}^{(n)}(\vec{z}) := \frac{p_{\Theta}(\vec{z} \mid \vec{x}^{(n)})}{\sum_{\vec{z}' \in \Phi^{(n)}} p_{\Theta}(\vec{z}' \mid \vec{x}^{(n)})} = \frac{p_{\Theta}(\vec{x}^{(n)} \mid \vec{z})\, p_{\Theta}(\vec{z})}{\sum_{\vec{z}' \in \Phi^{(n)}} p_{\Theta}(\vec{x}^{(n)} \mid \vec{z}')\, p_{\Theta}(\vec{z}')} \quad (4)$$

while $q_{\Phi}^{(n)}(\vec{z}) = 0$ for $\vec{z} \notin \Phi^{(n)}$. The subsets $\Phi = \{\Phi^{(n)}\}_{n=1}^{N}$ are the variational parameters. Centrally for this work, truncated posteriors allow for a specific alternative reformulation of the bound. The reformulation recombines the entropy term of the original form (Eq. (2)) with the first expectation value into a single term, and is given by (see [14,17,26] for details):

$$\mathcal{F}(\Phi, \Theta) = \sum_n \log \left(\sum_{\vec{z} \in \Phi^{(n)}} p_{\Theta}(\vec{x}^{(n)} \mid \vec{z})\, p_{\Theta}(\vec{z}) \right). \quad (5)$$

Thanks to the simplified form of the bound, the variational parameters $\Phi^{(n)}$ of the encoding model can now be sought using direct discrete optimization procedures. More concretely, because of the specific form of Eq. (5), pairwise comparisons of joint probabilities are sufficient to maximize the lower bound: if we update the set $\Phi^{(n)}$ for a given $\vec{x}^{(n)}$ by replacing a state $\vec{z}^{\mathrm{old}} \in \Phi^{(n)}$ with a state $\vec{z}^{\mathrm{new}} \notin \Phi^{(n)}$, then $\mathcal{F}(\Phi, \Theta)$ increases if and only if:

$$\log\left(p_\Theta(\vec{x}^{(n)}, \vec{z}^{\text{new}})\right) > \log\left(p_\Theta(\vec{x}^{(n)}, \vec{z}^{\text{old}})\right). \tag{6}$$

To obtain intuition for the pairwise comparison, consider the form of $\log(p_\Theta(\vec{x}, \vec{z}))$ when inserting the binary VAE defined by Eq. (1). Eliding terms that do not depend on $\vec{z}$ we obtain:

$$\widetilde{\log p_\Theta}(\vec{x}, \vec{z}) = -\|\vec{x} - \vec{\mu}(\vec{z}, W)\|^2 - 2\sigma^2 \sum_h \tilde{\pi}_h z_h, \tag{7}$$

where $\tilde{\pi}_h = \log\left((1 - \pi_h)/\pi_h\right)$. The expression assumes an even more familiar form if we restrict ourselves for a moment to sparse priors with $\pi_h = \pi < \frac{1}{2}$, i.e., $\tilde{\pi}_h = \tilde{\pi} > 0$. The criterion defined by Eq. (6) then becomes:

$$\|\vec{x}^{(n)} - \vec{\mu}(\vec{z}^{\text{new}}, W)\|^2 + 2\sigma^2\tilde{\pi}\,|\vec{z}^{\text{new}}| < \|\vec{x}^{(n)} - \vec{\mu}(\vec{z}^{\text{old}}, W)\|^2 + 2\sigma^2\tilde{\pi}\,|\vec{z}^{\text{old}}|, \tag{8}$$

where $|\vec{z}| = \sum_{h=1}^H z_h$ and $2\sigma^2\tilde{\pi} > 0$. Such functions are routinely encountered in sparse coding or compressive sensing [16]: for each set $\Phi^{(n)}$, we seek those states $\vec{z}$ that are reconstructing $\vec{x}^{(n)}$ well while being sparse ($\vec{z}$ with few non-zero bits). For VAEs, $\vec{\mu}(\vec{z}, W)$ is a DNN and as such much more flexible in matching the distribution of observables $\vec{x}$ than can be expected from linear mappings. Furthermore, criteria like Eq. (8) usually emerge for maximum a-posteriori (MAP) training in sparse coding [43]. In contrast to MAP, however, here we seek a *population* of states $\vec{z}$ in $\Phi^{(n)}$ for each data point. It is a consequence of the reformulated lower bound defined by Eq. (5) that it remains optimal to evaluate joint probabilities (as for MAP) although the constructed population of states $\Phi^{(n)}$ can capture (unlike MAP training) rich posterior structures.

Evolutionary Search. But how can new states $\vec{z}^{\text{new}}$ that optimize $\Phi^{(n)}$ be found efficiently in high-dimensional latent spaces? While blind random search for states $\vec{z}$ can in principle be used, it is not efficient; and adaptive search space approaches [17,26] are only defined for mixture models. However, a recently suggested combination of truncated variational optimization with evolutionary optimization (EVO; [14]) is more generally defined for models with discrete latents, and does only require the efficient computation of joint probabilities $p_\Theta(\vec{x}, \vec{z})$. It can consequently be adapted to the VAEs considered here.

EVO optimization interprets the sets $\Phi^{(n)}$ of Eq. 4 as populations of binary genomes $\vec{z}$, and we can here adapt it by using Eq. (7) in order to assign to each $\vec{z} \in \Phi^{(n)}$ a fitness for evolutionary optimization. For the concrete updates, we use for each EVO iteration $\Phi^{(n)}$ as initial parent pool. We then apply the following genetic operators in sequence to suggest candidate states $\vec{z}^{\text{new}}$ to update the $\Phi^{(n)}$ based on Eq. (6) (see Fig. S3 for an illustration and Sec. S1.2 and [14] for further details): Firstly, *parent selection* stochastically picks states from the parent pool. Subsequently, each of these states undergoes *mutation* which flips one or more entries of the bit vectors. Offspring diversity can be further increased by crossover operations. Using the *children* generated this way as the new parent pool, the procedure is repeated giving birth to multiple *generations* of candidate states. Finally, we update $\Phi^{(n)}$ by substituting individuals with low fitness with candidates with higher fitness according to Eq. (6). The whole procedure can be seen as an evolutionary algorithm (EA) with perfect memory or very strong elitism (individuals with higher fitness never drop out of the gene pool). Note that the improvement of the variational lower bound depends on generating as many as possible *different* children with high fitness over the course of training.

We point out that the EAs optimize each $\Phi^{(n)}$ independently, which allows for distributed execution s.t. the technique can be efficiently applied to large datasets in conjunction with stochastic or batch gradient descent on the model parameters Θ. The approach is, at the same time, memory intensive, i.e., all sets $\Phi^{(n)}$ need to be kept in memory (details in Sec. S1.1). Furthermore, we point out the we here optimize variational parameters $\Phi^{(n)}$ of the encoding model which is fundamentally different from the approach of Hajewski & Oliveira [24] (who use EAs to optimize DNN architectures of otherwise conventionally optimized VAEs with continuous latents).

Optimization of the Decoding Model. Using the previously described encoding model $q_{\Phi}^{(n)}(\vec{z})$, we can compute the gradient of Eq. (2) w.r.t. the decoder weights W which results in (see Sec. S1 for details):

$$\vec{\nabla}_W \mathcal{F}(\Phi, \Theta) = -\frac{1}{2\sigma^2} \sum_n \sum_{\vec{z} \in \Phi^{(n)}} q_{\Phi}^{(n)}(\vec{z}) \; \vec{\nabla}_W \|\vec{x}^{(n)} - \vec{\mu}(\vec{z}, W)\|^2. \tag{9}$$

The right-hand-side has salient similarities to standard gradient ascent for VAE decoders. Especially the familiar gradient of the mean squared error (MSE) shows that, e.g., standard automatic differentiation tools can be applied. However, the decisive difference is represented by the weighting factors $q_{\Phi}^{(n)}(\vec{z})$. Considering Eq. (4), we require all $\vec{z} \in \Phi^{(n)}$ to be passed through the decoder DNN in order to compute the $q_{\Phi}^{(n)}(\vec{z})$. As all states of $\Phi^{(n)}$ anyway have to be passed through the decoder for the MSE term of Eq. (9), the overall computational complexity is not higher than an estimation of the gradient with samples instead of states in $\Phi^{(n)}$ (but we use many states per $\Phi^{(n)}$, compare Tab. S1).

To complete the decoder optimization, update equations for variance σ^2 and prior parameters $\vec{\pi}$ can be computed in closed-form (compare, e.g., [57]) and are given by:

$$\sigma^2 = \frac{1}{DN} \sum_n \sum_{\vec{z} \in \Phi^{(n)}} q_{\Phi}^{(n)}(\vec{z}) \, \|\vec{x}^{(n)} - \vec{\mu}(\vec{z}, W)\|^2,$$

$$\vec{\pi} = \frac{1}{N} \sum_n \sum_{\vec{z} \in \Phi^{(n)}} q_{\Phi}^{(n)}(\vec{z}) \, \vec{z}. \tag{10}$$

The full training procedure for binary VAEs is summarized in Algorithm 1. We refer to the binary VAE trained with this procedure as *Truncated Variational Autoencoder* (TVAE) because of the applied truncated posteriors[1].

---

[1] Source code available at https://github.com/tvlearn.

Algorithm 1. Training Truncated Variational Autoencoders (TVAE)

Initialize model parameters $\Theta = (\vec{\pi}, W, \sigma^2)$
Initialize each $\Phi^{(n)}$ with S distinct latent states
repeat
    for all batches in dataset **do**
        for sample n in batch **do**
            $\Phi^{\text{new}} = \Phi^{(n)}$
            for all generations **do**
                $\Phi^{\text{new}} = \text{mutation}\,(\text{selection}\,(\Phi^{\text{new}}))$
                $\Phi^{(n)} = \Phi^{(n)} \cup \Phi^{\text{new}}$
            end for
            Define new $\Phi^{(n)}$ by selecting the S fittest elements in $\Phi^{(n)}$ using Eq. (6)
        end for
        Use Adam to update W using Eq. (9)
    end for
    Use Eq. (10) to update $\vec{\pi}$, σ^2
until parameters Θ have sufficiently converged

3   Numerical Experiments

TVAE can flexibly learn prior parameters $\vec{\pi}$, and if low values for the π_h are obtained (which will be the case), the code is sparse. The prototypical application domain to study sparse codes is image patch data [20,43]. We consequently use such data to investigate sparsity, scalability and efficiency on benchmarks. For all numerical experiments, we employ fully connected DNNs $\vec{\mu}(\vec{z}; W)$ for the decoder (compare Fig. S4); the exact network architectures and activations used are listed in Tab. S1. The DNN parameters are optimized based on Eq. (9) using mini-batches and the Adam optimizer (details in Sec. S2.1).

Verification and Scalability. After first verifying that the procedure can recover generating parameters using ground-truth data (see Sec. S2.2), we trained TVAE on $N = 100,000$ whitened image patches of $D = 16 \times 16$ pixels [25] using two different decoder architectures, namely a shallow, linear decoder with $H = 300$ binary latents, and second, a deep non-linear decoder with a 300-300-256 architecture (i.e., $H = 300$ binary latents and two hidden layers with 300 and 256 units, respectively; details in Sec. S2.3). For both linear and non-linear TVAE, we observed a sparse encoding with on average $\frac{\sum_h \pi_h}{H} = \frac{20.3}{300}$ and $\frac{\sum_h \pi_h}{H} = \frac{28.5}{300}$ active latents across data points, respectively. We observed sparse codes also when we varied the parameter initialization and further modified the decoder DNN architecture. As long as decoder DNNs were of small to intermediate size, we observed efficient scalability to large latent spaces (we went up to $H = 1,000$). Compared to linear decoders, the main additional computational cost is given by passing the latent states in the $\Phi^{(n)}$ sets through the decoder DNN instead of just through a linear mapping. The sets of states (i.e., the bitvectors in $\Phi^{(n)}$) could be kept small, at size $S = |\Phi^{(n)}| = 64$, such that $N \times (|\Phi^{(n)}| + |\Phi^{(n)}_{\text{new}}|)$ states had to be evaluated per epoch. This compares to $N \times M$ states that would be used for standard VAE training (given M samples are drawn per data point). In contrast to standard VAE training, the sets $\Phi^{(n)}$ have to be remembered across iterations. For very large datasets, the additional $\mathcal{O}(N \times |\Phi^{(n)}| \times H)$ memory demand can be distributed over compute nodes, however.

Denoising - Controlled Conditions. Due to its non-amortized encoding model, the computational load of TVAE increases more strongly with data points compared to amortized training. Consequently, tasks such as disentanglement of features using high-dimensional input data, large DNNs, and small latent spaces are not a regime where the approach can be applied efficiently. With this in mind, we focused on tasks with relatively few data for which an as effective as possible optimization is required, and for which advantages of a direct optimization can be expected. As one such task, we here considered 'zero-shot' image denoising. To apply TVAE in a 'zero-shot' setting (in which no additional information besides the noisy image is available, e.g., [28,59]), we trained the model on overlapping patches extracted from a given noisy image and subsequently applied the learned encoding to estimate non-noisy image pixels (details in Sec. S2.4). In general, denoising represents a canonical benchmark for evaluating image patch models, and approaches exploiting sparse encodings have shown to be particularly well suited (compare, e.g., [42,56,68]). The 'zero-shot' setting has recently become popular also because the application of conventional DNN-based approaches has shown to be challenging (see discussion in Sec. S2.4).

One denoising benchmark, which allows for an extensive comparison to other methods is the House image. Standard benchmark settings for this image make use of additive Gaussian white noise with standard deviations $\sigma \in \{15, 25, 50\}$ (Fig. 1 A). First, consider the comparison in Fig. 1 C where all models used the same patch size of $D = 8 \times 8$ pixels and $H = 64$ latent variables (details in Sec. S2.4). Figure 1 C lists the different approaches in terms of the standard measure of peak signal-to-noise ratio (PSNR). Values for MTMKL [61] and GSC [56] were taken from the respective original publications (which both established new state-of-the-art results when first published); for EBSC [14], we produced PSNRs ourselves by running publicly available source code (cf. Sec. S2.4). As can be observed, TVAE significantly improves performance for high noise levels; the approach is able to learn the best data representation for denoising and establishes new state-of-the-art results in this controlled setting (i.e., fixed D and H). The decoder DNN of TVAE provides the decisive performance advantage: TVAE significantly improves performance compared to EBSC (which can be considered as an approach with a shallow, linear decoding model), confirming that the high lower bounds of TVAE on natural images (compare Fig. S9) translate into improved performance on a concrete benchmark. For $\sigma = 25$ and $\sigma = 50$, TVAE also significantly improves on MTMKL, and GSC, which are both based on a spike-and-slab sparse coding (SSSC) model (also compare [20]).

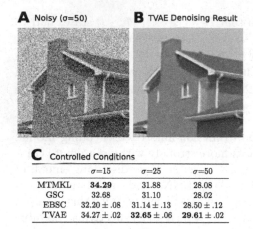

A Noisy (σ=50)   **B** TVAE Denoising Result

D Different Optimized Hyperparameters

	$\sigma=15$	$\sigma=25$	$\sigma=50$
N2V*	31.86 ± .37	29.34 ± .37	24.24 ± .57
S2S	34.86 ± .13	32.87 ± .08	29.95 ± .26
GSVAE-B	29.39 ± .02	28.32 ± .28	25.73 ± .14
GSVAE-C	30.94 ± .17	29.83 ± .04	26.75 ± .02
VLAE	34.47 ± .05	32.44 ± .05	29.16 ± .07
MTMKL	34.29	31.88	28.08
GSC	33.78	32.01	28.35
S5C	33.50	32.08	28.35
EBSC	33.66 ± .04	32.40 ± .07	28.98 ± .06
ES3C	**34.90** ± .04	**33.15** ± .06	29.83 ± .07
TVAE	34.27 ± .02	32.65 ± .06	**29.98** ± .05
N2V†	33.91	32.10	28.94
KSVD	34.32	32.15	27.95
WNNM	**35.13**	**33.22**	**30.33**
BM3D	34.94	32.86	29.37
EPLL	34.17	32.17	29.12
BDGAN	34.57	33.28	30.61
DPDNN	**35.40**	**33.54**	**31.04**

C Controlled Conditions

	$\sigma=15$	$\sigma=25$	$\sigma=50$
MTMKL	**34.29**	31.88	28.08
GSC	32.68	31.10	28.02
EBSC	32.20 ± .08	31.14 ± .13	28.50 ± .12
TVAE	34.27 ± .02	**32.65** ± .06	**29.61** ± .02

Fig. 1. Denoising results for House. **C** compares PSNRs (in dB) obtained with different 'zero-shot' models using a fixed patch size and number of latents (means and standard deviations were computed over three runs with independent noise realizations, see text for details). **D** lists PSNRs for different algorithms with different optimized hyperparameters. The top category only requires the noisy image ('zero-shot' setting). The middle uses additional information such as noise level (KSVD, WNNM, BM3D) or additional noisy images with matched noise level (N2V†). The bottom three algorithms use large clean datasets. The highest PSNR per category is marked bold, and the overall highest PSNR is bold and underlined. **B** depicts the denoised image obtained with TVAE for $\sigma = 50$ in the best run (PSNR = 30.03 dB).

Despite the less flexible Bernoulli prior, the decoder DNN of TVAE provides the highest PSNR values for high noise levels.

Denoising - Uncontrolled Conditions. To extend the comparison, we next evaluated denoising performance without controlling for equal conditions, i.e., we also included approaches in our comparison that use large image datasets and/or different patch sizes for training (including multi-scale and whole image processing). Note that different approaches may employ very different sets of hyperparameters that can be optimized for denoising performance (e.g., patch and dictionary sizes for sparse coding approaches, or network and training scheme hyper-parameters for DNN approaches). By allowing for comparison in this less controlled setting, we can compare to a number of recent approaches including large DNNs trained on clean data and training schemes specifically targeted to noisy training data. See Fig. 1 D for an extensive PSNR overview with results for other algorithms cited from their corresponding original publications if not stated otherwise. PSNRs for S5C originate from [55], GSVAE-B, EBSC and ES3C from [14], and WNNM and EPLL from [67]. For Noise2Void (N2V; [36]), Self2Self (S2S; [50]), GSVAE-C [29], and VLAE [47], we produced results ourselves by applying publicly available source code (details in Sec. S2.4). Note that the best performing approaches in Fig. 1 D were trained on noiseless data: EPLL

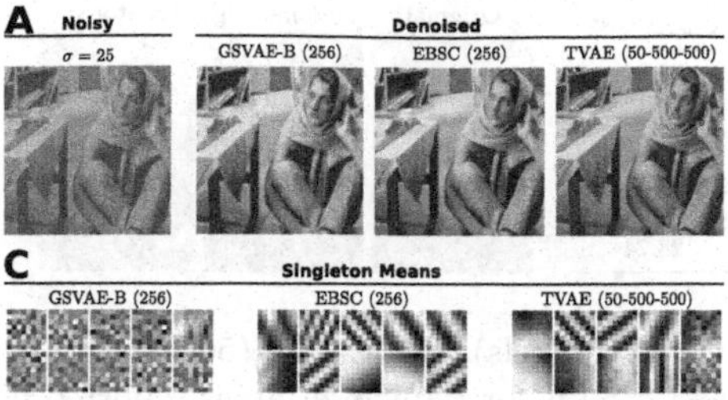

A

Noisy | Denoised

$\sigma = 25$ | GSVAE-B (256) | EBSC (256) | TVAE (50-500-500)

C

Singleton Means

GSVAE-B (256) | EBSC (256) | TVAE (50-500-500)

B

Model	Sparsity	Lower Bound	PSNR
GSVAE-B (256)	$\frac{128}{256}$	39.02	25.06
GSVAE-B (256-256-512)	$\frac{128}{256}$	38.68	25.06
GSVAE-C ((32×8)-256-512)	n/a	45.66	26.18
EBSC (256)	$\frac{1.5}{256}$	43.85	28.37
TVAE (256-256-512)	$\frac{2.3}{256}$	46.86	29.96
TVAE (50-500-500)	$\frac{2.6}{50}$	**47.98**	**30.03**
VLAE (50-500-500)	n/a	45.93*	28.83
ES3C (256)	$\frac{2.0}{256}$	**46.09**	**29.69**

Fig. 2. Data encodings and denoising results for Barbara obtained with generative model approaches and different decoding models. In **B**, approaches with binary (top) and continuous (bottom) latents are separated. EBSC and ES3C are considered as using a shallow, linear decoder. Listed are best performances of several runs of each algorithm. *VLAE uses importance sampling-based log-likelihood estimation. **C** compares decoder outputs for singleton (i.e., one-hot) input vectors. See Sec. S2.4 for details.

[71], BDGAN [70] and DPDNN [12] all make use of clean training data (typically hundreds of thousands of data points or more). EPLL, KSVD [15], WNNM [23] and BM3D [10] leverage a-priori noise level information (these algorithms use the ground-truth noise level of the test image as input parameter). As noisy data is very frequently occurring, lifting the requirement of clean data has been of considerable recent interest with approaches such as Noise2Noise (N2N; [37]), N2V, and S2S having received considerable attention.

Considering Fig. 1 D, first note that TVAE consistently improves PSNRs of N2V, also when comparing to a variant trained on external data with matched noise level (N2V[†] in Fig. 1 D). At high noise level ($\sigma = 50$), PSNRs of TVAE represent state-of-the-art performance in the 'zero-shot' category (Fig. 1 D, top); compared to methods which exploit additional a-priori information (Fig. 1 D, middle and bottom), the denoising performance of TVAE (at high noise level) is improved only by WNNM, BDGAN and DPDNN. At lower noise levels, TVAE still performs competitively in the 'zero-shot' setting, yet highest PSNRs are obtained by other methods (S2S and ES3C). Figure 1 D reveals that TVAE can improve on two competing VAE approaches, namely GSVAE (which uses Gumbel-softmax-based optimization for discrete latents) and VLAE (which uses continuous latents and Gaussian posterior approximations). For more systematic comparison, we applied the VAE approaches using identical decoder architectures and identical patch sizes (details in Sec. S2.4). As striking difference between the approaches, we observed GSVAE to learn a significantly denser encoding compared to TVAE. Furthermore, we observed that the sparse encodings of TVAE resulted in strong performance not only in terms of denoising PSNR but also in terms of lower bounds (see Fig. 2).

Inpainting. Finally, we applied TVAE to 'zero-shot' inpainting tasks. For TVAE, the treatment of missing data is directly available given the probabilistic formulation of the model. Concretely, when evaluating log-joint probabilities of a datapoint, missing values are treated as unknown observables (details in Sec. S2.5). In contrast, amortized approaches need to specify how the deterministic encoder

	House 50%		Castle 50%	Castle 80%	Original	50% missing	Restored
Papyan et al.	34.58	MTMKL	n/a	28.94			
BPFA	38.02	BPFA	36.45	29.12			
DIP	39.16	ES3C	**38.23**	**29.66**			
ES3C	**39.59**	TVAE	37.33	28.93			
TVAE	38.56	PLE	**38.34**	**30.07**			
		IRCNN	n/a	28.74			

Fig. 3. Inpainting results for House (50% missing pixels) and Castle (50% and 80% missing; top group lists 'zero-shot' approaches). PSNR for Papyan et al. as reported in [64]. Depicted is TVAE's restoration for 50% missing pixels (sparsity $\frac{\sum_h \pi_h}{H} = \frac{10.56}{512}$).

DNNs should treat missing values. Figure 3 evaluates performance of TVAE on two standard inpainting benchmarks with randomly missing pixels. Methods compared to include MTMKL, BPFA [69], ES3C, the method of Papyan et al. [46], DIP [64], PLE [66], and IRCNN [6]. PLE uses the noise level as a-priori information, and IRCNN is trained on external clean images. On House, TVAE improves the performance of Papyan et al. and BPFA; highest PSNRs for this benchmark are obtained by DIP (which, in contrast to TVAE, is not permutation invariant and uses large U-Nets) and ES3C (which is based on a SSSC model and EVO-based training). On Castle, PSNRs of TVAE are higher in comparison to SSSC-based BPFA (for 50% missing pixels) and IRCNN.

4   Discussion

We investigated a novel approach built upon Evolutionary Variational Optimization [14] to train VAEs with binary latents. Compared to all previous optimizations suggested for VAEs with discrete latents, the approach followed here differs the most substantially from conventional VAE training. While all other VAEs maintain amortization and reparameterization as key elements, the TVAE approach instead uses a direct and non-amortized optimization. Recent work using elementary generative models such as mixtures and shallow models [14, 17] have made considerations of direct VAE optimization possible for intermediately large scales. A conceptual advantage of the here developed approach is its concise formulation (compare Fig. S2) with fewer algorithmic elements, fewer hyperparameters and fewer model parameters (e.g., no parameters of encoder DNNs). Functional advantages of the approach are its avoidance of an amortization gap (e.g., [9, 30]), its ability to learn sparse codes, and its generality (it does not use a specific posterior model, and can be applied to other noise models, for instance). However, non-amortized approaches do in general have the disadvantage of a lower computational efficiency: an optimization of variational parameters for each data point is more costly (Tab. S3). Conventional amortized approaches (for discrete or continuous VAEs) are consequently preferable for large-scale data sets and for the optimization of large, intricate DNNs. There are, however, alternatives such as transformers (which can use >150M parameters) or diffusion nets, which both are considered to perform more strongly than VAEs for large-scale settings and density modeling ([7, 31] for recent comparisons).

At the same time, direct discrete optimization can be feasible and can be advantageous. For image patch data, for instance, we showed that TVAEs with intermediately large decoder DNNs perform more strongly than Gumbel-softmax VAEs (GSVAE), and TVAEs are also outperforming a recent continuous VAE baseline (VLAE; Figs. 1 and 2). The stronger performance of TVAE is presumably, at least in part, due to the approach not being subject to an amortization gap, due to it avoiding factored variational distributions, and, more generally, due to the emerging sparse codes being well suited for modeling image patch data. In comparison, the additional methods to treat discrete latents in GSVAE seem to result in dense codes with significantly lower performance than TVAE. Compared to GSVAE, the VLAE approach, which uses standard non-sparse (i.e. Gaussian) latents, is more competitive on the benchmarks we considered. The reason is presumably that VLAE's continuous latents are able to better capture component intensities in image patches. This advantage does not outweigh the advantages of sparse codes learned by TVAE, however. If sparse codes and continuous latents are combined, the example of ES3C shows that strong performances can be obtained (Figs. 1, 2 and 3). For the here considered binary latents, however, a linear decoder (compare EBSC) is much inferior to a deep decoder (Figs. 1, 2 and S9), which suggests future work on VAEs with more complex, sparse priors if the goal is to improve 'zero-shot' denoising and inpainting. Dense codes are notably not necessarily disadvantageous for image data. On the contrary, for datasets with many images of single objects like CIFAR, the dense codes of GSVAE and also of VLAE are, in terms of ELBO values, similar or better compared to TVAE (Tab. S4). The suitability of sparse versus dense encoding consequently seems to highly depend on the data, and here we confirm the suitability of sparse codes for image patches. In addition to learning sparse codes, direct optimization can have further advantages compared to conventional training. One such advantage is highlighted by the inpainting task: in contrast to other (continuous or discrete) VAEs, it is not required to additionally specify how missing data shall be treated by an encoder DNN (compare Sec. S2.5).

We conclude that direct discrete optimization can, depending on the data and task, serve as an alternative for training discrete VAEs. In a sense, the approach can be considered more brute-force than conventional amortized training: direct optimization is slower but at scales at which it can be applied, it is more effective. To our knowledge, the approach is also the first training method for discrete VAEs not using gradient optimization of encoder models, and can thus contribute to our understanding of how good representations can be learned by different approaches.

Acknowledgments. We acknowledge funding by the German Research Foundation (DFG) under grant SFB 1330/1&2 (B2), ID 352015383 (JD), and the H4a cluster of excellence EXC 2177/1, ID 390895286 (JL), and by the German Federal Ministry of Education and Research (BMBF) through a Wolfgang Gentner scholarship (awarded to EG, ID 13E18CHA). Furthermore, computing infrastructure support is acknowledged through the HPC Cluster CARL at UOL (DFG, grant INST 184/157-1 FUGG) and the HLRN Alliance, grant ID nim00006.

References

1. Bengio, Y., Lamblin, P., Popovici, D., Larochelle, H.: Greedy layer-wise training of deep networks. In: NeurIPS (2007)
2. Bengio, Y., Léonard, N., Courville, A.: Estimating or propagating gradients through stochastic neurons for conditional computation. arXiv:1308.3432 (2013)
3. Berliner, A., Rotman, G., Adi, Y., Reichart, R., Hazan, T.: Learning discrete structured variational auto-encoder using natural evolution strategies. In: ICLR (2022)
4. Bingham, E., et al.: PYRO: deep universal probabilistic programming. JMLR **20**(1), 973–978 (2019)
5. Bowman, S., Vilnis, L., Vinyals, O., Dai, A.M., Jozefowicz, R., Bengio, S.: Generating Sentences from a Continuous Space. In: CoNLL (2016)
6. Chaudhury, S., Roy, H.: Can fully convolutional networks perform well for general image restoration problems? In: IAPR International Conference MVA (2017)
7. Child, R., Gray, S., Radford, A., Sutskever, I.: Generating long sequences with sparse transformers. arXiv:1904.10509 (2019)
8. Cong, Y., Zhao, M., Bai, K., Carin, L.: GO gradient for expectation-based objectives. In: ICLR (2019)
9. Cremer, C., Li, X., Duvenaud, D.: Inference suboptimality in variational autoencoders. In: ICML (2018)
10. Dabov, K., Foi, A., Katkovnik, V., Egiazarian, K.: Image denoising by sparse 3D transform-domain collaborative filtering. IEEE Trans. Image Proc. **16**(8), 2080–2095 (2007)
11. Dimitriev, A., Zhou, M.: CARMS: categorical-antithetic-reinforce multi-sample gradient estimator. In: NeurIPS (2021)
12. Dong, W., Wang, P., Yin, W., Shi, G., Wu, F., Lu, X.: Denoising prior driven deep neural network for image restoration. TPAMI **41**(10), 2305–2318 (2019)
13. Dong, Z., Mnih, A., Tucker, G.: Coupled gradient estimators for discrete latent variables. In: NeurIPS (2021)
14. Drefs, J., Guiraud, E., Lücke, J.: Evolutionary Variational Optimization of Generative Models. JMLR **23**(21), 1–51 (2022)
15. Elad, M., Aharon, M.: Image denoising via sparse and redundant representations over learned dictionaries. IEEE Trans. Image Proc. **15**, 3736–3745 (2006)
16. Eldar, Y.C., Kutyniok, G.: Compressed Sensing: Theory and Applications. Cambridge University Press, Cambridge (2012)
17. Exarchakis, G., Oubari, O., Lenz, G.: A sampling-based approach for efficient clustering in large datasets. In: CVPR (2022)
18. Fajtl, J., Argyriou, V., Monekosso, D., Remagnino, P.: Latent bernoulli autoencoder. In: ICML (2020)
19. Foerster, J., Farquhar, G., Al-Shedivat, M., Rocktäschel, T., Xing, E., Whiteson, S.: DiCE: the infinitely differentiable monte Carlo estimator. In: ICML (2018)
20. Goodfellow, I.J., Courville, A., Bengio, Y.: Scaling up spike-and-slab models for unsupervised feature learning. TPAMI **35**(8), 1902–1914 (2013)
21. Goodfellow, I., et al.: Generative adversarial nets. In: NeurIPS (2014)
22. Grathwohl, W., Choi, D., Wu, Y., Roeder, G., Duvenaud, D.: Backpropagation through the void: optimizing control variates for black-box gradient estimation. In: ICLR (2018)
23. Gu, S., Zhang, L., Zuo, W., Feng, X.: Weighted nuclear norm minimization with application to image denoising. In: CVPR (2014)

24. Hajewski, J., Oliveira, S.: An evolutionary approach to variational autoencoders. In: CCWC (2020)
25. van Hateren, J.H., van der Schaaf, A.: Independent component filters of natural images compared with simple cells in primary visual cortex. Proc. Roy. Soc. London Ser. B **265**, 359–66 (1998)
26. Hirschberger, F., Forster, D., Lücke, J.: A variational EM acceleration for efficient clustering at very large scales. TPAMI (2022)
27. Hughes, M.C., Sudderth, E.B.: Fast learning of clusters and topics via sparse posteriors. arXiv:1609.07521 (2016)
28. Imamura, R., Itasaka, T., Okuda, M.: Zero-shot hyperspectral image denoising with separable image prior. In: ICCV Workshops (2019)
29. Jang, E., Gu, S., Poole, B.: Categorical reparameterization with gumbel-softmax. In: ICLR (2017)
30. Kim, Y., Wiseman, S., Miller, A., Sontag, D., Rush, A.: Semi-amortized variational autoencoders. In: ICML (2018)
31. Kingma, D.P., Salimans, T., Poole, B., Ho, J.: Variational diffusion models. In: NeurIPS (2021)
32. Kingma, D.P., Welling, M.: Auto-encoding variational bayes. In: ICLR (2014)
33. Kiran, B., Thomas, D., Parakkal, R.: An overview of deep learning based methods for unsupervised and semi-supervised anomaly detection in videos. J. Imaging **4**(2), 36 (2018)
34. Kool, W., van Hoof, H., Welling, M.: Estimating gradients for discrete random variables by sampling without Replacement. In: ICLR (2020)
35. van Krieken, E., Tomczak, J.M., Teije, A.T.: Storchastic: a framework for general stochastic automatic differentiation. In: NeurIPS (2021)
36. Krull, A., Buchholz, T.O., Jug, F.: Noise2Void - learning denoising from single noisy images. In: CVPR (2019)
37. Lehtinen, J., et al.: Noise2Noise: learning image restoration without clean data. In: ICML (2018)
38. Liu, R., Regier, J., Tripuraneni, N., Jordan, M., Mcauliffe, J.: Rao-Blackwellized stochastic gradients for discrete distributions. In: ICML (2019)
39. Lorberbom, G., Gane, A., Jaakkola, T., Hazan, T.: Direct Optimization through arg max for discrete variational auto-encoder. In: NeurIPS (2019)
40. Maaløe, L., Sønderby, C.K., Sønderby, S.K., Winther, O.: Auxiliary deep generative models. In: ICML (2016)
41. Maddison, C.J., Mnih, A., Teh, Y.W.: The concrete distribution: a continuous relaxation of discrete random variables. In: ICLR (2017)
42. Mairal, J., Elad, M., Sapiro, G.: Sparse representation for color image restoration. IEEE Trans. Image Proc. **17**(1), 53–69 (2008)
43. Olshausen, B., Field, D.: Emergence of simple-cell receptive field properties by learning a sparse code for natural images. Nature **381**, 607–9 (1996)
44. van den Oord, A., Vinyals, O., Kavukcuoglu, K.: Neural discrete representation learning. In: NeurIPS (2017)
45. Paiton, D.M., Frye, C.G., Lundquist, S.Y., Bowen, J.D., Zarcone, R., Olshausen, B.A.: Selectivity and robustness of sparse coding networks. J. Vis. **20**(12), 10–10 (2020)
46. Papyan, V., Romano, Y., Sulam, J., Elad, M.: Convolutional dictionary learning via local processing. In: ICCV (2017)
47. Park, Y., Kim, C., Kim, G.: Variational laplace autoencoders. In: ICML (2019)
48. Paulus, M.B., Maddison, C.J., Krause, A.: Rao-blackwellizing the straight-through gumbel-softmax gradient estimator. In: ICLR (2021)

49. Potapczynski, A., Loaiza-Ganem, G., Cunningham, J.P.: Invertible Gaussian reparameterization: revisiting the gumbel-softmax. In: NeurIPS (2020)
50. Quan, Y., Chen, M., Pang, T., Ji, H.: Self2Self With Dropout: learning self-supervised denoising from single image. In: CVPR (2020)
51. Rezende, D.J., Mohamed, S., Wierstra, D.: Stochastic backpropagation and approximate inference in deep generative models. In: ICML (2014)
52. Roberts, A., Engel, J., Raffel, C., Hawthorne, C., Eck, D.: A hierarchical latent vector model for learning long-term structure in music. In: ICML (2018)
53. Rolfe, J.T.: Discrete variational autoencoders. In: ICLR (2017)
54. Schulman, J., Heess, N., Weber, T., Abbeel, P.: Gradient estimation using stochastic computation graphs. In: NeurIPS (2015)
55. Sheikh, A.S., Lücke, J.: Select-and-sample for spike-and-slab sparse coding. In: NeurIPS (2016)
56. Sheikh, A.S., Shelton, J.A., Lücke, J.: A truncated EM approach for spike-and-slab sparse coding. JMLR **15**, 2653–2687 (2014)
57. Shelton, J., Bornschein, J., Sheikh, A.S., Berkes, P., Lücke, J.: Select and sample - a model of efficient neural inference and learning. In: NeurIPS (2011)
58. Shelton, J.A., Gasthaus, J., Dai, Z., Lücke, J., Gretton, A.: GP-Select: accelerating EM using adaptive subspace preselection. Neural Comp. **29**(8), 2177–2202 (2017)
59. Shocher, A., Cohen, N., Irani, M.: "Zero-Shot" super-resolution using deep internal learning. In: CVPR (2018)
60. Sulam, J., Muthukumar, R., Arora, R.: Adversarial robustness of supervised sparse coding. In: NeurIPS (2020)
61. Titsias, M.K., Lázaro-Gredilla, M.: Spike and slab variational inference for multi-task and multiple kernel learning. In: NeurIPS (2011)
62. Tomczak, J.M., Welling, M.: VAE with a VampPrior. In: AISTATS (2018)
63. Tonolini, F., Jensen, B.S., Murray-Smith, R.: Variational sparse coding. In: UAI (2020)
64. Ulyanov, D., Vedaldi, A., Lempitsky, V.: Deep image prior. In: CVPR (2018)
65. Williams, R.J.: Simple statistical gradient-following algorithms for connectionist reinforcement learning. Mach. Learn. **8**(3), 229–256 (1992)
66. Yu, G., Sapiro, G., Mallat, S.: Solving inverse problems with piecewise linear estimators: from gaussian mixture models to structured sparsity. IEEE Trans. Image Proc. **21**(5), 2481–2499 (2012)
67. Zhang, K., Zuo, W., Chen, Y., Meng, D., Zhang, L.: Beyond a Gaussian Denoiser: residual learning of deep CNN for image denoising. IEEE Trans. Image Proc. **26**(7), 3142–3155 (2017)
68. Zhou, M., Chen, H., Paisley, J., Ren, L., Sapiro, G., Carin, L.: Non-Parametric Bayesian dictionary learning for sparse image representations. In: NeurIPS (2009)
69. Zhou, M., et al.: Nonparametric Bayesian dictionary learning for analysis of noisy and incomplete images. IEEE Trans. Image Proc. **21**(1), 130–144 (2012)
70. Zhu, S., Xu, G., Cheng, Y., Han, X., Wang, Z.: BDGAN: image blind denoising using generative adversarial networks. In: PRCV (2019)
71. Zoran, D., Weiss, Y.: From learning models of natural image patches to whole image restoration. In: ICCV (2011)

Scalable Adversarial Online Continual Learning

Tanmoy Dam[1], Mahardhika Pratama[2(✉)], MD Meftahul Ferdaus[3],
Sreenatha Anavatti[1], and Hussein Abbas[1]

[1] SEIT, University of New South Wales, Canberra, Australia
t.dam@student.adfa.edu.au, {s.anavatti,h.abbas}@adfa.edu.au
[2] STEM, University of South Australia, Adelaide, Australia
dhika.pratama@unisa.edu.au
[3] ATMRI, Nanyang Technological University, Singapore, Singapore
mdmeftahul.ferdaus@ntu.edu.sg

Abstract. Adversarial continual learning is effective for continual learning problems because of the presence of feature alignment process generating task-invariant features having low susceptibility to the catastrophic forgetting problem. Nevertheless, the ACL method imposes considerable complexities because it relies on task-specific networks and discriminators. It also goes through an iterative training process which does not fit for online (one-epoch) continual learning problems. This paper proposes a scalable adversarial continual learning (SCALE) method putting forward a parameter generator transforming common features into task-specific features and a single discriminator in the adversarial game to induce common features. The training process is carried out in meta-learning fashions using a new combination of three loss functions. SCALE outperforms prominent baselines with noticeable margins in both accuracy and execution time.

Keywords: Continual learning · Lifelong learning · Incremental learning

1 Introduction

Continual learning (CL) has received significant attention because of its importance in improving existing deep learning algorithms to handle long-term learning problems. Unlike conventional learning problems where a deep model is presented with only a single task at once, a continual learner is exposed to a sequence of different tasks featuring varying characteristics in terms of different distributions or different target classes [9]. Since the goal is to develop a never-ending learning algorithm which must scale well to possibly infinite numbers of tasks,

---

T. Dam, M. Pratama and MD. M. Ferdaus—Equal Contribution.

Supplementary Information The online version contains supplementary material available at https://doi.org/10.1007/978-3-031-26409-2_23.

it is impossible to perform retraining processes from scratch when facing new tasks. The CL problem prohibits the excessive use of old data samples and only a small quantity of old data samples can be stored in the memory.

The CL problem leads to two major research questions. The first question is how to quickly transfer relevant knowledge of old tasks when embracing a new task. The second problem is how to avoid loss of generalization of old tasks when learning a new task. The loss of generalization power of old tasks when learning a new task is known as a catastrophic forgetting problem [9, 20] where learning new tasks catastrophically overwrites important parameters of old tasks. The continual learner has to accumulate knowledge from streaming tasks and achieves improved intelligence overtime.

There exists three common approaches for continual learning [20]: memory-based approach [16], structure-based approach [26], regularization-based approach [14]. The regularization-based approach makes use of a regularization term penalizing important parameters of old tasks from changing when learning new tasks. Although this approach is computationally light and easy to implement, this approach does not scale well for a large-scale CL problem because an overlapping region across all tasks are difficult to obtain. The structure-based approach applies a network growing strategy to accommodate new tasks while freezing old parameters to prevent the catastrophic forgetting problem. This approach imposes expensive complexity if the network growing phase is not controlled properly or the structural learning mechanism is often done via computationally expensive architecture search approaches thus being infeasible in the online continual learning setting. The memory-based approach stores a small subset of old samples to be replayed along with new samples to handle the catastrophic forgetting problem. Compared to the former two approaches, this approach usually betters the learning performance. The underlying challenge of this approach is to keep a modest memory size. SCALE is categorized as a memory-based approach here where a tiny episodic memory storing old samples is put forward for experience replay mechanisms.

The notion of adversarial continual learning (ACL) is proposed in [10]. The main idea is to utilize the adversarial learning strategy [12,13] to extract task-aligned features of all tasks deemed less prone to forgetting than task-specific features. It offers disjoint representations between common features and private features to be combined as an input of multi-head classifiers. The main bottleneck of this approach lies in expensive complexities because private features are generated by task-specific networks while common features are crafted by the adversarial game played by task-specific discriminators. In addition, ACL is based on an iterative training mechanism which does not fit for online (single-epoch) continual learning problems.

This paper proposes scalable adversarial continual learning (SCALE) reducing the complexity of ACL significantly via a parameter generator network and a single discriminator. The parameter generator network produces scaling and shifting parameters converting task-invariant features produced by the adversarial learning mechanism to task-specific features [21,22]. Our approach does not need to store task-specific parameters rather the parameter generator network predicts these parameters leading to private features. Production of private features are carried out with two light-weight operations, scaling and shifting. The parameter generator is trained in the meta-learning way using the validation loss of the base network, i.e., feature extractor and classifier. The meta-learning strategy is done by two data partitions: training set and validation set portraying both new and old concepts. The training set updates the base network while the validation set trains the parameter generator. Our approach distinguishes itself from [22] where the adversarial training approach is adopted to produce task-invariant features and we do not need to construct two different memories as per [22]. Unlike ACL, the adversarial game is played by a single discriminator without any catastrophic problem while still aligning the features of all tasks well.

SCALE outperforms prominent baselines with over 1% margins in accuracy and forgetting index while exhibiting significant improvements in execution times. The ablation study, memory analysis and sensitivity analysis further substantiate the advantages of SCALE for the online (one-epoch) continual learning problems. This paper offers four major contributions: 1) a new online continual learning approach, namely SCALE, is proposed; 2) our approach provides a scalable adversarial continual learning approach relying only on a single parameter generator for feature transformations leading to task-specific features and a single discriminator to induce task-invariant features; 3) the training process is done in the meta-learning manner using a new combination of three loss functions: the cross-entropy loss function, the DER++ loss function [4] and the adversarial loss function [10]. Although the adversarial loss function already exists in [10], the adversarial game is done differently here using the concept of BAGAN [18] rather than that the gradient reversal strategy [10,12]; 4) All source codes, data and raw numerical results are made available in https://github.com/TanmDL/SCALE to help further studies.

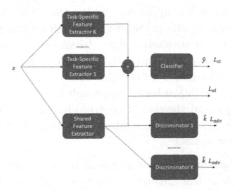

Fig. 1. Structure of ACL based on task-specific feature extractors and discriminators.

2 Related Works

Regularization-based Approach relies on a penalty term in the loss function preventing important parameters of old tasks from significant deviations. The L2-regularization strategy is combined with the parameter importance matrix indicating the significance of network parameters. Different strategies are proposed to construct the parameter importance matrix: Elastic Weight Consolidation (EWC) makes use of the Fisher Importance Matrix (FIM) [14], Synaptic Intelligence (SI) utilizes accumulated gradients [32], Memory Aware Synapses (MAS) adopts unsupervised and online criteria [1]. online EWC (oEWC) puts forward an online version of EWC using Laplace approximation [28]. Learning without Forgetting (LWF) utilizes the knowledge distillation (KD) approach to match between current and previous outputs. The regularization strategy is better performed in the neuron level rather than in the synaptic level [19] because of the hierarchical nature of the deep neural network. [17] follows the same principle as [19] and goes one step further using the concept of inter-task similarity. Such approach allows a node to be shared across related tasks. Another attempt to improve scalability of regularization-based approaches also exists in [5] where the projection concept is put forward to induce wide local optimum regions. The regularization-based approach heavily depends on the task-IDs and the task-boundaries.

Structure-based Approach offers different philosophies where new tasks are handled by adding new network components while isolating old components to avoid the catastrophic forgetting problem. The pioneering approach is the progressive neural network (PNN) [26] where a new network column is integrated when handling a new task. PNN incurs expensive structural complexities when dealing with a long sequence of tasks. [31] puts forward a network growing condition based on a loss criterion with the selective retraining strategy. The concept of neural architecture search (NAS) is proposed in [15] to select the best action when observing new tasks. Similar approach is designed in [30] but with the use of Bayesian optimization approach rather than the NAS concept. These approaches are computationally prohibitive and call for the presence of task IDs and boundaries. [3,23] put forward a data-driven structural learning for unsupervised continual learning problems where hidden clusters, nodes and layers dynamically grow and shrink. The key difference between the two approaches lies in the use of regularization-based approach in [3] via the Knowledge Distillation (KD) strategy and the use of centroid-based experience replay in [23]. The data-driven structural learning strategy does not guarantee optimal actions when dealing with new tasks.

Memory-based Approach utilizes a tiny memory storing a subset of old data samples. Old samples of the memory are interleaved with current samples for experience replay purposes to cope with the catastrophic forgetting problem. iCaRL exemplifies such approach [24] where the KD approach is performed with the nearest exemplar classification strategy. GEM [16] and AGEM [7] make use of the memory to identify the forgetting cases. HAL [6] proposes the idea of anchor samples maximizing the forgetting metric and constructed in the meta-learning manner. DER [4] devises the dark knowledge distillation and successfully achieves improved performances with or without the task IDs. CTN is proposed in [22] using the feature transformation concept of [21] and integrates the controller network trained in the meta-learning fashion. ACL [10] is categorized as the memory-based approach where a memory is used to develop the adversarial game. However, ACL imposes considerable complexities because of the use of task-specific feature extractors and discriminators. We offer an alternative approach here where private features are induced by the feature transformation strategy of [21] and the adversarial game is played by one and only one discriminator. Compared to [22], SCALE integrates the adversarial learning strategy to train the shared feature extractor generating common features being robust to the catastrophic forgetting problem and puts forward a new combination of three loss functions.

3   Problem Formulation

Continual learning (CL) problem is defined as a learning problem of sequentially arriving tasks $T_1, T_2, ..., T_k, k \in \{1, ..., K\}$ where K denotes the number of tasks unknown in practise. Each task carries triplets $T_k = \{x_i, y_i, t_i\}_{i=1}^{N_k}$ where N_k stands for a task size. $x_i^k \in \mathcal{X}_k$ denotes an input image while $y_i^k \in \mathcal{Y}_k$, $y_i^k = [l_1, l_2, ..., l_m]$ labels a class label and t_i^k stands for a task identifier (ID). The goal of CL problem is to build a continual learner $f_\phi(g_\theta(.))$ performing well on already seen tasks where $g_\theta(.)$ is the feature extractor and $f_\phi(.)$ is the classifier. This paper focuses on the online (one-epoch) task-incremental learning and domain-incremental learning problems [29] where each triplet of any tasks $\{x_i, y_i, t_i\} \curvearrowleft T_k$ is learned only in a single epoch. The task-incremental learning problem features disjoint classes of each task, i.e., Suppose that L_k and $L_{k'}$ stand for label sets of the $k - th$ task and the $k' - th$ task, $\forall k, k' L_k \cap L_{k'} = \emptyset$. The domain-incremental learning problem presents different distributions or domains of each task $P(X, Y)_k \neq P(X, Y)_{(k+1)}$ while still retaining the same target classes for each task. That is, a multi-head configuration is applied for the task-incremental learning problem where an independent classifier is created for each task $f_{\phi_k}(.)$. The domain-incremental learning problem is purely handled with a single head configuration $f_\phi(.)$. The CL problem prohibits the retraining process from scratch $\frac{1}{K} \sum_{k=1}^{K} \mathcal{L}_k, \mathcal{L}_k \triangleq \mathbb{E}_{(x,y) \sim \mathcal{D}_k}[l(f_\phi(g_\theta(x)), y)]$. The learning process is only supported by data samples of the current task T_k and a tiny memory $\mathcal{M}_{k-1}$ containing old samples of previously seen tasks to overcome the catastrophic forgetting problem.

4   Adversarial Continual Learning (ACL)

Figure 1 visualizes the adversarial continual learning method [10] comprising four parts: shared feature extractor, task-specific feature extractors, task-specific discriminator and multi-head classifiers. The shared feature extractor generates task-invariant features while the task-specific feature extractors offer private features of each task. The task-specific discriminator predicts the task's origins while the multi-head classifiers produce final predictions. The training process is governed by three loss functions: the classification loss, the adversarial loss and the orthogonal loss. The classification loss utilizes the cross-entropy loss function affecting the multi-head classifiers, the shared feature extractor and the task-specific feature extractors. The adversarial loss is carried out in the min-max fashion between the shared feature extractor and the task-specific discriminators. The gradient reversal layer is implemented when adjusting the feature extractor thus converting the minimization problem into the maximization problem. That is, the shared feature extractor is trained to fool the task-specific discriminators and eventually generates the task-invariant features. The orthogonal loss ensures clear distinctions between the task-specific features by the shared feature extractor and the private features by the task-specific feature extractors.

The task-specific discriminator is excluded during the testing phase and the inference phase is performed by feeding concatenated features of the private features and the common features to the multi-head classifiers producing the final outputs. ACL incurs high complexity because of the application of the task-specific feature extractors and the task-specific discriminators. We offer a parameter generator here generating scaling and shifting parameters to perform feature transformation. Hence, the task-specific features are generated with low overheads without loss of generalization, while relying only on a single discriminator to play the adversarial game inducing aligned features. In addition, ACL relies on an iterative training procedure violating the online continual learning requirements whereas SCALE fully runs in the one-epoch setting.

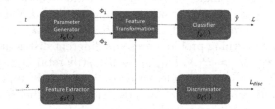

Fig. 2. SCALE consists of the parameter generator, the feature extractor, the multi-head classifiers (the single-head classifier in the domain-incremental learning problem) and the discriminator. The parameter generator generates the scaling and shifting coefficients converting the common features into the task-specific features. The task-specific features and the task-invariant features are combined and feed the classifier. The training process is controlled by the classification loss, the DER++ loss and the adversarial loss. The training process of the parameter generator is carried out in the meta-learning fashion minimizing the three loss functions. The single discriminator is updated by playing an adversarial game using the cross entropy loss and the DER++ loss.

5   Learning Policy of SCALE

The learning procedure of SCALE is visualized in Fig. 2 and Algorithm 1 where it comprises four blocks: the feature extractor $g_\theta(.)$, the parameter generator $P_\varphi(.)$, the multi-head classifiers $f_{\phi_k}(.)$ (the single-head classifier in the domain-incremental learning problem) and the single discriminator $D_\xi(.)$. The feature extractor extracts the task-invariant features enabled by the adversarial learning mechanism with the discriminator predicting the task IDs. Unlike ACL where task-specific features are produced by task-specific feature extractors, SCALE benefits from the feature transformation strategy with the scaling parameters Φ_1 and the shifting parameters Φ_2 produced by the parameter generator. The scaling and shifting parameters modify the common features into the task-specific features. The classifier receives aggregated features and thus delivers the final predictions. Since the scaling and shifting parameters assure distinct task-specific features of those common features, the orthogonal loss is removed. SCALE replaces the task-specific discriminators in ACL with only a single discriminator.

5.1   Feature Transformation

SCALE does not deploy any task-specific parameters violating the fixed architecture constraint [22] rather the parameter generator produces the scaling and shifting parameters thereby reducing its complexity significantly. We adopt similar idea of [21,22] where the scaling and shifting parameters creates the task-specific features via the feature transformation procedure as follows:

$$\tilde{g}_\theta(x) = \frac{\Phi_1}{||\Phi_1||_2} \odot g_\theta(x) + \frac{\Phi_2}{||\Phi_2||_2} \tag{1}$$

where $\odot$ denotes the element-wise multiplication. Φ_1, Φ_2 are the scaling and shifting parameters generated by the parameter generator $P_\varphi(t) = \{\Phi_1, \Phi_2\}$ taking the task IDs as input features with an embedding layer to produce low-dimensional features. This implies the parameter generator network φ to produce the scaling and shifting parameters Φ_1, Φ_2. A residual connection is implemented to linearly combine the shared and private features:

$$g_\theta(x) = \tilde{g}_\theta(x) + g_\theta(x) \tag{2}$$

We follow the same structure as [22] where the feature transformation strategy is implemented per layer with one parameter generator per layer. It is implemented for all intermediate layers except for the classifier in the case of multi-layer perceptron network while it is only applied to the last residual layer for convolutional neural network, thus only utilizing a single parameter generator network. A nonlinear activation function $s(.)$ is usually applied before feeding the combined features to the classifier $f_{\phi_k}(s(g_\theta(.)))$.

5.2   Loss Function

The loss function of SCALE consists of three components: the cross-entropy (CE) loss function, the dark-experience replay++ (DER++) loss function [4], and the adversarial loss function [10]. The CE loss function focuses on the current task and the previous tasks simultaneously while the DER++ loss function concerns on the past tasks thus distinguishing the second part of the DER++ loss with the CE loss function. The adversarial loss function is designed to align features of all tasks. Suppose that $o = f_\phi(g_\theta(.))$ stands for the output logits or the pre-softmax responses and $l(.)$ labels the cross-entropy loss function, the loss function of SCALE is expressed:

$$\mathcal{L} = \underbrace{\mathbb{E}_{(x,y)\sim\mathcal{D}_k\cup\mathcal{M}_{k-1}}[l(o,y)]}_{L_{CE}} + \underbrace{\mathbb{E}_{(x,y)\sim\mathcal{M}_{k-1}}[\lambda_1||o-h||_2 + \lambda_2 l(o,y)]}_{L_{DER++}} +$$

$$\underbrace{\mathbb{E}_{(x,y)\sim\mathcal{D}_k\cup\mathcal{M}_{k-1}}[\lambda_3 l(D_\xi(g_\theta(x)),t)]}_{L_{adv}} \tag{3}$$

where $h = f_\phi(g_\theta(.))_{k-1}$ is the output logit generated by a previous model, i.e., before seeing the current task. $\lambda_1, \lambda_2, \lambda_3$ are trade-off constants. The second term of L_{DER++}, $l(o,y)$, prevents the problem of label shifts ignored when only checking the output logits without the actual ground truth. The three loss functions are vital where the absence of one term is detrimental as shown in our ablation study.

5.3   Meta-training Strategy

The meta-training strategy [11, 27] is implemented here to update the parameter generator $P_\varphi(.)$ subject to the performance of the base network $f_\phi(g_\theta(.))$. This strategy initiates with creation of two data partitions: the training set $\mathcal{T}^k_{train}$ and the validation set $\mathcal{T}^k_{val}$ where both of them comprise the current data samples and the memory samples $\mathcal{T}^k \cup \mathcal{M}_{k-1}$. The meta-learning strategy is formulated as the bi-level optimization problem using the inner loop and the outer loop [22] as follows:

$$Outer : \min_\varphi \mathbb{E}_{(x,y)\sim\mathcal{T}^k_{val}}[\mathcal{L}]$$

$$Inner : s.t \quad \{\phi^*, \theta^*\} = \arg\min_{\phi,\theta} \mathbb{E}_{(x,y)\sim\mathcal{T}^k_{train}}[\mathcal{L}] \tag{4}$$

where $\mathcal{L}$ denotes the loss function of SCALE as formulated in (3). From (4), the parameter generator and the classifier are trained jointly. Because of the absence of ground truth of the scaling and shifting coefficients, our objective is to find the parameters of the parameter generator φ that minimizes the validation loss of the base network. This optimization problem is solvable with the stochastic gradient descent (SGD) method where it first tunes the parameters of the classifier in the inner loop:

$$\{\phi, \theta\} = \{\phi, \theta\} - \alpha \sum_{(x,y)\in\mathcal{T}^k_{train}} \nabla_{\{\phi,\theta\}}[\mathcal{L}] \tag{5}$$

where α is the learning rate of the inner loop. Once obtaining updated parameters of the base network, the base network is evaluated on the validation set and results in the validation loss. The validation loss is utilized to update the parameter generator:

$$\varphi = \varphi - \beta \sum_{(x,y) \in T_{val}^k} \nabla_\varphi [\mathcal{L}] \tag{6}$$

where β is the learning rate of the outer loop. Every outer loop (6) involves the inner loop . Both inner and outer loops might involve few gradient steps as in [22] but only a single epoch is enforced in SCALE to fit the online continual learning requirements.

5.4   Adversarial Training Strategy

The adversarial training strategy is applied here where it involves the feature extractor $g_\theta(.)$ and the discriminator $D_\xi(.)$. The goal is to generate task-invariant features, robust against the catastrophic forgetting problem. The discriminator and the feature extractor play a minimax game where the feature extractor is trained to fool the discriminator by generating indistinguishable features while the discriminator is trained to classify the generated features by their task labels [10]. The adversarial loss function L_{adv} is formulated as follows:

$$L_{adv} = \min_g \max_D \sum_{k=0}^{K} \mathbb{I}_{k=t^k} \log(D_\xi(g_\theta(x))) \tag{7}$$

where the index $k = 0$ corresponds to a fake task label associated with a Gaussian noise $\mathcal{N}(\mu, \Sigma)$ while $\mathbb{I}_{k=t^k}$ denotes an indicator function returning 1 only if $k = t^k$ occurs, i.e., t^k is the task ID of a sample x. The feature extractor is trained to minimize (7) while the discriminator is trained to maximize (7). Unlike [10] using the gradient reversal concept in the adversarial game, the concept of BAGAN [18] is utilized where the discriminator to trained to associate a data sample to either a fake task label $k = 0$ or one of real task labels $k = 1, .., K$ having its own output probability or soft label. A generator role is played by the feature extractor. The discriminator is trained with the use of memory as with the base network to prevent the catastrophic forgetting problem where its loss function is formulated:

$$L_{disc} = L_{adv} + L_{DER++} \tag{8}$$

where L_{DER++} is defined as per (3) except that the target attribute is the task labels rather than the class labels. Unlike [22] using two memories, we use a single memory shared across the adversarial training phase and the meta-training phase.

6   Experiments

The advantage of SCALE is demonstrated here and is compared with recently published baselines. The ablation study, analyzing each learning component, is

---

Algorithm 1. Learning Policy of SCALE

Input: continual dataset $\mathcal{D}$, learning rates μ, β, α, iteration numbers $n_{in} = n_{out} = n_{ad} = 1$

Output: parameters of the base learner $\{\phi, \theta\}$, parameters of the parameter generator φ, parameters of the discriminator ξ

for $k = 1$ to K **do**
   for $n_1 = 1$ to n_{out} **do**
      for $n_2 = 1$ to n_{in} **do**
         Update base learner parameters $\{\theta, \phi\}$ using (4)
      end for
      Update parameter generator parameters φ using (6)
   end for
   for $n_3 = 1$ to n_{ad} **do**
      Update discriminator parameters ξ minimizing (8)
   end for
   $\mathcal{M}_k = \mathcal{M}_{k-1} \cup B_k$ /*Update memory/*
end for

Table 1. Experimental Details

Datasets	#Tasks	#classes/task	#training/task	#testing/task	Dimensions
PMNIST	23	10	1000	1000	$1 \times 28 \times 28$
SCIFAR-10	20	5	2500	500	$3 \times 32 \times 32$
SCIFAR-10	5	2	10000	2000	$3 \times 32 \times 32$
SMINIIMAGENET	20	5	2400	600	$3 \times 84 \times 84$

provided along with the memory analysis studying the SCALE's performances under different memory budgets. All codes, data and raw numerical results are placed in https://github.com/TanmDL/SCALE to enable further studies.

6.1 Datasets

Four datasets, namely Permutted MNIST (PMNIST), Split CIFAR100 (SCIFAR100), Split CIFAR10 (SCIFAR10) and Split MiniImagenet (SMINIIMAGENET), are put forward to evaluate all consolidated algorithms. The PMNIST features a domain-incremental learning problem with 23 tasks where each task characterizes different random permutations while the rests focus on the task-incremental learning problem. The SCIFAR100 carries 20 tasks where each task features 5 distinct classes. As with the SCIFAR100, the SMINIIMAGENET contains 20 tasks where each task presents disjoint classes. The SCIFAR10 presents 5 tasks where each task features 2 mutually exclusive classes. Our experimental details are further explained in Table 1.

6.2 Baselines

SCALE is compared against five strong baselines: GEM [16], MER [25], ER-Ring [8], MIR [2] and CTN [22]. The five baselines are recently published and outperform other methods as shown in [22]. All algorithms are memory-based approaches usually performing better than structure-based approach, regularization-based approach [22]. No comparison is done against ACL [10] because ACL is not compatible under the online (one-epoch) continual learning problems due to its iterative characteristics. Significant performance deterioration is observed in ACL under the one-epoch setting. All baselines are recently published, thus representing state-of-the art results. All algorithms are executed in the same computer, a laptop with 1 NVIDIA RTX 3080 GPU having 16 GB RAM and 16 cores Intel i-9 processor having 32 GB RAM, to ensure fairness and their source codes are placed in https://github.com/TanmDL/SCALE.

6.3 Implementation Notes

Source codes of SCALE are built upon [16,22] and our experiments adopt the same network architectures for each problem to assure fair comparisons. A two hidden layer MLP network with 256 nodes in each layer is applied for PMNIST and a reduced ResNet18 is applied for SCIFAR10/100 and SMINIIMAGENET. The hyper-parameter selection of all consolidated methods is performed using the grid search approach in the first three tasks as with [7] to comply to the online learning constraint. Hyper-parameters of all consolidated algorithms are detailed in **the supplemental document**. Numerical results of all consolidated algorithms are produced with the best hyper-parameters. Since the main focus of this paper lies in the online (one-epoch) continual learning, all algorithms run in one epoch. Our experiments are repeated five times using different random seeds and the average results across five runs are reported. Two evaluation metrics, averaged accuracy [16] and forgetting measure [7] are used to evaluate all consolidated methods. Since all consolidated algorithms make use of a memory, the memory budget is fixed to 50 per tasks.

6.4 Numerical Results

The advantage of SCALE is demonstrated in Table 2 where it outperforms other consolidated algorithms with significant margins. In SMINIIMAGENET, SCALE beats CTN in accuracy with over 1.5% gap and higher than that for other algorithms, i.e., around 10% margin. It also shows the smallest forgetting index compared to other algorithms with over 1% improvement to the second best approach, CTN. The same pattern is observed in the SCIFAR100 where SCALE exceeds CTN by almost 2% improvement in accuracy and shows improved performance in the forgetting index by almost 2% margin. Other algorithms perform poorly compared to SCALE where the accuracy margin is at least over 9%. and the forgetting index margin is at least over 5%. In pMNIST, SCALE beats its counterparts with at least 2% gap in accuracy while around 2% margin is

Table 2. Numerical results of consolidated algorithms across the four problems. All methods use the same backbone network and 50 memory slots per task

Method	SMINIIMAGENET		SCIFAR100	
	ACC(↑)	FM(↓)	ACC(↑)	FM(↓)
GEM	54.50 ± 0.93	7.40 ± 0.89	60.22 ± 1.07	9.04 ± 0.84
MER	53.38 ± 1.74	10.96 ± 1.57	59.48 ± 1.31	10.44 ± 1.11
MIR	54.92 ± 2.29	8.96 ± 1.68	61.26 ± 0.46	9.06 ± 0.63
ER	54.62 ± 0.80	9.50 ± 1.09	60.68 ± 0.57	9.70 ± 0.97
CTN	63.42 ± 1.18	3.84 ± 1.26	67.62 ± 0.76	6.20 ± 0.97
SCALE	$\mathbf{64.96 \pm 1.10}$	$\mathbf{2.60 \pm 0.60}$	$\mathbf{70.24 \pm 0.76}$	$\mathbf{4.16 \pm 0.48}$
Method	pMNIST		SCIFAR10	
	ACC(↑)	FM(↓)	ACC(↑)	FM(↓)
GEM	71.10 ± 0.47	10.16 ± 0.41	75.9 ± 1.3	12.74 ± 2.85
MER	68.70 ± 0.35	12.04 ± 0.30	79.4 ± 1.51	9.7 ± 1.07
MIR	71.90 ± 0.49	11.74 ± 0.34	79 ± 1.16	9.28 ± 0.91
ER	74.76 ± 0.56	9.06 ± 0.58	79.76 ± 1.26	8.68 ± 1.49
CTN	78.70 ± 0.37	5.84 ± 0.36	83.38 ± 0.8	5.68 ± 1.43
SCALE	$\mathbf{80.70 \pm 0.46}$	$\mathbf{2.90 \pm 0.27}$	$\mathbf{84.9 \pm 0.91}$	$\mathbf{4.46 \pm 0.4}$

observed in the forgetting index. SCALE is also the best-performing continual learner in the SCIFAR10 where it produces the highest accuracy with 1.5% difference to CTN and the lowest forgetting index with about 1% gap to CTN. Numerical results of Table 2 are produced from five independent runs under different random seeds.

6.5   Memory Analysis

This section discusses the performances of consolidated algorithms, MER, MIR, CTN, SCALE under different memory budgets $|\mathcal{M}_k| = 50, 100, 150, 200$ per task. GEM and ER are excluded here because ER performs similarly to MER and MIR while GEM is usually worse than other algorithms. The memory analysis is carried out in the SCIFAR100 and in the SMINIIMAGENET. Our numerical results are visualized in Fig. 2(a) for the SCIFAR100 and in Fig. 2(B) for the SMINIIMAGENET. It is obvious that SCALE remains superior to other algorithms under varying memory budgets in the SCIFAR100 where the gap is at least 1% to CTN as the second best algorithm across all memory configurations. In SMINIIMAGENET problem, SCALE outperforms other algorithms with the most noticeable gap in $|\mathcal{M}_k| = 50$ presenting the hardest case. The gap with CTN becomes close when increasing the memory slots per tasks but still favours SCALE. Note that the performances of SCALE and CTN is close to the joint training (upper bound) with increased memory slots in the SMINIIMAGENET, i.e., no room for further performance improvement is possible.

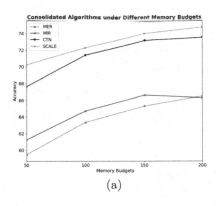
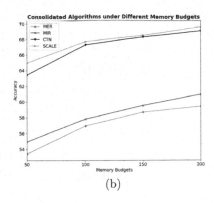

(a)   (b)

Fig. 3. Consolidated Algorithms under Different Memory Budgets in case of (a) SCI-FAR100, and (b) SMINIIMAGENET

Table 3. Ablation Study: Different Learning Configurations of SCALE

Method	SCIFAR10		SCIFAR100	
	ACC($\uparrow$)	FM($\downarrow$)	ACC($\uparrow$)	FM($\downarrow$)
A	83.66 ± 1.42	6.26 ± 1.23	68.58 ± 1.88	5.88 ± 2.06
B	76.08 ± 2.83	16.32 ± 3.57	45.32 ± 2.09	27.68 ± 1.95
C	81.68 ± 1.22	7.12 ± 1.63	66.64 ± 1.71	5.76 ± 0.78
SCALE	$\mathbf{84.9 \pm 0.91}$	$\mathbf{4.46 \pm 0.40}$	$\mathbf{70.24 \pm 0.76}$	$\mathbf{4.16 \pm 0.48}$

6.6   Ablation Study

This section discusses the advantage of each learning component of SCALE where it is configured into three settings: (A) SCALE with the absence of adversarial learning strategy meaning that the meta-training process is carried out only with the CE loss function and the DER++ loss function while removing any adversarial games; (B) SCALE with the absence of DER++ loss function meaning that the meta-training process is driven by the CE loss function and the adversarial loss function while the adversarial game in (8) is undertaken without the DER++ loss function; (C) SCALE with the absence of parameter generator network meaning that no task-specific features are generated here due to no feature transformation approaches. Table 3 reports our numerical results across two problems: SCIFAR10 and SCIFAR100.

Configuration (A) leads to drops in accuracy by about 2% and increases in forgetting by about 2% for SCIFAR10 and SCIFAR100. These facts confirm the efficacy of the adversarial learning strategy to boost the learning performances of SCALE. Such strategy allows feature's alignments of all tasks extracting common features, being robust to the catastrophic forgetting problem. Configuration (B) results in major performance degradation in both accuracy and forgetting index across SCIFAR10 and SCIFAR100, i.e., 10% drop in accuracy for SCIFAR10 and

Table 4. Execution Times of All Consolidated Algorithms across All Problems

Dataset	Methods	Execution Times
pMNIST	SCALE	41.33
	CTN	52.01
	ER	26.28
	MIR	**23.56**
	MER	26.80
	GEM	28.53
SCIFAR10	SCALE	**142.1**
	CTN	358.5
	ER	250.86
	MIR	242.52
	MER	257.44
	GEM	151.96
SCIFAR100	SCALE	**108.33**
	CTN	321.87
	ER	211.94
	MIR	218.24
	MER	222.39
	GEM	314.24
SMINIIMAGENET	SCALE	**193.73**
	CTN	298.33
	ER	290.27
	MIR	254.20
	MER	261.17
	GEM	540.46

25% drop in accuracy for SCIFAR100; 12% increase in forgetting for SCIFAR10 and 23% increase in forgetting for SCIFAR100. This finding is reasonable because the DER++ loss function is the major component in combatting the catastrophic forgetting problem. Configuration (C) leaves SCALE without any task-specific features, thus causing drops in performances. 3% drop in accuracy is observed for SCIFAR10 while 4% degradation in accuracy is seen for SCIFAR100. The same pattern exists for the forgetting index where 3% increase in forgetting occurs for SCIFAR10 and 1.5% increase in forgetting happens for SCIFAR100. Our finding confirms the advantage of each learning component of SCALE where it contributes positively to the overall performances.

Table 5. Sensitivity Analysis of Hyper-parameters in SCIFAR100

Hyper-parameters	ACC ($\uparrow$)	FM($\downarrow$)
$\lambda_1, \lambda_2 = 3, \lambda_3 = 0.9$	69.5 ± 0.54	5.5 ± 0.44
$\lambda_1, \lambda_2 = 3, \lambda_3 = 0.3$	69.34 ± 1.04	5.12 ± 0.53
$\lambda_1, \lambda_2 = 3, \lambda_3 = 0.09$	69.26 ± 1.22	5.2 ± 1.08
$\lambda_1, \lambda_2 = 3, \lambda_3 = 0.03$	69.88 ± 0.98	4.82 ± 0.55
$\lambda_1, \lambda_2 = 1, \lambda_3 = 0.9$	69.2 ± 0.76	5.16 ± 0.85
$\lambda_1, \lambda_2 = 1, \lambda_3 = 0.3$	69.24 ± 0.84	5.44 ± 1.19
$\lambda_1, \lambda_2 = 1, \lambda_3 = 0.09$	69.8 ± 0.73	4.88 ± 0.90
$\lambda_1, \lambda_2 = 1, \lambda_3 = 0.03$	$\mathbf{70.24 \pm 0.76}$	$\mathbf{4.16 \pm 0.48}$

6.7 Execution Times

Execution times of all consolidated algorithms are evaluated here because it is an important indicator in the online continual learning problems. Table 4 displays execution times of consolidated algorithms across all problems. The advantage of SCALE is observed in its low running times compared to other algorithms in three of four problems except in the pMNIST. SCALE demonstrates significant improvements by almost 50% speed-up from CTN in realm of execution times because it fully runs in the one-epoch setting whereas CTN undergoes few gradient steps in the inner and outer loops. Note that both SCALE and CTN implement the parameter generator network. This fact also supports the adversarial learning approach of SCALE, absent in CTN, where it imposes negligible computational costs but positive contribution to accuracy and forgetting index as shown in our ablation study. Execution times of SCALE are rather slow in pMNIST problem because the parameter generator is incorporated across all intermediate layers in the MLP network. The execution times significantly improves when using the convolution structure because the parameter generator is only implemented in the last residual block. SCALE only relies on one and only discriminator to produce aligned features while private features are generated via parameter generator networks.

7 Sensitivity Analysis

Sensitivity of different hyper-parameters, $\lambda_1, \lambda_2, \lambda_3$, are analyzed here under the SCIFAR100 where these hyper-parameters control the influence of each loss function (3). Other hyper-parameters are excluded from our sensitivity analysis because they are standard hyper-parameters of deep neural networks where their effects have been well-studied in the literature. Note that the hyper-parameter sensitivity is a major issue in the online learning context because of time and space constraints for reliable hyper-parameter searches. Specifically, we select $\lambda_1, \lambda_2 = 1$ and $\lambda_1, \lambda_2 = 3$, while varying $\lambda_3 = [0.03, 0.09, 0.3, 0.9]$. Table 5 reports our numerical results.

It is observed that SCALE is not sensitive to different settings of hyper-parameters. That is, there does not exist any significant gaps in performances compared to the best hyper-parameters as applied to produce the main results in Table 2, $\lambda_1, \lambda_2 = 1, \lambda_3 = 0.03$, the gaps are less than $< 1\%$. Once again, this finding confirms the advantage of SCALE for deployments in the online (one-epoch) continual learning problem.

8   Conclusion

This paper presents an online (one-epoch) continual learning approach, scalable adversarial continual learning (SCALE). The innovation of SCALE lies in one and only one discriminator in the adversarial games for the feature alignment process leading to robust common features while making use of the feature transformation concept underpinned by the parameter generator to produce task-specific (private) features. Private features and common features are linearly combined with residual connections where aggregated features feed the classifier for class inferences. The training process takes place in the strictly one-epoch meta-learning fashion based on a new combination of the three loss functions. Rigorous experiments confirm the efficacy of SCALE beating prominent algorithms with noticeable margins ($>1\%$) in accuracy and forgetting index across all four problems. Our memory analysis favours SCALE under different memory budgets while our ablation study demonstrates the advantage of each learning component. In addition, SCALE is faster than other consolidated algorithms in 3 of 4 problems and not sensitive to hyper-parameter selections. Our future work is devoted to continual time-series forecasting problems.

Acknowledgements. M. Pratama acknowledges the UniSA start up grant. T. Dam acknowledges UIPA scholarship from the UNSW PhD program.

References

1. Aljundi, R., Babiloni, F., Elhoseiny, M., Rohrbach, M., Tuytelaars, T.: Memory aware synapses: learning what (not) to forget. In: ECCV (2018)
2. Aljundi, R., et al.: Online continual learning with maximally interfered retrieval. In: NeurIPS (2019)
3. Ashfahani, A., Pratama, M.: Unsupervised continual learning in streaming environments. IEEE Trans. Neural Netw. Learn. Syst. PP (2022)
4. Buzzega, P., Boschini, M., Porrello, A., Abati, D., Calderara, S.: Dark experience for general continual learning: a strong, simple baseline. In: NeurIPS (2020)
5. Cha, S., Hsu, H., Calmon, F., Moon, T.: CPR: classifier-projection regularization for continual learning. In: ICLR (2020)
6. Chaudhry, A., Gordo, A., Dokania, P., Torr, P.H.S., Lopez-Paz, D.: Using hindsight to anchor past knowledge in continual learning. In: AAAI (2021)
7. Chaudhry, A., Ranzato, M., Rohrbach, M., Elhoseiny, M.: Efficient lifelong learning with a-gem. In: ICLR (2018)
8. Chaudhry, A., et al.: On tiny episodic memories in continual learning. arXiv: Learning (2019)

9. Chen, Z., Liu, B.: Lifelong machine learning. In: Synthesis Lectures on Artificial Intelligence and Machine Learning (2016)
10. Ebrahimi, S., Meier, F., Calandra, R., Darrell, T., Rohrbach, M.: Adversarial continual learning. In: Vedaldi, A., Bischof, H., Brox, T., Frahm, J.-M. (eds.) ECCV 2020. LNCS, vol. 12356, pp. 386–402. Springer, Cham (2020). https://doi.org/10.1007/978-3-030-58621-8_23
11. Finn, C., Abbeel, P., Levine, S.: Model-agnostic meta-learning for fast adaptation of deep networks. In: ICML (2017)
12. Ganin, Y., et al.: Domain-adversarial training of neural networks. J. Mach. Learn. Res. **17**(59), 1–35 (2016)
13. Goodfellow, I., et al.: Generative adversarial nets. In: NIPS (2014)
14. Kirkpatrick, J., et al.: Overcoming catastrophic forgetting in neural networks. Proc. Nat. Acad. Sci. **114**, 3521–3526 (2017)
15. Lai Li, X., Zhou, Y., Wu, T., Socher, R., Xiong, C.: Learn to grow: a continual structure learning framework for overcoming catastrophic forgetting. In: ICML (2019)
16. Lopez-Paz, D., Ranzato, M.: Gradient episodic memory for continual learning. In: NIPS (2017)
17. Mao, F., Weng, W., Pratama, M., Yapp, E.: Continual learning via inter-task synaptic mapping. Knowl. Based Syst. **222**, 106947 (2021)
18. Mariani, G., Scheidegger, F., Istrate, R., Bekas, C., Malossi, A.C.I.: Bagan: data augmentation with balancing GAN. ArXiv abs/1803.09655 (2018)
19. Paik, I., Oh, S., Kwak, T., Kim, I.: Overcoming catastrophic forgetting by neuron-level plasticity control. In: AAAI (2019)
20. Parisi, G.I., Kemker, R., Part, J.L., Kanan, C., Wermter, S.: Continual lifelong learning with neural networks: a review. Neural Netw. Official J. Int. Neural Netw. Soc. **113**, 54–71 (2019)
21. Perez, E., Strub, F., de Vries, H., Dumoulin, V., Courville, A.C.: FiLM: visual reasoning with a general conditioning layer. In: AAAI (2018)
22. Pham, Q H., Liu, C., Sahoo, D., Hoi, S.C.H.: Contextual transformation networks for online continual learning. In: ICLR (2021)
23. Pratama, M., Ashfahani, A., Lughofer, E.: Unsupervised continual learning via self-adaptive deep clustering approach. ArXiv abs/2106.14563 (2021)
24. Rebuffi, S.A., Kolesnikov, A., Sperl, G., Lampert, C.H.: iCaRL: incremental classifier and representation learning. In: 2017 IEEE Conference on Computer Vision and Pattern Recognition (CVPR), pp. 5533–5542 (2017)
25. Riemer, M., et al.: Learning to learn without forgetting by maximizing transfer and minimizing interference. In: ICLR (2019)
26. Rusu, A.A., et al.: Progressive neural networks. ArXiv abs/1606.04671 (2016)
27. Schmidhuber, J.: Evolutionary principles in self-referential learning. on learning now to learn: the meta-meta-meta...-hook. diploma thesis, Technische Universitat Munchen, Germany (14 May 1987). http://www.idsia.ch/juergen/diploma.html
28. Schwarz, J., et al.: Progress & compress: a scalable framework for continual learning. In: ICML (2018)
29. van de Ven, G.M., Tolias, A.: Three scenarios for continual learning. ArXiv abs/1904.07734 (2019)
30. Xu, J., Ma, J., Gao, X., Zhu, Z.: Adaptive progressive continual learning. IEEE Trans. Pattern Anal. Mach. Intell. **44**, 6715–6728 (2021)
31. Yoon, J., Yang, E., Lee, J., Hwang, S.J.: Lifelong learning with dynamically expandable networks. In: ICLR (2017)
32. Zenke, F., Poole, B., Ganguli, S.: Continual learning through synaptic intelligence. Proc. Mach. Learn. Res. **70**, 3987–3995 (2017)

Fine-Grained Bidirectional Attention-Based Generative Networks for Image-Text Matching

Zhixin Li$^{(\boxtimes)}$, Jianwei Zhu, Jiahui Wei, and Yufei Zeng

Guangxi Key Lab of Multi-source Information Mining and Security,
Guangxi Normal University, Guilin 541004, China
`lizx@gxnu.edu.cn`

Abstract. In this paper, we propose a method called BiKA (Bidirectional Knowledge-assisted embedding and Attention-based generation) for the task of image-text matching. It mainly improves the embedding ability of images and texts from two aspects: first, modality conversion, we build a bidirectional image and text generation network to explore the positive effect of mutual conversion between modalities on image-text feature embedding; then is relational dependency, we built a bidirectional graph convolutional neural network to establish the dependency between objects, introduce non-Euclidean data into image-text fine-grained matching to explore the positive effect of this dependency on fine-grained embedding of images and texts. Experiments on two public datasets show that the performance of our method is significantly improved compared to many state-of-the-art models.

Keywords: Cross-modal retrieval · Graph convolutional network · Knowledge embedding · Cross-attention · Attentional generative network

1 Introduction

The human brain will spontaneously switch between modalities. For example, when people hear a sentence or see a text, we will associate related images in their brains. Similarly, when people see a picture or a video, our brains will spontaneously construct their language expression. As shown in Fig. 1, these spontaneous modal transformation behaviors help us understand the world. In the field of deep learning, there are similar tasks, such as *Image Caption* and *Text to Image* (T2I) generation. Recently, both modal transformation tasks have made remarkable progress. The cross-modal retrieval task aims to capture the association between different modalities. Applying modal transformations to the cross-modal retrieval task should enhance the expressive ability of model embedding. Based on this idea, we propose a bidirectional attentional generative network. While performing image-text matching, additional generative tasks are introduced to constrain the expression ability of modal embedding vectors.

M.-R. Amini et al. (Eds.): ECML PKDD 2022, LNAI 13715, pp. 390–406, 2023.
https://doi.org/10.1007/978-3-031-26409-2_24

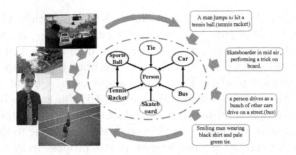

Fig. 1. Establish fine-grained dependency structure and bidirectional generative structure between objects

Traditional deep learning models such as LSTM and CNN have achieved great results on Euclidean data (such as language, image, video, etc.). However, there are certain limitations in the processing of non-Euclidean data. Kipf et al. [12] proposed Graph Convolutional Neural networks (GCNs) to process graph data. There are problems with data sparseness and a lack of key information for image and text retrieval tasks. The text description is usually very short and cannot describe the corresponding image in detail. For example, the text description in Fig. 1 "A man jumps to hit a tennis ball" does not express key information such as environment, tennis rackets, etc.

Moreover, a static image at a specific moment cannot fully express the state of objects at the current moment. For example, in Fig. 1, due to the high-speed motion, the image capture of a tennis ball is blurred and cannot be captured by the deep model. These problems limit the performance of retrieval. In response to this problem, we proposed a knowledge-assisted embedding method, combining traditional depth models with non-Euclidean data, and modeled the dependencies between objects. As shown in Fig. 1, by constructing the dependency relationship between "Sports Ball" and "Tennis Racket," text and image can be better matched.

Our main contributions are as follows:

- We propose a knowledge-assisted embedding network, which allows external knowledge to be integrated into the local embedding of images and text, which strengthens regional-level text and image embedding.
- We propose a location relationship embedding method, which embeds multiple location relationship information of multiple entities in an image into its encoding.
- We introduce a bidirectional attentional generative network to optimize the expression of global and local features of images and text.

2 Related Work

Bottom-up attention [1] refers to detecting salient regions at the object level, similar to spontaneous bottom-up attention in the human visual system (for example, the foreground in an image is generally more concerned than the background). Karpathy et al. [11] used R-CNN [5] to obtain The region-level embedding of the image calculates the similarity between the image and the text by aggregating the similarity of all regions and all words. Lee et al. [13] proposed a stacked cross-attention model that uses both bottom-up attention and conventional attention models. The model aligns each segment with all other segments from other modalities to train the model. The most advanced performance is achieved on several benchmark datasets for cross-modal retrieval, but the amount of calculation for inference becomes enormous due to the fine-grained matching method. Gu et al. [6] and Peng et al. [19] propose to incorporate generative models into cross-view feature embedding learning. Li et al. [14] used GCN to obtain image region features with enhanced semantic relations, at the same time, which incorporated generative models to enhance the expressive ability of image networks. Chen et al. [2] proposed an iterative matching method for cross-modal image text retrieval to deal with semantic complexity and proposed an iterative matching method of cyclic attention memory to refine the alignment relationship between images and text. Ji et al. [10] proposed a Stepwise Hierarchical Alignment Network (SHAN), which decomposes image-text matching into a multi-step cross-modal reasoning process. Compared with our models, these models either only consider coarse-grained matching between images and text or only consider fine-grained matching between image regions and words. We combine these two matching methods.

The regional features obtained by bottom-up attention may lack the expression of positional relationship. Hu et al. [9] introduced geometric attention to object detection for the first time, which uses bounding box coordinates and size to infer the weight of the relationship between the object pairs. The closer the two bounding boxes, the more similar the size, the stronger their relationship. Herdade et al. [8] changed the structure of Transformer [21] to embed the positional relationship of the region of interest. We get three kinds of relationship matrices from the image: the area matrix, distance matrix, and azimuth matrix. We combine these three relational matrices to obtain feature representations of embedded positional relations.

Chen et al. [4] used GCN for multi-label image recognition and achieved good performance by constructing a directed graph to model the relationship between objects. Li et al. [15] used a multi-head attention mechanism on the text and image side to enhance feature extraction. Chen et al. [3] used this structure in object detection tasks to make the generated bounding box more reasonable. We also use a similar structure, but we use GCN to assist in embedding text and images. We explore the relationship between objects to solve the insufficient expressive ability of text and the image itself and embed this relationship into the image and text. In local embedding, better feature expression is obtained.

3   Proposed Method

3.1   Framework

The purpose of knowledge-assisted embedding in a cross-modal retrieval network is to improve the performance of cross-media retrieval by exploring the semantics and dependencies of objects in two modalities. We designed an integrated framework based on the baseline model SCAN [13]. Unlike the method that only considers visual and textual features, the knowledge-assisted embedding cross-modal retrieval network explicitly considers the semantics and dependencies of the objects in the two modalities. Explore this dependency in images and text. The structure of the frame is shown in Fig. 2. The critical problem faced by the knowledge-assisted embedding cross-modal retrieval network is how to learn the dependency relationship between objects and embed it into the local feature vector of each modal. We use two graphs to model the dependencies between objects. Specifically, we designed a GCN-based network to get the dependency information between objects.

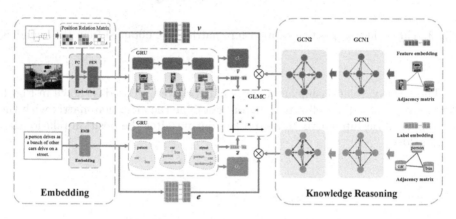

Fig. 2. The structure of BiKA, where G_I and G_T are the text generation network and the image generation network, respectively.

We propose a framework divided into two modules: a global semantic network and two external knowledge embeddings. The global semantic network extracts the global semantic information of images and texts; the external knowledge embedding module constructs a directed graph on the dataset, where the nodes on the text end are vectors represented by the word embedding of the label. The nodes on the image end are vectors pooled by the average of the corresponding classification features representation. Then we use GCN to obtain a higher-level semantic representation. The semantic representation obtained on the image side is multiplied by the encoding point of the image local feature to obtain the image local label encoding. The semantic representation obtained on the text side is multiplied by the encoding point of the image local feature to obtain Text local label encoding, and finally, we introduce the attention model to establish the alignment relationship between two local label encodings.

3.2   Positional Embedding

We pay attention to the bounding box from the bottom up to calculate three kinds of matrices, namely the orientation matrix O, the distance matrix D, and the area matrix A, as shown in Fig. 3.

Fig. 3. Three kinds of position relationship matrices are obtained from the bounding box.

The position relationship embedding network we proposed is shown in Fig. 4. We modify it based on self-attention and combine the attention matrix with the position relationship matrix to obtain the output after the position relationship is embedded:

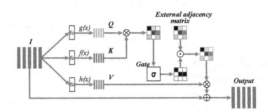

Fig. 4. Positional Relationship Embedding framework in BiKA.

$$v = \sum_{i=0}^{m} \sum_{j=0}^{n} (\sigma(Q_i^T K_i) \odot E_j) V_i + I. \tag{1}$$

in which I represents the local features of the input image, m represents the number of layers of the network, $\sigma(x)$ represents the Sigmoid gate structure, and its function is represented by $\sigma(x) = 1/(1 + e^{-x})$, E_j represents the external position relationship matrix, we have defined three position relationship matrices, namely the orientation relation matrix O, distance relationship Matrix D, and area matrix A. The distance matrix is symmetric, and the area matrix is diagonal. For the i-th and j-th image local embedding vectors, the three types of matrices are defined as follows:

$$O_{i,j} = \arctan(\frac{\bar{h}_i - \bar{h}_j}{\bar{w}_i - \bar{w}_j}),$$

$$D_{i,j} = D_{j,i} = \frac{\|W, H\|_2 - \|(\bar{h}_i - \bar{h}_j), (\bar{w}_i - \bar{w}_j)\|_2}{\|W, H\|_2}, \tag{2}$$

$$A_{i,i} = \frac{h_i}{H} \cdot \frac{w_i}{W},$$

where W and H represent the original image's horizontal and vertical pixel numbers, respectively, w and h represent the target bounding box's horizontal and vertical pixel numbers, respectively. $\bar{w}$ represents the horizontal coordinate representing the center of the target bounding box, and $\bar{h}$ represents the vertical coordinate representing the center of the target bounding box.

3.3   Object Relational Reasoning

The key problem faced by object relational networks is how to effectively capture the global semantic relations in labels. We use graphs to model global semantic relationships in labels, explore the topology of implicit global knowledge, and we construct an adjacency matrix for labels. Specifically, we design a fGCN-based network to obtain global semantic relation information.

We first define the GCN. The objective of GCN is to learn a function $f(D, A)$ on the graph $\mathcal{G}$, D is the input eigenvector, and A is the adjacency matrix. The feature expression obtained by the i-th level GCN is $D_i \in \mathbb{R}^{n \times d}$, where n is the number of graph nodes and d is the depth of the graph. The output of the next layer of GCN is $D_{i+1} \in \mathbb{R}^{n \times d'}$. Each GCN layer can be written as $D_{i+1} = f(D_i, A)$. After applying the convolutional operation, $f(D, A)$ can be expressed as $D_{i+1} = r(\hat{A} D_i W_i)$.

The nodes in GCN transmit information through edges, that is, to construct the adjacency matrix of the graph. How to construct the adjacency matrix is the key to modeling with graphs. We process the dataset with an object detection model, and the classification result corresponding to the output of each image is obtained. First, the number of occurrences of each classification label in the data set is calculated to obtain the matrix $T \in \mathbb{R}^{N \times N}$, where N is the number of class labels, and both dimensions of the matrix correspond to class labels $C \in \mathbb{R}^{1 \times N}$. $T_{i,j}$ represents the number of co-occurrences of categories C_i and C_j, and the conditional probability $P(C_i|C_j) = T_{j,i}/T_{j,j}$ represents the probability of the occurrence of the i-th category when the j-th category appears. For example, when a bicycle appears, there is a high probability that someone will appear next to it, and when someone appears, the probability of a bicycle appearing will be much smaller. This is the relationship between people and bicycles.

However, there are two serious problems with the above simple correlation analysis. First, the co-occurrence pattern between one label and other labels may exhibit a long-tailed distribution, where some numerically small co-occurrence times may be due to noise. Second, due to the different datasets in the training and testing phases, the number of simultaneous occurrences in training and testing is inconsistent. Therefore, We binarize the relation matrix T. Specifically, the noise is filtered by setting a threshold $\hbar$. The binary relationship matrix is

$$A_{ij} = \begin{cases} 0, & \text{if } P(C_i|C_j) < \hbar \\ 1, & \text{if } P(C_i|C_j) \geq \hbar \end{cases}, \tag{3}$$

The binary adjacency matrix brings a new problem, node features may be over-smoothed, so that nodes from different scenes (for example, nodes composed of categories related to kitchen and living room) may become indistinguishable. To alleviate this problem, we use the following reweighting scheme:

$$A'_{ij} = \begin{cases} p/\sum_{j=1}^{C} A_{ij}, & \text{if } i \neq j \\ 1 - p, & \text{if } i = j \end{cases}, \tag{4}$$

where A' is the reweighted relation matrix. The weights of the node itself and other related nodes are determined by p. When p is close to 1, the characteristics of the node itself are not considered. On the other hand, when p is close to 0, the neighbor information is easily ignored. After applying the obtained weighted relational adjacency matrix A to a multi-layer GCN, we get:

$$D_{i+1} = r(\hat{A}' D_i W_i) \tag{5}$$

where the matrix $\hat{A}'$ is the normalized version of the correlation matrix A'; W is the transformation matrix to be learned by the i-th layer GCN, D_i is the input node feature of the i-th layer; r is the nonlinear operation, and we use the LeakyReLU activation function.

In our proposed method, We apply GCN to both image and text side, using two two-layer GCN networks to capture the dependencies between objects to aid feature embedding for text and images.

3.4   Bidirectional Attentional Generative Network

The structure of the two generative models is shown in Fig. 5, in which G_I generates a sequence of sentences to make the generated sequence as similar as possible to the corresponding text description; The purpose of G_T is to generate a realistic image with similar semantics to the corresponding real image. In addition, they all use the attention mechanism of the Encoder-Decoder framework to focus on the input sequence when generating a specific word or image region instead of relying only on the context vector to complete the generative task.

Attentional Text Generative Network. We use the Encoder-Decoder structures in [25], we use GRU as the main structure of the encoder and decoder, using the Soft Attention mechanism, which is shown in the upper part of the Fig. 5. Different from the traditional encoder-decoder structure, after applying attention, the state of each step of the encoder will be considered when each word in the target sequence is generated. When the decoder generates each word y_i, the

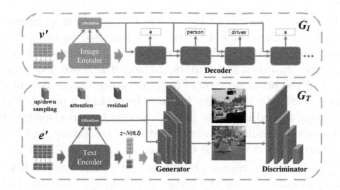

Fig. 5. Proposed bidirectional generative network's structure, in which e' and v' are the output of each step of the text encoder and image encoder in Fig. 2.

hidden layer state of the encoder per step is considered, and the current time step is output more accurately. After the application of attention, the process of the decoder generates the target sequence has become as follows:

$$y_1 = f(c_1)$$
$$y_i = f(c_i, y_1, ..., y_{i-1})$$

(6)

where c_i corresponds to the attention probability distribution of each word in the input sequence. In general, according to the widely used Soft Attention mechanism, c_i is weighted and obtained through all hidden layers in the encoder, which is represented as $c_j = \sum_{i=0}^{I-1} \alpha_{j,i} v_i'$, where $\alpha_{j,i} = \exp(r_{j,i}) / \sum_{k=0}^{T-1} \exp(r_{j,k})$, $r_{j,i} = h_j^T v_i'$, I represents the number of image region features. The weight of attention is obtained by comparing the hidden layer status of the current time decoder and all hidden layers in the encoder.

Attentional Generative Adversarial Network. For the text side, the embedding vector of the text is combined with the attention mechanism, and the feature map of the final size is obtained through multiple upsampling. The image of a specific size is generated by the generator and provided to the discriminator for discrimination. Specifically, for the global feature vector $\bar{e}$ and the local feature vector e obtained by the text encoder, there are the following formulas:

$$h_0 = F_0\left(z, \bar{e}'\right),$$
$$\hat{h} = \sum_{i=1}^{n} F_i(h_{i-1}, F_i^{att}(e', h_{i-1})),$$
$$\hat{x} = G(h),$$

(7)

in which z is a noise vector, which is usually sampled from the standard normal distribution, $\bar{e}'$ is the hidden layer of the last step of the G_T encoder in Fig. 5. F_i represents n feature transformation networks used to sample the feature map, $\hat{h}$

represents the feature map output by the last layer, $G(h)$ is a convolution kernel with a size of 3×3 convolutional neural network, used to generate RGB color images. In our model, n is 3, and the generated image size is 64×64.

Following [26], F_{att} has two inputs, namely the word features e and the hidden state h of the previous layer. First, the word feature is converted into a semantic space similar to the image features by adding a perceptron layer, i.e., $e' = We + b$. Then, according to the image's hidden features (query), the context vector is calculated for each region-level feature of the image. For j-th area-level features, the word context vector is expressed as $\hat{c}_j = \sum_{i=0}^{T-1} \beta_{j,i} e'_i$, where $\beta_{j,i} = \exp(s_{j,i})/\sum_{k=0}^{T-1} \exp(s_{j,k})$, represents the weight of the word when the model generates the j-th subregion of the image $s_{j,i} = h_j^T e'_i$. Then the image feature of the context matrix is generated by $F^{att}(e', h) = (\hat{c}_0, \hat{c}_1, ..., \hat{c}_{N-1})$. Finally, the next stage image is generated by fusing image features and corresponding context features.

3.5 Global and Local Multi-modal Cross-Attention

The text encoder uses the GRU structure. We use the hidden layer of each time step as the local vector of the text, all local features are represented by e, and the last time step hides the layer state $\bar{e}$ as a global embedded vector. Similarly, the local feature vector v and global feature vector $\bar{v}$ can be obtained.

Aiming at local and global embedding results, we propose Global and Local Multi-modal Cross-attention (GLMC). GLMC requires four inputs: image local features v, image global features $\bar{v}$, Word feature e and sentence feature $\bar{e}$, the output is the similarity score of the image feature-text pair at the regional level. We can get the similarity relationship between a certain image area and the corresponding word embedding: $s_{ij} = v_i^T e_j/\|v_i\| \|e_j\|, i \in [1, k], j \in [1, n]$, in which $k = 36$ corresponds to the regional features of 36 images obtained by bottom-up attention, and n is the number of tokens in the sentence. Then we normalize the calculated cosine similarity to get the similarity score

$$\bar{s}_{ij} = relu(s_{ij})/\sqrt{\sum_{i=1}^{n} relu(s_{ij})^2}, \tag{8}$$

where $relu(x) = max(0, x)$. In addition, we constructed a global attention model to calculate the regional context vector of each word (querry): $c_i = \sum_{j=1}^{n} \alpha_{ij} e_j$, where $\alpha_{ij} = \exp(\lambda_1 \bar{s}_{ij})/\sum_{j=1}^{n} \exp(\lambda_1 \bar{s}_{ij})$, λ_1 is the inverse temperature coefficient of the softmax function, which is used to adjust the smoothness of the attention distribution. Inspired by the minimum classification error formula in the speech recognition task [7], we use the following formula to calculate the similarity between image I and the sentence T:

$$R(I, T) = \log\left(\sum_{i=1}^{k} \exp(\lambda_2 R(v_i, c_i))\right)^{(1/\lambda_2)}, \tag{9}$$

in which λ_2 is a magnification factor, which determines how much the importance of the most relevant word and regional context pair is magnified. When $\lambda_2 \to \infty$, $S(I,T)$ approaches $\max_{i=1}^k R(v_i, c_i)$.

For a batch of image sentence pairs (I_i, T_i), according to the posterior probability of sentence T_i matching image I_i is calculated as follows: $P(T_i \mid I_i) = \exp(R(I_i, T_i))/\sum_{j=1}^M \exp(R(I_i, T_j))$, where M represents the number of image-text pairs in each batch. We define the loss function as the negative logarithmic posterior probability of an image that matches its corresponding text description:

$$\mathcal{L}_r = -\sum_{i=1}^M \log P(T_i \mid I_i) - \sum_{i=1}^M \log P(I_i \mid T_i). \tag{10}$$

We define $R(I,T) = (\bar{v}^T \bar{e})/(\|\bar{v}\|\|\bar{e}\|)$ for the global text and image embedding, from this, the global matching loss $\mathcal{L}_g$ can be obtained. Finally, the objective function of GLMC is defined as:

$$\mathcal{L}_{GLMC} = \mathcal{L}_r + \lambda_3 \mathcal{L}_g, \tag{11}$$

in which λ_3 is used to adjust the weight between the global loss $\mathcal{L}_g$ and the local loss $\mathcal{L}_r$.

3.6   Object Function

BiKA's objective function is mainly divided into three parts, namely matching loss $\mathcal{L}_{GLMC}$, image-to-text generation loss $\mathcal{L}_{cap}$, and text-to-image generation loss $\mathcal{L}_{gen}$, we define the final objective function of the BiKA network as

$$\mathcal{L} = \mathcal{L}_{GLMC} + \mathcal{L}_{cap} + \mathcal{L}_{gen} \tag{12}$$

For the captions generation part, the visual representation obtained at each step of the model should be able to generate sentences, making it close to the ground truth captions. Specifically, we use an attention-based encoder-decoder framework. We maximize the log-likelihood of the predicted output sentence. The loss function is defined as $\mathcal{L}_{cap} = -\sum_{t=1}^{l_i} \log p(y_t^i \mid y_{t-1}^i, V_i; \theta_i)$, where l^i is the length of the output word sequence $Y_i = (y_1^i, ... y_l^i)$ in the i-th step. θ_i is the parameter of the sequence-to-sequence model. The generator loss $\mathcal{L}_{gen}$ is defined as

$$\mathcal{L}_{gen} = \underbrace{-\frac{1}{2}\mathbb{E}_{x \sim p_G}[\log D(x)]}_{\text{uncondition}} \underbrace{-\frac{1}{2}\mathbb{E}_{x \sim p_G}[\log D(x, \bar{e})]}_{\text{condition}} \tag{13}$$

in which the unconditional loss is used to make the generated image more realistic, and the conditional loss is used to make the generated image more match the original text description. In contrast to the training of the generator, the discriminator is trained to judge the authenticity of the input by minimizing a defined cross-entropy loss

$$\mathcal{L}_D = \underbrace{-\frac{1}{2}\mathbb{E}_{x\sim p_{\text{data}}}\left[\log D\left(x\right)\right] - \frac{1}{2}\mathbb{E}_{x\sim p_G}\left[\log\left(1 - D\left(x\right)\right)\right]}_{\text{uncondition}}$$

$$\underbrace{-\frac{1}{2}\mathbb{E}_{x\sim p_{\text{data}}}\left[\log D\left(x, \bar{e}\right)\right] - \frac{1}{2}\mathbb{E}_{x\sim p_G}\left[\log\left(1 - D\left(x, \bar{e}\right)\right)\right]}_{\text{condition}}.$$

$$(14)$$

4   Experiments

We use two popular datasets: MS-COCO [16], and Flickr30k [27] datasets, and we evaluated our method on these two datasets. The division of the model is provided by [13]. The training set of MS-COCO has 113287 images, the validation set and the test set each have 5000 images, and each image corresponds to 5 captions. The training set of Flickr30k has 2800 images, the validation set and the test set each have 5000 images, and each image corresponds to 5 captions. We set $\hbar$ in Eq. (3) to be 0.4 and p in Eq. (4) to be 0.2.

4.1   Quantitative Analysis

Table 1 summarizes the performance comparison results of the MS-COCO 1K test set. We can see that our BiKA model achieves the best performance, surpassing all baselines. Our method has the same backbone as SCAN. On R-1, the relative gains from image to text and text to image are 6.7% and 6.8%, respectively; on R-5, the relative gains from image to text and text to image are 1.8% and 2.1%, respectively; on R-sum, the relative gain is 2.7%. In the MS-COCO 5k test set, these gains are 8.1%, 4.4%, 1.4%, 2.3%, 2.4%. Table 2 summarizes the performance comparison results of the Flickr30K 1K test set. We can see that the relative gains of image-to-text and text-to-image on R-1 between our method and SCAN are 11.6% and 12.8%, respectively. The relative gains of image-to-text and text-to-image on R-5 are 9.4% and 10.2%, respectively. The relative gains from image to text and text to image on R-10 are 1.4% and 6.1%, respectively. On R-sum, the relative gain is 5.4%. In addition, in comparison with the new method [2, 10, 14, 22–24], our model still has a relatively significant advantage. Therefore, it can be seen from the above comparison that the BiKA model is effective in assisting embedding.

4.2   Qualitative Analysis

We demonstrate some examples of fine-grained image-text matching in Fig. 6, some text retrieval examples of a given image query sentence in Fig. 7, and a given image in Fig. 8 A search example of a text query image.

As shown in Fig. 6, we visualize fine-grained text matching. It can be seen that when the word is "men," we mainly focus on the two riders in the image. When the word is "horses," we also focus on the image. Two horses and two riders embody the dependency between "horse" and "person." When there is a horse in the image, there is usually someone nearby, but when there is a person in the image, the horse will probably not appear.

Table 1. Quantitative experimental results on MS-COCO test set.

Method	Caption Retrieval			Image Retrieval			
	R-1	R-5	R-10	R-1	R-5	R-10	R-sum
MSCOCO-1K							
SCAN [13]	72.7	94.8	98.4	58.8	88.4	94.8	507.9
CAMP [24]	72.3	94.8	98.3	58.5	87.9	95.0	506.8
BFAN [17]	74.9	95.2	/	59.4	88.4	/	/
PFAN [23]	76.5	96.3	**99.0**	61.6	89.6	95.2	518.2
VSRN [14]	76.2	94.8	98.2	**62.8**	89.7	95.1	516.8
SGM [22]	73.4	93.8	97.8	57.5	87.3	94.3	504.1
IMRAM [2]	76.7	95.6	98.5	61.7	89.1	95.0	516.6
SHAN [10]	76.8	96.3	98.7	62.6	89.6	**95.8**	519.8
BiKA	**77.6**	**96.5**	98.6	**62.8**	**90.3**	**95.8**	**521.6**
MSCOCO-5K							
SCAN [13]	50.4	82.2	90.0	38.6	69.3	80.4	410.9
CAMP [24]	50.1	82.1	89.7	39.0	68.9	80.2	410.0
SGM [22]	50.0	79.3	87.9	35.3	64.9	76.5	393.9
VSRN [14]	53.0	81.1	89.4	**40.5**	70.6	**81.1**	415.7
IMRAM [2]	53.7	83.2	91.0	39.7	69.1	79.8	416.5
BiKA	**54.5**	**83.4**	**91.4**	40.2	**70.9**	80.7	**421.1**

Table 2. Quantitative experimental results on Flickr30k test set.

Method	Caption Retrieval			Image Retrieval			
	R-1	R-5	R-10	R-1	R-5	R-10	R-sum
SCAN [13]	67.4	90.3	95.8	48.6	77.7	85.2	465.0
CAMP [24]	68.1	89.7	95.2	51.5	77.1	85.3	466.9
BFAN [17]	68.1	91.4	/	50.8	78.4	/	/
PFAN [23]	70.0	91.8	95.0	50.4	78.7	86.1	472.0
VSRN [14]	71.3	90.6	96.0	54.7	81.8	88.2	482.6
SGM [22]	71.8	91.7	95.5	53.5	79.6	86.5	478.6
IMRAM [2]	74.1	93.0	96.6	53.9	79.4	87.2	484.2
SHAN [10]	74.6	**93.5**	96.9	**55.3**	81.3	88.4	489.9
BiKA	**75.2**	91.6	**97.4**	54.8	**82.5**	**88.6**	**490.1**

4.3 GCN Parameter Analysis

Layer Number Analysis. As shown in Fig. 9(a), we have done experimental comparisons for the effect of the number of GCN layers on the model performance. It can be seen from the figure that the performance of the model does

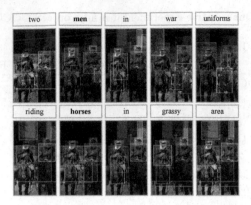

Fig. 6. Image text matching examples, showing the contribution of each image region to the similarity of the current word-image region. The brightness represents the strength of the similarity, and we select the image area with the similarity greater than a certain threshold to highlight it with a bounding box.

Fig. 7. A few qualitative examples of text retrieval for a given image query. Error results are highlighted in red and marked with an "×." reasonable mismatch are black. (Color figure online)

not improve with the deepening of the GCN. The model can get the best performance when the number of GCN layers is 2.

Node Number Analysis. The number of different nodes may have a greater impact on the performance of the model. We use Faster-RCNN [20] to predict the dataset, obtain the co-occurrence matrix between objects, and sort by the appearance frequency of the objects, according to the order of appearance frequency. We select 40, 80, 120, 160, 200, 240, 280 objects respectively, and obtain the word embeddings of the corresponding words through the pre-trained GloVe model and use them as graph nodes, the co-occurrence matrix between them as the adjacency matrix of the graph. The experimental results are shown in Fig. 9(b). It can be seen from the figure that when the number of nodes is small, as the number of nodes increases, the performance of the model is also enhanced, but when the number of nodes reaches 160 or more, the performance of the model no longer increases, even to a certain extent. The weakening. Therefore, we select 160 nodes as GCN graph nodes to assist text and image embedding.

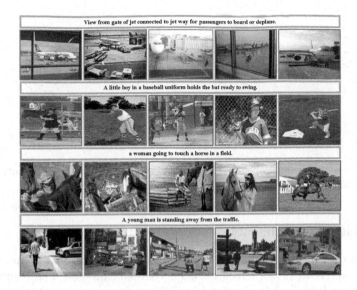

Fig. 8. A few qualitative examples of image retrieval for a given text query. Correct results are highlighted with a green border, and incorrect results are highlighted with a red border. A reasonable mismatch is highlighted with a blue border. (Color figure online)

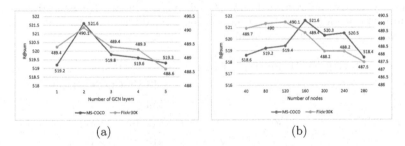

Fig. 9. The effect of GCN parameters on performance, where (a) represents the effect of the number of GCN layers, and (b) is the effect of the number of nodes on performance.

4.4 Visualization

In Fig. 10, we use t-SNE [18] to visualize the output of the last layer of GCN. It can be seen that the semantics of nodes show clustering patterns. Specifically, in the visualization on the image side, objects are more likely to appear at the same time in the same scene, For example, "baseball" and "baseball glove," "keyboard" and "mouse," and so on. They all have similar semantics. The same characteristics are also shown in the text-side GCN output visualization. Our method is effective in modeling the dependencies between objects.

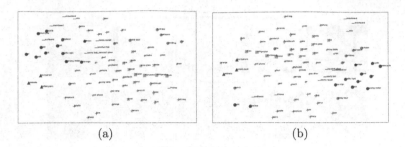

(a)   (b)

Fig. 10. Visual representations of GCN output, where (a) represents the output on the image side, and (b) represents the output on the text side.

4.5   Ablation Experiments

We perform ablation experiments to demonstrate the gain of our modification on mutual retrieval performance between text and images. The results of the corresponding four models are shown in Table 3. It can be seen that our changes to the model are effective, in which the effect of GCN is significant.

Table 3. Comparison of ablation experiment results. "Bi-Gen" represents the bidirectional generative network in Fig. 5. "GCN" means knowledge-assisted embedding structure. "Pos-Emb" means position relation embedding module, and "Cross-Att" means fine-grained text image matching.

Bi-Gen	GCN	Pos-Emb	Cross-Att	R-sum	
				MS-COCO	Flickr30k
	✓			514.0	481.2
		✓		512.3	471.2
			✓	512.2	472.3
	✓	✓		516.2	482.7
	✓		✓	516.1	483.3
		✓	✓	514.3	483.1
	✓	✓	✓	518.6	487.9
✓	✓	✓	✓	521.6	490.1

5   Conclusion

In this paper, we proposed BiKA, which uses external knowledge to assist text and image embedding, models the dependencies between objects, and aggregates the information of each node through the GCN network. At the same time, we explored a new positional relationship embedding structure, which makes the distance information, orientation information between objects, and the area information of each object embedded in the feature vector, which enhances the expressive ability of the model. In addition, we propose a bidirectional generative structure to constrain the extraction of features through the mutual generation of images and text. Our work has promising results on two public datasets. We have done sufficient experimental analysis and provided the visualization effect of node characteristics, which proves the robustness of our method. We conducted ablation experiments to prove that each structural change is effective.

Acknowledgments. This work is supported by National Natural Science Foundation of China (Nos. 61966004, 61866004), Guangxi Natural Science Foundation (No. 2019GXNSFDA245018), Guangxi "Bagui Scholar" Teams for Innovation and Research Project, Guangxi Talent Highland Project of Big Data Intelligence and Application, and Guangxi Collaborative Innovation Center of Multi-source Information Integration and Intelligent Processing.

References

1. Anderson, P., et al.: Bottom-up and top-down attention for image captioning and visual question answering. In: Proceedings of the IEEE Conference on Computer Vision and Pattern Recognition, pp. 6077–6086 (2018)
2. Chen, H., Ding, G., Liu, X., Lin, Z., Liu, J., Han, J.: IMRAM: iterative matching with recurrent attention memory for cross-modal image-text retrieval. In: Proceedings of the IEEE Conference on Computer Vision and Pattern Recognition, pp. 12655–12663 (2020)
3. Chen, S., Li, Z., Tang, Z.: Relation R-CNN: a graph based relation-aware network for object detection. IEEE Signal Process. Lett. **27**, 1680–1684 (2020)
4. Chen, Z.M., Wei, X.S., Wang, P., Guo, Y.: Multi-label image recognition with graph convolutional networks. In: Proceedings of the IEEE Conference on Computer Vision and Pattern Recognition, pp. 5177–5186 (2019)
5. Girshick, R., Donahue, J., Darrell, T., Malik, J.: Rich feature hierarchies for accurate object detection and semantic segmentation. In: Proceedings of the IEEE Conference on Computer Vision and Pattern Recognition, pp. 580–587 (2014)
6. Gu, J., Cai, J., Joty, S.R., Niu, L., Wang, G.: Look, imagine and match: improving textual-visual cross-modal retrieval with generative models. In: Proceedings of the IEEE Conference on Computer Vision and Pattern Recognition, pp. 7181–7189 (2018)
7. He, X., Deng, L., Chou, W.: Discriminative learning in sequential pattern recognition. IEEE Signal Process. Mag. **25**(5), 14–36 (2008)
8. Herdade, S., Kappeler, A., Boakye, K., Soares, J.: Image captioning: transforming objects into words. arXiv preprint arXiv:1906.05963 (2019)
9. Hu, H., Gu, J., Zhang, Z., Dai, J., Wei, Y.: Relation networks for object detection. In: Proceedings of the IEEE Conference on Computer Vision and Pattern Recognition, pp. 3588–3597 (2018)
10. Ji, Z., Chen, K., Wang, H.: Step-wise hierarchical alignment network for image-text matching. arXiv preprint arXiv:2106.06509 (2021)
11. Karpathy, A., Fei-Fei, L.: Deep visual-semantic alignments for generating image descriptions. In: Proceedings of the IEEE Conference on Computer Vision and Pattern Recognition, pp. 3128–3137 (2015)

12. Kipf, T.N., Welling, M.: Semi-supervised classification with graph convolutional networks. arXiv preprint arXiv:1609.02907 (2016)
13. Lee, K.H., Chen, X., Hua, G., Hu, H., He, X.: Stacked cross attention for image-text matching. In: Proceedings of the European Conference on Computer Vision, pp. 201–216 (2018)
14. Li, K., Zhang, Y., Li, K., Li, Y., Fu, Y.: Visual semantic reasoning for image-text matching. In: Proceedings of the IEEE/CVF International Conference on Computer Vision, pp. 4654–4662 (2019)
15. Li, Z., Xie, X., Ling, F., Ma, H., Shi, Z.: Matching images and texts with multi-head attention network for cross-media hashing retrieval. Eng. Appl. Artif. Intell. **106**, 104475 (2021)
16. Lin, T.-Y., et al.: Microsoft COCO: common objects in context. In: Fleet, D., Pajdla, T., Schiele, B., Tuytelaars, T. (eds.) ECCV 2014. LNCS, vol. 8693, pp. 740–755. Springer, Cham (2014). https://doi.org/10.1007/978-3-319-10602-1_48
17. Liu, C., Mao, Z., Liu, A.A., Zhang, T., Wang, B., Zhang, Y.: Focus your attention: a bidirectional focal attention network for image-text matching. In: Proceedings of the 27th ACM International Conference on Multimedia, pp. 3–11 (2019)
18. Van der Maaten, L., Hinton, G.: Visualizing data using T-SNE. J. Mach. Learn. Res. **9**(11), 2579–2605 (2008)
19. Peng, Y., Qi, J.: CM-GANs: cross-modal generative adversarial networks for common representation learning. ACM Trans. Multimed. Comput. Commun. Appl. **15**(1), 1–24 (2019)
20. Ren, S., He, K., Girshick, R., Sun, J.: Faster R-CNN: towards real-time object detection with region proposal networks. Adv. Neural. Inf. Process. Syst. **28**, 91–99 (2015)
21. Vaswani, A., et al.: Attention is all you need. Adv. Neural. Inf. Process. Syst. **30**, 5998–6008 (2017)
22. Wang, S., Wang, R., Yao, Z., Shan, S., Chen, X.: Cross-modal scene graph matching for relationship-aware image-text retrieval. In: Proceedings of the IEEE/CVF Winter Conference on Applications of Computer Vision, pp. 1508–1517 (2020)
23. Wang, Y., et al.: Position focused attention network for image-text matching. arXiv preprint arXiv:1907.09748 (2019)
24. Wang, Z., Liu, X., Li, H., Sheng, L., Yan, J., Wang, X., Shao, J.: CAMP: cross-modal adaptive message passing for text-image retrieval. In: Proceedings of the IEEE International Conference on Computer Vision, pp. 5764–5773 (2019)
25. Xu, K., et al.: Show, attend and tell: neural image caption generation with visual attention. In: Proceedings of the International Conference on Machine Learning, pp. 2048–2057. PMLR (2015)
26. Xu, T., et al.: AttnGAN: fine-grained text to image generation with attentional generative adversarial networks. In: Proceedings of the IEEE Conference on Computer Vision and Pattern Recognition, pp. 1316–1324 (2018)
27. Young, P., Lai, A., Hodosh, M., Hockenmaier, J.: From image descriptions to visual denotations: new similarity metrics for semantic inference over event descriptions. Trans. Assoc. Comput. Linguist. **2**, 67–78 (2014)

Computer Vision

Learnable Masked Tokens for Improved Transferability of Self-supervised Vision Transformers

Hao Hu$^{(\boxtimes)}$, Federico Baldassarre, and Hossein Azizpour

KTH Royal Institute of Technology, Stockholm, Sweden
{haohu,fedbal,azizpour}@kth.se

Abstract. Vision transformers have recently shown remarkable performance in various visual recognition tasks specifically for self-supervised representation learning. The key advantage of transformers for self supervised learning, compared to their convolutional counterparts, is the reduced inductive biases that makes transformers amenable to learning rich representations from massive amounts of unlabelled data. On the other hand, this flexibility makes self-supervised vision transformers susceptible to overfitting when fine-tuning them on small labeled target datasets. Therefore, in this work, we make a simple yet effective architectural change by introducing new learnable masked tokens to vision transformers whereby we reduce the effect of overfitting in transfer learning while retaining the desirable flexibility of vision transformers. Through several experiments based on two seminal self-supervised vision transformers, SiT and DINO, and several small target visual recognition tasks, we show consistent and significant improvements in the accuracy of the fine-tuned models across all target tasks.

Keywords: Vision transformer · Transfer learning · Computer vision

1 Introduction

Deep learning on small datasets usually relies on transferring a model that is pretrained on a large-scale source task [25]. Recent concurrent advancements in transformers [1] and self-supervised pretraining [9–12] have made self-supervised Vision Transformers (ViTs) a viable alternative to supervised pretraining of Convolutional Networks (ConvNets) [5–7]. Mainly based on self-attention [1,4] and multi-layer perceptron, ViTs have shown improved performance over the state-of-the-art ConvNets on large datasets [2,3,62,64] while retaining computational efficiency [23,24]. Considering that collecting large volumes of unlabeled data is becoming increasingly easier, a practical approach for transfer learning would be to pretrain ViTs with self-supervision and then fine-tune them on the downstream task with a small amount of labeled data.

This work is partially supported by KTH Digital Futures and Wallenberg AI, Autonomous Systems and Software Program (WASP).

The supremacy of ViTs for self-supervised learning over ConvNets can be attributed to the reduced inductive biases of ViTs which facilitates learning from the abundance of unlabelled data that is commonly available for self-supervised learning. However, this comes at a cost. That is, such flexibility of ViTs makes a fully-fledged fine-tuning of them on small target datasets susceptible to overfitting. This is due to the fact that the dense self-attention among image patches in ViTs is more likely to find spurious patterns in small datasets. This makes the locality and sparsity inductive biases of ConvNets, in contrast to ViTs, crucial for fine-tuning on small amount of labelled data.

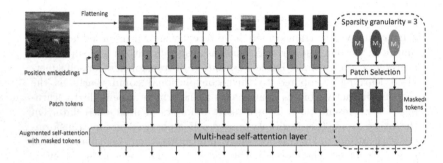

Fig. 1. An overview of the vision transformer with masked tokens. The part in the dashed rectangle is the proposed structural augmentation.

Therefore, in this paper, we aim at alleviating overfitting of fine-tuning self-supervised ViTs on small, domain-specific target sets while preserving their flexibility when learning from large unlabeled data. To this purpose, we propose *masked tokens*, a simple and flexible structural augmentation for self-attention layers. Each masked token aggregates a selected *subset* of patches to draw out sparse informative patterns. By varying the subset size from small to large, masked tokens encode the spatial information at different sparsity levels. We augment a self-attention layer by adding all the masked tokens to regular ones such that its output contains not only dependencies between patches, but also among different sparsity levels. Furthermore, we employ a data-driven method and two regularization techniques to *learn* the patch selection function for each masked token that can select patches with the most informative sparse patterns. The introduced sparsity makes the fine-tuning less prone to overfitting while the learnt selectivity retains the benefits of ViTs. Importantly, the proposed masked tokens are trained to encode details from local regions, reminiscent of the locality bias in the convolutional layers but with two key differences that the locality (i) can be learnt and (ii) can happen at various levels.

We summarize our contributions as below.

- We mitigate the overfitting of fine-tuned self-supervised ViTs by integrating sparsity and locality biases of ConvNets through masked tokens.

- We propose data-driven mechanisms to dynamically select the local region individually for each masked token and at different sparsity levels.
- We conduct extensive experiments on two self-supervised ViTs and various target tasks which show effectiveness of learnable masked tokens for ViTs.

2   Vision Transformers with Learnable Masked Tokens

2.1   Background: Vision Transformers

Given an input image $\mathbf{I} \in \mathrm{R}^{H \times W}$, it is divided into T non-overlapping patches $\{\mathbf{p}^i \in \mathrm{R}^{P \times P}\}_T$ and flattened into a sequence, where $T = \lceil \frac{HW}{P^2} \rceil$. A transformer [1] consists of L identical blocks with residual connections [33,34]. Each block processes the input patch sequence $\{\mathbf{p}^i\}$ as

$$\mathbf{Z}_0 = [\mathbf{h}_{cls}; \mathcal{F}(\mathbf{p}^1) + \mathbf{e}^1; \cdots ; \mathcal{F}(\mathbf{p}^T) + \mathbf{e}^T], \tag{1}$$

$$\mathbf{Z}'_l = \mathrm{MSA}(\mathrm{LN}(\mathbf{Z}_{l-1})) + \mathbf{Z}_{l-1}, \tag{2}$$

$$\mathbf{Z}_l = \mathrm{MLP}(\mathrm{LN}(\mathbf{Z}'_l)) + \mathbf{Z}'_l, \tag{3}$$

$$\mathbf{y} = \mathtt{softmax}(\mathrm{MLP}(\mathbf{Z}^0_L)), \tag{4}$$

where MSA, MLP, LN and $\mathtt{softmax}(\cdot)$ respectively indicate multi-head self-attentions, MLP with GELU, layer normalization and softmax. $\mathcal{F}(\cdot)$ is a convolutional feature extractor and $\{\mathbf{e}^i\}_T$ are position embeddings. Further, a learnable class token $\mathbf{h}_{cls} \in \mathrm{R}^d$ is used at the beginning of the sequence to globally represent entire image by taking an attention-weighted sum of every patch.

We augment ViT blocks by introducing *masked tokens*, which can alleviate overfitting by masking out redundant regions of input images. Given the sequence $\mathbf{Z}_l \backslash \mathbf{h}_{cls} = [\mathbf{z}^1_l; \cdots ; \mathbf{z}^T_l]$ of input tokens for layer $l+1$, we construct N masked tokens $\{\mathbf{s}^j_l\}_N$ via selecting and aggregating a subset of patch tokens for each $\mathbf{s}^j_l$ using a selection function $\mathcal{G}(\cdot, \cdot)$. More specifically, the subset $\mathbf{S}^j_l \subset \mathbf{Z}_l$ of tokens selected for $\mathbf{s}^j_l$ can be presented as

$$\mathbf{S}^j_l = \{\mathbf{z}^{i_1}_l, \cdots , \mathbf{z}^{i_{M_j}}_l\}, M_j = \lceil j \cdot \frac{T}{N} \rceil, \tag{5}$$

where M_j defines the sparsity level of $\mathbf{s}^j_l$ indicating the size of the informative sub regions for encoding. Then $\mathbf{s}^j_l$ can be produced by aggregating $\mathbf{S}^j_l$ using any permutation-invariant pooling. We use mean pooling in this work. Finally, the generated masked tokens for this layer are appended at the end of $\mathbf{Z}_l$:

$$\tilde{\mathbf{Z}}_l = [\mathbf{h}_{cls}; \mathbf{z}^1_l; \cdots ; \mathbf{z}^T_l; \mathbf{s}^1_l; \cdots ; \mathbf{s}^N_l]. \tag{6}$$

This augmented $\tilde{\mathbf{Z}}_l$ is then fed to the MSA layer instead of the original $\mathbf{Z}_l$. For two consecutive augmented MSA layers[1] l and $l+1$, where masked tokens $\mathbf{s}^j_l$

---

[1] We omit LN and MLP layers in between for convenience.

and self-attention output $\mathbf{z}_l^{T+j}$ are both there, we combine these two parts using a weighted summation as

$$\tilde{\mathbf{s}}_l^j = \alpha \mathbf{s}_l^j + (1 - \alpha)\mathbf{z}_l^{T+j}, \tag{7}$$

where α is a hyper-parameter set to 0.2 as default. In such cases, $\tilde{\mathbf{s}}_l^j$ will replace $\mathbf{s}_l^j$ in Eq. (6) as masked tokens.

Figure 1 illustrates the workflow of an augmented self-attention. We name the number of masked tokens N as the *sparsity granularity* since it spans the sparsity level, from sparse to dense, that masked tokens cover. It is also worth noting that despite the fact that a single masked token can only encode information for a region, with multi-head MSA layers, it can be extended to multiple regions.

2.2   Learning the Selection Function

We propose a data-driven approach to learn the patch selection function $\mathcal{G}(\cdot, \cdot)$ such that it can choose the most informative patches for masked tokens. We reformulate it as a corresponding ranking problem, where each masked token $\mathbf{s}_l^j$ takes the top M_j patch tokens based on ranking scores $\mathbf{o}_l^j = \{o_l^{i,j}\}_T$. To obtain $\mathbf{o}_l^j$, we define a set of new parameters $\{\mathbf{w}_l^j \in R^d\}_N$ to dot-product with each $\mathbf{z}_l^i$, whose score $o_l^{i,j}$ can be computed as

$$o_l^{i,j} = (\mathbf{z}_l^i)^\intercal \mathbf{w}_l^j. \tag{8}$$

We name $\mathbf{w}_l^j$ as *Masked Query Embedding* (MQE), it can be seen as a learned query that selects the (masked) tokens. It is worth mentioning that similar to position embedding, when N is changed, $\mathbf{w}_l^j$ can be interpolated to match the new sparsity granularity. Now the selection function $\mathcal{G}(\cdot, \cdot)$ can be further defined as

$$\mathcal{G}(\mathbf{z}_l^{1:T}, M_j; \mathbf{w}_l^j) = \mathtt{argsort}(\mathbf{o}_l^j)\mathbf{z}_l^{1:T}|_{1:M_j}, \tag{9}$$

where $\mathtt{argsort}(\cdot)$ returns a $T \times T$ matrix whose rows are one-hot vectors, indicating the location of i-th largest value at the i-th row. $\cdot|_{1:M_j}$ means it takes only the top M_j rows as the output.

To overcome the discrete nature of $\mathtt{argsort}(\cdot)$, we approximate it by a differentiable relaxation named SoftSort [14], denoted by $\mathtt{SS}(\cdot)$:

$$\mathtt{SS}(\mathbf{o}_l^j) = \mathtt{softmax}\left(\frac{|\mathtt{sort}(\mathbf{o}_l^j)\mathbf{1}^\intercal - \mathbf{1}(\mathbf{o}_l^j)^\intercal|}{\tau}\right), \tag{10}$$

where $\mathtt{softmax}(\cdot)$ is applied row-wise. $\mathtt{sort}(\cdot)$ returns a sorted input. $|\cdot|$ takes element-wise absolute value and τ is a temperature set to 0.1 by default. By replacing $\mathtt{argsort}$ with $\mathtt{SS}$, $\mathbf{w}_l^j$ can be learnt jointly with other network weights.

2.3   Regularizations on Masked Tokens

We additionally introduce two regularizations that directly work on masked tokens to stabilize the training. First, to avoid masked tokens collapsing due to overlapping patches [11,52], we add a linear classifier $\mathcal{C}(\cdot)$ at the top of the last layer to identify the sparsity index j associated to each masked token features. This is trained with the cross-entropy loss $\mathcal{L}_{\text{sparse}}$:

$$\mathcal{L}_{\text{sparse}} = -\frac{1}{N} \sum_{j=1}^{N} \delta_j^{\mathsf{T}} \log(\mathcal{C}(\mathbf{z}_L^{T+j})), \qquad (11)$$

where δ_j is the one-hot vector with one at element j. Conversely, we use contrastive loss [13], to have masked tokens of the same image with maximal similarities to each other. Specifically, given a batch of images $\{\mathbf{I}_k\}_K$, we consider any pair of the form $(\mathbf{z}_{k_1}^{T+j_1}, \mathbf{z}_{k_1}^{T+j_2})$ as a positive pair[2], and the rest as negative pairs. We compute the contrastive loss $\mathcal{L}_{\text{con}}$ between the positive pairs like

$$\mathcal{L}_{\text{con}}(\mathbf{z}_{k_1}^{T+j_1}, \mathbf{z}_{k_1}^{T+j_2}) = -\frac{\exp\left(\mathbf{cs}(\mathbf{z}_{k_1}^{T+j_1}, \mathbf{z}_{k_1}^{T+j_2})\right)}{\exp\left(\mathbf{cs}(\mathbf{z}_{k_1}^{T+j_1}, \mathbf{z}_{k_1}^{T+j_2})\right) + \sum_{j,k \neq k_1}^{N \cdot (K-1)} \exp\left(\mathbf{cs}(\mathbf{z}_{k_1}^{T+j_1}, \mathbf{z}_k^{T+j})\right)}, \qquad (12)$$

and the total contrastive loss $\mathcal{L}_{\text{total\_con}}$ as

$$\mathcal{L}_{\text{total\_con}} = \frac{1}{K \cdot N \cdot (N-1)} \sum_{k=1}^{K} \sum_{j_1=1}^{N} \sum_{j_2 \neq j_1}^{N} \log \mathcal{L}_{con}(\mathbf{z}_k^{T+j_1}, \mathbf{z}_k^{T+j_2}), \qquad (13)$$

here $\mathbf{cs}(\cdot)$ is a cosine-similarity function.

3   Experiments

In this section we evaluate the effectiveness of learning masked tokens on various image benchmarks. This aim of the proposed modifications is to improve the transferability of self-supervised ViTs. Thus, we mainly focus on fine-tuning pretrained models on small datasets. We only consider image-level classification tasks to simplify the architectural choices of the backbone.

3.1   Configurations

Baselines. We apply two state-of-the-art self-supervised ViTs as our pretraining schemes and baselines: SiT [8] and DINO [9].

- **SiT** [8] replaces the class token $\mathbf{h}_{cls}$ with two tokens, namely a rotation token $\mathbf{h}_{rot}$ and a contrastive token $\mathbf{h}_{contr}$, such that it can be trained by predicting image rotations [15] and maximizing the similarity between positive pairs [13]. Furthermore, it features another regularization task where corrupted inputs are reconstructed via inpainting.

---

[2] We remove the layer index l, and replace it with the image index k for convenience.

Table 1. Top-1 accuracy (%) for linear evaluations on CIFAR datasets. All the baseline performance are reported from [8].

Method	Backbone	CIFAR-10	CIFAR-100
DeepCluster [41]	ResNet-32	43.31	20.44
RotationNet [15]	ResNet-32	62.00	29.02
Deep InfoMax [42]	ResNet-32	47.13	24.07
SimCLR [12]	ResNet-32	77.02	42.13
Rel. Reasoning [43]	ResNet-32	74.99	46.17
Rel. Reasoning [43]	ResNet-56	77.51	47.90
SiT [8]	ViT-B/16	81.20	55.97
MT SiT (ours)	ViT-B/16	**81.98**	**57.18**

- **DINO** [9] takes a self-distillation paradigm by simultaneously updating the teacher with an exponential moving average and encouraging the student to have similar outputs as the teacher. Such objective is further optimised using multi-crops augmentation to ensure consistency between different scales.

Implementations. We implement our proposed augmentations based on their officially released codes in PyTorch. Our pilot studies show that augmenting many layers with masked tokens will show diminishing return. Thus, unless specified otherwise, we use a single ViT variant by replacing the MSA layers in the last four blocks with the augmented ones, and set the default sparsity granularity to 4. To reduce the computational cost we only do the token selection for the first of the four augmented blocks. For both baselines, we refer to their augmented ones with the prefix "MT". All experiments are done with 8 Nvidia A100 GPUs.

3.2 In-domain Transfer Learning

We first present the results using the SiT-based [8] pretraining on three datasets: CIFAR-10/100 [17] and STL-10 [19]. For a fair comparison, we follow the same experimental protocols as [8] including random seeds, hyper-parameters and data augmentations. In this way, we first train the model on the entire dataset using SiT losses, then fine-tune the model on a fully labeled subset. Since both source and target are from the same domain, we refer it as **In-domain Transfer Learning** (IdTL) in the rest of the paper. ViT-B/16 will be our default backbone.

IdTL for CIFAR-10 and CIFAR-100

Linear Evaluation. We first report the linear evaluation results in Table 1 to make sure that masked tokens won't degrade the pretrained features due to additional model complexities. As we can see, MT SiT can outperform ConvNet-based methods by significant margins of 4.47 percentage points on CIFAR-10 and 9.28 on CIFAR-100. However, such gains become smaller when compared

Table 2. Top-1 accuracy (%) of IdTL on CIFAR datasets. Referred to as 'few-shot' in [8].

Method	1%	10%	25%
CIFAR-10			
[43]	76.55	80.14	85.30
SiT [8]	74.78	87.16	92.90
MT SiT (ours)	**82.52**	**92.23**	**95.60**
CIFAR-100			
[43]	**46.10**	49.55	54.44
SiT [8]	27.50	53.72	67.58
MT SiT (ours)	24.51	**61.39**	**72.69**

with SiT, with only 0.78 for CIFAR-10 and 1.21 percentage point for CIFAR-100. This supports the assumption that the architectural change of introducing masked tokens would not make overfitting worse.

Fine-Tuning. Following [8], we fine-tune the MT SiT on subsets with different percentage of available labels. From Table 2, we can observe significant improvements over the SiT baseline in most cases. More specifically, we achieve 7.74, 5.07 and 2.70 percentage point improvements on CIFAR-10 with only 1%, 10% and 25% of labels. Moreover, in most cases MT SiT can achieve higher performance gain over the vanilla SiT when fine-tuning labels become much less, indicating the positive effects for reducing overfitting brought by masked tokens. On the other hand, while we can find similar improvements on CIFAR-100 with 10% and 25% labels, MT SiT performs worse than the SiT baseline and has a nearly 20 percentage point gap with the ConvNet baseline [43]. We argue that this is due to too few training samples to learn meaningful patterns on the target set. In such cases, fine-tuning can have a high variance and furthermore attentions between masked tokens may put an overly strong emphasis on localities, causing the drop in transferability.

IdTL for STL-10. Now we consider the STL-10 [19] dataset, which contains $100,000$ unlabeled and $5,000$ labeled training images. Thus, compared to CIFAR, it almost doubles the pretraining size while keeping the target set small. We directly fine-tune our models with all training labels without further dividing them into various subsets.

Fine-Tuning. Table 3 summarizes the fine-tuning results for STL-10. Similar to the CIFAR, MT SiT consistently outperforms the SiT and other ConvNet baselines with a small 1.84 percentage points margin, showing the relative effectiveness of involving masked tokens for fine-tuning. Moreover, the experiments on three popular benchmarks, so far, suggest that ViTs could benefit from masked tokens for IdTL on small datasets.

Table 3. IdTL comparisons with SOTAs on STL-10 dataset.

Method	Backbone	Fine-tuning (%)
Exemplars [37]	Conv-3	72.80
Artifacts [38]	Custom	80.10
ADC [39]	ResNet-34	56.70
Invariant Info Clustering [40]	ResNet-34	88.80
DeepCluster [41]	ResNet-34	73.37
RotationNet [15]	ResNet-34	83.22
Deep InfoMax [42]	AlexNet	77.00
Deep InfoMax [42]	ResNet-34	76.03
SimCLR [12]	ResNet-34	89.31
Relational Reasoning [43]	ResNet-34	89.67
SiT [8]	ViT-B/16	93.02
MT SiT (ours)	ViT-B/16	**94.86**

Table 4. Ablation studies of pretraining the MT SiT with different components on the STL-10 dataset.

Method	MT	$\mathcal{L}_{\text{sparse}}$	$\mathcal{L}_{\text{con}}$	$\mathcal{G}(\cdot,\cdot)$	Linear	Fine tuning
SiT [8]	-	-	-	-	78.58	93.02
MT SiT (ours)	✓				71.95	94.22
	✓	✓			68.77	93.89
	✓		✓		69.47	94.00
	✓			✓	77.71	94.44
	✓	✓	✓		78.75	94.78
	✓	✓	✓	✓	**78.99**	**94.86**

Ablation Study. We further perform ablation studies on STL-10 to understand how each component affects the performance. The corresponding results are listed in Table 4. Although the fine-tuning accuracy can be boosted by any of the individual components, randomly selecting patches (as opposed to $\mathcal{G}$), pretraining with no or partial regularizers has produced worse self-supervised features than the complete model. Therefore, all proposed components seem to be important for achieving the best performance.

Visualization. This additional qualitative study investigates how the learnt masked regions are spatially distributed by visualizing selected patches of each masked token at the first augmented block. We randomly sample 10 examples from STL-10 and highlight the positions of selected patches using different colors for each masked token in Fig. 2. Overall, in most cases, the majority of patches in the same sparsity level are spatially close to each other, forming local clusters that

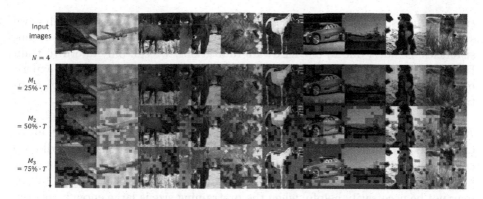

Fig. 2. Visualized examples from STL-10 datasets showing selected patch tokens at the first augmented MSA layer for each sparsity level using the learnt selection.

Table 5. Accuracy on ImageNet-1K. We list both results that reported from [9] (top) and trained by ourselves (bottom) to assure comparison is fair. 'FT' means 'fine-tune'.

Method	Arch	KNN	Linear	FT-10%	FT-20%
BYOL [11]	ViT-S	66.6	71.4	N/A	N/A
MoCov2 [10]	ViT-S	64.4	72.7	N/A	N/A
SwAV [16]	ViT-S	66.3	73.5	N/A	N/A
DINO [9]	ViT-S	73.3	76.0	N/A	N/A
DINO (300 epochs)	ViT-S/16	73.06	75.83	58.48	68.46
MT DINO (4 masked tokens)	ViT-S/16	73.10	75.93	57.71	68.69
MT DINO (28 masked tokens)	ViT-S/16	73.08	75.90	**60.26**	**70.02**

cover multiple small regions. This lends evidence that masked tokens can indeed encode local information from the informative sub-regions at various sparsity levels in the image. On the other hand, it is also surprising to see that low-level tokens tend to select the patches lying outside of the main interested object in many cases, which is slightly counter-intuitive. We conjecture about this observation as the usefulness of the associated contexts for reducing overfitting. While global attentions are pretrained to focus on the interested object due to large pretraining samples, the secondary contents also becomes informative and complimentary when the training size decreases during the fine-tuning. Low-level masked tokens are flexible enough to be tuned to capture such information.

IdTL for ImageNet-1K. Here we consider a larger pretraining source i.e., ImageNet-1K [18], which serves as a fundamental pretraining source for many small datasets. Since SiT [8] doesn't report ImageNet-pretrained results, we switch to another state-of-the-art baseline DINO [9] to avoid any setting inconsistencies. Besides, due to our hardware limitations, we can only afford to train DINO and MT DINO using ViT-small (ViT-S) [2]. Similar to previous experiments, we follow the same protocol provided by the official baseline implementation.

KNN and Linear Classification. Like [9], we do KNN ($K = 20$) and linear classification for self-supervised features first, whose results are in the middle two columns in Table 5. Compared with the baseline, MT DINO does not show clear improvements for either KNN or linear evaluation, implying that masked tokens may not be necessarily helpful when the pretraining size is large enough, as even the most informative localities are likely to be modelled by global attentions.

Fine-Tuning with Increased Number of Masked Tokens. We further inspect the fine-tuning on two subsets of ImageNet-1K with only 10% and 20% labels, and report their accuracy in the last two columns of Table 5. Surprisingly, we initially find MT DINO with default number of scale tokens are outperformed by the baseline on 10% labeled subset with a 0.77 percentage point margin. We then increase the sparsity granularity up to 28 and find the performances are boosted by 2.55 and 1.33 for each subset. We speculate that as the dataset size grows, there are enough samples for the global patch tokens to model some sparse and local patterns, therefore, more masked tokens are needed to become complementary in addition to the standard tokens. Thus, a proper sparsity granularity is also important. Moreover, comparing with Sect. 3.2, the performance gain significantly drops, implying masked tokens may become less effective as dataset size increases. This also coincides with [2] that ViTs may beat ConvNets as training set size grows.

Costs for Introducing Masked Tokens. Here we briefly discuss the additional model complexity and time consumption added by masked tokens. It is easy to see that the only new model weights are MQE for the first augmented block, bringing around 1%(4/384) more parameters than any projections of a self-attention layer in our implementation. Thus, the computational overhead of masked token is quite negligible. Meanwhile, the inference time using 4 and 28 tokens increases 1 and 5 s respectively on the entire ImageNet validation set, showing that the extra computational costs don't affect the ViT's efficiency too much.

3.3   Cross-Domain Transfer Learning

We now conduct experiments of transferring ImageNet-pretrained ViTs to various domain-specific datasets, which is closer to the mainstream transfer learning applications. Compared to Sect. 3.2, the target datasets exhibit a significant domain shift from the source, making them more challenging. Thus, we refer to such tasks as **Cross-domain Transfer Learning (CdTL)** in contrast to IdTL. We

Table 6. CdTL performance comparing with SOTA baselines for fine-grained recognition on CUB200-2011 dataset. The input size is 448. Baselines that outperform MT DINO are underlined.

Method	Backbone	Supervised pretraining?	Accuracy (%)
RA-CNN [30]	VGG-19	✓	85.30
ResNet-50 [6]	ResNet-50	✓	85.50
M-CNN [31]	VGG-16	✓	85.70
GP-256 [45]	VGG-16	✓	85.80
MaxEnt [46]	DenseNet161	✓	86.60
DFL-CNN [47]	ResNet-50	✓	87.40
Nts-Net [32]	ResNet-50	✓	87.50
Cross-X [50]	ResNet-50	✓	87.70
DCL [49]	ResNet-50	✓	87.80
CIN [48]	ResNet-101	✓	88.10
ViT [2]	ViT-B/16	✓	90.80
TransFG [51]	ViT-B/16	✓	91.70
DINO [9]	ViT-S/16	✗	86.47
MT DINO	ViT-S/16	✗	86.68
MT DINO (28 masked tokens)	ViT-S/16	✗	**87.38**

continue using ViT-S/16 [2] as the backbone and DINO [9] for self-supervised pretraining on ImageNet-1K.

Comparison with the State-of-the-Art. We compare the MT DINO with DINO and other related baselines on four small datasets from three different domains, CUB-200-2011 birds [20] for fine-grained recognition, SoybeanLocal and Cotton80 [21] for ultra fine-grained recognition, and COVID-CT [22] for medical imagery-based diagnosis.

Fine-Grained Classifications (FG). Table 6 lists the accuracy for MT DINO and other state-of-the-art baselines on CUB birds dataset. MT DINO can achieve ~ 1 percentage point improvement over vanilla DINO with 28 masked tokens and a comparable result with most ConvNet baselines. It is worth emphasising that computational cost prevents us from getting higher performance by either pretraining on larger datasets or using larger backbones. Besides, those outperforming baselines (underlined in the table) are achieved by extra mechanisms such as *fully-supervised* pretraining on larger datasets using more powerful backbones [2,51], or fine-tuned with FG-specific losses [32,48–50]. We believe this does not undermine the effectiveness of masked tokens.

Ultra Fine-Grained Classifications (UFG). Comparing to the FG, UFG requires more subtle details to distinguish its categories, effectively rendering the available

Table 7. CdTL performances for UFG datasets. Same as [21], the input size is set to 384.

Method	Backbone	Supervised pretraining?	Soybean Local	Cotton80
Nts-Net [32]	ResNet-50	✓	42.67	51.67
ADL [35]	ResNet-50	✓	34.67	43.75
Cutmix [36]	ResNet-50	✓	26.33	45.00
MoCov2 [10]	ResNet-50	✗	32.67	45.00
BYOL [11]	ResNet-50	✗	33.17	52.92
SimCLR [12]	ResNet-50	✗	37.33	51.67
ViT [2]	ViT-B/16	✓	39.33	51.25
BeiT [3]	ViT-B/16	✓	38.67	**53.75**
TransFG [51]	ViT-B/16	✓	38.67	45.84
DINO [9]	ViT-S/16	✗	<u>41.33</u>	49.58
MT DINO	ViT-S/16	✗	41.17	51.67
MT DINO (28 masked tokens)	ViT-S/16	✗	**43.33**	**53.75**

Table 8. CdTL performances for COVID-CT dataset.

Method	Backbone	Supervised pretraining?	Accuracy (%)
DenseNet [44]	DenseNet-169	✓	84.65
DINO	ViT-S/16	✗	83.25
MT DINO	ViT-S/16	✗	82.76
MT DINO (28 masked tokens)	ViT-S/16	✗	**85.22**

data even smaller. Table 7 shows the comparison with multiple ViT and ConvNet baselines on SoybeanLocal and Cotton80 datasets, which only have 600 and 240 fine-tuning samples for each. It is encouraging to see that MT DINO performs 2.00 and 4.17 percentage points better than DINO with 28 masked tokens on the two datasets respectively, and outperforms most baselines. Especially, MT DINO can improve over ViT baselines [2,3,51] that use more powerful backbones and supervised pretraining, demonstrating the usefulness of masked tokens for reducing overfitting. Similar to ImageNet results, more scale tokens help improve the fine-tuning performance.

Medical Imagery-Based Diagnosis. We conduct domain-specific transferability experiments on COVID-CT dataset [22], which contains only 425 samples for fine-tuning. The results are shown in Table 8. Similar to the UFG, we achieve a better performance than baselines with increased masked token number than the default case, which provides corroborates to our assumption that higher performance can be achieved with more masked tokens.

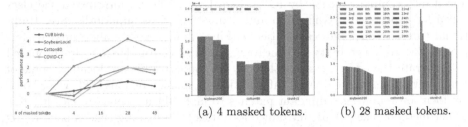

(a) 4 masked tokens.   (b) 28 masked tokens.

Fig. 3. Performance variations versus masked tokens numbers (sparsity granularity) on CdTL tasks.

Fig. 4. Average attention values of classification head and each masked tokens across different datasets and token numbers.

The Impact of Sparsity Granularity. Inspired by previous observations *w.r.t.* the number of masked tokens, we further study the relationship between the performance and the masked tokens in two more experiments.

Granularity vs. Performance. We plot the line chart in Fig. 3 to show the performance gains with different masked token numbers on all four CdTL datasets. Overall, despite a few exceptions, the performance increases as masked token number grows for all datasets, confirming that more sparsity levels can yield better results. However, after a certain point, the performance begins to drop as number continue growing. This is expected as too many masked tokens can carry many overlapped patches, leading to a higher chance of overfitting on smaller sets. Thus, it is not always good to keep a large masked token number.

Attentions for Masked Tokens. We additionally compute the attention values between the class token $\mathbf{h}_{cls}$ and each masked tokens, and visualize their means across multi-heads and samples for UFG and COVID-CT datasets in Fig. 4. Basically, the patterns of attention are similar when masked token number is small, where attentions are uniformly distributed across each masked token. As token number increases, these patterns act differently for each dataset. For Cotton80, the class token has more dependencies with both the low and high sparsity levels than the mid level, while such dependencies tend to decrease from low to high for SoybeanLocal and COVID-CT as their attention values drop when the sparsity level goes higher. Especially on COVID-CT, tokens with the lower levels have significantly higher attention than others, indicating the class token relies more on the lower sparsity level information.

4   Related Works

Vision Transformers. Inspired by works in NLP [56,57], transformers are introduced into computer vision by iGPT [55]. Later, ViT [2] introduced the class token for supervised classification and demonstrated its superiority over traditional CovNets on large-scale datasets. Since it may yield suboptimal performance due to venerability of overfitting, works such as [58–64] are proposed to moderate the effect by strengthening the inductive bias of locality. Some of which try to aggregate spatial information in smaller regions [58], where others focus on removing redundant patches to highlight informative ones [62–64]. Other methods like [59–61] introduce localities by reshaping tokens back to 2D grids and forward them to a convolution kernel just like ConvNets.

Self-supervised Learning. Numerous techniques are introduced to train a visual model in a self-supervised fashion. Some earlier works do this by predicting patch orders [26], image rotations [15], or colorization [27]. Recently, contrastive-based methods have become increasingly popular [10,12,28,29,52], which augment the input image into multiple views and optimise the model by maximizing the similarity between positive pairs. To prevent from collapsing, [10,12,28] propose to increase the number of informative negative pairs by constructing large memory banks or batches, while other works [11,52] build non-gradient-based targets without explicitly involving negative pairs. Besides, a few methods focus on clustering-based training [16,53,54], or using transformers as backbones [8,9].

5   Conclusions

We tackle the problem of alleviating overfitting for fine-tuned self-supervised ViTs on small, domain-specific datasets. We introduce masked tokens, which mask out redundant regions by aggregating a subset of informative patch tokens. Defined by their sparsity levels, multiple masked tokens encode different sub regions of input images with sizes from small to large. With the proposed patch selection and regularizations, masked tokens can be trained to determine most interesting encoding regions in a data-driven manner. Via integrating masked tokens with self-attentions, we augment ViTs with sparsity and locality biases without altering their core structures. We conduct extensive experiments on various datasets and have found that masked tokens can more effectively capture local secondary contents, which can be complimentary to the standard global attention. Thus with a proper number of masked tokens, an augmented ViT is more amenable to small sets, and retains capabilities of learning rich representations when training sets grow larger.

Acknowledgements. The project was partially funded by KTH Digital Futures and the Wallenberg AI, Autonomous Systems and Software Program (WASP) funded by the Knut and Alice Wallenberg Foundation.

References

1. Vaswani, A., et al.: Attention is all you need. Advances In: Neural Information Processing Systems, pp. 5998–6008 (2017)
2. Dosovitskiy, A., et al.: An image is worth 16x16 words: transformers for image recognition at scale. ArXiv Preprint ArXiv:2010.11929. (2020)
3. Touvron, H., Cord, M., Douze, M., Massa, F., Sablayrolles, A. & Jégou, H. Training data-efficient image transformers & distillation through attention. In: International Conference on Machine Learning, pp. 10347–10357 (2021)
4. Bahdanau, D., Cho, K. & Bengio, Y.: Neural machine translation by jointly learning to align and translate. ArXiv Preprint ArXiv:1409.0473. (2014)
5. Simonyan, K., Zisserman, A.: Very deep convolutional networks for large-scale image recognition. ArXiv Preprint ArXiv:1409.1556. (2014)
6. He, K., Zhang, X., Ren, S., Sun, J.: Deep residual learning for image recognition. In: Proceedings of The IEEE Conference on Computer Vision and Pattern Recognition, pp. 770–778 (2016)
7. Huang, G., Liu, Z., Van Der Maaten, L., Weinberger, K.: Densely connected convolutional networks. In: Proceedings of The IEEE Conference on Computer Vision and Pattern Recognition, pp. 4700–4708 (2017)
8. Atito, S., Awais, M., Kittler, J.: SIT: Self-supervised vision transformer. ArXiv Preprint ArXiv:2104.03602 (2021)
9. Caron, M., Touvron, H., Misra, I., Jégou, H., Mairal, J., Bojanowski, P., Joulin, A.: Emerging properties in self-supervised vision transformers. ArXiv Preprint ArXiv:2104.14294 (2021)
10. Chen, X., Fan, H., Girshick, R., He, K.: Improved baselines with momentum contrastive learning. ArXiv Preprint ArXiv:2003.04297 (2020)
11. Grill, J., et al.: Bootstrap your own latent: A new approach to self-supervised learning. ArXiv Preprint ArXiv:2006.07733 (2020)
12. Chen, T., Kornblith, S., Norouzi, M., Hinton, G.: A simple framework for contrastive learning of visual representations. In: International Conference on Machine Learning, pp. 1597–1607 (2020)
13. Park, T., Efros, A.A., Zhang, R., Zhu, J.-Y.: Contrastive learning for unpaired image-to-image translation. In: Vedaldi, A., Bischof, H., Brox, T., Frahm, J.-M. (eds.) ECCV 2020. LNCS, vol. 12354, pp. 319–345. Springer, Cham (2020). https://doi.org/10.1007/978-3-030-58545-7_19
14. Prillo, S., Eisenschlos, J.: SoftSort: a continuous relaxation for the argsort operator. International Conference on Machine Learning, pp. 7793–7802 (2020)
15. Gidaris, S., Singh, P., Komodakis, N.: Unsupervised representation learning by predicting image rotations. ArXiv Preprint ArXiv:1803.07728 (2018)
16. Caron, M., Misra, I., Mairal, J., Goyal, P., Bojanowski, P., Joulin, A.: Unsupervised learning of visual features by contrasting cluster assignments. ArXiv Preprint ArXiv:2006.09882 (2020)
17. Krizhevsky, A., Hinton, G., Others Learning multiple layers of features from tiny images. (Citeseer 2009
18. Deng, J., Dong, W., Socher, R., Li, L., Li, K., Fei-Fei, L. ImageNet: a large-scale hierarchical image database. In: 2009 IEEE Conference on Computer Vision and Pattern Recognition, pp. 248–255 (2009)
19. Coates, A., Ng, A., Lee, H.: An analysis of single-layer networks in unsupervised feature learning. In: Proceedings of the Fourteenth International Conference on Artificial Intelligence and Statistics, pp. 215–223 (2011)

20. Welinder, P., Branson, S., Mita, T., Wah, C., Schroff, F., Belongie, S., Perona, P.: Caltech-UCSD birds 200 (California Institute of Technology, 2010)

21. Yu, X., Zhao, Y., Gao, Y., Yuan, X., Xiong, S.: Benchmark platform for ultra-fine-grained visual categorization beyond human performance. In: Proceedings Of The IEEE/CVF International Conference on Computer Vision, pp. 10285–10295 (2021)

22. Zhao, J., Zhang, Y., He, X., Xie, P.: Covid-CT-dataset: a CT scan dataset about Covid-19. ArXiv Preprint ArXiv:2003.13865 490 (2020)

23. Brown, T., et al.: Language models are few-shot learners. ArXiv Preprint ArXiv:2005.14165 (2020)

24. Lepikhin, D., et al.: Gshard: scaling giant models with conditional computation and automatic sharding. ArXiv Preprint ArXiv:2006.16668 (2020)

25. Pan, S., Yang, Q.: A survey on transfer learning. IEEE Trans. Knowl. Data Eng. **22**, 1345–1359 (2009)

26. Noroozi, M., Favaro, P.: unsupervised learning of visual representations by solving jigsaw puzzles. In: Leibe, B., Matas, J., Sebe, N., Welling, M. (eds.) ECCV 2016. LNCS, vol. 9910, pp. 69–84. Springer, Cham (2016). https://doi.org/10.1007/978-3-319-46466-4_5

27. Zhang, R., Isola, P., Efros, A.A.: Colorful image colorization. In: Leibe, B., Matas, J., Sebe, N., Welling, M. (eds.) ECCV 2016. LNCS, vol. 9907, pp. 649–666. Springer, Cham (2016). https://doi.org/10.1007/978-3-319-46487-9_40

28. Wu, Z., Xiong, Y., Yu, S., Lin, D.: Unsupervised feature learning via non-parametric instance discrimination. In: Proceedings Of The IEEE Conference on Computer Vision and Pattern Recognition, pp. 3733–3742 (2018)

29. Oord, A., Li, Y., Vinyals, O.: Representation learning with contrastive predictive coding. ArXiv Preprint ArXiv:1807.03748 (2018)

30. Zheng, H., Fu, J., Mei, T., Luo, J.: Learning multi-attention convolutional neural network for fine-grained image recognition. In: Proceedings Of The IEEE International Conference on Computer Vision, pp. 5209–5217 (2017)

31. Wei, X., Xie, C., Wu, J., Shen, C.: Mask-CNN: localizing parts and selecting descriptors for fine-grained bird species categorization. Pattern Recogn. **76**, 704–714 (2018)

32. Nawaz, S., Calefati, A., Caraffini, M., Landro, N., Gallo, I.: Are these birds similar: Learning branched networks for fine-grained representations. In: 2019 International Conference on Image and Vision Computing New Zealand (IVCNZ), pp. 1–5 (2019)

33. Wang, Q., Li, B., Xiao, T., Zhu, J., Li, C., Wong, D., Chao, L.: Learning deep transformer models for machine translation. ArXiv Preprint ArXiv:1906.01787 (2019)

34. Baevski, A., Auli, M.: Adaptive input representations for neural language modeling. ArXiv Preprint ArXiv:1809.10853 (2018)

35. Choe, J.., Shim, H.: Attention-based dropout layer for weakly supervised object localization. In: Proceedings Of The IEEE/CVF Conference On Computer Vision And Pattern Recognition, pp. 2219–2228 (2019)

36. Yun, S., Han, D., Oh, S., Chun, S., Choe, J., Yoo, Y.: CutMix: regularization strategy to train strong classifiers with localizable features. In: Proceedings Of The IEEE/CVF International Conference On Computer Vision, pp. 6023–6032 (2019)

37. Dosovitskiy, A., Springenberg, J., Riedmiller, M., Brox, T.: Discriminative unsupervised feature learning with convolutional neural networks. Adv. Neural. Inf. Process. Syst. **27**, 766–774 (2014)

38. Jenni, S., Favaro, P.: Self-supervised feature learning by learning to spot artifacts. In: Proceedings of The IEEE Conference on Computer Vision And Pattern Recognition, pp. 2733–2742 (2018)

39. Haeusser, P., Plapp, J., Golkov, V., Aljalbout, E., Cremers, D.: Associative deep clustering: training a classification network with no labels. In: German Conference On Pattern Recognition, pp. 18–32 (2018)

40. Ji, X., Henriques, J., Vedaldi, A.: Invariant information clustering for unsupervised image classification and segmentation. In: Proceedings Of The IEEE/CVF International Conference On Computer Vision, pp. 9865–9874 (2019)

41. Caron, M., Bojanowski, P., Joulin, A., Douze, M.: Deep clustering for unsupervised learning of visual features. In: Ferrari, V., Hebert, M., Sminchisescu, C., Weiss, Y. (eds.) Computer Vision – ECCV 2018. LNCS, vol. 11218, pp. 139–156. Springer, Cham (2018). https://doi.org/10.1007/978-3-030-01264-9_9

42. Hjelm, R., et al.: Learning deep representations by mutual information estimation and maximization. ArXiv Preprint ArXiv:1808.06670 (2018)

43. Patacchiola, M., Storkey, A.: Self-supervised relational reasoning for representation learning. ArXiv Preprint ArXiv:2006.05849 (2020)

44. He, X., Yang, X., Zhang, S., Zhao, J., Zhang, Y., Xing, E., Xie, P.: Sample-efficient deep learning for COVID-19 diagnosis based on CT scans. Medrxiv (2020)

45. Wei, X., Zhang, Y., Gong, Y., Zhang, J., Zheng, N.: Grassmann pooling as compact homogeneous bilinear pooling for fine-grained visual classification. In: Ferrari, V., Hebert, M., Sminchisescu, C., Weiss, Y. (eds.) ECCV 2018. LNCS, vol. 11207, pp. 365–380. Springer, Cham (2018). https://doi.org/10.1007/978-3-030-01219-9_22

46. Dubey, A., Gupta, O., Raskar, R., Naik, N.: Maximum-entropy fine-grained classification. ArXiv Preprint ArXiv:1809.05934 (2018)

47. Wang, Y., Morariu, V., Davis, L.: Learning a discriminative filter bank within a CNN for fine-grained recognition. In: Proceedings Of The IEEE Conference on Computer Vision And Pattern Recognition, ,pp. 4148–4157 (2018)

48. Gao, Y., Han, X., Wang, X., Huang, W., Scott, M.: Channel interaction networks for fine-grained image categorization. In: Proceedings Of The AAAI Conference On Artificial Intelligence. **34**, 10818–10825 (2020)

49. Chen, Y., Bai, Y., Zhang, W., Mei, T.: Destruction and construction learning for fine-grained image recognition. In: Proceedings Of The IEEE/CVF Conference on Computer Vision and Pattern Recognition, pp. 5157–5166 (2019)

50. Luo, W., et al,: Cross-X learning for fine-grained visual categorization. In: Proceedings Of The IEEE/CVF International Conference On Computer Vision, pp. 8242–8251 (2019)

51. He, J., et al.:TransFG: a Transformer Architecture for fine-grained recognition. ArXiv Preprint ArXiv:2103.07976 (2021)

52. Chen, X., He, K.: Exploring simple Siamese representation learning. In: Proceedings Of The IEEE/CVF Conference on Computer Vision and Pattern Recognition, pp. 15750–15758 (2021)

53. Asano, Y., Rupprecht, C.., Vedaldi, A.: Self-labelling via simultaneous clustering and representation learning. ArXiv Preprint ArXiv:1911.05371 (2019)

54. Li, J., Zhou, P., Xiong, C., Hoi, S..: Prototypical contrastive learning of unsupervised representations. ArXiv Preprint ArXiv:2005.04966 (2020)

55. Chen, M., Radford, A., Child, R., Wu, J., Jun, H., Luan, D., Sutskever, I.: Generative pretraining from pixels. In: International Conference On Machine Learning, pp. 1691–1703 (2020)

56. Devlin, J., Chang, M., Lee, K.,Toutanova, K.: BERT: pre-training of deep bidirectional transformers for language understanding. ArXiv Preprint ArXiv:1810.04805 (2018)
57. Radford, A., Narasimhan, K., Salimans, T., Sutskever, I.: Improving language understanding by generative pre-training (2018)
58. Yuan, L., et al.: Tokens-to-token vit: Training vision transformers from scratch on ImageNet. ArXiv Preprint ArXiv:2101.11986 (2021)
59. Yuan, K., Guo, S., Liu, Z., Zhou, A., Yu, F., Wu, W.:Incorporating convolution designs into visual transformers. ArXiv Preprint ArXiv:2103.11816 (2021)
60. Li, Y., Zhang, K., Cao, J., Timofte, R., Van Gool, L.: Localvit: Bringing locality to vision transformers. ArXiv Preprint ArXiv:2104.05707 (2021)
61. Hudson, D., Zitnick, C.: Generative adversarial transformers. ArXiv Preprint ArXiv:2103.01209 (2021)
62. Rao, Y., Zhao, W., Liu, B., Lu, J., Zhou, J., Hsieh, C.: Dynamicvit: efficient vision transformers with dynamic token sparsification. Adv. Neural. Inf. Process. Syst. **34**, 13937–13949 (2021)
63. Liang, Y., Ge, C., Tong, Z., Song, Y., Wang, J., Xie, P.: Not all patches are what you need: Expediting vision transformers via token reorganizations. ArXiv Preprint ArXiv:2202.07800 (2022)
64. Tang, Y., Han, K., Wang, Y., Xu, C., Guo, J., Xu, C., Tao, D.: Patch slimming for efficient vision transformers. In: Proceedings Of The IEEE/CVF Conference On Computer Vision And Pattern Recognition, pp. 12165–12174 (2022)

Rethinking the Misalignment Problem in Dense Object Detection

Yang Yang[1,2], Min Li[1,2(✉)], Bo Meng[1,3], Zihao Huang[1,2], Junxing Ren[1,2], and Degang Sun[1,2]

[1] Institute of Information Engineering, Chinese Academy of Sciences, Beijing, China
{yangyang1995,limin,renjunxing,sundegang}@iie.ac.cn
[2] School of Cyber Security, University of Chinese Academy of Sciences, Beijing, China
[3] Beijing Institute of Technology, Beijing, China

Abstract. Object detection aims to localize and classify the objects in a given image, and these two tasks are sensitive to different object regions. Therefore, some locations predict high-quality bounding boxes but low classification scores, and some locations are quite the opposite. A misalignment exists between the two tasks, and their features are spatially entangled. In order to solve the misalignment problem, we propose a plug-in **S**patial-disentangled and **T**ask-aligned operator (SALT). By predicting two task-aware point sets that are located in each task's sensitive regions, SALT can reassign features from those regions and align them to the corresponding anchor point. Therefore, features for the two tasks are spatially aligned and disentangled. To minimize the difference between the two regression stages, we propose a **S**elf-distillation regression (SDR) loss that can transfer knowledge from the refined regression results to the coarse regression results. On the basis of SALT and SDR loss, we propose SALT-Net, which explicitly exploits task-aligned point-set features for accurate detection results. Extensive experiments on the MS-COCO dataset show that our proposed methods can consistently boost different state-of-the-art dense detectors by ∼2 AP. Notably, SALT-Net with Res2Net-101-DCN backbone achieves 53.8 AP on the MS-COCO *test-dev*.

Keywords: Object detection · Misalignment problem · Spatial disentanglement

1 Introduction

The main goal of object detection contains two tasks, one is to give the accurate location of the object in an image (i.e., regression), and the other is to predict the category of the object (i.e., classification). During the inference step, the regression and classification results predicted from the same location are paired together as the detection result. Then the NMS algorithm is usually applied to remove redundant detection results by taking the classification scores as the

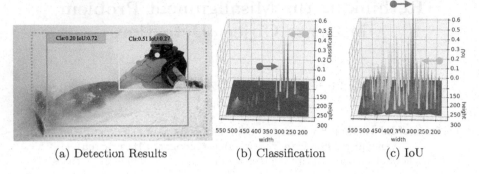

(a) Detection Results   (b) Classification   (c) IoU

Fig. 1. Illustration of the spatial misalignment of classification and regression. In (a), the blue box denotes the ground truth, and the other boxes are the detection results of ATSS [34]. The two points are the locations where the detection results are predicted. (b) and (c) are the distributions of classification and IoU scores over all image pixels, and "IoU" denotes the intersection over union between the predicted box and the ground truth. (Color figure online)

ranking keywords. For the same instance, the detection result with a high classification score will be kept, while others are filtered out. However, the natures of these two tasks are so distinct that they require features from different object locations [22]. As shown in Fig. 1, the classification and regression quality (i.e., IoU) scores from the same location can be quite different. Classification focus on the salient part of the object (e.g., the head of the person), while regression is sensitive to the whole object, especially for its border part. *Therefore, the prediction distributions of the two tasks are misaligned.* The detection result with a high classification score can have low-quality regression prediction and vice versa.

We model the prediction qualities of the two tasks as two Discrete distributions. Therefore, the goal of solving the misalignment problem is *bridging the gap between these two distributions (i.e., minimizing the distance of their peak positions).*

CNN-based dense detectors utilize a coupled or decoupled head to conduct classification and regression. As illustrated in Fig. 2(a), the coupled head predicts the classification and regression results based on the shared features [15,19,20]. As a result, the coupled head structure introduces feature conflicts between the two tasks and makes them compromise each other. To solve this problem, the decoupled head structure [28] is proposed and has been widely adopted in recent years [16,22,23]. As shown in Fig. 2(b), the decoupled head utilizes two parrel sub-networks to perform regression and classification, respectively. This could alleviate the conflict problem by reducing the shared parameters. However, the point features (i.e., the two orange points) that predict the detection result still share the identical receptive field. In conclusion, *both the coupled and decoupled heads predict the classification and regression results from the spatially identical and entangled features.* Considering the difference in their spatial sensitivity, the

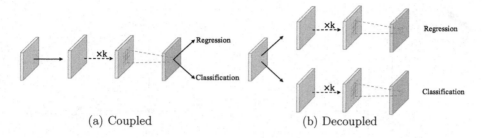

Fig. 2. Illustration of the entangled features in the coupled and decoupled heads.

entangled features inevitably make a location prefer one task over the other one, thereby compounding the misalignment problem.

In this paper, we propose a plug-in operator to address the misalignment problem: the **S**patial-disentangled and **T**ask-aligned operator (SALT). The first stage of our network is the coarse regression predictions made by a simple Dirac delta decoder [34]. After that, SALT predicts two sets of spatial-disentangled points to represent each task's sensitive regions, respectively. Then we use bilinear interpolation to reassign features from those regions to the corresponding anchor point. In the second stage, SALT utilizes spatial-disentangled and task-aligned features to make refined predictions with a General distribution decoder [12]. Therefore, a single anchor point can obtain accurate regression and classification predictions simultaneously. Feature reassignment can bring the peak positions of the two Discrete distributions closer so that SALT can weaken the impact of the misalignment problem.

In order to minimize the difference between the first and second stage predictions, we also propose a novel **S**elf-**d**istillation **r**egression (SDR) loss, making the coarse predictions learn from the refined predictions. As a result, the final performance got improved without any extra inference cost.

1. We propose an operator that can generate spatial-disentangled and task-aligned features for regression and classification, respectively.
2. The proposed operator can be easily plugged into most dense object detectors and bring a considerable improvement of ~2 AP.
3. Our proposed SDR loss can also boost the overall performance in an inference cost-free fashion.
4. Without bells and whistles, our best single-scale model (Res2Net-101-DCN) yields 51.5 AP on the COCO *test-dev* set, which is very competitive results among dense object detectors.

2   Related Work

Misalignment: Dense detectors, such as IoU-aware [27], FCOS [23] and PAA [10] apply an extra branch to predict the regression confidence and combine it

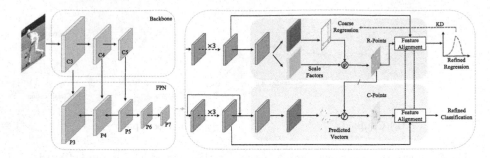

Fig. 3. Architecture of SALT-Net. Our proposed architecture consists of a backbone, an FPN (P3-P7), and two subnetworks for classification and regression, respectively. "φ" and "γ" denote Eqs. (2) and (5), respectively. "/" denotes the gradient flow detachment. "KD" denotes our proposed self-distillation approach. "R-points" and "C-Points" denote the regression-aware and classification-aware points, respectively.

with the classification confidence as the detection score. Different from previous methods, GFL [13] and VFNet [31] propose a joint representation format by merging the regression confidence and classification result to eliminate the inconsistency between training and inference. TOOD [6] proposes a prediction alignment method that predicts the offset between each location and the best anchor and then readjusts the prediction results. Guided Anchoring [25], RefineDet [35], and SRN [3] learn an offset field for the preset anchor and then utilize a feature adaption module to extract features from the refined anchors. RepPoints [29] and VFNet [31] utilize the deformable convolution [4] to extract accurate point feature. However, all the aforementioned methods extract features for regression and classification from the same locations, without considering their spatial preference. That is, the features for these tasks are spatially entangled, which leads to inferior performance.

Self-distillation: Model distillation [7] usually refers to transferring knowledge from a pre-trained heavy teacher network to a compact student network. DML [36] provides a new paradigm that a pre-trained teacher is no longer needed and all the student counterparts are trained simultaneously in a cooperative peer-teaching manner. Following this paradigm, many self-distillation approaches [8,14,30,32] are proposed for classification knowledge transfer learning. However, transferring regression knowledge of object detection has been proven to be difficult [9,26], as different locations of an image have different contributions to the regression task. LGD [33] is the only self-distillation approach for general object detection, which proposes an intra-object knowledge mapper that generates a better feature pyramid and then performs distillation with feature imitation. This approach provides performance gains but also introduces too many auxiliary layers.

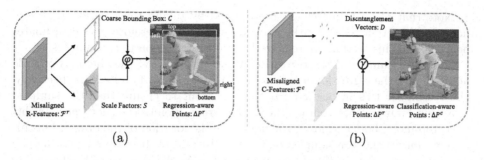

Fig. 4. (a) is the illustration of the regression-aware points. "φ" and the white box denotes Eq. (2) and the coarse bounding box prediction, respectively. (b) is the illustration of the spatial disentanglement method. "γ" denotes Eq. (5).

3   Proposed Approach

In this section, we first detail the proposed operator SALT. Then we introduce our self-distillation approach that enables the first-stage decoder to learn from the second-stage decoder. Finally, we introduce the loss function of SALT-Net.

3.1   SALT: Regression-Aware Points

As shown in Fig. 4(a), given the misaligned regression features $\mathcal{F}^r$ from the last layer of the regression tower (i.e., the $3\times$ convolutions shown in Fig. 3), SALT first predicts the coarse bounding box $\mathcal{C}$ with the Dirac delta decoder, as in [23,34]. The coarse bounding box is represented by the top-left corner and its width and height (i.e., $(xmin, ymin, w, h)$).

Then SALT predicts the scale Factors $\mathcal{S}$ that measures the normalized distances between the top-left corner of the coarse bounding box and the regression-sensitive regions (i.e., regression-aware points $\Delta\mathcal{P}^r$). Scale factors $\mathcal{S}$ and the coarse bounding box $\mathcal{C}$ are obtained by only two convolution layers, i.e.:

$$\begin{cases} \mathcal{C} = \delta(conv_c(\mathcal{F}^r)) \\ \mathcal{S} = \sigma(conv_s(\mathcal{F}^r)) \end{cases} \tag{1}$$

where σ and δ are Sigmoid and ReLU, respectively. $\mathcal{C} \in \mathbb{R}^{H \times W \times 4}$, $\mathcal{S} \in \mathbb{R}^{H \times W \times (2N-4)}$ and N is the number of the regression-aware points. Then the location of i-th regression-aware point Δp_i can be obtained with Eq. (2):

$$\begin{cases} x_i = x_{\min} + w * \mathcal{S}_{ix} \\ y_i = y_{\min} + h * \mathcal{S}_{iy} \end{cases} \tag{2}$$

where $(xmin, ymin)$ is the location of the top-left corner of the coarse bounding box $\mathcal{C}$, and $(\mathcal{S}_{ix}, \mathcal{S}_{iy})$ are scale factors that measure the normalized distance between the i-th point and top-left corner. Therefore, the location of the regression-aware points $\Delta\mathcal{P}^r$ can be represented by:

$$\Delta\mathcal{P}^r = \{\Delta p_i^r\}_{i=1}^N \tag{3}$$

Note that all coordinates are represented by taking the location that predicts the detection result as the coordinate origin. Therefore, the coordinates mentioned in this section are relative locations, not absolute coordinates.

The total number of the regression-aware points is N, the channels of scale factors $\mathcal{S}$ and the regression-aware points $\Delta\mathcal{P}^r$ are $2N-4$ and $2N$, respectively. The reason for this inconstancy is that we want to ensure that the sampled regression-aware points contain the four extreme points (i.e., left-most, right-most, top-most, bottom-most), which encode the location of the object. On this account, four points are sampled on the four bounds of the coarse bounding box (i.e., the green points in Fig. 4 (a)), respectively. As the location of the bounding box has been predicted, four axial coordinates of the extreme points are preset and do not need to be learned (i.e., $xmin, ymin, xmin + w, ymin + h$).

3.2 SALT: Classification-Aware Points

Regression and classification are sensitive to different areas of the object. For this reason, extracting features from the regression-interested-locations hinders the detection performance. Therefore, SALT contains a spatial disentangle module to guide the classification branch to generate a set of classification-aware points. As shown in Fig. 4(b), the regression-aware points act as the shape hypothesis of the object to be classified. In other words, we take the regression-aware points as a point-set anchor for predicting the classification-aware points.

Similar to the scale factors, this module also consists of only one convolution layer. As shown in Fig. 4(b), given the feature map $\mathcal{F}^c$ from the last layer of the classification tower, the disentanglement vectors $\mathcal{D}$ are obtained by:

$$\mathcal{D} = \delta(conv_d(\mathcal{F}^c)) \tag{4}$$

With the regression-aware points $\Delta\mathcal{P}^r$ taken as the point-set anchor, we propose two functions to generate the classification-aware points, as illustrated by Eqs. (5) and (6). We choose Eq. (5) as the final prediction strategy. Details and analysis can be found in Sect. 4.2.

$$\Delta\mathcal{P}^c = e^{\mathcal{D}} \cdot \Delta\mathcal{P}^r \tag{5}$$

$$\Delta\mathcal{P}^c = \mathcal{D} + \Delta\mathcal{P}^r \tag{6}$$

To make sure the learning process of classification and regression are independent of each other. The gradient flow of the regression-aware points $\Delta\mathcal{P}^r$ is detached from the classification branch. $\Delta\mathcal{P}^r$ only serves as the prior knowledge in this module. Therefore, the supervision of the classification task does not affect the learning of regression-aware points.

3.3 SALT: Feature Alignment

The regression-aware and classification-aware points are located in each task's sensitive regions, and they are spatially misaligned. Therefore, we aggregate

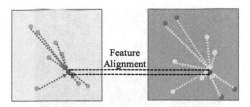

Fig. 5. Illustration of the feature alignment method. The red point denotes the location that predicts the detection result. (Color figure online)

features from those regions to the same anchor point (shown in Fig. 5). Given the learned point set $\Delta \mathcal{P} = \{\Delta p_i\}_{i=1}^{N}$, we use the bilinear interpolation to make $\Delta \mathcal{P}$ differentiable. Let $\{p_i\}_{i=1}^{N}$ be the sampling window of a regular convolution grid, where N is the number of grid points. The new irregular sampling locations can be represented by Eq. (7), and the bilinear interpolation is formulated as Eq. (8),

$$\hat{\mathcal{P}} = \{p + p_i + \Delta p_i \mid i = 1, \ldots, N\} \tag{7}$$

$$\hat{\mathcal{F}}(p) = \sum_{\hat{p}} G(\hat{p}, p) \cdot \mathcal{F}(\hat{p}) \tag{8}$$

where $\mathcal{F}(\cdot)$ and $\hat{\mathcal{F}}(\cdot)$ are the input and output feature maps, and $G(\cdot, \cdot)$ is the bilinear interpolation kernel. $\hat{p} \in \hat{\mathcal{P}}$, and p is the location that predicts the detection result.

The aligned task features are extracted from the locations of the task-aware points, and then they are used for classification and regression refinement. Different from the first stage, the second regression stage utilizes the General distribution decoder [13] that outputs the discrete representation of the bounding box.

3.4   Self-distillation

As Fig. 6 shows, the bottom boundary of the handcrafted annotation is inaccurate and ambiguous, which can misguide and hurt the training process. However, the network's localization prediction results sometimes provide better and clearer regression targets that are easier for the network to learn. For this reason, we propose a self-distillation regression loss (SDR) that could transfer regression knowledge from the refined predictions to the coarse ones.

As Eq. 9 shows, R_1, R_2, and y are the output of the stage-one, stage-two decoders, and the classification score. IoU_1 and IoU_2 denote the Intersection over Union between the ground truth label and the corresponding regression results, and $GIoU$ is the Generalized Intersection over Union as in [21]. As the stage-two decoder is fed with better features, we take its outputs as the regression upper bound of the stage-one decoder. That is, we utilize the integral results from

Fig. 6. The red and green boxes are the ground truth and the bounding box prediction made by our stage-two decoder, respectively. (Color figure online)

the Discrete distribution decoder as the soft target for the Dirac delta decoder. Here, $y \cdot IoU_2$ denotes the confidence score of the refined regression result, and its gradient is detached. Thus, SDR loss pays more attention to the high-confidence prediction results. Notably, SDR loss only penalizes the network when the predictions of the stage-two decoder are better than the stage-one decoder (i.e., $IoU_2 > IoU_1$). Our proposed SDR loss enables the coarse predictions to learn from the refined results and bridges the gap between them. Better stage-one predictions lead to better stage-two predictions and promote the training process into positive circulation.

$$\mathcal{L}_{SDR}(R_1, R_2, y) = \begin{cases} y \cdot IoU_2 \cdot (1 - GIoU(R_1, R_2)), & \text{if } IoU_2 > IoU_1 \\ 0 & \text{otherwise} \end{cases} \quad (9)$$

3.5 Loss Function

The proposed SALT-Net is optimized in an end-to-end fashion, and both the coarse and the refined detection stages utilize ATSS [34] as the positive and negative targets assignment strategy. The training loss of SALT-Net is defined as follows:

$$\begin{aligned} \mathcal{L} = &\frac{1}{N_{\text{pos}}} \sum_z \lambda_0 \mathcal{L}_{\mathcal{Q}} \\ &+ \frac{1}{N_{\text{pos}}} \sum_z \mathbf{1}_{\{c_z^* > 0\}} (\lambda_1 \mathcal{L}_{\mathcal{R}_1} + \lambda_2 \mathcal{L}_{\mathcal{R}_2} + \lambda_3 \mathcal{L}_{\mathcal{D}} + \lambda_4 \mathcal{L}_{\mathcal{SDR}}) \end{aligned} \quad (10)$$

where $\mathcal{L}_{\mathcal{Q}}$ is the Quality Focal loss [13] for the classification task. $\mathcal{L}_{\mathcal{R}_1}$ and $\mathcal{L}_{\mathcal{R}_2}$ are both $GIoU$ loss [21], one for the coarse bounding box prediction and the other for the refined regression result. $\mathcal{L}_{\mathcal{D}}$ is the Distribution Focal Loss [13] for optimizing the general distribution representation of the bounding box, and $\mathcal{L}_{\mathcal{SDR}}$ is the proposed self-distillation loss. $\lambda_0 \sim \lambda_4$ are the hyperparameters used to balance different losses, and they are set as $1, 1, 2, 0.5$, and 1, respectively.

Table 1. Ablation study of SALT on the COCO val2017 split. S_1 and S_2 denote the stage-one and stage-two regression results. "R-Points" and "C-Points" denote the regression-aware and Classification-aware points, respectively. "P-anchor" denotes utilizing the regression-aware points as the point-set anchor for generating the Classification-aware points. "skip" denotes the skip connection of the classification tower, as shown in Fig. 3.

Method	R-Points	C-Points	P-anchor	skip	AP	AP_{50}	AP_{75}	AP_S	AP_M	AP_L
baseline										
S_1 [34]					39.9	58.5	43.0	22.4	43.9	52.7
S_2 [12]					40.9	58.3	44.4	23.9	44.7	53.5
S_2	✓				41.3	58.7	44.9	23.3	45.0	54.2
S_2	✓	✓			41.6	59.1	45.6	23.7	45.4	54.7
S_2	✓	✓	✓		42.1	59.6	45.6	24.8	45.4	55.5
S_2	✓	✓	✓	✓	**42.5**	**60.1**	**46.2**	**25.1**	**45.9**	**56.4**
S_1	✓	✓	✓	✓	41.3	58.6	44.9	23.2	45.1	54.2

Fig. 7. Visualization of the regression-aware (upper row) and classification-aware (lower row) points. Different task-aware points are located on the different areas of the object, and their sensitive regions (i.e., the bounding boxes) are distinct.

N_{pos} denotes the number of selected positive samples, and z denotes all the locations on the pyramid feature maps. $\mathbf{1}_{\{c_z^* > 0\}}$ is the indicator function, being 1 if $c_z^* > 0$ and 0 otherwise.

4 Experiments

Figure 3 presents the network of our proposed SALT-Net. We take state-of-the-art dense detectors ATSS [34] and GFLv2 [12] as our baseline, and they serve as the stage-one and stage-two decoders, respectively. Our SALT-Net is evaluated on the challenging MS-COCO benchmark [17]. Following the common practice, we use the COCO train2017 split (115K images) as the training set and the COCO val2017 split (5K images) for the ablation study. To compare with state-of-the-art detectors, we report the COCO AP on the *test-dev* split (20K images) by uploading the detection result to the MS-COCO server.

Table 2. Spatial disentanglement strategies. "*exp*" and "+" denotes utilizing Equation (5) and (6) to generate the classification-aware points, respectively.

Method	AP	AP_{50}	AP_{75}	AP_S	AP_M	AP_L
w/+	42.3	60.1	46.1	24.7	**46.0**	56.1
w/*exp*	**42.5**	**60.1**	**46.2**	**25.1**	45.9	**56.4**

Table 3. Performance of implementing our proposed approach in popular dense detectors.

Method	AP	AP_{50}	AP_{75}
FCOS	38.6	57.2	41.7
SALT-FCOS	**40.9** (+2.3)	**59.3**	**44.2**
RepPoints w/ GridF	37.4	58.9	39.7
SALT-RepPoints	**39.0** (+1.6)	**60.5**	**41.6**

4.1   Performance of SALT's Component Parts

To validate the effectiveness of different component parts of our proposed operator SALT, we gradually add the proposed modules to the baseline. As shown in Table 1, the second and third rows are the baseline performances of the stage-one and stage-two decoders, respectively. Note that the stage-one decoder utilizes joint representation of IoU and classification scores instead of its original centerness branch, as in [13]. The baseline performances of the two stages are 39.9 AP and 40.9 AP, respectively.

As presented in the fourth row, the first experiment investigates the effect of implementing the regression-aware points. Therefore, SALT only predicts the scale factors $\mathcal{S}$ for generating the regression-aware points. Both subnetworks utilize aligned features from the locations of the regression-aware points for the refined detection results. The AP is improved to 41.3, which indicates that the aligned features do improve the detection accuracy, even though features for the two tasks are still spatially entangled.

As shown in the fifth row, to test the effect of spatial disentanglement, SALT predicts the disentanglement vectors $\mathcal{D}$ for generating the classification-aware points. Note that these points are learned without the regression-aware points acting as the point-set anchor (i.e., $\Delta\mathcal{P}^c = \mathcal{D}$), yet the AP is still boosted to 41.6. These classification-aware points are located in different regions from the regression-aware points, and higher accuracy is obtained (41.6 vs. 41.3). Therefore, spatial disentanglement does raise the detection performance by eliminating their spatial feature conflicts.

The sixth row shows the performance when taking the regression-aware points as the point-set anchor for generating the classification-aware points. It can be observed that a notable performance gain is achieved (i.e., 0.5 AP improvement). That thereby proves the effectiveness of utilizing the regression-

Table 4. The effect of SDR loss

Method	SDR	AP	AP_{50}	AP_{75}	AP_S	AP_M	AP_L
S_1		41.3	58.6	44.9	23.2	45.1	54.2
S_2		42.5	60.1	46.2	25.1	45.9	56.4
S_1	✓	**42.1**(+0.8)	60.5	45.9	24.8(+1.6)	45.7	54.9
S_2	✓	**42.8**(+0.3)	60.6	46.7	25.1	46.4	56.0

aware points as the shape hypothesis and the importance of task disentanglement. Figure 7 is the visualization of task-aware points and their sensitive regions. This figure indicates that classification and regression are sensitive to different locations of the object, which also gives the interpretability of spatial disentanglement.

As shown in the seventh row, the long-range skip connection (i.e., the residual connection on the classification tower) can also bring a considerable performance boost and gain 0.4 AP. Note that the overall performance has been improved by 1.6 AP and 2.9 AP_L compared with the strong baseline. Finally, the last row indicates that the coarse regression results with the refined classification results can also improve the baseline performance by 1.4 AP.

4.2 The Selection of Spatial Disentanglement Strategies

We propose two disentanglement functions to generate the classification-aware points, as illustrated in Eqs. (5) and (6). In Eq. (5), the disentanglement vector set $\mathcal{D}$ is taken as the exponent, whereas $\mathcal{D}$ is directly aggregated with $\Delta\mathcal{P}^r$ in Eq. (6). As illustrated in Table 2, the "*exp*" strategy performs better than that of "+." The reason is that predicting log-space transforms (i.e., $\mathcal{D} = \ln\frac{\Delta\mathcal{P}^c}{\Delta\mathcal{P}^r}$), instead of directly predicting the distance (i.e., $\mathcal{D} = \Delta\mathcal{P}^c - \Delta\mathcal{P}^r$), prevents unstable gradients during training. Therefore, it is easier to be learned.

4.3 Generality of SALT

Our proposed SALT can act as a plug-in operator for dense detectors. Therefore, we plug SALT into popular detectors [23,29], to validate its generality. As shown in Table 3, the performance gain is 2.3 AP on FCOS, which is a considerable improvement. Compared with RepPoints, our SALT-RepPoints performs better than it and gains 1.6 AP. One can see that SALT can significantly improve the accuracy of different detectors, which demonstrates its generality.

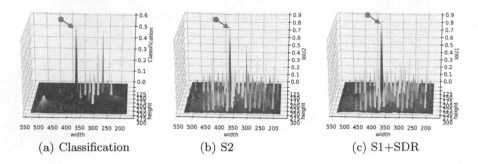

Fig. 8. Prediction distributions (Fig. 1) after applying SALT and SDR loss. The locations of their distribution peaks are identical.

4.4 Self-distillation Regression Loss

The baseline for this ablation study is the best model of Table 1. Here, both stages utilize the refined classification scores as the NMS ranking keywords. As Table 4 shows, after applying the SDR loss to the SALT-Net, the performance of both stages got improved. The performance gain of the stage-one decoder is an absolute 0.8 AP score. Notably, the performance on small objects has been improved by 1.6 AP, which is a relatively large margin compared with the strong baseline. Furthermore, the improvement of the stage-one decoder also brings positive feedback to the stage-two decoder and leads to the highest performance of our SALT-Net (i.e., 42.8 AP).

4.5 Evaluations for Task-Alignment of SALT-Net

Figure 8(a) and (b) are the distributions of the refined detection results when implementing SALT, whereas Fig. 1(b) and (c) are the original coarse prediction distributions made by the stage-one decoder. The green arrows point to the distribution peaks, and one can see that they are spatially aligned (i.e., at the same location). Therefore, the detection result with the highest classification score also has the best regression result, and the misalignment gap is bridged. Figure 8(c) is the IoU distribution of the stage-one decoder after applying the SDR loss. Its quality distributions become very close to the stage-two decoder (i.e., Fig. 8(b)), which proves the effectiveness of the regression knowledge transfer. In Fig. 9, the qualitative results show that SALT can align the regression and detection tasks and thereby suppress some low IoU but high classification score results.

4.6 Comparisons with State-of-the-Arts

The multi-scale training strategy (i.e., input images are resized from [400, 1333] to [960, 1333]) and the 2 × schedule [1] are adopted as they are commonly used strategies in state-of-the-art methods. GFLV2 only applies DCN on the last two stages of the backbone, whereas the common practices [24,31] usually apply it on

Table 5. SALT-Net vs. State-of-the-art Detectors. All test results are reported on the COCO *test-dev* set. "DCN^2, DCN"-applying Deformable Convolutional Network [38] on the last two and three stages of the backbone, respectively. "DCN^P"-applying DCN on both the backbone and the FPN. "*" indicates applying SDR loss and "†" indicates test-time augmentations, including horizontal flip and multi-scale testing.

Method	Backbone	Epoch	AP	AP$_{50}$	AP$_{75}$	AP$_S$	AP$_M$	AP$_L$
Multi-stage								
GuidedAnchor [25]	R-50	12	39.8	59.2	43.5	21.8	42.6	50.7
DCNV2 [38]	X-101-32x8d-DCN	24	44.5	65.8	48.4	27.0	48.5	58.9
BorderDet [18]	X-101-64x4d-DCN	24	48.0	67.1	52.1	29.4	50.7	60.5
RepPointsV2 [2]	X-101-64x4d-DCN	24	49.4	68.9	53.4	30.3	52.1	62.3
TSD† [22]	SE154-DCN	24	51.2	71.9	56.0	33.8	54.8	64.2
VFNet [31]	X-101-32x8d-DCN	24	50.0	68.5	54.4	30.4	53.2	62.9
LSNet [5]	R2-101-DCN^P	24	51.1	70.3	55.2	31.2	54.3	65.1
One-stage								
CornerNet [11]	HG-104	200	40.5	59.1	42.3	21.8	42.7	50.2
SAPD [37]	X-101-32x8d-DCN	24	46.6	66.6	50.0	27.3	49.7	60.7
ATSS [34]	X-101-32x8d-DCN	24	47.7	66.5	51.9	29.7	50.8	59.4
GFL [13]	X-101-32x8d-DCN	24	48.2	67.4	52.6	29.2	51.7	60.2
FCOS-imprv [24]	X-101-32x8d-DCN	24	44.1	63.7	47.9	27.4	46.8	53.7
PAA [10]	X-101-64x4d-DCN	24	49.0	67.8	53.3	30.2	52.8	62.2
GFLV2 [12]	R-50	24	44.3	62.3	48.5	26.8	47.7	54.1
GFLV2 [12]	X-101-32x8d-DCN^2	24	49.0	67.6	53.5	29.7	52.4	61.4
TOOD [6]	X-101-64x4d-DCN	24	51.1	69.4	55.5	31.9	54.1	63.7
SALT-Net*	R-50	24	46.1	64.0	50.3	28.0	49.5	57.2
SALT-Net	X-101-32x8d-DCN^2	24	49.8	68.5	54.2	30.6	53.2	62.6
SALT-Net	X-101-32x8d-DCN	24	50.2	68.8	54.9	31.2	53.4	63.1
SALT-Net	R2-101-DCN^2	24	51.1	69.7	55.7	32.3	54.5	64.0
SALT-Net*	R2-101-DCN	24	51.5	70.0	56.2	32.1	55.1	64.8
SALT-Net*†	R2-101-DCN	24	**53.8**	**71.1**	**59.9**	**36.3**	**56.9**	**65.1**

the last three stages. Therefore, for a fair comparison, the results of the proposed method with both settings are reported. As Table 5 shows, our model achieves a 46.1 AP with ResNet-50, which outperforms other state-of-the-art methods with heavier backbones (e.g., FCOS with X-101-32x8d-DCN). With test-time augmentations and R2-101-DCN as the backbone, our best model achieves a 53.8 AP, which is a very competitive result among dense object detectors.

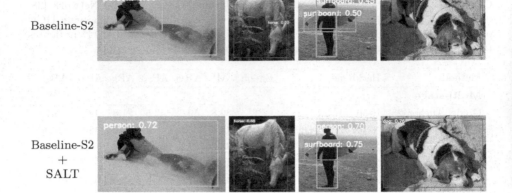

Fig. 9. Comparisons between the stage-two baseline decoder and our SALT-Net.

5   Conclusion

In this work, we presented SALT, a simple yet effective plug-in operator that can solve the misalignment problem between regression and classification. Our new-fashioned framework can disentangle classification and regression from the spatial dimension by extracting features from each task's sensitive locations and aligning them to the same anchor point. We also proposed SDR loss to transfer the regression knowledge from the stage-two decoder to the stage-one decoder. The refined detection results also received positive feedback by improving the coarse regression results, and the final performance improved in an inference cost-free fashion. Extensive experiments showed that SALT could considerably raise the performance of various dense detectors, and SALT-Net showed promising results among the state-of-the-art dense detectors.

References

1. Chen, K., et al.: Mmdetection: open mmlab detection toolbox and benchmark. arXiv preprint arXiv:1906.07155 (2019)
2. Chen, Y., Zhang, Z., Cao, Y., Wang, L., Lin, S., Hu, H.: Reppoints v2: verification meets regression for object detection. In: 33rd Proceedings of Conference on Advances in Neural Information Processing Systems(2020)
3. Chi, C., Zhang, S., Xing, J., Lei, Z., Li, S.Z., Zou, X.: Selective refinement network for high performance face detection. In: AAAI, vol. 33, pp. 8231–8238 (2019)
4. Dai, J., et al.: Deformable convolutional networks. In: ICCV, pp. 764–773 (2017)
5. Duan, K., Xie, L., Qi, H., Bai, S., Huang, Q., Tian, Q.: Location-sensitive visual recognition with cross-iou loss. arXiv preprint arXiv:2104.04899 (2021)
6. Feng, C., Zhong, Y., Gao, Y., Scott, M.R., Huang, W.: TOOD: task-aligned one-stage object detection. In: ICCV., pp. 3490–3499. IEEE Computer Society (2021)
7. Hinton, G., Vinyals, O., Dean, J., et al.: Distilling the knowledge in a neural network. arXiv preprint arXiv:1503.02531 2(7) (2015)

8. Ji, M., Shin, S., Hwang, S., Park, G., Moon, I.C.: Refine myself by teaching myself: Feature refinement via self-knowledge distillation. In: CVPR, pp. 10664–10673 (2021)

9. Kang, Z., Zhang, P., Zhang, X., Sun, J., Zheng, N.: Instance-conditional knowledge distillation for object detection. In: NeurIPS (2021)

10. Kim, K., Lee, H.S.: Probabilistic anchor assignment with IoU prediction for object detection. In: Vedaldi, A., Bischof, H., Brox, T., Frahm, J.-M. (eds.) ECCV 2020. LNCS, vol. 12370, pp. 355–371. Springer, Cham (2020). https://doi.org/10.1007/978-3-030-58595-2_22

11. Law, H., Deng, J.: Cornernet: Detecting objects as paired keypoints. In: ECCV, pp. 734–750 (2018)

12. Li, X., Wang, W., Hu, X., Li, J., Tang, J., Yang, J.: Generalized focal loss v2: learning reliable localization quality estimation for dense object detection. In: CVPR, pp. 11632–11641 (2021)

13. Li, X., et al.: Generalized focal loss: Learning qualified and distributed bounding boxes for dense object detection. In: NeurIPS (2020)

14. Li, Z., Li, X., Yang, L., Yang, J., Pan, Z.: Student helping teacher: teacher evolution via self-knowledge distillation. arXiv preprint arXiv:2110.00329 (2021)

15. Lin, T.Y., Dollár, P., Girshick, R., He, K., Hariharan, B., Belongie, S.: Feature pyramid networks for object detection. In: CVPR, pp. 2117–2125 (2017)

16. Lin, T.Y., Goyal, P., Girshick, R., He, K., Dollár, P.: Focal loss for dense object detection. In: ICCV, pp. 2980–2988 (2017)

17. Lin, T., et al.: Microsoft COCO: common objects in context. In: Fleet, D., Pajdla, T., Schiele, B., Tuytelaars, T. (eds.) ECCV 2014. LNCS, vol. 8693, pp. 740–755. Springer, Cham (2014). https://doi.org/10.1007/978-3-319-10602-1_48

18. Qiu, H., Ma, Y., Li, Z., Liu, S., Sun, J.: BorderDet: border feature for dense object detection. In: Vedaldi, A., Bischof, H., Brox, T., Frahm, J.-M. (eds.) ECCV 2020. LNCS, vol. 12346, pp. 549–564. Springer, Cham (2020). https://doi.org/10.1007/978-3-030-58452-8_32

19. Redmon, J., Divvala, S., Girshick, R., Farhadi, A.: You only look once: Unified, real-time object detection. In: CVPR, pp. 779–788 (2016)

20. Redmon, J., Farhadi, A.: Yolov3: an incremental improvement. arXiv preprint arXiv:1804.02767 (2018)

21. Rezatofighi, H., Tsoi, N., Gwak, J., Sadeghian, A., Reid, I., Savarese, S.: Generalized intersection over union: a metric and a loss for bounding box regression. In: CVPR, pp. 658–666 (2019)

22. Song, G., Liu, Y., Wang, X.: Revisiting the sibling head in object detector. In: CVPR, pp. 11563–11572 (2020)

23. Tian, Z., Shen, C., Chen, H., He, T.: FCOS: fully convolutional one-stage object detection. In: ICCV, pp. 9627–9636 (2019)

24. Tian, Z., Shen, C., Chen, H., He, T.: FCOS: a simple and strong anchor-free object detector. IEEE Trans. Pattern Anal. Mach. Intell. (2020)

25. Wang, J., Chen, K., Yang, S., Loy, C.C., Lin, D.: Region proposal by guided anchoring. In: CVPR, pp. 2965–2974 (2019)

26. Wang, T., Yuan, L., Zhang, X., Feng, J.: Distilling object detectors with fine-grained feature imitation. In: CVPR, pp. 4933–4942 (2019)

27. Wu, S., Li, X., Wang, X.: IOU-aware single-stage object detector for accurate localization. Image Vis. Comput. **97**, 103911 (2020)

28. Wu, Y., Chen, Y., Yuan, L., Liu, Z., Wang, L., Li, H., Fu, Y.: Rethinking classification and localization for object detection. In: CVPR, pp. 10186–10195 (2020)

29. Yang, Z., Liu, S., Hu, H., Wang, L., Lin, S.: Reppoints: point set representation for object detection. In: ICCV, pp. 9657–9666 (2019)
30. Yao, A., Sun, D.: Knowledge transfer via dense cross-layer mutual-distillation. In: Vedaldi, A., Bischof, H., Brox, T., Frahm, J.-M. (eds.) ECCV 2020. LNCS, vol. 12360, pp. 294–311. Springer, Cham (2020). https://doi.org/10.1007/978-3-030-58555-6_18
31. Zhang, H., Wang, Y., Dayoub, F., Sunderhauf, N.: Varifocalnet: An iou-aware dense object detector. In: CVPR. pp. 8514–8523 (2021)
32. Zhang, L., Song, J., Gao, A., Chen, J., Bao, C., Ma, K.: Be your own teacher: Improve the performance of convolutional neural networks via self distillation. In: ICCV, pp. 3713–3722 (2019)
33. Zhang, P., Kang, Z., Yang, T., Zhang, X., Zheng, N., Sun, J.: Lgd: label-guided self-distillation for object detection. arXiv preprint arXiv:2109.11496 (2021)
34. Zhang, S., Chi, C., Yao, Y., Lei, Z., Li, S.Z.: Bridging the gap between anchor-based and anchor-free detection via adaptive training sample selection. In: CVPR, pp. 9759–9768 (2020)
35. Zhang, S., Wen, L., Bian, X., Lei, Z., Li, S.Z.: Single-shot refinement neural network for object detection. In: CVPR, pp. 4203–4212 (2018)
36. Zhang, Y., Xiang, T., Hospedales, T.M., Lu, H.: Deep mutual learning. In: CVPR, pp. 4320–4328 (2018)
37. Zhu, C., Chen, F., Shen, Z., Savvides, M.: Soft anchor-point object detection. In: Vedaldi, A., Bischof, H., Brox, T., Frahm, J.-M. (eds.) ECCV 2020. LNCS, vol. 12354, pp. 91–107. Springer, Cham (2020). https://doi.org/10.1007/978-3-030-58545-7_6
38. Zhu, X., Hu, H., Lin, S., Dai, J.: Deformable convnets v2: More deformable, better results. In: CVPR. pp. 9308–9316 (2019)

No More Strided Convolutions or Pooling: A New CNN Building Block for Low-Resolution Images and Small Objects

Raja Sunkara and Tie Luo$^{(\boxtimes)}$

Computer Science Department, Missouri University of Science and Technology,
Rolla, MO 65409, USA
{rs5cq,tluo}@mst.edu

Abstract. Convolutional neural networks (CNNs) have made resounding success in many computer vision tasks such as image classification and object detection. However, their performance degrades rapidly on tougher tasks where images are of low resolution or objects are small. In this paper, we point out that this roots in a defective yet common design in existing CNN architectures, namely the use of *strided convolution* and/or *pooling layers*, which results in a loss of fine-grained information and learning of less effective feature representations. To this end, we propose a new CNN building block called *SPD-Conv* in place of each strided convolution layer and each pooling layer (thus eliminates them altogether). SPD-Conv is comprised of a *space-to-depth* (SPD) layer followed by a *non-strided* convolution (Conv) layer, and can be applied in most if not all CNN architectures. We explain this new design under two most representative computer vision tasks: object detection and image classification. We then create new CNN architectures by applying SPD-Conv to YOLOv5 and ResNet, and empirically show that our approach significantly outperforms state-of-the-art deep learning models, especially on tougher tasks with low-resolution images and small objects. We have open-sourced our code at https://github.com/LabSAINT/SPD-Conv.

1 Introduction

Since AlexNet [18], convolutional neural networks (CNNs) have excelled at many computer vision tasks. For example in image classification, well-known CNN models include AlexNet, VGGNet [30], ResNet [13], etc.; while in object detection, those models include the R-CNN series [9,28], YOLO series [4,26], SSD [24], EfficientDet [34], and so on. However, all such CNN models need "good quality" inputs (fine images, medium to large objects) in both training and inference. For example, AlexNet was originally trained and evaluated on 227×227 clear images, but after reducing the image resolution to 1/4 and 1/8, its classification accuracy drops by 14% and 30%, respectively [16]. The similar observation was made on VGGNet and ResNet too [16]. In the case of object detection, SSD suffers from a remarkable mAP loss of 34.1 on 1/4 resolution images or equivalently 1/4 smaller-size objects, as demonstrated in [11]. In fact, small object detection is a very challenging task because smaller objects inherently have lower resolution, and also limited context information for a model to learn from. Moreover, they often (unfortunately) co-exist with large objects in the same image, which (the large ones) tend to dominate the feature learning process, thereby making the small objects undetected.

In this paper, we contend that such performance degradation roots in a defective yet common design in existing CNNs. That is, the use of strided convolution and/or pooling, especially in the earlier layers of a CNN architecture. The adverse effect of this design usually does not exhibit because most scenarios being studied are "amiable" where images have good resolutions and objects are in fair sizes; therefore, there is plenty of *redundant* pixel information that strided convolution and pooling can *conveniently skip* and the model can still learn features quite well. However, in tougher tasks when images are blurry or objects are small, the lavish assumption of redundant information no longer holds and the current design starts to suffer from loss of fine-grained information and poorly learned features.

To address this problem, we propose a new building block for CNN, called *SPD-Conv*, in substitution of (and thus eliminate) strided convolution and pooling layers altogether. SPD-Conv is a *space-to-depth* (SPD) layer followed by a *non-strided* (i.e., vanilla) convolution layer. The SPD layer downsamples a feature map X but retains all the information in the *channel* dimension, and thus there is no information loss. We were inspired by an image transformation technique [29] which rescales a raw image before feeding it into a neural net, but we substantially generalize it to downsampling *feature maps* inside and throughout the entire network; furthermore, we add a non-strided convolution operation after each SPD to reduce the (increased) number of channels using learnable parameters in the added convolution layer. Our proposed approach is both *general* and *unified*, in that SPD-Conv (i) can be applied to most if not all CNN architectures and (ii) replaces both strided convolution and pooling the same way. In summary, this paper makes the following contributions:

1) We identify a defective yet common design in existing CNN architectures and propose a new building block called SPD-Conv in lieu of the old design. SPD-Conv downsamples feature maps without losing learnable information, completely jettisoning strided convolution and pooling operations which are widely used nowadays.
2) SPD-Conv represents a general and unified approach, which can be easily applied to most if not all deep learning based computer vision tasks.
3) Using two most representative computer vision tasks, object detection and image classification, we evaluate the performance of SPD-Conv. Specifically, we construct YOLOv5-SPD, ResNet18-SPD and ResNet50-SPD, and evaluate them on COCO-2017, Tiny ImageNet, and CIFAR-10 datasets in comparison with several state-of-the-art deep learning models. The results demonstrate

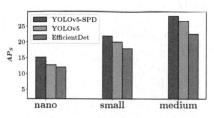

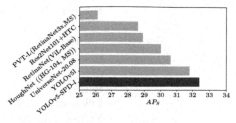

(a) Nano, small, and medium models.   (b) Large-scale models.

Fig. 1. Comparing AP for small objects (AP_S). "SPD" indicates our approach.

significant performance improvement in AP and top-1 accuracy, especially on small objects and low-resolution images. See Fig. 1 for a preview.

4) SPD-Conv can be easily integrated into popular deep learning libraries such as PyTorch and TensorFlow, potentially producing greater impact. Our source code is available at https://github.com/LabSAINT/SPD-Conv.

The rest of this paper is organized as follows. Section 2 presents background and reviews related work. Section 3 describes our proposed approach and Sect. 4 presents two case studies using object detection and image classification. Section 5 provides performance evaluation. This paper concludes in Sect. 6.

2 Preliminaries and Related Work

We first provide an overview of this area, focusing more on object detection since it subsumes image classification.

Current state-of-the-art object detection models are CNN-based and can be categorized into one-stage and two-stage detectors, or anchor-based or anchor-free detectors. A two-stage detector firstly generates coarse region proposals and secondly classifies and refines each proposal using a head (a fully-connected network). In contrast, a one-stage detector skips the region proposal step and runs detection directly over a dense sampling of locations. Anchor-based methods use *anchor boxes*, which are a predefined collection of boxes that match the widths and heights of objects in the training data, to improve loss convergence during training. We provide Table 1 that categorizes some well-known models.

Generally, one-stage detectors are faster than two-stage ones and anchor-based models are more accurate than anchor-free ones. Therefore, later in our case study and experiments we focus more on one-stage and anchor-based models, i.e., the first cell of Table 1.

A typical one-stage object detection model is depicted in Fig. 2. It consists of a CNN-based *backbone* for visual feature extraction and a detection *head* for predicting class and bounding box of each contained object. In between, a *neck* of extra layers is added to combine features at multiple scales to produce semantically strong features for detecting objects of different sizes.

Table 1. A taxonomy of OD models.

Model	Anchor-based	Anchor-free
One-stage	Faster R-CNN [27], SSD [24], RetinaNet [21], EfficientDet [34], YOLO [4,14,26,36]	FCOS [35], CenterNet [7], DETR [5], YOLOX [8]
Two-stage	R-CNN [10], Fast R-CNN [9]	RepPoints, CenterNet2

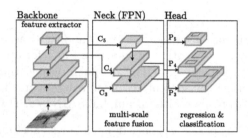

Fig. 2. A one-stage object detection pipeline.

2.1 Small Object Detection

Traditionally, detecting both small and large objects is viewed as a multi-scale object detection problem. A classic way is *image pyramid* [3], which resizes input images to multiple scales and trains a dedicated detector for each scale. To improve accuracy, SNIP [31] was proposed which performs *selective backpropagation* based on different object sizes in each detector. SNIPER [32] improves the efficiency of SNIP by only processing the context regions around each object instance rather than every pixel in an image pyramid, thus reducing the training time. Taking a different approach to efficiency, Feature Pyramid Network (FPN) [20] exploits the multi-scale features inherent in convolution layers using lateral connections and combine those features using a top-down structure. Following that, PANet [22] and BiFPN [34] were introduced to improve FPN in its feature information flow by using shorter pathways. Moreover, SAN [15] was introduced to map multi-scale features onto a scale-invariant subspace to make a detector more robust to scale variation. All these models unanimously use strided convolution and max pooling, which we get rid of completely.

2.2 Low-Resolution Image Classification

One of the early attempts to address this challenge is [6], which proposes an end-to-end CNN model by adding a super-resolution step before classification. Following that, [25] proposes to transfer fine-grained knowledge acquired from high-resolution training images to low-resolution test images. However, this approach

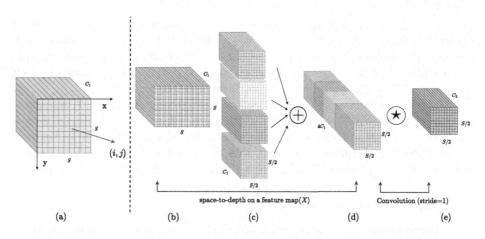

Fig. 3. Illustration of SPD-Conv when $scale = 2$ (see text for details).

requires high-resolution training images corresponding to the specific application (e.g., the classes), which are not always available.

This same requirement of high-resolution training images is also needed by several other studies such as [37]. Recently, [33] proposed a loss function that incorporate attribute-level separability (where attribute means fine-grained, hierarchical class labels) so that the model can learn class-specific discriminative features. However, the fine-grained (hierarchical) class labels are difficult to obtain and hence limit the adoption of the method.

3   A New Building Block: SPD-Conv

SPD-Conv is comprised of a Space-to-depth (SPD) layer followed by a non-strided convolution layer. This section describes it in detail.

3.1   Space-to-depth (SPD)

Our SPD component generalizes a (raw) image transformation technique [29] to downsampling feature maps inside and throughout a CNN, as follows.

Consider any intermediate feature map X of size $S \times S \times C_1$, slice out a sequence of sub feature maps as

$$f_{0,0} = X[0:S:scale, 0:S:scale], f_{1,0} = X[1:S:scale, 0:S:scale], \ldots,$$
$$f_{scale-1,0} = X[scale-1:S:scale, 0:S:scale];$$
$$f_{0,1} = X[0:S:scale, 1:S:scale], \ldots, f_{scale-1,1} = X[scale-1:S:scale, 1:S:scale];$$
$$\vdots$$
$$f_{0,scale-1} = X[0:S:scale, scale-1:S:scale], f_{1,scale-1}, \ldots,$$
$$f_{scale-1,scale-1} = X[scale-1:S:scale, scale-1:S:scale].$$

In general, given any (original) feature map X, a sub-map $f_{x,y}$ is formed by all the entries $X(i,j)$ that $i+x$ and $j+y$ are divisible by $scale$. Therefore, each sub-map downsamples X by a factor of $scale$. Figure 3(a)–(c) give an example when $scale = 2$, where we obtain four sub-maps $f_{0,0}, f_{1,0}, f_{0,1}, f_{1,1}$ each of which is of shape $(\frac{S}{2}, \frac{S}{2}, C_1)$ and downsamples X by a factor of 2.

Next, we concatenate these sub feature maps along the channel dimension and thereby obtain a feature map X', which has a reduced spatial dimension by a factor of $scale$ and an increased channel dimension by a factor of $scale^2$. In other words, SPD transforms feature map $X(S, S, C_1)$ into an intermediate feature map $X'(\frac{S}{scale}, \frac{S}{scale}, scale^2 C_1)$. Figure 3(d) gives an illustration using $scale = 2$.

3.2   Non-strided Convolution

After the SPD feature transformation layer, we add a non-strided (i.e., stride=1) convolution layer with C_2 filters, where $C_2 < scale^2 C_1$, and further transforms $X'(\frac{S}{scale}, \frac{S}{scale}, scale^2 C_1) \rightarrow X''(\frac{S}{scale}, \frac{S}{scale}, C_2)$. The reason we use non-strided convolution is to retain all the discriminative feature information as much as possible. Otherwise, for instance, using a 3×3 filer with stride=3, feature maps will get "shrunk" yet each pixel is sampled only once; if stride=2, *asymmetric sampling* will occur where even and odd rows/columns will be sampled different times. In general, striding with a step size greater than 1 will cause *non-discriminative loss* of information although at the surface, it appears to convert feature map $X(S, S, C_1) \rightarrow X''(\frac{S}{scale}, \frac{S}{scale}, C_2)$ too (but without X').

4   How to Use SPD-Conv: Case Studies

To explain how to apply our proposed method to redesigning CNN architectures, we use two most representative categories of computer vision models: object detection and image classification. This is without loss of generality as almost all CNN architectures use strided convolution and/or pooling operations to downsample feature maps.

4.1   Object Detection

YOLO is a series of very popular object detection models, among which we choose the latest YOLOv5 [14] to demonstrate. YOLOv5 uses CSPDarknet53 [4] with a SPP [12] module as its backbone, PANet [23] as its neck, and the YOLOv3 head [26] as its detection head. In addition, it also uses various data augmentation methods and some modules from YOLOv4 [4] for performance optimization. It employs the cross-entropy loss with a sigmoid layer to compute objectness and classification loss, and the CIoU loss function [38] for localization loss. The CIoU loss takes more details than IoU loss into account, such as edge overlapping, center distance, and width-to-height ratio.

YOLOv5-SPD. We apply our method described in Sect. 3 to YOLOv5 and obtain YOLOv5-SPD (Fig. 4), simply by replacing the YOLOv5 stride-2 convolutions with our SPD-Conv building block. There are 7 instances of such replacement because YOLOv5 uses five stride-2 convolution layers in the backbone to

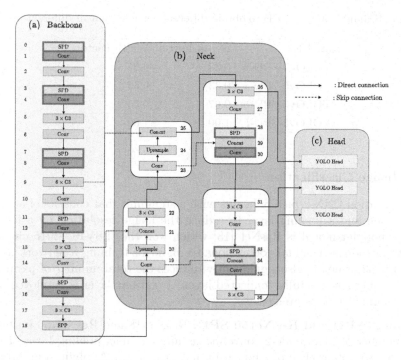

Fig. 4. Overview of our YOLOv5-SPD. Red boxes are where the replacement happens. (Color figure online)

downsample the feature map by a factor of 2^5, and two stride-2 convolution layers in the neck. There is a concatenation layer after each strided convolution in YOLOv5 neck; this does not alter our approach and we simply keep it between our SPD and Conv.

Scalability. YOLOv5-SPD can suit different application or hardware needs by easily scaling up and down in the same manner as YOLOv5. Specifically, we can simply adjust (1) the number of filters in every non-strided convolution layer and/or (2) the repeated times of C3 module (as in Fig. 4), to obtain different versions of YOLOv5-SPD. The first is referred to as *width scaling* which changes the original width n_w (number of channels) to $\lceil n_w \times width\_factor \rceil_8$ (rounded off to the nearest multiple of 8). The second is referred to as *depth scaling* which changes the original depth n_d (times of repeating the C3 module; e.g., 9 as in $9 \times$ C3 in Fig. 4) to $\lceil n_d \times depth\_factor \rceil$. This way, by choosing different width/depth factors, we obtain *nano, small, medium,* and *large* versions of YOLOv5-SPD as shown in Table 2, where factor values are chosen the same as YOLOv5 for the purpose of comparison in our experiments later.

Table 2. Scaling YOLOv5-SPD to obtain different versions that fit different use cases.

Models	Depth_Factor	Width_Factor
YOLOv5-SPD-n	0.33	0.25
YOLOv5-SPD-s	0.33	0.50
YOLOv5-SPD-m	0.67	0.75
YOLOv5-SPD-l	1.00	1.00

4.2 Image Classification

A classification CNN typically begins with a stem unit that consists of a stride-2 convolution and a pooling layer to reduce the image resolution by a factor of four. A popular model is ResNet [13] which won the ILSVRC 2015 challenge. ResNet introduces residual connections to allow for training a network as deep as up to 152 layers. It also significantly reduces the total number of parameters by only using a single fully-connected layer. A softmax layer is employed at the end to normalize class predictions.

ResNet18-SPD and ResNet50-SPD. ResNet-18 and ResNet-50 both use a total number of four stride-2 convolutions and one max-pooling layer of stride 2 to downsample each input image by a factor of 2^5. Applying our proposed building block, we replace the four strided convolutions with SPD-Conv; but on the other hand, we simply remove the max pooling layer because, since our main target is low-resolution images, the datasets used in our experiments have rather small images (64×64 in Tiny ImageNet and 32×32 in CIFAR-10) and hence pooling is unnecessary. For larger images, such max-pooling layers can still be replaced the same way by SPD-Conv. The two new architectures are shown in Table 3.

5 Experiments

This section evaluates our proposed approach SPD-Conv using two representative computer vision tasks, object detection and image classification.

5.1 Object Detection

Dataset and Setup. We use the COCO-2017 dataset [1] which is divided into `train2017` (118,287 images) for training, `val2017` (5,000 images; also called `minival`) for validation, and `test2017` (40,670 images) for testing. We use a wide range of state-of-the-art baseline models as listed in Tables 4 and 5. We report the standard metric of average precision (AP) on `val2017` under different IoU thresholds [0.5:0.95] and object sizes (small, medium, large). We also report the AP metrics on `test-dev2017` (20,288 images) which is a subset of `test2017` with accessible labels. However, the labels are not publicly released but one needs

Table 3. Our ResNet18-SPD and ResNet50-SPD architecture.

Layer name	ResNet18-SPD	ResNet50-SPD
spd1	**SPD-Conv**	
conv1	3×3 kernel, 64 output channels	
conv2	$\begin{bmatrix} 3 \times 3, 64 \\ 3 \times 3, 64 \end{bmatrix} \times 2$	$\begin{bmatrix} 1 \times 1, 64 \\ 3 \times 3, 64 \\ 1 \times 1, 256 \end{bmatrix} \times 3$
spd2	**SPD-Conv**	
conv3	$\begin{bmatrix} 3 \times 3, 128 \\ 3 \times 3, 128 \end{bmatrix} \times 2$	$\begin{bmatrix} 1 \times 1, 128 \\ 3 \times 3, 128 \\ 1 \times 1, 512 \end{bmatrix} \times 4$
spd3	**SPD-Conv**	
conv4	$\begin{bmatrix} 3 \times 3, 256 \\ 3 \times 3, 256 \end{bmatrix} \times 2$	$\begin{bmatrix} 1 \times 1, 256 \\ 3 \times 3, 256 \\ 1 \times 1, 1024 \end{bmatrix} \times 6$
spd4	**SPD-Conv**	
conv5	$\begin{bmatrix} 3 \times 3, 512 \\ 3 \times 3, 512 \end{bmatrix} \times 2$	$\begin{bmatrix} 1 \times 1, 512 \\ 3 \times 3, 512 \\ 1 \times 1, 2048 \end{bmatrix} \times 3$
fc (fully conn.)	Global avg. pooling + fc(no. of classes) + softmax	

to submit all the *predicted* labels in JSON files to the `CodaLab COCO Detection Challenge` [2] to retrieve the evaluated metrics, which we did.

Training. We train different versions (nano, small, medium, and large) of YOLOv5-SPD and all the baseline models on `train2017`. Unlike most other studies, we *train from scratch without using transfer learning*. This is because we want to examine the *true learning capability* of each model without being disguised by the rich feature representation it inherits via transfer learning from ideal (high quality) datasets such as ImageNet. This was carried out on our own models (*-SPD-n/s/m/l) and on all the existing YOLO-series models (v5, X, v4, and their scaled versions like nano, small, large, etc.). The other baseline models still used transfer learning because of our lack of resource (training from scratch consumes an enormous amount of GPU time). However, note that this simply means that *those baselines are placed in a much more advantageous position* than our own models as they benefit from high quality datasets.

We choose the SGD optimizer with momentum 0.937 and a weight decay of 0.0005. The learning rate linearly increases from 0.0033 to 0.01 during three warm-up epochs, followed by a decrease using the Cosine decay strategy to a final value of 0.001. The *nano* and *small* models are trained on four V-100 32 GB GPU with a batch size of 128, while *medium* and *large* models are trained with batch size 32. CIoU loss [38] and cross-entropy loss are adopted for objectness and classification. We also employ several data augmentation techniques to mitigate overfitting and improve performance for *all* the models; these techniques include

Table 4. Comparison on MS-COCO validation dataset (val2017).

Model	Backbone	Image size	AP	AP_S (small obj.)	Params (M)	Latency (ms) (batch_size=1)
YOLOv5-SPD-*n*	-	640 × 640	31.0	16.0 (+13.15%)	2.2	7.3
YOLOv5n	-	640 × 640	28.0	14.14	1.9	6.3
YOLOX-Nano	-	640 × 640	25.3	-	0.9	-
YOLOv5-SPD-*s*	-	640 × 640	**40.0**	**23.5 (+11.4%)**	8.7	7.3
YOLOv5s	-	640 × 640	37.4	21.09	7.2	6.4
YOLOX-S	-	640 × 640	39.6	-	9.0	9.8
YOLOv5-SPD-*m*	-	640 × 640	**46.5**	**30.3 (+8.6%)**	24.6	8.4
YOLOv5m	-	640 × 640	45.4	27.9	21.2	8.2
YOLOX-M	-	640 × 640	46.4	-	25.3	12.3
YOLOv5-SPD-*l*	-	640 × 640	48.5	**32.4 (+1.8%)**	52.7	10.3
YOLOv5l	-	640 × 640	49.0	31.8	46.5	10.1
YOLOX-L	-	640 × 640	**50.0**	-	54.2	14.5
Faster R-CNN	R50-FPN	–	40.2	24.2	42.0	-
Faster R-CNN+	R50-FPN	–	42.0	26.6	42.0	-
DETR	R50	-	42.0	20.5	41.0	-
DETR-DC5	ResNet-101	800 × 1333	44.9	23.7	60.0	-
RetinaNet	ViL-Small-RPB	800 × 1333	44.2	28.8	35.7	-

(i) photometric distortions of hue, saturation, and value, (ii) geometric distortions such as translation, scaling, shearing, fliplr and flipup, and (iii) multi-image enhancement techniques such as mosaic and cutmix. Note that augmentation is not used at inference. The hyperparameters are adopted from YOLOv5 without re-tuning.

Results
Table 4 reports the results on val2017 and Table 5 reports the results on test-dev. The AP_S, AP_M, AP_L in both tables mean the AP for small/medium/ large *objects*, which should not be confused with *model* scales (nano, small, medium, large). The image resolution 640 × 640 as shown in both tables is not considered high in object detection (as opposed to image classification) because the resolution on the actual objects is much lower, especially when the objects are small.

Results on val2017. Table 4 is organized by model scales, as separated by horizontal lines (the last group are large-scale models). In the first category of nano models, our YOLOv5-SPD-*n* is the best performer in terms of both AP and AP_S: its AP_S is 13.15% higher than the runner-up, YOLOv5n, and its overall AP is 10.7% higher than the runner-up, also YOLOv5n.

In the second category, small models, our YOLOv5-SPD-*s* is again the best performer on both AP and AP_S, although this time YOLOX-S is the second best on AP.

In the third, medium model category, the AP performance gets quite close although our YOLOv5-SPD-*m* still outperforms others. On the other hand, our

Table 5. Comparison on MS-COCO test dataset (test-dev2017).

Model	ImgSize	Params (M)	AP	AP$_{50}$	AP$_{75}$	AP$_S$ (small obj.)	AP$_M$	AP$_L$
YOLOv5-SPD-n	640 × 640	2.2	**30.4**	48.7	32.4	**15.1(+19%)**	33.9	37.4
YOLOv5n	640 × 640	1.9	28.1	45.7	29.8	12.7	31.3	35.4
EfficientDet-D0	512 × 512	3.9	33.8(Trf)	52.2	35.8	12.0	38.3	51.2
YOLOv5-SPD-s	640 × 640	8.7	**39.7**	59.1	43.1	**21.9(+9.5%)**	43.9	49.1
YOLOv5s	640 × 640	7.2	37.1	55.7	40.2	20.0	41.5	45.2
EfficientDet-D1	640 × 640	6.6	39.6	58.6	42.3	17.9	44.3	56.0
EfficientDet-D2	768 × 768	8.1	43.0(Trf)	62.3	46.2	22.5(Trf)	47.0	58.4
YOLOv5-SPD-m	640 × 640	24.6	**46.6**	65.2	50.8	**28.2(+6%)**	50.9	57.1
YOLOv5m	640 × 640	21.2	45.5	64.0	49.7	26.6	50.0	56.6
YOLOX-M	640 × 640	25.3	46.4	65.4	50.6	26.3	51.0	59.9
EfficientDet-D3	896 × 896	12.0	45.8	65.0	49.3	26.6	49.4	59.8
SSD512	512 × 512	36.1	28.8	48.5	30.3	-	-	-
YOLOv5-SPD-l	640 × 640	52.7	48.8	67.1	53.0	**30.0**	52.9	60.5
YOLOv5l	640 × 640	46.5	49.0	67.3	53.3	29.9	53.4	61.3
YOLOX-L	640 × 640	54.2	**50.0**	68.5	54.5	29.8	54.5	64.4
YOLOv4-CSP	640 × 640	52.9	47.5	66.2	51.7	28.2	51.2	59.8
PP-YOLO	608 × 608	52.9	45.2	65.2	49.9	26.3	47.8	57.2
YOLOX-X	640 × 640	99.1	51.2	69.6	55.7	31.2	56.1	66.1
YOLOv4-P5	896 × 896	70.8	51.8	70.3	56.6	33.4	55.7	63.4
YOLOv4-P6	1280 × 1280	127.6	54.5	72.6	59.8	36.8	58.3	65.9
RetinaNet (w/ SpineNet-143)	1280 × 1280	66.9	50.7	70.4	54.9	33.6	53.9	62.1

AP$_S$ has a larger winning margin (8.6% higher) than the runner-up, which is a good sign because SPD-Conv is especially advantageous for smaller objects and lower resolutions.

Lastly for large models, YOLOX-L achieves the best AP while our YOLOv5-SPD-l is only slightly (3%) lower (yet much better than other baselines shown in the bottom group). On the other hand, our AP$_S$ remains the highest, which echos SPD-Conv's advantage mentioned above.

Results on test-dev2017. As presented in Table 5, our YOLOv5-SPD-n is again the clear winner in the nano model category on AP$_S$, with a good winning margin (19%) over the runner-up, YOLOv5n. For the average AP, although it appears as if EfficientDet-D0 performed better than ours, that is because EfficientDet has almost double parameters than ours and was trained using high-resolution images (via transfer learning, as indicated by "Trf" in the cell) and AP is highly correlated with resolution. This training benefit is similarly reflected in the small model category too.

In spite of this benefit that other baselines receive, our approach reclaims its top rank in the next category, medium models, on both AP and AP$_S$. Finally in the large model category, our YOLOv5-SPD-l is also the best performer on AP$_S$, and closely matches YOLOX-L on AP.

Summary. It is clear that, by simply replacing the strided convolution and pooling layers with our proposed SPD-Conv building block, a neural net can significantly improves its accuracy, while maintaining the same level of parameter size. The improvement is more prominent when objects are small, which meets our goal well. Although we do not constantly notch the first position in all the cases, SPD-Conv is the only approach that *consistently* performs very well; it is only occasionally a (very close) runner-up if not performing the best, and is *always* the winner on AP$_S$ which is the chief metric we target.

Lastly, recall that we have adopted YOLOv5 hyperparameters without re-tuning, which means that our models will likely perform even better after dedicated hyperparameter tuning. Also recall that all the non-YOLO baselines (and PP-YOLO) were trained using transfer learning and thus have benefited from high quality images, while ours do not.

Visual Comparison. For a visual and intuitive understanding, we provide two real examples using two randomly chosen images, as shown in Fig. 5. We compare YOLOv5-SPD-m and YOLOv5m since the latter is the best performer among all the baselines in the corresponding (medium) category. Figure 5(a)(b) demonstrates that YOLOv5-SPD-m is able to detect the occluded giraffe which YOLOv5m misses, and Fig. 5(c)(d) shows that YOLOv5-SPD-m detects very small objects (a face and two benches) while YOLOv5m fails to.

5.2   Image Classification

Dataset and Setup. For the task of image classification, we use the Tiny ImageNet [19] and CIFAR-10 datasets [17]. Tiny ImageNet is a subset of the ILSVRC-2012 classification dataset and contains 200 classes. Each class has 500 training images, 50 validation images, and 50 test images. Each image is of resolution $64 \times 64 \times 3$ pixels. CIFAR-10 consists of 60,000 images of resolution $32 \times 32 \times 3$, including 50,000 training images and 10,000 test images. There are 10 classes with 6,000 images per class. We use the top-1 accuracy as the metric to evaluate the classification performance.

Training. We train our ReseNet18-SPD model on Tiny ImageNet. We perform random grid search to tune hyperparameters including learning rate, batch size, momentum, optimizer, and weight decay. Figure 6 shows a sample hyperparameter sweep plot generated using the wandb MLOPs. The outcome is the SGD optimizer with a learning rate of 0.01793 and momentum of 0.9447, a mini batch size of 256, weight decay regularization of 0.002113, and 200 training epochs. Next, we train our ResNet50-SPD model on CIFAR-10. The hyperparameters are adopted from the ResNet50 paper, where SGD optimizer is used with an initial learning rate 0.1 and momentum 0.9, batch size 128, weight decay regularization 0.0001, and 200 training epochs. For both ReseNet18-SPD and ReseNet50-SPD, we use the same decay function as in ResNet to decrease the learning rate as the number of epochs increases.

(a) Purple boxes: YOLOv5m predictions. (b) Green boxes: YOLOv5-SPD-*m* predictions.

(c) Purple boxes: YOLOv5m predictions. (d) Green boxes: YOLOv5-SPD-*m* predictions.

Fig. 5. Object detection examples from `val2017`. Blue boxes indicate the ground truth. Red arrows highlight the differences.

Testing. The accuracy on Tiny ImageNet is evaluated on the validation dataset because the ground truth in the test dataset is not available. The accuracy on CIFAR-10 is calculated on the test dataset.

Results. Table 6 summarizes the results of top-1 accuracy. It shows that our models, ResNet18-SPD and ResNet50-SPD, clearly outperform all the other baseline models.

Finally, we provide in Fig. 7 a visual illustration using Tiny ImageNet. It shows 8 examples misclassified by ResNet18 and correctly classified by ResNet18-SPD. The common characteristics of these images is that the resolution is low and therefore presents a challenge to the standard ResNet which loses fine-grained information during its strided convolution and pooling operations.

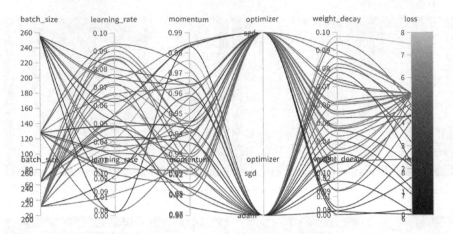

Fig. 6. Hyperparameter tuning in image classification: a sweep plot using `wandb`.

Table 6. Image classification performance comparison.

Model	Dataset	Top-1 accuracy (%)
ResNet18-SPD	Tiny ImageNet	**64.52**
ResNet18	Tiny ImageNet	61.68
Convolutional Nystromformer for Vision	Tiny ImageNet	49.56
WaveMix-128/7	Tiny ImageNet	52.03
ResNet50-SPD	CIFAR-10	**95.03**
ResNet50	CIFAR-10	93.94
Stochastic Depth	CIFAR-10	94.77
Prodpoly	CIFAR-10	94.90

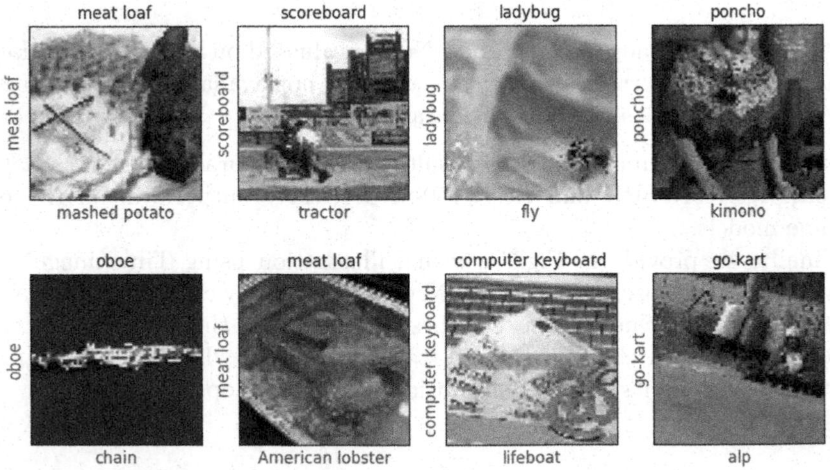

Fig. 7. Green labels: ground truth. Blue labels: ResNet18-SPD predictions. Red labels: ResNet-18 predictions. (Color figure online)

6   Conclusion

This paper identifies a common yet defective design in existing CNN architectures, which is the use of strided convolution and/or pooling layers. It will result in the loss of fine-grained feature information especially on low-resolution images and small objects. We then propose a new CNN building block called SPD-Conv that eliminates the strided and pooling operations altogether, by replacing them with a space-to-depth convolution followed by a non-strided convolution. This new design has a big advantage of downsampling feature maps while retaining the discriminative feature information. It also represents a general and unified approach that can be easily applied to perhaps any CNN architecture and to strided conv and pooling the same way. We provide two most representative use cases, object detection and image classification, and demonstrate via extensive evaluation that SPD-Conv brings significant performance improvement on detection and classification accuracy. We anticipate it to widely benefit the research community as it can be easily integrated into existing deep learning frameworks such as PyTorch and TensorFlow.

References

1. COCO dataset (2017). https://cocodataset.org
2. CodaLab COCO detection challenge (bounding box) (2019). https://competitions.codalab.org/competitions/20794
3. Adelson, E.H., Anderson, C.H., Bergen, J.R., Burt, P.J., Ogden, J.M.: Pyramid methods in image processing. RCA Eng. **29**(6), 33–41 (1984)
4. Bochkovskiy, A., Wang, C.Y., Liao, H.Y.M.: YOLOV4: optimal speed and accuracy of object detection. arXiv preprint arXiv:2004.10934 (2020)
5. Carion, N., Massa, F., Synnaeve, G., Usunier, N., Kirillov, A., Zagoruyko, S.: End-to-end object detection with transformers. In: Vedaldi, A., Bischof, H., Brox, T., Frahm, J.-M. (eds.) ECCV 2020. LNCS, vol. 12346, pp. 213–229. Springer, Cham (2020). https://doi.org/10.1007/978-3-030-58452-8_13
6. Chevalier, M., Thome, N., Cord, M., Fournier, J., Henaff, G., Dusch, E.: LR-CNN for fine-grained classification with varying resolution. In: 2015 IEEE International Conference on Image Processing (ICIP), pp. 3101–3105. IEEE (2015)
7. Duan, K., Bai, S., Xie, L., Qi, H., Huang, Q., Tian, Q.: CenterNet: Keypoint triplets for object detection. In: Proceedings of the IEEE/CVF International Conference on Computer Vision, pp. 6569–6578 (2019)
8. Ge, Z., Liu, S., Wang, F., Li, Z., Sun, J.: Yolox: exceeding yolo series in 2021. arXiv preprint arXiv:2107.08430 (2021)
9. Girshick, R.: Fast R-CNN. In: Proceedings of the IEEE International Conference on Computer Vision, pp. 1440–1448 (2015)
10. Girshick, R., Donahue, J., Darrell, T., Malik, J.: Rich feature hierarchies for accurate object detection and semantic segmentation. In: Proceedings of the IEEE Conference on Computer Vision and Pattern Recognition, pp. 580–587 (2014)
11. Haris, M., Shakhnarovich, G., Ukita, N.: Task-driven super resolution: object detection in low-resolution images. In: Mantoro, T., Lee, M., Ayu, M.A., Wong, K.W., Hidayanto, A.N. (eds.) ICONIP 2021. CCIS, vol. 1516, pp. 387–395. Springer, Cham (2021). https://doi.org/10.1007/978-3-030-92307-5_45

12. He, K., Zhang, X., Ren, S., Sun, J.: Spatial pyramid pooling in deep convolutional networks for visual recognition. IEEE Trans. Pattern Anal. Mach. Intell. **37**(9), 1904–1916 (2015)
13. He, K., Zhang, X., Ren, S., Sun, J.: Deep residual learning for image recognition. In: IEEE/CVF Conference on Computer Vision and Pattern Recognition, pp. 770–778 (2016)
14. Jocher, G., et al.: https://github.com/ultralytics/yolov5 (2021). Released version available at the time of evaluation: 12 October 2021
15. Kim, Y., Kang, B.-N., Kim, D.: SAN: learning relationship between convolutional features for multi-scale object detection. In: Ferrari, V., Hebert, M., Sminchisescu, C., Weiss, Y. (eds.) ECCV 2018. LNCS, vol. 11209, pp. 328–343. Springer, Cham (2018). https://doi.org/10.1007/978-3-030-01228-1_20
16. Koziarski, M., Cyganek, B.: Impact of low resolution on image recognition with deep neural networks: an experimental study. Int. J. Appl. Math. Comput. Sci. **28**(4) (2018)
17. Krizhevsky, A., Nair, V., Hinton, G.: Cifar-10 (canadian institute for advanced research) http://www.cs.toronto.edu/kriz/cifar.html
18. Krizhevsky, A., Sutskever, I., Hinton, G.E.: ImageNet classification with deep convolutional neural networks. In: NeurIPS, vol. 25 (2012)
19. Le, Y., Yang, X.: Tiny ImageNet visual recognition challenge. CS 231N **7**, 3 (2015)
20. Lin, T.Y., Dollár, P., Girshick, R., He, K., Hariharan, B., Belongie, S.: Feature pyramid networks for object detection. In: Proceedings of the IEEE Conference on Computer Vision and Pattern Recognition, pp. 2117–2125 (2017)
21. Lin, T.Y., Goyal, P., Girshick, R., He, K., Dollár, P.: Focal loss for dense object detection. In: IEEE ICCV, pp. 2980–2988 (2017)
22. Liu, S., Qi, L., Qin, H., Shi, J., Jia, J.: Path aggregation network for instance segmentation. In: Proceedings of the IEEE Conference on Computer Vision and Pattern Recognition, pp. 8759–8768 (2018)
23. Liu, S., Qi, L., Qin, H., Shi, J., Jia, J.: Path aggregation network for instance segmentation. In: Proceedings of the IEEE Conference on Computer Vision and Pattern Recognition (CVPR) (June 2018)
24. Liu, W., et al.: SSD: single shot multibox detector. In: Leibe, B., Matas, J., Sebe, N., Welling, M. (eds.) ECCV 2016. LNCS, vol. 9905, pp. 21–37. Springer, Cham (2016). https://doi.org/10.1007/978-3-319-46448-0_2
25. Peng, X., Hoffman, J., Stella, X.Y., Saenko, K.: Fine-to-coarse knowledge transfer for low-res image classification. In: 2016 IEEE International Conference on Image Processing (ICIP), pp. 3683–3687. IEEE (2016)
26. Redmon, J., Farhadi, A.: Yolov3: an incremental improvement. arXiv preprint arXiv:1804.02767 (2018)
27. Ren, S., He, K., Girshick, R., Sun, J.: Faster R-CNN: towards real-time object detection with region proposal networks. Adv. Neural. Inf. Process. Syst. **28**, 91–99 (2015)
28. Ren, S., He, K., Girshick, R., Sun, J.: Faster R-CNN: towards real-time object detection with region proposal networks. IEEE Trans. Pattern Anal. Mach. Intell. **39**(6), 1137–1149 (2016)
29. Sajjadi, M.S., Vemulapalli, R., Brown, M.: Frame-recurrent video super-resolution. In: Proceedings of the IEEE Conference on Computer Vision and Pattern Recognition, pp. 6626–6634 (2018)
30. Simonyan, K., Zisserman, A.: Very deep convolutional networks for large-scale image recognition. arXiv preprint arXiv:1409.1556 (2014)

31. Singh, B., Davis, L.S.: An analysis of scale invariance in object detection - snip. In: IEEE CVPR, pp. 3578–3587 (2018)
32. Singh, B., Najibi, M., Davis, L.S.: Sniper: efficient multi-scale training. In: 31st Proceedings of Conference on Advances in Neural Information Processing Systems (2018)
33. Singh, M., Nagpal, S., Vatsa, M., Singh, R.: Enhancing fine-grained classification for low resolution images. In: 2021 International Joint Conference on Neural Networks (IJCNN), pp. 1–8. IEEE (2021)
34. Tan, M., Pang, R., Le, Q.V.: Efficientdet: scalable and efficient object detection. In: Proceedings of the IEEE/CVF Conference on Computer Vision and Pattern Recognition, pp. 10781–10790 (2020)
35. Tian, Z., Shen, C., Chen, H., He, T.: FCOS: fully convolutional one-stage object detection. In: Proceedings of the IEEE/CVF International Conference on Computer Vision, pp. 9627–9636 (2019)
36. Wang, C.Y., Bochkovskiy, A., Liao, H.Y.M.: scaled-yolov4: scaling cross stage partial network. In: Proceedings of the IEEE/CVF Conference on Computer Vision and Pattern Recognition, pp. 13029–13038 (2021)
37. Wang, Z., Chang, S., Yang, Y., Liu, D., Huang, T.S.: Studying very low resolution recognition using deep networks. In: Proceedings of the IEEE Conference on Computer Vision and Pattern Recognition, pp. 4792–4800 (2016)
38. Zheng, Z., et al.: Distance-IOU loss: Faster and better learning for bounding box regression. In: Proceedings of the AAAI Conference on Artificial Intelligence, vol. 34, pp. 12993–13000 (2020)

SAViR-T: Spatially Attentive Visual Reasoning with Transformers

Pritish Sahu[✉], Kalliopi Basioti[✉], and Vladimir Pavlovic[✉]

Department of Computer Science, Rutgers University, Piscataway, NJ, USA
pritish.sahu@rutgers.edu, kib21@scarletmail.rutgers.edu,
vladimir@cs.rutgers.edu

Abstract. We present a novel computational model, SAViR-T, for the family of visual reasoning problems embodied in the Raven's Progressive Matrices (RPM). Our model considers explicit spatial semantics of visual elements within each image in the puzzle, encoded as spatio-visual tokens, and learns the intra-image as well as the inter-image token dependencies, highly relevant for the visual reasoning task. Token-wise relationship, modeled through a transformer-based SAViR-T architecture, extract group (row or column) driven representations by leveraging the group-rule coherence and use this as the inductive bias to extract the underlying rule representations in the top two row (or column) per token in the RPM. We use this relation representations to locate the correct choice image that completes the last row or column for the RPM. Extensive experiments across both synthetic RPM benchmarks, including RAVEN, I-RAVEN, RAVEN-FAIR, and PGM, and the natural image-based "V-PROM" demonstrate that SAViR-T sets a new state-of-the-art for visual reasoning, exceeding prior models' performance by a considerable margin.

Keywords: Abstract visual reasoning · Raven's progressive matrices · Transformer

1 Introduction

Human abstract reasoning is the analytic process aimed at decision-making or solving a problem [1]. In the realm of visual reasoning, humans find it advantageous, explicitly or implicitly, to break down an image into well-understood low-level concepts before proceeding with the reasoning task, e.g., examining object properties or counting objects. These low-level concepts are combined to form high-level abstract concepts in a list of images, enabling relational reasoning functions such as assessing the increment of the object count, changes in the type of object, or object properties, and subsequently applying the acquired knowledge to unseen scenarios. However, replicating such reasoning processes in machines is particularly challenging [12].

Supplementary Information The online version contains supplementary material available at https://doi.org/10.1007/978-3-031-26409-2_28.

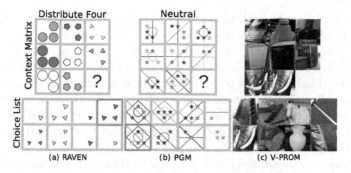

Fig. 1. Three RPM examples from the (a) RAVEN, (b) PGM, and (c) V-PROM datasets. The highlighted green box shows the correct image in the choice list. Solving these RPM requires identifying the underlying rules applied to image attributes or lines along a row or column and, the image with the best fit from the choice list to the rules is the correct answer. RAVEN example (left), the rules governing are "distribute_three" on {position, color }(*Color figure online*) and "progression" on {type, size}. The rules governing PGM(middle) are "OR" on object type line and "XOR" on the position of shapes. And V-PROM (right) example is "And". (Color figure online)

A popular test format of abstract reasoning in the visual domain is the Raven's Progressive Matrix (RPM), developed on Spearman's work on human general intelligence [23]. The test is designed as an incomplete 3×3 matrix, with each matrix element being an image and the bottom right location left empty, c.f., Fig. 1. Every image can contain one or more objects or lines characterized by the attributes of shape, color, scale, rotation angle, and, holistically, the counts of items or their variability (all circles, all pentagons, both pentagons, and circles, etc.). The top two rows or columns follow a certain unknown rule applied to the attributes; the task is to pick the correct image from an unordered set of choices, satisfying the same constraints. For example in Fig. 1, we present three instances of RPM, RAVEN [32], PGM [2], and V-PROM [25] datasets, where the first eight images are denoted as context images, and below them is the set of choice images.

Classical computational models for solving RPMs, built upon access to symbolic attribute representations of the images [5,15–17], are incapable of adapting to unseen domains. The success of deep models in other computer vision tasks made it possible to exploit the feature representation and relational learning concepts in visual reasoning [18]. Initial studies [2,32] using widely popular neural network architecture such as ResNet [9] and LSTM [10] failed in solving general reasoning tasks. These models aim to discover underlying rules by mapping the eight context images to each choice image. Modeling the reasoning network [3,11,33] to mimic the human reasoning process has led to a large performance gain. All recent works utilize an encoding mechanism to extract the features/attributes of single or groups of images, followed by a reasoning model that learns the underlying rule from the extracted features to predict scores for images in the choice list. The context features are contrasted against each choice features to elucidate the best image in the missing location. However, these models make use of holistic image representations, which ignore the important local, spatially contextualized features. Because typical reasoning patterns use intra and inter-image object-level relationships, holistic representations are likely to lead to suboptimal reasoning performance of the models that rely on it.

In this work, we focus on using local, spatially-contextualized features accompanied by an attention mechanism to learn the rule constraint within and across groups (rows or columns). We use a bottom-up and top-down approach to the visual encoding and the reasoning process. From the bottom-up, we address how a set of image regions are associated with each other via the self-attention mechanism from visual transformers. Specifically, instead of extracting a traditional holistic feature vector on image-level , we constrain semantic visual tokens to attend to different image patches. The top-down process is driven to solve visual reasoning tasks that predict an attention distribution over the image regions. To this end, we propose "SAViR-T", Spatially Attentive Visual Reasoning with Transformers, that naturally integrates the attended region vectors with abstract reasoning. Our reasoning network focuses on entities of interest obtained from the attended vector since the irrelevant local areas have been filtered out. Next, the reasoning task determines the Principal Shared rules in the two complete groups (typically, the top two rows) of the RPM per local region, which are then fused to provide an integrated rule representation. We define a similarity metric to compare the extracted rule representation with the rules formed in the last row when placing each choice at the missing location. The choice with the highest score is predicted as the correct answer. Our contributions in this work are three-fold:

- We propose a novel abstract visual reasoning model, SAViR-T, using spatially-localized attended features for reasoning tasks. SAViR-T accomplishes this using: the Backbone Network responsible for extracting a set of image region encodings; the Visual Transformer performing self-attention on the tokenized feature maps; and finally Reasoning Network, which elucidates the rules governing the puzzle over rows-columns to predict the solution to the RPM.
- SAViR-T automatically learns to focus on different semantic regions of the input images, addressing the problem of extracting holistic feature vectors per image, which may omit critical objects at finer visual scales. Our approach is generic because it is suitable for any configuration of the RPM problems without the need to modify the model for different image structures.
- We drastically improved the reasoning accuracy over all RPM benchmarks, echoed in substantial enhancement on the "3×3 Grid", "Out-In Single", and "Out Single, In Four Distribute" configurations for RAVEN and I-RAVEN, with strong accuracy gains in the other configurations. We show an average improvement of $2 - 3\%$ for RAVEN-type and PGM datasets. Performance improvement of SAViR-T on V-PROM, a natural image RPM benchmark, significantly improves by 10% over the current state-of-the-art models.

2 Related Works

2.1 Abstract Visual Reasoning

RPM is a form of non-verbal assessment for human intelligence with strong roots in cognitive science [2,5,32]. It measures an individual's eductive ability, i.e., the ability to find patterns in the apparent chaos of a set of visual scenes [23]. RPM consists of a context matrix of 3×3 that has eight images and a missing image at the last row last column. The participant has to locate the correct image from a choice set of size eight. In the early stages, RPM datasets were created manually, and the popular computational models solved them using hand-crafted feature representations [19], or access to symbolic representations [15]. It motivated the need for large-scale RPM datasets and the requirement for efficient reasoning models that utilized minimal prior knowledge. The first automatic RPM generation [28] work was based on using first-order logic, followed by two large-scale RPM datasets RAVEN [32] and Procedurally Generated Matrices (PGM) [2]. However, the RAVEN dataset contained a hidden shortcut solution where a model trained on the choice set only can achieve better performance than many state-of-the-art models. The reason behind this behavior is rooted in the creation of the choice set. Given the correct image, the distractor images were derived by randomly changing only one attribute. In response, two modified versions of the dataset, I-RAVEN by SRAN [11] and RAVEN-FAIR [3], were proposed to remove the shortcut solution and increase the difficulty levels of the distractors. Both the works, devised algorithm to generate a different set of distractors and provide evidence through experiments to claim the non-existence of any shortcut solution. The first significant advancement in RPM was by Wild Relational Network (WReN) [2], which applies the relation network of [24] multiple times to solve the abstract reasoning problem. LEN [34] learns to reason using a triplet of images in a row or column as input to a variant of the relation network. This work empirically supports improvements in performance using curriculum and reinforcement learning frameworks. CoPINet [33] suggests a contrastive learning algorithm to learn the underlying rules from given images. SRAN [11] designs a hierarchical rule-aware framework that learns rules through a series of steps of learning image representation, followed by row representation, and finally learning rules by pairing rows.

2.2 Transformer in Vision

Transformers [26] for machine translation have become widely adopted in numerous NLP tasks [7,14,22]. A transformer consists of self-attention layers added along with MLP layers. The self-attention mechanism plays a key role in drawing out the global dependencies between input and output. Grouped with the non-sequential processing of sentences, transformers demonstrate superiority in large-scale training scenarios compared to Recurrent Neural Networks. It avoids a drop in performance due to long-term dependencies. Recently, there has been a steady influx of visual transformer models to various vision tasks: image classification [6], object detection [4], segmentation [29], image generation [21], video processing [31], VQA [13]. Among them, Vision Transformer (ViT) [8] designed for image classification has closed the gap with the performance provided by any state-of-the-art models (e.g., ImageNet, ResNet) based on convolution. Similar to the word sequence referred to as tokens required by transformers in NLP, ViT splits an image into patches and uses the linear embeddings of these patches as an input sequence. Our idea is closely related to learning intra-sequence relationships via self-attention.

3 Method

Before presenting our reasoning model, we provide a formal description of the RPM task in Sect. 3.1 designed to measure abstract reasoning. The description articulates the condition required in a reasoning model to solve RPM questions successfully. In Sect. 3.2, we describe the three major components of our SAViR-T: the backbone network, the visual transformer, and the reasoning network. Our objective is based on using local feature maps as tokens for rule discovery among visual attributes along rows or columns to solve RPM questions.

3.1 Raven's Progressive Matrices

Given a list of observed images in the form of Raven's Progressive Matrix ($\mathcal{M}$) referred to as the context of size 3×3 with a missing final element at $\mathcal{M}_{3,3}$, where $\mathcal{M}_{i,j}$ denotes the j-th image at i-th row. In Fig. 2, for ease of notation, we refer to the images in $\mathcal{M}$ as index locations 1 through 16 as formatted in the dataset, where the first eight form the context matrix and the rest belong to the choice list. The task of a learner is to solve the context $\mathcal{M}$ by finding the best-fit answer image from an unordered set of choices $\mathcal{A} = \{a_1, \ldots, a_8\}$. The images in an RPM can be decomposed into attributes, objects, and the object count. Learning intra-relationship between these visual components will guide the model to form a stronger inter-relationship between images constrained by rules. The learner needs to locate objects in each image, extract their visual attributes such as color, size, and shape, followed by inferring the rules "r" such as "constant", "progression", "OR", etc., that satisfies the attributes among a list of images. Usually, the rules "r" are applied row-wise or column-wise according to [5] on the decomposed visual elements. Since an RPM is based on a set of rules applied either row-wise or column-wise, the learner needs to pick the shared rules between the top two rows or columns. Among the choice list, the correct image from $\mathcal{A}$, when placed at the missing location in the last row or column, will satisfy these shared rules.

3.2 Our Approach: SAViR-T

Our method consists of three sub-modules: (i) a Backbone Network, (ii) a Visual Transformer (VT), and (iii) a Reasoning Network, trained end to end. Please

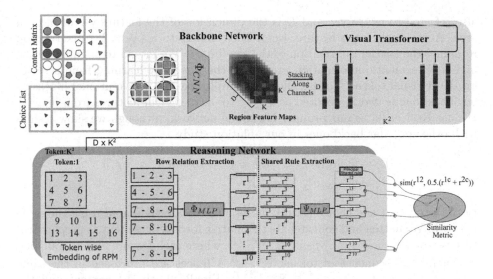

Fig. 2. SAViR-T consists of **Backbone Network, Visual Transformer, Reasoning Network**. Each image in $\mathcal{M}$ given to the Backbone Network (Φ_{CNN}) to extract the "Feature Maps", $f \in \mathbb{R}^{D \times K \times K}$. Visual Transformer attends on the features of local image patches and returns the attended vectors $\hat{f}$ of each image. The Reasoning Network functions per patch (depicted as K parallel layers) for the entire context and choice attended vectors. Per patch, we start with group (row or column) rule extraction r^i via Φ_{MLP}, followed by Shared rule extraction r^{ij} via Ψ_{MLP}. The Principal Shared rule r^{12} is compared against extracted Shared rule for choice a, $\frac{1}{2}(r^{1c} + r^{2c})$ ("c" is the choice row index for choice image "a"). The choice image with max similarity score is predicted as the answer.

refer Fig. 2 for an illustration of our model. First, we process the images in $\mathcal{M}$ via several convolutional blocks referred to as the backbone network. The output feature map is given as to the visual transformer to extract the attended visual embedding. The attended embedding is given to the reasoning network to discover the embedded rule representation in RPM. A scoring function is used to rank the choice images by comparing its row or column representation (extracted by placing it in the missing location) with the rule representation and predicting the index with the highest similarity as the correct choice. We leverage the strength of convolutions, which learns location invariant low-level neighborhood structures and visual transformer to relate to the higher-order semantic concepts. We treat each local region as a token separately in reasoning and apply a fusion function to recover the hidden rules that point the model to the correct answer.

Backbone Network. The backbone network receives as input an image from the context ($\mathcal{M}$) or choice list ($\mathcal{A}$) of size $\mathbb{R}^{C \times H \times W}$. The extracted feature map (f_{ij}, f_a) is of dimension $\mathbb{R}^{D \times K \times K}$ where $K \times K$ is the number of image regions also referred to as tokens, and D is the dimension of the feature vector of each area. Accordingly, each feature vector corresponds to a $(\frac{W}{K} \times \frac{H}{K})$ pixel region retaining the spatial information of the raw image. We intend to summarize the high-level semantic information present in the image by learning from a set of low-level visual tokens. To this effect, we employ several convolutional blocks as our backbone network, denoted as Φ_{CNN}, to extract local features from images. We use ResNet [9] as our primary backbone network, although we show results with other popular backbones in our ablation study.

$$f_{ij} = \Phi_{\text{CNN}}(\mathcal{M}_{ij}), \quad i,j = 1,\ldots,3 \quad \text{and} \quad i,j \neq 3 \tag{1}$$

$$f_a = \Phi_{\text{CNN}}(a), \quad \forall a \in \mathcal{A} \tag{2}$$

Both the context and choice feature representation are generated using the same network. We flatten and concatenate the matrix format in context to prepare feature vector $F_{\mathcal{M}} = [f_{11}, \ldots, f_{32}] \in \mathbb{R}^{8 \times K^2 \times D}$, where each feature map is reshaped into a K^2-tall sequence of tokens, $\mathbb{R}^{K^2 \times D}$. Choices are processed in the same manner, $F_{\mathcal{A}} = [f_{a_1}, \ldots, f_{a_8}] \in \mathbb{R}^{8 \times K^2 \times D}$. Finally, both the context and the choice representations are concatenated as $F = [F_{\mathcal{M}}, F_{\mathcal{A}}]$, with $[\cdot, \cdot]$ denoting the concatenation operator.

Visual Transformer. To learn the concepts responsible for reasoning, we seek to model the interactions between local regions of an image as a bottom-up process, followed by top-down attention to encoder relations over the regions. We adopt Visual Transformer [8], which learns the attention weights between tokens to focus on relationally-relevant regions within images. The transformer is composed of "Multi-head Self-Attention" (MSA) mechanism followed by a multi-layer perceptron (MLP). Both are combined together in layers $l = 1, \ldots, L$ to form the transformer encoder. A layer normalization layer (LN) and residual connection are added before and after every core component. The interactions between the tokens generate an attention map for each layer and head. Below we describe the steps involved in learning the attended vectors.

Reasoning Model. Human representations of space are believed to be hierarchical [20], with objects parsed into parts and grouped into part constellations. To mimic this, SAViR-T generates weighted representations over local regions in the image described above. Our reasoning module combines the inductive bias in an RPM and per-patch representations to learn spatial relations between images. We start by translating these attended region vectors obtained from RPM above into within-group relational reasoning instructions, expressed in terms of the row representations r_k^i via Φ_{MLP}. These representations hold knowledge of the rules that bind the images in the i-th row. We realize ten row representations, including eight possible last rows, where each choice is replaced at the missing location (similarly for the columns). We define function Ψ_{MLP} which retrieves the common across-group rule representations r_k^{ij}, given the pair (r_k^i, r_k^j). The maximum similarity score between the last-row rules based on the Choice list,

$\{r_k^{1c}, r_k^{2c}\}, \forall c = 3, \ldots, 10$, and the extracted Principal Shared rule from the top two rows r_k^{12} indicates the correct answer.

Row Relation Extraction Module. Since our image encoding is prepared region-wise, we focus on these regions independently to detect patterns maintaining the order of RPM, i.e., row-wise (left to right) or column-wise (top to bottom). We restructure the resulting output of a RPM from the transformer $\hat{F}$ as $\mathbb{R}^{K^2 \times 16 \times D}$, where the emddeddings of each local region k for an image at index n in the RPM is denoted $\hat{f}_k^n$ for $n = 1, 2, \ldots, 16^1$ and $k = 1, \ldots, K^2$. We collect these embeddings for every row and column as triplets $(\hat{f}_k^1, \hat{f}_k^2, \hat{f}_k^3)$, $(\hat{f}_k^4, \hat{f}_k^5, \hat{f}_k^6)$, $(\hat{f}_k^7, \hat{f}_k^8, \hat{f}_k^a)$, $a = 9, \ldots, 16$. Similarly, the triplets formed from columns are $(\hat{f}_k^1, \hat{f}_k^4, \hat{f}_k^7)$, $(\hat{f}_k^2, \hat{f}_k^5, \hat{f}_k^8)$, $(\hat{f}_k^3, \hat{f}_k^6, \hat{f}_k^a)$. Each row and column is concatenated along the feature dimension and processed through a relation extraction function Φ_{MLP}:

$$r_k^1 = \Phi_{\mathrm{MLP}}(\hat{f}_k^1, \hat{f}_k^2, \hat{f}_k^3), \quad c_k^1 = \Phi_{\mathrm{MLP}}(\hat{f}_k^1, \hat{f}_k^4, \hat{f}_k^7)$$
$$r_k^2 = \Phi_{\mathrm{MLP}}(\hat{f}_k^4, \hat{f}_k^5, \hat{f}_k^6), \quad c_k^2 = \Phi_{\mathrm{MLP}}(\hat{f}_k^2, \hat{f}_k^5, \hat{f}_k^8) \qquad (3)$$
$$r_k^3 = \Phi_{\mathrm{MLP}}(\hat{f}_k^7, \hat{f}_k^8, \hat{f}_k^a), \quad c_k^3 = \Phi_{\mathrm{MLP}}(\hat{f}_k^3, \hat{f}_k^6, \hat{f}_k^a).$$

The function Φ_{MLP} has two-fold aims: (a) it seeks to capture common properties of relational reasoning along a row/column for each local region; (b) it updates the attention weights to cast aside the irrelevant portions that do not contribute to rules.

Shared Rule Extraction Module. The common set of rules conditioned on visual attributes in the first two rows of a RPM is the Principal Shared rule set which the last row has to match to select the correct image in the missing location. The top two rows will contain these common sets of rules, possibly along with the rules unique to their specific rows. Given r_k^i, r_k^j, the goal of function Ψ_{MLP} is to extract these Shared rules between the pairs of rows/columns,

$$r_k^{ij} = \Psi_{\mathrm{MLP}}(r_k^i, r_k^j), \quad c_k^{ij} = \Psi_{\mathrm{MLP}}(c_k^i, c_k^j), \quad rc_k^{ij} = [r_k^{ij}, c_k^{ij}]. \qquad (4)$$

Similar to Φ_{MLP}, function Ψ_{MLP} aims to elucidate the Shared relationships between any pair i, j of rows/columns. Those relationships should hold across image patches, thus we fuse rc_k^{ij} obtained for $k = 1, \ldots, K^2$ regions via averaging, leading to the Shared rule embedding $rc^{ij} = \frac{1}{K^2} \sum_k rc_k^{ij}$ for the entire image.

3.3   Training and Inference

Given the principal Shared rule embedding rc^{12} from the top two row pairs, a similarity metric is a function of closeness between $\mathrm{sim}(rc^{12}, rc^a)$, where $rc^a = \frac{1}{2}(rc^{1c} + rc^{2c}), \forall c = 3, \ldots, 10$ is the average of the Shared rule embeddings among the choice a and the top two rows, and $\mathrm{sim}(\cdot, \cdot)$ is the inner product between

---

^{1} $16 = 8 + 8$ for eight Context and eight Choice images of an RPM.

rc^{12} and the average Shared rule embedding. The similarity score will be higher for the correct image is placed at the last row compared to the wrong choice,

$$a^* = \arg\max_a \mathrm{sim}(rc^{12}, rc^a). \tag{5}$$

We use cross-entropy as our loss function to train SAViR-T end-to-end. To bolster generalization property of our model, two types of augmentation were adapted from [33]: (i) shuffle the order of the top two rows or columns, as the resulting change will not affect the final solution since the rules remain unaffected; and (ii) shuffling the index location of the correct image in the unordered set of choice list. After training our model, we can use SAViR-T to solve new RPM problems (i.e., during testing) by applying (5).

4   Experiments

We study the effectiveness of our proposed SAViR-T for solving the challenging RPM questions, specifically focusing on PGM [2], RAVEN [32], I-RAVEN [11], RAVEN-FAIR [3], and V-PROM [25]. Details about the datasets can be found in the Supplementary. Next, we describe the experimental details of our simulations, followed by the results of these experiments and the performance analysis of obtained results, including an ablation study of our SAViR-T.

4.1   Experimental Settings

We trained SAViR-T for 100 epochs on all three RAVEN datasets, 50 epochs for PGM and 100 epochs for V-PROM, where each RPM is scaled to $16 \times 224 \times 224$. For V-PROM, similar to [25], we use the features extracted from the pre-trained ResNet-101 before the last pooling layer; i.e., dimension of $2048 \times 7 \times 7$. To further reduce the complexity in case of V-PROM, we use an MLP layer to derive $512 \times 7 \times 7$ feature vectors; we must mention that this MLP becomes part of SAViR-T's training parameters. We use the validation set to track model performance during the training process and use the best validation checkpoint to report the accuracy on the test set. For SAViR-T, we adopt ResNet-18 as our backbone for results in Table 1 and Table 2. Our transformer depth is set to one for all datasets and the counts of heads set to 3 for RAVEN datasets, to 6 for PGM and to 5 for V-PROM. Finally, in our reasoning module, we use a two-layer MLP, Φ_{MLP}, and a four-layer MLP, Ψ_{MLP} with a dropout of 0.5 applied to the last layer. As the RAVEN dataset is created by applying rules row-wise, we set the column vector c_k^{ij} in (4) as zero-vector while training our model. No changes are made while training for PGM, as the tuple (rule, object, attribute) can be applied either along the rows or columns.

4.2   Performance Analysis

Table 1 summarizes the performance of our model and other baselines on the test set of RAVEN and I-RAVEN datasets. We report the scores from [35] for

Table 1. Model performance (%) on RAVEN / I-RAVEN.

Method	Acc	Center	2×2Grid	3×3Grid	L-R	U-D	O-IC	O-IG
LSTM [10]	13.1/18.9	13.2/26.2	14.1/16.7	13.7/15.1	12.8/14.6	12.4/16.5	12.2/21.9	13/21.1
WReN [2]	34.0/23.8	58.4/29.4	38.9/26.8	37.7/23.5	21.6/21.9	19.7/21.4	38.8/22.5	22.6/21.5
ResNet+DRT [32]	59.6/40.4	58.1/46.5	46.5/28.8	50.4/27.3	65.8/50.1	67.1/49.8	69.1/46.0	60.1/34.2
LEN [34]	72.9/39.0	80.2/45.5	57.5/27.9	62.1/26.6	73.5/44.2	81.2/43.6	84.4/50.5	71.5/34.9
CoPINet [33]	91.4/46.1	95.1/54.4	77.5/36.8	78.9/31.9	99.1/51.9	99.7/52.5	98.5/52.2	91.4/42.8
SRAN [11]	56.1[†]/60.8	78.2[†]/78.2	44.0[†]/50.1	44.1[†]/42.4	65.0[†]/70.1	61.0[†]/70.3	60.2[†]/68.2	40.1[†]/46.3
DCNet [36]	93.6/49.4	97.8/57.8	81.7/34.1	**86.7**/35.5	99.8/58.5	**99.8**/60	**99.0**/57.0	91.5/42.9
SCL [30]	91.6/95.0	**98.1**/99.0	91.0/96.2	82.5/89.5	96.8/97.9	96.5/97.1	96.0/97.6	80.1/87.7
SAViR-T (Ours)	**94.0/98.1**	97.8/**99.5**	**94.7**/**98.1**	83.8/**93.8**	97.8/**99.6**	98.2/**99.1**	97.6/**99.5**	88.0/**97.2**
Human	84.41/-	95.45/-	81.82/-	79.55/-	86.36/-	81.81/-	86.36/-	81.81/-

[†] indicates our evaluation of the baseline in the absence of published results.

Table 2. Test accuracy of different models on PGM.

	LSTM [10]	ResNet	CoPINet [33]	WReN [2]	MXGNet [27]	LEN [34]	SRAN [11]	DCNet [36]	SCL [30]	SAViR-T (Ours)
Acc	35.8	42.0	56.4	62.6	66.7	68.1	71.3	68.6	88.9	**91.2**

Table 3. Test accuracy of different models on V-PROM.

Method	RN [25]	DCNet [36]	SRAN [11]	SAViR-T (Ours)
Acc	52.8[††]	30.4[†]	40.8[†]	**62.6**

[†] indicates our evaluation of the baseline in the absence of published results.
[††] indicates our evaluation of the baseline (reported result in [25] was 51.2).

I-RAVEN on LEN, COPINet, and DCNet. We also report the performance of humans on the RAVEN dataset [32]; there is no reported human performance on I-RAVEN. Overall, our SAViR-T achieves superior performance among all baselines for I-RAVEN and a strong performance, on average, on RAVEN. Our method performs similar to DCNet for RAVEN with a slight improvement of 0.4%. We notice DCNet has better performance over ours by a margin of 1.4%–3.5% over "3 × 3 Grid", "L-R", "U-D", "O-IC" and "O-IG" , while we show 13% improvement for "2 × 2 Grid". For I-RAVEN, the average test accuracy of our model improves from 95% (SCL) to 98.13% and shows consistent improvement over all configurations across all models. The most significant gain, spotted in "3 × 3 Grid" and "O-IG" is expected since our method learns to attend to semantic spatio-visual tokens. As a result, we can focus on the smaller scale objects present

Table 4. Test accuracy of different models on RAVEN-FAIR.

Method	ResNet [9]	LEN [34]	COPINet [33]	DCNet [36]	MRNet [3]	SAViR-T (Ours)
Acc	72.5	78.3	91.4	54.5†	96.6	**97.4**

† indicates our evaluation of the baseline in the absence of published results.

in these configurations, which is essential since the attributes of these objects define the rules of the RPM problem. In Table 1, DCNet achieves better accuracy for "3 × 3", "L-R", "U-D", "O-IC", and "O-IG" for RAVEN but significantly lower accuracy for I-RAVEN, **suggesting DCNet exploits the shortcut in RAVEN**. See our analysis in Table 5 that supports this observation.

We also observe that our reasoning model shows more significant improvements on I-RAVEN than RAVEN. This is because the two datasets differ in the selection process of the negative choice set. The wrong images differ in only a single attribute from the right panel for RAVEN, while in I-RAVEN, they differ in at least two characteristics. The latter choice set prevents models from deriving the puzzle solution by only considering the available choices. At the same time, this strategy also helps the classification problem (better I-RAVEN scores) since now the choice set images are more distinct. In Table 4, we also report the test scores on the RAVEN-FAIR dataset against several baselines. Similar to the above, our model achieves the best performance.

Table 2 reports performance of SAViR-T and other models trained on the neutral configuration in the PGM dataset. Our model improves by 2.3% over the best baseline model (SCL). PGM dataset is 20 times larger than RAVEN, and the applied rule can be present either row-wise or column-wise. Additionally, PGM contains "line" as an object type, increasing the complexity compared to RAVEN datasets. Even under these additional constraints, SAViR-T is able to improve RPM solving by mimicking the reasoning process.

In Table 3 we report the performance on the V-PROM dataset, made up of natural images. The background signal for every image in the dataset can be considered as noise or a distractor. In this highly challenging benchmark, our SAViR-T shows a major improvement of over 8% over the Relation Network (RN) reported in [25] (51.2 reported in [25] and 52.83 for our evaluation of the RN model –since the code is not available–). Since the V-PROM dataset is the most challenging one, we define the margin Δ for each testing sample to better understand the performance of SAViR-T:

$$\Delta = \mathrm{sim}(rc^{12}, rc^{a^*}) - \max_{a \neq a^*} \mathrm{sim}(rc^{12}, rc^a),$$

where a^* indicates the correct answer among eight choices. The model is confident and correctly answers the RPM question for $\Delta \gg 0$). When $\Delta \approx 0$, the model is uncertain about its predictions, $\Delta < 0$ indicating incorrect and $\Delta > 0$ correct uncertain predictions. Finally, the model is confident but incorrectly answers for $\Delta \ll 0$).

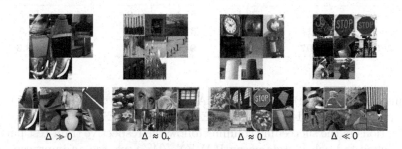

Fig. 3. V-PROM RPM examples (trained on SAViR-T) for 4-cases of Δ inorder from correct prediction with strong certainty to incorrect prediction with strong certainty.

In Fig. 3 we present examples from the V-PROM testing set. In the Supplementary we present additional detailed analysis of these results. The first example has a $\Delta > 100$, which means that the trained SAViR-T is very confident about its prediction. Next, we move to examples with Δ very close to zero, either negative or positive. We visualize two such examples in Fig. 3, in the second (correctly classified RPM) and the third (misclassified) puzzles. The second is a counting problem where the first image in a row has x objects and the next two images in the same row y (i.e., first row $x = 7, y = 2$). Some images are distorted and/or blurred after the pre-processing required to use the pre-trained ResNet-101, which makes the recognition and "counting" of objects difficult. The second example is an "And" rule on object attributes. The first row contains circle objects and the second and third cylinders. The wrongly selected image (depicted in a red bounding box) contains as well cylinder objects. In these two cases, the model is uncertain about which image is solving the RPM.

Lastly, we study examples which SAViR-T very confidently misclassifies. Specifically, we picked the worst six instances of the testing set. One of these examples is depicted in the last puzzle of Fig. 3. All examples belong to the "And rule" for object attributes like the case $\Delta \gg 0$. In the depicted example, "And" rule of the last row refers to "Players"; where in both cases, they are playing "Tennis". The problem is that images five and eight depict "Players" playing "Tennis", resulting in a controversial situation. Although the model misclassifies the "Player" in the second choice, it is reasonable to choose either the second, fifth, or eighth images as the correct image in the puzzle.

Table 5 presents our cross-dataset results between models trained and tested on different RAVEN-based datasets. Since all three datasets only differ in the manner their distractors in the Choice set were created but are identical in the Context, a model close to the generative process should be able to pick the correct image irrespective of how difficult the distractors are. In the first two

Table 5. Results of cross-dataset evaluation. Top row indicates the training set, next the test set.

Model training set	RAVEN		I-RAVEN		RAVEN-F	
Evaluation Dataset	I-RAVEN	RAVEN-F	RAVEN	RAVEN-F	RAVEN	I-RAVEN
SRAN	72.8	78.5	57.1	71.9	54.7	60.5
DCNet	14.7	27.4	37.9	51.7	57.8	46.5
SAViR-T (Ours)	**97**	**97.5**	**95.1**	**97.7**	**94.7**	**88.3**

columns, we train SAViR-T with the RAVEN dataset and measure the trained model performance on I-RAVEN, RAVEN-FAIR (RAVEN-F) testing sets; other combinations in the succeeding columns follow the respective train-test patterns. As was expected, when training on RAVEN (94%), we see an improvement on both I-RAVEN (97%) and RAVEN-FAIR (97.5%) test since the latter datasets have more dissimilar choice images, helping the reasoning problem. This increase in performance can be seen for SRAN from 56.1% to 72.8% and 78.5% respectively. Since DCNet (93.6%) utilizes the short solution, the accuracy drops to 14.7% and 27.4% respectively. For the same reasons, when training on I-RAVEN (98.8%), our model shows a drop in RAVEN (95.1%) performance; for RAVEN-FAIR (97.7%), the performance remains close to the one on the training dataset.

4.3   Ablation Study

Exploiting Shortcut Solutions. As shown in SRAN [11], any powerful model that learns by combining the extracted features from the choices is capable of exploiting the shortcut solution present in the original RAVEN. In a context-blind setting, a model trained only on images in the RAVEN choice list should predict randomly. However, context-blind {ResNet, CoPINet, DCNet} models attain 71.9%, 94.2% and 94.1% test accuracy respectively. We train and report the accuracy for context-blind DCNet and reported the scores in [11] for ResNet, CoPINet. Similarly, we investigate our SAViR-T in a context-blind setting. We remove the reasoning module and use the extracted attended choice vectors from the visual transformer, passed through an MLP, to output an eight-dimensional logit vector. After training for 100 epochs, our model performance remained at 12.2%, similar to the random guess of $1/8 = 12.5\%$, suggesting that our Backbone with the Visual Transformer does not contribute towards finding a shortcut. Thus, our semantic tokenized representation coiled with the reasoning module learns rules from the context to solve the RPM questions.

Does SAViR-T Learn Rules? The rules in RPMs for the I-RAVEN dataset are applied row-wise. However, these rules can exist in either rows or columns. We evaluate performance on two different setups to determine if our model can discover the rule embeddings with no prior knowledge of whether the rules were applied row-wise or column-wise. In our first setup, we train SAViR-T with the prior knowledge of row-wise rules in RPMs. We train our second model by preparing rule embeddings on both row and column and finally concatenating them to predict the correct answer. Our model performance drop was only 3.8% from 98.1% to 94.3%. In the case of SRAN [11], the reported drop in performance was 1.2% for the above setting. Overall, this indicates that our model is capable of ignoring the distraction from the column-wise rule application.

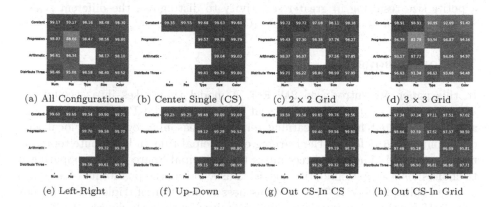

(a) All Configurations (b) Center Single (CS)   (c) 2 × 2 Grid   (d) 3 × 3 Grid

(e) Left-Right   (f) Up-Down   (g) Out CS-In CS   (h) Out CS-In Grid

Fig. 4. Testing set classification accuracy for the SAViR-T trained on all configurations based on each I-RAVEN rule (constant, progression, arithmetic, and distribute three) and used attribute (number, position, type, size, color of the objects). From left to right, Fig. 4a presents the classification accuracy for all configurations, Figure 4b for "Center Single", Fig. 4c for 2 × 2 Grid, and Fig. 4d for 3 × 3 Grid in the top row. Similarly in the bottom row, Fig. 4e for left right, Fig. 4f for the up down, Fig. 4g for out single, in center single and finally Fig. 4h for out single, in 2 × 2 Grid.

Figure 4 presents the I-RAVEN performance on the test set for SAViR-T when trained on the I-RAVEN dataset. The eight different heatmap images correspond to the setting with all configurations (Fig. 4a) and individual configurations from "Center Single"(Fig. 4b) to "Out-In-Grid" (Fig. 4h). The row dimension in each heatmap corresponds to the RPM rules used in the puzzles (Constant, Progression, Arithmetic, and Distribute Three). In I-RAVEN, each rule is associated with an attribute. Therefore, in the columns, we identify the characteristics of different objects, such as their "Number", "Position," "Type," "Size," and "Color". The Blank cells in the heatmap e.g., "(Arithmetic, Type)", indicate non-existence for that (rule, attribute) pair combination in the dataset.

The most challenging combination of (rule, attribute) is a progression with the position. In this setup, the different objects progressively change position on the 2 × 2 and 3 × 3 grids. The models fail to track this change well. To further understand this drop in performance, we performed extended experiments, investigating the difference in attributes between the correct image and the predicted (misclassified) image (more details in the Appendix). We notice that for the 2 × 2 grid configuration, the predicted differences are only in one attribute, primarily the "Position". This means the distractor objects are of the same color, size, type, and number as the correct one but have different positions inside the grid. The second group of misclassified examples differs in the "Type" of the present objects. Therefore, we can conclude the model finds it challenging to track the proper position of the entities and their type for some examples. Again, in the 3×3 grid, "Position" is the main differentiating attribute, but the "Size" attribute follows it; this makes sense since, in the 3×3 grid, the objects have a petite size resulting in greater sensitivity to distinguish the different scales. Similar behavior is observed in the O-IG configuration.

5   Conclusions

In this paper, we introduced SAViR-T, a model that takes into account the visually-critical spatial context present in image-based RPM. By partitioning an image into patches and learning relational reasoning over these local windows, our SAViR-T fosters the learning of Principal rule and attribute representations in RPM. The model recognizes the Principal Shared rule, comparing it to choices via a simple similarity metric, thus avoiding the possibility of finding a shortcut solution. SAViR-T shows robustness to injection of triplets that disobey the RPM formation patterns, e.g., when trained with both choices of row- and column-wise triplets on the RPM with uniquely, but unknown, row-based rules. We are the first to provide extensive experiments results on all three RAVEN-based datasets, PGM, and the challenging natural image-based V-PROM, which suggests that SAViR-T outperforms all baselines by a significant margin except for RAVEN where we match their accuracy.

References

1. Apps, J.N.: Abstract Thinking. In: Loue, S.J., Sajatovic, M. (eds.) Encyclopedia of Aging and Public Health. Springer, Boston, MA (2006). https://doi.org/10.1007/978-0-387-33754-8_2
2. Barrett, D., Hill, F., Santoro, A., Morcos, A., Lillicrap, T.: Measuring abstract reasoning in neural networks. In: International Conference on Machine Learning, pp. 511–520. PMLR (2018)
3. Benny, Y., Pekar, N., Wolf, L.: Scale-localized abstract reasoning. In: Proceedings of the IEEE/CVF Conference on Computer Vision and Pattern Recognition, pp. 12557–12565 (2021)
4. Carion, N., Massa, F., Synnaeve, G., Usunier, N., Kirillov, A., Zagoruyko, S.: End-to-end object detection with transformers. In: European Conference on Computer Vision, pp. 213–229. Springer, Cham (2020)
5. Carpenter, P.A., Just, M.A., Shell, P.: What one intelligence test measures: a theoretical account of the processing in the raven progressive matrices test. Psychol. Rev. **97**(3), 404 (1990)
6. Chen, M.,et al.: Generative pretraining from pixels. In: International Conference on Machine Learning, pp. 1691–1703. PMLR (2020)
7. Devlin, J., Chang, M.W., Lee, K., Toutanova, K.: BERT: pre-training of deep bidirectional transformers for language understanding. arXiv preprint arXiv:1810.04805 (2018)

8. Dosovitskiy, A., et al.: An image is worth 16×16 words: transformers for image recognition at scale. arXiv preprint arXiv:2010.11929 (2020)

9. He, K., Zhang, X., Ren, S., Sun, J.: Deep residual learning for image recognition. In: Proceedings of the IEEE Conference on Computer Vision and Pattern Recognition, pp. 770–778 (2016)

10. Hochreiter, S., Schmidhuber, J.: Long short-term memory. Neural Comput. **9**(8), 1735–1780 (1997)

11. Hu, S., Ma, Y., Liu, X., Wei, Y., Bai, S.: Stratified rule-aware network for abstract visual reasoning. arXiv preprint arXiv:2002.06838 (2020)

12. Lake, B.M., Ullman, T.D., Tenenbaum, J.B., Gershman, S.J.: Building machines that learn and think like people. Behav. brain sciences **40** (2017)

13. Li, L.H., Yatskar, M., Yin, D., Hsieh, C.J., Chang, K.W.: Visualbert: A simple and performant baseline for vision and language. arXiv preprint arXiv:1908.03557 (2019)

14. Liu, Y., et al.: Zettlemoyer, L., Stoyanov, V.: Roberta: a robustly optimized BERt pretraining approach. arXiv preprint arXiv:1907.11692 (2019)

15. Lovett, A., Forbus, K., Usher, J.: Analogy with qualitative spatial representations can simulate solving raven's progressive matrices. In: Proceedings of the Annual Meeting of the Cognitive Science Society. vol. 29 (2007)

16. Lovett, A., Forbus, K., Usher, J.: A structure-mapping model of raven's progressive matrices. In: Proceedings of the Annual Meeting of the Cognitive Science Society. vol. 32 (2010)

17. Lovett, A., Tomai, E., Forbus, K., Usher, J.: Solving geometric analogy problems through two-stage analogical mapping. Cogn. Sci. **33**(7), 1192–1231 (2009)

18. Małkiński, M., Mańdziuk, J.: A review of emerging research directions in abstract visual reasoning. arXiv preprint arXiv:2202.10284 (2022)

19. McGreggor, K., Goel, A.: Confident reasoning on raven's progressive matrices tests. In: Proceedings of the AAAI Conference on Artificial Intelligence. vol. 28 (2014)

20. Palmer, S.E.: Hierarchical structure in perceptual representation. Cogn. Psychol. **9**(4), 441–474 (1977)

21. Parmar, N., Vaswani, A., Uszkoreit, J., Kaiser, L., Shazeer, N., Ku, A., Tran, D.: Image transformer. In: International Conference on Machine Learning, pp. 4055–4064. PMLR (2018)

22. Radford, A., Narasimhan, K., Salimans, T., Sutskever, I.: Improving language understanding with unsupervised learning (2018)

23. Raven, J.C., Court, J.H.: Raven's Progressive Matrices and Vocabulary Scales, vol. 759. Oxford pyschologists Press, Oxford (1998)

24. Santoro, A., et al.: A simple neural network module for relational reasoning. In: 30th Proceedings of Conference on Advances in Neural Information Processing Systems (2017)

25. Teney, D., Wang, P., Cao, J., Liu, L., Shen, C., van den Hengel, A.: V-PROM: a benchmark for visual reasoning using visual progressive matrices. In: Proceedings of the AAAI Conference on Artificial Intelligence, vol. 34, pp. 12071–12078 (2020)

26. Vaswani, A., et al.: Attention is all you need. In: 30th Proceedings of Advances in Neural Information Processing Systems (2017)

27. Wang, D., Jamnik, M., Lio, P.: Abstract diagrammatic reasoning with multiplex graph networks. arXiv preprint arXiv:2006.11197 (2020)

28. Wang, K., Su, Z.: Automatic generation of raven's progressive matrices. In: Twenty-Fourth International Joint Conference on Artificial Intelligence (2015)

29. Wang, Y., et al.: End-to-end video instance segmentation with transformers. In: Proceedings of the IEEE/CVF Conference on Computer Vision and Pattern Recognition, pp. 8741–8750 (2021)
30. Wu, Y., Dong, H., Grosse, R., Ba, J.: The scattering compositional learner: discovering objects, attributes, relationships in analogical reasoning. arXiv preprint arXiv:2007.04212 (2020)
31. Zeng, Y., Fu, J., Chao, H.: Learning Joint spatial-temporal transformations for video inpainting. In: Vedaldi, A., Bischof, H., Brox, T., Frahm, J.-M. (eds.) ECCV 2020. LNCS, vol. 12361, pp. 528–543. Springer, Cham (2020). https://doi.org/10.1007/978-3-030-58517-4_31
32. Zhang, C., Gao, F., Jia, B., Zhu, Y., Zhu, S.C.: Raven: A dataset for relational and analogical visual reasoning. In: Proceedings of the IEEE/CVF Conference on Computer Vision and Pattern Recognition, pp. 5317–5327 (2019)
33. Zhang, C., Jia, B., Gao, F., Zhu, Y., Lu, H., Zhu, S.C.: Learning perceptual inference by contrasting. In: 32nd Proceedings of Advances in Neural Information Processing Systems (2019)
34. Zheng, K., Zha, Z.J., Wei, W.: Abstract reasoning with distracting features. In: 32nd Advances in Neural Information Processing Systems (2019)
35. Zhuo, T., Huang, Q., Kankanhalli, M.: Unsupervised abstract reasoning for raven's problem matrices. IEEE Trans. Image Process. 30, 8332–8341 (2021)
36. Zhuo, T., Kankanhalli, M.: Effective abstract reasoning with dual-contrast network. In: International Conference on Learning Representations (2020)

A Scaling Law for Syn2real Transfer: How Much Is Your Pre-training Effective?

Hiroaki Mikami[1], Kenji Fukumizu[1,2], Shogo Murai[1], Shuji Suzuki[1],
Yuta Kikuchi[1], Taiji Suzuki[3,4], Shin-ichi Maeda[1], and Kohei Hayashi[1(✉)]

[1] Preferred Networks, Inc., Tokyo, Japan
{mhiroaki,murai,ssuzuki,kikuchi,ichi,hayasick}@preferred.jp
[2] The Institute of Statistical Mathematics, Tokyo, Japan
fukumizu@ism.ac.jp
[3] The University of Tokyo, Tokyo, Japan
taiji@mist.i.u-tokyo.ac.jp
[4] AIP-RIKEN, Tokyo, Japan

Abstract. Synthetic-to-real transfer learning is a framework in which a synthetically generated dataset is used to pre-train a model to improve its performance on real vision tasks. The most significant advantage of using synthetic images is that the ground-truth labels are automatically available, enabling unlimited expansion of the data size without human cost. However, synthetic data may have a huge domain gap, in which case increasing the data size does not improve the performance. How can we know that? In this study, we derive a simple scaling law that predicts the performance from the amount of pre-training data. By estimating the parameters of the law, we can judge whether we should increase the data or change the setting of image synthesis. Further, we analyze the theory of transfer learning by considering learning dynamics and confirm that the derived generalization bound is consistent with our empirical findings. We empirically validated our scaling law on various experimental settings of benchmark tasks, model sizes, and complexities of synthetic images.

1 Introduction

The success of deep learning relies on the availability of large data. If the target task provides limited data, the framework of transfer learning is preferably employed. A typical scenario of transfer learning is to pre-train a model for a similar or even different task and fine-tune the model for the target task. However, the limitation of labeled data has been the main bottleneck of supervised pre-training. While there have been significant advances in the representation capability of the models and computational capabilities of the hardware, the size and the diversity of the baseline dataset have not been growing as fast [57]. This

---

H. Mikami, K. Fukumizu, K. Hayashi—Equal contribution.

Supplementary Information The online version contains supplementary material available at https://doi.org/10.1007/978-3-031-26409-2_29.

is partially because of the sheer physical difficulty of collecting large datasets from real environments (e.g., the cost of human annotation).

In computer vision, *synthetic-to-real (syn2real) transfer* is a promising strategy that has been attracting attention [9,12,22,29,44,56,61]. In syn2real, images used for pre-training are synthesized to improve the performance on real vision tasks. By combining various conditions, such as 3D models, textures, light conditions, and camera poses, we can synthesize an infinite number of images with ground-truth annotations. Syn2real transfer has already been applied in some real-world applications. Teed and Deng [59] proposed a simultaneous localization and mapping (SLAM) system that was trained only with synthetic data and demonstrated state-of-the-art performance. The object detection networks for autonomous driving developed by Tesla was trained with 370 million images generated by simulation [36].

The performance of syn2real transfer depends on the similarity between synthetic and real data. In general, the more similar they are, the stronger the effect of pre-training will be. On the contrary, if there is a significant gap, increasing the number of synthetic data may be completely useless, in which case we waste time and computational resources. A distinctive feature of syn2real is that we can control the process of generating data by ourselves. If a considerable gap exists, we can try to regenerate the data with a different setting. But how do we know that? More specifically, in a standard learning setting without transfer, a "power law"-like relationship called a *scaling law* often holds between data size and generalization errors [35,53]. Is there such a rule for pre-training?

In this study, we find that the generalization error on fine-tuning is explained by a simple scaling law,

$$\text{test error} \simeq Dn^{-\alpha} + C, \tag{1}$$

where coefficient $D > 0$ and exponent $\alpha > 0$ describe the convergence speed of pre-training, and constant $C \geq 0$ determines the lower limit of the error. We refer to α as *pre-training rate* and C as *transfer gap*. We can predict how large the pre-training data should be to achieve the desired accuracy by estimating the parameters α, C from the empirical results. Additionally, we analyze the dynamics of transfer learning using the recent theoretical results based on the neural tangent kernel [50] and confirm that the above law agrees with the theoretical analysis. We empirically validated our scaling law on various experimental settings of benchmark tasks, model sizes, and complexities of synthetic images.

Our contributions are summarized as follows.

- From empirical results and theoretical analysis, we elicit a law that describes how generalization scales in terms of data sizes on pre-training and fine-tuning.
- We confirm that the derived law explains the empirical results for various settings in terms of pre-training/fine-tuning tasks, model size, and data complexity (e.g., Fig. 1). Furthermore, we demonstrate that we can use the estimated parameters in our scaling law to assess how much improvement we can expect from the pre-training procedure based on synthetic data.
- We theoretically derive a generalization bound for a general transfer learning setting and confirm its agreement with our empirical findings.

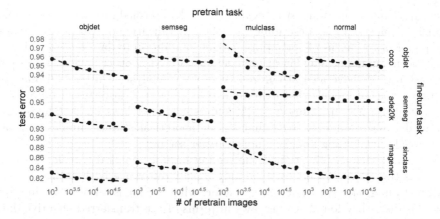

Fig. 1. Empirical results of syn2real transfer for different tasks. We conducted four pre-training tasks: object detection (`objdet`), semantic segmentation (`semseg`), multi-label classification (`mulclass`), surface normal estimation (`normal`), and three fine-tuning tasks for benchmark datasets: object detection for MS-COCO, semantic segmentation for ADE20K, and single-label classification (`sinclass`) for ImageNet. The y-axis indicates the test error for each fine-tuning task. Dots indicate empirical results and dashed lines indicate the fitted curves of scaling law (1). For more details, see Sect. 4.2.

2 Related Work

Pre-training for Visual Tasks. Many empirical studies show that the performance at a fine-tuning task scales with pre-training data (and model) size. For example, Huh et al. [32] studied the scaling behavior on ImageNet pre-trained models. Beyond ImageNet, Sun et al. [57] studied the effect of pre-training with pseudo-labeled large-scale data and found a logarithmic scaling behavior. Similar results were observed by Kolesnikov et al. [38].

Syn2real Transfer. The utility of synthetic images as supervised data for computer vision tasks has been continuously studied by many researchers [9,12,14, 22,29,31,43–45,56,61]. These studies found positive evidence that using synthetic images is helpful to the fine-tuning task. In addition, they demonstrated how data complexity, induced by e.g., light randomization, affects the final performance. For example, Newell and Deng [45] investigated how the recent self-supervised methods perform well as a pre-training task to improve the performance of downstream tasks. In this paper, following this line of research, we quantify the effects under the lens of the scaling law (1).

Neural Scaling Laws. The scaling behavior of generalization error, including some theoretical works [e.g., 3], has been studied extensively. For modern neural networks, Hestness et al. [28] empirically observed the power-law behavior of generalization for language, image, and speech domains with respect to the training size. Rosenfeld et al. [53] constructed a predictive form for the power-law in terms of data and model sizes. Kaplan et al. [35] pushed forward this direction in the language domain, describing that the generalization of transformers obeys the power law in terms of a compute budget in addition to data and model sizes. Since then, similar scaling laws have been discovered in other data domains [25]. Several authors have also attempted theoretical analysis. Hutter [33] analyzed a simple class of models that exhibits a power-law $n^{-\beta}$ in terms of data size n with arbitrary $\beta > 0$. Bahri et al. [5] addressed power laws under four regimes for model and data size. Note that these theoretical studies, unlike ours, are concerned with scaling laws in a non-transfer setting.

Hernandez et al. [27] studied the scaling laws for general transfer learning, which is the most relevant to this study. A key difference is that they focused on fine-tuning data size as a scaling factor, while we focus on pre-training data size. Further, they found scaling laws in terms of the transferred effective data, which is converted data amount necessary to achieve the same performance gain by pre-training. In contrast, Eq. (1) explains the test error with respect to the pre-training data size directly at a fine-tuning task. Other differences include task domains (language vs. vision) and architectures (transformer vs. CNN).

Theory of Transfer Learning. Theoretical analysis of transfer learning has been dated back to decades ago [7] and has been pursued extensively. Among others, some recent studies [16,42,62] derived an error bound of a fine-tuning task in the multi-task scenario based on complexity analysis; the bound takes an additive form $O(An^{-1/2} + Bs^{-1/2})$, where n and s are the data size of pre-training and fine-tuning, respectively, with coefficients A and B. Neural network regression has been also discussed with this bound [62]. In the field of domain adaptation, error bounds have been derived in relation to the mismatch between source and target input distributions [1,19]. They also proposed algorithms to adopt a new data domain. However, unlike in this study, no specific learning dynamics has been taken into account. In the area of hypothesis transfer learning [18,64], among many theoretical works, Du et al. [17] has derived a risk bound for kernel ridge regression with transfer realized as the weights on the training samples. The obtained bound takes a similar form to our scaling law. However, the learning dynamics of neural networks initialized with a pre-trained model has never been explored in this context.

3   Scaling Laws for Pre-training and Fine-tuning

The main obstacle in analyzing the test error is that we have to consider interplay between the effects of pre-training and fine-tuning. Let $L(n, s) \geq 0$ be the test error of a fine-tuning task with pre-training data size n and fine-tuning data size s. As the simplest case, consider a fine-tuning task without pre-training ($n = 0$), which boils the transfer learning down to a standard learning setting. In this case, the prior studies of both classical learning theory and neural scaling laws

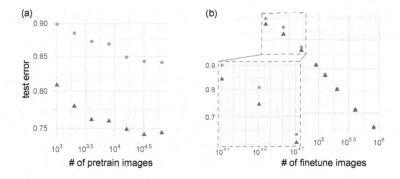

Fig. 2. Scaling curves with different (a) pre-training size and (b) fine-tuning size.

tell us that the test error decreases polynomially[1] with the fine-tuning data size s, that is, $L(0, s) = Bs^{-\beta} + \mathcal{E}$ with decay rate $\beta > 0$ and irreducible loss $\mathcal{E} \geq 0$. The irreducible loss $\mathcal{E}$ is the inevitable error given by the best possible mapping; it is caused by noise in continuous outputs or labels. Hereafter we assume $\mathcal{E} = 0$ for brevity.

3.1   Induction of Scaling Law with Small Empirical Results

To speculate a scaling law, we conducted preliminary experiments.[2] We pre-trained ResNet-50 by a synthetic classification task and fine-tuned by ImageNet. Figure 2(a) presents the log-log plot of error curves with respect to pre-training data size n, where each shape and color indicates a different fine-tuning size s. It shows that the pre-training effect diminishes for large n. In contrast, Fig. 2(b) presents the relations between the error and the fine-tuning size s with different n. It indicates the error drops straight down regardless of n, confirming the power-law scaling with respect to s. The above observations and the fact that $L(0, s)$ decays polynomially are summarized as follows.

Requirement 1. $\lim_{s \to \infty} L(n, s) = 0$.

Requirement 2. $\lim_{n \to \infty} L(n, s) = \text{const}$.

Requirement 3. $L(0, s) = Bs^{-\beta}$.

Requirements 1 and 3 suggest the dependency of n is embedded in the coefficient $B = g(n)$, i.e., the pre-training and fine-tuning effects interact multiplicatively. To satisfy Requirement 2, a reasonable choice for the pre-training effect is $g(n) =$

---

[1] For classification with strong low-noise condition, it is known that the decay rate can be exponential [49]. However, we focus only on the polynomial decay without such strong condition in this paper.

[2] The results are replicated from Appendix C.2; see the subsection for more details.

$n^{-\alpha} + \gamma$; the error decays polynomially with respect to n but has a plateau at γ. By combining these, we obtain

$$L(n, s) = \delta(\gamma + n^{-\alpha})s^{-\beta}, \tag{2}$$

where $\alpha, \beta > 0$ are decay rates for pre-training and fine-tuning, respectively, $\gamma \geq 0$ is a constant, and $\delta > 0$ is a coefficient. The exponent β determines the convergence rate with respect to fine-tuning data size. From this viewpoint, $\delta(\gamma + n^{-\alpha})$ is the coefficient factor to the power law. The influence of the pre-training appears in this coefficient, where the constant term $\delta\gamma$ comes from the irreducible loss of the pre-training task and $n^{-\alpha}$ expresses the effect of pre-training data size. The theoretical consideration in Section E.5 suggests that the rates α and β can depend on both the target functions of pre-training and fine-tuning as well as the learning rate.

3.2   Theoretical Deduction of Scaling Law

Next, we analyze the fine-tuning error from a purely theoretical point of view. To incorporate the effect of pre-training that is given as an initialization, we need to analyze the test error during the training with a given learning algorithm such as SGD. We apply the recent development by [50] to transfer learning. The study successfully analyzes the generalization of neural networks in the dynamics of learning, showing it achieves minmax optimum rate. The analysis uses the framework of the reproducing kernel Hilbert space given by the neural tangent kernel [34].

For theoretical analysis of transfer, it is important to formulate a task similarity between pre-training and fine-tuning. If the tasks were totally irrelevant (e.g., learning MNIST to forecast tomorrow's weather), pre-training would have no benefit. Following Nitanda and Suzuki [50], for simplicity of analysis, we discuss only a regression problem with square loss. We assume that a vector input x and scalar output y follow $y = \phi_0(x)$ for pre-training and $y = \phi_0(x) + \phi_1(x)$ for fine-tuning, where we omit the output noise for brevity; the task types are identical sharing the same input-output form, and task similarity is controlled by ϕ_1.

We analyze the situation where the effect of pre-training remains in the fine-tuning even for large data size ($s \to \infty$). More specifically, the theoretical analysis assumes a regularization term as the ℓ_2-distance between the weights and the initial values, and a smaller learning rate than constant in the fine-tuning. Hence we control how the pre-training effect is preserved through the regularization and learning rate. Other assumptions made for theoretical analysis concern the model and learning algorithm; a two-layer neural network having M hidden units with continuous nonlinear activation[3] is adopted; for optimization, the averaged SGD [51], an online algorithm, is used for a technical reason.

The following is an informal statement of the theoretical result. See Appendix E for details. We emphasize that our result holds not only for syn2real transfer but also for transfer learning in general.

---

[3] ReLU is not included in this class, but we can generalize this condition; see [50].

Theorem 1 (Informal). *Let $\hat{f}_{n,s}(x)$ be a model of width M pre-trained by n samples $(x_1, y_1), \ldots, (x_n, y_n)$ and fine-tuned by s samples $(x'_1, y'_1), \ldots, (x'_s, y'_s)$ where inputs $x, x' \sim p(x)$ are i.i.d. with the input distribution $p(x)$ and $y = \phi_0(x)$ and $y' = \varphi(x') = \phi_0(x') + \phi_1(x')$. Then the generalization error of the squared loss $L(n, s) = |\hat{f}_{n,s}(x) - \varphi(x)|^2$ is bounded from above with high probability as*

$$E_x L(n, s) \leq A_1(c_M + A_0 n^{-\alpha})s^{-\beta} + \varepsilon_M. \tag{3}$$

ε_M and c_M can be arbitrary small for large M; A_0 and A_1 are constants; the exponents α and β depend on ϕ_0, ϕ_1, $p(x)$, and the learning rate of fine-tuning.

The above bound (3) shows the correspondence with the empirical derivation of the full scaling law (2). Note that the approximation error ε_M is omitted in (2).

We note that the derived bound takes a multiplicative form in terms of the pre-training and fine-tuning effects, which contrasts with the additive bounds such as $An^{-1/2} + Bs^{-1/2}$ [62]. The existing studies consider the situation where a part of a network (e.g., backbone) is frozen during fine-tuning. Therefore, the error of pre-training is completely preserved after fine-tuning, and both errors appear in an additive way. This means that the effect of pre-training is irreducible by the effect of fine-tuning, and vice versa. In contrast, our analysis deals with the case of re-optimizing the entire network in fine-tuning. In that case, the pre-trained model is used as initial values. As a result, even if the error in pre-training is large, the final error can be reduced to zero by increasing the amount of fine-tuning data.

3.3 Insights and Practical Values

The form of the full scaling law (2) suggests that there are two scenarios depending on whether fine-tuning data is big or small. In "big fine-tune" regime, pre-training contributes relatively little. By taking logarithm, we can separate the full scaling law (2) into the pre-training part $u(n) = \log(n^{-\alpha} + \gamma)$ and the fine-tuning part $v(s) = -\beta \log s$. Consider to increase n by squaring it. Since the pre-training part cannot be reduced below $\log(\gamma)$ as

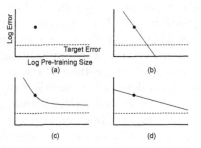

Fig. 3. Pre-training scenarios.

$u(n) > u(n^2) > \log(\gamma)$, the relative improvement $(u(n^2) - u(n))/v(s)$ becomes infinitesimal for large s. Figure 2(b) confirms this situation. Indeed, prior studies provide the same conclusion that the gain from pre-training can easily vanish [24,45] or a target task accuracy even degrade [67] if we have large enough fine-tuning data.

The above observation, however, does not mean pre-training is futile. Dense prediction tasks such as depth estimation require pixel-level annotations, which critically limits the number of labeled data. Pre-training is indispensable in such "small fine-tune" regime. Based on this, we hereafter analyze the case where the fine-tuning size s is fixed. By eliminating s-dependent terms in (2), we obtain a simplified law (1) by setting $D = \delta s^{-\beta}$ and $C = \delta \gamma s^{-\beta}$. After several evaluations, these parameters including α can be estimated by the nonlinear least squares method (see also Sect. 4.1).

As a practical benefit, the estimated parameters of the simplified law (1) bring a way to assess syn2real transfer. Suppose we want to solve a classification task that requires at least 90% accuracy with limited labels. We generate some number of synthetic images and pre-train with them, and we obtain 70% accuracy as Fig. 3(a). How can we achieve the required accuracy? It depends on the parameters of the scaling law. The best scenario is (b)—transfer gap C is low and pre-training rate α is high. In this case, increasing synthetic images eventually leads the required accuracy. In contrast, when transfer gap C is larger than the required accuracy (c), increasing synthetic images does not help to solve the problem. Similarly, for low pre-training rate α (d), we may have to generate tremendous amount of synthetic images that are computationally infeasible. In the last two cases, we have to change the rendering settings such as 3D models and light conditions to improve C and/or α, rather than increasing the data size. The estimation of α and C requires to compute multiple fine-tuning processes. However, the estimated parameters tell us whether we should increase data or change the data generation process, which can reduce the total number of trials and errors.

4   Experiments

4.1   Settings

For experiments, we employed the following transfer learning protocol. First, we pre-train a model that consists of backbone and head networks from random initialization until convergence, and we select the best model in terms of the validation error of the pre-training task. Then, we extract the backbone and add a new head to fine-tune all the model parameters. For notations, the task names of object detection, semantic segmentation, multi-label classification, single-label classification, and surface normal estimation are abbreviated as `objdet`, `semseg`, `mulclass`, `sinclass`, and `normal`, respectively. The settings for transfer learning are denoted by arrows. For example, `objdet→semseg` indicates that a model is pre-trained by object detection, and fine-tuned by semantic segmentation. All the results including Fig. 1 are shown as log-log plots. The details of pre-training, fine-tuning, and curve fitting are described in Appendix A.1.

4.2   Scaling Law Universally Explains Downstream Performance for Various Task Combinations

Figure 1 shows the test errors of each fine-tuning task and fitted learning curves with (1), which describes the effect of pre-training data size n for all combinations of pre-training and fine-tuning tasks. The scaling law fits with the empirical fine-tuning test errors with high accuracy in most cases.

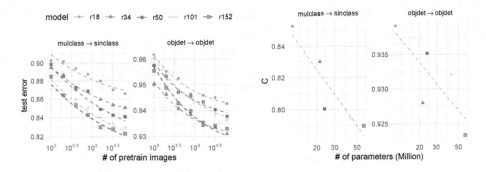

Fig. 4. Effect of model size. Best viewed in color. **Left**: The scaling curves for mulclass→sinclass and objdet→objdet cases. The meanings of dots and lines are the same as those in Fig. 1. **Right**: The estimated transfer gap C (y-axis) versus the model size (x-axis) in log-log scale. The dots are estimated values, and the lines are linear fittings of them.

4.3   Bigger Models Reduce the Transfer Gap

We compared several ResNet models as backbones in mulclass→sinclass and objdet→objdet to observe the effects of model size. Figure 4(left) shows the curves of scaling laws for the pre-training data size n for different sizes of backbone ResNet-x, where $x \in \{18, 34, 50, 101, 152\}$. The bigger models attain smaller test errors. Figure 4(right) shows the values of the estimated transfer gap C. The results suggest that there is a roughly power-law relationship between the transfer gap and model size. This agrees with the scaling law with respect to the model size shown by Hernandez et al. [27].

4.4   Scaling Law Can Extrapolate for More Pre-training Images

We also evaluated the extrapolation ability of the scaling law. We increased the number of synthetic images from the original size ($n = 64,000$) to 1.28 million, and see how the fitted scaling law predicts the unseen test errors where $n > 64,000$. As a baseline, we compared the power-law model, which is equivalent to the derived scaling law (1) with $C = 0$. Figure 5(left) shows the extrapolation results in objdet→objdet setting, which indicates the scaling law follows the saturating trend in regions with large pre-training sizes for all models, while the power-law model fails to capture it. The prediction errors is numerically shown in Fig. 5(right), which again shows our scaling law achieves better prediction performance.

4.5   Data Complexity Affects both Pre-training Rate and Transfer Gap

We examined how the complexity of synthetic images affects fine-tuning performance. We controlled the following four rendering parameters: *Appearance*:

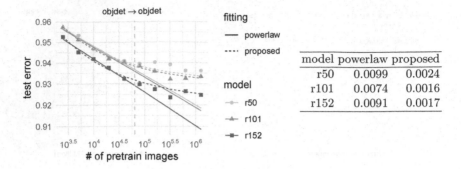

Fig. 5. Ability to extrapolate. **Left:** The solid lines represent the fitted power law and the dashed curves represent the fitted scaling law (1), in which the laws were fitted using the empirical errors where the pre-training size n was less than 64,000 (the first five dots). The vertical dashed line indicates where $n = 64,000$. **Right:** The root-mean-square errors between the laws and the actual test errors in the area of extrapolation (the last five dots).

Number of objects in each image; `single` or `multiple` (max 10 objects). *Light*: Either an area and point light is `randomized` or `fixed` in terms of height, color, and intensity. *Background*: Either the textures of floor/wall are `randomized` or `fixed`. *Object texture*: Either the 3D objects used for rendering contain texture (`w/`) or not (`w/o`). Indeed, the data complexity satisfies the following ordered relationships: `single` < `multiple` in *appearance*, `fix` < `random` in *light* and *background*, and `w/o` < `w/` in *object texture*[4]. To quantify the complexity, we computed the negative entropy of the Gaussian distribution fitted to the last activation values of the backbone network. For this purpose, we pre-trained ResNet-50 as a backbone with MS-COCO for 48 epochs and computed the empirical covariance of the last activations for all the synthetic data sets.

The estimated parameters are shown in Fig. 6, which indicates the following (we discuss the implications of these results further in Sect. 5.1).

- Data complexity controlled by the rendering settings correlates with the negative entropy, implying the negative entropy expresses the actual complexity of pre-training data.
- Pre-training rate α correlates with data complexity. The larger complexity causes slower rates of convergence with respect to the pre-training data size.
- Transfer gap C mostly correlates negatively with data complexity, but not for *object texture*.

As discussed in Sect. 4.1, we have fixed the value of D to avoid numerical instability, which might cause some bias to the estimates of α. We postulate, however, the value of D depends mainly on the fine-tuning task and thus has a fixed value for different pre-training data complexities. This can be inferred

---

[4] The object category of `w/o` is a subset of `w/`, and `w/` has a strictly higher complexity than `w/o`.

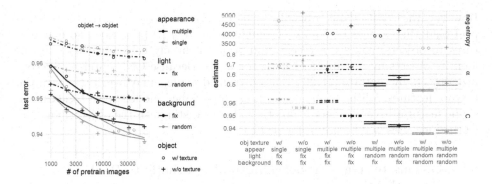

Fig. 6. Effect of synthetic image complexity. Best viewed in color. **Left**: Scaling curves of different data complexities. **Right**: Estimated parameters. The error bars represent the standard error of the estimate in least squares.

from the theoretical analysis in Appendix E.5: the exponent β in the main factor $s^{-\beta}$ of D does not depend on the pre-training data distribution but only on the fine-tuning task or the pre-training true mapping. Thus, the values of D should be similar over the different complexities, and the correlation of α preserves.

5   Conclusion and Discussion

In this paper, we studied how the performance on syn2real transfer depends on pre-training and fine-tuning data sizes. Based on the experimental results, we found a scaling law (1) and its generalization (2) that explain the scaling behavior in various settings in terms of pre-training/fine-tuning tasks, model sizes, and data complexities. Further, we present the theoretical error bound for transfer learning and found our theoretical bound has a good agreement with the scaling law.

5.1   Implication of Complexity Results in Sect. 4.5

The results of Sect. 4.5 has two implications. First, data complexity (i.e., the diversity of images) largely affects the pre-training rate α. This is reasonable because if we want a network to recognize more diverse images, we need to train it with more examples. Indeed, prior studies [5,55] observed that α is inversely proportional to the intrinsic dimension of the data (e.g., dimension of the data manifold), which is an equivalent concept of data complexity.

Second, the estimated values of the transfer gap C suggest that increasing the complexity of data is generally beneficial to decrease C, but not always. Figure 6(right) shows that increasing complexities in terms of *appearance*, *light*, and *background* reduces the transfer gap, which implies that these rendering operations are most effective to cover the fine-tuning task that uses real images.

However, the additional complexity in *object texture* works negatively. We suspect that this occurred because of *shortcut learning* [20]. Namely, adding textures to objects makes the recognition problem falsely easier because we can identify objects by textures rather than shapes. Because CNNs prefer to recognize objects by textures [21,26], the pre-trained models may overfit to learn the texture features. Without object textures, pre-trained models have to learn the shape features because there is no other clue to distinguish the objects, and the learned features will be useful for real tasks.

5.2   Lessons to Transfer Learning and Synthetic-to-Real Generalization

Our results suggest the transfer gap C is the most crucial factor for successful transfer learning because C determines the maximum utility of pre-training. Large-scale pre-training data can be useless when C is large. In contrast, if C is negligibly small, the law is reduced essentially to $n^{-\alpha}$, which tells that the volume of pre-training data is directly exchanged to the performance of fine-tuning tasks. Our empirical results suggest two strategies for reducing C: 1) Use bigger models and 2) fill the domain gap in terms of the decision rule and image distribution. For the latter, existing techniques such as domain randomization [60] would be helpful.

5.3   Limitations

We have not covered several directions in this paper. In theory, we assume several conditions that may not fit with the actual setting; the additive fine-tuning model $\phi_0(x) + \phi_1(x)$ in Theorem 1 does not address the transfer to different type of tasks, and the distributional difference of the inputs (synthetic versus real) is not considered. We analyzed only ASGD as the optimization and the effect of the choice is not fully clarified. In spite of these theoretical simplifications, our analysis has revealed the important aspects of the transfer learning as discussed in Sect. 3. In the experiments, although our theory is justified, we have not investigated the case when a pre-training dataset is not synthetic but real. These topics are left for future work.

Acknowledgments. We thank Daisuke Okanohara, Shoichiro Yamaguchi, Takeru Miyato, Katsuhiko Ishiguro for valuable comments and discussions in the early stage of this study. We also thank Masanori Koyama and Kenta Oono for reading the draft and providing detailed feedback. TS was partially supported by JSPS KAKENHI (18H03201, 20H00576), and JST CREST. KF was partially supported by JST CREST JPMJCR2015.

References

1. Acuna, D., Zhang, G., Law, M.T., Fidler, S.: F-domain-adversarial learning: theory and algorithms. arXiv:2106.11344 (2021)

2. Allen-Zhu, Z., Li, Y., Liang, Y.: Learning and generalization in overparameterized neural networks, going beyond two layers. CoRR abs/1811.04918 (2018)
3. Amari, S., Fujita, N., Shinomoto, S.: Four types of learning curves. Neural Comput. **4**(4), 605–618 (1992)
4. Arora, S., Du, S., Hu, W., Li, Z., Wang, R.: Fine-grained analysis of optimization and generalization for overparameterized two-layer neural networks. In: Proceedings of the 36th International Conference on Machine Learning, vol. 97, pp. 322–332 (2019)
5. Bahri, Y., Dyer, E., Kaplan, J., Lee, J., Sharma, U.: Explaining neural scaling laws. arXiv:2102.06701 (2021)
6. Bartlett, P.L., Foster, D.J., Telgarsky, M.J.: Spectrally-normalized margin bounds for neural networks. In: Advances in Neural Information Processing Systems (2017)
7. Baxter, J.: A model of inductive bias learning. J. Artif. Intell. Res. **12**, 149–198 (2000)
8. Bolya, D., Zhou, C., Xiao, F., Lee, Y.J.: YOLACT: real-time instance segmentation. In: Proceedings of the IEEE/CVF International Conference on Computer Vision, pp. 9157–9166 (2019)
9. Borrego, J., Dehban, A., Figueiredo, R., Moreno, P., Bernardino, A., Santos-Victor, J.: Applying domain randomization to synthetic data for object category detection. arXiv:1807.09834 (2018)
10. Caponnetto, A., De Vito, E.: Optimal rates for regularized least-squares algorithm. Found. Comput. Math. **7**(3), 331–368 (2007)
11. Chen, L.C., Papandreou, G., Schroff, F., Adam, H.: Rethinking atrous convolution for semantic image segmentation. arXiv:1706.05587 (2017)
12. Chen, W., et al.: Contrastive syn-to-real generalization. arXiv:2104.02290 (2021)
13. Denninger, M., et al.: Blenderproc. arXiv:1911.01911 (2019)
14. Devaranjan, J., Kar, A., Fidler, S.: Meta-Sim2: unsupervised learning of scene structure for synthetic data generation. In: Vedaldi, A., Bischof, H., Brox, T., Frahm, J.-M. (eds.) ECCV 2020. LNCS, vol. 12362, pp. 715–733. Springer, Cham (2020). https://doi.org/10.1007/978-3-030-58520-4_42
15. Du, S., Lee, J., Li, H., Wang, L., Zhai, X.: Gradient descent finds global minima of deep neural networks. In: Proceedings of the 36th International Conference on Machine Learning, vol. 97, pp. 1675–1685 (2019)
16. Du, S.S., Hu, W., Kakade, S.M., Lee, J.D., Lei, Q.: Few-shot learning via learning the representation, provably. arXiv:2002.09434 (2020)
17. Du, S.S., Koushik, J., Singh, A., Poczos, B.: Hypothesis transfer learning via transformation functions. In: Advances in Neural Information Processing Systems, vol. 30 (2017)
18. Fei-Fei, L., Fergus, R., Perona, P.: One-shot learning of object categories. IEEE Trans. Pattern Anal. Mach. Intell. **28**(4), 594–611 (2006)
19. Ganin, Y., et al.: Domain-adversarial training of neural networks. J. Mach. Learn. Res. **17**(1), 2096-2030 (2016)
20. Geirhos, R., et al.: Shortcut learning in deep neural networks. Nat. Mach. Intell. **2**(11), 665–673 (2020)
21. Geirhos, R., Rubisch, P., Michaelis, C., Bethge, M., Wichmann, F.A., Brendel, W.: ImageNet-trained CNNs are biased towards texture; increasing shape bias improves accuracy and robustness. arXiv:1811.12231 (2018)
22. Georgakis, G., Mousavian, A., Berg, A.C., Kosecka, J.: Synthesizing training data for object detection in indoor scenes. arXiv:1702.07836 (2017)
23. Goyal, P., et al.: Accurate, large minibatch SGD: training ImageNet in 1 hour. arXiv:1706.02677 (2017)

24. He, K., Girshick, R., Dollár, P.: Rethinking imagenet pre-training. arXiv:1811.08883 (2018)
25. Henighan, T., et al.: Scaling laws for autoregressive generative modeling. arXiv:2010.14701 (2020)
26. Hermann, K.L., Chen, T., Kornblith, S.: The origins and prevalence of texture bias in convolutional neural networks. arXiv:1911.09071 (2019)
27. Hernandez, D., Kaplan, J., Henighan, T., McCandlish, S.: Scaling laws for transfer. arXiv:2102.01293 (2021)
28. Hestness, J., et al.: Deep learning scaling is predictable, empirically. arXiv:1712.00409 (2017)
29. Hinterstoisser, S., Pauly, O., Heibel, H., Marek, M., Bokeloh, M.: An annotation saved is an annotation earned: using fully synthetic training for object instance detection. arXiv:1902.09967 (2019)
30. Hodaň, T., et al.: BOP challenge 2020 on 6D object localization. In: Bartoli, A., Fusiello, A. (eds.) ECCV 2020. LNCS, vol. 12536, pp. 577–594. Springer, Cham (2020). https://doi.org/10.1007/978-3-030-66096-3_39
31. Hodaň, T., et al.: Photorealistic image synthesis for object instance detection. In: 2019 IEEE International Conference on Image Processing (ICIP), pp. 66–70. IEEE (2019)
32. Huh, M., Agrawal, P., Efros, A.A.: What makes imagenet good for transfer learning? arXiv:1608.08614 (2016)
33. Hutter, M.: Learning curve theory. arXiv:2102.04074 (2021)
34. Jacot, A., Gabriel, F., Hongler, C.: Neural tangent kernel: convergence and generalization in neural networks. In: Advances in Neural Information Processing Systems, vol. 31, pp. 8571–8580. Curran Associates, Inc. (2018)
35. Kaplan, J., et al.: Scaling laws for neural language models. arXiv:2001.08361 (2020)
36. Karpathy, A.: Tesla AI day (2021). https://www.youtube.com/watch?v=j0z4FweCy4M
37. Kirkpatrick, J., et al.: Overcoming catastrophic forgetting in neural networks. Proc. Natl. Acad. Sci. U.S.A. **114**(13), 3521–3526 (2017)
38. Kolesnikov, A., et al.: Big transfer (bit): general visual representation learning. arXiv:1912.11370 (2019)
39. Lin, T.Y., Dollár, P., Girshick, R., He, K., Hariharan, B., Belongie, S.: Feature pyramid networks for object detection. In: Proceedings of the IEEE Conference on Computer Vision and Pattern Recognition, pp. 2117–2125 (2017)
40. Lin, T.-Y., et al.: Microsoft COCO: common objects in context. In: Fleet, D., Pajdla, T., Schiele, B., Tuytelaars, T. (eds.) ECCV 2014. LNCS, vol. 8693, pp. 740–755. Springer, Cham (2014). https://doi.org/10.1007/978-3-319-10602-1_48
41. Long, J., Shelhamer, E., Darrell, T.: Fully convolutional networks for semantic segmentation. In: Proceedings of the IEEE Conference on Computer Vision and Pattern Recognition, pp. 3431–3440 (2015)
42. Maurer, A., Pontil, M., Romera-Paredes, B.: The benefit of multitask representation learning. J. Mach. Learn. Res. **17**(81), 1–32 (2016)
43. Mousavi, M., Khanal, A., Estrada, R.: AI playground: unreal engine-based data ablation tool for deep learning. In: Bebis, G., et al. (eds.) ISVC 2020. LNCS, vol. 12510, pp. 518–532. Springer, Cham (2020). https://doi.org/10.1007/978-3-030-64559-5_41
44. Movshovitz-Attias, Y., Kanade, T., Sheikh, Y.: How useful is photo-realistic rendering for visual learning? arXiv:1603.08152 (2016)
45. Newell, A., Deng, J.: How useful is self-supervised pretraining for visual tasks? arXiv:2003.14323 (2020)

46. Neyshabur, B., Bhojanapalli, S., McAllester, D., Srebro, N.: Exploring generalization in deep learning. In: Advances in Neural Information Processing Systems, vol. 30, pp. 5947–5956 (2017)
47. Neyshabur, B., Tomioka, R., Srebro, N.: Norm-based capacity control in neural networks. In: Proceedings of the 28th Conference on Learning Theory, pp. 1376–1401 (2015)
48. Nitanda, A., Chinot, G., Suzuki, T.: Gradient descent can learn less over-parameterized two-layer neural networks on classification problems (2020)
49. Nitanda, A., Suzuki, T.: Stochastic gradient descent with exponential convergence rates of expected classification errors. In: Proceedings of the Twenty-Second International Conference on Artificial Intelligence and Statistics. Proceedings of Machine Learning Research, vol. 89, pp. 1417–1426 (2019)
50. Nitanda, A., Suzuki, T.: Optimal rates for averaged stochastic gradient descent under neural tangent kernel regime. In: International Conference on Learning Representations (2021)
51. Polyak, B.T., Juditsky, A.B.: Acceleration of stochastic approximation by averaging. SIAM J. Control. Optim. 30(4), 838–855 (1992)
52. Ren, S., He, K., Girshick, R., Sun, J.: Faster R-CNN: towards real-time object detection with region proposal networks. IEEE Trans. Pattern Anal. Mach. Intell. 39(6), 1137–1149 (2016)
53. Rosenfeld, J.S., Rosenfeld, A., Belinkov, Y., Shavit, N.: A constructive prediction of the generalization error across scales. arXiv:1909.12673 (2019)
54. Russakovsky, O., et al.: ImageNet large scale visual recognition challenge. Int. J. Comput. Vis. 115(3), 211–252 (2015). https://doi.org/10.1007/s11263-015-0816-y
55. Sharma, U., Kaplan, J.: A neural scaling law from the dimension of the data manifold. arXiv:2004.10802 (2020)
56. Su, H., Qi, C.R., Li, Y., Guibas, L.J.: Render for CNN: viewpoint estimation in images using CNNs trained with rendered 3D model views. In: Proceedings of the IEEE International Conference on Computer Vision, pp. 2686–2694 (2015)
57. Sun, C., Shrivastava, A., Singh, S., Gupta, A.: Revisiting unreasonable effectiveness of data in deep learning era. In: Proceedings of the IEEE International Conference on Computer Vision, pp. 843–852 (2017)
58. Suzuki, T.: Fast generalization error bound of deep learning from a kernel perspective. In: Proceedings of the Twenty-First International Conference on Artificial Intelligence and Statistics, vol. 84, pp. 1397–1406 (2018)
59. Teed, Z., Deng, J.: Droid-SLAM: deep visual slam for monocular, stereo, and RGB-D cameras. arXiv:2108.10869 (2021)
60. Tobin, J., Fong, R., Ray, A., Schneider, J., Zaremba, W., Abbeel, P.: Domain randomization for transferring deep neural networks from simulation to the real world. arXiv:1703.06907 (2017)
61. Tremblay, J., et al.: Training deep networks with synthetic data: bridging the reality gap by domain randomization. arXiv:1804.06516 (2018)
62. Tripuraneni, N., Jordan, M.I., Jin, C.: On the theory of transfer learning: the importance of task diversity. arXiv:2006.11650 (2020)
63. Wei, C., Ma, T.: Improved sample complexities for deep neural networks and robust classification via an all-layer margin. In: International Conference on Learning Representations (2020)
64. Yang, J., Yan, R., Hauptmann, A.G.: Cross-domain video concept detection using adaptive SVMs. In: Proceedings of the 15th ACM International Conference on Multimedia, MM 2007, pp. 188–197. Association for Computing Machinery, New York (2007)

65. Zhou, B., Zhao, H., Puig, X., Fidler, S., Barriuso, A., Torralba, A.: Scene parsing through ADE20K dataset. In: Proceedings of the IEEE Conference on Computer Vision and Pattern Recognition, pp. 633–641 (2017)
66. Zhou, B., et al.: Semantic understanding of scenes through the ADE20K dataset. arXiv:1608.05442 (2016)
67. Zoph, B., et al.: Rethinking pre-training and self-training. arXiv:2006.06882 (2020)

Submodular Meta Data Compiling
for Meta Optimization

Fengguang Su, Yu Zhu, Ou Wu$^{(\boxtimes)}$, and Yingjun Deng

National Center for Applied Mathematics, Tianjin University, Tianjin, China
{fengguangsu,yuzhu,wuou,yingjun.deng}@tju.edu.cn

Abstract. The search for good hyper-parameters is crucial for various deep learning methods. In addition to the hyper-parameter tuning on validation data, meta-learning provides a promising manner for optimizing the hyper-parameters, referred to as meta optimization. In all existing meta optimization methods, the meta data set is directly given or constructed from training data based on simple selection criteria. This study investigates the automatic compiling of a high-quality meta set from training data with more well-designed criteria and the submodular optimization strategy. First, a theoretical analysis is conducted for the generalization gap of meta optimization with a general meta data compiling method. Illuminated by the theoretical analysis, four criteria are presented to reduce the gap's upper bound. Second, the four criteria are cooperated to construct an optimization problem for the automatic meta data selection from training data. The optimization problem is proven to be submodular, and the submodular optimization strategy is employed to optimize the selection process. An extensive experimental study is conducted, and results indicate that our compiled meta data can yield better or comparable performances than the data compiled with existing methods.

Keywords: Hyper-parameter optimization · Meta optimization · Generalization gap · Submodular optimization · Selection criteria

1 Introduction

Hyper-parameters have a considerable effect on the final performance of a model in machine learning. In shallow learning, cross-validation is usually leveraged to search (near) optimal hyper-parameters; in deep learning, due to the high time consumption of cross-validation, an independent validation set is constructed, and the hyper-parameters with the best performance are selected as the final hyper-parameters. In both strategies, the hyper-parameters are searched in a pre-defined grid. Recently, meta-learning has provided an effective manner to directly optimize the hyper-parameters instead of the grid search in existing strategies. Various hyper-parameters, such as learning rates [1], weights of noisy

---

Supplementary Information The online version contains supplementary material available at https://doi.org/10.1007/978-3-031-26409-2_30.

M.-R. Amini et al. (Eds.): ECML PKDD 2022, LNAI 13715, pp. 493–511, 2023.
https://doi.org/10.1007/978-3-031-26409-2_30

or imbalanced samples [3–5], pseudo labels [6–8], and others inside particular methods [9,10], have been optimized via meta-learning on an additional small meta data set. Meta-learning based hyper-parameter optimization is called meta optimization.

In meta optimization, an independent meta data set is required, and ideally the meta set is unbiased. For example, in meta semantic data augmentation [10], which applies meta optimization for the covariance matrix, the meta data set in an experimental run contains a certain number of images independent of the training set. Although the leveraged meta set is claimed to be unbiased, no "unbiased" standard is provided. Most existing studies directly assume that an independent and high-quality meta set is ready for training. However, independent meta data do not usually exist. Recently, Zhang and Pfister [11] combine two criteria to compile meta data from training data with a simple greedy selection strategy. Initial promising results are reported in their study. However, their utilized criteria are still simple and may be insufficient in meta data compiling. This study proposes a new effective method for compiling meta data only from the corresponding training data. First, the generalization gap is analyzed for compiled meta data. Based on the upper bound of the gap, we analyze the characteristics that meta data should meet. Four selection criteria are then obtained: cleanness, balance, diversity, and uncertainty. The submodular optimization strategy [12] is leveraged to optimize the selection process with the criteria. Experiments on the two typical meta optimization scenarios, namely, imbalance learning and noisy label learning, are performed to verify the effectiveness of our method. The main contributions are summarized as follows:

- The expected generalization gap of the meta optimization is inferred when the ideal (i.e., not unbiased) meta set is not given, and the employed meta set is constructed through a meta data compiling method. This gap facilitates the understanding and explanation of the performances of meta optimization with different meta data compiling methods. Moreover, the gap provides theoretical guidance for automatic meta data construction.
- A new meta data compiling method is proposed to select meta data from training data for meta optimization. In our method, four sophisticated criteria are considered illuminated by the gap, and the submodular optimization strategy is introduced to solve the optimal subset selection with the fused criteria. Extensive experiments indicate our compiled meta data yield better accuracies in typical meta optimization scenarios than existing strategies.

2   Related Work

This section briefly introduces meta optimization, meta data compiling, and submodular optimization in machine learning.

2.1   Meta Optimization

Meta optimization is the instantiation of meta-learning [13,14], which optimizes the target hyper-parameters by minimizing the learning error on meta data.

Compared with the grid search, meta optimization is more efficient and has theoretical advantages over traditional cross-validation [15]. Let T and S be the training and (unbiased) meta sets, respectively. Let Θ and μ be the model parameters and hyper-parameters, respectively. Given μ, an optimal $\Theta^*(\mu)$ can be subsequently obtained as follows:

$$\Theta^*(\mu) = \arg\min_{\Theta} \mathcal{L}_T(\Theta, \mu), \tag{1}$$

where $\mathcal{L}$ is the loss. The optimal hyper-parameters μ^* can thus be obtained by minimizing the loss on the meta set S:

$$\mu^* = \arg\min_{\mu} \mathcal{L}_S(\Theta^*(\mu)). \tag{2}$$

Meta optimization has been widely used in various scenarios, such as imbalance learning and noisy label learning.

2.2 Meta Data Compiling

In existing studies, meta data are assumed to be given in advance, and no standard for selecting meta data is provided and discussed. Take meta optimization as an example in imbalance to illustrate how the meta data are compiled in nearly all existing studies. The benchmark data set CIFAR10 [16] contains 50,000 training samples on ten balanced categories. In the experiments, a balanced subset of 50,000 images is used as the independent meta set. Then, the rest of the images are used to build imbalanced training set by different category-wise probabilities.

Unfortunately, the above simulation process is infeasible in real applications. A promising solution is to define a set of "unbiased" criteria and then select meta data from training data. So far, only one recent study [11] has investigated this technical line. However, only the "cleanness" criterion and the "balance" criterion are considered. For the balance criterion in [11], if the number of samples for a certain class is not enough, the authors simply repeat the samples to attain balance. Theoretical guidance for how to compile meta data is still lacking up till now. This study attempts to construct guidance with a theoretical basis.

2.3 Submodular Optimization

Submodular optimization provides an efficient framework to solve the NP-hard combination problem with fast greedy optimization. A submodular optimization instance LtLG [17] can achieve linear time complexity in the data size, which is independent of the cardinality constraint in expectation. Submodular optimization has been widely used in text summarization, sensor placement, and speech recognition [18]. Joseph et al. [18] proposed an effective submodular optimization-based method to construct a mini-batch in DNN training. Significant improvements in convergence and accuracy with submodular mini-batches have been observed.

When more sophisticated criteria are considered in automatic meta data compiling, the optimizing is very likely to become NP-hard, and simple greedy strategies are ineffective. Submodular optimization provides an effective solution.

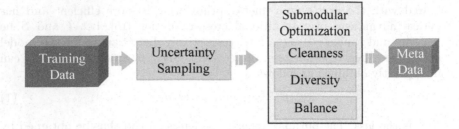

Fig. 1. Overview of our submodular meta data compiling. Our method is called submodular optimization-based meta data compiling (denoted as SOMC for briefly).

3   Methodology

Figure 1 illustrates the proposed submodular compiling process for the meta data set. The theoretical analysis for meta data construction is conducted firstly. And then the meta data selection method is described.

3.1   Theoretical Analysis for Meta Data Construction

Ideally, the distribution of samples in a compiled meta set equals that of testing samples. Bao et al. [15] infer a generalization gap for the meta optimization associated with independent ideal (i.e., unbiased) meta data. Let X be the sample space. Let p^{tr} and p^{me} be the distributions of training and meta data, respectively. Let T be a set of n training samples, and S_m^{me} be a set of m meta samples. Let $R(\mathcal{A}(T, S_m^{me}), p^{me}) = E_{x \sim p^{me}}[l(\mathcal{A}(T, S_m^{me}), x)]$ be the expected risk for the learning on the meta set S^{me}, where $l(\cdot, x)$ is the loss on x, $\mathcal{A}$ is a meta optimization method and $\mathcal{A}(T, S_m^{me})$ is the learned hyper-parameters and model with the training set T and meta set S_m^{me}. Let $\hat{R}(\mathcal{A}(T, S_m^{me}), S_m^{me}) = \frac{1}{m} \sum_{x \in S_m^{me}} l(\mathcal{A}(T, S_m^{me}), x)$ be the empirical risk for the learning on the meta set S_m^{me}.

The involved meta optimization method is assumed to be β-uniformly stable [15]. That is, for a randomized meta optimization algorithm $\mathcal{A}$, if for two arbitrary compiled meta sets S_m^{me} and $S_m'^{me}$ such that they differ in at most one sample, then $\forall\, T \in X^n, \forall\, x \in X$, we have

$$|E_{\mathcal{A}}[l(\mathcal{A}(T, S_m^{me}), x) - l(\mathcal{A}(T, S_m'^{me}), x)]| \le \beta. \tag{3}$$

The generalization gap is defined as

$$gap(T, S_m^{me}) = R(\mathcal{A}(T, S_m^{me}), p^{me}) - \hat{R}(\mathcal{A}(T, S_m^{me}), S_m^{me}). \tag{4}$$

The expected generalization gap satisfies [15]

$$|E_{\mathcal{A}, T, S_m^{me}}[gap(T, S_m^{me})]| \le \beta. \tag{5}$$

We infer the expected generalization gap when a meta set is not ideal and constructed from training data, including ours method. As the involved meta optimization method is not changed in our study, the β-uniform stability is still assumed. Let S_m^{sme} be the compiled meta set consisting of m samples. Let P_m^{me} and P_m^{sme} be the distributions of S_m^{me} and S_m^{sme}, respectively.

Definition 1. *The distance between two distributions P_m^{me} and P_m^{sme} is defined as follow:*

$$d(P_m^{me} \| P_m^{sme}) = \int_{S \in X^m} |P_m^{me}(S) - P_m^{sme}(S)| dS. \tag{6}$$

For brevity, $d(P_m^{me} \| P_m^{sme})$ is denoted as d_m. If the two distributions are identical, then d_m is zero. Let $\hat{R}(\mathcal{A}(T, S_m^{sme}), S_m^{sme}) = \frac{1}{m} \sum_{x \in S_m^{sme}} l(\mathcal{A}(T, S_m^{sme}), x)$ be the empirical risk for the learning on our compiled meta set S_m^{sme}. We first define the generalization gap for S_m^{sme} as follows:

$$gap(T, S_m^{me}, S_m^{sme}) = R(\mathcal{A}(T, S_m^{me}), p^{me}) - \hat{R}(\mathcal{A}(T, S_m^{sme}), S_m^{sme}). \tag{7}$$

We obtain the theorem for the expectation of the above generalization gap as follows:

Theorem 1. *Suppose a randomized meta optimization algorithm $\mathcal{A}$ is β-uniformly stable on meta data in expectation, then we have*

$$|E_{\mathcal{A},T,S_m^{me},S_m^{sme}}[gap(T, S_m^{me}, S_m^{sme})]| \leq \beta + bd_m, \tag{8}$$

where b is the upper bound of the losses of samples in the whole space (following the assumption in [15]), and $d_m = d(P_m^{me} \| P_m^{sme})$.

The proof of Theorem 1 is presented in the supplementary material[1]. Compared with the expected generalization gap for independent (ideal) meta sets given in Eq. (5), our expected generalization gap for automatically compiled meta sets contains an additional term bd_m. Naturally, an ideal criterion for S_m^{sme} should make sure both β and bd_m as small as possible. Note that $\beta = \frac{2cL^2}{m} [\frac{1}{\kappa}((\frac{Ns(l)}{2cL^2})^\kappa - 1) + 1]$ (Theorem 2 in [15]), where m, c, L, γ, κ, and N remain unchanged and only $s(l) = b - a$ (the range of the loss) may change in terms of different meta data selection criteria. As $a \to 0$ when the cross-entropy loss is used, only b and d_m affect the upper bound of the gap (i.e., the value of the right-side of (8)). Consequently, we explore the selection criteria according to the minimization of both d_m and b, separately[2]. First, we have the following conclusion.

Corollary 1. *The optimal selected meta data distribution $P_m^{sme}(S)$ should satisfy that $d_m = 0$, i.e., $P_m^{sme}(S) = P_m^{me}(S)$.*

---

[1] The supplementary material is uploaded to https://github.com/ffgg11/Submodular-Meta-Data-Compiling-for-Meta-Optimization.
[2] The value of b affects both β and bd_m, while d_m only affects bd_m.

Accordingly, it is inappropriate to select training data uniformly at random as meta data as the training data in many scenarios (e.g., imbalance learning) is biased against the true meta data. In practice, the true distribution of meta data is unknown. However, two requirements [3–6, 8–10] are usually assumed to be met for an arbitrary meta data set:

- **Cleanness**. As meta data are assumed to be drawn from the true distribution without observation noises, meta data should be as clean as possible.
- **Balance**. The balance over categories is usually taken as a prior in previous studies utilizing meta optimization. This study also inherits this assumption.

According to Corollary 1, to reduce the value of d_m, cleanness and balance should be leveraged as two selection criteria in our meta data compiling. We will show that cleanness and balance may also reduce the value of b in the succeeding discussion.

As d_m cannot be guaranteed to be zero only with the two criteria mentioned above, b should also be as small as possible. The value of b is determined by both the ideal yet unknown meta set (actually the true distribution of meta data) and the compiled meta set (actually the underlying distribution of our compiled meta data). Considering that the ideal meta set is not given and our selection criteria do not affect the distribution of the ideal meta set, the ideal meta set can be ignored in the discussion for the reduction of b. To reduce the value of b, the following selection criteria are beneficial:

- **Cleanness**. If there are noisy samples in the compiled meta set, then the losses of clean samples will be larger as noisy samples usually damage the generalization ability [38]. Therefore, keeping the compiled meta data as clean as possible will also reduce the value of b in a high probability.
- **Balance**. Even though the balance prior does not hold in a specific learning task, the balance over categories may reduce the maximum loss of the samples of tail categories [5]. For this consideration, balance is still useful.
- **Uncertainty**. Pagliardini et al. [37] show that adding more samples with high uncertainty will increase the classification margin. Accordingly, the maximum loss may also be reduced if the meta data are noisy-free. Indeed, uncertainty sampling [30–32] is prevalent in sample selection in active learning. It is proven to be more data-efficient than random sampling [34].
- **Diversity**. Diversity can be seen as the balance prior for the samples within a category. This balance prior may also reduce the maximum loss of each category. The maximum loss may subsequently be reduced. Indeed, diversity-aware selection has other merits. Madan et al. [36] find that using the same amount of training data, increasing the number of in-distribution combinations (i.e., data diversity) also significantly improves the generalization ability to out-of-distribution data.

According to the above considerations, two more criteria, namely, uncertainty and diversity[3], are also considered in addition to the cleanness and balance

---

[3] Indeed, Ren et al. [29] revealed that the uncertainty and the diversity criteria are usually used together to improve the model performance in deep active learning.

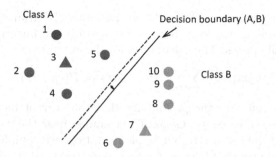

Fig. 2. An illustrative example of the four selection criteria. There are two classes and five samples per class with a decision boundary. The samples $\{3, 7\}$ are those with noisy labels. The cleanness criterion prefers the samples $\{4, 6\}$ to $\{3, 7\}$. The balance criterion promotes to select the samples $\{4, 5, 6, 10\}$ instead of $\{2, 4, 5, 6\}$. Diversity prefers samples $\{6, 8, 10\}$ to $\{8, 9, 10\}$. The uncertainty criterion promotes the selection of samples $\{4, 5\}$ instead of $\{1, 2\}$.

criteria. Figure 2 illustrates the roles of each of the four summarized selection criteria in terms of the reduction of b. There are two classes of points with a decision boundary between them. There are two noisy samples, 3 and 7. First, if cleanness is not considered, then the noisy samples $\{3, 7\}$ may appear in the meta set. It is highly possible that the losses of clean samples near $\{3\}$ and $\{7\}$ in the whole space are relatively high. Second, if the balance criterion is not considered, the losses of the samples in the tail categories are high to a certain extent. For example, if we choose the samples $\{1, 2, 4, 5, 6\}$ as meta data, then the losses of the samples near $\{9, 10\}$ in Class B will become high with a high probability. Third, if the diversity criterion is not considered, then the samples $\{8, 9, 10\}$ may be chosen. The samples around the sample $\{6\}$ may have higher loss values. Finally, if the uncertainty criterion is not considered (e.g., if $\{4, 5\}$ are not selected, the decision boundary will move in the direction of the dotted line.), then the classification margin will decrease [37]. Consequently, the losses of the samples near $\{4, 5\}$ will increase. Further, uncertainty can avoid selecting too many clean samples with small losses through Eq. (10), and thus can improve the update efficiency of meta optimization. Based on the above analysis, if any of the four summarized criteria are ignored, the losses of samples in specific local regions of the whole space will increase. As a result, b will increase.

3.2 Details of the Four Selection Criteria

This subsection describes how the four selection criteria are applied in the meta data compiling from a given training set. Considering that the training sizes in deep learning tasks are usually large, it is inappropriate to run all four selection criteria on each training sample. Therefore, sampling will firstly be performed to reduce the size of candidate samples fed to other criteria.

Uncertainty Criterion. Let Θ be the current model parameter. The output entropy of a training sample x_i is used to measure the uncertainty of x_i. Let C be the set of all classes. The calculation for the out entropy of x_i is as follows:

$$u(x_i) = -\sum_{c \in C} P(c|x_i, \Theta) \log P(c|x_i, \Theta), \tag{9}$$

where $P(c|x_i, \Theta)$ refers to the probability that the current model predicts the sample x_i as the c-th category. Generally, a sample near the decision boundary has a high uncertainty score. In our implementation, we sample the data based on the normalized uncertainty score for each class, respectively. That is, the sampling probability of x_i is $u(x_i)/\sum_{j:y_j=y_i} u(x_j)$. More details about uncertainty sampling can be found in Algorithm 1 and the experimental implementation details in supplementary materials.

Cleanness Criterion. This criterion aims to select data with clean labels or clean features. Many metrics can be used to judge the noisy degree of a sample, including loss (prediction) [11], loss variance [2], gradient norm [20], etc. Considering that the loss metric is the most widely used, this study also adopts it. The cleanness degree of a set is defined as follows:

$$\mathcal{C}(S) = \sum_{x_i \in S} c(x_i) = \sum_{x_i \in S} P(y_i|x_i, \Theta), \tag{10}$$

where y_i is the label of x_i, and Θ is the model parameter(s). If y_i is a noisy label or x_i has non-trivial noisy features, then $P(y_i|x_i, \Theta)$ is usually small during training.

Balance Criterion. Imbalance can cause the model to have a good performance on the head categories but poor performance on the tail ones. Let n_c^s be the number of meta samples of the c-th category, and m be the total number of meta samples. The balance score of a subset is formulated as follows:

$$\mathcal{B}(S) = \prod_{c \in C} I(\lfloor \frac{m}{|C|} \rfloor \leq n_c^s \leq \lceil \frac{m}{|C|} \rceil), \tag{11}$$

where C is the category set. When $\mathcal{B}(S) = 1$, the subset is balanced.

Diversity Criterion. The criterion selects samples with different features by considering the relationship among samples. The following approach is utilized to measure the diversity of a subset. Given $\phi(\cdot, \cdot)$ to be any distance metric between the two data points, a larger value of the minimum distance among points would imply more diversity in the subset.

$$\mathcal{D}(S) = \sum_{x_i \in S} \min_{x_j \in S: i \neq j} \phi(\tilde{x}_i, \tilde{x}_j), \tag{12}$$

where $\tilde{x}_i$ is the output of the final feature encoding layer of x_i. This score is dependent on the choice of distance metric. In our implementation, Euclidean distance ($\|\tilde{x}_i - \tilde{x}_j\|_2$) is employed according to the performances of different distance metrics reported in [18].

3.3 Submodular Optimization

The four criteria are cooperated to construct an optimization problem for the final meta data compiling. As previously described, the uncertainty criterion is first utilized to reduce the candidate training data. The diversity and cleanness criteria are then combined as follows:

$$\mathcal{F}(S) = \lambda \mathcal{D}(S) + (1 - \lambda)\mathcal{C}(S), \tag{13}$$

where λ is a hyper-parameter. Let T be the candidate training data which is passed through the uncertainty criterion. Consequently, an optimal meta set of size m is selected by solving the following optimization problem:

$$S^* = \arg\max_{S \subseteq T} \mathcal{F}(S)$$
$$s.t. \quad |S| \leq m; \quad \mathcal{B}(S) = 1 \tag{14}$$

The maximization of Eq. (14) is a NP-hard problem as the total diversity score in Eq. (12) cannot be factorized into the sum of diversity scores of each sample. The simple greedy method leveraged in [11] is inapplicable. Hence, to conduct an efficient and effective maximization, the submodular optimization manner is leveraged.

Submodular optimization guarantees a solution for a submodular objective function which is at least (in the worst case) $1 - 1/e$ of the optimal solution [21], where e is the base of the natural logarithm. Further, some fast submodular optimization algorithms such as LtLG [17] have been put forward. An optimization problem can be solved with submodular optimization if its objective function is submodular and monotonically non-decreasing. Therefore, to apply submodular optimization, we have two lemmas for the objective function.

Lemma 1. $\mathcal{F}(S)$ in Eq. (13) is submodular.

Lemma 2. $\mathcal{F}(S)$ in Eq. (13) is monotonically non-decreasing.

According to Lemmas 1 and 2, the submodular optimization technique can be used directly to solve Eq. (14). Inspired by the general submodular optimization framework SMDL [18], our method consists of three main processes shown in Algorithm 1. First, a training subset T is obtained based on uncertainty sampling and is randomly partitioned into K disjoint subsets. Secondly, a subset is further generated from each of the K subset by maximizing the marginal gain $\mathcal{F}(a|S) = \mathcal{F}(\{a\} \cup S) - \mathcal{F}(S)$. Lastly, the subsets are merged to generate the final meta data set by considering the margin gain maximization and the balance constraint.

The time complexity of our proposed submodular optimization-based meta data compiling (SOMC) is $O((|T| + Km)md)$, where d is the feature dimension. In practice, Algorithm 1 can be implemented in parallel and the time complexity becomes approximately $O((|T|/K + Km)md)$. When m is large, the time consumption can be significantly reduced by first compiling a batch of small meta sets and then merging them as the final meta set S. The entire algorithmic steps and more details are presented in the supplementary material.

Algorithm 1. SOMC

Input: Training set T, $u(x_i)$, $i = 1, \cdots, |T|$, m, K, λ, and $\mathcal{F}(\cdot)$ in Eq. (13).
Output: Meta data set S

 1: $S \longleftarrow \emptyset$;
 2: Obtain a subset (still marked as T) of size $\frac{|T|}{2}$ based on uncertainty sampling;
 3: Partition T into K disjoint sets $T_1, T_2, ..., T_K$;
 4: Generate a subset S_k (m samples) from T_k using LtLG [17], $k = 1, \cdots, K$;
 5: $\tilde{S} \longleftarrow \bigcup_{k=1}^{K} S_k$;
 6: While $|S| < m$
 7:    Select a sample $(x^*, y^*) \in \tilde{S} \setminus S$ using LtLG;
 8:    If $n_{y^*} \leq \lceil \frac{m}{|C|} \rceil - 1$
 9:        $S \longleftarrow \{(x^*, y^*)\} \cup S$;
10: Return S.

---

4   Experiments

This section evaluates the performance of SOMC in benchmark image classification corpora, including CIFAR [16], ImageNet-LT [22], iNaturelist [23], and Clothing1M [24]. Details of these corpora and the source code are provided in the supplementary material.

4.1   Evaluation on CIFAR10 and CIFAR100

Nearly all existing meta optimization studies utilize independent meta sets, and thus they should be compared. In this part, the independent meta data used in existing studies are replaced by the data compiled by our SOMC. In addition, the only existing automatic meta data selection method FSR [11] is also compared. FSR only uses cleanness and balance to select meta data in the training set. Indeed, FSR also contains multiple data augmentation tricks and a novel meta optimization method. For a fair comparison, only the module of meta data compiling of FSR (denoted as "FSRC") is compared in this experiment.

Results on Imbalance Classification. Following [25], we use CIFAR10 and CIFAR100 to build imbalance training sets by varying imbalance factors $\mu \in \{200, 100, 50, 20, 10\}$, namely, CIFAR10-LT and CIFAR100-LT. The original balanced test sets are still used. The concrete hyper-parameters setting is described in the supplementary material. The average accuracy of the three repeated runs is recorded for each method. The meta set for all existing studies contains ten images for each category. However, the numbers of images in some tail categories in CIFAR10-LT and CIFAR100-LT are less than ten. Thus, data augmentation techniques are used to generate more candidates for the successive meta image selection for these categories for our SOMC and FSRC. ResNet-32 [26] is used as the base network. The parameter λ of our SOMC is searched

Table 1. Test top-1 accuracy (%) of ResNet-32 on CIFAR10-LT and CIFAR100-LT under different imbalance settings.

Data set	CIFAR10-LT					CIFAR100-LT				
Imbalance factor	200	100	50	20	10	200	100	50	20	10
Base model (CE)	65.87	70.14	74.94	82.44	86.18	34.70	38.46	44.02	51.06	55.73
MCW+100/1000 meta images (CE)	70.66	76.41	80.51	86.46	**88.85**	39.31	43.35	48.53	55.62	59.58
MCW+**FSRC** (CE)	72.34	77.65	81.31	86.25	88.02	38.53	44.21	49.72	55.98	60.17
MCW+**SOMC** (CE)	**73.71**	**79.24**	**82.34**	**86.98**	88.67	**39.95**	**45.97**	**51.28**	**57.32**	**61.11**
MetaSAug +100/1000 meta images (CE)	76.16	**80.48**	83.52	87.20	88.89	42.27	46.97	51.98	57.75	61.75
MetaSAug+**FSRC** (CE)	75.41	79.28	82.87	86.81	88.37	42.53	47.02	51.61	57.87	61.35
MetaSAug+**SOMC** (CE)	**76.25**	80.25	**83.61**	**87.43**	**89.02**	**43.32**	**48.03**	**52.36**	**58.52**	**61.88**
MCW+100/1000 meta images (FL)	74.43	78.90	82.88	86.10	88.37	39.34	44.70	50.08	55.73	59.59
MCW+**FSRC** (FL)	74.57	79.23	83.06	86.22	88.59	39.67	44.85	50.35	55.89	59.87
MCW+**SOMC** (FL)	**75.26**	**80.17**	**83.65**	**86.52**	**88.84**	**40.26**	**45.96**	**51.13**	**56.67**	**60.35**
MetaSAug+100/1000 meta images (FL)	75.73	80.25	83.04	**86.95**	88.61	40.42	45.95	51.57	57.65	61.17
MetaSAug+**FSRC** (FL)	75.12	79.87	82.52	85.99	88.21	39.77	45.86	51.22	57.25	60.84
MetaSAug+**SOMC** (FL)	**76.01**	**80.44**	**83.41**	86.77	**88.87**	**40.69**	**46.90**	**51.99**	**57.81**	**61.65**
MCW+100/1000 meta images (LDAM)	77.23	80.00	82.23	84.37	87.40	39.53	44.08	49.16	52.38	58.00
MCW+**FSRC** (LDAM)	76.85	79.97	82.04	85.12	88.03	40.25	44.83	49.79	53.34	59.46
MCW+**SOMC** (LDAM)	**77.69**	**80.43**	**82.86**	**85.74**	**88.51**	**41.37**	**45.73**	**50.62**	**54.29**	**60.30**
MetaSAug+100/1000 meta images (LDAM)	76.42	80.43	83.72	87.32	**88.77**	42.87	**48.29**	52.18	57.65	61.37
MetaSAug+**FSRC** (LDAM)	75.89	79.93	83.21	86.72	87.93	42.69	47.43	51.65	57.54	61.35
MetaSAug+**SOMC** (LDAM)	**76.56**	**80.61**	**83.96**	**87.45**	88.57	**43.48**	48.17	**52.56**	**58.43**	**61.93**

Table 2. Test top-1 accuracy (%) on CIFAR10 and CIFAR100 of WRN-28-10 with varying noise rates under uniform noise.

Data set	CIFAR10			CIFAR100		
Noise rate	0%	40%	60%	0%	40%	60%
Base model (CE)	95.60 ± 0.22	68.07 ± 1.23	53.12 ± 3.03	79.95 ± 1.26	51.11 ± 0.42	30.92 ± 0.33
MSLC+1000 meta images	95.42 ± 0.07	**91.54 ± 0.15**	**87.27 ± 0.27**	80.75 ± 0.11	**71.83 ± 0.24**	**65.37 ± 0.53**
MSLC+**FSRC**	95.23 ± 0.17	88.15 ± 0.31	81.84 ± 0.33	80.49 ± 0.23	67.86 ± 0.14	59.63 ± 0.42
MSLC+**SOMC**	**95.65 ± 0.05**	89.38 ± 0.13	83.56 ± 0.27	**81.36 ± 0.31**	68.75 ± 0.29	61.03 ± 0.17
MWNet+1000 meta images	94.52 ± 0.25	89.27 ± 0.28	84.07 ± 0.33	78.76 ± 0.24	67.73 ± 0.26	58.75 ± 0.11
MWNet+**FSRC**	95.03 ± 0.23	88.78 ± 0.16	84.26 ± 0.17	79.95 ± 0.08	67.88 ± 0.25	59.37 ± 0.28
MWNet+**SOMC**	**95.69 ± 0.09**	**89.81 ± 0.13**	**85.16 ± 0.12**	**80.68 ± 0.32**	**68.63 ± 0.14**	**60.65 ± 0.19**

Table 3. Test top-1 accuracy (%) of ResNet-32 on CIFAR10 and CIFAR100 with varying noise rates under flip noise.

Data set	CIFAR10			CIFAR100		
Noise rate	0%	20%	40%	0%	20%	40%
Base model (CE)	**92.89±0.32**	76.83±2.30	70.77±2.31	70.50±0.12	50.86±0.27	43.01±1.16
MSLC+1000 meta images	92.75±0.15	**91.67±0.19**	**90.23±0.13**	70.37±0.31	**67.59±0.06**	**65.02±0.21**
MSLC+**FSRC**	92.46±0.13	89.78±0.32	88.61±0.27	70.29±0.21	64.97±0.19	61.15±0.46
MSLC+**SOMC**	92.83±0.09	91.13±0.21	89.55±0.25	**70.82±0.15**	66.33±0.11	62.58±0.28
MWNet+1000 meta images	92.04±0.15	90.33±0.61	87.54±0.23	70.11±0.33	64.22±0.28	58.64±0.47
MWNet+**FSRC**	92.42±0.12	90.65±0.36	87.25±0.41	70.52±0.11	65.26±0.12	59.47±0.22
MWNet+**SOMC**	**93.06±0.06**	**91.37±0.11**	**88.65±0.26**	**71.39±0.31**	**66.69±0.11**	**60.34±0.19**

in $\{0.3, 0.5, 0.7\}$, and K is searched in $\{2, 5\}$. More details are described in the supplementary file.

Two meta optimization methods, namely, MCW [5] and MetaSAug [10], are leveraged. Partial results on the early representative method MWNet [4] are shown in the supplementary file. The original study of both methods provides source codes and meta sets on the above data sets. Our experimental results are obtained directly on these codes and hyper-parameter settings.

The classification accuracies of the three meta optimization methods with independent meta sets, FSRC, and our proposed SOMC on CIFAR10-LT and CIFAR100-LT are shown in Table 1. The results are organized into three distinct groups according to the adopted loss functions (i.e., Cross-entropy (CE), Focal loss (FL), and LDAM). SOMC can construct more effective meta data only from training data than both independent meta sets and FSRC in nearly all the cases.

Results on Noisy Labels Learning. Two typical types of corrupted training labels are constructed: 1) **Uniform noise.** The label of each sample is independently changed to a random class with probability p. 2) **Flip noise.** The label of each sample is independently flipped to similar classes with total probability p. Details are described in the supplementary file. Two typical meta optimization methods, MSLC [6] and MWNet [4], are used. In previous studies, the meta data for these two methods consist of absolutely clean images. These clean images will be replaced by the compiled images with our SOMC. ResNet-32 [26] and WRN-28-10 [27] are used as the base network. The hyper-parameters setting is presented in the supplementary file.

The classification results under different noise rates are shown in Tables 2 and 3. Our method outperforms the independent meta data in MWNet. As the noise ratio increases, SOMC degrades more than the independent meta data in MSLC. It is reasonable to require independent clean meta data in the case of a high noise rate. Our method SOMC consistently outperforms FSRC under different noise rates on both sets.

4.2   Evaluation of Large Data Sets

Four large data sets, iNaturalist2017 (iNat2017), iNaturalist2018 (iNat2018), ImageNet-LT, and Clothing1M are used. The former three are leveraged for imbalance learning, while the last is for noisy label learning. MCW and MetaSAug are utilized for ImageNet-LT, INat 2017 and 2018. MSLC and MWNet are utilized for Clothing1M. The experimental settings, including the hyper-parameters, are presented in the supplementary material.

Tables 4 and 5 show the results of the competing methods on iNaturalist data sets and ImageNet-LT. Although 25445 (for iNat2017), 16284 (for iNat2018), and 10000 (for ImageNet-LT) independent meta data are used for MCW and MetaSAug, their performances are worse than those of meta data compiled by our SOMC. FSRC yields the lowest accuracies among the three meta data construction methods for iNat2017 and 2018. In addition, MetaSAug+SOMC with the pre-trained BBN [28] yields the highest top-1 accuracy for iNat2017 and 2018. For ImageNet-LT, SOMC still yields the best results.

Table 6 shows the results on Clothing1M. It can be seen that SOMC achieves better results than 7000 independent meta data and compiled meta data by FSRC on MWNet. For MSLC, compared with 7000 independent meta data, SOMC still achieves comparable results. However, FSRC yields the worst results.

Table 4. Test top-1 accuracy (%) on iNaturalist 2017 and 2018.

Method	iNat2017	iNat2018
Base model (CE)	56.79	65.76
MCW+**25445/16284 meta images**	59.38	67.55
MCW+**FSRC**	58.76	67.52
MCW+**SOMC**	**60.47**	**68.89**
MetaSAug+**25445/16284 meta images**	63.28	68.75
MetaSAug+**FSRC**	62.59	68.28
MetaSAug+**SOMC**	63.53	69.05
MetaSAug+**SOMC** with BBN model	**65.34**	**70.66**

4.3   Discussion

The supplementary file provides more details (including results and analysis) on the issues discussed in this part. The above comparisons suggest that the meta data compiled by our SOMC are more effective than the independent meta data in most cases (except the cases of high noise rate when MSLC is used) and those compiled by FSRC in nearly all cases. This conclusion can be explained by Theorem 1. We calculate the upper bounds of the test losses of the models corresponding to the three meta data compiling methods, namely,

Table 5. Test top-1 accuracy (%) on ImageNet-LT.

Method	ImageNet-LT
Base model (CE)	38.88
MCW+**10000 meta images**	44.92
MCW+**FSRC**	45.05
MCW+**SOMC**	**45.97**
MetaSAug+**10000 meta images**	46.21
MetaSAug+**FSRC**	45.77
MetaSAug+**SOMC**	**46.68**

Table 6. Test top-1 accuracy (%) on on Clothing1M.

Method	Clothing1M
Base model (CE)	68.94
MWNet+**7000 meta images**	73.72
MWNet+**FSRC**	73.01
MWNet+**SOMC**	**73.89**
MSLC+**7000 meta images**	**74.02**
MSLC+**FSRC**	73.23
MSLC+**SOMC**	73.67

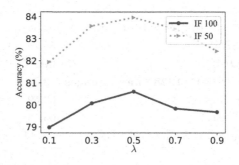

Fig. 3. Effect of λ on CIFAR-10-LT under the different imbalance factors (IF) based on MetaSAug+**SOMC** (LDAM).

FSRC, IDMD (independent meta data), and our method SOMC, respectively. Figure 4 shows the recorded values. SOMC does achieve the minimum upper bound of test losses (i.e., b) among the three methods. This is consistent with the theoretical analysis in Sect. 3.1 that the four criteria mainly aim to reduce d_m and b. More comparisons of the upper bounds of test losses are presented in the supplementary file.

There are two important hyper-parameters, namely, λ and K, in SOMC. They are tuned with grid search in the experiments. Nevertheless, the performances are usually satisfactory when $\lambda \in \{0.3, 0.5\}$ (shown in Fig. 3) and $K = 5$. In all the experiments, the parameter m in our SOMC equals the size of independent meta data used in existing studies for a fair comparison. In addition, the time cost of SOMC is recorded.

An ablation study is conducted for the importance of each criterion in SOMC. The results on imbalance learning (ResNet-32) are shown in Table 7. Removing each criterion causes a performance drop. This result indicates that each of the four criteria is useful in SOMC.

We use different backbone networks (i.e., ResNet-50, ResNet-101, and ResNet-152 [26]). The results indicate that our method still achieves competitive performances. Comparisons with more competing methods and settings are conducted in the supplementary file.

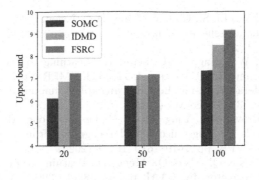

Fig. 4. The upper bounds of test losses on CIFAR10-LT for the three meta data compiling methods under different imbalance factors (IF) based on MetaSAug (LDAM).

Table 7. Ablation study of MetaSAug+**SOMC** using CE loss on CIFAR-100-LT.

Imbalance factor	200	100	50
SOMC w/o Uncertainty	42.19	47.14	51.29
SOMC w/o Diversity	41.21	46.23	50.21
SOMC w/o Cleanness	41.37	46.42	50.13
SOMC w/o Balance	40.09	45.59	49.52
SOMC	**43.32**	**48.03**	**52.36**

5 Conclusions

This study has investigated the automatic compiling of meta data from training data for meta optimization. A theoretical analysis is firstly conducted for the generalization gap for automatic meta data compiling methods, and theoretical guidance for the construction of meta data is obtained. Four sophisticated selection criteria, namely, cleanness, balance, diversity, and uncertainty, are summarized to reduce the upper bound of the generalization gap. These criteria are cooperated to construct an objective function for optimal subset selection from training data. The submodular optimization technique is leveraged to search for the optimal subset. Extensive experiments on six benchmark data sets verify the effectiveness and competitive performance of the proposed method compared with SOTA competing methods.

Acknowledgements. This study is supported by NSFC 62076178, TJF 19ZXAZNG X00050, and Zhijiang Fund 2019KB0AB03.

References

1. Saxena, S., Vyas, N., DeCoste, D.: Training with data dependent dynamic learning rates. arXiv preprint arXiv:2105.13464 (2021)

2. Shin, W., Ha, J.W., Li, S., Cho, Y., Song, H., s Kwon, S.: Which strategies matter for noisy label classification? insight into loss and uncertainty. arXiv preprint arXiv:2008.06218 (2020)

3. Ren, M., Zeng, W., Yang, B., Urtasun, R.: Learning to reweight examples for robust deep learning. In: ICML, pp. 4334–4343. PMLR, Stockholm (2018)

4. Shu, J., et al.: Meta-weight-net: Learning an explicit mapping for sample weighting. In: NeurIPS, Vol. 32. Vancouver (2019)

5. Jamal, M.A., Brown, M., Yang, M.H., Wang, L., Gong, B.: Rethinking class-balanced methods for long-tailed visual recognition from a domain adaptation perspective. In: CVPR, pp. 7610–7619. IEEE (2020)

6. Wu, Y., Shu, J., Xie, Q., Zhao, Q., Meng, D.: Learning to purify noisy labels via meta soft label corrector. In: AAAI, pp. 10 388–10 396. AAAI Press, SlidesLive (2021)

7. Zhang, Z., Zhang, H., Arik, S. O., Lee, H., Pfister, T.: Distilling effective supervision from severe label noise. In: CVPR, pp. 9294–9303. IEEE (2020)

8. Zheng, G., Awadallah, A. H., Dumais, S.: Meta label correction for noisy label learning. In: AAAI, AAAI Press, SlidesLive (2021)

9. Mai, Z., Hu, G., Chen, D., Shen, F., Shen, H.T.: Metamixup: learning adaptive interpolation policy of mixup with metalearning. IEEE Trans. Neural Netw. Learn. Syst. **33**, 3050 –3064 (2021)

10. Li, S. et al.: Metasaug: meta semantic augmentation for long-tailed visual recognition. In: CVPR, pp. 5212–5221. IEEE, Nashville (2021)

11. Zhang, Z., Pfister, T.: Learning fast sample re-weighting without reward data. In: ICCV, pp. 725–734. IEEE, Montreal (2021)

12. Mirzasoleiman, B., Karbasi, A., Sarkar, R., Krause, A.: Distributed submodular maximization: Identifying representative elements in massive data. In: NeurIPS, vol. 26. Lake Tahoe (2013)

13. Thrun, S., Pratt, L.: Learning to Learn: Introduction and Overview. In: Thrun, S., Pratt, L. (eds) Learning to Learn, pp. 3–17. Springer, Boston (1998)..https://doi.org/10.1007/978-1-4615-5529-2_1

14. Finn, C., Abbeel, P., Levine, S.: Model-agnostic meta-learning for fast adaptation of deep networks. In: ICML, pp. 1126–1135. PMLR, Sydney (2017)

15. Bao, F., Wu, G., Li, C., Zhu, J., Zhang, B.: Stability and Generalization of Bilevel Programming in Hyperparameter Optimization. In: NeurIPS, vol. 34 (2021)

16. Krizhevsky, A., Hinton, G.: Learning multiple layers of features from tiny images. In: Handbook of Systemic Autoimmune Diseases, 1st edn, vol. 4. Elsevier Science; (2009)

17. Mirzasoleiman, B., Badanidiyuru, A., Karbasi, A., Vondrák, J., Krause, A.: Lazier than lazy greedy. In: AAAI, pp. 1812–1818. AAAI Press, Austin Texas (2015)

18. Joseph, K.J., Singh, K., Balasubramanian, V.N.: Submodular batch selection for training deep neural networks. In: IJCAI, pp. 2677–2683 (2019)

19. Xiao, Y., Wang, W.Y.: Quantifying uncertainties in natural language processing tasks. In: AAAI, vol. 33, pp. 7322–7329. AAAI Press, Hawaii (2019)

20. Paul, M., Ganguli, S., Dziugaite, G.K.: Deep Learning on a Data Diet: Finding Important Examples Early in Training. In: NeurIPS, vol. 34 (2021)

21. Nemhauser, G.L., et al.: An analysis of approximations for maximizing submodular set functions-I. Math. Program. **14**(1), 265–294 (1978)

22. Liu, Z., Miao, Z., Zhan, X., Wang, J., Gong, B., Yu, S.X.: Large-scale long-tailed recognition in an open world. In: CVPR, pp. 2537–2546. IEEE, California (2019)

23. Van Horn, G. et al.: The inaturalist species classification and detection dataset. In: CVPR, pp. 8769–8778. IEEE, Salt Lake City (2018)

24. Xiao, T., Xia, T., Yang, Y., Huang, C., Wang, X.: Learning from massive noisy labeled data for image classification. In: CVPR, pp. 2691–2699. IEEE (2015)
25. Cui, Y., Jia, M., Lin, T.Y., Song, Y., Belongie, S.: Class-balanced loss based on effective number of samples. In: CVPR, pp. 9268–9277. IEEE, California (2019)
26. He, K., Zhang, X., Ren, S., Sun, J.: Deep residual learning for image recognition. In: CVPR, pp. 770–778. IEEE, Las Vegas (2016)
27. Zagoruyko, S., Komodakis, N.: Wide residual networks. In: BMVC, pp. 87.1-87.12. BMVA Press, York (2016)
28. Zhou, B. et al.: BBN: Bilateral-branch network with cumulative learning for long-tailed visual recognition. In: CVPR, pp. 9719–9728. IEEE (2020)
29. Ren, P., et al.: A survey of deep active learning. ACM Comput. Surv. **54**(9), 1–40 (2021)
30. Beluch, W. H. et al.: The power of ensembles for active learning in image classification. In: CVPR, pp. 9368–9377. IEEE, Salt Lake City (2018)
31. Joshi, A.J., Porikli, F., Papanikolopoulos, N.: Multi-class active learning for image classification. In: CVPR, pp. 2372–2379. IEEE, Florida (2009)
32. Lewis, D.D., Gale, W.A.: A sequential algorithm for training text classifiers. In: SIGIR, pp. 3–12. Springer, London (1994). https://doi.org/10.1007/978-1-4471-2099-5_1
33. Ranganathan, H., Venkateswara, H., Chakraborty, S., Panchanathan, S: Deep active learning for image classification. In: ICIP, pp. 3934–3938. IEEE, Beijing (2017)
34. Mussmann, S., Liang, P.: On the relationship between data efficiency and error for uncertainty sampling. In: ICML, pp. 3674–3682. PMLR, Stockholm (2018)
35. Lam, C.P., Stork, D. G.: Evaluating classifiers by means of test data with noisy labels. In: IJCAI, pp. 513–518. ACM, San Francisco (2003)
36. Madan, S., et al.: When and how do CNNs generalize to out-of-distribution category-viewpoint combinations? Nat. Mach. Intell. **4**(2), 146–153 (2022)
37. Pagliardini, M. et al.: Improving generalization via uncertainty driven perturbations. arXiv preprint arXiv:2202.05737 (2022)
38. Algan, G., Ulusoy, I.: image classification with deep learning in the presence of noisy labels: a survey. Knowl. Based Syst. **215**(3), 106771 (2019)

Supervised Contrastive Learning for Few-Shot Action Classification

Hongfeng Han[1,3] ![ID], Nanyi Fei[1,3] ![ID], Zhiwu Lu[2,3(✉)] ![ID], and Ji-Rong Wen[2,3] ![ID]

[1] School of Information, Renmin University of China, Beijing, China
[2] Gaoling School of Artificial Intelligence, Renmin University of China, Beijing, China
[3] Beijing Key Laboratory of Big Data Management and Analysis Methods, Beijing, China
{hanhongfeng,feinanyi,luzhiwu,jrwen}@ruc.edu.cn

Abstract. In a typical few-shot action classification scenario, a learner needs to recognize unseen video classes with only few labeled videos. It is critical to learn effective representations of video samples and distinguish their difference when they are sampled from different action classes. In this work, we propose a novel supervised contrastive learning framework for few-shot video action classification based on spatial-temporal augmentations over video samples. Specifically, for each meta-training episode, we first obtain multiple spatial-temporal augmentations for each video sample, and then define the contrastive loss over the augmented support samples by extracting positive and negative sample pairs according to their class labels. This supervised contrastive loss is further combined with the few-shot classification loss defined over a similarity score regression network for end-to-end episodic meta-training. Due to its high flexibility, the proposed framework can deploy the latest contrastive learning approaches for few-shot video action classification. The extensive experiments on several action classification benchmarks show that the proposed supervised contrastive learning framework achieves state-of-the-art performance.

Keywords: Few-shot learning · Contrastive learning · Action classification

1 Introduction

Recently, the metric-based meta-learning paradigm has led to great advances in few-shot learning (FSL) and become the mainstream [7,10,36]. Following such a paradigm, FSL models are typically trained via two learning stages [21]: (1) They are first trained on base classes to learn visual representations, acquiring transferable visual analysis abilities. (2) During the second stage, the models learn to classify novel classes that are unseen before by using only a few labelled samples per novel class. Similar to FSL, contrastive learning (CL) [21] is also deployed to address the labelled data-hungry problem. Specifically, CL is defined as unsupervised or self-supervised learning. The target of CL is to obtain better visual representations to transfer the learned knowledge to downstream tasks

M.-R. Amini et al. (Eds.): ECML PKDD 2022, LNAI 13715, pp. 512–528, 2023.
https://doi.org/10.1007/978-3-031-26409-2_31

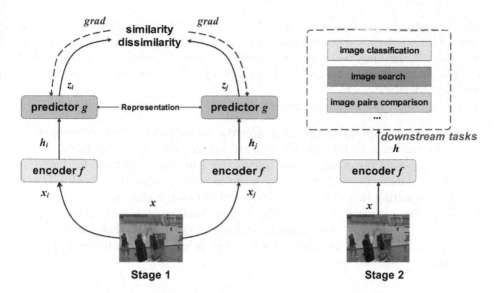

Fig. 1. A typical contrastive learning framework for unsupervised image representation learning. Specifically, two image views x_i and x_j are generated from the same family of image augmentations $(x_i, x_j \sim \Sigma)$. A CNN-based encoder network f along with the projection head g is applied to represent each sample effectively. After the network parameters are trained based on a contrastive loss, the projection head g is put away, and only the encoder network f and representations h_i/h_j are used for downstream tasks.

such as image classification [8,29]. As illustrated in Fig. 1, a classic CL framework [12–14,25] also follows the two learning stages (similar to metric-based meta-learning): (1) An encoder named f and a predictor named g are first trained with constructed positive and negative sample pairs; (2) The learned latent embeddings h_i/h_j are further adapted to downstream tasks of interest. Therefore, it is natural and indispensable to combine FSL and CL.

However, for few-shot action classification, the integration of CL and FSL is extremely challenging because of the complicated video encoding methods. Specifically, two typical methods are widely used: (1) Extracting frame features and then aggregating them. For example, combined with long-short term memory (LSTM), 2D Convolutional Neural Networks (CNNs) are often used for video encoding [5,20,32,40]); (2) Directly extracting spatial-temporal features using 3D CNNs [9,18,30,38,39] or their variants. For both video encoding practices, the high-level semantic contexts among video frames are difficult to be aligned either spatially or temporally [4,6].

In this work, we thus propose a novel supervised contrastive learning framework to make a closer integration of CL and FSL for few-shot action classification. Specifically, we first obtain multiple spatial-temporal augmentations from each video sample for each meta-training episode. Further, we define a supervised contrastive loss over the augmented support samples by constructing positive and negative pairs based on their class labels. Finally, the contrastive loss is combined with a few-shot classification loss defined over a similarity score regression network for the end-to-end episodic meta-training. In addition, the proposed framework can deploy the latest CL methods for few-shot action classification with high flexibility.

In summary, the major contributions of this paper are three-fold:

(1) We devise a spatial-temporal augmentation method to generate different augmentations, facilitating CL to learn better video representations.
(2) We propose a novel supervised contrastive learning framework for few-shot action classification. A similarity score network is shared by both CL and FSL, resulting in a closer integration of the two paradigms.
(3) Extensive experiments on three benchmarks (i.e., HMDB51 [27], UCF101 [34], and Something-Something-V2 [22]) show that the proposed supervised contrastive learning framework achieves state-of-the-art performance.

2   Related Work

Few-Shot Learning for Action Classification. Few-shot learning (FSL) approaches are often divided into two main categories: (1) The goal of gradient-based approaches [1,19,28,31] is to achieve rapid learning on a new task with a limited number of gradient update steps while simultaneously avoiding overfitting (which can happen when few labelled samples are used). (2) Metric-based approaches [2,4,6,21,42] first extract image/video features and then measure the distances/similarities between an embedded query sample and embedded support samples. It is essential to measure the distances in the latent space to determine the class label of query samples. We examine the simplicity and adaptability of the metric-based meta-learning framework in this paper. But note that our proposed video augmentation methods and supervised contrastive learning strategy are also compatible with other few-shot classification solutions.

Contrastive Learning. Contrastive learning (CL) is now a relatively new paradigm for unsupervised or self-supervised learning for visual representations, and it has shown some promising results [8,12–15,23,25,26,29]. It is customary for CL methods to learn representations by optimizing the degree to which multiple augmented views of the same data sample agree with one another. This is accomplished by suffering a contrastive loss in the latent embedding space. For example, SimCLR [12] achieves the highest level of agreement possible between various augmented views of the same data sample by obtaining representations and employing a contrastive loss while operating in the latent space. It comes with an improved version called SimCLR v2 [13] that explores larger-sized ResNet models, boosts the performance of the non-linear network (multiple-layer perception, MLP), and incorporates the memory mechanism. Momentum Contrast (MoCo) [25] approach creates a dynamic dictionary using a queue structure and a moving-averaged encoder. It allows for the construction of an extensive and

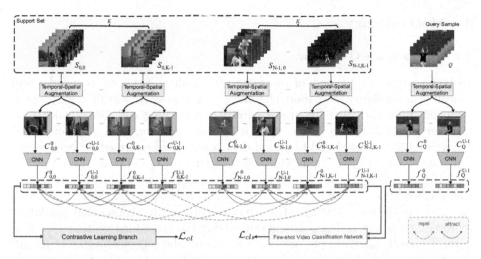

Fig. 2. Architecture of our proposed few-shot action classification framework boosted by supervised contrastive learning. A set of effective spatial-temporal augmentation methods are utilized to generate various video clips (views), which are subsequently fed into the feature extractor (3D CNN) to obtain semantic representation vectors. All these sampled video semantic vectors $f_{i,j}^r$ ($i \in \{0, 1, \cdots, N-1\}, j \in \{0, 1, \cdots, K-1\}, r \in \{0, 1, \cdots, U-1\}$) from the support set are exploited to train a similarity measurement network $\mathcal{M}$ in a supervised way with the contrastive learning loss $\mathcal{L}_{cl}$. Furthermore, $f_{i,j}^r$ together with the representation vectors $f_Q^r, (r \in \{0, 1, \cdots, U-1\})$ of the augmented views of query samples are used to train downstream few-shot classification tasks with softmax loss $\mathcal{L}_{cls}$.

consistent dictionary on-the-fly, which enables unsupervised contrastive learning to take place more easily. In this second version [14], the authors apply an MLP-based projection head and more kinds of data augmentation methods to establish strong representations. By performing a stop-gradient operation on one of the two encoder branches, SimSiam [15] is able to optimize the degree to which two augmentations of the same image are similar to one another, which allows it to obtain more meaningful representations even when none of the relevant factors (negative sample pairs, larger batch sizes, or momentum encoders) are present. In this paper, we also evaluate our proposed few-shot video action classification framework with the latest/mainstream CL methods, verifying the flexibility and the independence of our method.

3   Methodology

3.1   Framework Overview

To increase the effectiveness of representation ability of the video encoder and measure the similarity score more effectively via contrastive learning, we propose a unified framework that integrates contrastive learning and few-shot learning together in Fig. 2. For an N-way K-shot few-shot episode, video augmentations considering both spatial and temporal dimensions are performed for each video. Concretely, for the j-th input video ($j \in \{0, 1, \cdots, K - 1\}$) in the i-th class ($i \in \{0, 1, \cdots, N - 1\}$) in the support set (i.e., $S_{i,j}$), we obtain U augmented views/video clips $C_{i,j}^r$ ($r \in \{0, 1, \cdots, U - 1\}$). Subsequently, these views with diversity are then followed by a CNN-based feature extractor so that the latent representations can be learned, and outputting the embedded vectors $f_{i,j}^r$. Similarly, for each query sample, we can also obtain the representations of its different augmented views, denoted as f_Q^r ($r \in \{0, 1, \cdots, U - 1\}$). Since we have label information for the support set, on the basis of the class labels, we are able to generate positive and negative sample pairs for the purpose of engaging in contrastive learning. That is, two latent vectors belonging to the same class are considered as a positive pair, while they are negative to each other if they come from different classes. In the N-way K-shot scenario with U augmentations, we can generate $N \times U \times (U - 1)$ positive pairs and $U^2 \times N(N-1)/2$ negative ones in total (as is illustrated in the dash-lined frame in Fig. 2). Then two branches are extended with the latent vectors: the contrastive learning branch and the few-shot classification branch. The positive and negative pairs are used to train the feature extractor with supervised learning as the input for the contrastive learning branch.

With the loss function defined as $\mathcal{L}_{cl}$, contrastive learning aims to facilitate the feature extractor to generate more discriminative representations, which make positive samples close and negative ones far away in the high-dimensional latent space. As for a few-shot classification scenario, we make use of the mean representation of the K shots for each class (denoted as a prototype) as the class center for the nearest-neighbor search. And a similarity measurement neural network $\mathcal{M}$ is intended to regress the distances between both query samples and prototypes, with the classification softmax loss defined as $\mathcal{L}_{cls}$.

3.2   Supervised Contrastive Learning

For each of the U data augmentation methods, we adopt a combination of temporal and spatial augmentations. The spatial one is the same across all U augmentations, i.e., we perform a random crop in each selected frame (as is shown in Fig. 3(f)). As for the temporal augmentations, we use $U = 5$ methods to provide a diversity of visual representations: uniform sampling, random sampling, speedup sampling, slow-down sampling, and Gaussian sampling. The augmented video clips (views) are further exploited to generate positive and negative sample pairs related to contrastive learning.

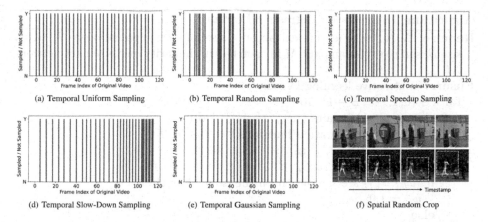

(a) Temporal Uniform Sampling   (b) Temporal Random Sampling   (c) Temporal Speedup Sampling

(d) Temporal Slow-Down Sampling   (e) Temporal Gaussian Sampling   (f) Spatial Random Crop

Fig. 3. Demonstration of temporal-spatial view augmentations for an input original video with $D = 117$ frames and sampling $T = 32$ frames: (a-e) temporal sampling using uniform, random, speed-up, slow-down, and Gaussian methods, respectively; (f) random spatial crop for each selected frame.

(1) **For uniform sampling**, let $\mathcal{I}(\sigma)$ denote the frame index of the selected σ-th frame ($\sigma \in \{0, 1, \cdots, T-1\}$, and T is the quantity of selected frames) from the original input video, which follows the distribution defined as:

$$\mathcal{I}(\sigma) \sim \mathcal{U}(0, D), \tag{1}$$

where D represents the total number of the original input video sample, and $\mathcal{U}$ is the uniform distribution.

(2) **For random sampling**, we directly obtain T frames by independently sampling T times from the original video without any replacement or sorting.

(3) **As for speed-up or slow-down sampling**, we are motivated by the observation that sometimes meaningful behaviors happen at the front/end along the time dimension in the original video, but which may be ignored by the uniform/random sampling method. The sampled frame $\mathcal{I}(\sigma)$ in both speedup and slow-down cases are defined as:

$$\frac{d\mathcal{I}(\sigma)}{d\sigma} = v, \quad \mathcal{I}(0) = 0, \quad \mathcal{I}(T) = D, \tag{2}$$

where v is the sampling velocity which is positive for speedup sampling while negative for the slow-down case. Note that the initial state $\mathcal{I}(0) = 0$ and $\mathcal{I}(T) = D$ limits the range of the sampled index. Speedup sampling samples more frames at the beginning of the input video, and slow-down sampling focus more on frames at the tail.

(4) **Gaussian sampling**, with slow-down as its first half part and speedup as second half, i.e., it samples most intensively at the middle of a given video sample. Its sampling formulation is the same with Eq. (2) but the border

---

Algorithm 1 Supervised Contrastive Learning (SCL)

Require: N, K, video feature extractor g, a set of view augmentations $\mathcal{T}$, batch size B, sampled pair amount M in each batch, similarity measurement network $\mathcal{M}$

Ensure: contrastive learning loss $\mathcal{L}_{cl}$

  for $b \in \{0, 1, \cdots, B-1\}$ **do**

    for sampled video pairs $\{(\boldsymbol{x}_{b,l}, \boldsymbol{x}'_{b,l})\}_{l=0}^{M-1}$ **do**

    Draw two augmentation functions $t \sim \mathcal{T}$, $t' \sim \mathcal{T}$;

    $C_{b,l}$, $C'_{b,l} = t(\boldsymbol{x}_{b,l})$, $t'(\boldsymbol{x}'_{b,l})$; # clip generation

    $f_{b,l}$, $f'_{b,l} = g(C_{b,l})$, $g(C'_{b,l})$; # representation

    if $(\boldsymbol{x}_{b,l}, \boldsymbol{x}'_{b,l})$ are sampled from the same class **then**

      $y_{b,l} = 1.0$;

    else

      $y_{b,l} = 0.0$;

    end if

    end for

  end for

  for $b \in \{0, 1, \cdots, B-1\}$, $l \in \{0, 1, \cdots, M-1\}$ **do**

    $d_{b,l} = 1.0 - \mathcal{M}(f_{b,l}, f'_{b,l})$; # pairwise distance

  end for

Update video clip representation network g and similarity measurement network $\mathcal{M}$ by minimizing $\mathcal{L}_{cl}$.

---

state should be initialized as $\mathcal{I}(0) = 0$, $\mathcal{I}(T/2) = D/2$ for the first half and $\mathcal{I}(T/2) = D/2$, $\mathcal{I}(T) = D$ for the second half. Figure 3(a–e) illustrate five examples with the same input video sample ($D = 117$) for the five augmentations, respectively ($T = 32$, $v = 4$).

The supervised contrastive learning (SCL) algorithm is summarized in **Algorithm** 1, where the similarity measurement network $\mathcal{M}$ is also shared in the few-shot classification branch, which is used to reflect the distance within each positive/negative pair (the details of $\mathcal{M}$ are described in Sect. 3.3). We follow the contrastive loss function $\mathcal{L}_{cl}$ used in [11,16,24,35,44], which is defined as:

$$\mathcal{L}_{cl} = -\frac{1}{BM} \sum_{b=0}^{B-1} \sum_{l=0}^{M-1} y_{b,l} d_{b,l}^2 + (1 - y_{b,l}) \max(m - d_{b,l}, 0)^2, \qquad (3)$$

where M is the total constructed positive and negative pairs with a single minibatch, and $d_{b,l}$ is the distance between two samples of the l-th pair in the b-th input episode, and $y_{b,l}$ is the corresponding ground truth label ($y_{b,l} = 1$ if the pair consists of two views generated from the same class and $y_{b,l} = 1$ otherwise). Note that m is a margin that defines a radius, and the negative pairs affect the loss only when the distance is within this radius.

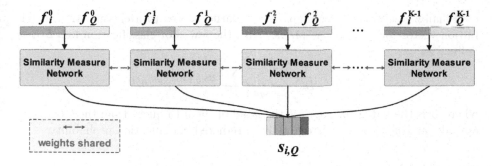

Fig. 4. The schematic illustration of the few-shot action classification process. For the r-th augmented view in the i-th class, the class prototype f_i^r is obtained by averaging the latent representations $f_{i,j}^r$ along the shot dimension j. Together with each query sample's augmented view f_Q^r, the prototype-query pairs are fed into the same similarity measurement network $\mathcal{M}$ which is also used in supervised contrastive learning (see Fig. 2) to obtain the final similarity score vector $\mathbf{s}_{i,Q}$.

3.3 Few-Shot Classification

The integration process related to contrastive learning and few-shot learning is reflected in two aspects: (1) The supervised contrastive learning loss is combined with the few-shot classification loss during training. (2) There exists a similarity measurement network $\mathcal{M}$ that is shared across the few-shot classification and the contrastive learning branch to measure the latent distance/similarity between two given augmented views. To exploit the few shots in the support set, we follow Prototypical Network [33] and summarize all shots' latent representations $f_{i,j}^r$ ($i \in \{0, 1, \cdots, N-1\}, j \in \{0, 1, \cdots, K-1\}, r \in \{0, 1, \cdots, U-1\}$) by computing their average response:

$$f_i^r = \frac{1}{K} \sum_{j=0}^{K-1} f_{i,j}^r. \tag{4}$$

Figure 4 illustrates the few-shot action classification network. For all the augmented views for a specific class in the support set, the class prototypes f_i^r ($i \in \{0, 1, \cdots, N-1\}, r \in \{0, 1, \cdots, U-1\}$) are only concerned with the query sample f_Q^r coming from the same augmentation. The similarity measurement network $\mathcal{M}$ is then utilized to predict the similarity score $s_{i,Q}^r$ between two input views:

$$s_{i,Q}^r = \mathcal{M}(f_i^r, f_Q^r). \tag{5}$$

It is worth mentioning that the similarity score vector $\mathbf{s}_{i,Q}$ of all views is further weighted by a linear layer $\mathbf{w} \in \mathbb{R}^{1 \times U}$, to obtain the final predicted similarity score $s_{i,Q}$ between the i-th class prototype and the query sample:

$$s_{i,Q} = \mathbf{w} \cdot \mathbf{s}_{i,Q}, \tag{6}$$

where $\mathbf{s}_{i,Q} = [s_{i,Q}^0, s_{i,Q}^1, \cdots, s_{i,Q}^{U-1}]^T$.

Finally, a softmax layer maps N similarity scores to a classification distribution vector for each query sample. And the few-shot classification loss $\mathcal{L}_{cls}$ is defined as:

$$\mathcal{L}_{cls} = -\frac{1}{BQ}\sum_{b=0}^{B-1}\sum_{q=0}^{Q-1}\sum_{i=0}^{N-1} y_{b,q,i}\log(\hat{y}_{b,q,i}), \qquad (7)$$

where B is the batch size, $y_{b,q,i}$ is the label of the q-th query from the b-th input episode, and $\hat{y}_{b,q,i}$ is the corresponding predicted classification probability.

3.4 Total Learning Objective

We incorporate supervised contrastive learning to the few-shot classification task by adding an auxiliary loss $\mathcal{L}_{cl}$, i.e., the final weighted loss $\mathcal{L}$ is constructed as:

$$\mathcal{L} = \mathcal{L}_{cls} + \alpha\mathcal{L}_{cl}, \qquad (8)$$

where α is the balance hyper-parameter.

4 Experiments

4.1 Datasets and Settings

Datasets. In this paper, the proposed supervised contrastive learning framework is evaluated the performance on three different action recognition datasets: HMDB51 [27], UCF101 [34] and Sth-Sth-V2 [22]. **HMDB51** totally contains 6,766 videos distributed in 51 action categories. **UCF101** has included 13,320 videos covering 101 different action-based categories. **Sth-Sth-V2** includes 220,847 videos with 174 different classes. For UCF101 and Sth-Sth-V2, we follow the same splits as in OTAM [6], and they are randomly sampling 64 classes for meta training, 12 classes for meta validation, and 24 classes for meta testing, respectively. For HMDB51, we randomly select 32/6/13 classes for meta training, validation, and testing.

Configuration. It is considered the few-shot scenarios with $N = 5$ and $K = 1, 3, 5$. In each episode, we randomly select N categories, each consisting of K samples as the support set and select another video for each class as the query sample. We train our model over 2,000 episodes and check that the validation set matches an early stopping criterion for every 128 episodes. We use Adam optimizer, and the learning rate is set to 0.001. Furthermore, the average classification accuracies are reported by evaluating 500 and 1000 episodes in the meta-validation and meta-test split, respectively.

3D Backbones. To better demonstrate the generalizability of the proposed framework, we perform extensive experiments with 5 different video feature extraction backbones: C3D [38], R(2+1)D [39], P3D [30], I3D [41] and Slow-Fast [18]. All backbones are trained with the input size of 224×224. The input clip length for C3D, R(2+1)D, P3D, I3D, and SlowFast are 16, 16, 16, 32, and 40 frames, respectively. The global average pooling layer in 3D backbones are

Table 1. Comparison to state-of-the-art video action classification approaches on the HMDB51, UCF101, and Sth-Sth-V2 datasets. All backbones are trained from scratch. Accuracy (%) are reported on average over 1,000 episodes. Note that Neg./Pos. pairs ratio is configured as 2.5.

Methods	Backbone	HMDB51 [27]			UCF101 [34]			Sth-Sth-V2 [22]		
		1-shot	3-shot	5-shot	1-shot	3-shot	5-shot	1-shot	3-shot	5-shot
ARN [43]	C3D	45.53	53.60	59.82	66.60	78.40	84.48	33.44	38.80	45.74
TARN [3]	C3D	66.52	73.30	75.50	85.40	86.72	93.40	38.43	44.54	48.63
ProtoGAN [17]	C3D	35.41	49.89	52.90	61.73	75.89	79.70	33.90	40.72	44.68
FAN [37]	C3D	69.90	71.48	78.20	77.56	87.62	90.80	37.20	43.32	45.82
OTAM [6]	C3D	64.63	79.80	81.90	88.12	91.07	92.10	39.60	47.10	52.30
TAV [4]	C3D	71.30	78.42	83.80	87.90	92.30	92.26	39.40	46.60	49.92
Ours (w/o SCL)	C3D	70.04	77.62	80.51	86.00	90.60	91.20	34.75	41.75	46.28
Ours (full)	C3D	**75.78**	**86.89**	**89.84**	**92.19**	**94.96**	**95.31**	**41.42**	**49.22**	**53.12**

remained, and the dimensions of the final clip representation vectors are 4096, 2048, 2048, 2048, and 2304, respectively. All the backbones are trained from scratch. As for the similarity measurement network $\mathcal{M}$, it consists of 5 fully connected layers with 1024, 1024, 512, 512, and 1 neuron.

Contrastive Learning Loss. With contrastive learning enabled, its loss $\mathcal{L}_{cl}$ contributes to the final loss with $\alpha = 1.0$. For the 5-way few-shot action classification scenario, the maximum numbers of generated positive and negative pairs are 100 and 250, respectively. Different positive and negative ratios can be achieved via masking the selected pairs. The margin parameter m in Equation (3) is configured to 0.75 in our work. That is, the distance between two clips of a negative pair is expected to be larger than it.

4.2 Main Results

Comparison to State-of-the-Art. In this paper, we evaluate our proposed architecture with supervised contrastive learning against the action classification methods on *HMDB51* [27], *UCF101* [34] and *Sth-Sth-V2* [22] datasets. Frame-level feature extraction based on 2D CNN and then aggregating them together as the video descriptor is used in original OTAM [6]. For a fair comparison, we change its backbone to C3D to extract feature vectors (each video is split into 16 segments, and each contains 16 frames (clip length)). As for TAV [4], we also re-implement it and replace its 2D backbone with the C3D model, which is then combined with the original temporal structure filter (TSF). For ARN [43], TARN [3], ProtoGAN [17] and FAN [37], we follow the original configurations. The only difference between them and our re-implementation versions is that we train all 3D backbones from scratch rather than use pre-trained weights (such as Kinetics-400) since there inevitably exists a category overlap between mainstream pre-trained models and our evaluation datasets. In Table 1, we summarizes the classification accuracy over 1/3/5 shot(s):

Table 2. Comparison with different contrastive learning approaches on the HMDB51, UCF101, and Sth-Sth-V2 datasets. All contrastive learning methods adopt the C3D model (trained from scratch) as their backbones (clip length is 16) to extract video feature vectors. Mean accuracies (%) are reported over 1, 000 episodes. Note that Neg./Pos. pair ratio is configured as 2.5.

Methods	HMDB51 [27]			UCF101 [34]			Sth-Sth-V2 [22]		
	1-shot	3-shot	5-shot	1-shot	3-shot	5-shot	1-shot	3-shot	5-shot
FSL	70.04	77.62	80.51	86.00	90.60	91.20	34.75	41.75	46.28
FSL+SCL (MoCo [4])	74.26	78.12	83.22	88.74	91.20	92.51	37.54	44.85	48.96
FSL+SCL (MoCov2 [14])	74.88	85.90	88.60	91.19	93.86	94.30	39.40	48.60	52.04
FSL+SCL (SimCLR [12])	72.32	81.60	84.10	88.90	91.08	92.48	36.90	45.72	49.28
FSL+SCL (SimCLRv2 [13])	74.92	85.20	89.28	91.16	93.66	94.37	40.06	48.90	53.00
FSL+SCL (SimSiam [15])	74.90	85.41	89.17	91.12	93.73	94.70	41.29	48.34	52.92
FSL+SCL (ours)	**75.78**	**86.89**	**89.84**	**92.19**	**94.96**	**95.31**	**41.42**	**49.22**	**53.12**

(1) With supervised contrastive learning disabled, our proposed few-shot classification architecture achieves better performance than ARN [43], ProtoGAN [17] on all three datasets and achieves competitive performance w.r.t. TARN [3] and FAN [37]. However, it performs weaker than OTAM [6] and TAV [4] because both OTAM and TAV mine the temporal alignment information between query and support samples in the latent space, which benefits the subsequent distance measurement and classification.

(2) With supervised contrastive learning enabled, we achieve better classification accuracy in all cases, surpassing prior methods with a significant margin. It illustrates that the auxiliary SCL loss can boost the representation ability and similarity score measurement capacity, resulting in improved final classification accuracy.

(3) Sth-Sth-V2 is much more difficult than HMDB51 and UCF101, as we can observe that the classification results on Sth-Sth-V2 are much lower than those on HMDB51/UCF101 with supervised contrastive learning enabled. Improving classification results on a complex dataset is much more difficult than on simple ones. The difficulty of Sth-Sth-V2 can be further explained by the diversity of samples in each category. For example, the category "putting something onto something" on Sth-Sth-V2 contains many different types of video clips. Almost all labels are general descriptions rather than actions with concrete object names (e.g., not like "putting a cup onto a table"?. The general descriptions increase the classification difficulty significantly.

Contrastive Learning Framework Evaluation. The proposed few-shot action classification architecture with supervised contrastive learning is designed not only for high efficient video representations, but also for pairwise similarity score regression. Therefore, it can adopt other mainstream contrastive learning

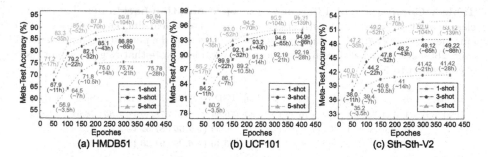

Fig. 5. Model convergence analysis of our proposed supervised contrastive learning algorithm for a few-shot action classification task. Experiments are performed in AWS *ml.g4dn.16xlarge* EC2 instance (64 vCPU and 256G RAM).

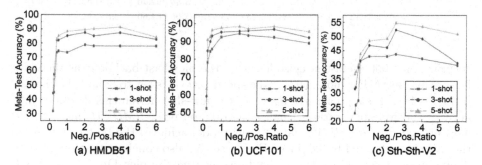

Fig. 6. Comparison of different negative/positive pair ratios for contrastive learning on HMDB51, UCF101, and Sth-Sth-V2 datasets with few-shot action classification. SlowFast is adopted (the speed ratio $\alpha = 8$, and the channel ratio $\beta = 1/8$) as the backbone (clip length is 40).

methods. To demonstrate its generalization ability, MoCo [4], MoCov2 [14], Sim-CLR [12], SimCLRv2 [13] and SimSiam [15] are compared with our supervised contrastive learning algorithm. The batch size B is 128, and all these models are trained up to 400 epochs. Table 2 shows the few-shot action classification results. As shown in Fig. 5, we show the model convergence curves and training time cost using our SCL algorithm. Experimental results demonstrate that adding the supervised contrastive learning branch indeed improves the few-shot action classification performance. Furthermore, since our proposed SCL algorithm considers an additional similarity network $\mathcal{M}$, it achieves competitive performance boosting.

4.3 Further Evaluations

Different Pos./Neg. Pair Ratios. In the experiment, we evaluate the influence of negative/positive pair ratio in contrastive learning. We configure the ratio to 0.2, 0.25, 0.4, 0.5, 1.0, 2.0, 2.5, 4.0, 6.0 and Fig. 6 plots the average

Table 3. Comparison of different video representation backbones. The average classification accuracy (%) with supervised contrastive learning enabled over 1,000 episodes are reported. The values in parentheses represent the percentage improvements over a baseline model with contrastive learning disabled. Note that Neg./Pos. pair ratio is configured to 2.5.

Dataset	K	C3D ([38])	R(2+1)D ([39])	P3D ([30])	I3D ([41])	SlowFast ([18])
HMDB51	1-shot	75.78 (+5.74)	76.22 (+4.78)	76.84 (+5.40)	78.80 (+3.80)	78.91 (+2.10)
	3-shot	86.89 (+9.27)	85.82 (+6.09)	86.30 (+7.60)	87.52 (+4.40)	87.50 (+3.26)
	5-shot	89.84 (+9.33)	90.02 (+8.24)	90.40 (+6.29)	91.38 (+6.10)	91.41 (+5.32)
UCF101	1-shot	92.19 (+6.19)	92.60 (+5.80)	93.90 (+6.70)	94.60 (+4.65)	94.53 (+3.74)
	3-shot	94.96 (+4.36)	95.00 (+5.96)	96.38 (+5.92)	96.88 (+5.46)	96.88 (+5.28)
	5-shot	95.31 (+4.11)	96.48 (+3.70)	97.96 (+4.26)	98.50 (+4.80)	98.44 (+3.90)
Sth-Sth-V2	1-shot	41.42 (+6.67)	42.69 (+6.10)	43.50 (+3.29)	43.74 (+2.10)	43.75 (+2.80)
	3-shot	49.22 (+7.47)	51.00 (+7.32)	52.28 (+3.50)	52.40 (+2.46)	52.34 (+2.45)
	5-shot	53.12 (+6.84)	53.18 (+5.43)	53.93 (+3.00)	54.60 (+2.65)	54.78 (+1.58)

accuracy on 1,000 meta-test episodes using the SlowFast backbone as the video feature extractor. For more details, the speed ratio α is set to 8, and the channel ratio β is 1/8. It is a poor performance of the few-shot action classification when negative/positive pair ratio is smaller than 0.5 on both *HMDB51* and *UCF101* datasets. On the *Sth-Sth-V2* dataset, our model achieves the best results when the ratio is configured to 2.5. From Fig. 6, we can also conclude that unlike Sim-Siam, our proposed SCL indeed depends on negative samples. One reason is that: not only the video representations are improved (i.e., more discriminative) by contrastive learning, but also the distances between video clips that are essential for few-shot classification are explicitly learned by the contrastive learning loss.

Influence of Different Backbones. To evaluate the generalisability of our proposed framework, we further integrate different video feature extraction back-bones. In Table 3, we summarize the few-shot action classification accuracies respectively based on C3D [38], R(2+1)D [39], P3D [30], I3D [41] and Slow-Fast [18] (the speed ratio $\alpha = 8$, and the channel ratio $\beta = 1/8$) with supervised contrastive learning enabled. The performance improvements are also given in parentheses compared to a simple model with a single few-shot classification branch without contrastive learning. It is clear to see that: (1) For all cases on three different datasets, the proposed framework achieves better results with the supervised contrastive learning branch enabled, which demonstrates the effectiveness as well as the potential for generalization of the methodology that we have developed, i.e., contrastive learning indeed improves the video representation capacity and benefits the distance measurement for classification. (2) The performance improvements are less significant for high-capacity video extraction backbones such as I3D and SlowFast.

Table 4. Comparison of different combinations of spatial-temporal augmentations on HMDB51 with C3D as the backbone. Few-shot classification accuracies (%) are reported over 1,000 episodes. Note that Neg./Pos. pair ratio is configured to 2.5.

Augmentation Method	1-shot	3-shot	5-shot
Uniform Samp. (US)	75.00	84.28	87.40
Random Samp. (RS)	74.84	84.10	87.26
Speedup Samp. (SS)	70.42	81.28	83.40
Slow-Down Samp. (SDS)	71.30	82.00	83.36
Gaussian Samp. (GS)	72.60	83.90	86.45
US+RS	74.89	85.27	88.31
US+RS+SS	75.18	85.63	88.99
US+RS+SS+SDS	75.34	86.74	89.70
US+RS+SS+SDS+GS	**75.78**	**86.89**	**89.84**

Effect of Spatial-Temporal Augmentations. To evaluate the effect of spatial-temporal augmentation methods, we combine different temporal sampling methods with the spatial random crop. In Table 4, we report the performance on HMDB51 with the C3D backbone. We can observe from Table 4 that uniform sampling and random sampling can achieve better performance than speedup, slow-down, or gaussian sampling, which because uniform and random sampling usually obtain the temporal information across the whole time dimension, while for speedup, slow-down, and gaussian sampling, they pay more attention to the beginning, the end and the middle of the video along the time dimension, respectively. Furthermore, combining all these sampling methods together and using learnable weights (attentive) to get the final similarity score (see Fig. 3) will help us mine the video features better.

5   Conclusions

This paper proposes a general few-shot action classification framework powered by supervised contrastive learning, where contrastive learning is deployed to improve the representation quality of videos and a similarity score network is shared by both contrastive learning and few-shot learning to make a closer integration of the two paradigms. Besides, five spatial-temporal video augmentation methods are designed for generating various video sample views in the N-way K-shot few-shot classification scenarios. The significantly improvements achieved by our proposed framework in few-shot action classification is mainly due to: (1) The auxiliary supervised contrastive learning loss makes the video representations more discriminative. (2) The distance measurement between clips is reflected by the similarity score more precisely thanks to a shared similarity score measurement network in both few-shot classification and contrastive learning branches. Importantly, our proposed framework shows strong generalization abilities when different video representation backbones are used. Our proposed framework also has highly flexibility as it can achieve competitive performance when other mainstream contrastive learning approaches are integrated.

Acknowledgements. We would like to thank the anonymous reviewers. This work was supported in part by National Natural Science Foundation of China (61976220 and 61832017), and Beijing Outstanding Young Scientist Program (BJJWZYJH012019100020098).

References

1. Andrychowicz, M., et al.: Learning to learn by gradient descent by gradient descent. In: Advances in Neural Information Processing Systems, pp. 3981–3989 (2016)
2. Ben-Ari, R., Nacson, M.S., Azulai, O., Barzelay, U., Rotman, D.: TAEN: temporal aware embedding network for few-shot action recognition. In: IEEE Conference on Computer Vision and Pattern Recognition, pp. 2786–2794 (2021)
3. Bi, M., Zou, G., Pat, I.: TARN: temporal attentive relation network for few-shot and zero-shot action recognition. arXiv preprint arXiv:1907.09021 (2019)
4. Bo, Y., Lu, Y., He, W.: Few-shot learning of video action recognition only based on video contents. In: IEEE Winter Conference on Applications of Computer Vision, pp. 584–593 (2020)
5. Brattoli, B., Buchler, U., Wahl, A.S., Schwab, M.E., Ommer, B.: LSTM self-supervision for detailed behavior analysis. In: IEEE Conference on Computer Vision and Pattern Recognition, pp. 6466–6475 (2017)
6. Cao, K., Ji, J., Cao, Z., Chang, C.Y., Niebles, J.C.: Few-shot video classification via temporal alignment. In: IEEE Conference on Computer Vision and Pattern Recognition, pp. 10615–10624 (2020)
7. Careaga, C., Hutchinson, B., Hodas, N., Phillips, L.: Metric-based few-shot learning for video action recognition. arXiv preprint arXiv:1909.09602 (2019)
8. Caron, M., Misra, I., Mairal, J., Goyal, P., Bojanowski, P., Joulin, A.: Unsupervised learning of visual features by contrasting cluster assignments. In: Advances in Neural Information Processing Systems, pp. 9912–9924 (2020)
9. Carreira, J., Zisserman, A.: Quo vadis, action recognition? A new model and the Kinetics dataset. In: IEEE Conference on Computer Vision and Pattern Recognition, pp. 4724–4733 (2017)
10. Chen, J., Zhan, L.M., Wu, X.M., Chung, F.l.: Variational metric scaling for metric-based meta-learning. In: AAAI Conference on Artificial Intelligence, pp. 3478–3485 (2020)
11. Chen, J., et al.: DASNet: dual attentive fully convolutional Siamese networks for change detection in high-resolution satellite images. IEEE J. Sel. Topics Appl. Earth Observ. Remote Sens. **14**, 1194–1206 (2020)
12. Chen, T., Kornblith, S., Norouzi, M., Hinton, G.: A simple framework for contrastive learning of visual representations. In: International Conference on Machine Learning, pp. 1597–1607 (2020)
13. Chen, T., Kornblith, S., Swersky, K., Norouzi, M., Hinton, G.E.: Big self-supervised models are strong semi-supervised learners. In: Advances in Neural Information Processing Systems, pp. 22276–22288 (2020)
14. Chen, X., Fan, H., Girshick, R., He, K.: Improved baselines with momentum contrastive learning. arXiv preprint arXiv:2003.04297 (2020)

15. Chen, X., He, K.: Exploring simple Siamese representation learning. In: IEEE Conference on Computer Vision and Pattern Recognition, pp. 15750–15758 (2021)
16. Choi, H., Som, A., Turaga, P.: AMC-loss: angular margin contrastive loss for improved explainability in image classification. In: IEEE Conference on Computer Vision and Pattern Recognition Workshops, pp. 3659–3666 (2020)
17. Dwivedi, S.K., Gupta, V., Mitra, R., Ahmed, S., Jain, A.: ProtoGAN: towards few shot learning for action recognition. In: IEEE International Conference on Computer Vision Workshop, pp. 1308–1316 (2019)
18. Feichtenhofer, C., Fan, H., Malik, J., He, K.: SlowFast networks for video recognition. In: IEEE International Conference on Computer Vision, pp. 6201–6210 (2019)
19. Finn, C., Abbeel, P., Levine, S.: Model-agnostic meta-learning for fast adaptation of deep networks. In: International Conference on Machine Learning, pp. 1126–1135 (2017)
20. Gammulle, H., Denman, S., Sridharan, S., Fookes, C.: Two stream LSTM: a deep fusion framework for human action recognition. In: IEEE Winter Conference on Applications of Computer Vision, pp. 177–186 (2017)
21. Gidaris, S., Bursuc, A., Komodakis, N., Pérez, P., Cord, M.: Boosting few-shot visual learning with self-supervision. In: IEEE International Conference on Computer Vision, pp. 8059–8068 (2019)
22. Goyal, R., et al.: The "something something" video database for learning and evaluating visual common sense. In: IEEE International Conference on Computer Vision, pp. 5842–5850 (2017)
23. Grill, J.B., et al.: Bootstrap your own latent-a new approach to self-supervised learning. In: Advances in Neural Information Processing Systems, pp. 21271–21284 (2020)
24. Hadsell, R., Chopra, S., LeCun, Y.: Dimensionality reduction by learning an invariant mapping. In: IEEE Conference on Computer Vision and Pattern Recognition, pp. 1735–1742 (2006)
25. He, K., Fan, H., Wu, Y., Xie, S., Girshick, R.: Momentum contrast for unsupervised visual representation learning. In: IEEE Conference on Computer Vision and Pattern Recognition, pp. 9726–9735 (2020)
26. Henaff, O.: Data-efficient image recognition with contrastive predictive coding. In: International Conference on Machine Learning, pp. 4182–4192 (2020)
27. Kuehne, H., Jhuang, H., Garrote, E., Poggio, T., Serre, T.: HMDB: a large video database for human motion recognition. In: International Conference on Computer Vision, pp. 2556–2563 (2011)
28. Nichol, A., Achiam, J., Schulman, J.: On first-order meta-learning algorithms. arXiv preprint arXiv:1803.02999 (2018)
29. van den Oord, A., Li, Y., Vinyals, O.: Representation learning with contrastive predictive coding. arXiv preprint arXiv:1807.03748 (2018)
30. Qiu, Z., Yao, T., Mei, T.: Learning spatio-temporal representation with pseudo-3D residual networks. In: IEEE International Conference on Computer Vision, pp. 5534–5542 (2017)
31. Ravi, S., Larochelle, H.: Optimization as a model for few-shot learning. In: International Conference on Learning Representations (2017)
32. Si, C., Chen, W., Wang, W., Wang, L., Tan, T.: An attention enhanced graph convolutional LSTM network for skeleton-based action recognition. In: IEEE Conference on Computer Vision and Pattern Recognition, pp. 1227–1236 (2019)
33. Snell, J., Swersky, K., Zemel, R.: Prototypical networks for few-shot learning. In: Advances in Neural Information Processing Systems, pp. 4077–4087 (2017)

34. Soomro, K., Zamir, A.R., Shah, M.: A dataset of 101 human action classes from videos in the wild. Center Res. Comput. Vision **2**(11) (2012)
35. Sun, Y., Chen, Y., Wang, X., Tang, X.: Deep learning face representation by joint identification-verification. In: Advances in Neural Information Processing Systems, pp. 1988–1996 (2014)
36. Sung, F., Yang, Y., Zhang, L., Xiang, T., Torr, P.H.S., Hospedales, T.M.: Learning to compare: relation network for few-shot learning. In: IEEE Conference on Computer Vision and Pattern Recognition, pp. 1199–1208 (2018)
37. Tan, S., Yang, R.: Learning similarity: feature-aligning network for few-shot action recognition. In: International Joint Conference on Neural Networks, pp. 1–7 (2019)
38. Tran, D., Bourdev, L., Fergus, R., Torresani, L., Paluri, M.: Learning spatiotemporal features with 3D convolutional networks. In: IEEE International Conference on Computer Vision, pp. 4489–4497 (2015)
39. Tran, D., Wang, H., Torresani, L., Ray, J., LeCun, Y., Paluri, M.: A closer look at spatiotemporal convolutions for action recognition. In: IEEE Conference on Computer Vision and Pattern Recognition, pp. 6450–6459 (2018)
40. Tsunoda, T., Komori, Y., Matsugu, M., Harada, T.: Football action recognition using hierarchical LSTM. In: IEEE Conference on Computer Vision and Pattern Recognition Workshops, pp. 155–163 (2017)
41. Wang, X., Girshick, R., Gupta, A., He, K.: Non-local neural networks. In: IEEE Conference on Computer Vision and Pattern Recognition, pp. 7794–7803 (2018)
42. Wu, Z., Li, Y., Guo, L., Jia, K.: PARN: position-aware relation networks for few-shot learning. In: IEEE International Conference on Computer Vision, pp. 6658–6666 (2019)
43. Zhang, H., Zhang, L., Qi, X., Li, H., Torr, P.H.S., Koniusz, P.: Few-shot action recognition with permutation-invariant attention. In: Vedaldi, A., Bischof, H., Brox, T., Frahm, J.-M. (eds.) ECCV 2020. LNCS, vol. 12350, pp. 525–542. Springer, Cham (2020). https://doi.org/10.1007/978-3-030-58558-7_31
44. Zhang, J., Shih, K.J., Elgammal, A., Tao, A., Catanzaro, B.: Graphical contrastive losses for scene graph parsing. In: IEEE Conference on Computer Vision and Pattern Recognition, pp. 11527–11535 (2019)

A Novel Data Augmentation Technique for Out-of-Distribution Sample Detection Using Compounded Corruptions

Ramya Hebbalaguppe[1,2]([✉]), Soumya Suvra Ghosal[1], Jatin Prakash[1],
Harshad Khadilkar[2], and Chetan Arora[1]

[1] IIT Delhi, New Delhi, India
ramya.s.hebbalaguppe@cse.iitd.ac.in
[2] TCS Research, Mumbai, India
https://github.com/cnc-ood

Abstract. Modern deep neural network models are known to erroneously classify out-of-distribution (OOD) test data into one of the in-distribution (ID) training classes with high confidence. This can have disastrous consequences for safety-critical applications. A popular mitigation strategy is to train a separate classifier that can detect such OOD samples at test time. In most practical settings OOD examples are not known at train time, and hence a key question is: *how to augment the ID data with synthetic OOD samples for training such an OOD detector?* In this paper, we propose a novel Compounded Corruption (CnC) technique for the OOD data augmentation. One of the major advantages of CnC is that it does not require any hold-out data apart from training set. Further, unlike current state-of-the-art (SOTA) techniques, CnC does not require backpropagation or ensembling at the test time, making our method much faster at inference. Our extensive comparison with 20 methods from the major conferences in last 4 years show that a model trained using CnC based data augmentation, significantly outperforms SOTA, both in terms of OOD detection accuracy as well as inference time. We include a detailed post-hoc analysis to investigate the reasons for the success of our method and identify higher relative entropy and diversity of CnC samples as probable causes. Theoretical insights via a piece-wise decomposition analysis on a two-dimensional dataset to reveal (visually and quantitatively) that our approach leads to a tighter boundary around ID classes, leading to better detection of OOD samples.

Keywords: OOD detection · Open Set recognition · Data augmentation

1 Introduction

Deep neural network (DNN) models generalize well when the test data is independent and identically distributed (IID) with respect to training data [42]. However, the condition is difficult to enforce in the real world due to distributional drifts, covariate shift, and/or adversarial perturbations. A *reliable* system based on a DNN model must be able

Supplementary Information The online version contains supplementary material available at https://doi.org/10.1007/978-3-031-26409-2_32.

M.-R. Amini et al. (Eds.): ECML PKDD 2022, LNAI 13715, pp. 529–545, 2023.
https://doi.org/10.1007/978-3-031-26409-2_32

to detect an OOD sample, and either abstain from making any decision on such samples, or flag them for human intervention. We assume that the in-distribution (ID) samples belong to one of the K known classes, and club all OOD samples into a new class called a *reject*/OOD class. We do not attempt to identify which specific class (unseen label) the unknown sample belongs to. Our goal is to build a classifier to accurately detect OOD samples as the $(K + 1)^{\text{th}}$ OOD class, with an objective to reject samples belonging to any novel class.

Most techniques for OOD detection assume the availability of validation samples from the OOD set for tuning model hyper-parameters [2,19,31,33]. Based on the samples, the techniques either update the model weights so as to predict lower scores for the OOD samples, or try to learn correlation between activations and the output score vector [31]. Such approaches have limited utility as in most practical scenarios, either the OOD samples are not available, or cover a tiny fraction of OOD sample space. Yet, other class of techniques learn the threshold on the uncertainty of the output score using deep ensembling [28] or MC dropout [9]. Understandably, OOD detection capability of these techniques suffer when the samples from a different OOD domain are presented.

The other popular class of OOD detectors do not use representative samples from OOD domain, but generate them synthetically [17,36,37]. The synthetic samples can be used to train any of the earlier mentioned SOTA models in lieu of the real OOD samples. This obviates the need for any domain specific OOD validation set. Such methods typically use natural corruptions (e.g. blur, noise, and geometric transformations etc.) or adversarial perturbations to generate samples near decision boundary of a classifier. This class also have limited accuracy on real OOD datasets, as the synthetic images generated in such a way are visually similar/semantically similar to the ID samples, and the behavior of a DNN when shown natural OOD images much farther (in terms of ℓ_2 distance in RGB space) from the ID samples still remains unknown.

Recent theoretical works towards estimating or minimizing open set loss recommend training with OOD samples covering as much of the probable input space as possible. For example, [24] show that a piece-wise DNN model shatters the input space into a polyhedral complex, and prove that empirical risk of a DNN model in a region of input space scales inversely with the density of training samples lying inside the polytope corresponding to the region. Similarly, [8] show that under an unknown OOD distribution, the best way to minimize the open set loss is by choosing OOD samples uniformly from the support set in the input space. Encouraged by such theoretical results, we propose a data augmentation technique which does not focus on generating samples visually similar to the ID samples but synthesizing OOD samples in two key regions of the input space: (i) finely distributed at the boundary of ID classes, and (ii) coarsely distributed in the inter-ID sample space (See Sect. 3.3 for details). We list the key contributions:

1. We propose a novel data augmentation strategy, Compounded Corruptions (CnC) for OOD detection. Unlike contemporary techniques [12,19,31,33] the proposed approach does not need a separate OOD train or validation dataset.
2. Unlike SOTA techniques which detect OOD samples by lowering the confidence of ID classes [1,18,31,35], we classify OOD samples into a separate reject class. We show empirically that our approach leads to clearer separation between ID and OOD samples in the embedding space (Fig. 4).

3. Our method does not require any input pre-processing at the test time, or a second forward pass with perturbation/noise. This makes it significantly faster in inference as compared to the other SOTA methods [22,33].
4. Visualization and analysis of our results indicate that finer granularity of the polyhedral complex around the ID regions learnt by a model is a good indicator of performance of a OOD data augmentation technique. Based on our analysis, we also recommend higher entropy and diversity of generated OOD samples as good predictors for OOD detection performance.

2   Related Work

Our approach is a hyper-parameter-free OOD detection technique, which does not need access to a validation OOD dataset. We review contemporary works below.

Hyper-parameter Tuning Using OOD Data. This class comprises of OOD detection methods that fine-tune hyper-parameters on a validation set. ODIN [33] utilizes temperature scaling with input perturbations using the OOD validation dataset to tune hyper-parameters for calibrating the neural networks. However, hyper-parameters tuned with one OOD dataset may not generalize to other datasets. Lee et al. [31] propose training a logistic regression detector on the Mahalanobis distance vectors calculated between test images' feature representations and class conditional Gaussian distribution at each layer.

Retraining a Model Using OOD Data. G-ODIN [22] decompose confidence score along with modified input pre-processing for detecting OOD, whereas ATOM [2] essentially makes a model robust to the small perturbations, and hard negative mining for OOD samples. MOOD [34] introduce multi-level OOD detection based on the complexity of input data, and exploit simpler classifier for faster OOD inference.

Using a Pre-trained Model's Score for OOD Detection. Hendrycks and Gimpel [18] use maximum confidence scores from a softmax output to detect OOD. Liu et al. [35] use energy as a scoring function for OOD detection without tuning hyper-parameters. Shastry and Oore [41] leverage p^{th}-order Gram matrices to identify anomalies between activity patterns and the predicted class. Blundell et al. [1] focus on a closed world assumption which forces a DNN to choose from one of the ID classes, even for the OOD data. *OpenMax* estimates the probability of an input being from an unknown class using a Weibull distribution. G-OpenMax [10] explicitly model OOD samples and report findings on small datasets like MNIST.

OOD Detection Using Uncertainty Estimation. OOD samples can be rejected by thresholding on the uncertainty measure. Graves et al. [11], Wen et al. [46] propose anomaly detection based on stochastic Bayesian inference. Gal et al. [9] propose MC-dropout to measure uncertainty of a model using multiple inferences. Deep Ensembles [28] use multiple networks trained independently to improve uncertainty estimation.

Data Augmentation for OOD Detection. This line of research augments the training set to improve OOD detection. Data augmentations like flipping and cropping generate samples that can be easily classified by a pre-trained classifier. Generative techniques based on VAEs, and GANs try to synthesize data samples near the decision boundary [7,30,32,39,40,45,47]. Other data augmentation strategies do not directly target OOD detection, but domain generalization: SaliencyMix [44], CutOut [6], GridMask [3], AugMix [20], RandomErase [52], PuzzleMix [26], RandAugment [4], SuperMix [5]. Mixup [51] generates new data through convex combination of training samples and labels to improve DNN generalization. CutMix [48] which generates samples by replacing an image region with a patch from another training image. The approach is not directly suitable for OOD detection, as the generated samples lie on the line joining the training samples, and may not cover the large input space [8,24].

3   Proposed Approach

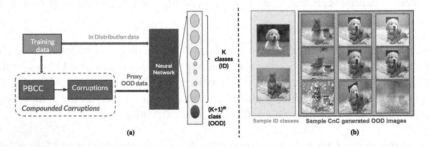

Fig. 1. Creating augmented data samples using Compounded Corruptions (CnC). Pane (a) shows block diagram of the training procedure: first we take a patch based convex combination (PBCC) of patches chosen from image pair belonging to $\binom{K}{2}$ labels; second, we apply corruptions on the data points obtained using PBCC. This proxy OOD data is then used to train a $(K+1)$ way classifier, where, first K classes correspond to the ID classes and $(K+1)^{th}$ class contains synthesized OOD samples corresponding to reject/OOD class. Pane (b) shows CnC synthesized sample images from cat and dog classes. Intuitively, CnC gives two knobs for generating OOD samples: a coarse exploration ability through linear combination of two ID classes achieved through PBCC operation, and a finer warping capability through corruption of these images. The order of the two operations (PBCC before corruption) is important, as we show later.

3.1 Problem Formulation

We consider a training set, $\mathcal{D}_{\text{in}}^{\text{train}}$, consisting of N training samples: $(x_n, y_n)_{n=1}^N$, where samples are drawn independently from a probability distribution: $\mathcal{P}_{X,Y}$. Here, $X \in \mathcal{X}$ is a random variable defined in the image space, and $Y \in \mathcal{Y} = \{1, \ldots, K\}$ represents its label. Traditionally, a classifier $f_\theta : \mathcal{X} \to \mathcal{Y}$ is trained on in-distribution samples drawn from a marginal distribution $\mathcal{P}_X$ of X derived from the joint distribution $\mathcal{P}_{X,Y}$. Let θ refers to model parameters and $\mathcal{Q}_X$ be another distinct data distribution defined on the image space $\mathcal{X}$. During testing phase, input images are drawn from a conditional mixture distribution $\mathcal{M}_{X|Z}$ where $Z \in \{0, 1\}$, such that $\mathcal{M}_{X|Z=0} = \mathcal{P}_X$, and $\mathcal{M}_{X|Z=1} = \mathcal{Q}_X$. We define all $\mathcal{Q}_X \nsim \mathcal{P}_X$ as OOD distributions, and Z is a latent (binary) variable to denote ID if $Z = 0$ and OOD if $Z = 1$.

One possible approach to detecting an OOD sample is if confidence of f_θ for a given input is low for all elements of $\mathcal{Y}$. However, we use an alternative approach where we learn to map OOD samples generated using our technique to an additional label $(K+1)$. Given any two ID samples $x_1, x_2 \sim \mathcal{P}_X$, we generate the synthetic data using the CnC operation $C(x_1, x_2) : \mathcal{X} \times \mathcal{X} \to \mathcal{X}$. We then define an extended label set $\mathcal{Y}^+ = \{1, \ldots, K+1\}$, and train a classifier f_θ^+ over $\mathcal{Y}^+$. The goal is to train f_θ^+ to implicitly build an estimate $\hat{Z}$ of Z, such that the output of f_θ^+ is $(K+1)$ if $\hat{Z} = 1$, and one of the elements of $\mathcal{Y}$ if $\hat{Z} = 0$.

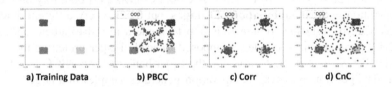

| a) Training Data | b) PBCC | c) Corr | d) CnC |

Fig. 2. Intuition with an illustrative plot of OOD synthesis on a toy dataset with four ID classes. Each sample is in $\mathbb{R}^2$. Consider $\mathbf{p}_1 = (x_1, y_1)$, and $\mathbf{p}_2 = (x_2, y_2)$ to be the two input samples belonging to distinct classes 1 and 2, then $\mathbf{p}_3 = (x_3, y_3)$ is the geometric convex combination of $\mathbf{p}_1$ and $\mathbf{p}_2$ such that: $\mathbf{p}_3 = \lambda\mathbf{p}_1 + (1-\lambda)\mathbf{p}_2$, $0 \le \lambda \le 1$. (a) training data corresponding to 4 distinct classes; Synthesised OOD points are in red; (b) PBCC generates OOD points through a convex combination of ID points from different classes in $\binom{4}{2}$ ways, whereas corruptions depicted in (c) can generate OOD points around each cluster. Observe that points generated by CnC spans wider OOD space including inter-ID-cluster area and outside the convex hull of ID points. (Color figure online)

3.2 Synthetic OOD Data Generation

Our synthetic sample generation strategy consists of following two steps.

Step 1: Patch Based Convex Combination (PBCC). We generate synthetic samples by convex combination of two input images. Let $x \in \mathbb{R}^{W \times H \times C}$, and y denote a training image and its label respectively. Here, W, H, C denote width, height, channels of the image respectively. A new sample, $\tilde{x}$, is generated by a convex combination of two training samples (x_A, y_A), and (x_B, y_B):

$$\tilde{x} = \mathbf{M} \odot x_A + (\mathbf{1} - \mathbf{M}) \odot x_B. \tag{1}$$

Here, x_A and x_B do not belong to a same class $(y_A \neq y_B)$, and $\mathbf{M} \in \{0, 1\}^{W \times H}$ denotes a rectangular binary mask that indicates which region to drop, or use from the two images. $\mathbf{1}$ is a binary mask filled with ones, and $\odot$ is element-wise multiplication. To sample $\mathbf{M}$, we first sample the bounding box coordinates $\mathbf{B} = (r_x, r_y, r_w, r_h)$, indicating the top-left coordinates, and width, and height of the box. The region $\mathbf{B}$ in x_A is cut-out and filled in with the patch cropped from $\mathbf{B}$ of x_B. The coordinates of $\mathbf{B}$ is uniformly sampled according to: $r_x \sim \mathrm{U}(0, W), r_w = W\sqrt{1 - \lambda}$ and similarly, $r_y \sim \mathrm{U}(0, H), r_h = H\sqrt{1 - \lambda}$. Here, $\lambda \in [0, 1]$ denotes the crop area ratio, and is fixed at different values for generating random samples. The cropping mask $\mathbf{M}$ is generated by filling zeros within the bounding box $\mathbf{B}$ and ones outside. We generate the samples by choosing each pair of labels in $\binom{K}{2}$ ways, and then randomly selecting input images corresponding to the chosen labels. This generates OOD samples spread across various inter-class regions in the embedding space. For ablation on range of λ to ensure that a large number of OOD samples are generated outside the ID clusters see supplementary. We label all generated samples as that of the $(K + 1)^{\text{th}}$ reject class.

PBCC and CutMix [48]: Note that PBCC and CutMix [48] both rely on the same basic operation **convex combination of images**, but for two very different objectives. Whereas, CutMix uses the combination step to guide a model to attend on less discriminative parts of objects e.g. leg as opposed to head of a person letting the network generalize better on object detection. On the other hand, we use PBCC as a first step for OOD data generation, where the operation generates samples in a large OOD space between a pair of classes in $\binom{K}{2}$ ways.

PBCC Shortcomings: Note that PBCC performs a convex combination of the two ID images belonging to two distinct classes. Hence, unlike adversarial perturbations, it is able to generate sample points far from the ID points in the RGB space. However, still it can generate samples from only within the convex hull of the ID points corresponding to all classes. Thus, as we show in our ablation studies, sample generated using this step alone are insufficient to train a good OOD detector. Below we show how to improve upon the shortcoming of PBCC.

Step 2: Compounded Corruptions. We aim to address the above shortcomings by using corruptions on top of PBCC generated samples, thus increasing the sample density in inter-class regions as well as generating samples outside the convex hull. We reason that such compounded corruptions increase the spread of the augmented data to a much wider region. Thus, a reasoning based on "per sample" generalisation error bound from [24]: [Fig. 1, Eq. 11] could be utilized for our problem. [24] constructs an input-dependent generalization error bound by analysing the subfunction membership of each input, and show that generalisation error bound improves with smoother training sample density (as defined by number of samples in each region). Intuitively, corruptions over PBCC produces a smoother approximation of ID classes with a finer fit at the ID class boundary. A detailed analysis is given in Sect. 3.3. To give an intuitive understanding, Fig. 2 shows visualizations of the generated OOD samples in red using a 4 class toy dataset in two dimensions.

Hendrycks et al. [17] benchmark robustness of a DNN using 15 algorithmically generated image corruptions that mimic natural corruptions. Each corruption severity ranges from 1 to 5 based on the intensity of corruption, where 5 is most severe. The corruptions can be seen as perturbing a sample point in its local neighborhood, while remaining in the support space of the probability distribution of valid images. We apply these corruptions on the samples generated using PBCC step described earlier. Together, PBCC, and corruptions, allow us to generate a synthetic sample far from, and outside the convex hull of ID samples. At the same time, unlike pure random noise images, the process maintains plausibility of the generated samples. Specifically we apply following corruptions: Gaussian noise, Snow, Fog, Contrast, Shot noise/Poisson noise, Elastic transform, JPEG compression, and blur such as Defocus, Motion etc.

Figure 1 gives a pictorial overview of the overall proposed scheme with a few OOD image samples generated by our approach. CnC formulates the problem as $(K + 1)$ class classification which improves the model representation of underlying distribution, and at the same time improves DNN calibration as seen in Sect. 5.2. Please see Suppl. for the precise steps of our algorithm.

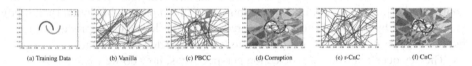

(a) Training Data   (b) Vanilla   (c) PBCC   (d) Corruption   (e) r-CnC   (f) CnC

Fig. 3. Visualization of trained classifiers as a result of OOD augmentation. A ReLU type DNN is trained on the two-dimensional half-moon data set shown in (a). The shattered neural networks [16] show that CnC has the tightest fit around the ID regions, as measured by the area of the (white colored) polytopes in which no training ID point is observed but a network predicts a point in that region as ID. The measured areas for such polytopes are (b) **Vanilla training without data augmentation**: 5.65, (c) **PBCC**: 8.20, (d) **Corruption**: 0.40, (e) **r-CnC**: 5.66, (f) **CnC**: 0.37. Note: [24] state that the more densely supported a polytope is by the training set, the more reliable the network is in that region. Hence, the samples declared ID in the regions where no ID sample is observed may actually be OOD with high probability. We observe that PBCC/r-CnC/Vanilla, all predict ID in many such polytopes. Note: r-CnC we reverse the order of PBCC and corruptions **Best viewed at 200%**

3.3   CnC Analysis via Polyhedral Decomposition of Input Space

While we validate the improved performance of CnC in Sect. 5, in this section we seek to provide a plausible explanation for the CnC's performance. We draw inspiration from theoretical support provided in recent work by [24] who formally derive and empirically test prediction unreliability for ReLU based neural networks.

Consider a ReLU network with n inputs and m neurons in total. [24] show that parameters of a trained model partition the input space into a polyhedral complex (PC) consisting of individual *convex* polytopes (also called *activation regions* in [24]). See Fig. 3 for an example with a 2D input space. Each possible input corresponds to a unique state (active or inactive) of each of the m ReLU neurons, and the interior of each polytope corresponds to a unique combination of states of all m neurons. Thus a trained network behaves linearly in the interior of corresponding polytopes. Each edge in the PC corresponds to the state flip of a single neuron (active to inactive, or vice versa).

For the purpose of classification based on the final layer activation, a key corollary from [24] is that *the decision boundary between two classes must be a straight line within a polytope, and can only turn at the vertices*. This is an immediate consequence of the observation that the decision boundary is the locus along which the two highest activations (most probable labels) in the output layer remain equal to each other. This implies that smaller polytopes near the decision boundary are needed for finer control over the boundary between training samples from different classes. Note also that the authors in [24]: [Eq. (11)] infer that (paraphrased) "the more a subfunction (polytope) is surrounded by samples with accurate predictions, the lower its empirical error and bound on generalization gap, and thus the lower its expected error bound".

The key question from OOD detection perspective is, how do we force a network to create tighter polytopes at the ID class decision boundaries? We believe the answer is to distribute a large number of the augmented samples (over which we have control) with contrasting OOD and ID labels all around each ID region, forcing the decision boundary to form a tight bounding surface. At the same time, we must also retain a good fraction of the augmented samples in the open space between ID classes, which can be covered by relatively large polytopes (recall that the maximum number of polytopes is bounded by the number of neurons, and thus small polytopes in one region may need to be traded off by larger polytopes in another region). Neglecting the inter-ID space entirely would run the risk of creating very large polytopes in this region, which increases the empirical error bound ([24]: [Eqs. (5) and (11), large subfunctions have low probability mass and hence higher error bound. Refer Supplementary for further details.]. CnC lets us achieve this dual objective by using compounding to sample the space between ID classes, and corruption to pepper the immediate neighborhoods around ID classes (especially for λ values near 0 and 1).

In Fig. 3, we show polyhedral complex corresponding to the DNN models trained on two-dimensional half-moon dataset [16,25], and OOD samples generated using various techniques. The first plot shows the input space with training samples from two ID classes (green and yellow semicircles). The learnt polytope structure for vanilla uses a neural network of size $[2, 32, 32, 2]$, while the remaining three plots use $[2, 32, 32, 3]$ (with an additional *reject*/OOD class).

Recall from Fig. 2 that PBCC produces samples sparsely between the ID classes, but not around the ID class boundaries. Pure corruptions produce samples only near and on ID classes, but not in the inter-ID space. On the other hand, CnC produces samples both near the ID boundaries as well as in the inter-ID space. In Fig. 3, we define any polytope that is fully or partially (decision boundary crosses through it) classified as ID, as an "ID classified polytope" and mark it in white color. *Visually, we can see that the white polytopes occupy a smaller total area when we compare Vanilla to CnC, with the actual values noted in the caption. This indicates that the CnC produces the tightest approximation of OD classes in our example, which in turn leads to better OOD detection.* Though we show for two-dimensional data, we posit that the same generalizes to higher dimensional input data as well, and is the reason for success of CnC based OOD detection.

CnC and Robustness to Adversarial Attacks: Note that, small polytopes in the input space partitioned by a DNN may also provide better safety against black box adversarial attacks as suggested by [16,25]. This is because the black box adversarial attacks extrapolate the gradients based upon a particular test sample. Since the linearity of the output, and thus the gradients is only valid inside a polytope, smaller polytopes near the ID or in the OOD region makes it difficult for an adversary to extrapolate an output to a large region. However, since adversarial robustness is not the focus of this paper, we do not further explore this direction.

3.4 Training Procedure

We train a $(K + 1)$ class classifier network f_θ^+, where first K classes correspond to the multi-classification ID classes, and the $(K + 1)^{th}$ class label indicates the OOD class. Our training objective takes the form:

$$\mathcal{L} = \underset{\theta}{\text{minimize}} \quad \mathbb{E}_{(x,y)\sim D_{\text{in}}^{\text{train}}}[\mathcal{L}_{\text{CE}}(x, y; f_\theta^+(x))]$$
$$+ \alpha \cdot \mathbb{E}_{(x,y)\sim D_{pbcc}^{corr}}[\mathcal{L}_{\text{CE}}(x, K + 1; f_\theta^+(x))], \tag{2}$$

where $\mathcal{L}_{\text{CE}}$ is the cross entropy loss, $f_\theta^+(x)$ denotes the softmax output of neural network for an input sample x. We use $\alpha = 1$ in our experiments based on the ablation study reported in the supplementary material. For above experiments setup we set the ratio of IID:OOD training points as $1 : 1$.

3.5 Inference

After training, we obtain a trained model F^+. We use $F^+(x)[K + 1]$ as the OOD score of x during testing, and define an OOD detector $D(x)$ as:

$$D(x) = \begin{cases} 0, & \text{if } F^+(x)[K + 1] > \delta \\ 1, & \text{if } F^+(x)[K + 1] \leq \delta \end{cases} \tag{3}$$

where, $D(x) = 0$ indicates an OOD prediction, and $D(x) = 1$ implies an ID sample prediction. δ is a threshold such that TPR, i.e., fraction of ID images correctly classified as ID is 95%. For images which are characterized as ID by $D(x)$, the labels are given as:.

$$\hat{y} = \arg\max_{i \in 1,\dots,K} F^+(x)_i \tag{4}$$

4 Dataset and Evaluation Methodology

In-Distribution Datasets: For ID samples, we use SVHN (10 classes) [38], CIFAR-10 (10 classes), CIFAR-100 (100 classes) [27] containing images of size 32 × 32. We also use TinyImageNet (200 classes) [29] containing images of resolution 64 × 64 images. Out-of-Distribution Datasets: For comparison, we use the following OOD datasets: TinyImageNet-crop (TINc), TinyImageNet-resize (TINr), LSUN-crop (LSUNc), LSUN-resize (LSUNr), iSUN, SVHN. Evaluation Metrics: We compare the performance of various approaches using TNR@TPR95, AUROC and Detection Error. See Suppl. for description on evaluation metrics.

5 Experiments and Results

To show that our data augmentation is effective across different feature extractors, we train using both DenseNet-BC [23] and ResNet-34 [14]. DenseNet has 100 layers with growth rate of 12. WideResNet [49] models have the same training configuration as [35].

5.1 Comparison with State-of-the-art

OOD Detection Performance: Table 1 shows comparison of CnC with recent state-of-the-art. The numbers indicate averaged OOD detection performance on 6 datasets as mentioned in Sect. 4 (TinyImagenet, TinyImageNet-crop (TINc), TinyImageNet-resize (TINr), LSUN-crop (LSUNc), LSUN-resize (LSUNr), iSUN, SVHN) with more details included in the supplementary. We would like to emphasize that CnC does not need any validation OOD data for fine-tuning. But ODIN [33] and Mahalanobis [31] require OOD data for fine-tuning the hyper-parameters; the hyper-parameters for ODIN and Mahalanobis methods [31,33] are set by validating on 1K images randomly sampled from the test set $\mathcal{D}_{\text{in}}^{\text{test}}$. Table 1 clearly shows that CnC outperforms the existing methods.

Comparison with Other Data Generation Methods: Table 2 shows how CnC fairs against recent OOD data generation methods. In each case we train a $(K + 1)$ way classier where first K classes correspond to ID and $(K + 1)^{\text{th}}$ class comprised of OOD data generated by corresponding method. As seen from the table, CnC outperforms the recent data augmentation schemes.

Table 1. Comparison of competing OOD detectors. TIN: TinyImageNet, and RN50: ResNet50, WRN: WideResNet-40-2 Values are averaged over all OOD benchmark datasets. We give individual dataset-wise results in the supplementary. Note that ATOM [2], and OE [19] require large image datasets like 80-Million Tiny Images [43] as representative of OOD samples. However, CnC synthesises its own OOD dataset using the ID training data. CnC models were trained using the same configuration as defined by OE [19] and EBO [35] paper, with the exception that CnC did not use any external auxiliary OOD dataset like [43] in training. CnC reasults are averaged on 3 evaluation runs.

$\mathcal{D}_{\text{in}}^{\text{train}}$	Method	TNR@TPR95 ↑	AUROC ↑	DetErr ↓	ID Acc. ↑
CIFAR-10 DenseNet-BC	MSP (ICLR'17) [18]	56.1	93.5	12.3	95.3
	ODIN (ICLR'18) [33]	92.4	98.4	5.8	95.3
	Maha (NeurIPS'18) [31]	83.9	93.5	10.2	95.3
	Gen-ODIN (CVPR'20) [22]	94.0	98.8	5.4	94.1
	Gram Matrices (ICML'20) [41]	96.4	99.3	3.6	95.3
	ATOM (ECML'21) [2]	98.3	99.2	1.2	94.5
	CnC (Proposed)	**98.4 ± 0.8**	**99.5 ± 1.2**	**2.7 ± 0.2**	94.7
CIFAR-100 DenseNet-BC	MSP (ICLR'17) [18]	21.7	75.2	31.4	77.8
	ODIN (ICLR'18) [33]	61.7	90.6	16.7	77.8
	Gen-ODIN (CVPR'20) [22]	86.5	97.4	8.0	74.6
	Maha (NeurIPS'18) [31]	68.3	92.8	13.4	77.8
	Gram Matrices(ICML'20) [41]	88.8	97.3	7.3	77.8
	ATOM (ECML'21) [2]	67.7	93	5.6	75.9
	CnC (Proposed)	**97.1 ± 1.4**	**98.5 ± 0.4**	**4.6 ± 0.6**	76.8
TIN RN50	MSP (ICLR'17) [18]	53.15	85.3	22.1	57.0
	ODIN (ICLR'18) [33]	68.5	93.7	12.3	57.0
	CnC (Proposed)	**97.8 ± 0.8**	**99.6 ± 0.2**	**2.1 ± 0.2**	60.5
C-10 WRN	OE (ICLR'19) [19]	93.23	98.64	5.32	94.8
	EBO (NeurIPS'20) [35]	**96.7**	99.0	**3.83**	95.2
	CnC (Proposed)	96.2 ± 1.5	**99.02 ± 0.1**	4.5 ± 0.8	94.3
C-100 WRN	OE (ICLR'19) [19]	47.35	86.02	21.24	75.6
	EBO (NeurIPS'20) [35]	54.0	86.65	19.7	75.7
	CnC (Proposed)	**97.6 ± 0.9**	**99.5 ± 0.1**	**2.2 ± 0.3**	75.1

Table 2. Comparison with other synthetic data generation methods. We consider CIFAR10 as ID. The values are averaged over all OOD benchmarks. We have used DenseNet [23] as the architecture for all methods trained for $(K + 1)$ class classification. Samples obtained through the listed data augmentation schemes were assumed to be of $(K+1)^{\text{th}}$ class. Observe that CnC has superior OOD detection performance. We report average and standard deviation of CnC trained models computed over 3 runs.

Data augmentation methods	TNR (95% TPR) ↑	AUROC ↑	Detection Err ↓
Mixup (ICLR'18) [51]	60.6	90.9	15.5
CutOut (arXiV'17) [6]	80.8	94.8	10
CutMix (ICCV'19) [48]	83.2	92.7	8.6
GridMask (arXiV'20) [3]	50.3	79.1	23.6
SaliencyMix (ICLR'21) [44]	85.3	95.7	8.0
AugMix (ICLR'20) [20]	81.3	94.6	11.2
RandomErase (AAAI'20) [52]	41.9	68.1	24.2
Corruptions (ICLR'19) [17]	98.0	99.4	2.8
PuzzleMix (ICML'20) [26]	66.8	84.1	15.2
RandAugment (NeurIPS'20) [4]	89.5	97.9	4.7
Fmix (ICLR'21) [13]	73	90.3	12.6
Standard Gaussian Noise	71.5	93.2	11.7
CnC (Proposed)	**98.4 ± 0.8**	**99.5 ± 1.2**	**2.7 ± 0.2**

Table 3. Detecting domain shift using CnC. A model trained with CnC data on CIFAR-100 as the ID using DenseNet-BC [23] feature extractor can successfully detect the domain shift when observing ImageNet-R at the test time.

Method	TNR@0.95TPR	AUROC	DetErr
MSP (ICLR'17) [18]	24.4	80.1	26.5
ODIN (ICLR'18) [33]	46.0	88.6	18.9
Gen-ODIN (CVPR'20) [22]	45.0	88.7	18.8
Mahalanobis (NeurIPS'18) [31]	14.0	56.2	41.6
Gram Matrices (ICML'20) [41]	35.0	81.5	25.8
CnC (Proposed)	**60.0**	**91.6**	**15.7**

Table 4. Using entropy/diversity of synthesized data to predict quality of OOD detection. Please refer to text for more details.

Method	TNR@ 0.95TPR	AUROC	DetErr	Mean diversity	Mean entropy
	↑	↑	↓	↑	↑
PBCC	93.7	98.6	6.2	2.30	0.33
Corruptions	95.5	97.4	3.5	2.68	0.38
CnC	**98.3**	**99.6**	**2.6**	**3.40**	**0.80**

5.2   Other Benefits of CnC

Detecting Domain Shift as OOD: We analyze if a model trained with CnC augmented data can detect non-semantic domain shift, i.e. images with the same label but different distribution. For the experiments we use a model trained using CIFAR-100 as ID, and ImageNet-O/ImageNet-R/Corrupted-ImageNet [21] as the OOD. While testing, we downsample the images from ImageNet-O, ImageNet-R and TinyImageNet-C to a size of 32 × 32. Table 3 shows results on ImageNet-R OOD dataset. We outperform the next best technique by 14% on TNR@0.95TPR, 2.9% in AUROC, 3.1% in detection error. See supplementary for results on ImageNet-O and Corrupted ImageNet.

Model Calibration. Another benefit of training with CnC is model calibration on ID data as well. A classifier is said to be calibrated if the confidence probabilities matches the empirical frequency of correctness [12,15], hence a crucial to measure of trust in classification models. Tables in the supplementary show the calibration error for a model trained on CIFAR-10, and CIFAR-100 as the ID data, with CnC samples as the $(K + 1)^{\text{th}}$ class. Note that the calibration error is measured only for the ID test samples. We compare the error for a similar model, trained using only ID train data, and calibrated using temperature scaling (TS) [12].

Time Efficiency. For applications demanding real-time performance, it is crucial to have low latency in systems using DNN for inference. Supplementary reports the competative performance of our method.

5.3   Ablation Studies

Rationale for Design choice of K vs. (K+1) Classifier. We empirically verify having a separate class helps in better optimization/learning during training a model using CnC augmentation. Figure 4 shows the advantages of using a $(K + 1)$ way classifier as compared to standard K class training with better ID-OOD separation. Supplementary material details the advantage of CnC with ACET [16] (CVPR'19) for uncertainty quantification on a half-moon dataset.

Recommendation for a Good OOD Detector. We performed detailed comparison of various configurations of our technique to understand the quantitative scores which can predict the quality of an OOD detector. For the experiment we keep the input images used same across configs, PBCC and corruptions applied are also fixed to remove any kind of randomness. We use ResNet34 as feature extractor for all methods. CIFAR-10 is used as ID dataset and TinyImageNet-crop as OOD dataset. We observe that the quality of OOD detection improves as the diversity, and entropy of the synthesized data increases (Table 4). Here, entropy is computed as the average entropy of the predicted probability vectors by the K class model for the synthesized data. We adapt data diversity from Zhang et al. [50] to measure diversity of OOD data. Refer supplementary for Algorithm for diversity computation.

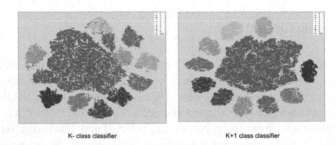

K- class classifier   K+1 class classifier

Fig. 4. We show sample t-SNE plots for K Vs. $(K + 1)$ classifiers, where CIFAR-10 is used as ID and SVHN is used as OOD(marked in red). The K-class classifier uses temperature scaling (TS) [12], where T is tuned on SVHN test set. On the other hand, the $(K + 1)$ class classifier uses SVHN data for $(K + 1)^{\text{th}}$ class during training. The visualization shows that the OOD data (marked in red) is better separated in a $(K+1)$-class classifier as compared to a K-class classifier (Color figure online)

Limitations of CnC Data Augmentation: Introduction of additional synthetic data indeed increases training time. For e.g., training a model with CnC data on TinyIm-ageNet dataset takes 10 min 23 s/epoch, whereas without CnC data it takes 5 min 30 s/epoch on the same Nvidia V100 GPU. Performance gain the overhead of training time can be discounted as inference time remains same. We assume the absence of adversarial intentions in this approach, Our method fails when tested against L_∞ norm bounded perturbed image. In future we intend to look at OOD detection using CnC variants for non-visual domains.

6   Conclusions

We have introduced Compounded Corruptions(CnC), a novel data augmentation technique for OOD detection in image classifiers. CnC outperforms all the SOTAOOD detectors on standard benchmark datasets tested upon. The major benefit of CnC over SOTA is absence of OOD exposure requirement for training or validation. We also show additional results for robustness to distributional drift, and calibration for CnC trained models. CnC requires just one inference pass at the test time, and thus has much faster inference time compared to SOTA. Finally, we also recommend high diversity and entropy of the synthesized data as good measures to predict quality of OOD detection using it.

Acknowledgements. Thanks to Subhashis Banerjee, Rahul Narain, Lokender Tiwari, Sandeep Subramanian for insightful comments.

References

1. Bendale, A., Boult, T.E.: Towards open set deep networks. In: Proceedings of the IEEE CVPR, pp. 1563–1572 (2016)
2. Chen, J., Li, Y., Wu, X., Liang, Y., Jha, S.: ATOM: robustifying out-of-distribution detection using outlier mining. In: Oliver, N., Pérez-Cruz, F., Kramer, S., Read, J., Lozano, J.A. (eds.) ECML PKDD 2021. LNCS (LNAI), vol. 12977, pp. 430–445. Springer, Cham (2021). https://doi.org/10.1007/978-3-030-86523-8_26
3. Chen, P., Liu, S., Zhao, H., Jia, J.: GridMask data augmentation (2020)
4. Cubuk, E.D., Zoph, B., Shlens, J., Le, Q.: RandAugment: practical automated data augmentation with a reduced search space. In: Larochelle, H., Ranzato, M., Hadsell, R., Balcan, M.F., Lin, H. (eds.) Advances in NeurIPS (2020)
5. Dabouei, A., Soleymani, S., Taherkhani, F., Nasrabadi, N.M.: SuperMix: supervising the mixing data augmentation. In: Proceedings of the IEEE/CVF CVPR, pp. 13794–13803 (2021)
6. DeVries, T., Taylor, G.W.: Improved regularization of convolutional neural networks with cutout (2017)
7. Du, X., Wang, Z., Cai, M., Li, Y.: VOS: learning what you don't know by virtual outlier synthesis. In: ICLR (2022)
8. Fang, Z., Lu, J., Liu, A., Liu, F., Zhang, G.: Learning bounds for open-set learning. In: International Conference on Machine Learning, pp. 3122–3132. PMLR (2021)
9. Gal, Y., Ghahramani, Z.: Dropout as a Bayesian approximation: representing model uncertainty in deep learning. In: ICML (2016)
10. Ge, Z., Demyanov, S., Chen, Z., Garnavi, R.: Generative OpenMax for multi-class open set classification. In: BMVC (2017)
11. Graves, A.: Practical variational inference for neural networks. In: Shawe-Taylor, J., Zemel, R.S., Bartlett, P.L., Pereira, F., Weinberger, K.Q. (eds.) Advances in Neural Information Processing Systems, vol. 24, pp. 2348–2356 (2011)
12. Guo, C., Pleiss, G., Sun, Y., Weinberger, K.Q.: On calibration of modern neural networks. In: ICML, pp. 1321–1330 (2017)
13. Harris, E., Marcu, A., Painter, M., Niranjan, M., Prügel-Bennett, A., Hare, J.: FMix: enhancing mixed sample data augmentation (2021)
14. He, K., Zhang, X., Ren, S., Sun, J.: Identity mappings in deep residual networks. In: Leibe, B., Matas, J., Sebe, N., Welling, M. (eds.) ECCV 2016. LNCS, vol. 9908, pp. 630–645. Springer, Cham (2016). https://doi.org/10.1007/978-3-319-46493-0_38
15. Hebbalaguppe, R., Prakash, J., Madan, N., Arora, C.: A stitch in time saves nine: a train-time regularizing loss for improved neural network calibration. In: Proceedings of the IEEE/CVF CVPR, pp. 16081–16090 (2022)
16. Hein, M., Andriushchenko, M., Bitterwolf, J.: Why ReLU networks yield high-confidence predictions far away from the training data and how to mitigate the problem. In: Proceedings of the IEEE/CVF CVPR, pp. 41–50 (2019)
17. Hendrycks, D., Dietterich, T.: Benchmarking neural network robustness to common corruptions and perturbations. In: ICLR (2018)
18. Hendrycks, D., Gimpel, K.: A baseline for detecting misclassified and out-of-distribution examples in neural networks. In: ICLR (2017)
19. Hendrycks, D., Mazeika, M., Dietterich, T.: Deep anomaly detection with outlier exposure. In: International Conference on Learning Representations (ICLR) (2019)

20. Hendrycks*, D., Mu*, N., Cubuk, E.D., Zoph, B., Gilmer, J., Lakshminarayanan, B.: Aug-Mix: a simple method to improve robustness and uncertainty under data shift. In: International Conference on Learning Representations (2020)
21. Hendrycks, D., Zhao, K., Basart, S., Steinhardt, J., Song, D.: Natural adversarial examples. In: Proceedings of the IEEE/CVF CVPR, pp. 15262–15271 (2021)
22. Hsu, Y.C., Shen, Y., Jin, H., Kira, Z.: Generalized ODIN: detecting out-of-distribution image without learning from out-of-distribution data. In: Proceedings of the IEEE/CVF CVPR, pp. 10951–10960 (2020)
23. Huang, G., Liu, Z., Van Der Maaten, L., Weinberger, K.Q.: Densely connected convolutional networks. In: Proceedings of the IEEE Conference on Computer Vision and Pattern Recognition, pp. 4700–4708 (2017)
24. Ji, X., Pascanu, R., Hjelm, R.D., Vedaldi, A., Lakshminarayanan, B., Bengio, Y.: Predicting unreliable predictions by shattering a neural network. CoRR abs/2106.08365 (2021). https://arxiv.org/abs/2106.08365
25. Jordan, M., Lewis, J., Dimakis, A.G.: Provable certificates for adversarial examples: fitting a ball in the union of polytopes. In: 33rd Conference on Neural Information Processing Systems (NeurIPS) (2019)
26. Kim, J.H., Choo, W., Song, H.O.: Puzzle mix: exploiting saliency and local statistics for optimal mixup. In: ICML (2020)
27. Krizhevsky, A., Hinton, G., et al.: Learning multiple layers of features from tiny images (2009)
28. Lakshminarayanan, B., Pritzel, A., Blundell, C.: Simple and scalable predictive uncertainty estimation using deep ensembles. In: Advances in NeurIPS, pp. 6402–6413 (2017)
29. Le, Y., Yang, X.: Tiny ImageNet visual recognition challenge. CS 231N 7(7), 3 (2015)
30. Lee, K., Lee, H., Lee, K., Shin, J.: Training confidence-calibrated classifiers for detecting out-of-distribution samples. arXiv preprint arXiv:1711.09325 (2017)
31. Lee, K., Lee, K., Lee, H., Shin, J.: A simple unified framework for detecting out-of-distribution samples and adversarial attacks. In: Advances in NeurIPS, vol. 31 (2018)
32. Li, D., Chen, D., Goh, J., Ng, S.K.: Anomaly detection with generative adversarial networks for multivariate time series. In: ACM KDD (2018)
33. Liang, S., Li, Y., Srikant, R.: Enhancing the reliability of out-of-distribution image detection in neural networks. In: ICLR (2018)
34. Lin, Z., Roy, S.D., Li, Y.: Mood: multi-level out-of-distribution detection. In: Proceedings of the IEEE/CVF CVPR, pp. 15313–15323 (2021)
35. Liu, W., Wang, X., Owens, J., Li, Y.: Energy-based out-of-distribution detection. In: Advances in Neural Information Processing Systems (NeurIPS) (2020)
36. Mohseni, S., Pitale, M., Yadawa, J., Wang, Z.: Self-supervised learning for generalizable out-of-distribution detection. In: AAAI, pp. 5216–5223, April 2020
37. Neal, L., Olson, M., Fern, X., Wong, W.-K., Li, F.: Open set learning with counterfactual images. In: Ferrari, V., Hebert, M., Sminchisescu, C., Weiss, Y. (eds.) ECCV 2018. LNCS, vol. 11210, pp. 620–635. Springer, Cham (2018). https://doi.org/10.1007/978-3-030-01231-1_38
38. Netzer, Y., Wang, T., Coates, A., Bissacco, A., Wu, B., Ng, A.Y.: Reading digits in natural images with unsupervised feature learning. In: NeurIPS 2021 (2011)
39. Perera, P., Nallapati, R., Xiang, B.: OCGAN: one-class novelty detection using GANs with constrained latent representations. In: IEEE/CVF CVPR (2019)
40. Ramírez Rivera, A., Khan, A., Bekkouch, I.E.I., Sheikh, T.S.: Anomaly detection based on zero-shot outlier synthesis and hierarchical feature distillation. IEEE Trans. Neural Netw. Learn. Syst. 33(1), 281–291 (2022)
41. Sastry, C.S., Oore, S.: Detecting out-of-distribution examples with gram matrices. In: International Conference on Machine Learning, pp. 8491–8501. PMLR (2020)

42. Simonyan, K., Zisserman, A.: Very deep convolutional networks for large-scale image recognition. arXiv preprint arXiv:1409.1556 (2014)
43. Torralba, A., Fergus, R., Freeman, W.T.: 80 million tiny images: a large data set for nonparametric object and scene recognition. IEEE Trans. PAMI **30**, 1958–1970 (2008)
44. Uddin, S., Monira, M.S., Shin, W., Chung, T., Bae, S.H.: SaliencyMix: a saliency guided data augmentation strategy for better regularization. In: ICLR (2021)
45. Wang, W., Wang, A., Tamar, A., Chen, X., Abbeel, P.: Safer classification by synthesis. arXiv preprint arXiv:1711.08534 (2017)
46. Wen, Y., Vicol, P., Ba, J., Tran, D., Grosse, R.: FlipOut: efficient pseudo-independent weight perturbations on mini-batches. In: ICLR (2018)
47. Xiao, Z., Yan, Q., Amit, Y.: Likelihood regret: an out-of-distribution detection score for variational auto-encoder. In: Advances in Neural Information Processing Systems (2020)
48. Yun, S., Han, D., Oh, S.J., Chun, S., Choe, J., Yoo, Y.: CutMix: regularization strategy to train strong classifiers with localizable features. In: Proceedings of the IEEE/CVF ICCV, pp. 6023–6032 (2019)
49. Zagoruyko, S., Komodakis, N.: Wide residual networks. In: BMVC (2016)
50. Zhang, C., Öztireli, C., Mandt, S., Salvi, G.: Active mini-batch sampling using repulsive point processes. In: AAAI, vol. 33, pp. 5741–5748 (2019)
51. Zhang, H., Cisse, M., Dauphin, Y.N., Lopez-Paz, D.: MixUp: beyond empirical risk minimization. In: ICLR (2018)
52. Zhong, Z., Zheng, L., Kang, G., Li, S., Yang, Y.: Random erasing data augmentation. In: AAAI, vol. 34, pp. 13001–13008 (2020)

Charge Own Job: Saliency Map and Visual Word Encoder for Image-Level Semantic Segmentation

Yuhui Guo⬤, Xun Liang(✉), Hui Tang, Xiangping Zheng, Bo Wu,
and Xuan Zhang

Renmin University of China, Beijing, China
{yhguo,xliang,huitang,xpzheng,wubochn,zhangxuanalex}@ruc.edu.cn

Abstract. Significant advances in weakly-supervised semantic segmentation (WSSS) methods with image-level labels have been made, but they have several key limitations: incomplete object regions, object boundary mismatch, and co-occurring pixels from non-target objects. To address these issues, we propose a novel joint learning framework, namely **S**aliency **M**ap and **V**isual **W**ord **E**ncoder (**SMVWE**), which employs two weak supervisions to generate the high-quality pseudo labels. Specifically, we develop a visual word encoder to encode the localization map into semantic words with a learnable codebook, making the network generate localization maps containing more semantic regions with the encoded fine-grained semantic words. Moreover, to obtain accurate object boundaries and eliminate co-occurring pixels, we design a saliency map selection mechanism with the pseudo-pixel feedback to separate the foreground from the background. During joint learning, we fully utilize the cooperation relationship between semantic word labels and saliency maps to generate high-quality pseudo-labels, thus remarkably improving the segmentation accuracy. Extensive experiments demonstrate that our proposed method better tackles above key challenges of WSSS and obtains the state-of-the-art performance on the PASCAL VOC 2012 segmentation benchmark.

Keywords: Weakly-supervised semantic segmentation · Saliency map · Visual word encoder · Pseudo labels

1 Introduction

Semantic segmentation aims to predict pixel-wise classification results on images, which is one important and challenging task of computer vision. With the development of deep learning, a variety of Convolutional Neural Network (CNN) based semantic segmentation methods [7,8] have achieved promising successes. However, they require a large number of training images annotated with pixel-level labels, which is both expensive and time-consuming. Thus, various weakly supervised semantic segmentation (WSSS) methods have attracted increasing interest of researchers. Most existing WSSS studies adopt image-level labels as the weak supervision of the segmentation model, in which a segmentation network is trained on images with less comprehensive annotations that are cheaper

© The Author(s), under exclusive license to Springer Nature Switzerland AG 2023
M.-R. Amini et al. (Eds.): ECML PKDD 2022, LNAI 13715, pp. 546–561, 2023.
https://doi.org/10.1007/978-3-031-26409-2_33

to obtain than pixel-level labels. The image-level WSSS methods usually perform semantic segmentation through generated pseudo-labels as weak supervision. In general, using a classification network to generate class activation maps (CAM) [46] containing object localization maps, which can be as initial pseudo labels to achieve the semantic segmentation performance [4,35]. However, the classification network has the ability to classifity, which does not locate the integral extents of target objects, leading to the generated CAM that typically only cover the most discriminative parts of target objects. Thus, during the process of producing pseudo labels, WSSS will be confronted with the following key challenges: i) the extents of the target objects can not be covered completely [46], ii) the localization map is unable to obtain accurate object boundaries [22], and iii) the localization map contains co-occurring pixels between target objects and the background [23]. These three aspects in pseudo labels are important to the final semantic segmentation performance [4,35].

Recently, many WSSS methods have been proposed to focus on tackling these issues. According to different issues, existing methods can be divided into three categories. To address the incomplete object region issue of pseudo-labels, researchers utilize the pixel-affinity based strategy [1,2] or erasing strategy [10,22,25] to enlarge the receptive field and discover more discriminative parts for target objects. However, they only focus on the object coverage extents, and neglect that accurate object boundaries are benefit for semantic annotation. Thus, in order to address the object boundary mismatch issue, researchers propose to use the idea of explicitly exploring object boundaries from training images [9,13] to keep coincidence of segmentation and boundaries. Due to some co-occurring pixels exist in between the foreground and the background [11], these methods still lack of the clue to explore the correlation between the foreground and the background, thus they are unable to correctly separate the foreground from the background. In order to alleviate the co-occurring pixels issue between the foreground and the background, most existing WSSS methods use the saliency map [15,19,23,26,34,36–38] to induce processing the background, reducing the computation burden of the segmentation model and helping the segmentation model distinguish coincident pixels of non-target objects from a target object. However, these WSSS methods directly utilize the saliency maps from off-the-shelf saliency detection models as the clue of co-occurring pixels, which is easy to separate the foreground from the background, but such a way is not beneficial to that the segmentation model generates self-saliency maps, leading to a not end-to-end manner training process.

In this paper, our goal is to overcome these challenges of WSSS with image-level labels by improving the performance of the localization map generated by the classification network. For this purpose, we propose a novel joint learning method for WSSS, namely saliency map and visual word encoder (SMVWE), to simultaneously learn semantic word labels and saliency maps. As shown in Fig. 1, we design a visual word encoder to help the classification network learn the semantic word labels, leading to that the generated localization map could cover more integral semantic extents of target objects. Due to the image-level

WSSS task is unable to directly use the semantic word labels, we use an unsupervised way to generate their vector representations in each forward pass, i.e., each semantic word in a trainable codebook utilizes the manhattan distance to encode the feature maps from the classification network. In such a way, it alleviates the sparse object region problem, but does not separate their boundaries from the background effectively. Thus, we design a saliency map selection mechanism to address inaccurate object boundaries and co-occurring pixels among objects, where the saliency maps from off-the-shelf saliency detection models are used as pseudo-pixel feedback. Specifically, the classification network based on image-level labels performs semantic segmentation for L target object classes and one background class, thus generating L foreground localization maps and one background localization map to represent the saliency maps. To obtain accurate object boundaries and discard the co-occurring pixels, we compare our generated saliency maps with off-the-shelf groundtruth saliency maps by a saliency loss, producing more effective saliency maps to improve the quality of final pseudo labels. Moreover, we also use the multi-label classification losses containing the image-level label prediction and the semantic word label prediction, which combine with the saliency loss to optimize our proposed model, thus generating higher-quality pseudo-labels for training the semantic segmentation network.

In summary, our main contributions are three folds:

- We propose a novel joint learning framework for WSSS, namely saliency map and visual word encoder (SMVWE), which learns from pseudo-pixel feedback by combining two weak supervisions, thereby effectively preventing the localization map from producing wrong attention regions.
- We develop a visual word encoder to generate semantic word labels. By enforcing the classification network to learn the generated semantic word labels, more object extents could be discovered, thus alleviating the sparse object region problem.
- We design a saliency map selection mechanism to separate the foreground from the background, which could capture precise object boundaries and discard co-occurring pixels of non-target objects, remarkably improving the quality of pseudo-labels for training semantic segmentation networks.

2   Related Work

2.1   Weakly-Supervised Semantic Segmentation

Existing weakly-supervised semantic segmentation methods using image-level labels mainly focus on two types of algorithms, including single- and multi-stage methods. Single-stage methods [17,27,30,31] could achieve the semantic segmentation of images through a high-speed and simple end-to-end process. For example, RRM [43] proposes an end-to-end network to mine reliable and tiny regions and use them as ground-truth labels, then combining a dense energy loss to optimize the segmentation network. SSSS [3] adopts local consistency, semantic fidelity, and completeness as guidelines, proposing a segmentation-based network and a self-supervised training scheme to solve the sparse object region problem for WSSS. Though these methods are effective for semantic segmentation, they barely achieve high-quality pseudo-labels to improve the segmentation accuracy.

Moreover, existing multi-stage methods generally perform the following three steps: (i) generate an initial localization map to localize the target objects; (ii) improve the initial localization map as the pseudo labels; and (iii) using generated pseudo-labels to train the segmentation network. Recently, many approaches [19,23,34] are devoted to alleviate the incomplete object region problem during generating pseudo-labels process. For example, adversarial erasing methods [18,36] help the classification network learn non-salient regions features and expand activation maps through erasing the most discriminative part of CAMs. Instead of using the erasing scheme, SEAM [35] proposes the consistency regularization on generated CAMs from various transformed images, and designs a pixel correlation module to exploit the context appearance information, leading to further improvement on CAMs consistency for semantic segmentation. ScE [4] proposes to iteratively aggregate image features, helping the network learn non-salient object parts, hence improving the quality of the initial localization maps. To improve the network training, MCOF [34] mines common object features from the initial localization and expands object regions with the mined features, then using saliency maps to refine the object regions as supervision to train the segmentation network. Similarly, the DSRG approach [19] proposes to train a semantic segmentation network starting from the discriminative regions and progressively increase the pixel-level supervision using the seeded region growing strategy. Moreover, MCIS [32] proposes to learn the cross-image semantic relations to mine the comprehensive object pattern and uses the co-attention to exploit context from other related images, thus improving localization maps to benefit the semantic segmentation learning. In this work, we also focus on semantic segmentation with image-level supervision and aim to improve the quality of the initial pseudo labels.

2.2 Saliency Detection

Saliency detection (SD) methods generate the saliency map that separates the foreground objects from the background in an image, which is benefit for many computer vision tasks. Most existing WSSS [15,26,36–38] methods have greatly benefited from SD that exploits the saliency map as the background cues of pseudo-labels. For example, the MDC method [38] uses CAMs of a classification network with different dilated convolutional rates to find object regions, and uses saliency maps to find background regions for training a segmentation model. STC [37] trains an initial segmentation network using the saliency maps of simple images, and uses the image-level annotations as supervision information to improve the initial segmentation network. Moreover, some methods [5,40] integrate class-agnostic saliency priors into the attention mechanism and utilize class-specific attention cues as an additional supervision to boost the segmentation performance. SSNet [42] jointly solves WSSS and SD using a single network, and makes full use of segmentation cues from saliency annotations to improve

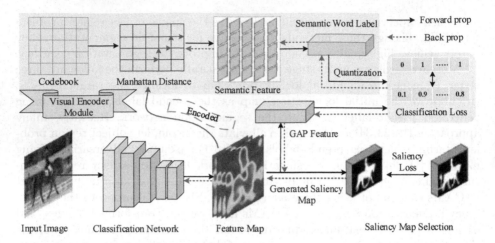

Fig. 1. Overview of the proposed method. We develop a visual encoder module to encode the feature map from the classification network into semantic words with a learnable codebook, covering more object regions. Moreover, we design a saliency map selection mechanism to separate the foreground from the background. The proposed model is jointly trained based on the classification loss and the saliency loss.

the segmentation performance. Different from these saliency-guided methods, our SMVWE method generates self-saliency maps using localization maps and utilizes off-the-shelf saliency maps as their pseudo-pixel feedback, while most existing methods directly use the off-the-shelf saliency map to guide the generation process of the pseudo labels, which is not benefit to tackle the co-occurring pixel problem.

3   Proposed Method

3.1   Motivation

Our SMVWE mainly focus on these two comprehensive information containing the target object location from the localization map and the boundary information from the saliency map. Firstly, we explore more fine-grained labels in the training procedure, namely semantic word labels, to supervise the classification network, making the network discover more semantic regions, thus the generated localization map could be more accurate for covering the object parts. Then, we employ the saliency map as pseudo-pixel feedback to the localization maps from both the foreground objects and the background. Next, we will explain how SMVWE can tackle the sparse object coverage, inaccurate object boundary and co-occurring pixel problems in image-level WSSS.

The image-level WSSS task is unable to directly use the semantic word labels, so we use an unsupervised way to generate their vector representations in each forward pass, i.e., each semantic word in a trainable codebook utilizes the manhattan distance to encode the feature maps from the classification network. In such a way, it alleviates the sparse object region problem, and improves the accuracy of the generated localization map.

To tackle the inaccurate object boundary and co-occurring pixel problems, we first use the $L + 1$ localization maps encoded by the semantic word labels to generate the foreground object map and the background map, then these generated saliency map are evaluated by a saliency loss using off-the-shelf saliency maps, addressing the boundary mismatch and assigning the co-occurring pixels of non-target objects to the background. Thus, our method can better separate the foreground objects from the background.

Lastly, the objective function of SMVWE is formulated with three parts: two multi-label classification losses from semantic word labels and image-level labels respectively, and the saliency loss from the generation process of the saliency map. By jointly training the three objectives, we can combine the localization map encoded by semantic word labels with the saliency map to generate higher-quality pseudo labels.

3.2   Semantic Word Learning

The localization map generated from the classification network only covers the most discriminative extents of objects. The reason is that the goal of the classification network is essentially classification ability, not localization map generation. Thus, we propose a visual word encoder (VWE) module to enforce the classification network to cover integral object regions via the semantic word labels.

Due to only image-level labels in the WSSS task can be employed to annotate pixels in images, no extra labels are available. For this reason, we employ the codebook to encode the extracted convolutional feature map $M \in R^{C \times H \times W}$ to specific semantic words, where C denotes the channels, W and H denote width and height, respectively. Then, the manhattan distance is used to measure the similarity between the pixel at position i in M and the j-th word in codebook $B \in R^{N \times K}$, where N is the number of words and K is the feature dimension. The similarity matrix D can be formulated as below:

$$D_{ij} = manhattan(M_i, B_j) = |M_i - B_j| \qquad (1)$$

After obtained D, we use $softmax$ to normalize row-wise, then computing the j-th word in codebook B represents the semantic probability of the i-th pixel in feature map M.

$$P_{ij} = softmax(D_i) = \frac{exp(D_{ij})}{\sum_{n=1}^{N} exp(D_{in})} \qquad (2)$$

The semantic word Z_i with the maximum probability will be denoted the semantic word label for M_i, where the index of the maximum value in the i-th row of P_{ij} is denoted as:

$$Z_i = argmax P_{ij} \qquad (3)$$

Then, we use a N-dimensional vector $\boldsymbol{z}^{word}$ to denote the semantic word label of the image $\boldsymbol{I}$, where $z_j^{word} = 1$ if the j-th word is in $\boldsymbol{Z}$, and $z_j^{word} = 0$, otherwise. $\boldsymbol{z}^{word}$ will make the classification network discover more semantic extents of target objects during the training procedure.

If employing the histogram distributions of each semantic word generated by counting their frequencies to represent the feature map, it will lead to non-continuities and make the training process intractable [28]. Thus, we compute the soft frequency of the j-th word by accumulating the probabilities in $\boldsymbol{P}$:

$$e_j^{word} = \frac{1}{H \cdot W} \sum_{i=1}^{H \cdot W} P_{ij} \tag{4}$$

where e_j^{word} denotes the appearance frequency of the j-th word in $\boldsymbol{M}$. As shown in Fig. 1, $\boldsymbol{e}^{word}$ will model the mapping relations between semantic words and image-level labels. Moreover, inspired by [28], we will set the codebook $\boldsymbol{B}$ as a trainable parameter, which makes it could be learned automatically via the back propagated gradients.

3.3 Saliency Map Feedback

In WSSS, utilizing the saliency map is a common practice to better provide the information of object boundaries. Different from existing methods that make full use of the off-the-shelf saliency map as a part of their feature maps, our method generates the saliency maps using the foreground localization map and the background localization map, where the off-the-shelf saliency map is only used as the pseudo-pixel feedback by a saliency loss.

First, generating a foreground map $\boldsymbol{F}_{fg} \in R^{H \times W}$ by aggregating the localization maps of target objects, and performing the inversion of a background map $\boldsymbol{F}_{bg} \in R^{H \times W}$ generated by the background localization map to represent the foreground map. Then, we use $\boldsymbol{F}_{fg}$ and $\boldsymbol{F}_{bg}$ to generate the saliency map $\boldsymbol{F}_s$.

$$\boldsymbol{F}_s = (1 - \mu)\boldsymbol{F}_{fg} + \mu(1 - \boldsymbol{F}_{bg}) \tag{5}$$

where $\mu \in [0, 1]$ is a hyper-parameter to adjust a weighted sum of the foreground map and the inversion of the background map.

Moreover, our method also addresses the saliency bias during generating the foreground map and the background map. Because the saliency detection model obtains the saliency map via different datasets, the saliency bias is inevitable. Thus, we introduce an overlapping ratio strategy [42] between the localization map and the saliency map to address this issue, i.e., the i-th localization map F_i is overlapped with the groundtruth saliency map F_s' more than $\delta\%$, which is classified as the foreground, otherwise the background. The foreground map and the background map are represented as follows:

$$F_{fg} = \sum_{i=1}^{L} z_i \cdot F_i \cdot 1[\phi(F_i, F'_s) > \delta] \tag{6}$$

$$F_{bg} = \sum_{i=1}^{L} z_i \cdot F_i \cdot 1[\phi(F_i, F'_s) \leq \delta] \tag{7}$$

where $z_i \in R^L$ is the binary image-level label and $\phi(F_i, F'_s)$ is used to compute the overlapping ratio between F_i and F'_s. We first use C_i and C_s to represented the binarized maps corresponding to F_i and F'_s respectively. For example, at the pixel Q in F, $C_N(Q) = 1$ if $F_N(Q) > 0.5$; $C_N(Q) = 0$, otherwise. Then, using $\phi(F_i, F'_s) = |C_i \cap C_s|/|C_i|$ to compute the overlapping ratio $\delta\%$ between F_i and F'_s.

3.4   Jointly Learning of Pseudo Label Generation

Our method generates the pseudo labels by two comprehensive information, i.e., semantic word encoding and saliency map, they respectively focus on different issues in WSSS task. To tackle sparse object region problem, we train the classification network on the localization map M through predicting the semantic word label z^{word}, where the global average pooling is used to compute the semantic word score $s^{word} = conv(f^{gap}, W^{word})$, and W^{word} denotes the weight matrix. We use the multi-label soft margin loss [29] to compute the classification loss for semantic words as follows:

$$L_{cls}(s^{word}, z^{word}) = \frac{1}{L} \sum_{i=1}^{L} [z_i^{word} log \frac{exp(s_i^{word})}{1 + exp(s_i^{word})} + (1 - z_i^{word}) log \frac{1}{1 + exp(s_i^{word})}] \tag{8}$$

where z^{word} is obtained by Eq. 3, L is the number of image classes.

To model the mapping relations between semantic words and image classes, we use an 1×1 conv layer with weight matrix W^{w2i} to transfer the semantic word frequency e^{word} into the class probability space, where the predicted score and the ground-truth image label are denoted by p^{w2i} and z^{img}, respectively. Thus, the loss function $L_{cls}(p^{w2i}, z^{img})$ is formulated as the same form as Eq. 8.

Then, we utilize the saliency map to tackle inaccurate object boundaries and co-occurring pixels, where the average pixel-level distance between the ground-truth saliency map F'_s and the generated saliency map F_s is employed to calculate the saliency loss.

$$L_{sal} = \frac{1}{H \cdot W} \left\| F'_s - F_s \right\|^2 \tag{9}$$

where F'_s is obtained from the off-the-shelf saliency detection model PFAN [45] trained on DUTS dataset [33].

The overall loss of our proposed method is finally represented as the sum of the aforementioned loss terms.

$$L_{total} = L_{cls}(s^{word}, z^{word}) + L_{cls}(s^{w2i}, z^{img}) + L_{sal} \tag{10}$$

where L_{sal} mainly focuses on updating the parameters of L target object classes and one background class, while L_{cls} only evaluates the label prediction for L target object classes, excluding the background class.

4   Experiments

4.1   Experimental Setup

Datasets and Evaluation Criteria. Following previous works [21,42], we evaluate the proposed method on the PASCAL VOC 2012 semantic segmentation benchmark [12]. PASCAL VOC 2012 consists of 21 classes, i.e., 20 foreground objects and the background. Following the common practice in semantic segmentation, we use the augmented training set with 10,582 images [16], validation set with 1,449 images and testing set with 1,456 images. For all experiments, the mean Intersection-over-Union (mIoU) is used as the evaluation criteria.

Implementation Details. The ResNet38 [39] is employed as the backbone network to extract feature maps. The classification network is trained via the SGD optimizer with a batch size of 4. Besides, we set the initial learning rate to 0.01 and decrease the learning rate every iteration with a polynomial decay strategy. The number of semantic words is set to 256. The images are randomly rescaled to 448×448. For the segmentation networks, we adopt DeepLab-LargeFOV (V1) [6] and DeepLab-ASPP (V2) [7], where VGG16 and ResNet101 are their backbone networks, i.e., VGG16 based DeepLab-V1 and DeepLab-V2, and ResNet101 based DeepLab-V1 and DeepLab-V2.

4.2   Ablation Study and Analysis

To validate the effectiveness of our proposed method, we conduct several experiments to analyze the effect of each component in the proposed method. For all experiments in this section, we adopt the DeepLab-V1 with VGG-16 as the segmentation network and measure the mIoU on the VOC 2012 validation set.

Dealing with Sparse Object Region
To validate whether the proposed VWE can cover more object regions in the input images reasonably, we compute the mIoU of the semantic word labels on the PASCAL VOC 2012 validation set. As shown in Table 1, it shows that the codebook can distinguish different semantic words reasonably and the proposed VWE can work effectively for encoding different objects of an image. Compared with existing methods, our VWE module can obtain higher performance on

Table 1. Comparison with representative methods on the sparse object region problem. The best three results are in red, blue and green, respectively.

Method	bkg	aero	bike	bird	boat	bottle	bus	car	cat	chair	cow	table
AffinityNet [2]	88.2	68.2	30.6	81.1	49.6	61.0	77.8	66.1	75.1	29.0	66.0	40.2
MCOF [34]	87.0	78.4	29.4	68.0	44.0	67.3	80.3	74.1	82.2	21.1	70.7	28.2
SSNet [42]	90.0	77.4	37.5	80.7	61.6	67.9	81.8	69.0	83.7	13.6	79.4	23.3
SEAM [35]	88.8	68.5	33.3	85.7	40.4	67.3	78.9	76.3	81.9	29.1	75.5	48.1
CIAN [14]	88.2	79.5	32.6	75.7	56.8	72.1	85.3	72.9	81.7	27.6	73.3	39.8
Ours (VWE)	89.2	75.7	31.1	82.4	66.1	61.7	87.5	77.8	82.8	32.3	81.4	34.5
Ours (SMVWE)	90.8	77.9	31.6	89.4	56.9	57.8	86.4	77.9	82.9	32.3	76.9	52.5

Method	dog	horse	mbk	person	plant	sheep	sofa	train	tv	mIoU
AffinityNet [2]	80.4	62.0	70.4	73.7	42.5	70.7	42.6	68.1	51.6	58.4
MCOF [34]	73.2	71.5	67.2	53.0	47.7	74.5	32.4	71.0	45.8	60.3
SSNet [42]	78.0	75.3	71.4	68.1	35.2	78.2	32.5	75.5	48.0	63.3
SEAM [35]	79.9	73.8	71.4	75.2	48.9	79.8	40.9	58.2	53.0	64.5
CIAN [14]	76.4	77.0	74.9	66.8	46.6	81.0	29.1	60.4	53.3	64.3
Ours (VWE)	77.4	77.6	76.7	75.1	51.2	78.7	42.7	71.8	59.6	**65.4**
Ours (SMVWE)	80.7	80.3	81.8	74.3	44.5	80.7	54.7	68.8	60.5	67.5

Table 2. Comparison with representative methods on the inaccurate object boundary problem using the SBD set of the VOC 2012 validation set.

Method	Recall (%)	Precision (%)	F1-score (%)
CAM [46]	22.3	35.8	27.5
SEAM [35]	40.2	45.0	42.5
BES [9]	45.5	46.4	45.9
Our SMVWE	**62.3**	**76.5**	**69.4**

most objects for semantic segmentation, and brings an improvement of 0.9% (65.4% vs 64.5%) compared to the state-of-the-art method [35]. Thus, under the supervision of the generated semantic word labels, our proposed method can cover more object extents, which effectively addresses the sparse object-region problem and improves the performance of the localization map.

Dealing with Inaccurate Boundary and Co-occurring Pixel Inaccurate Boundary Problem. To evaluate the boundary quality of pseudo-labels, our method compares with representative methods [9,35,46] by using the SBD set of the VOC 2012 validation set, where the SBD set containing boundary annotations is benefit to test the boundary quality of pseudo labels through the

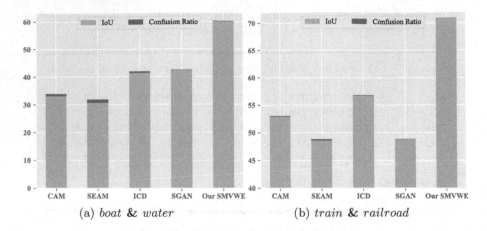

(a) *boat & water*   (b) *train & railroad*

Fig. 2. Comparison with representative methods on the co-occurring pixel problem. The lower confusion ratio denotes the better, and the higher IoU denotes the better.

Laplacian edge detector [9]. As shown in Table 2, we use the evaluation metrics of recall, precision, and F1-score to demonstrate that our method remarkably outperforms other methods. Figure 3 shows our some visualization results, which validate that our method works well on tackling the object boundary mismatch problem (Table 3).

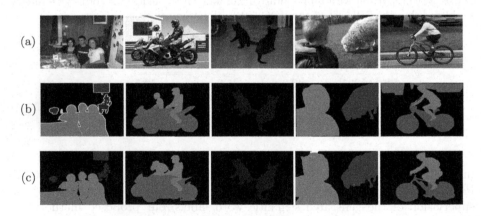

Fig. 3. Qualitative segmentation results on PASCAL VOC 2012 validation set. (a) Original images, (b) groundtruth and (c) our SMVWE. Segmentation results are predicted by ResNet101 based DeepLab-V2 segmentation network.

Co-occurring Pixel Problem. To measure the ability of our method on addressing the co-occurring pixels problem, we compare the performance of our method with representative methods (i.e., CAM [46], SEAM [35], ICD [13],

Table 3. Performance comparisons of our method with state-of-the-art WSSS methods on PASCAL VOC 2012 dataset. All results are based on VGG16. S means the saliency map is used for existing methods and ours.

Methods	S	val (%)	test (%)
Segmentation network: DeepLab-V1 (VGG-16)			
GAIN [25]	✓	55.3	56.8
MCOF [34]	✓	60.3	59.6
AffinityNet [2]	✗	58.4	60.5
SeeNet [18]	✓	61.1	60.8
OAA [20]	✓	63.1	62.8
RRM [43]	✗	60.7	61.0
ICD [13]	✓	64.0	63.9
BES [9]	✗	60.1	61.1
DRS [21]	✓	63.5	64.5
NSRM [41]	✓	65.5	65.3
Ours (SMVWE)	✓	67.5	67.2
Segmentation network: DeepLab-V2 (VGG-16)			
DSRG [19]	✓	59.0	60.4
FickleNet [24]	✓	61.2	61.9
Split and Merge [44]	✓	63.7.	64.5
SGAN [40]	✓	64.2	65.0
Ours (SMVWE)	✓	68.2	68.1

SGAN [40]) by IoU and confusion ratio evaluation criteria, where the lower confusion ratio denotes the better, and the higher IoU denotes the better. The IoU measures how much the target classes are predicted correctly, and the confusion ratio measures how much the co-occurring non-target class is incorrectly predicted as the target class.

As shown in Fig. 2, we use two co-occurring pairs, i.e. *boat* with *water*, *train* with *railroad*, to compare our method with existing methods. Our method markedly outperforms other methods on the IoU evaluation criteria. Moreover, compared to other methods, only SGAN [40] method has a same lower confusion ratio with ours. For the following reasons, CAM [46] only captures the most discriminative region of target objects; SEAM [35] and ICD [13] both ignore the co-occurring pixels between target objects and non-target objects, while our method proposes a semantic word labels to discover more object regions, and designs a saliency map selection mechanism to obtain accurate object boundaries and discard the co-occurring pixels of non-target objects. Thus, our method generates higher-quality pseudo labels to perform the semantic segmentation task.

Table 4. Performance comparisons of our method with state-of-the-art WSSS methods on PASCAL VOC 2012 dataset. All results are based on ResNet101. S means the saliency map is used for existing methods and ours.

Methods	S	val (%)	test (%)
Segmentation Network : DeepLab-V1 (ResNet-101)			
MCOF [34]	✓	60.3	61.2
SeeNet [18]	✓	63.1	62.8
AffinityNet [2]	✗	61.7	63.7
FickleNet [24]	✓	64.9	65.3
OAA [20]	✓	65.2	65.2
RRM [43]	✗	66.3	65.5
ICD [13]	✓	67.8	68.0
DRS[21]	✓	66.5	67.5
Ours (SMVWE)	✓	70.1	69.6
Segmentation Network : DeepLab-V2 (ResNet-101)			
DSRG [19]	✓	61.4	63.2
BES [9]	✗	65.7	66.6
SGAN [40]	✓	67.1	67.2
DRS [21]	✓	70.4	70.7
Ours (SMVWE)	✓	71.3	71.5

4.3   Comparison with State-of-the-Arts

We compare our SMVWE method with state-of-the-art WSSS methods using only image-level labels. As shown in Table 4, our method remarkably outperforms other methods on the same VGG16 backbone. Noting that our performance improvement does not rely on a larger network structure and is superior to other existing methods based on a more powerful backbone (i.e. ResNet101 in Table 5). Because our method mainly relies on the cooperation of visual word encoder and saliency map selection strategy, which generates better pseudo labels for the semantic segmentation task. As shown in Table 5, our method achieves a new state-of-the-art performance (71.3% on validation set and 71.5% on test set) with the ResNet101 based DeepLab-V2 segmentation network. Figure 3 visualizes our semantic segmentation results on the validation set. These results show that our method can obtain more integral object regions and accurate object boundaries, and discard co-occurring pixels between target objects and the background.

5   Conclusion

In this paper, we proposed a saliency map and visual word encoder (SMVWE) method for image-level semantic segmentation. Particularly, we explored more fine-grained semantic word labels to supervise the classification network, making the generated localization map could cover more integral object regions. Moreover, we designed a saliency map selection mechanism to separate the foreground from the background, where the saliency maps were used as pseudo-pixel feedback. By joint learning of visual word encoder and saliency map feedback, our SMVWE successfully tackles the sparse object regions, boundary mismatch and co-occurring pixels problems, thus producing higher-quality pseudo labels for WSSS task. Extensive experiments demonstrate the superiority of our proposed method, and achieve the state-of-the-art performance using only image-level labels.

Acknowledgements. This work was supported by the National Natural Science Foundation of China (62072463, 71531012), and the National Social Science Foundation of China (18ZDA309).

References

1. Ahn, J., Cho, S., Kwak, S.: Weakly supervised learning of instance segmentation with inter-pixel relations. In: CVPR, pp. 2209–2218 (2019)
2. Ahn, J., Kwak, S.: Learning pixel-level semantic affinity with image-level supervision for weakly supervised semantic segmentation. In: CVPR, pp. 4981–4990 (2018)
3. Araslanov, N., Roth, S.: Single-stage semantic segmentation from image labels. In: CVPR, pp. 4252–4261 (2020)
4. Chang, Y., Wang, Q., Hung, W., Piramuthu, R., Tsai, Y., Yang, M.: Weakly-supervised semantic segmentation via sub-category exploration. In: CVPR, pp. 8988–8997 (2020)
5. Chaudhry, A., Dokania, P.K., Torr, P.H.S.: Discovering class-specific pixels for weakly-supervised semantic segmentation. In: BMVC (2017)
6. Chen, L., Papandreou, G., Kokkinos, I., Murphy, K., Yuille, A.L.: Semantic image segmentation with deep convolutional nets and fully connected CRFs. In: ICLR (2015)
7. Chen, L., Papandreou, G., Kokkinos, I., Murphy, K., Yuille, A.L.: DeepLab: semantic image segmentation with deep convolutional nets, atrous convolution, and fully connected CRFs. IEEE Trans. Pattern Anal. Mach. Intell. **40**(4), 834–848 (2018)
8. Chen, L.-C., Zhu, Y., Papandreou, G., Schroff, F., Adam, H.: Encoder-decoder with atrous separable convolution for semantic image segmentation. In: Ferrari, V., Hebert, M., Sminchisescu, C., Weiss, Y. (eds.) ECCV 2018. LNCS, vol. 11211, pp. 833–851. Springer, Cham (2018). https://doi.org/10.1007/978-3-030-01234-2_49
9. Chen, L., Wu, W., Fu, C., Han, X., Zhang, Y.: Weakly supervised semantic segmentation with boundary exploration. In: Vedaldi, A., Bischof, H., Brox, T., Frahm, J.-M. (eds.) ECCV 2020. LNCS, vol. 12371, pp. 347–362. Springer, Cham (2020). https://doi.org/10.1007/978-3-030-58574-7_21
10. Choe, J., Lee, S., Shim, H.: Attention-based dropout layer for weakly supervised single object localization and semantic segmentation. IEEE Trans. Pattern Anal. Mach. Intell. **43**(12), 4256–4271 (2021)
11. Choe, J., Oh, S.J., Lee, S., Chun, S., Akata, Z., Shim, H.: Evaluating weakly supervised object localization methods right. In: CVPR, pp. 3130–3139 (2020)

12. Everingham, M., Eslami, S.M.A., Gool, L.V., Williams, C.K.I., Winn, J.M., Zisserman, A.: The pascal visual object classes challenge: a retrospective. Int. J. Comput. Vis. **111**(1), 98–136 (2015)
13. Fan, J., Zhang, Z., Song, C., Tan, T.: Learning integral objects with intra-class discriminator for weakly-supervised semantic segmentation. In: CVPR, pp. 4282–4291 (2020)
14. Fan, J., Zhang, Z., Tan, T., Song, C., Xiao, J.: CIAN: cross-image affinity net for weakly supervised semantic segmentation. In: AAAI, pp. 10762–10769 (2020)
15. Fan, R., Hou, Q., Cheng, M.-M., Yu, G., Martin, R.R., Hu, S.-M.: Associating inter-image salient instances for weakly supervised semantic segmentation. In: Ferrari, V., Hebert, M., Sminchisescu, C., Weiss, Y. (eds.) ECCV 2018. LNCS, vol. 11213, pp. 371–388. Springer, Cham (2018). https://doi.org/10.1007/978-3-030-01240-3_23
16. Hariharan, B., Arbelaez, P., Bourdev, L.D., Maji, S., Malik, J.: Semantic contours from inverse detectors. In: ICCV, pp. 991–998 (2011)
17. Hong, S., Oh, J., Lee, H., Han, B.: Learning transferrable knowledge for semantic segmentation with deep convolutional neural network. In: CVPR, pp. 3204–3212 (2016)
18. Hou, Q., Jiang, P., Wei, Y., Cheng, M.: Self-erasing network for integral object attention. In: NeurIPS, pp. 547–557 (2018)
19. Huang, Z., Wang, X., Wang, J., Liu, W., Wang, J.: Weakly-supervised semantic segmentation network with deep seeded region growing. In: CVPR, pp. 7014–7023 (2018)
20. Jiang, P., Hou, Q., Cao, Y., Cheng, M., Wei, Y., Xiong, H.: Integral object mining via online attention accumulation. In: ICCV, pp. 2070–2079 (2019)
21. Kim, B., Han, S., Kim, J.: Discriminative region suppression for weakly-supervised semantic segmentation. In: AAAI, pp. 1754–1761 (2021)
22. Kim, D., Cho, D., Yoo, D.: Two-phase learning for weakly supervised object localization. In: ICCV, pp. 3554–3563 (2017)
23. Kolesnikov, A., Lampert, C.H.: Seed, expand and constrain: three principles for weakly-supervised image segmentation. In: Leibe, B., Matas, J., Sebe, N., Welling, M. (eds.) ECCV 2016. LNCS, vol. 9908, pp. 695–711. Springer, Cham (2016). https://doi.org/10.1007/978-3-319-46493-0_42
24. Lee, J., Kim, E., Lee, S., Lee, J., Yoon, S.: FickleNet: weakly and semi-supervised semantic image segmentation using stochastic inference. In: CVPR, pp. 5267–5276 (2019)
25. Li, K., Wu, Z., Peng, K., Ernst, J., Fu, Y.: Tell me where to look: guided attention inference network. In: CVPR, pp. 9215–9223 (2018)
26. Oh, S.J., Benenson, R., Khoreva, A., Akata, Z., Fritz, M., Schiele, B.: Exploiting saliency for object segmentation from image level labels. In: CVPR, pp. 5038–5047 (2017)
27. Papandreou, G., Chen, L., Murphy, K.P., Yuille, A.L.: Weakly-and semi-supervised learning of a deep convolutional network for semantic image segmentation. In: ICCV, pp. 1742–1750 (2015)
28. Passalis, N., Tefas, A.: Learning bag-of-features pooling for deep convolutional neural networks. In: IEEE International Conference on Computer Vision, ICCV 2017, Venice, Italy, 22–29 October 2017, pp. 5766–5774 (2017)
29. Paszke, A., et al.: PyTorch: an imperative style, high-performance deep learning library. In: NeurIPS, pp. 8024–8035 (2019)
30. Pinheiro, P.H.O., Collobert, R.: From image-level to pixel-level labeling with convolutional networks. In: CVPR, pp. 1713–1721 (2015)

31. Roy, A., Todorovic, S.: Combining bottom-up, top-down, and smoothness cues for weakly supervised image segmentation. In: CVPR, pp. 7282–7291 (2017)
32. Sun, G., Wang, W., Dai, J., Gool, L.V.: Mining cross-image semantics for weakly supervised semantic segmentation. In: ECCV, vol. 12347, pp. 347–365 (2020)
33. Wang, L., et al.: Learning to detect salient objects with image-level supervision. In: CVPR, pp. 3796–3805 (2017)
34. Wang, X., You, S., Li, X., Ma, H.: Weakly-supervised semantic segmentation by iteratively mining common object features. In: CVPR, pp. 1354–1362 (2018)
35. Wang, Y., Zhang, J., Kan, M., Shan, S., Chen, X.: Self-supervised equivariant attention mechanism for weakly supervised semantic segmentation. In: CVPR, pp. 12272–12281 (2020)
36. Wei, Y., Feng, J., Liang, X., Cheng, M., Zhao, Y., Yan, S.: Object region mining with adversarial erasing: a simple classification to semantic segmentation approach. In: CVPR, pp. 6488–6496 (2017)
37. Wei, Y., et al.: STC: a simple to complex framework for weakly-supervised semantic segmentation. IEEE Trans. Pattern Anal. Mach. Intell. **39**(11), 2314–2320 (2017)
38. Wei, Y., Xiao, H., Shi, H., Jie, Z., Feng, J., Huang, T.S.: Revisiting dilated convolution: a simple approach for weakly- and semi-supervised semantic segmentation. In: CVPR, pp. 7268–7277 (2018)
39. Wu, Z., Shen, C., van den Hengel, A.: Wider or deeper: revisiting the ResNet model for visual recognition. Pattern Recognit. **90**, 119–133 (2019)
40. Yao, Q., Gong, X.: Saliency guided self-attention network for weakly and semi-supervised semantic segmentation. IEEE Access **8**, 14413–14423 (2020)
41. Yao, Y., et al.: Non-salient region object mining for weakly supervised semantic segmentation. In: CVPR, pp. 2623–2632 (2021)
42. Yu, Z., Zhuge, Y., Lu, H., Zhang, L.: Joint learning of saliency detection and weakly supervised semantic segmentation. In: ICCV, pp. 7222–7232 (2019)
43. Zhang, B., Xiao, J., Wei, Y., Sun, M., Huang, K.: Reliability does matter: an end-to-end weakly supervised semantic segmentation approach. In: AAAI, pp. 12765–12772 (2020)
44. Zhang, T., Lin, G., Liu, W., Cai, J., Kot, A.: Splitting vs. Merging: mining object regions with discrepancy and intersection loss for weakly supervised semantic segmentation. In: Vedaldi, A., Bischof, H., Brox, T., Frahm, J.-M. (eds.) ECCV 2020. LNCS, vol. 12367, pp. 663–679. Springer, Cham (2020). https://doi.org/10.1007/978-3-030-58542-6_40
45. Zhao, T., Wu, X.: Pyramid feature attention network for saliency detection. In: CVPR, pp. 3085–3094 (2019)
46. Zhou, B., Khosla, A., Lapedriza, À., Oliva, A., Torralba, A.: Learning deep features for discriminative localization. In: CVPR, pp. 2921–2929 (2016)

Understanding Adversarial Robustness of Vision Transformers via Cauchy Problem

Zheng Wang🆔 and Wenjie Ruan$^{(\boxtimes)}$🆔

University of Exeter, Exeter EX4 4PY, UK
{zw360,W.Ruan}@exeter.ac.uk

Abstract. Recent research on the robustness of deep learning has shown that *Vision Transformers (ViTs)* surpass the *Convolutional Neural Networks (CNNs)* under some perturbations, e.g., natural corruption, adversarial attacks, etc. Some papers argue that the superior robustness of ViT comes from the segmentation on its input images; others say that the *Multi-head Self-Attention (MSA)* is the key to preserving the robustness [30]. In this paper, we aim to introduce a principled and unified theoretical framework to investigate such argument on ViT's robustness. We first theoretically prove that, unlike *Transformers* in *Natural Language Processing*, ViTs are *Lipschitz* continuous. Then we theoretically analyze the adversarial robustness of ViTs from the perspective of *Cauchy Problem*, via which we can quantify how the robustness propagates through layers. We demonstrate that the first and last layers are the critical factors to affect the robustness of ViTs. Furthermore, based on our theory, we empirically show that unlike the claims from existing research, MSA only contributes to the adversarial robustness of ViTs under weak adversarial attacks, e.g., *FGSM*, and surprisingly, MSA actually comprises the model's adversarial robustness under stronger attacks, e.g., *PGD attacks*. We release our code via https://github.com/TrustAI/ODE4RobustViT.

Keywords: Adversarial robustness · Cauchy problem · Vision Transformer

1 Introduction

Since *Transformers* have been transplanted from *Natural Language Processing (NLP)* to *Computer Vision (CV)*, great potential has been revealed by *Vision Transformers* for various CV tasks [19]. It is so successful that some papers even argue that CNNs are just a special case of ViTs [9]. Recently, the robustness of

This work is supported by Partnership Resource Fund (PRF) on Towards the Accountable and Explainable Learning-enabled Autonomous Robotic Systems from UK EPSRC project on Offshore Robotics for Certification of Assets (ORCA) [EP/R026173/1].

Supplementary Information The online version contains supplementary material available at https://doi.org/10.1007/978-3-031-26409-2_34.

ViTs has been studied, for example, some research showed that ViT has superior robustness than CNNs under natural corruptions [31]. Very recently, some researchers have also begun to investigate the robustness of ViTs against *adversarial perturbations* [26].

However, existing research on adversarial robustness for ViTs mainly focuses on *adversarial attacks*. The main idea is to adopt the attacks on CNNs to ViTs, e.g., *SAGA* [26] and *IAM-UAP* [18]. Meanwhile, some pioneering studies demonstrate that ViTs are more robust than CNNs against *adversarial patch attacks*, arguing that the *dynamic receptive field* of MSA is the key factor to its superior robustness [30]. On the other hand, some others argue that the tokenization of ViTs plays an essential role in adversarial robustness [1]. While some researchers say the patch embedding method is a critical factor to contribute the adversarial robustness of ViTs [28]. However, most existing works concerning the superior robustness of ViTs are purely based on empirical experiments in an ad-hoc manner. A principled and unified theoretical framework that can quantify the adversarial robustness of ViT is still lacking in the community.

In our paper, instead of analyzing the robustness of *Vision Transformer* purely based on empirical evidence, a theoretical framework has been proposed to examine whether MSA contributes to the robustness of ViTs. Inspired by the fact that ViTs and ResNets share a similar structure of residual additions, we show that, ViTs, under certain assumptions, can be regarded as a *Forward Euler* approximation of the underlying *Ordinary Differential Equations (ODEs)* defined as

$$\frac{d\boldsymbol{x}}{dt} = \mathcal{F}(\boldsymbol{x}, t).$$

With this approximation, each block in transformer can be modeled as the nonlinear function $\mathcal{F}(\boldsymbol{x})$. Based on the assumption that function $\mathcal{F}(\boldsymbol{x})$ is Lipschitz continuous, we then can theoretically bridge the adversarial robustness with the *Cauchy Problem* by first-order *Taylor expansion* of $\mathcal{F}(\boldsymbol{x})$. With the proposed theoretical framework, this paper is able to quantify how robustness is changing among each block in ViTs. We also observe that the first and last layers are vital for the robustness of ViTs.

Furthermore, according to our theoretical and empirical studies, different to the existing claim made by Naseer et al. [30] that MSA in ViTs strengthens the robustness of ViTs against patch attacks. We show that MSA in ViTs is *not always* improving the model's adversarial robustness. Its strength to enhance the robustness is minimal and even comprises the adversarial robustness against strong L_p norm adversarial attacks. In summary, the key contributions of this paper are listed below.

1. To our knowledge, this is the first work to formally bridge the gap between the robustness problem of ViTs and the *Cauchy problem*, which provides a principled and unified theoretical framework to quantify the robustness of transformers.
2. We theoretically prove that ViTs are Lipschitz continuous on vision tasks, which is an important requisite to building our theoretical framework.

3. Based on our proposed framework, we observe that the first and last layers in the encoder of ViTs are the most critical factors to affect the robustness of the transformers.
4. Unlike existing claims, surprisingly, we observe that MSA can only improve the robustness of ViTs under weak attacks, e.g., *FGSM attack*, and it even comprises the robustness of ViTs under strong attacks, e.g., *PGD attack*.

2 Related Work

2.1 Vision Transformers and Its Variants

To the best of our knowledge, the first work using the transformer to deal with computer vision tasks is done by Carion et al. [6], since then, it has quickly become a research hotspot, though it has to be pre-trained on a larger dataset to achieve comparable performance due to its high complexity and lack of ability to encode local information. To reduce the model complexity, *DeiT* [36] leverages the *Knowledge Distillation* [17] technique, incorporating information learned by *Resnets* [15]; *PvT* [37] and *BoTNet* [34] adopt more efficient backbones; *Swim Transformer* [24] and *DeepViT* [38] modifies the MSA. Other variants, e.g., *TNT*, *T2T-ViT*, *CvT*, *LocalViT* and *CeiT* manage to incorporate local information to the ViTs [19].

2.2 Robustness of Vision Transformer

Many researchers focus on the robustness of ViTs against *natural corruptions* [16] and empirically show that ViTs are more robust than CNNs [31]. The adversarial robustness of ViTs has also been empirically investigated. Compared with CNNs and MLP-Mixers under different attacks, it claims that for most of the white-box attacks, some black-box attacks, and Universal Adversarial Perturbations (UAPs) [29], ViTs show superior robustness [30]. However, ViTs are more vulnerable to simple FGSM attacks [5]. The robustness of variants of ViTs is also investigated and shown that the local window structure in Swim-ViT harms the robustness and argues that positional embedding and the *completeness/compactness* of heads are crucial for performance and robustness [27].

However, the reason for the superior robustness of ViTs is rarely investigated. Most of the research concentrate on frequency analysis [31]. Benz et al. argue that shift-invariance property [4] harms the robustness of CNNs. Naseer et al. say the flexible receptive field is the key to learning more shape information which strengthens the robustness of ViTs by studying the severe occlusions [30]. And yet Mao et al. argue that ViTs are still overly reliant on the texture, which could harm their robustness against out-of-distribution data [27]. Qin et al. investigate the robustness from the perspective of robust features and argue ViTs are insensitive to patch-level transformation, which is considered as non-robust features [32].

2.3   Deep Neural Network via Dynamic Point of View

The connection between differential equations and neural networks is first introduced by S. Grossberg [14] to describe a continuous additive RNN model. After ResNet had been proposed, new relations appeared that regard forward prorogation as Euler discretization of the underlying ODEs [33]. And many variants of ResNets can also be analyzed in the framework of numerical schemes for ODEs, e.g., PolyNet, FracalNet, RevNet and LMResNet [25]. Instead of regarding neural networks as discrete methods, Neural ODE has been proposed [7], replacing the ResNet with its Underlying ODEs, and the parameters are calculated by a black-box ODE solver. However, E. Dupont et al. [11] argue that neural ODEs hardly learn some representations. In addition to ODEs, PDEs and even SDEs are also involved in analyzing the Neural Network [35].

3   Preliminaries

The original ViTs are generally composed of *Patch Embedding, Transformer Block* and *Classification Head*. We follow the definition from [10]. Let $x \in \mathbf{R}^{H \times W \times C}$ stands for the input image. Hence, Each image is divided equally into a sequence of $N = HW/P^2$ patches, and each one is denoted as $x_p \in \mathbf{R}^{N \times (P^2 \cdot C)}$.

$$z_0 = [x_{class}, x_p^1 E, x_p^2 E, ..., x_p^N E] + E_{pos},$$
$$z_l' = MSA(LN(z_{l-1})) + z_{l-1},$$
$$z_l = MLP(LN(z_l')) + z_l',$$
$$y = LN(z_L^0),$$

where $E \in \mathbb{R}^{P^2 \cdot C \times D}, E_{pos} \in \mathbb{R}^{(N+1) \times D}$ and $l = 1, 2, ..., L$. LN denotes *Layer Normalization*, *MSA* is *Multihead Self-Attention* and *MLP* represents *Multilayer Perceptron*. MSA is the concatenation of *Self-Attentions (SA)* before linear transformation by $W^{(O)} \in \mathbb{R}^{D \times D}$ defined by

$$MHA := (SA_1 \ SA_2 \ ... \ SA_H) \, W^{(O)},$$

where H is the number of heads and SA is defined by

$$SA := PzW^{(V)} = softmax\left(zW^{(Q)}W^{(K)T}z^T\right)W^{(V)},$$

where $W^{(Q)}, W^{(K)}, W^{(V)} \in \mathbb{R}^{D \times (D/H)}$, and $z \in \mathbb{R}^{N \times D}$ defines the inputs of transformers.

4   Theoretically Analysis

4.1   Vision Transformers are Lipschitz

To model the adversarial robustness to Cauchy Problem, we first prove that ViTs are Lipschitz functions. Unlike the conclusion drawn by Kim et al. [21]

that Dot-product self-attention is not Lipschitz, it can be proved that Vision Transformers are Lipchitz continuous since inputs are bounded between $[0, 1]$. We follow the same definition from [21] that a function $f : \mathcal{X} \to \mathcal{Y}$ is called Lipschitz continuous if $\exists K \geq 0$ such that $\forall \boldsymbol{x} \in \mathcal{X}, \boldsymbol{y} \in \mathcal{Y}$ we have

$$d_{\mathcal{Y}}(f(\boldsymbol{x}), f(\boldsymbol{x}_0)) \leq K d_{\mathcal{X}}(\boldsymbol{x}, \boldsymbol{x}_0), \tag{1}$$

where $(\mathcal{X}, d_{\mathcal{X}}), (\mathcal{Y}, d_{\mathcal{Y}})$ are given metric spaces, and given p-norm distance, the Lipschitz constant K is given by

$$Lip_p(f) = \sup_{\boldsymbol{x} \neq \boldsymbol{x}_0} \frac{\|f(\boldsymbol{x}) - f(\boldsymbol{x}_0)\|_p}{\|\boldsymbol{x} - \boldsymbol{x}_0\|_p}. \tag{2}$$

Similar to the analysis by Kim et al. [20], since Linear transformation by $W^{(V)}$ is Lipchitz and does not impact our analysis, we will drop it and focus on the non-linear part of $P\boldsymbol{z}$.

Since Patch embeddings are conducted by convolutional operations and the classification heads are fully connected layers, they are Lipschitz continuous [20]. Therefore as long as the transformer blocks are Lipchitz continuous, ViTs are Lipschitz continuous because the composite Lipchitz functions, i.e., $f \circ g$, are also Lipschitz continuous [12]. To this end, we have the Theorem 1.

Theorem 1. *(Transformer Blocks in ViTs are Lipschitz continuous) Given vision transformer block with trained parameters $\boldsymbol{w}$ and convex bounded domain $\mathcal{Z}_{l-1} \subseteq \mathbb{R}^{N \times D}$, we show that the transformer block $\mathcal{F}_l : \mathcal{Z}_{l-1} \to \mathbb{R}^{N \times D}$ mapping from z_{l-1} to z_l is Lipchitz function for all $l = 1, 2, ..., L$.*

Proof. For simplicity, we only prove the case that the number of heads H and the dimension of patch embedding D are all equal to 1. The general case can be found in Appendix.

Because the composition of the transformer block includes an MLP layer that is Lipchitz continuous, as argued by Kim et al. [20], it is the non-linear part of MSA that need to be proved Lipchitz continuous. We formulated the non-linear part as mapping $f : \mathcal{Z} \to \mathbb{R}^{N \times 1}$ shown in Eq. (3)

$$f(\boldsymbol{z}) = softmax(a\boldsymbol{z}\boldsymbol{z}^T)\boldsymbol{z} = P\boldsymbol{z} = \begin{pmatrix} p_1(\boldsymbol{u}_1) & \cdots & p_N(\boldsymbol{u}_1) \\ \vdots & \ddots & \vdots \\ p_1(\boldsymbol{u}_N) & \cdots & p_N(\boldsymbol{u}_N) \end{pmatrix} \boldsymbol{z} \tag{3}$$

where $a = W^{(Q)}W^{(K)T} \in \mathbb{R}$, $\boldsymbol{z} \in \mathcal{Z}$ which is a bounded convex set and belongs to $\mathbb{R}^{N \times 1}$, P is defined by *softmax* operator. Each row in P defines a discrete probability distribution. Therefore P can be regarded as the *transition matrix* for a finite discrete *Markov Chain* with $z_1, ..., z_n$ as observed value for random valuables. Since f has continuous first deviates, *Mean Value Inequality* can be used to find Lipchitz constant. Let $\boldsymbol{z}, \boldsymbol{z}_0 \in \mathcal{Z}$ and $\|\cdot\|_p$ denote the p-norm distance for vectors and *induced norm* for matrices. Specifically, when $p = 2$, the *induce norm* coincides with *spectral norm*, then we have

$$\|f(z) - f(z_0)\|_2 \leq \|J_f(\xi)\|_2 \|(z - z_0)\|_2, \tag{4}$$

where $\xi \in \mathcal{Z}$ is on the line through x and x_0, and $J_f(\cdot)$ denotes the *Jacobian* of f. As long as the Jacobian J_f is bounded for $\mathcal{Z}$, f is Lipschitz continuous. The Jacobian J_f is shown in Eq. (5) (see detail in Appendix).

$$J_f(z) = a\{diag(z)Pdiag(z) - Pdiag(z)diag(\mu) + diag(\sigma^2)\} + P, \tag{5}$$

where $\mu = Pz$ define the mean vector for the Finite Markov Chain and the *variance* are defined by

$$\sigma^2 = \begin{pmatrix} \sum_{k=1}^{N} p_k(u_1)x_k^2 - \left(\sum_{k=1}^{N} p_k(u_1)x_k\right)^2 \\ \vdots \\ \sum_{k=1}^{N} p_k(u_N)x_k^2 - \left(\sum_{k=1}^{N} p_k(u_N)x_k\right)^2 \end{pmatrix} = \begin{pmatrix} \sigma_1^2 \\ \vdots \\ \sigma_N^2 \end{pmatrix}, \tag{6}$$

Since every component on the right-hand-side in (5) is bounded since z is bounded. We conclude that $J_f(z)$ is also bounded, therefore the Lipchitz continuous.

Remark 1. The use of the Mean Value Theorem requires the domain $\mathcal{Z}$ to be convex, however as long as $\mathcal{Z}$ is bounded, we can always find a larger convex set $\mathcal{Z}' \supseteq \mathcal{Z}$.

Remark 2. Different from the conclusion drawn by Kim et al. [20] that the transformer is not Lipschitz continuous, ViTs are Lipschitz continuous due to the bounded input.

4.2 Model Adversarial Robustness as Cauchy Problem

Since there exists the *Residual Structure* in the *Transformer Encoder*, just like *ResNet*, which can be formulated as *Euler Method* [25], the *forward propagation* through Transformer Encoder can also be regarded as a *Forward Euler Method* to approximate the underlying *Ordinary Differential Equations (ODEs)*.

Let $f : \mathcal{X} \to \mathcal{Y}$ denote the ViTs, where $\mathcal{X} \subseteq \mathbb{R}^n$ denotes the input space and $\mathcal{Y} = \{1, 2, ..., C\}$ refers to the labels, and $\mathcal{F}_i, i = 1, ..., L$ denote the basic blocks. Notice that for simplicity, let $\mathcal{F}_1(x_0; w_0) + x_0$ refer to the patch embedding and $\mathcal{F}_L(x_{L-1}; w_{L-1}) + x_{L-1}$ be the classification head, the rest are transformer blocks. Hence, the forward propagation can be described in Eq. (7).

$$\begin{cases} x_k & = \mathcal{F}_k(x_{k-1}; w_{k-1}) + x_{k-1}, k = 1, ..., L \\ y_{logit} & = softmax(LP(x_L)) \\ y & = \arg\max_{\mathcal{Y}} y_{logit}, \end{cases} \tag{7}$$

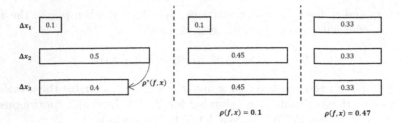

Fig. 1. Illustration of $\rho^\star(f,x)$. For better illustration L_1-norm is taken while calculating the $\rho(f,x)$.

where $x_0 \in \mathcal{X}$, $LP(\cdot)$ stands for *Linear Projection*, y_{logit} shows the likelihood for each class and $y \in \mathcal{Y}$ denote the classification result. As argued by Liao, et al. [23], the Transformer blocks in Eq. (7) can be regarded as *Forward Euler* approximation of the underlying ODE shown below.

$$\frac{d}{dt}x(t) = \mathcal{F}(x,t), t \in [t_0, T] \tag{8}$$

where $\mathcal{F}(\cdot)$ corresponds to the basic blocks in ViTs and $t \in [t_0, T]$ refers to the continuous indexing of those blocks.

The backward-propagation of Eq. (8) can be regarded as an estimation problem for parameters w of given boundary conditions defined by $\mathcal{X}$ and $\mathcal{Y}$, which leads to Neural ODEs [7].

Before the main theorem that models the adversarial robustness as Cauchy problem, we first define the adversarial robustness metrics. Given neural network f, and the fixed input $x \in \mathcal{X}$, the *local Adversarial Robustness* proposed by Bastani et al. [3] is defined as

$$\rho(f,x) \stackrel{def}{=} \inf\{\epsilon > 0 | \exists \hat{x} : \|\hat{x} - x\| \le \epsilon, f(\hat{x}) \neq f(x)\},$$

where $\|\cdot\|$ defines the general L_p norm. Usually, p is taken as $1, 2$ and ∞. The adversarial robustness is defined as the minimum radius that the classifier can be perturbed from their original corrected result. As illustrated in Fig. 1, considering the fact that even in the final layer $\Delta x(T)_1 < \Delta x(T)_2$, it is still possible that $softmax(LP(x(T) + \Delta x(T)_1))$ has been perturbed but $softmax(LP(x(T) + \Delta x(T)_2))$ is not, we use the minimal distortion to define the robustness as

$$\rho^\star(f,x) \stackrel{def}{=} \inf_{\|\hat{x}(T) - x(T)\|} \rho(f,x), \tag{9}$$

where $\hat{x}(T) - x(T) = \Delta x(T)$.

Lemma 1. *(Existence and Uniqueness for the Solution of Underlying ODE) Since the continuous mapping $\mathcal{F}$ defined in ODE (8) satisfies the Lipschitz condition on $z \in \mathcal{Z}$ for $t \in [t_0, T]$ as claimed in Theorem 1, where $\mathcal{Z}$ is a bounded closed convex set. There exists and only exists one solution for the underlying ODE defined in (8).*

Lemma 2. *(Error Bound for Forward Euler approximation) Given Forward Euler approximation shown in Eq. (7) and its underlying ODE in Eq. (8). Let $K > 0$ denotes the Lipschitz constant for the underlying ODE, and $\|\hat{\mathcal{F}}(x,t) - \mathcal{F}(x,t)\| \leq \delta$, hence the error of solution is given by*

$$\|\Delta x\| \leq \frac{\delta}{K}(e^{K|t-t_0|} - 1)$$

Since $\mathcal{F}(x,t)$ is continuous, δ can be arbitrary small as long as step for Euler approximation is small enough, namely, neural network is deep enough. The proof of Lemma 1 and 2 can be found in [8].

Theorem 2. *Let f and g be two neural networks defined in Eq. (7), which have the underlying ODEs as shown in Eq. (8), and denote the basic blocks of g as $\mathcal{G}_k, k = 1, ..., L'$ with its corresponding ODE defined as $\mathcal{G}$ to show the difference. Given point $x \in X$ and robustness metric $\rho^*(\cdot)$ defined in (9), classifier f is more robust than g, such that*

$$\rho^*(f, x) \leq \rho^*(g, x), \tag{10}$$

if $\forall t \in [t_0, T]$

$$\sigma_{max}(J_{\mathcal{F}}(t)) \leq \sigma_{max}(J_{\mathcal{G}}(t)) \tag{11}$$

where $J_f(t)$ and $J_g(t)$ refers to the Jacobian of the basic blocks $\mathcal{F}$ and $\mathcal{G}$ w.r.t. x and $\sigma_{max}(\cdot)$ denotes the largest singular value.

Proof. Consider 2 solutions $x(t), \hat{x}(t)$ of ODE defined in (8) such that for $\epsilon > 0$

$$\|\hat{x}(t_0) - x(t_0)\|_2 \leq \epsilon$$

and let $\Delta x(t) = \hat{x}(t) - x(t), t \in [t_0, T]$ hence

$$\frac{d}{dt}\Delta x = \mathcal{F}(\hat{x}, t) - \mathcal{F}(x, t) = J_{\mathcal{F}}(t)\Delta x + r_{\mathcal{F}}(\Delta x), \tag{12}$$

where $r_{\mathcal{F}}(\Delta x)$ is the residual of *Taylor Expansion* of $\mathcal{F}$ w.r.t. x, such that $\|r_{\mathcal{F}}(\Delta x)\| = \mathcal{O}(\|\Delta x\|^2)$ [2]. Instead of Δx, $\|\Delta x\|_2$ is more of our interest, hence

$$\frac{d}{dt}\|\Delta x\|_2 \leq \|\frac{d}{dt}\Delta x\|_2 \leq \|J_{\mathcal{F}}(t)\|_2\|\Delta x\|_2 + \mathcal{O}(\|\Delta x\|_2^2), \tag{13}$$

since $\|\Delta x(t_0)\|_2 = 0$ is trivial, we assume $\|\Delta x(t_0)\|_2 > 0$. And because there exist unique solution for the ODE system, we have $\|\Delta x(t)\|_2 > 0, t \in [t_0, T]$ therefore Eq. (13) becomes

$$\frac{1}{\|\Delta x\|_2}\frac{d}{dt}\|\Delta x\|_2 \leq \|J_{\mathcal{F}}(t)\|_2 + \mathcal{O}(\|\Delta x\|_2). \tag{14}$$

After integral of the both sides from t_0 to T we have

$$\int_{t_0}^{T} \frac{1}{\|\Delta x\|_2} d\|\Delta x\|_2 \leq \int_{t_0}^{T} \|J_{\mathcal{F}}(t)\|_2 + M\epsilon dt,$$

where $M > 0$ is a given large number. The integral for $[t_0, T]$ is given by

$$\|\Delta x(T)\|_2 \leq \epsilon e^{\int_{t_0}^{T} \|J_{\mathcal{F}}(t)\|_2 dt + (T-t_0)M\epsilon}. \tag{15}$$

It is obvious that the perturbed output of neural network $\Delta x(T)$ is actually bounded by the right-hand-side of Eq. (15) which is determined by the $\|J_{\mathcal{F}}(t)\|_2, t \in [t_0, T]$, namely the largest *singular value* of $J_{\mathcal{F}}(t)$, denoted as $\sigma_{max}(J_{\mathcal{F}}(t))$. The rest of the proof is simple, since if $\forall t \in [t_0, T]$ (11) holds and $(T - t_0)M\epsilon$ is negligible, we have

$$\|\Delta x_{\mathcal{F}}(T)\|_2 \leq \epsilon e^{\int_{t_0}^{T} \|J_{\mathcal{F}}(t)\|_2 dt} \leq \epsilon e^{\int_{t_0}^{T} \|J_{\mathcal{G}}(t)\|_2 dt},$$

therefore for any $\|\Delta x_{\mathcal{F}}(t_0)\|_2 \leq \rho^*(g, x)$ the classification result will also not change for f, hence the Eq. (10).

Remark 3. Theorem 2 is particularly useful for adversarial perturbation since the approximation in Eq. (15) relies on the narrowness of ϵ. If it is too large, the first-order approximation may fail.

Remark 4. Theorem 2 assumes that the approximation error induced in Lemma 2 is small enough to neglect. For very shallow models, e.g., ViT-S_1, ViT-S_2, the relation is violated, as is shown in Table 2.

5   Empirical Study

In order to verify the proposed theorem and find out whether self-attention indeed contributes to the adversarial robustness of ViTs, we replace the self-attention with a 1-D convolutional layer, as shown in Fig. 2. And we name the modified model *CoViT*, which stands for *Convolutional Vision Transformer*. We use Average Pooling instead of the classification token since the classification token can only learn the nearest few features rather than the whole feature maps for CoViT.

Both ViTs and CoViTs are trained from sketches without any pertaining to ensure that they are comparable. *Sharpness-Aware Minimization (SAM)* [13] optimizer is used throughout the experiments to ensure adequate clear accurate.

5.1   Configuration and Training Result

The configurations for Both ViTs and CoViTs have an input resolution of 224 and embedding sizes of 128 and 512. The use of a smaller embedding size of 128 is to calculate the maximum singular value exactly. An upper bound is calculated

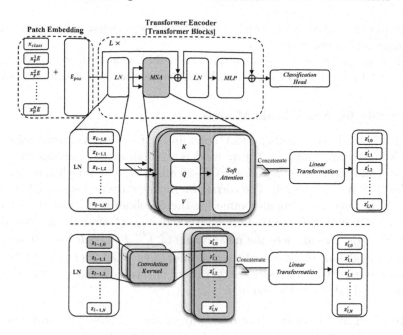

Fig. 2. Illustration of ViT and CoViTs. After *Patch Embedding*, the *Transformer Encoder* is composed of $L\times$ *Transformer Blocks*, of which in each K, Q and V stands for *Key*, *Query* and *Value* are computed as linear projection from former tokens z_{l-1}, hence *Self-Attention* is calculated as $softmax(\frac{QK^T}{\sqrt{D}})V$. In order to better understand whether self-attention indeed contributes to adversarial robustness, it is replaced by *1-D convolution layers* where different kernel are used and the intermediates, denoted by z_l^*, are generated before concatenation and linearly projecting to z_l'. The kernel size can be different for each convolutional projection.

instead for models with a larger embedding size since the exact calculation is intractable. We change the number of heads for ViTs, the kernels for CoViTs, the depth, and the patch size for the experiment. All the models are divided into four groups: S, M, L, T, standing for *Small*, *Medium*, *Large*, and *Tiny* of parameter size. The tiny model uses an embedding size of 128. The detailed configuration is shown in Appendix.

All the models are trained on CIFAR10, and the base optimizer for SAM is SGD with the One-cycle learning scheduler of maximum learning rate equals to 0.1. In order to have a better performance, augmentations, including *Horizontal Flipping, Random Corp* and *Color Jitter*, are involved during training. We resize the image size to the resolution of 224×224. The model with an embedding size of 512 is trained by 150 epochs, and the tiny model with an embedding size of 128 is trained by 300 to achieve adequate performance. The performance of models within the same group is similar, and the shallow networks, e.g., ViT-S_1, ViT-S_2, CoViT-S_1, CoViT-S_2, are harder to train and may need extra training to be converged. This may be due to the optimizer used, i.e., SAM, since SAM will try to find shallow-wide optima instead of a deep-narrow one, which requires a stronger model capacity.

The experiments are conducted on Nvidia RTX3090 with python 3.9.7, and realized by PyTorch 1.9.1. *Torchattacks* [20] is used for adversarial attacks.

5.2   Study for Small Scale Models

In order to find out whether MSA contributes to the adversarial robustness of ViTs and verify Theorem 2, tiny models with an embedding size of 128 are employed and attacked by L_2-norm PGD-20 and CW. The threshed of successful attacks for CW is set to 260. The corresponding average and standard deviation of the exact maximum singular value for the Jacobian is calculated over layers and images to indicate the overall magnitude of $\sigma_{max}(t)$ over the interval $[t_0, T]$. In other words, we calculate the mean value of $\int_{t_0}^{T} \|J_{\mathcal{F}}(t)\|_2 dt$ for 500 images to indicate the global robustness of the classifier. The PGD-20(L_2) and CW share the same setting with large-scale experiments in Table 2, except that the total iteration for PGD is 20 instead of 7.

Verification of Theorem. The result, as shown in Table 1, generally matches our theoretic analysis since the most robust model has the lowest average maximum singular value. It is worth mentioning that the smaller value of $\bar{\sigma}_{max}$ cannot guarantee stronger robustness for ViTs in Table 1, since the standard deviations of σ_{max} are much larger, e.g., 11.66, than that of CoViTs.

Contribution of MSA. Another observation is that CoViTs are generally more robust than ViTs. In other words, without enough embedding capacity, Self-Attention could even hurt both the robustness and generalization power. In addition, increasing the models' depth will enhance both generalization power and robustness.

Distribution of Maximum Singular Value in Each Layer. In order to know which layer contributes most to the non-robustness, the distribution of σ_{max} is calculated. The layer that has the highest value of σ_{max} may dominate the robustness of the network. As is shown in Fig. 3, maximum singular values for the CoViTs are much concentrated around the means, reflecting more stable results for classification. And the maximum singular values for the first and last layer of all tiny models are significantly higher than that for in-between layers, indicating that the first and last layers in the transformer encoder are crucial for adversarial robustness.

5.3   Contribution of MSA to Robustness for Large Scale Models

We attack both ViTs and CoViTs with FGSM, PGD, and CW for large-scale models and compare the robust accuracy. And since it is intractable to compute exact maximum singular value for the matrix of size $(128 \cdot 512) \times (128 \cdot 512)$, an upper bound of maximum singular value is calculated as

Table 1. Attack result and the average maximum singular value for tiny model. All the models have embedding size of 128 with different depth and head or kernels. The cleaning accuracy and the robust accuracy for PGD-20 and CW attack are shown in the Table. The mean of exact maximum singular value σ_{max} for over layers and 500 input images are calculated with standard deviation shown in square. The highest accuracy and lowest maximum singular value are marked in bold.

Net-name	Depth	#Head/ Kernel	Clear Acc.	PGD-20(L_2)	CW(L_2)	$\bar{\sigma}_{max}$
ViT-T_1	4	1	0.819	0.397	0.0305	10.45(4.125)
ViT-T_2	4	4	0.820	0.407	0.039	18.18(11.66)
CoViT-T_1	4	$K3$	0.849	0.492	**0.083**	9.25(0.822)
CoViT-T_2	4	$4 \times K3$	**0.852**	0.492	0.036	**9.186(1.088)**
ViT-T_3	8	1	0.836	0.442	0.040	7.17(1.52)
ViT-T_4	8	4	0.834	0.463	0.048	10.03(2.73)
CoViT-T_3	8	$K3$	**0.860**	0.514	**0.076**	**6.413(0.562)**
CoViT-T_4	8	$4 \times K3$	0.860	**0.515**	0.065	6.662(0.386)

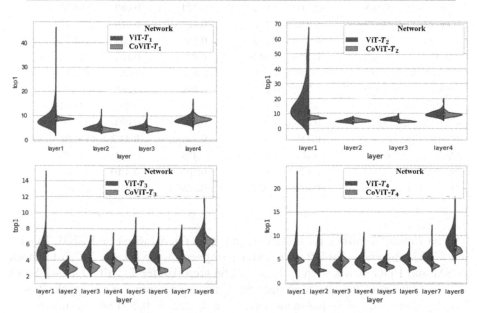

Fig. 3. Violin plot for maximum singular value for each layer of the ViTs/CoViTs. The y-axis shows the maximum singular value.

$$\|J\|_2 \leq \left(\|J\|_1 \|J\|_\infty \right)^{\frac{1}{2}}, \tag{16}$$

which is used as an approximation to the maximum singular value of the Jacobian. $\|J\|_1$ and $\|J\|_\infty$ denotes L_1 and L_∞ induced norm for the Jacobian. The mean value for 50 images is taken.

Table 2. Summary of attacking results and corresponding estimated largest singular value. The attacks are employed for both ViTs and CoViTs with FGSM, PGD-7 and CW, and the robust accuracy are shown for each attack. The models with patch size 32×32 are marked with $*$. $\|J\|_1$ and $\|J\|_\infty$ are the L_1 and L_∞ norm respectively. The highest accuracy and lowest estimated maximum singular value are marked in bold.

	CLEAN ACC.	FGSM	PGD-7(L_∞)	PGD-7(L_2)	CW(L_2)	$(\|J\|_1\|J\|_\infty)^{\frac{1}{2}}$
ViT-S_1	0.676	0.213	0.135	0.267	0.059	812.69
ViT-S_2	0.739	**0.273**	0.162	**0.348**	0.067	1003.20
CoViT-S_1	0.734	0.254	**0.173**	0.341	**0.144**	242.78
CoViT-S_2	0.737	0.244	0.163	0.328	0.143	**206.78**
ViT-S_3	0.847	0.369	0.221	0.444	0.053	296.48
ViT-S_4	0.863	**0.392**	**0.240**	**0.448**	0.065	462.12
CoViT-S_3	0.882	0.320	0.179	0.413	**0.104**	**146.48**
CoViT-S_4	0.876	0.306	0.170	0.401	0.088	150.02
CoViT-S_5	0.868	0.341	0.192	0.424	0.082	163.50
ViT-M_1	0.877	0.415	0.267	0.467	0.049	236.21
ViT-M_2	0.861	0.415	0.260	0.461	0.053	294.06
$*$ViT-M_3	0.853	**0.478**	**0.356**	**0.519**	0.103	139.23
CoViT-M_1	0.881	0.336	0.185	0.422	0.051	93.21
CoViT-M_2	0.882	0.337	0.197	0.417	0.086	109.79
CoViT-M_3	0.870	0.337	0.194	0.424	0.072	131.94
CoViT-M_4	0.875	0.357	0.208	0.427	0.093	99.57
$*$CoViT-M_5	0.861	0.416	0.303	0.480	**0.152**	**78.70**
$*$ViT-L	0.848	0.461	0.347	0.499	0.094	111.38
$*$CoViT-L_1	0.867	0.443	0.333	0.505	**0.140**	**59.54**
$*$CoViT-L_2	0.853	**0.466**	**0.357**	**0.528**	0.096	37.26

The *Robust Accuracy* for both ViTs and CoViTs attacked by FGSM, PGD-7 and CW is shown in Table 2. For better comparison, we set $\epsilon = 2/225$ for both FGSM and PDG attack with L_∞ norm. The step size for L_∞ PGD attack is set to $\alpha = 2/255$ and it is iterated only for 7 times to represent the weak attack. The L_2 norm PGD-7 is parameterized by $\epsilon = 2, \alpha = 0.2$. The parameters set for stronger CW attack is that $c = 1$, adversarial confidence level *kappa* = 0, learning rate for Adam [22] optimizer in CW is set to 0.01 and the total iteration number is set to 100.

As is shown in Table 2, for weak attacks, i.e., FGSM and PGD-7, ViTs are generally exhibiting higher robust accuracy within the same group of similar parameter sizes with only a few exceptions. Also, both for ViTs and CoViTs, the robustness is strengthened as the model becomes deeper with more parameters.

For a stronger CW attack, the result is almost reversed, CoViT model shows significantly better robustness and agrees with the approximation of the maximum singular value for the Jacobian. The ability of Self-Attention to avoid perturbed pixels is compromised as the attacking becomes stronger. And it seems that the *translation invariance* of CNNs has more defensive power against strong attacks. In addition, a larger patch size always induces better adversarial robustness for both ViTs and CoViTs.

6   Conclusion

This paper first proves that ViTs are Lipschitz continuous for vision tasks, then we formally bridge up the local robustness of transformers with the Cauchy problem. We theoretically proved that the maximum singular value determines local robustness for the Jacobian of each block. Both small-scale and large-scale experiments have been conducted to verify our theories. With the proposed framework, we open the black box of ViTs and study how robustness changes among layers. We found that the first and last layers impede the robustness of ViTs. In addition, unlike existing research that argues MSA could boost robustness, we found that the defensive power of MSA in ViT only works for the large model under weak adversarial attacks. MSA even compromises the adversarial robustness under strong attacks.

7   Discussion and Limitation

The major limitations in this paper are embodied by the several approximations involved. The first one is the approximation of the underlying ODEs to the forward propagation of neural networks with a residual addition structure. As is shown in Lemma 2, the approximation is accurate only when the neural networks are deep enough, and it is hard to know what depth is enough, given the required error bound. One possible way to make it accurate is to consider the *Difference Equation*, which is a discrete parallel theory to ODEs. The second one is the approximation of the second-order term in Eq. (14). For small-size inputs, we can say that the L_2-norm of perturbations of adversarial examples is smaller enough so that the second term is negligible. However, the larger inputs may inflate the L_2-norm of perturbations since simply up sampling could result in a larger L_2-norm. Therefore, including the second term or choosing a better norm should be considered. The third approximation is shown in Eq. (16). Since the size of the Jacobian depends on the size of the input image, making it impossible to directly calculate the singular value of the Jacobian for larger images, hence, we use an upper bound instead, which inevitably compromises the validation of the experiment. Moreover, since the adversarial attack can only get the upper bound of the minimal perturbations, it is also an approximation of the local robustness, as shown in Table 2.

In the experimental part, we only take into account for the small to moderate size models because it is necessary to rule out the influence of pre-training, and we have to admit that the calculation for the singular value of Jacobian w.r.t. inputs of too large size is hardly implemented.

References

1. Aldahdooh, A., Hamidouche, W., Deforges, O.: Reveal of vision transformers robustness against adversarial attacks. arXiv preprint arXiv:2106.03734 (2021)
2. Ascher, U.M., Mattheij, R.M., Russell, R.D.: Numerical solution of boundary value problems for ordinary differential equations. SIAM (1995)
3. Bastani, O., Ioannou, Y., Lampropoulos, L., Vytiniotis, D., Nori, A., Criminisi, A.: Measuring neural net robustness with constraints. Adv. Neural. Inf. Process. Syst. **29**, 2613–2621 (2016)
4. Benz, P., Ham, S., Zhang, C., Karjauv, A., Kweon, I.S.: Adversarial robustness comparison of vision transformer and MLP-mixer to CNNs. arXiv preprint arXiv:2110.02797 (2021)
5. Bhojanapalli, S., Chakrabarti, A., Glasner, D., Li, D., Unterthiner, T., Veit, A.: Understanding robustness of transformers for image classification. In: CVF International Conference on Computer Vision, ICCV, vol. 9 (2021)
6. Carion, N., Massa, F., Synnaeve, G., Usunier, N., Kirillov, A., Zagoruyko, S.: End-to-end object detection with transformers. In: Vedaldi, A., Bischof, H., Brox, T., Frahm, J.-M. (eds.) ECCV 2020. LNCS, vol. 12346, pp. 213–229. Springer, Cham (2020). https://doi.org/10.1007/978-3-030-58452-8_13
7. Chen, R.T., Rubanova, Y., Bettencourt, J., Duvenaud, D.K.: Neural ordinary differential equations. In: Advances in Neural Information Processing Systems, vol. 31 (2018)
8. Coddington, E.A., Levinson, N.: Theory of Ordinary Differential Equations. Tata McGraw-Hill Education, New York (1955)
9. Cordonnier, J.B., Loukas, A., Jaggi, M.: On the relationship between self-attention and convolutional layers. In: International Conference on Learning Representations (2020)
10. Dosovitskiy, A., et al.: An image is worth 16×16 words: transformers for image recognition at scale. In: International Conference on Learning Representations (2021)
11. Dupont, E., Doucet, A., Teh, Y.W.: Augmented neural odes. In: Advances in Neural Information Processing Systems, vol. 32 (2019)
12. Federer, H.: Geometric Measure Theory. Springer, Cham (2014). https://doi.org/10.1007/978-3-642-62010-2
13. Foret, P., Kleiner, A., Mobahi, H., Neyshabur, B.: Sharpness-aware minimization for efficiently improving generalization. arXiv preprint arXiv:2010.01412 (2020)
14. Grossberg, S.: Recurrent neural networks. Scholarpedia **8**(2), 1888 (2013)
15. He, K., Zhang, X., Ren, S., Sun, J.: Deep residual learning for image recognition. Corr abs/1512.03385 (2015)
16. Hendrycks, D., Zhao, K., Basart, S., Steinhardt, J., Song, D.: Natural adversarial examples. In: CVPR (2021)
17. Hinton, G., Vinyals, O., Dean, J.: Distilling the knowledge in a neural network. arXiv preprint arXiv:1503.02531 (2015)
18. Hu, H., Lu, X., Zhang, X., Zhang, T., Sun, G.: Inheritance attention matrix-based universal adversarial perturbations on vision transformers. IEEE Signal Process. Lett. **28**, 1923–1927 (2021)
19. Khan, S., Naseer, M., Hayat, M., Zamir, S.W., Khan, F.S., Shah, M.: Transformers in vision: a survey. ACM Comput. Surv. (CSUR) **54**(10s), 1–41 (2021)
20. Kim, H.: TorchAttacks: a PyTorch repository for adversarial attacks. arXiv preprint arXiv:2010.01950 (2020)

21. Kim, H., Papamakarios, G., Mnih, A.: The Lipschitz constant of self-attention. In: International Conference on Machine Learning, pp. 5562–5571. PMLR (2021)
22. Kingma, D.P., Ba, J.: Adam: a method for stochastic optimization. arXiv preprint arXiv:1412.6980 (2014)
23. Liao, Q., Poggio, T.: Bridging the gaps between residual learning, recurrent neural networks and visual cortex. arXiv preprint arXiv:1604.03640 (2016)
24. Liu, Z., et al.: Swin transformer: hierarchical vision transformer using shifted windows. In: Proceedings of the IEEE/CVF International Conference on Computer Vision, pp. 10012–10022 (2021)
25. Lu, Y., Zhong, A., Li, Q., Dong, B.: Beyond finite layer neural networks: Bridging deep architectures and numerical differential equations. In: International Conference on Machine Learning, pp. 3276–3285. PMLR (2018)
26. Mahmood, K., Mahmood, R., Van Dijk, M.: On the robustness of vision transformers to adversarial examples. In: Proceedings of the IEEE/CVF International Conference on Computer Vision, pp. 7838–7847 (2021)
27. Mao, C., Jiang, L., Dehghani, M., Vondrick, C., Sukthankar, R., Essa, I.: Discrete representations strengthen vision transformer robustness. arXiv preprint arXiv:2111.10493 (2021)
28. Mao, X., et al.: Towards robust vision transformer. In: Proceedings of the IEEE/CVF Conference on Computer Vision and Pattern Recognition, pp. 12042–12051 (2022)
29. Moosavi-Dezfooli, S.M., Fawzi, A., Fawzi, O., Frossard, P.: Universal adversarial perturbations. In: Proceedings of the IEEE Conference on Computer Vision and Pattern Recognition, pp. 1765–1773 (2017)
30. Naseer, M.M., Ranasinghe, K., Khan, S.H., Hayat, M., Shahbaz Khan, F., Yang, M.H.: Intriguing properties of vision transformers. Adv. Neural. Inf. Process. Syst. **34**, 23296–23308 (2021)
31. Paul, S., Chen, P.Y.: Vision transformers are robust learners. arXiv preprint arXiv:2105.07581 (2021)
32. Qin, Y., Zhang, C., Chen, T., Lakshminarayanan, B., Beutel, A., Wang, X.: Understanding and improving robustness of vision transformers through patch-based negative augmentation. arXiv preprint arXiv:2110.07858 (2021)
33. Ruseckas, J.: Differential equations as models of deep neural networks. arXiv preprint arXiv:1909.03767 (2019)
34. Srinivas, A., Lin, T.Y., Parmar, N., Shlens, J., Abbeel, P., Vaswani, A.: Bottleneck transformers for visual recognition. In: Proceedings of the IEEE/CVF Conference on Computer Vision and Pattern Recognition, pp. 16519–16529 (2021)
35. Sun, Q., Tao, Y., Du, Q.: Stochastic training of residual networks: a differential equation viewpoint. arXiv preprint arXiv:1812.00174 (2018)
36. Touvron, H., Cord, M., Douze, M., Massa, F., Sablayrolles, A., Jégou, H.: Training data-efficient image transformers & distillation through attention. In: International Conference on Machine Learning, pp. 10347–10357. PMLR (2021)
37. Wang, W., et al.: Pyramid vision transformer: a versatile backbone for dense prediction without convolutions. In: Proceedings of the IEEE/CVF International Conference on Computer Vision, pp. 568–578 (2021)
38. Zhou, D., et al.: DeepViT: towards deeper vision transformer. arXiv preprint arXiv:2103.11886 (2021)

Meta-learning, Neural Architecture Search

Automatic Feature Engineering Through Monte Carlo Tree Search

Yiran Huang$^{(\boxtimes)}$ [ID], Yexu Zhou [ID], Michael Hefenbrock [ID], Till Riedel [ID], Likun Fang [ID], and Michael Beigl [ID]

Telecooperation Office, Karlsruhe Institute of Technology, Karlsruhe, Germany
{yhuang,zhou,hefenbrock,riedel,fang,michael}@teco.edu

Abstract. The performance of machine learning models depends heavily on the feature space and feature engineering. Although neural networks have made significant progress in learning latent feature spaces from data, compositional feature engineering through nested feature transformations can reduce model complexity and can be particularly desirable for interpretability. To find suitable transformations automatically, state-of-the-art methods model the feature transformation space by graph structures and use heuristics such as ϵ-greedy to search for them. Such search strategies tend to become less efficient over time because they do not consider the sequential information of the candidate sequences and cannot dynamically adjust the heuristic strategy. To address these shortcomings, we propose a reinforcement learning-based automatic feature engineering method, which we call Monte Carlo tree search Automatic Feature Engineering (mCAFE). We employ a surrogate model that can capture the sequential information contained in the transformation sequence and thus can dynamically adjust the exploration strategy. It balances exploration and exploitation by Thompson sampling and uses a Long Short Term Memory (LSTM) based surrogate model to estimate sequences of promising transformations. In our experiments, mCAFE outperformed state-of-the-art automatic feature engineering methods on most common benchmark datasets.

Keywords: Data mining · Feature engineering · Monte Carlo tree search · Reinforce learning

1 Introduction

In many applications, the success of machine learning is often attributed to the experience of experts who use not only the best-fitting algorithms but also extensive domain knowledge. This domain knowledge is often reflected in the pre-processing of raw data: it is transformed step-by-step so that it can be optimally processed by an automated machine learning pipeline. Most of this heuristic search performed by an expert is commonly referred to as feature engineering. Due to limited human resources but ever-growing computing capabilities, automating this search process is becoming increasingly attractive.

© The Author(s), under exclusive license to Springer Nature Switzerland AG 2023
M.-R. Amini et al. (Eds.): ECML PKDD 2022, LNAI 13715, pp. 581–598, 2023.
https://doi.org/10.1007/978-3-031-26409-2_35

Feature engineering can be understood as a combinatorial optimization that attempts to maximize the utility of a subsequent optimization step, i.e., fitting the model. By employing explicit feature engineering, as opposed to deep learning (e.g., Long Short Term Memory (LSTM) [10] in [1]), we obtain a tighter control over the model space.

Furthermore, good feature engineering can increase the robustness (generalizability to unknown data) and interpretability (predictability of decisions based on input features) of the overall machine learning architecture.

However, there are several challenges in automatically searching for useful features made up of sequences of atomic mathematical transformations such as addition (add), logarithm (log), or a sine function (sin). First, the search space grows large very quickly, as the number of possible transformation sequences grows exponentially with their length and the atomic transformations allowed. Second, evaluating a potentially promising transformation sequence can be time-consuming, as it requires training and evaluation of a machine learning model. Both of these features make the search challenging and require methods that search the space efficiently.

To address these challenges, Cognito [6] models the exploration of the transformation space with a transformation tree and explores the tree with some handcrafted heuristic traversal strategies such as depth-first, global traversal, or balanced traversal. Furthermore, the recently proposed reinforcement-based approach [7] applies a Q-learning algorithm and approximates the Q value with linear approximation to automate feature engineering.

While these methods achieve good results, our hypothesis is that they can be significantly improved by addressing two aspects, namely:

- **Choice of search hyperparameters and dynamic adaptation of the heuristic strategy**: A serious problem with an approach that relies entirely on guidance is the tendency to fall into local optima. Strategies like ϵ-greedy and Upper Confidence Bound (UCB) [17] can mitigate this problem, however, both need careful tuning of the initial hyperparameters that also control the dynamic adaptation of their search strategy.
- **Sequential information of the composite transformations**: New features can be transformations of existing features. Such compositions are sensitive to the order in which the atomic transformations are applied. State of the art feature engineering methods approximate the performance of a given transformation sequence with a linear model [7] or a deep convolutional neural network [8], which do not exploit the sequential information (order) contained in the composite transformation.

To address these shortcomings, we present a novel algorithm called Monte Carlo tree search for Automatic Feature Engineering (mCAFE). We choose Thompson sampling as an automatically adjusting selection policy, in combination with an LSTM network to capture the sequential information in the feature transformation sequences, while the main structure follows a Monte Carlo Tree Search (MCTS) [9].

Our contributions can be summarized as 1) we leverage Thompson sampling to guide the exploration, thus avoiding the parameters initialization and strategy dynamic adjustment problem. 2) we utilize sequential information of composed transformations by training an LSTM-based surrogate model for predicting the expected reward of a transformation sequence to a given dataset. 3) we evaluated the algorithm on common benchmark datasets (see [7]) and achieved improvements on most of them.

2   Related Work

In recent years, a large number of research results have emerged for domain-specific feature engineering. The work of [2] investigates how to share information through feature engineering in multi-task learning tasks, and [3] tries to find suitable features to improve the class separation. However, less new research has been done on feature engineering applicable to all data types. FCTree, proposed in [12], uses the original and constructed features as the splitting point to partition the data through a decision tree. It constructs local features where the local error is high and the features constructed so far are not well predicted. FEADIS [13] uses a random combination of mathematical functions, including ceiling, modulus, sin, and feature selection methods to construct new features. Of these, features are then selected greedily and added it to the original features. The Data Science Machine (DSM) [4] applies transformations to all features at once. Then, feature selection and model optimization are performed on the generated dataset. A similar procedure was also applied in [5]. In contrast, ExploreKit [14] increases the constructed features iteratively. To overcome the exponential growth of the feature space, ExploreKit uses a novel machine learning-based feature selection approach to predict the usefulness of new candidate features. Similarly, Cognito [6] introduced the notion of a tree-like exploration of the transformation space. Through a few handcrafted heuristics traversal strategies, such as depth-first and global-first strategy, Cognito can efficiently explore the set of available transformations. However, several factors, such as episode budget constraints, are beyond the consideration of the strategy. As an improvement, a reinforcement learning-based feature engineering method was proposed in [7] to explore the available feature engineering choices under a given budget. Finally, LFE [15] considers each feature individually and predicts the best transformation of each feature through the learning-based method. However, none of these methods takes the order of the transformations of the features into account. More recently, a graph-based method was proposed in [8] that guides the exploration of the transformation space with a deep neural network.

3   Methodology

We model the feature engineering problem as a classic episode-based reinforcement learning problem consisting of an agent interacting with the environment.

The search starts from the initial state representing the original dataset $D_0 \in \mathcal{D}$, where $\mathcal{D}$ denotes the state space. From D_0, a transformation $t \in \mathcal{T}$ (action) can be chosen to transform the dataset (all the features contained in the dataset) according to t. The new state D' is then obtained by the concatenation of the data of the old state D with $t(D)$, i.e., $D' = [D, t(D)]$. Through this, the new state contains all information (data) from the previous states, which can be seen as a Markov property. Finally, for each state D, a machine learning model can be trained on D to obtain its n fold cross-validation performance. However, since we seek to obtain the best sequence of length L, we further define a feature engineering pipeline, as an ordered sequence $(t_1, \cdots, t_i, \cdots, t_L)$ consisting of L transformations. The i-th entry of the sequence denotes the decision in the i-th step, i.e., the transformation to apply to the data in order to generate new features. Overall, the environment can be summarized with a 3-tuple $(\mathcal{D}, \mathcal{T}, r)$, denoting the state space $\mathcal{D}$, the transformation (action) space $\mathcal{T}$ and the rewards $r \in \{0, 1\}$. The reward thereby expresses whether a transformation pipeline of length L improved over the best performing model found so far.

Monte-Carlo Tree Search (MCTS) defines a class sampling-based tree search algorithms used to find optimal decisions in vast search domains and has been successfully applied to related problems like feature subset selection [25]. To deal with huge search spaces, MCTS models the search space as a tree structure and explores the tree iteratively. It gradually favors the most promising regions in the search tree given an arbitrary evaluation function.

Evidently, our search space of feature transformations can span such a tree, which allows the application of MCTS to find a feature set that contains features constructed by an optimal transformation pipeline on the original dataset. We discuss the construction of the tree in the following, alongside the selection policy (Thompson sampling) and a surrogate model-based (LSTM) expansion policy. Finally, we will outline the overall mCAFE algorithm. In contrast to problems like feature subset selection, the ordering of the nodes inside the tree is of critical importance in our case.

3.1 The Transformation Tree

We illustrate the reinforcement task with a transformation tree of maximum depth L, in which each node represents a state (dataset), each edge represents an action (transformation) and each path from the root to a leaf node represents a feature engineering pipeline. Additionally, each edge in the tree is associated with a distribution, which shows the mean success (reward $= 1$) probability of taking the action at its parent state. The nodes in the tree are divided into two categories: (1) *root node* D_0 is the initial state for each pipeline and represents the original dataset; (2) *derived nodes* D_i, where $i > 0$, has only one parent node D_j, $i > j \geq 0$ and the connecting edge responds to the action $t \in \mathcal{T}$ applied to the parent node, i.e., $D_i = [D_j, t(D_j)]$. In this way, we translate the feature engineering problem into a problem of exploring the transformation tree to find the node that maximizes the expected reward.

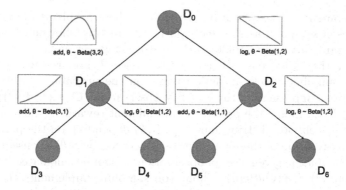

Fig. 1. Representation of feature transformation with a tree structure. Here, each node corresponds to a state D and each edge corresponds to a transformation (action). The distributions on the edges display the distribution over the mean success (reward = 1) probability when taking the action in the parent state.

Figure 1 shows a full transformation tree for a pipeline of $L = 2$ and two available actions $\mathcal{T} = \{\log, \text{add}\}$. Each node in the tree is a candidate dataset for the feature engineering problem. For example, the derived nodes D_4 and D_5, represent

$$D_4 = \{D_0, \text{add}(D_0), \log(D_0), \log(\text{add}(D_0))\},$$
$$D_5 = \{D_0, \log(D_0), \text{add}(D_0), \text{add}(\log(D_0))\}.$$

Note that, although the transformations in D_4 and D_5 are the same, the resulting dataset is not identical due to the order in which the transformations are applied.

We can find the optimal node by traversing this tree. However, the complexity of this task grows exponentially as L and the number of available transforms $|\mathcal{T}|$ becomes larger. Since traversing all possible nodes of the tree is prohibitive, mCAFE focuses on optimizing the selection policy π_s and expansion policy π_e to reduce the number of evaluations required to find a good transformation sequence.

3.2   The Selection Policy

The selection policy π_s determines the balance between exploration and exploitation. It guides the selection for known parts of the MCTS. The UCB and ϵ-greedy are the two most commonly used selection policies, for which also strong theoretical guarantees on the regret[1] can be proven. While they have proven successful in various reinforcement learning settings, they are not ideal for the application of feature engineering. This is mainly due to their requirement to explicitly define the exploration and exploitation trade-off through ϵ in ϵ-greedy and λ in the UCB. Additionally, ϵ greedy does not adapt the trade-off dynamically

---

[1] The amount we lose for not selecting optimal action in each state.

but always pursues ϵ % exploration. To address these problems, we make use of the Thompson sampling as the selection policy. In the following, we introduce Thompson sampling and adapt it to the feature engineering case.

Consider the state space $\mathcal{D}$, the action space $\mathcal{T}$ and rewards $r \in \{0,1\}$. Thompson sampling selects an action based on the probability of it being the optimal action. Representing the set of N observations $\mathcal{O} = \{(r,t,D)\}^N$, where $D \in \mathcal{D}, t \in \mathcal{T}$, we model the probability of different rewards of each action with a parametric likelihood distribution $p(r|t, D, \theta)$ depending on the parameters θ. The prior distribution of these parameters is denoted by $p(\theta)$. Consequently, the posterior distribution given a set of observations $\mathcal{O}$ can be calculated using Bayes rule, i.e., $p(\theta|\mathcal{O}) \propto p(\mathcal{O}|\theta)p(\theta)$. Thompson sampling implements the selection policy π_s by sampling a parameter θ from the posterior distribution $p(\theta|\mathcal{O})$, and taking the action that maximizes the expected reward. Hence,

$$\pi_s(D) = \underset{t \in \mathcal{T}}{\mathrm{argmax}} \quad \mathbb{E}\left[r|t, D, \theta\right] \quad \text{where} \quad \theta \sim p(\theta|\mathcal{O}). \tag{1}$$

Since, in the case of feature engineering, each state $D \in \mathcal{D}$ satisfies the Markov property, we can simplify the problem of which action to take on state D, to whether taking the action $t \in \mathcal{T}$ leads to a performance improvement. This can be modeled as a classic Bernoulli bandit problem, where the variable $\theta = (\theta_1, \theta_2, \cdots)$ denotes the expected values of a Bernoulli random variable expressing the probability of taking the selected action in given a state (and obtaining a reward of one). The distribution of the parameter θ_t can be modeled through a beta distribution

$$p(\theta_t|\alpha, \beta) = \frac{\Gamma(\alpha + \beta)}{\Gamma(\alpha)\Gamma(\beta)}\theta_t^{\alpha-1}(1 - \theta_t)^{\beta-1},$$

where Γ is the Gamma function. $\frac{\Gamma(\alpha+\beta)}{\Gamma(\alpha)\Gamma(\beta)}$ serves as a normalisation constant that ensures the integration of the density function over (0,1) is 1. The parameters α and β control the shape of the distribution and the mean of the distribution is $\frac{\alpha}{\alpha+\beta}$. It denotes the expectation that taking the corresponding action will lead to performance improvement. The higher α, the larger the mean and therefore the probability of the action to be selected. On the other hand, the larger β, the lower the probability.

The beta distribution is conjugate to the Bernoulli distribution (i.e., the posterior distribution $p(\theta|\mathcal{O})$ inherits the functional form the prior distribution $p(\theta)$). Given an observed sample $O = (r, t, D)$, the posterior distribution of the parameters θ is given by

$$\theta_{t'} \sim \text{Beta}\left(\alpha + \mathbb{1}_{r=1,t'=t}, \beta + \mathbb{1}_{r=0,t'=t}\right), \quad t' \in \mathcal{T}. \tag{2}$$

The parameter α is incremented when the action led to an improvement in performance. Otherwise, the parameter β is incremented. In this view, α represents the number of successes in the Bernoulli trial and β represents the number of failures. Furthermore, the support of the beta distribution is $(0, 1)$,

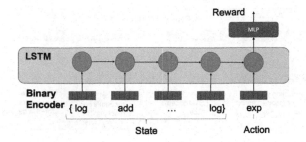

Fig. 2. The Surrogate network consist of 2 LSTM layers of size 32 and a two fully connected layers of size 32 with a ReLU activation function.

independent of the parameterization. This ensures that there is always a nonzero probability for each action to be selected. Consequently, there is always a nonzero probability to take each path in the tree.

Figure 1 shows an example of the tree representation. Each edge in the tree maintains a beta distribution $\text{Beta}(\alpha, \beta)$. By comparing the two transformations on D_0 with the same β, we can see that the higher the value of α the more the distribution is shifted towards sampling larger values (higher probabilities of success). In each step, an edge is selected based on the sampling result. This ensures the priority of high-quality edges while also allowing inferior edges to be selected occasionally. By using Thompson sampling as the selection policy, we avoid choosing hyperparameters to balance the exploration and exploitation trade-off. In contrast, the trade-off is adjusted dynamically through the posterior distribution of the parameter θ, which is updated along with the observation. Even though α and β represent hyperparameters, their choice is arguably more intuitive as $\alpha = \beta = 1$ describes a uniform distribution.

The requirement to construct and sample from a beta distribution for each action may rise efficiency concerns, as this process is slow compared to an ϵ-greedy selection. However, this is not an issue for feature engineering as in each episode, the selection phase takes little time compared to the other phases of the algorithm. This will be further explored in Sect. 4 (see Table 1).

3.3 The Expansion Policy

The selection policy π_s guides the selection of actions in parts of the search space that have been explored. Outside of the explored search space and beyond the leaves of the MCTS, the expansion policy π_e guides the selection of the actions t. It expands the child nodes to the tree and selects the one with the maximum expectation reward (Q value) as the next exploration candidate

$$\begin{aligned} \pi_e(D) &= \operatorname*{argmax}_{t \in \mathcal{T}} \ \mathbb{E}\left[r|t, D, \theta\right] \\ &= \operatorname*{argmax}_{t \in \mathcal{T}} \ \hat{Q}(D, t). \end{aligned} \tag{3}$$

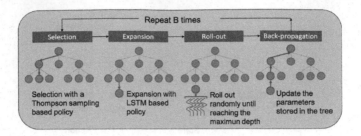

Fig. 3. The mCAFE framework: each iteration (episode) includes four phases: selection, expansion, roll-out, back-propagation. B is the number of iterations.

Since the state space $\mathcal{D}$ is huge and it is infeasible to calculate the expectation directly, mCAFE models it with a surrogate network $\hat{Q}(D, t)$ as shown in Fig. 2. This network takes the selected action t and the action sequence, which was used to generate the leaf node state D as input, and outputs the expectation reward of taking this action at the state. Considering the characteristic input and order information in the sequence, the surrogate network consists of three parts, namely a binary encoder which takes an action as input and outputs a binary code, one LSTM layer with a hidden size of 32 to deal with different lengths of the input sequence and capture their sequential information, and a fully connected layer with an input size of 32 and ReLU activation function to map the LSTM output to the expectation.

Since each edge in the tree maintains a beta distribution, we collect training data from all the existing edges in the tree and update the surrogate model after each iteration (episode). With the help of the surrogate model $\hat{Q}(D, t)$, the expansion policy can be defined as selecting the action t that maximizes the expectation reward predicted by the surrogate model.

3.4   The mCAFE Algorithm

The mCAFE applies MCTS to explore the target space, while the selection policy gradually biases the actions taken towards the more promising regions of search space in order to find the optimal sequence of actions. It follows the general MCTS scheme, where the main four phases have been modified as follows (Fig. 3):

Selection. Starting from the root node, mCAFE selects the child node according to the selection policy π_s iteratively until it reaches a leaf node.

Expansion. In a leaf node of the transformation tree, all the available child nodes are expanded to the tree. One of these nodes is selected to explore according to the expansion policy π_e.

Roll-out. Instead of the performance of the current node, we are interested in whether the expectation performance of its descendant nodes has been better than the best performance so far. To achieve this, mCAFE combines the n-folds

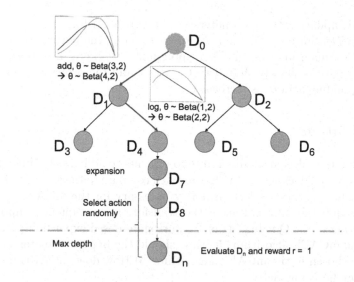

Fig. 4. Example of an episode of mCAFE. The beta distributions of the edges in the selected path are displayed next to the corresponding edge. Blue denotes the distribution before back-propagation and orange after back-propagation. (Color figure online)

cross-validation and the general Roll-out process by the following: Assuming that the current node is of depth l in the MCTS tree, mCAFE completes the feature engineering pipeline by sampling $L - l$ transformations from $\mathcal{T}$ randomly with replacement, where L is the predefined length of the pipeline (transformation sequence). This process is repeated n times to get n different pipelines (transformation sequences), where n is the number of iterations in n fold cross validation. A reward of $r = 1$ is returned if the mean evaluation score of the transformation sequences is higher, else $r = 0$.

Back-propagation. The reward from the roll-out process is back propagated along the path from the node selected in the expansion process to the root node in the tree, updating the parameters α, β in each edge on the path with the update rule (see Sect. 3.2).

The algorithm stops after the computational budget is exhausted, e.g. the algorithm stops when the number of episodes reaches 100 in the experiment.

Figure 4 shows an example of an episode of the mCAFE algorithm. Starting from the root node D_0, it selects explored nodes according to the selection policy π_s until reaching the leaf node D_4. Then an unexplored node D_7 is selected and expanded to the tree according to the expansion policy π_e. If the depth of the current node l (expanded node D_7) is smaller than the predefined pipeline length L (max depth), an action is selected according to the random policy and applied to the current node to create a new node, which is regarded as the new current node. This process is repeated until the depth of the new node l is larger than the pipeline length L. Finally, the current node is evaluated and its reward is back-propagated, updating the parameters of the beta distributions along the path from D_0 to D_7. Since $r = 1$, the α of all edges along the path are incremented, while the β remain unchanged. Correspondingly, the beta distribution of each edge in the path is slightly shifted to the right, and the probability of selecting the corresponding actions is increased.

4   Evaluation

In this section, we design six different experiments to address the following questions: 1) How well does the mCAFE approach compare to the state-of-the-art [7]? 2) Is the sampling-based selection policy necessary for the mCAFE algorithm? 3) Is the sequential information of the transformation sequence important for the prediction of the Q value? 4) Is the surrogate-based expansion policy necessary for the mCAFE algorithm? 5) How should the hyperparameter L (pipeline length) be chosen in the mCAFE algorithm? 6) How does mCAFE perform for different predictive models?

For the first five experiments, we use the same benchmarks as [7]. For this, we tried to reproduce this previous work. Some datasets were removed from the experiment since either the results of the base model differ considerably from those in [7], e.g., 'Amazon Employ' and 'Whine Quality Red'. Additionally, 'Wine Quality White', 'Higgs Boson', 'SVMGuide3', 'Bikeshare DC' were removed as they displayed a different dataset size compared to the one cited in [7]. To overcome this problem for future work, we published our code and datasets at https://github.com/HuangYiran/MonteCarlo-AFE.git.

We run the last experiment on the Automatic Machine Learning (AutoML) benchmark datasets [24]. We keep the same hyperparameter setting as in the first four experiments for both our work and the baseline.

In the experiments, we use episode budgets instead of time budgets for the following three reasons. 1) Different from some other optimization tasks, the time spent on candidate evaluation for feature engineering tasks dominates the overall time spent. This time is inevitable for all the evaluation-oriented optimization methods when dealing with feature engineering tasks. Table 1 shows the average percentage of time taken by the mCAFE for each step in the first 20 episodes. The roll-out phase, which consist of random transformation selection and candidate evaluation, takes up an average of 97% of the overall time. 2) The run time varies greatly across datasets. It is influenced by the size of the data and the sensitivity of the data to different transformations. 3) The algorithm implementation and operating environment have a significant impact on the run time.

For the first five experiments, we use the random forest model of the sklearn package (version 0.24) with default parameters and an episode budget of $B = 100$ as in [7], in order to make the result more comparable.

We set the pipeline length to $L = 4$ according to the result of the third experiment and all the beta distributions are initialized with $(1, 1)$ for a uniform prior. To reduce the computation time, we sub-sample to the dataset with a large number of data points. For the sub-sampling, up to 10^4 data points are considered.

Table 1. Average percentage of time for each process in the first 20 episodes.

Dataset	Size		Time spent in percentage (%)			
	Rows	Feat	Selection	Expansion	Roll-out	Back-propagation
SpecFact	267	44	0.01	0.04	96.94	3.01
PimaIndian	768	8	0.02	0.05	97.29	2.66
Lymphography	148	18	0.01	0.04	96.98	2.97
Ionosphere	351	34	0.01	0.06	96.37	3.56
AP-omentum-ovary	275	10936	0.01	0.02	98.57	1.40
SpamBase	4601	57	0.01	0.01	98.74	1.24

To ensure comparability, we did not tune any hyperparameters of the feature engineering algorithms to suit a concrete data set or prediction model (which we changed for the last experiment). Considering the imbalanced datasets, we apply the F1-score to assess the classification performance and use 1 - RAE (Relative Absolute Error) as in [7] as the metric for the regression task. All performances are obtained under 5-folds cross-validation, which also means the parameter n in roll out process is set to 5.

In the experiment, we used the transformation functions $T = \{$Log, Exp, Square, Sin, Cos, TanH, Sigmoid, Abs, Negative, Radian, K-term, Difference, Add, Minus, Product, Div, NormalExpansion, Aggregation, Normalization, Binning$\}$.

4.1 Performance of mCAFE

We evaluate the improvement of mCAFE algorithm in comparison with the following methods, namely the *original dataset* (Base), a *Reinforcement-Based Model (RBM)* with discount factor 0.99, learning rate 0.05 and B 100, a *tree-heuristic model (Cognito)* with global search heuristic for 100 nodes, *random selection* selecting a transformation from the available transformation set and applying it to one or more features in the original dataset. If the addition of the new features leads to an improved performance, we keep the new feature. This process is repeated 100 times to get the final dataset.

We summarized the performance of the methods in Table 2. It can be seen that, mCAFE achieves the best score in all the regression datasets against the reinforcement-based model and achieved superior results on most of the classification datasets. However, mCAFE performs worse than the reinforcement based model in two datasets on 'Credit Default' and 'SpamBase'. Among them, the difference on 'SpamBase' is not significant. From these results, we conclude that the proposed method performs better than the state-of-the-art automatic feature engineering approaches.

4.2 Ablation Study

The proposed selection strategy and extension strategy are the most important components supporting the performance of the algorithm. To verify their importance, we designed two ablation experiments.

Table 2. Comparing performance of without feature engineering (Base), reinforcement-based model (RBM) [7], Cognito [6], random selection and mCAFE in 100 episodes using 15 open source datasets. Classification (C) tasks are evaluated with F1-score and regression (R) tasks are evaluated with (1-relative absolute error).

Dataset	C/R	Rows	Feat.	Base	RBM	Cognito	Random	mCAFE
SpecFact	C	267	44	0.686	0.788	0.790	0.748	**0.855 ± 0.036**
PimaIndian	C	768	8	0.721	0.756	0.732	0.709	**0.773 ± 0.026**
German Credit	C	1001	21	0.661	0.724	0.662	0.655	**0.764 ± 0.026**
Lymphography	C	148	18	0.832	0.895	0.849	0.680	**0.967 ± 0.016**
Ionosphere	C	351	34	0.927	0.941	0.941	0.934	**0.962 ± 0.014**
Credit Default	C	30000	25	0.797	**0.831**	0.799	0.766	0.796 ± 0.006
AP-omentum-ovary	C	275	10936	0.615	0.820	0.758	0.710	**0.831 ± 0.036**
SpamBase	C	4601	57	0.955	**0.961**	0.959	0.937	0.953 ± 0.016
Openml_618	R	1000	50	0.428	0.589	0.532	0.428	**0.743 ± 0.015**
Openml_589	R	1000	25	0.542	0.687	0.644	0.571	**0.776 ± 0.018**
Openml_616	R	500	50	0.343	0.559	0.450	0.343	**0.622 ± 0.010**
Openml_607	R	1000	50	0.380	0.647	0.629	0.411	**0.803 ± 0.010**
Opemml_620	R	1000	25	0.524	0.683	0.583	0.524	**0.765 ± 0.012**
Openml_637	R	500	50	0.313	0.585	0.582	0.313	**0.637 ± 0.021**
Openml_586	R	1000	25	0.547	0.704	0.647	0.549	**0.783 ± 0.020**

Selection Policy. We apply the traditional UCB with ϵ-greedy policy as selection policy in the mCAFE algorithm (mCAFE-ucb) and compare its performance with the proposed model, which uses Thompson sampling based selection policy (mCAFE-ts). The parameter λ of UCB is set to 1.412 as proposed in [23], the ϵ is set to 0.1 while the mCAFE algorithm keeps the same setting as the last experiment. Performance of the classification task is measured with F1-score and regression task is measured with (1- relative absolute error). Figure 5 divides the results of the comparison into four categories. 1) mCAFE-ts gets better result: the result performance measured is higher than with mCAFE-ucb. 2) mCAFE-ts is faster: the number of episodes needed to obtain the same result is larger on MCAFE-ucb. 3) Tie: mCAFE-ucb obtains the same result and requires the similar number of epochs (difference smaller than 5). 4) mCAFE-ucb gets better result: mCAFE-ucb obtains the same results and requires a smaller number of episodes than mCAFE). The result in Fig. 5 demonstrates the importance of the selection strategy. mCAFE achieved better performance on 64.7% of the datasets and tied on 13.3%.

Expansion Policy. In the expansion process, we use an LSTM neural network to approximate the expectation reward (Q value) of taking an action, since it can capture the sequential information of the transformation sequence. To prove that this information is important for the Q value prediction, we designed an experiment to compare the performance of using MLP and LSTM as the surrogate model in mCAFE.

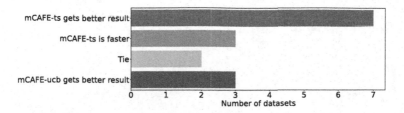

Fig. 5. Comparing the performance between mCAFE-ucb and mCAFE.

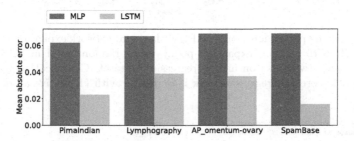

Fig. 6. Comparing the performance of MLP and LSTM model in predicting the Q value.

To make the trained models comparable, the MLP model here contains two 76 dimension hidden layers so that it has a similar number of parameters as the LSTM surrogate model mentioned above. We use the mean absolute error as the evaluation criterion. A smaller value indicates a better model. Both models are trained with 100 epochs. We can see from Fig. 6 that, the LSTM obtains significantly better results than the MLP model in all datasets.

To evaluate its contribution to mCAFE, we compare the performance of the following three models, namely mCAFE with LSTM-based expansion policy, mCAFE with random expansion policy, mCAFE with greedy expansion policy, which always expand the best action explored.

All three models used the same initial parameters as the last experiment. Each model is evaluated 10 times on each dataset. The performance of the models on the regression datasets is displayed with the box plot in Fig. 7. We can see that mCAFE with neural network achieves best performances on all datasets except two, where mCAFE with random policy performs better on dataset 'Openml_618' and mCAFE with fixed expansion policy performs better on dataset 'Openml_586'. For all datasets except 'Openml_618' and 'Openml_586', mCAFE with neural network expansion policy also loses to mCAFE with random expansion policy on dataset 'AP-omentum-ovary'.

The main differences between these three expansion approaches are the usages of previous observations and the dispersion of the selected actions. mCAFE with a fixed expansion policy selects actions greedily according to the performance of the actions in the first layer. This selection process is stable, however, hinders the exploration of new transformations, which is likely the rea-

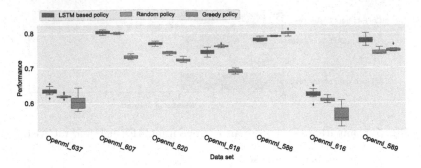

Fig. 7. Comparing performance of mCAFE with neural network expansion policy (with nn), mCAFE with random expansion policy (with random) and mCAFE with fix expansion policy (with fix) on all the regression dataset. Classification task is evaluated with F1-score and regression task is evaluated with (1-relative absolute error).

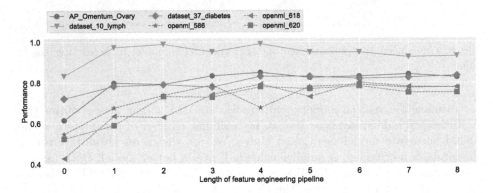

Fig. 8. Comparing performance of mCAFE with different maximum pipeline length on 3 classification datasets (F1-score) and 3 regression datasets (1-relative absolute error).

son for its failure in most cases. mCAFE with neural network expansion policy captures the performance information of previous observation and uses it in the prediction of the reward expectation of future actions.

4.3 Length of Feature Engineering Pipeline

The length of the feature engineering pipeline L determines the number of actions selected in each roll-out step, as well as the length of the final transformation sequence. It influences the performance of mCAFE algorithm not only on the final result but also on the time and memory consumption. In general, the larger L, the larger the time and memory consumption and, at the same time, the larger the number of features after the transformation. To achieve the best results with limited resources, we conducted an experiment to find a suitable parameter L by comparing the performance of the algorithm with different L values.

Table 3. The performances of mCAFE with different predictive models on AutoML benchmark dataset [24]. The improvements brought by the mCAFE are shown in the parentheses. Classification task is evaluated with F1-score and regression task is evaluated with (1-relative absolute error)

AutoML benchmark datasets	Base performance				Performance with mCafe			
	Rbf-svm	Linear-svm	Linear model	Decision tree	Rbf-svm	Linear-svm	Linear model	Decision tree
shuttle	0.901	0.996	0.859	**0.998**	0.996 (0.095)	0.998 (0.002)	0.996 (0.137)	**0.998** (0.001)
php2LgL9q	0.407	0.432	**0.503**	0.454	0.407 (0.000)	0.432 (0.000)	**0.503** (0.000)	0.454 (0.000)
phpyM5ND4	0.563	0.848	0.775	**0.920**	0.752 (0.189)	0.943 (0.095)	0.893 (0.118)	**0.959** (0.039)
phpvcoG8S	0.471	0.426	0.468	**0.572**	0.561 (0.090)	**0.579** (0.153)	0.529 (0.061)	0.578 (0.006)
phpQOf0wY	0.320	0.385	0.457	**0.698**	0.620 (0.300)	0.493 (0.225)	0.652 (0.225)	**0.698** (0.000)
phpnBqZGZ	0.016	**0.768**	0.700	0.747	0.467 (0.451)	**0.768** (0.000)	0.760 (0.060)	0.750 (0.003)
phpmPOD5A	**0.919**	0.749	0.869	0.912	**0.919** (0.000)	0.908 (0.159)	0.908 (0.039)	0.913 (0.001)
phpmcGu2X	0.930	**0.953**	0.941	0.854	**0.970** (0.040)	0.953 (0.000)	0.941 (0.000)	0.854 (0.001)
phpMawTba	0.650	0.589	0.716	**0.797**	0.802 (0.152)	0.787 (0.198)	0.787 (0.071)	**0.820** (0.023)
phpkIxskf	0.833	0.767	0.847	**0.877**	0.883 (0.050)	0.883 (0.116)	0.870 (0.023)	**0.892** (0.015)

Figure 8 shows the relationship between length L and the best performance displayed by the mCAFE algorithm for six datasets. $L = 0$ signifies the performance of the random forest model on the base dataset. We can see that some achieve good results with $L = 1$, however, increasing L can further improve its performance. Most of the datasets reach the maximum performance with $L = 4$, while a small fraction shows a lower performance. This may be due to the random selection in the starting process. The performance on 'Dataset_10_lymph' is worse for higher L, which is probably due to overfitting. From this experiment, we can conclude that the optimal L depends on the dataset. However, $L = 4$ should be a suitable choice in most cases.

4.4 Performances of mCAFE on Different Predictive Models

Different predictive models differ in their performance and sensitivity to mCAFE on the same dataset. To test this conjecture, we tested the performance of the mCAFE with the following predictive models on the AutoML benchmark datasets separately, namely Rbf-svm, Linear-svm, Linear model, Decision tree.

Table 3 summarizes the results of the experiment. We can see that mCAFE brings performance improvements to most of the datasets. The value of feature engineering is more prominent for linear and svm models. It is worth noting that although the performance of each model on the original dataset varies greatly, the performance obtained after mCAFE tends to be close.

5 Conclusion and Future Work

In this paper, we show that existing automatic feature engineering methods can be significantly improved by building upon two simple observations. Our results suggest that feature engineering should make use of sequence information, incorporating composite transformations into the surrogate model. In addition, a suitable selection policy should be chosen. The proposed novel MCTS-based framework uses an LSTM neural network for the expansion policy to explore the search space efficiently. Furthermore, Thompson sampling is employed to address the trade-off between exploration and exploitation in the selection policy. Through this, we manage to obtain superior results to state-of-the-art methods for automatic feature engineering on the majority of commonly used benchmarks. We believe that further improvements could be made to the algorithms by adding transformations that might also reduce redundant and irrelevant feature during the construction.

Acknowledgements. This work was partially funded by the Ministry of The Ministry of Science, Research and the Arts Baden-Wuerttemberg as part of the SDSC-BW and by the German Ministry for Research as well as by Education as part of SDI-C (Grant 01IS19030A).

References

1. Zhou, Y., Hefenbrock, M., Huang, Y., Riedel, T., Beigl, M.: Automatic remaining useful life estimation framework with embedded convolutional LSTM as the backbone. In: Dong, Y., Mladenić, D., Saunders, C. (eds.) ECML PKDD 2020. LNCS (LNAI), vol. 12460, pp. 461–477. Springer, Cham (2021). https://doi.org/10.1007/978-3-030-67667-4_28

2. Guo, P., Deng, C., Xu, L., Huang, X., Zhang, Yu.: Deep multi-task augmented feature learning via hierarchical graph neural network. In: Oliver, N., Pérez-Cruz, F., Kramer, S., Read, J., Lozano, J.A. (eds.) ECML PKDD 2021. LNCS (LNAI), vol. 12975, pp. 538–553. Springer, Cham (2021). https://doi.org/10.1007/978-3-030-86486-6_33

3. Schelling, B., Bauer, L.G.M., Behzadi, S., Plant, C.: Utilizing structure-rich features to improve clustering. In: Hutter, F., Kersting, K., Lijffijt, J., Valera, I. (eds.) ECML PKDD 2020. LNCS (LNAI), vol. 12457, pp. 91–107. Springer, Cham (2021). https://doi.org/10.1007/978-3-030-67658-2_6

4. Kanter, J.M., Veeramachaneni, K.: Deep feature synthesis: towards automating data science endeavors. In: DSAA 2015. IEEE (2015). https://doi.org/10.1109/DSAA.2015.7344858

5. Lam, H.T., Thiebaut, J.M., Sinn, M., Chen, B., Mai, T., Alkan, O.: One button machine for automating feature engineering in relational databases (2017). arXiv https://doi.org/10.48550/arXiv.1706.00327

6. Khurana, U., Turaga, D., Samulowitz, H., Parthasrathy, S. Cognito: automated feature engineering for supervised learning. In: ICDMW 2016. IEEE (2016). https://doi.org/10.1109/ICDMW.2016.0190

7. Khurana, U., Samulowitz, H., Turaga, D.: Feature engineering for predictive modeling using reinforcement learning. In: AAAI 2018. PKP (2018). https://doi.org/10.1609/aaai.v32i1.11678

8. Zhang, J., Hao, J., Fogelman-Soulié, F., Wang, Z.: Automatic feature engineering by deep reinforcement learning. In: AAMAS 2019, ACM, ACM DL (2019). https://doi.org/10.5555/3306127.3332095

9. Coulom, R.: Efficient selectivity and backup operators in Monte-Carlo tree search. In: van den Herik, H.J., Ciancarini, P., Donkers, H.H.L.M.J. (eds.) CG 2006. LNCS, vol. 4630, pp. 72–83. Springer, Heidelberg (2007). https://doi.org/10.1007/978-3-540-75538-8_7

10. Graves, A.: Long short-term memory. In: Supervised Sequence Labelling with Recurrent Neural Networks. Studies in Computational Intelligence, vol. 385. Springer, Berlin, Heidelberg (2012). https://doi.org/10.1007/978-3-642-24797-2_4

11. Markovitch, S., Rosenstein, D.: Feature generation using general constructor functions. Mach. Learn. **49**, 59–98 (2002). https://doi.org/10.1023/A:1014046307775

12. Fan, W., Zhong, E., Peng, J., Verscheure, O., Zhang, K., Ren, J., et al.: Generalized and heuristic-free feature construction for improved accuracy. SDM (2010). https://doi.org/10.1137/1.9781611972801.55

13. Dor, O., Reich, Y.: Strengthening learning algorithms by feature discovery. Inf. Sci. **189**, 176–190. Elsevier (2012). https://doi.org/10.1016/j.ins.2011.11.039

14. Katz, G., Shin, E.C.R., Song, D.: Explorekit: automatic feature generation and selection. In: ICDM 2016. IEEE (2016). https://doi.org/10.1109/ICDM.2016.0123

15. Nargesian, F., Samulowitz, H., Khurana, U., Khalil, E.B., Turaga, D.S.: Learning feature engineering for classification. In: IJCAI 2017 (2017). https://doi.org/10.24963/ijcai.2017/352

16. Gelly, S., Silver, D.: Monte-Carlo tree search and rapid action value estimation in computer go. Artif. Intell. **175**(11), 1856–1875. Elsevier (2011). https://doi.org/10.1016/j.artint.2011.03.007

17. Auer, P.: Using confidence bounds for exploitation-exploration trade-offs. JMLR **3**, 397–422 (2002). ACM DL. https://doi.org/10.5555/944919.944941

18. Kocsis, L., Szepesvári, C.: Bandit based Monte-Carlo planning. In: Fürnkranz, J., Scheffer, T., Spiliopoulou, M. (eds.) ECML 2006. LNCS (LNAI), vol. 4212, pp. 282–293. Springer, Heidelberg (2006). https://doi.org/10.1007/11871842_29

19. Silver, D., Tesauro, G.: Monte-Carlo simulation balancing. In: Proceedings of the 26th Annual International Conference on Machine Learning (2009). https://doi.org/10.1145/1553374.1553495

20. Rimmel, A., Teytaud, F.: Multiple overlapping tiles for contextual Monte Carlo tree search. In: Di Chio, C., Cagnoni, S., Cotta, C., Ebner, M., Ekárt, A., Esparcia-Alcazar, A.I., Goh, C.-K., Merelo, J.J., Neri, F., Preuß, M., Togelius, J., Yannakakis, G.N. (eds.) EvoApplications 2010. LNCS, vol. 6024, pp. 201–210. Springer, Heidelberg (2010). https://doi.org/10.1007/978-3-642-12239-2_21

21. Fernández-Delgado, M., Cernadas, E., Barro, S., Amorim, D.: Do we need hundreds of classifiers to solve real world classification problems? JMLR **15**(1), 3133–3181 (2014). ACM DL https://doi.org/10.5555/2627435.2697065

22. Helmbold, D.P., Parker-Wood, A.: All-moves-as-first heuristics in Monte-Carlo go. In: IC-AI 2009. CiteSeer (2009). https://doi.org/10.1.1.183.7924

23. Coquelin, P.A., Munos, R.: Bandit algorithms for tree search (2007). arXiv preprint, https://doi.org/10.48550/arXiv.cs/0703062

24. Gijsbers, P., LeDell, E., Thomas, J., Poirier, S., Bischl, B., Vanschoren, J.: An open source AutoML benchmark (2019). arXiv preprint, https://doi.org/10.48550/arXiv.1907.00909

25. Gaudel, R., Sebag, M.: Feature selection as a one-player game. In: ICML 2010, ACM DL (2010). https://doi.org/10.5555/3104322.3104369

MRF-UNets: Searching UNet with Markov Random Fields

Zifu Wang$^{(\boxtimes)}$ and Matthew B. Blaschko

ESAT-PSI, KU Leuven, Leuven, Belgium
zifu.wang@kuleuven.be

Abstract. UNet [27] is widely used in semantic segmentation due to its simplicity and effectiveness. However, its manually-designed architecture is applied to a large number of problem settings, either with no architecture optimizations, or with manual tuning, which is time consuming and can be sub-optimal. In this work, firstly, we propose Markov Random Field Neural Architecture Search (MRF-NAS) that extends and improves the recent Adaptive and Optimal Network Width Search (AOWS) method [4] with (i) a more general MRF framework (ii) diverse M-best loopy inference (iii) differentiable parameter learning. This provides the necessary NAS framework to efficiently explore network architectures that induce loopy inference graphs, including loops that arise from skip connections. With UNet as the backbone, we find an architecture, MRF-UNet, that shows several interesting characteristics. Secondly, through the lens of these characteristics, we identify the sub-optimality of the original UNet architecture and further improve our results with MRF-UNetV2. Experiments show that our MRF-UNets significantly outperform several benchmarks on three aerial image datasets and two medical image datasets while maintaining low computational costs. The code is available at: https://github.com/zifuwanggg/MRF-UNets.

Keywords: Neural architecture search · Probabilistic graphical models · Semantic segmentation

1 Introduction

Neural architecture search (NAS) has greatly improved the performance on various vision tasks, for example, classification [4,8,30], object detection [10,23,34] and semantic segmentation [18,31,32,35] via automating the architecture design process. UNet [27] is widely adopted to a large number of problem settings such as aerial [9,28] and medical image segmentation [11,20,24] due to its simplicity and effectiveness, but either with no architecture optimization, or with simple manual tuning. A natural question is, can we improve its manually-designed architecture with NAS?

AOWS [4] is a resource-aware NAS method and it is able to find effective architectures that strictly satisfy resource constraints, e.g. latency or the number of floating-point operations (FLOPs). The main idea of AOWS is to model the search problem as parameter learning and maximum a posteriori (MAP)

© The Author(s), under exclusive license to Springer Nature Switzerland AG 2023
M.-R. Amini et al. (Eds.): ECML PKDD 2022, LNAI 13715, pp. 599–614, 2023.
https://doi.org/10.1007/978-3-031-26409-2_36

inference over a Markov Random Field (MRF). However, the main limitation of AOWS is the adoption of Viterbi inference. As a result, the approach is only applicable to simple tree-structured graphs where the architecture cannot have skip connections [14]. Skip connections are widely adopted in modern neural networks as they can ease the training of deep models via shortening effective paths [29]; skip connections have also played an important role in the success of semantic segmentation where an encoder and a decoder are connected by long skip connections to aggregate features at different levels, e.g. UNet.

Besides, MAP assignment over a weight-sharing network is usually sub-optimal due to the discrepancy between the one-shot super-network and stand-alone child-networks [36,42]. Restricting the search to a single MAP solution also results in the search method having high variance [25], so one usually needs to repeat the search process with different random seeds and hyper-parameters. Furthermore, parameter learning of AOWS is non-differentiable, and this disconnects AOWS from recent advances in the differentiable NAS community [7,25,37], despite its advantages in efficient inference.

The contributions of this paper are twofold. Firstly, we propose MRF-NAS, which extends and improves AOWS with (i) a more general framework that shows close connections with other NAS approaches and yields better representation capability, (ii) loopy inference algorithms so we can apply it to more complex search spaces, (iii) diverse M-best inference instead of a single MAP assignment to reduce the search variance and to improve search results and (iv) a novel differentiable parameter learning approach with Gibbs sampling and Long-Short-Burnin-Scheme (LSBS) to save on computational cost. With UNet as the backbone, we find an architecture, MRF-UNet, that shows several interesting characteristics. Secondly, through the lens of these characteristics, we identify the sub-optimality of the original UNet architecture and further improve our results with MRF-UNetV2. We show the effectiveness of our approach on three aerial image datasets: DeepGlobe Land, Road and Building [9] and two medical image datasets: CHAOS [20] and PROMISE12 [24]. Compared with the benchmarks, our MRF-UNets achieve superior performance while maintaining low computational cost.

2   Related Works

Neural architecture search (NAS) [25,44] is a technique for automating the design process of neural network architectures. The early attempt [44] trains a RNN with reinforcement learning and costs thousands of GPU hours. In order to reduce the search cost, one usually uses some proxy to infer an architecture's performance without training it from scratch. For example, the significance of learnable architecture parameters [7,15,18,25,31,32,35,37] or validation accuracy from a super-network with shared weights [3,4,6,13].

Resource-aware NAS [4,8,38,40] focuses on architectures that achieve good performance while satisfy resource targets such as FLOPs or latency. FBNet [8] inserts a differentiable latency term into the loss function to penalize networks that consume high latency. However, the found architectures are not guaranteed to strictly satisfy the constraints. AutoSlim [38] trains a slimmable network [39,41] as the super-network and applies a greedy heuristics to search for channel configurations under different FLOPs targets. AOWS [4] models resource-aware NAS as a constrained optimization problem which can then be solved via inference over a chain-structured MRF. Nevertheless, their method can only be applied to simple search spaces which do not include skip connections.

3   Preliminaries

3.1   Markov Random Field

For an arbitrary integer n, let $[n]$ be shorthand for $\{1, 2, ..., n\}$. We have a set of discrete variables $\boldsymbol{x} = \{x_i | i \in [n]\}$ and each x_i takes value in a finite label set $X_i = \{x_i^j | j \in [k_i]\}$. For a set $S \subseteq [n]$, we use x_S to denote $\{x_i | i \in S\}$, and $X_S = \times_{i \in S} X_i$, where $\times$ is the cartesian product.

A Markov Random Field (MRF) is an undirected graph $G = (V, E)$ over these variables, and equipped with a set of factors $\Phi = \{\phi_S | S \subseteq [n]\}$ where $\phi_S : X_S \to \mathbb{R}$, such that $V = [n]$ and an edge $e_{i,j} \in E$ when there exists some $\phi_S \in \Phi$ and $\{i, j\} \subseteq S$ [21]. It is common to employ a pairwise MRF where $\Phi = \{\phi_S | S \subseteq [n] \text{ and } |S| \leq 2\}$. A set of factors Φ explicitly defines a probabilistic distribution $\mathbb{P}_\Phi(\boldsymbol{x}) = \frac{1}{Z} \exp\left(\sum_S \phi_S(x_S)\right)$ where Z is the normalizing constant. The goal of MAP inference is to find an assignment $\boldsymbol{x}^*$ so as to maximize a real-valued energy function $\mathcal{E}(\boldsymbol{x})$:

$$\boldsymbol{x}^* = \underset{\boldsymbol{x} \in X_V}{\operatorname{argmax}} \mathcal{E}(\boldsymbol{x}) = \underset{\boldsymbol{x} \in X_V}{\operatorname{argmax}} \exp\left(\sum_S \phi_S(x_S)\right). \tag{1}$$

3.2   Diverse M-Best Inference

In MRFs, there exist optimization error (approximate inference), approximation error (limitations of the model, e.g. a pairwise MRF can only represent pairwise interactions), and estimation error (factors are learnt from a finite dataset). In the context of NAS, in order to reduce the cost of searching, we often resort to proxies [3,25] on a weight-sharing network and they can be inaccurate. Instead of giving all our hope to a single MAP solution, diverse M-best inference [2] aims to find a diverse set of highly probable solutions.

Given some dissimilarity function $\Delta(\boldsymbol{x}^p, \boldsymbol{x}^q)$ between two solutions and a dissimilarity target k^q, we denote $\boldsymbol{x}^1$ as the MAP, $\boldsymbol{x}^2$ the second-best solution and so on until $\boldsymbol{x}^m$ the mth-best solution. Then for each $2 \leq p \leq m$, we have the following constrained optimization problem

$$\boldsymbol{x}^p = \underset{\boldsymbol{x} \in X_V}{\operatorname{argmax}} \mathcal{E}(\boldsymbol{x}) \tag{2}$$

$$\text{s.t. } \Delta(\boldsymbol{x}^p, \boldsymbol{x}^q) \geq k^q \quad \text{for } q = 1, ..., p-1. \tag{3}$$

Therefore, we are interested in a diverse set of solutions $\{x^1, ..., x^p, ..., x^m\}$ such that each x^p maximizes the energy function, and is at least k^q-units away from each of the $p - 1$ previously found solutions. If we consider a pairwise MRF and choose Hamming distance as the dissimilarity function, we can turn Eq. (2) into a new MAP problem such that pairwise factors remain the same and unary factors become $\phi_i^p(x_i^j) = \phi_i(x_i^j) - \sum_{q=1}^{p-1} \lambda^q \cdot \mathbb{1}(x_i^{;q} = x_i^j)$ where λ^q is the Lagrange multiplier and $x_i^{;q}$ is the assignment of x_i in the q-th solution [2].

3.3 AOWS

Having introduced the notations for MRFs, here we illustrate how AOWS models the NAS problem as a MRF. In NAS, there is a neural network N that has n choice nodes, i.e. $x = \{x_i | i \in [n]\}$, and each node x_i can take some value from a label set X_i, e.g. kernel size $= 3$ or 5. Therefore, we can use x to represent the architecture of a neural network. Let $N(x)$ be a neural network whose architecture is x. Given some task-specific performance measurement $\mathcal{M}$, e.g. classification accuracy, and a resource measurement $\mathcal{R}$, e.g. latency or FLOPs, resource-aware NAS can be represented as a constrained optimization problem

$$\max_x \mathcal{M}(N(x)) \quad \text{s.t. } \mathcal{R}(N(x)) \le R_T \tag{4}$$

where R_T is the resource target. Consider the following Lagrangian relaxation of the problem

$$\min_\gamma \max_x \mathcal{M}(N(x)) + \gamma(\mathcal{R}(N(x)) - R_T) \tag{5}$$

with γ a Lagrange multiplier. If the inner maximization problem can be solved efficiently, then the minimization problem in Eq. (5) can be solved by binary search over γ since the objective is concave in γ [4,28].

The key idea of AOWS [4] is to model Eq. (5) as parameter learning and MAP inference over a pairwise MRF such that $\mathcal{M}(N(x)) = \sum_i \phi_i$ and $\mathcal{R}(N(x)) = \sum_{i,j} \phi_{i,j}$. For $\mathcal{M}(N(x))$, they assume $\phi_i(x_i^j) = -\frac{1}{|T_{i,j}|} \sum_t l(w|x_i^{(t)} = j)$ where $l(w|x_i^{(t)} = j)$ is the training loss when $x_i = j$ is sampled at iteration t, and $|T_{i,j}|$ is the total number of times x_i^j is sampled. For $\mathcal{R}(N(x))$, many resource models have a pairwise form. For example, FLOPs can be calculated exactly as pairwise sums; latency is usually modeled as a pairwise model due to sequential execution of the forward pass. Once these factors are known, the inner maximization problem can be solved efficiently via Viterbi inference.

4 MRF-NAS

Here we generalize the idea of AOWS [4] to a broader setting. We assume that there exists some non-decreasing mapping $\mathcal{F} : \mathbb{R} \to \mathbb{R}$ such that

$$\mathcal{M}(N(x)) = \mathcal{F}(\mathcal{E}(x)) \tag{6}$$

Therefore, we extend their framework and no longer require $\mathcal{M}(N(\boldsymbol{x})) = \mathcal{E}(\boldsymbol{x})$, but let $\mathbb{P}_\Phi$ be defined by a set of factors Φ such that $\mathbb{P}_\Phi(\boldsymbol{x}_1) \geq \mathbb{P}_\Phi(\boldsymbol{x}_2) \Rightarrow \mathcal{M}(N(\boldsymbol{x}_1)) \geq \mathcal{M}(N(\boldsymbol{x}_2))$. Then NAS becomes MAP inference over a set of properly defined factors

$$\boldsymbol{x}^* = \underset{x}{\operatorname{argmax}} \mathcal{M}(N(\boldsymbol{x})) = \underset{x}{\operatorname{argmax}} \mathcal{E}(\boldsymbol{x}). \tag{7}$$

For resource-aware NAS, we follow [4] to introduce another set of factors

$$\mathcal{R}(N(\boldsymbol{x})) = \mathcal{E}'(\boldsymbol{x}) = \exp \left(\sum_{i \in V} \phi_i'(x_i) + \sum_{(i,j) \in E} \phi_{i,j}'(x_i, x_j) \right) \tag{8}$$

and combine these two energy functions as in Eq. (5). As discussed in Sect. 3.3, many resource models can be represented exactly or approximately as a pairwise model. Following [4], we focus on latency. We can populate each element ϕ_i' and $\phi_{i,j}'$ through profiling the entire network on some target hardware and solving a system of linear equations.

Many existing methods show close connections to our formulation. For example, one-shot methods with weight sharing [3,6,13] define a single factor ϕ_V whose scope includes all nodes in the graph where $\phi_V(\boldsymbol{x})$ is the validation accuracy of the super-network evaluated with architecture $\boldsymbol{x}$. Their formulation imposes no factorization, and therefore the cardinality of $\phi_V(\boldsymbol{x})$ grows exponentially in the order of n, the number of nodes in the graph, which makes MAP inference impossible. On the contrary, AOWS [4] and differentiable NAS approaches [7,25,37] introduce a set of unary factors $\Phi = \{\phi_i | i \in V\}$ such that in AOWS, $\phi_i(x_i)$ is the averaged losses, and in differentiable NAS approaches, $\phi_i(x_i)$ is the learnable architecture parameter. With no higher order interaction, MAP inference deteriorates to marginal maximization whose solution can be derived easily. However, this model imposes strong local independence and greatly limits the representation capability of the underlying graphical model.

Since we model the resource measurement as a pairwise model, for the ease of joint inference in Eq. (5), we also consider $\mathcal{E}(\boldsymbol{x})$ to be pairwise

$$\mathcal{E}(\boldsymbol{x}) = \exp \left(\sum_{i \in V} \phi_i(x_i) + \sum_{(i,j) \in E} \phi_{i,j}(x_i, x_j) \right). \tag{9}$$

Moreover, compared with methods that only use unary terms, our pairwise model imposes weaker local independence and has more representation power. Although the inclusion of pairwise terms increases the number of learnable factors from $O(|X|)$ to $O(|X|^2)$, where $|X|$ is the cardinality of factors and is usually less than 20, the added overhead is negligible compared with the number in learnable parameters of modern neural networks.

4.1   Diverse M-Best Loopy Inference

The main limitation of AOWS [4] is the adoption of Viterbi inference, which is only applicable to simple chain graphs such as MobileNetV1 [16]. When the computational graph forms loops, i.e. it includes skip connections [14], we need to resort to loopy inference algorithms. Exact inference over a loopy graph can be very expensive, especially when the graph is densely connected. In the worst case, the complexity of exact inference can be exponential in n when the graph is fully connected. Therefore, sometimes we can only hope for approximate solutions. However, we find in practice that for realistic architectures, fast approximate inference yields excellent performance on par with exact inference. The difference between exact and approximate inference will be discussed in more detail in Sect. 6.1.

Furthermore, MAP assignment on a weight-sharing network is usually of poor quality, but we can still find architectures that achieve good performance by examining other top solutions [36,42]. Therefore, instead of a single MAP assignment $\boldsymbol{x}^*$, we use diverse M-best inference [2] to find a set of diverse solutions $\{\boldsymbol{x}^1, ..., \boldsymbol{x}^m\}$ so as to reduce the variance in the search phase.

Diverse M-best inference requires a set of balanced dissimilarity constraints, each with an associated Lagrange multiplier. In Eq. (2), it is crucial to choose a dissimilarity target k^q, and rather than searching via the diversity constraints, we can directly perform model selection on the Lagrange multiplier λ^q [2]. We find that the absolute value of ϕ_i can be very different across factors. Instead of using a single scalar value λ^q for all i, we set it to be a vector $\boldsymbol{\lambda}^q = (\lambda_1^q, ..., \lambda_i^q, ..., \lambda_n^q)$ such that

$$\lambda_i^q = \frac{\max_j \phi_i^q(x_i^j) - \min_j \phi_i^q(x_i^j)}{L}, \tag{10}$$

where $\phi_i^q(\cdot)$ is the modified unary factor. Then we can tune L instead.

4.2   Differentiable Parameter Learning

In the previous sections, we have discussed how to find optimal solutions if the factors in MRF are already known. Here we propose a differentiable approach to learn these factors so as to close the gap between AOWS and other differentiable NAS approaches. Following the formulation in [1], the goal of differentiable NAS is to maximize the following objective

$$- \mathbb{E}_{\boldsymbol{x} \sim \mathbb{P}_\Phi(\boldsymbol{x})}[l(\boldsymbol{w}^*|\boldsymbol{x})] \quad \text{s.t. } \boldsymbol{w}^* = \underset{\boldsymbol{w}}{\operatorname{argmin}} \, l(\boldsymbol{w}|\boldsymbol{x}), \tag{11}$$

where $l(\cdot)$ is some loss function and $\boldsymbol{w}$ encodes connection weights of the neural network. In order to make Eq. (11) differentiable, we can approximate it through Monte Carlo with n_{mc} samples and use the Gumbel-Softmax reparameterization trick [19] to smooth the discrete categorical distribution.

However, there is one more caveat in the aforementioned approach: to sample from the joint probability distribution $\mathbb{P}_\Phi(\boldsymbol{x})$. When we only have unary factors such as in [1], sampling from $\mathbb{P}_\Phi(\boldsymbol{x})$ is the same as independently sampling from the marginal probability distribution $\mathbb{P}_{\phi_i}(x_i)$:

$$\mathbb{P}_\Phi(\boldsymbol{x}) = \frac{1}{Z} \exp\left(\sum_i \phi_i(x_i) \right) = \prod_i \frac{1}{Z} \exp\left(\phi_i(x_i) \right) = \prod_i \mathbb{P}_{\phi_i}(x_i). \tag{12}$$

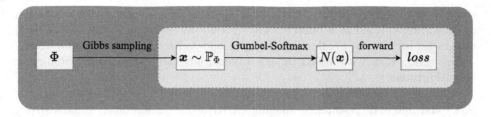

Fig. 1. Workflow of our differentiable parameter learning approach. Vanilla differentiable NAS methods [1] with Monte Carlo approximation and Gumbel-Softmax trick are shown in the blue rectangle. In our case, since the joint probability distribution $\mathbb{P}_\Phi(\boldsymbol{x})$ cannot be decomposed as the product of only unary terms, we need to perform an extra step of MCMC, e.g. Gibbs sampling.

When we have high-order interactions such as pairwise terms, sampling from the joint usually involves the use of Markov Chain Monte Carlo (MCMC) methods. Here we use Gibbs sampling for simplicity:

$$x_i^t \sim \mathbb{P}_\Phi(x_i|\boldsymbol{x}_{-i}) \tag{13}$$

where the distribution of x_i at t-th iteration is determined by $\boldsymbol{x}_{-i}$, all nodes except i. In an MRF, $\boldsymbol{x}_{-i}$ can be simplified to the Markov blanket of i. Since $\mathbb{P}_\Phi(x_i|\boldsymbol{x}_{-i})$ is just the product of several factors, if we apply the Gumbel-Softmax trick, the sampling process is differentiable with respect to these factors. A graphical illustration of the overall procedure is shown in Fig. 1.

After a burn-in period with n_{burnin} samples, Gibbs sampling will converge to the stationary distribution. The length of burn-in period is theoretically unknown and is often decided empirically. Gibbs sampling can be expensive because every time we update Φ, we will have a new $\mathbb{P}_\Phi$. Therefore, we need to re-enter the burn-in period just to draw n_{mc} samples where $n_{\text{mc}} \ll n_{\text{burnin}}$, and then update Φ again. In order to mitigate this problem, we propose a Long-Short-Burnin-Scheme (LSBS). Specifically, at the beginning of each epoch, we run a long burn-in period n_{long}, but at each iteration within that epoch, we only run a short burn-in period n_{short}. We can assume that $\mathbb{P}_{\Phi^t} \approx \mathbb{P}_{\Phi^{t+1}}$ since Φ will only change by a small amount. Starting from a sample $\boldsymbol{x}^t \sim \mathbb{P}_{\Phi^t}$, we can quickly transit to a sample $\boldsymbol{x}^{t+1} \sim \mathbb{P}_{\Phi^{t+1}}$ without running a long burnin period. As a result, as opposed to $n_{\text{long}} + n_{\text{mc}}$, we only need to draw $n_{\text{short}} + n_{\text{mc}}$ samples at each iteration, where $n_{\text{mc}} \approx n_{\text{short}} \ll n_{\text{long}}$.

5 Experiments

5.1 Datasets

We choose five semantic segmentation datasets with diverse contents. Specifically, DeepGlobe challenge [9] provides three aerial image datasets: Land, Road and Building. Land is a multi-class (urban, agriculture, rangeland, forest, water and barren) segmentation dataset and it contains 803 satellite images focusing on rural areas. Road is a binary segmentation dataset and it consists of 6226 images captured over Thailand, Indonesia, and India. Building is also a binary segmentation task with 10593 images taken from Las Vegas, Paris, Shanghai, and Khartoum.

CHAOS [20] is medical image dataset including both computed tomography (CT) and magnetic resonance imaging (MRI) scans of abdomen organs (liver, right kidney, left kidney and spleen). We only use MRI scans, and it has 20 cases and 1270 2D-slices. PROMISE [24] contains prostate MRI images, and it has 50 cases and 1377 2D-slices. For all datasets, only training sets are available. We split 60%/20%/20% of the training set as train/val/test set. For simplicity, we resize all images to 256×256.

5.2   Search Space

We use UNet [27] as the backbone. Generally speaking, UNet and other encoder-decoder networks usually contain three types of operations: Normal, Down and Up. We search for both size and width of the convolution kernel and summarize the search space in Table 1. For the original UNet that has 26 layers, our search space has about 4×10^{23} configurations in total.

For a fair comparison, for manually designed architectures, e.g. UNet [27], UNet++ [43] and BiO-Net [33], we use their templates and implement the Normal/Up/Down operations in the same way as our search space, but fix the kernel size to be 3. For automatically found architectures, e.g. NAS-UNet [32], MS-NAS [35], and BiX-Net [31], we do not make any modification and use their implementations directly. However, there exist discrepancies. For instance, we use transposed convolution for up-sampling while NAS-UNet [32] uses dilated transposed convolution and BiX-Net [31] uses bilinear interpolation.

Table 1. Search space with UNet as the backbone.

Type	Size	Width
Normal	3, 5	0.5, 0.75, 1.0, 1.25, 1.5
Down	3	0.5, 0.75, 1.0, 1.25, 1.5
Up	2	0.5, 0.75, 1.0, 1.25, 1.5

5.3   Implementation Details

In the search phase, we train a super-network using the sandwich rule [39] for $T = 50$ epochs. Initially, factors are not updated until the super-network is trained for a warmup period of 10 epochs. The learning rate of network weights starts from 0.0005 and is then decreased by a factor of $(1 - \frac{t}{T})^{0.9}$ at each epoch t. We use the Adam optimizer with weight decay of 0.0001. The learning rate of MRF factors is fixed at 0.0003 and we also use the Adam optimizer with the same weight decay. For the sampling, we use $n_{\text{long}} = 10000$, $n_{\text{short}} = 10$ and $n_{\text{mc}} = 1$. The temperature parameter τ in Gumbel-Softmax is fixed at 1. For inference, we choose $m = 5$ and $L = 10$ for diverse M-best inference, and the number of binary search iterations is $n_{\text{iter}} = 20$. For simplicity, we search on Deepglobe Land [9] and the found architectures are evaluated on other datasets. Our results can be improved by searching on each dataset individually. In the re-train phase, we use the same hyper-parameters as in the search phase, except that we train for $T = 100$ epochs. We use the same hyper-parameters for both architectures found by our methods as well as the baselines.

5.4   Computational Cost

The overhead of our method comes from Gibbs sampling and inference over a complex loopy graph. Since we run a long burn-in period n_{long} only at the start of each epoch, and a short burn-in period $n_{\text{short}} + n_{\text{mc}}$ at each training iteration, the cost of sampling is negligible compared with a forward-backward pass of the neural network. We will discuss the cost of loopy inference algorithms in Sect. 6.1, and since we use approximate inference algorithm as a default, the cost of inference is also minimal. In our experiments, the overhead takes up less than 2% of the total search time.

5.5   MRF-UNets Architecture

MRF-UNet shows several interesting characteristics that differ from the original UNet: (i) it has a larger encoder but a smaller decoder (ii) layers that are connected by the long skip connections are shallower (iii) layers that need to process these concatenated feature maps are wider and also have larger kernel size. As a result, there exists a bottleneck pattern in the encoder and an inverted bottleneck pattern in the decoder. Our observations show that the encoder and decoder do not need to be balanced as in the original UNet and many other encoder-decoder architectures [5, 26, 45]. They also demonstrate that the widely adopted "half resolution, double width" principle might be sub-optimal in an encoder-decoder network. Indeed, feature maps are concatenated by the long skip connections and are processed by the following layer, which form the most computationally extensive part in the whole network. Therefore, their widths should be smaller to reduce complexity. However, layers that need to process this rich information should be wider and have larger receptive fields.

Inspired by these observations, we propose MRF-UNetV2 to emphasize these characteristics. As shown in Fig. 2, MRF-UNetV2 has a simpler architecture that is easier for implementation, and we show that it can sometimes outperform MRF-UNet in Table 2 and Table 3. We note that a recent trend in NAS is to design a more and more complex search space to include as many candidates as possible [30], but it becomes difficult to interpret the search results. We hope that our observations can inspire practitioners when designing other encoder-decoder architectures.

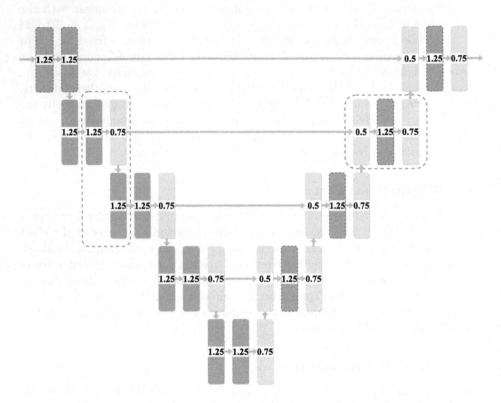

Fig. 2. Architecture of MRF-UNetV2. Numbers inside rectangles are width ratios to the original UNet. Rectangles surrounded by colored dashed lines use 5×5 kernels while others use 3×3 kernels. The gray dashed line on the left highlights an example of a bottleneck block in the encoder, and the gray dashed line on the right shows an example of an inverted bottleneck block in the decoder. (Color figure online)

5.6   Main Results

Our benchmarks include manually designed architectures: UNet [27], UNet++ [43], BiO-Net [33] and architectures found with NAS: NAS-UNet [32], MS-NAS [35], BiX-Net [31]. The main results are in Table 2 and Table 3. We set the latency target to 1.70ms which is the same as the original UNet, and we also include FLOPs for comparison. Our MRF-UNets outperform benchmarks over five datasets with diverse semantics, while require less computational resources.

Table 2. mIoU (%) on DeepGlobe challenge. mean ± std are computed over 5 runs.

Model	Land (%)	Road (%)	Building (%)	FLOPs (G)	Latency (ms)
UNet	58.41 ± 0.52	57.05 ± 0.13	75.12 ± 0.09	4.84	1.70
UNet++	59.53 ± 0.21	56.96 ± 0.08	74.27 ± 0.21	11.76	5.11
BiO-Net	58.10 ± 0.37	57.06 ± 0.22	74.29 ± 0.15	37.22	3.28
NAS-UNet	58.11 ± 0.91	57.73 ± 0.56	74.46 ± 0.19	30.44	4.25
MS-NAS	58.75 ± 0.68	57.34 ± 0.35	74.61 ± 0.17	24.28	3.96
BiX-Net	57.96 ± 0.94	57.74 ± 0.08	74.63 ± 0.12	13.28	1.87
MRF-UNet	**59.64 ± 0.44**	57.81 ± 0.17	75.50 ± 0.09	4.74	1.68
MRF-UNetV2	58.56 ± 0.25	**57.90 ± 0.23**	**75.84 ± 0.13**	4.66	1.70

Table 3. Dice scores (%) on CHAOS and PROMISE. mean ± std are computed over 5 runs.

Model	CHAOS (%)	PROMISE (%)	FLOPs (G)	Latency (ms)
UNet	91.16 ± 0.23	84.60 ± 0.68	4.84	1.70
UNet++	91.46 ± 0.15	86.29 ± 0.35	11.76	5.11
BiO-Net	91.80 ± 0.42	86.04 ± 0.77	37.22	3.28
NAS-UNet	91.30 ± 0.65	85.04 ± 0.90	30.44	4.25
MS-NAS	91.47 ± 0.35	85.42 ± 0.72	24.28	3.96
BiX-Net	91.22 ± 0.39	84.35 ± 0.91	13.28	1.87
MRF-UNet	92.03 ± 0.31	**86.76 ± 0.32**	4.74	1.68
MRF-UNetV2	**92.14 ± 0.24**	86.61 ± 0.36	4.66	1.70

6 Ablation Study

6.1 Exact vs. Approximate Loopy Inference

Without our loopy inference extension, AOWS fails on more complex loopy graphs. However, inference on a loopy MRF is a NP-hard problem [21], so we cannot always hope for exact solutions. Here we use Max-Product Clique Tree algorithm (MPCT) for exact solutions, and Max-Product Linear Programming (MPLP) for approximate inference [21]. We compare architectures found by MPCT and MPLP in Table 4. They usually obtain very similar solutions and their results are almost identical. The complexity of MPCT and in general of exact inference is $O(|X|^{|C|})$ where $|X|$ is the cardinality of factors and in our experiments it is 10, and $|C|$ is the size of the largest clique. Generally, $|C|$ increases when the graph is more densely connected, and in the worst case $|C| = n$ the number of nodes when the graph is fully connected, e.g. DenseNet [17]. In Fig. 3, we show the size of the largest clique vs. the number nodes for UNet [27], UNet+ [43] and UNet++ [43]. UNet++, being more densely connected, has a much larger clique size than UNet. The clique size of the original UNet is 5, and MPCT already takes several minutes on our MacBook Pro (14-

inch, 2021). Note that we need to repeat the inference for $m \times n_{\text{iter}} = 100$ times, and it will soon become infeasible if we want to apply MPCT on deeper UNet or UNet+/UNet++. Nevertheless, MPLP can converge within a few seconds. Since they usually find similar solutions, we use MPLP as a default.

Table 4. Evaluating architectures found by MPCT and MPLP on DeepGlobe Land. mean $\pm$ std are computed over 5 runs.

Algorithm	MPCT	MPLP
mIoU (%)	59.56 ± 0.57	$\mathbf{59.64 \pm 0.44}$

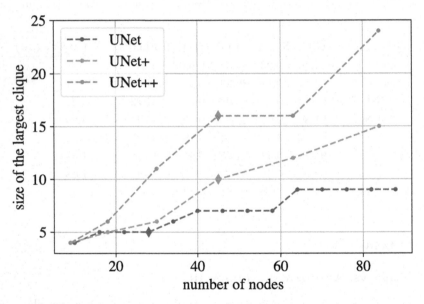

Fig. 3. Size of the largest clique vs. the number of nodes for UNet [27], UNet+ [43] and UNet++ [43]. Diamonds indicate the original architectures whose depth is 5.

6.2   Diverse Solutions

As shown in [36,42], there exists an inconsistency between the true rank of an architecture and its rank on a weight-sharing network, but we can still find architectures that are reasonably good by examining other top solutions. This motivates us to apply diverse M-best inference [2]. In Table 5, we show the results of diverse 5-best. The MAP solution is not the best quality and diverse M-best inference can greatly improve our results. It also helps reduce the variance in the search phase since we can evaluate m highly probable solutions at the same time, and the total computational cost is thus decreased from $m \times (\text{cost}_{\text{search}} + \text{cost}_{\text{eval}})$ to $\text{cost}_{\text{search}} + m \times \text{cost}_{\text{eval}}$.

Calibration [12] is critical for medical diagnosis and deep ensembles [22] have been shown to effectively reduce the calibration error. Compared with a deep ensemble that consists of the same architecture from different initialization, we can form an ensemble with a diverse set of solutions. As shown in Table 7, our diverse ensemble achieves a lower Expected Calibration Error (ECE) on CHAOS.

Should we further increase m if it is so helpful? The answer is no: further increasing m generally does not help us to find a better solution, and the majority of solutions are often of sub-optimal performance so it does not help reduce the variance. In Table 6, we show the result of diverse 10-best. We choose a higher $L = 20$ and expect that the best solution will come later, since it leads to a lower dissimilarity target in Eq. (10). Indeed, the best architecture is now the 5th one instead of the 3rd, but it does not show a better performance and many solutions are of similar quality. Therefore, we do not benefit from increasing m, while it adds the cost of both inference and evaluation.

Table 5. Evaluating architectures found by diverse 5-best on DeepGlobe Land ($L = 10$). mean ± std are computed over 5 runs.

Solution	1	2	3	4	5
mIoU (%)	57.37 ± 0.61	57.41 ± 0.32	**59.64 ± 0.44**	59.43 ± 0.46	56.59 ± 0.70

Table 6. Evaluating architectures found by diverse 10-best on DeepGlobe Land ($L = 20$). mean ± std are computed over 5 runs.

Solution	1	2	3	4	5
mIoU (%)	57.02 ± 0.53	58.37 ± 0.41	56.94 ± 0.55	59.37 ± 0.41	**59.56 ± 0.22**
Solution	6	7	8	9	10
mIoU (%)	57.86 ± 0.41	59.45 ± 0.92	57.82 ± 0.51	57.79 ± 0.48	57.84 ± 0.39

Table 7. Comparing deep ensemble [22] with our diverse deep ensemble on CHAOS. Lower is better. mean ± std are computed over 5 runs.

Model	Deep ensemble	Diverse deep ensemble
ECE (%)	0.7091 ± 0.0140	**0.6872 ± 0.0126**

6.3 Pairwise Formulation and Differentiable Parameter Learning

Except diverse M-best loopy inference, we make other two modifications: pairwise formulation and differentiable parameter learning. In Table 8, we compare the architectures found by our MRF-NAS vs. MRF-NAS without pairwise factors and MRF-NAS without differentiable parameter learning.

Table 8. Evaluating architectures found by A1: MRF-NAS w/o pairwise factors, A2: MRF-NAS w/o differentiable parameter learning and MRF-UNet on DeepGlobe Land. mean ± std are computed over 5 runs.

Architecture	A1	A2	MRF-UNet
mIoU (%)	59.17 ± 0.58	59.33 ± 0.31	$\mathbf{59.64} \pm \mathbf{0.44}$

7 Conclusion

In this paper, we propose MRF-NAS that extends and improves AOWS [4] with a more general framework based on a pairwise Markov Random Field (MRF) formulation, which leads to applying various statistical techniques for MAP optimization. With diverse M-best loopy inference algorithms and differentiable parameter learning, we find an architecture, MRF-UNet, with several interesting characteristics. Through the lens of these characteristics, we identify the sub-optimality of the original UNet and propose MRF-UNetV2 with a simpler architecture that can further improve our results. MRF-UNets, albeit requiring less computational resources, outperform several SOTA benchmarks over three aerial image datasets and two medical image datasets that contain diverse contents. This demonstrates that the found architectures are robust and effective.

Acknowledgements. We acknowledge support from the Research Foundation - Flanders (FWO) through project numbers G0A1319N and S001421N, and funding from the Flemish Government under the Onderzoeksprogramma Artificiële Intelligentie (AI) Vlaanderen programme.

References

1. Ardywibowo, R., Boluki, S., Gong, X., Wang, Z., Qian, X.: NADS: neural architecture distribution search for uncertainty awareness. In: ICML (2020)
2. Batra, D., Yadollahpour, P., Guzman-Rivera, A., Shakhnarovich, G.: Diverse M-best solutions in Markov random fields. In: Fitzgibbon, A., Lazebnik, S., Perona, P., Sato, Y., Schmid, C. (eds.) ECCV 2012. LNCS, vol. 7576, pp. 1–16. Springer, Heidelberg (2012). https://doi.org/10.1007/978-3-642-33715-4_1
3. Bender, G., Kindermans, P.J., Zoph, B., Vasudevan, V., Le, Q.: Understanding and simplifying one-shot architecture search. In: ICML (2018)
4. Berman, M., Pishchulin, L., Xu, N., Blaschko, M., Medioni, G.: AOWS: adaptive and optimal network width search with latency constraints. In: CVPR (2020)
5. Chaurasia, A., Culurciello, E.: LinkNet: exploiting encoder representations for efficient semantic segmentation. In: VCIP (2017)
6. Chu, X., Zhang, B., Xu, R.: FairNAS: rethinking evaluation fairness of weight sharing neural architecture search. In: ICCV (2021)
7. Chu, X., Zhou, T., Zhang, B., Li, J.: Fair DARTS: eliminating unfair advantages in differentiable architecture search. In: Vedaldi, A., Bischof, H., Brox, T., Frahm, J.-M. (eds.) ECCV 2020. LNCS, vol. 12360, pp. 465–480. Springer, Cham (2020). https://doi.org/10.1007/978-3-030-58555-6_28

8. Dai, X., et al.: FBNetV3: joint architecture-recipe search using predictor pretraining. In: CVPR (2021)
9. Demir, I., et al.: DeepGlobe 2018: a challenge to parse the earth through satellite images. In: CVPR Workshop (2018)
10. Ding, M., et al.: Learning versatile neural architectures by propagating network codes. In: ICLR (2022)
11. Eelbode, T., et al.: Optimization for medical image segmentation: theory and practice when evaluating with dice score or Jaccard index. TMI **39**, 3679–3690 (2020)
12. Guo, C., Pleiss, G., Sun, Y., Weinberger, K.Q.: On calibration of modern neural networks. In: ICML (2017)
13. Guo, Z., et al.: Single path one-shot neural architecture search with uniform sampling. In: Vedaldi, A., Bischof, H., Brox, T., Frahm, J.-M. (eds.) ECCV 2020. LNCS, vol. 12361, pp. 544–560. Springer, Cham (2020). https://doi.org/10.1007/978-3-030-58517-4_32
14. He, K., Zhang, X., Ren, S., Sun, J.: Deep residual learning for image recognition. In: CVPR (2016)
15. He, Y., Yang, D., Roth, H., Zhao, C., Xu, D.: DiNTS: differentiable neural network topology search for 3D medical image segmentation. In: CVPR (2021)
16. Howard, A.G., et al.: MobileNets: efficient convolutional neural networks for mobile vision applications. arXiv (2017)
17. Huang, G., Liu, Z., van der Maaten, L, Weinberger, K.Q.: Densely connected convolutional networks. In: CVPR (2017)
18. Huang, Z., Wang, Z., Yang, Z., Gu, L.: AdwU-Net: adaptive depth and width U-Net for medical image segmentation by differentiable neural architecture search. In: MIDL (2022)
19. Jang, E., Gu, S., Poole, B.: Categorical reparameterization with Gumbel-softmax. In: ICLR (2017)
20. Kavur, A.E., et al.: CHAOS challenge - combined (CT-MR) healthy abdominal organ segmentation. MIA (2021)
21. Koller, D., Friedman, N.: Probabilistic Graphical Models: Principles and Techniques. MIT Press, Cambridge (2009)
22. Lakshminarayanan, B., Pritzel, A., Blundell, C.: Simple and scalable predictive uncertainty estimation using deep ensembles. In: NeurIPS (2017)
23. Liang, T., Wang, Y., Tang, Z., Hu, G., Ling, H.: OPANAS: one-shot path aggregation network architecture search for object detection. In: CVPR (2021)
24. Litjens, G., et al.: Evaluation of prostate segmentation algorithms for MRI: the PROMISE12 challenge. MIA (2014)
25. Liu, H., Simonyan, K., Yang, Y.: DARTS: differentiable architecture search. In: ICLR (2019)
26. Milletari, F., Navab, N., Ahmadi, S.A.: V-Net: fully convolutional neural networks for volumetric medical image segmentation. In: 3DV (2016)
27. Ronneberger, O., Fischer, P., Brox, T.: U-Net: convolutional networks for biomedical image segmentation. In: Navab, N., Hornegger, J., Wells, W.M., Frangi, A.F. (eds.) MICCAI 2015. LNCS, vol. 9351, pp. 234–241. Springer, Cham (2015). https://doi.org/10.1007/978-3-319-24574-4_28
28. Srivastava, S., Berman, M., Blaschko, M.B., Tuia, D.: Adaptive compression-based lifelong learning. In: BMVC (2019)
29. Veit, A., Wilber, M., Belongie, S.: Residual networks behave like ensembles of relatively shallow networks. In: NeurIPS (2016)
30. Wan, X., Ru, B., Esperança, P.M., Li, Z.: On redundancy and diversity in cell-based neural architecture search. In: ICLR (2022)

31. Wang, X., et al.: BiX-NAS: searching efficient bi-directional architecture for medical image segmentation. In: de Bruijne, M., et al. (eds.) MICCAI 2021. LNCS, vol. 12901, pp. 229–238. Springer, Cham (2021). https://doi.org/10.1007/978-3-030-87193-2_22

32. Weng, Y., Zhou, T., Li, Y., Qiu, X.: NAS-Unet: neural architecture search for medical image segmentation. IEEE Access **7**, 44247–44257 (2019)

33. Xiang, T., Zhang, C., Liu, D., Song, Y., Huang, H., Cai, W.: BiO-Net: learning recurrent bi-directional connections for encoder-decoder architecture. In: Martel, A.L., et al. (eds.) MICCAI 2020. LNCS, vol. 12261, pp. 74–84. Springer, Cham (2020). https://doi.org/10.1007/978-3-030-59710-8_8

34. Xu, H., Yao, L., Zhang, W., Liang, X., Li, Z.: Auto-FPN: automatic network architecture adaptation for object detection beyond classification. In: ICCV (2019)

35. Yan, X., Jiang, W., Shi, Y., Zhuo, C.: MS-NAS: multi-scale neural architecture search for medical image segmentation. In: Martel, A.L., et al. (eds.) MICCAI 2020. LNCS, vol. 12261, pp. 388–397. Springer, Cham (2020). https://doi.org/10.1007/978-3-030-59710-8_38

36. Yang, A., Esperança, P.M., Carlucci, F.M.: NAS evaluation is frustratingly hard. In: ICLR (2020)

37. Ye, P., Li, B., Li, Y., Chen, T., Fan, J., Ouyang, W.: β-DARTS: beta-decay regularization for differentiable architecture search. In: CVPR (2022)

38. Yu, J., Huang, T.: AutoSlim: towards one-shot architecture search for channel numbers. In: NeurIPS Workshop (2019)

39. Yu, J., Huang, T.S.: Universally slimmable networks and improved training techniques. In: ICCV (2019)

40. Yu, J., et al.: BigNAS: scaling up neural architecture search with big single-stage models. In: Vedaldi, A., Bischof, H., Brox, T., Frahm, J.-M. (eds.) ECCV 2020. LNCS, vol. 12352, pp. 702–717. Springer, Cham (2020). https://doi.org/10.1007/978-3-030-58571-6_41

41. Yu, J., Yang, L., Xu, N., Yang, J., Huang, T.: Slimmable neural networks. In: ICLR (2019)

42. Yu, K., Sciuto, C., Jaggi, M., Musat, C., Salzmann, M.: Evaluating the search phase of neural architecture search. In: ICLR (2020)

43. Zhou, Z., Siddiquee, M.M.R., Tajbakhsh, N., Liang, J.: UNet++: redesigning skip connections to exploit multiscale features in image segmentation. TMI **39**, 1856–1867 (2020)

44. Zoph, B., Le, Q.V.: Neural architecture search with reinforcement learning. In: ICLR (2017)

45. Çiçek, Ö., Abdulkadir, A., Lienkamp, S.S., Brox, T., Ronneberger, O.: 3D U-Net: learning dense volumetric segmentation from sparse annotation. In: Ourselin, S., Joskowicz, L., Sabuncu, M.R., Unal, G., Wells, W. (eds.) MICCAI 2016. LNCS, vol. 9901, pp. 424–432. Springer, Cham (2016). https://doi.org/10.1007/978-3-319-46723-8_49

Adversarial Projections to Tackle Support-Query Shifts in Few-Shot Meta-Learning

Aroof Aimen[1]([✉]), Bharat Ladrecha[1], and Narayanan C. Krishnan[2]

[1] Indian Institute of Technology, Ropar, Punjab, India
{2018csz0001,2018csb1080}@iitrpr.ac.in
[2] Indian Institute of Technology, Palakkad, Kerala, India
ckn@iitpkd.ac.in

Abstract. Popular few-shot Meta-learning (ML) methods presume that a task's support and query data are drawn from a common distribution. Recently, Bennequin et al. [4] relaxed this assumption to propose a few-shot setting where the support and query distributions differ, with disjoint yet related meta-train and meta-test support-query shifts (SQS). We relax this assumption further to a more pragmatic SQS setting (SQS+) where the meta-test SQS is anonymous and need not be related to the meta-train SQS. The state-of-the-art solution to address SQS is transductive, requiring unlabelled meta-test query data to bridge the support and query distribution gap. In contrast, we propose a theoretically grounded inductive solution - Adversarial Query Projection (AQP) for addressing SQS+ and SQS that is applicable when unlabeled meta-test query instances are unavailable. AQP can be easily integrated into the popular ML frameworks. Exhaustive empirical investigations on benchmark datasets and their extensions, different ML approaches, and architectures establish AQP's efficacy in handling SQS+ and SQS.

Keywords: Meta-learning · Task · Support · Query · Projection · Shift

1 Introduction

Learning Deep neural networks (DNN) from limited training data is of increasing relevance due to its ability to mitigate the challenges posed by the costly data annotation process for various real-world problems. A popular framework for learning with limited training data is few-shot learning, i.e., learning a model from few shots (examples) of data. Meta-learning (ML) approaches for few-shot learning have proven to be robust at handling data scarcity [1,10,24,27]. A

A. Aimen and B. Ladrecha—Equal Contribution.

Supplementary Information The online version contains supplementary material available at https://doi.org/10.1007/978-3-031-26409-2_37.

typical ML setup follows an episodic training regimen. An episode or a task is an N-way K-shot learning problem, where N is the number of classes in a task and K is the number of examples per class. Each task comprises of a task-train data (support set) and task-test data (query set), containing disjoint examples from the same classes. Models are adapted separately for the tasks using the support set. The adapted model's loss on the query set is used to update the meta-model. The model meta-trained in this fashion extracts rich class discriminative features [14] that can quickly adapt to a new unseen test task.

The ML approach assumes that the meta-train and meta-test tasks are drawn from a common distribution. The shared distribution assumption prevents the use of meta-learned models in evolving test environments deviating from the training set. Recent ML works attempt at relaxing this assumption [26,29]. However, these ML approaches assume a common distribution inside the tasks, i.e., the task-train and task-test data come from the same distribution. But a distribution shift may exist between the support and query set because of the evolving or deteriorating nature of real-world objects or environments, differences in the data acquisition techniques from support to query sets, extreme data deficiency from one distribution, etc. Addressing support query shift (SQS) inside a task has gained attention very recently [4]. However, this pioneering work assumes the prior knowledge of SQS in the meta-test set and induces a related although disjoint SQS in the meta-train set. The model trained on such a meta-train set is accustomed to handle the SQS and, to some extent, becomes robust to the related unseen meta-test SQS.

In this paper, we consider, SQS+, a more generic SQS problem where the prior knowledge of the meta-test SQS is absent. We expect an unknown SQS in the meta-test set and therefore cannot induce any related SQS in the meta-train set. The earlier work on addressing SQS [4] is a limiting case of SQS+.

We illustrate the significance of SQS+ in Fig. 1 on a 5-way 5-shot problem: *Case a)* miniImagenet with No SQS [27] where both meta-train and meta-test sets do not contain SQS; *Case b)* miniImagenet with SQS [4] where meta-train and meta-test sets have related but disjoint SQS and *Case c)* miniImagenet with SQS+ *(ours)*

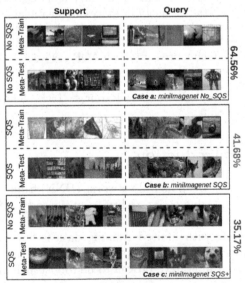

Fig. 1. Performance drop of a ProtoNet model due to SQS. **Case a)** No SQS in meta-train/test set **Case b)** Related but disjoint SQS in both meta-train/test sets **Case c)** Meta-train set lacks SQS, but meta-test set contains SQS.

where meta-train set lacks SQS, but meta-test set possesses SQS. The average performance of a meta-trained prototypical network (ProtoNet) [24] for the cases *(a)*, *(b)*, and *(c)* is 64.56%, 41.68%, and 35.17% respectively. The nearly 29% performance drop from *case a* to *case c* indicates that the naive ML model is vulnerable to SQS and cannot extrapolate its training experience to comparable scenarios. Bennequin et al. [4] initiated the research on SQS to address the problem specified in *case (b)*. We extend it to a more generic an challenging setting where there is no SQS during meta-training, but meta-test tasks may contain a distribution shift between the support and the query sets. The approximately 6% drop in the accuracy of the ProtoNets trained for settings *case b* and *case c* reinforces our challenging problem setting.

The solution proposed by Bennequin et al. [4] uses optimal transport (OT) to bridge the gap between support and query distributions, but assumes the availability of labeled and unlabelled query data during meta-training and testing respectively. While this solution can be adopted for our proposed problem, access to unlabelled query data during meta-test may be unrealistic in many real-world scenarios. Our solution to address the support query (SQ) shift problem - Adversarial Query Projection (AQP), does not require transduction during meta-testing. During meta-training, we induce a distribution shift between support and query sets by adversarially perturbing the query sets to create more "challenging" virtual query sets. New virtual tasks are constructed from the original support and virtual query sets. Due to the disparity between the initial and perturbed distributions, a distribution mismatch occurs between the support and query set of a virtual task. The adversarial perturbations are dynamic and adaptive, seeking to inhibit the model's learning. A model trained in such a setup performs well only if it learns to be resilient to the SQS in a task. As adversarial perturbations lack a static structure, the model is forced to learn various shift-invariant representations and thus becomes robust to various unknown distribution shifts. Overall, we make the following contributions:

- We propose, SQS+, a practical SQS setting for few-shot meta-learning. The shift between support and query sets during meta-testing is unknown while meta-training the model.
- We contribute to the FewShiftBed [4] realistic datasets for evaluating methods that address SQS and SQS+. In these datasets, meta-train data lacks SQS while meta-test data contains SQS.
- We design an inductive solution for tackling SQS+ using adversarial query projections (AQP). We theoretically justify the feasibility of meta-optimizing the model using adversarially projected query sets and verify the existence of an adversarial query projection for each query set. The AQP module is standalone and could be integrated with any few-shot ML episodic training regimen. We verify this capability by integrating AQP into ProtoNet and Matching Networks (MatchingNet).
- Exhaustive empirical investigation validates the effectiveness of the AQP on various settings and datasets, preventing a negative impact even in the absence of SQS.

2   Related Work

We segregate the discussion of the related work into approaches for cross-domain few-shot learning and tackling support-query shift in few-shot learning.

2.1   Cross-Domain Few-shot Learning (CDFSL)

Classical few-shot learning (FSL) [7,15] does not expect distribution shifts between train and test sets. Domain generalization approaches that generate examples from a fictitious hard domain through adversarial training [28] or synthesize virtual train and test domains to simulate a shift during the training process using a critic network [17] aim to encourage generalization on unseen target domain. Typical domain generalization setting assumes abundant training data and shared labels between train and test domains, which need not hold in a cross domain FSL setup. The early approaches to bridge the domain shifts in FSL relied on adaptive batch normalization [9] and batch spectral normalization [19]. Recent work [4] suggests limitations of batch normalization as a strategy for handling SQS. A common hypothesis among cross domain FSL approaches is that a model's over-reliance on the meta-train domain inhibits its generalizability to unseen test domains. While some cross domain FSL approaches relied on model's generalizability by enhancing diversity in the feature representations [25,26], others have tried ensembles [19], large margin enforcement [30], and adversarial perturbations [29]. Though these approaches handle domain discrepancy between meta-train and meta-test sets, they assume a common distribution over support and query sets. On the other hand, we focus on the scenarios where support and query distributions vary.

2.2   Support-Query Shift in FSL

Transductive meta-learning approaches that utilize unlabeled query data in the training process are effective baselines for handling SQS in FSL. Ren et al. [22] introduce a transductive prototypical network that refines the learned prototypes with cluster assignments of unlabelled query examples. Boudiaf et al. [6] induce transduction by maximizing the mutual information between query features and their predicted labels in conjunction with minimizing cross-entropy loss on the support set. Minimizing the entropy of the unlabeled query instance predictions during adaptation [8] also achieves a similar goal. Liu et al. [20] propose a graph based label propagation from the support to the unlabeled query set that exploits the data manifold properties to improve the efficiency of adaptation . Antoniou et al. [2] show that minimizing a parameterized label-free loss function that utilizes unlabelled query data during training can also bridge SQS. Inspired from learning invariant representations [3,11,12], Bennequin et al. [4] use Optimal Transport (OT) [21] during meta-training and meta-testing to address SQS. In contrast, we propose an inductive method to tackle SQS in few-shot meta-learning where access to the unlabelled meta-test query instances is not required.

Inductive approaches to tackle train-test domain shifts have relied on adversarial methods for data/task augmentations. Goldblum et al. [13] propose adversarial data augmentation for few-shot meta-learning and demonstrate the robustness of the model trained on augmented tasks to adversarial attacks at meta-test time. Wang et al. [29] bridge the shift between meta-train and meta-test domains by adversarial augmentation by constructing virtual tasks learned through adversarial perturbations. A model trained on such virtual tasks becomes resilient to meta-train and meta-test domain shifts. While adversarial perturbations are central to our approach, we use it to tackle a different problem, support query distribution shifts inside a task for few-shot meta-learning.

3 Methodology

3.1 Preliminaries

Notations. A typical ML setup has three phases - meta-train M, meta-validation M_v and meta-test M_t. A model is trained on M and evaluated on M_t. M_v is used for hyperparameter tuning and model selection. The dataset $(C, \mathcal{D})$ comprising of classes and domains is partitioned into $(C_M, \mathcal{D}_M)$, $(C_{M_v}, \mathcal{D}_{M_v})$, and $(C_{M_t}, \mathcal{D}_{M_t})$ corresponding to the phases M, M_v and M_t, respectively. Each phase is a collection of tasks and every task T_0 is composed of a support set T_{S_0} and a query set T_{Q_0}. The support set $T_{S_0} = \{\{x_k^c, y_k^c\}_{k=1}^K\}_{c=1}^N$ and query set $T_{Q_0} = \{\{x_q^{*c}, y_q^{*c}\}_{q=1}^Q\}_{c=1}^N$ contain (example x, label y) pairs from N-classes with K and Q examples per class, with the label of meta-test query instances being used only for evaluation.

The classical few-shot learning setup does not consider diverse domains. The tasks are sampled from a common distribution $\mathcal{T}_0$. A model meta-trained on tasks sampled from $\mathcal{T}_0$ learns representations that extend to the disjoint meta-test tasks from the same distribution. Given a task T_0 (support-query pair $\{T_{S_0}, T_{Q_0}\}$), few-shot learning learns a classifier ϕ using T_{S_0}, which correctly categorizes instances of T_{Q_0}. A model parameterized by θ is meta-trained on a collection of tasks sampled from $\mathcal{T}_0$ using a bi-level optimization procedure. First, θ is adapted on the tasks' support set T_{S_0} to obtain ϕ. Then ϕ is evaluated on the query set T_{Q_0} to estimate query loss L^*, which is used to update θ. The model is meta-trained according to the objective $\min_{\theta \in \Theta} \mathbb{E}_{T_{Q_0}}[L^*(\phi, T_{Q_0})]$, where $\phi \leftarrow \theta - \alpha \nabla_\theta L(\theta; T_{S_0})$; L and L^* are the losses of the model on the support and query sets respectively. Note that ML approaches such as ProtoNet [16] and MatchingNet [27] do not require adaptation, and hence $\theta = \phi$.

Support-Query Distribution Shift. In a classical few-shot learning setup, the domain is constant across M, M_v, M_t phases and within the tasks. So, in addition to a common distribution $\mathcal{T}_0$ over tasks, a shared distribution exists even at the task composition level, i.e., $\mathcal{T}_{S_0} = \mathcal{T}_{Q_0}$, where $\mathcal{T}_{S_0}$ and $\mathcal{T}_{Q_0}$ are the distributions on support and query sets respectively. A more pragmatic case is that of SQS, wherein a distribution mismatch occurs between the support and query sets within a task. Let $\mathcal{D}_M$ and $\mathcal{D}_{M_t}$ be the set of domains for the M and M_t phases. We skip M_v for convenience, but it follows the same characteristics as M and M_t. We define our version of the support query shift problem termed SQS+ illustrated in the Fig. 1 (*case c*) as follows.

Definition 1. *(SQS+) The support and query sets of every meta-train task come from the domain $\mathcal{D}_M$ and share a common distribution $\mathcal{T}_{S_0} = \mathcal{T}_{Q_0}$. Let $D_S^{M_t}, D_Q^{M_t} \in \mathcal{D}_{M_t}$ be the support and query domains for a meta-test task. The SQS+ setting is characterized by an unknown shift in the support and query domains of a meta-test task, $D_S^{M_t} \neq D_Q^{M_t}$ (introducing a shift in the support and query distributions $\mathcal{T}_{S_0} \neq \mathcal{T}_{Q_0}$), along with the standard SQS assumption of disjoint meta-train and meta-test domains - $\mathcal{D}_M \cap \mathcal{D}_{M_t} = \emptyset$.*

Bennequin et al. [4] identified the SQS problem, but assumed only a similar but disjoint SQS in the meta-train and meta-test datasets. A model learned on such a meta-train set is compelled to extract shift-invariant features during adaptation on the support set to reduce L^* on query sets. Although $\mathcal{D}_M$ and $\mathcal{D}_{M_t}$ are disjoint, they share a latent structure that facilitates learning of shift-invariant features on $\mathcal{D}_M$ that can be extended to $\mathcal{D}_{M_t}$. This makes the learned model impervious to SQS in the meta-test set. SQS+, on the other hand, is more general and challenging. We neither anticipate the occurrence of SQS in the meta-test set nor maintain a common structure between the meta-train and meta-test SQS's. Relaxing the shared structure constraint between $\mathcal{D}_M$ and $\mathcal{D}_{M_t}$ removes the need for prior access to the meta-test set (consequently its domains) to imbibe SQS in meta-train tasks. Hence, we tackle a more challenging problem of learning a resilient model for an unknown meta-test SQS.

A model trained using the classical ML objective has not encountered support and query set shifts during meta-training. Thus the learned representations are not shift-invariant, due to which the model does not generalize to the unknown meta-test SQS. Bennequin et al.'s [4] transductive optimal transport (OT)-based solution to bridge the gap between the support and query shifts could also be adopted SQS+. However, the solution needs access to unlabeled query sets during meta-training and meta-testing, which is unavailable in our setting. We propose an inductive adversarial query projection (AQP) strategy to address SQS+ that can also work in the vanilla SQS setting.

3.2 Adversarial Query Projection (AQP)

Without leveraging unlabelled meta-test query instances, our solution induces the hardest distribution shift for the meta-model's current state. For a task T_0, we simulate the worst distribution shift by adversarially perturbing its query set T_{Q_0} such that the model's query loss L^* maximizes. Let H be the task composition space, i.e., H is the distribution of support and query distributions such that $\mathcal{T}_{Q_0} \sim H$ and $\mathcal{T}_Q \sim H$. Let T_{Q_0} and T_Q be the samples belonging to $\mathcal{T}_{Q_0}$ and $\mathcal{T}_Q$ respectively (we occasionally denote $T_Q \sim H$ because $T_Q \sim \mathcal{T}_Q \sim H$, to improve readability). Also, let Θ be the parameter space with $\theta, \phi \sim \Theta$, and

$d : H \times H \rightarrow R_+$ be the distance metric that satisfies $d(T_{Q_0}, T_{Q_0}) = 0$ and $d(T_Q, T_{Q_0}) \geq 0$. We consider a Wasserstein ball B centered at T_{Q_0} with radius ρ denoted by $B_\rho(T_{Q_0})$ such that:

$$B_\rho(T_{Q_0}) = \{T_Q \in H : W_d(T_Q, T_{Q_0}) \leq \rho\}$$

where $W_d(T_Q, T_{Q_0}) = \inf\limits_{M \in \pi(T_Q, T_{Q_0})} \mathbb{E}_M [d(T_Q, T_{Q_0})]$ is the Wasserstein distance that measures the minimum transportation cost required to transform T_{Q_0} to T_Q, and $\pi(T_Q, T_{Q_0})$ denotes all joint distributions for (T_Q, T_{Q_0}) with marginals T_Q and T_{Q_0}.

AQP aims to find the most challenging query distribution T_Q for an original query distribution T_{Q_0} that lies within or on the Wasserstein ball $B_\rho(T_{Q_0})$. The hardest perturbation to the query distribution T_{Q_0} is the one that maximizes the model's query loss L^*. Updating the model using such difficult query distribution T_Q improves its generalizability. Further, the transformation of T_{Q_0} into T_Q induces a distributional disparity in a new virtual task comprising of the original support set from T_{S_0} and the projected query set from T_Q. A model adapted to such virtual tasks is compelled to extract the shift-invariant representations from $T_{S_0} \sim T_{S_0}$ transferable to $T_Q \sim T_Q$ to reduce the query loss L^*. As adversarial perturbations are adaptive to the model's state, they do not have a monotonic structure throughout the meta-training phase. The evolving augmentations expose the model to diverse SQS. A model meta-trained on such virtual tasks with different SQ shifts learns to extract diverse shift-invariant representations increasing the model's endurance to unknown meta-test SQS. The simultaneous restrain of T_Q to a Wasserstein ball radius ρ ensures T_Q does not deviate extensively from T_{Q_0}, and T_Q, T_{Q_0} share the label space, and $T_{Q_0}, T_Q \in H$ is maintained. Thus the newly-framed meta-objective is:

$$\min_{\theta \in \Theta} \quad \sup_{W_d(T_Q, T_{Q_0}) \leq \rho} \mathbb{E}_{(T_Q \sim T_Q)} [L^*(\phi, T_Q)] \tag{1}$$

where $\phi \leftarrow \theta - \alpha \nabla_\theta L(T_{S_0}; \theta)$. As Eq. 1 is intractable for an arbitrary ρ, we aim to convert this constrained optimization problem to an unconstrained optimization problem for a fixed penalty parameter $\gamma \geq 0$ as given below:

$$\min_{\theta \in \Theta} \sup_{T_Q} \{\mathbb{E}_{T_Q}[L^*(\phi, T_Q)] - \gamma W_d(T_Q, T_{Q_0})\} \tag{2}$$

We first show that the unconstrained objective is strongly concave and then define a shift robust surrogate, $\psi_\gamma(\phi, T_{Q_0})$, that is easy to optimize.

Theorem 1. *For the loss function $L^*(\phi, T_Q)$ smooth in T_Q, a distance metric $d : H \times H \rightarrow R_+$ convex in T_Q and a large penalty γ (by duality small ρ), the function $L^*(\phi; T_Q) - \gamma d(T_Q, T_{Q_0})$ is $\gamma - \mathcal{L}$ strongly concave for $\gamma \geq \mathcal{L}$.*

Proof. Deferred to the supplementary material.

We next define a robust surrogate inspired from Sinha et al. [23] for this unconstrained objective that is the dual of the minimax problem in Eq. 1.

Theorem 2. *Let* $L^* : \Theta \times H \to R$ *and* $d : H \times H \to R_+$ *be continuous. Let*
$\psi_\gamma(\phi; T_{Q_0}) = \sup_{T_Q \in H} \{L^*(\phi, T_Q) - \gamma d(T_Q, T_{Q_0})\}$ *be a shift robust surrogate. For*
any query set distribution T_Q *and any* $\rho > 0$,

$$\sup_{T_Q : W_d(T_Q, T_{Q_0}) \leq \rho} \mathbb{E}_{T_Q \sim T_Q} [L^*(\phi, T_Q)] = \inf_{\gamma \geq 0} \{\gamma\rho + \mathbb{E}_{T_{Q_0}}[\psi_\gamma(\phi; T_{Q_0})]\}$$

and for any $\gamma \geq 0$,

$$\sup_{T_Q} \{\mathbb{E}_{T_Q}[L^*(\phi, T_Q)] - \gamma W_d(T_Q, T_{Q_0})\} = \mathbb{E}_{T_{Q_0}}[\psi_\gamma(\phi; T_{Q_0})]$$

Using Theorem 2, we arrive at the following surrogate meta-objective:

$$\min_{\theta \in \Theta} \{\mathbb{E}_{T_{Q_0}}[\psi_\gamma(\phi; T_{Q_0})]\} \tag{3}$$

Thus, meta-optimizing the robust surrogate involves maximizing the loss L^* on adversarial query projections T_Q while simultaneously restraining T_Q to a ρ distance from T_{Q_0}. We now show the existence of the adversarial projection for an original query set using the results from [5,29].

Theorem 3. *Let* $L^* : \Theta \times H \to R$ *be* $\mathcal{L}$-*Lipshitz smooth and* $d(., T_{Q_0})$ *be a* μ-*strongly convex for each* $T_{Q_0} \in H$. *If* $\gamma > \dfrac{\mathcal{L}}{\mu}$ *then there exists a unique* $\hat{T}_Q$ *satisfying*

$$\hat{T}_Q = \arg \operatorname{Sup}_{T_Q \in H} \{L^*(\phi, T_Q) - \gamma d(T_Q, T_{Q_0})\} \tag{4}$$

and

$$\nabla_\theta \psi_\gamma(\phi, T_{Q_0}) = \nabla_\theta L^*(\theta; \hat{T}_Q) \tag{5}$$

Proof. Deferred to the supplementary material.

Remark 1. $L^*(\phi, T_Q) - \gamma d(T_Q, T_{Q_0})$ is a $\gamma - \mathcal{L}/\mu$ strongly concave function for $\gamma \geq \mathcal{L}/\mu$ and so $L^*(\phi, T_Q) - \gamma d(T_Q, T_{Q_0})$ admits one and only one unique maximizer $\hat{T}_Q$ ($\mu = 1$ for Euclidean distance).

Estimation of AQP. To find the adversarial query projection, we approximate Eq. 4 by employing gradient ascent with early stopping on the query set. We consider a task $T_0 = \{T_{S_0} \cup T_{Q_0}\}$ and let $\{X, Y\}$ and $\{X^*, Y^*\}$ be the set of all instance-label pairs in T_{S_0} and T_{Q_0}, respectively. We propose algorithm 1 to induce SQS in the meta-train tasks. The original query instances X^* initialize the worst-case query augmentations X_w^*. We perform an iterative gradient ascent on X^* using L^*, resulting in an augmented query set X_w^*. This augmented query set X_w^* has distributional disparity with original support set X. Early stopping by Adv_iter and initializing X_w^* with X^* regularizes $(-\gamma d(T_Q, T_{Q_0}))$ and ensures X_w^* does not deviate extensively from X^*. The algorithm returns a virtual task with original support X and projected query X_w^*, which is used to update θ.

Algorithm 1: Adversarial Query Projection $AQP(T_{S_0}, T_{Q_0})$

Input: Task Support and Query Sets - $(T_{S_0} = \{X, Y\}, T_{Q_0} = \{X^*, Y^*\})$,
model parameters ϕ
$X_w^* \leftarrow X^*$
for $i = 0$ *to Adv_iter* **do**
$\quad | \quad X_w^* \longleftarrow X_w^* + \eta \nabla_{X_w^*} L_i^*(\phi, X_w^*)$
end
$T_Q = \{X_w^*, Y^*\}$
return $(T_{S_0} \cup T_Q)$

---

4   Experiments and Results

We design experiments to investigate the challenging nature of our proposed
SQS+ benchmark and empirically validate the efficacy of the proposed AQP
over the state-of-the-art approach [4] to address SQS in inductive settings. We
consider Cifar 100, miniImagenet, tieredImagenet, FEMNIST, and their state-
of-the-art SQS variants for evaluation. We also demonstrate the AQP's efficiency
on our proposed datasets (introduced in Sect. 4.1 and elaborated in the supplementary material). We used Conv4 models [4] for Cifar 100, FEMNIST and their
variants, and ResNet-18 [16] for miniImagenet, tieredImagenet, and their extensions. We use 32×32 images for Cifar 100, 28×28 for FEMNIST, and 84×84 for
miniImagenet and tieredImagenet. We next present the implementation details,
followed by the contributions to FewShiftBed and empirical investigations.

4.1   Implementation Details

Following [4], we fix the meta-learning rate as 0.001 for all approaches (Ind_OT,
AQP), models (ProtoNet, MatchingNet), and datasets (Cifar 100, miniImagenet,
tieredImagenet, FEMNIST, and their variants) and learn the models for 60,000
episodes. We perform a grid search using Ray over 35 configurations for 12000
episodes to find the optimal hyper-parameters. The search space is shared for
all approaches, datasets, and models. The hyper-parameters of regularization in
Ind_OT and AQP's adversarial learning rate are sampled from log uniform distribution in the ranges [15, 50] and [0.001, 1.0], respectively. Further, the number of
iterations required to project data in AQP and Ind_OT iterations are randomly
sampled from ranges [2,9] and [500, 1500] (increments of 100), respectively. However, for Ind_OT, we obtained better results on default hyper-parameters than
tuned ones on miniImagenet and its SQS variants. So we fixed its parameters
as mentioned in [4] for all the cases. We report the hyper-parameters (learning
rate (η) and number of iterations (Adv_iter)) for AQP in Table 1.

Table 1. Hyperparameter details of AQP for different datasets and approaches.

Dataset	ProtoNet						MatchingNet					
	No SQS		SQS		SQS+		No SQS		SQS		SQS+	
	η	Adv_iter	η	Adv_iter	η	Adv_iter	η	Adv_iter	η	Adv_iter	η	Adv_iter
Cifar 100	22.0	4	31.0	3	22.0	4	22.0	4	31.0	3	32.0	2
miniImagenet	22.0	4	31.0	3	22.0	4	22.0	4	41.0	8	24.5	5
tieredImagenet	22.0	4	17.0	4	22.0	4	22.0	4	41.0	9	22.0	4
FEMNIST	22.0	4	16.5	2	24.0	2	22.0	4	25.8	5	30.0	8

4.2   Contributions to FewShiftBed

We make significant contributions to the FewShiftBed [4]. Firstly, we have created challenging datasets wherein SQS is present only at meta-test time (SQS+).
The SQS+ versions of Cifar 100, miniImagenet, and tieredImagenet datasets are
constructed from their SQS counterparts [4] by removing perturbations from
the meta-train datasets. Similarly, the SQS+ variant of FEMNIST also follows
its SQS counterpart, but the meta-train set contains alpha-numerals from users
randomly. We add these SQS+ versions of benchmark datasets to the testbed.
The perturbations applied to the tasks are entirely modular, i.e., a task may have
augmentation in support, query, both, or none. More details about the datasets
are available in the supplementary material. Secondly, we integrate our theoretically grounded inductive solution, Adversarial Query Projections (AQP), into
the testbed. The AQP implementation is standalone and can be integrated with
any episodic training regimen. We have successfully integrated AQP with ML
approaches like Prototypical and Matching networks [24,27]. Thirdly, we have
also added a hyperparameter optimization module that uses RAY [18] for tuning parameters. We believe these additions improve the usability and coverage
of FewShiftBed to study SQS. The modified FewShiftBed, which includes the
proposed solution, datasets, and experimental setup, is publicly available.[1]

4.3   Evaluation of SQS+

We first validate that SQS+ is more challenging than the SQS problem [4]. We
train Prototypical and Matching networks on Cifar 100, miniImagenet, tiered-
Imagenet, and FEMNIST on all three settings - No SQS, SQS, and SQS+. We
report the results in Table 2 and observe that for all the datasets, models trained
with both the approaches (Prototypical and Matching network) perform best
in the No SQS setting, followed by SQS and SQS+. In the classical few-shot
setting, meta-train and meta-test phases share the domain, due to which the
meta-knowledge is easily transferable across the phases. However, in SQS, each
task's support and query set represent different domains, but share a latent
structure, during the meta-train and meta-test phases. In SQS versions of Cifar
100, miniImagenet, and tieredImagenet, both meta-train and meta-test SQS are
characterized by different types of data perturbations. However, in FEMNIST's

---

[1] https://github.com/Few-Shot-SQS/adversarial-query-projection.

Table 2. Comparison of ML methods with their Ind_OT and AQP counterparts across Cifar 100, miniImagenet, tieredImagenet, FEMNIST datasets, and their SQS and SQS+ variants. The results are obtained on 5-way tasks with 5 support and 8 query instances per class except for FEMNIST and its variants, which contains only one support and one query instance per class. The ± represents the 95% confidence intervals over 2000 tasks. AQP outperforms classic, and Ind_OT-based ML approaches approximately on all datasets.

Method	Test Accuracy					
	No SQS	SQS	SQS+	No SQS	SQS	SQS+
	Cifar 100			miniImagenet		
ProtoNeT	48.07 ± 0.44	43.15 ± 0.48	40.59 ± 0.69	64.56 ± 0.42	41.68 ± 0.76	35.17 ± 0.78
Ind_OT+ ProtoNeT	48.62 ± 0.44	43.62 ± 0.49	41.74 ± 0.65	63.74 ± 0.42	39.84 ± 0.78	34.75 ± 0.80
AQP+ ProtoNeT	**48.70 ± 0.42**	**45.09 ± 0.46**	**45.06 ± 0.46**	**66.81 ± 0.42**	**42.65 ± 0.57**	**40.61 ±0.60**
MatchingNet	46.03 ± 0.42	39.89 ± 0.44	36.63 ± 0.45	59.68 ± 0.43	39.66± 0.54	35.40 ±0.52
Ind_OT+ MatchingNet	45.77 ± 0.42	40.82 ± 0.45	37.13 ± 0.47	59.64 ± 0.44	38.25± 0.54	33.22± 0.50
AQP+ MatchingNet	**46.53 ± 0.43**	**42.40 ± 0.46**	**41.26 ± 0.46**	**62.29 ± 0.42**	**42.32 ± 0.52**	**37.90 ± 0.53**
	tieredImagenet			FEMNIST		
ProtoNeT	**71.04 ± 0.45**	41.59 ± 0.57	38.57 ± 0.65	93.09 ± 0.51	84.36 ± 0.74	82.67 ± 0.77
Ind_OT+ ProtoNeT	69.56 ± 0.46	40.08 ± 0.56	35.81 ± 0.58	91.66 ± 0.55	79.64 ± 0.80	76.37 ± 0.84
AQP+ ProtoNeT	69.62 ± 0.45	**45.34 ± 0.60**	**40.94 ± 0.66**	**94.61 ± 0.45**	**85.92 ± 0.69**	**84.42 ± 0.74**
MatchingNet	67.85 ± 0.46	43.30 ± 0.56	37.57 ± 0.57	93.69 ± 0.49	85.88 ± 0.69	83.48 ± 0.74
Ind_OT+ MatchingNet	67.79 ± 0.46	44.27 ± 0.56	39.24 ± 0.59	**93.76 ± 0.48**	84.08 ± 0.71	83.09 ± 0.74
AQP+ MatchingNet	**68.40 ± 0.45**	**45.26 ± 0.56**	**39.39 ± 0.58**	93.69 +- 0.49	**87.24 ± 0.67**	**84.98 ± 0.72**

SQS variant, meta-train and meta-test SQS is induced due to different writers. A meta-model trained in this setup becomes partially resilient to the related but disjoint SQS during meta-testing. A common SQS structure across meta-train and meta-test sets may not exist. Thus, SQS+ datasets are more challenging, which is empirically validated by the baseline approach's poor performance.

4.4   Evaluation of AQP

We compare the efficiency of the proposed AQP and optimal transport (OT) based state-of-the-art solution for handling vanilla SQS and SQS+ on the benchmark datasets. A strong baseline for SQS+ is the inductive version of OT (Ind_OT), where we employ OT only in the meta-train phase to generate projected support sets using support and query instances of a task. We evaluate ProtoNet and Matching Networks versions of Ind_OT and AQP. Table 2 presents the results for this evaluation. We observe that the models learned on projected support data obtained by Ind_OT are less robust to both SQS and SQS+ than the models learned on AQP for all approaches and datasets. Hence, AQP is

better at addressing SQS+ (and SQS), when meta-test unlabeled query instances
are unavailable.

To inspect whether the proposed AQP negatively impacts the models' generalization in the absence of meta-test SQS, we evaluate the ML approaches
and their Ind_OT and AQP counterparts on classic datasets containing no support query shifts (No SQS). We observe from Table 2 that AQP does not lead
to degradation in the performance in the absence of SQS, instead improves the
generalizability of the model even when SQS is absent. We note that Ind_OT
sometimes deteriorates the model's performance when SQS is missing. AQP outperforms both classic methods and their Ind_OT versions in almost all cases.

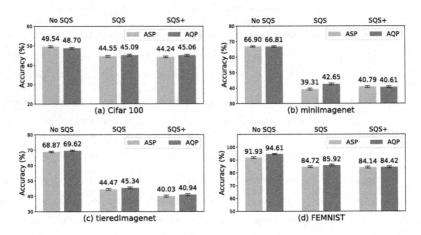

Fig. 2. Impact of adversarially projecting support and query data in a task on the
model's performance across No SQS and SQS and SQS+ variants of Cifar 100, mini-
Imagenet, tieredImagenet, and FEMNIST datasets.

Following [4], we used a Conv4 backbone for Cifar 100, FEMNIST and their
transformations, and a ResNet-18 [16] backbone for miniImagenet, tieredImagenet, and their variants. Thus, Table 2 not only shows the robustness of a
model trained via AQP on different SQ shifts but also its thoroughness across
architectures. We randomly projecting 25% of the tasks with AQP to reduce
the computational cost. Extending this idea to Ind_OT, resulted in a significant decline in the performance. We thus maintain the standard-setting [4] for
Ind_OT, wherein support sets of all the tasks are projected.

4.5   Ablations

We perform ablations to investigate the sensitivity of the proposed approach
to task characteristics (varying number of support and query shots) and design
choices (support vs query projections).

Ablation on Projections. We study the impact of adversarially perturbing support *vs* query set in a task and evaluate the model's (ProtoNet) performance across all settings and datasets. From Fig. 2 we observe perturbing query sets is empirically more meritorious in 9 out of 12 settings. We measure the model's generalizability from in-distribution support to out-of distribution (OOD) query set in a task by perturbing a query set. The magnitude of loss and hence gradients on the OOD query set is high, resulting in more meaningful meta-updates. As performance on the query set directly impacts the meta-update, the model's invariance to SQS is directly reflected in the meta-update. On the other hand, projecting support sets creates potent prototypes (robust adaptation) as adversarial projections distort the images. However, the meta-update may not be impactful due to the model's good performance on clean query images.

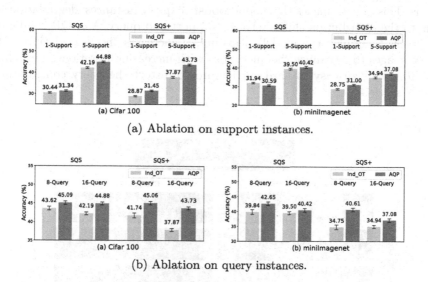

(a) Ablation on support instances.

(b) Ablation on query instances.

Fig. 3. Ablation on the number of support and query instances per class on SQS and SQS+ variants of Cifar 100 and miniImagenet datasets. In (a), we consider 5-way tasks with 1 and 5 support instances with 16 query instances. In (b), we vary query instances between 8 and 16 with 5 support instances per class.

Ablation on Support and Query Shots. We ablate the number of shots per class in the support and query sets, limited to Cifar 100 and miniImagenet datasets, to inspect the efficacy of our proposed AQP employing a ProtoNet. AQP outperforms Ind_OT when the number of query instances are fixed to 16 per class, and support shots per class vary from 1 to 5 (Fig. 3a). We also vary the number of query instances per class from 8 to 16 and observe that AQP surpasses Ind_OT with varying query instances (Fig. 3b).

4.6   Visual Analysis of AQP

We visualize the impact of AQP on the query instances across meta-training iterations. We train a Prototypical network in a 5-way 5-shot setting on the SQS+ version of miniImagenet for 150 epochs. Extended results on No SQS and SQS versions of miniImagenet are presented in the supplementary material. For better illustration, we fix one task and one instance per class and show the transformation in the query images over meta-train iterations (Fig. 4). The images in the top row are the original query set, the left column are the query images impacted by AQP with increasing iterations, and the right column represents the change mask (in the increasing order of iterations), which is the difference between the pixel intensities of the original image and its adversarially perturbed counterpart. We observe gradual increase in the distortions with increasing iterations. This in turn makes the model robust to query instances' degradation and thus to the distribution shifts between support and query. As AQP is adaptive and seeks to inhibit the model's learning, it increases the degradation in the query images to maximize the query loss with increasing iterations. This shows that following an easy to hard curriculum to distort the query contributes to AQP's success.

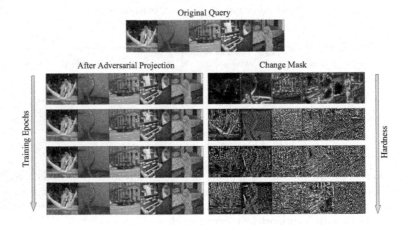

Fig. 4. Evolution of Adversarial Query Projections across training epochs for SQS+ version of miniImagenet.

However, this experiment also reflects the potential limitations of the proposed AQP. We evaluated AQP in the cases where SQS is characterized by the perturbations in data (SQS variants of Cifar 100, miniImagenet, and tieredImagenet), and for a small-realistic dataset (FEMNIST and its variants) where different writers characterize SQS. The masks (Fig. 4) reflect that AQP adds varying noise to distort the images, which may not resemble complex SQ shifts. Investigating AQP in more complex SQ shifts, e.g., real to sketch or caricature pictures, is part of our future work.

5   Conclusion and Future Directions

This paper proposes SQS+ - a more challenging distribution shift between the support and query sets of a task in a few-shot meta-learning setup. SQS+ includes an unknown SQ shift in the meta-test tasks, and empirical evidence suggests SQS+ is a complex problem than the prevalent SQS notion. We propose a theoretically grounded solution - Adversarial Query Projection (AQP) to address SQS+ without leveraging unlabelled meta-test query instances. Exhaustive experiments involving AQP on multiple benchmark datasets (Cifar 100, miniImagenet, tieredImagenet, and FEMNIST - their SQS and proposed SQS+ variants), different architectures, and ML approaches demonstrate its effectiveness. The future work lies in verifying the effectiveness of AQP in complex SQ shifts, e.g., shift from real to sketch images and creating datasets corresponding to these difficult SQ shifts, and integrating AQP with gradient and transductive ML approaches. We incorporate proposed AQP and SQS+ versions of Cifar 100, miniImagenet, tieredImagenet, and FEMNIST to FewShiftBed and make it publicly available to encourage research in this direction.

Acknowledgements. The resources provided by 'PARAM Shivay Facility' under the National Supercomputing Mission, Government of India at the Indian Institute of Technology, Varanasi are gratefully acknowledged.

References

1. Aimen, A., Sidheekh, S., Ladrecha, B., Krishnan, N.C.: Task attended meta-learning for few-shot learning. In: Fifth Workshop on Meta-Learning at the Conference on Neural Information Processing Systems (2021)
2. Antoniou, A., Storkey, A.J.: Learning to learn by self-critique. Adv. Neural Inf. Process. Syst. **32**, 1–11 (2019)
3. Ben-David, S., Blitzer, J., Crammer, K., Pereira, F.: Analysis of representations for domain adaptation. Adv. Neural Inf. Process. Syst. **19**, 1–8 (2006)
4. Bennequin, E., Bouvier, V., Tami, M., Toubhans, A., Hudelot, C.: Bridging few-shot learning and adaptation: new challenges of support-query shift. In: Joint European Conference on Machine Learning and Knowledge Discovery in Databases, pp. 554–569 (2021)
5. Bonnans, J.F., Shapiro, A.: Perturbation Analysis of Optimization Problems. Springer, Heidelberg (2013). https://doi.org/10.1007/978-1-4612-1394-9
6. Boudiaf, M., Ziko, I., Rony, J., Dolz, J., Piantanida, P., Ben Ayed, I.: Information maximization for few-shot learning. Adv. Neural Inf. Process. Syst. **33**, 2445–2457 (2020)
7. Chen, W., Liu, Y., Kira, Z., Wang, Y.F., Huang, J.: A closer look at few-shot classification. In: International Conference on Learning Representations (2019)
8. Dhillon, G.S., Chaudhari, P., Ravichandran, A., Soatto, S.: A baseline for few-shot image classification. In: International Conference on Learning Representations (2020)
9. Du, Y., Zhen, X., Shao, L., Snoek, C.G.: Metanorm: learning to normalize few-shot batches across domains. In: International Conference on Learning Representations (2020)

10. Finn, C., Xu, K., Levine, S.: Probabilistic model-agnostic meta-learning. Adv. Neural Inf. Process. Syst. **31**, 1–12 (2018)
11. Flamary, R., Courty, N., Tuia, D., Rakotomamonjy, A.: Optimal transport for domain adaptation. IEEE Trans. Pattern Anal. Mach. Intell. **1**, 1–40 (2016)
12. Ganin, Y., Lempitsky, V.: Unsupervised domain adaptation by backpropagation. In: International Conference on Machine Learning, pp. 1180–1189 (2015)
13. Goldblum, M., Fowl, L., Goldstein, T.: Adversarially robust few-shot learning: a meta-learning approach. Adv. Neural Inf. Process. Syst. **33**, 17886–17895 (2020)
14. Goldblum, M., Reich, S., Fowl, L., Ni, R., Cherepanova, V., Goldstein, T.: Unraveling meta-learning: understanding feature representations for few-shot tasks. In: International Conference on Machine Learning, pp. 3607–3616 (2020)
15. Guo, Y., et al.: A broader study of cross-domain few-shot learning. In: European Conference on Computer Vision, pp. 124–141 (2020)
16. Laenen, S., Bertinetto, L.: On episodes, prototypical networks, and few-shot learning. Adv. Neural Inf. Process. Syst. **34**, 24581–24592 (2021)
17. Li, Y., Yang, Y., Zhou, W., Hospedales, T.M.: Feature-critic networks for heterogeneous domain generalization. In: International Conference on Machine Learning, pp. 3915–3924 (2019)
18. Liaw, R., Liang, E., Nishihara, R., Moritz, P., Gonzalez, J.E., Stoica, I.: Tune: a research platform for distributed model selection and training. arXiv preprint arXiv:1807.05118 (2018)
19. Liu, B., Zhao, Z., Li, Z., Jiang, J., Guo, Y., Ye, J.: Feature transformation ensemble model with batch spectral regularization for cross-domain few-shot classification. arXiv preprint arXiv:2005.08463 (2020)
20. Liu, Y., et al.: Learning to propagate labels: transductive propagation network for few-shot learning. In: International Conference on Learning Representations (2019)
21. Peyré, G., Cuturi, M., et al.: Computational optimal transport: with applications to data science. Found. Trends® Mach. Learn. **11**, 355–607 (2019)
22. Ren, M., et al.: Meta-learning for semi-supervised few-shot classification. In: International Conference on Learning Representations (2018)
23. Sinha, A., Namkoong, H., Duchi, J.C.: Certifying some distributional robustness with principled adversarial training. In: International Conference on Learning Representations (2018)
24. Snell, J., Swersky, K., Zemel, R.S.: Prototypical networks for few-shot learning. Adv. Neural Inf. Process. Syst. **30**, 1–11 (2017)
25. Sun, J., Lapuschkin, S., Samek, W., Zhao, Y., Cheung, N., Binder, A.: Explanation-guided training for cross-domain few-shot classification. In: International Conference on Pattern Recognition, pp. 7609–7616 (2020)
26. Tseng, H., Lee, H., Huang, J., Yang, M.: Cross-domain few-shot classification via learned feature-wise transformation. In: International Conference on Learning Representations (2020)
27. Vinyals, O., Blundell, C., Lillicrap, T., Wierstra, D., et al.: Matching networks for one shot learning. Adv. Neural Inf. Process. Syst. **29**, 1–9 (2016)
28. Volpi, R., Namkoong, H., Sener, O., Duchi, J.C., Murino, V., Savarese, S.: Generalizing to unseen domains via adversarial data augmentation. Adv. Neural Inf. Process. Syst. **31**, 5339–5349 (2018)
29. Wang, H., Deng, Z.: Cross-domain few-shot classification via adversarial task augmentation. In: International Joint Conference on Artificial Intelligence, pp. 1075–1081 (2021)
30. Yeh, J.F., et al.: Large margin mechanism and pseudo query set on cross-domain few-shot learning. arXiv preprint arXiv:2005.09218 (2020)

Discovering Wiring Patterns Influencing Neural Network Performance

Aleksandra I. Nowak[1]([✉]) [iD] and Romuald A. Janik[2] [iD]

[1] Faculty of Mathematics and Computer Science, Jagiellonian University,
Łojasiewicza 6, 30-348 Kraków, Poland
nowak.aleksandrairena@gmail.com
[2] Institute of Theoretical Physics, Jagiellonian University, Łojasiewicza 11,
30-348 Kraków, Poland

Abstract. The search for optimal neural network architecture is a well-known problem in deep learning. However, as many algorithms have been proposed in this domain, little attention is given to the analysis of wiring properties that are beneficial or detrimental to the network performance. We take a step at addressing this issue by performing a massive evaluation of artificial neural networks with various computational architectures, where the diversity of the studied constructions is obtained by basing the wiring topology of the networks on different types of random graphs. Our goal is to investigate the structural and numerical properties of the graphs and assess their relation to the test accuracy of the corresponding neural networks. We find that none of the classical numerical graph invariants by itself allows to single out the best networks. Consequently, we introduce a new numerical graph characteristic, called *quasi-1-dimensionality*, which is able to identify the majority of the best-performing graphs.

Keywords: Deep learning · Artificial neural networks · Neural architectures · Network analysis · Image classification

1 Introduction

Over the recent years many different neural architectures have been proposed, varying from hand-engineered solutions [11,13,23] to very complicated, automatically generated patterns produced by Neural Architecture Search (NAS) algorithms [6,15,19,31]. However, in this vast panorama of searching methods and benchmarking data, little focus is placed upon analyzing what specific structural properties of the architectures are related to the performance of the network. Apart from studies revolving around residual connections [25] and the impact of width or depth of the network [23,29], we still lack an understanding of why certain wiring topologies work better than others. We believe that addressing this issue would not only increase our knowledge about deep learning systems but also provide guidelines and principles for constructing new, better neural network architectures. Moreover, gathering empirical data linking the

---

Supplementary Information The online version contains supplementary material available at https://doi.org/10.1007/978-3-031-26409-2_38.

graph structure of the information flow with the performance could contribute nontrivial benchmark data for the, yet to be developed, theory.

The aim of this paper is to perform a wide-ranging study of neural network architectures for which the wiring pattern between the blocks of operations is based on a variety of random graphs. We focus on analyzing the interrelation of the structure of the graph with the performance of the corresponding neural network in an image recognition task. This allows us to address some fundamental questions for deep neural networks such as to what extent does the performance of the network depends on the pattern of information flow encoded in its global architecture. Is the performance basically independent of the structure or can we identify *quantitatively* structural patterns which typically yield enhanced performance? The goal of this work is to identify and analyze the discriminative features of neural network architectures. We do not aim at constructing, nor searching for, an optimal architecture.

Another motivation is the observation that artificial neural networks typically have a quite rigid connectivity structure, yet in recent years significant advances in performance have been made through novel global architectural changes like ResNets, [11] or DenseNets [12]. This has been further systematically exploited in the field of Neural Architecture Search (see [6] for a review). Hence there is a definite interest in exploring a wide variety of possible global network structures. On the other hand, biological neural networks in the brain do not have rigid structures and some randomness is an inherent feature of networks that evolved ontogenetically [4]. Contrarily, we also do not expect these networks to be uniformly random [17]. Therefore, it is very interesting to investigate the interrelations of structural randomness and global architectural properties with the network's performance.

To this end, we explore a wide variety of neural network architectures for an image recognition task, constructed accordingly to wiring topologies defined by random graphs. This approach can efficiently produce many qualitatively different connectivity patterns by alternating only the random graph generators [28]. The nodes in the graph correspond to a simple convolutional computational unit, whose internal structure is kept fixed. Apart from that, we do not impose any restrictions on the overall structure of the neural network. In particular, the employed constructions allow for modeling arbitrary global (as well as local) connectivity. We investigate a very diversified set of graph architectures, which range from the quintessential random, scale-free, and small-world families, through some edge-direction sensitive constructions, to graphs based on fMRI data. Altogether we conduct an analysis of more than 1000 neural networks, each corresponding to a different directed acyclic graph[1]. Such a wide variety of graphs is crucial for our goal of analyzing the properties of the network architecture by studying various characteristics of the corresponding graph and examining their impact on the performance of the model.

We find that among more than 50 graph-theoretical properties tested by us in this study, none is able to distinguish, by itself, the best performing graphs. We

---

[1] The code is available at https://github.com/rmldj/random-graph-nn-paper.

are able to identify one group of the top graphs by introducing a new numerical graph criterion, which we refer to as *quasi-1-dimensionality*. This criterion captures graphs characterized by mostly local connections with a global elongated structure, providing guidelines for beneficial biases in the architectural design of neural networks.

2   Related Work

Neural Architecture Search. Studies undertaken over the recent years indicate a strong connection between the wiring of network layers and its generalization performance. For instance, ResNet introduced by [11], or DenseNet proposed in [12], enabled successful training of very large multi-layer networks, only by adding new connections between regular blocks of convolutional operations. The possible performance enhancement that can be gained by the change of network architecture has posed the question, whether the process of discovering the optimal neural network topology can be automatized. In consequence, many approaches to this Neural Architecture Search (NAS) problem were introduced over the recent years [6]. Among others, algorithms based on reinforcement learning [2,31], evolutionary techniques [19] or differentiable methods [15]. Large benchmarking datasets of the cell-operation blocks produced in NAS have been also proposed by [29] and extended by [5].

The key difference between NAS approaches an the present work is that we are not concentrating on directly optimizing the architecture of a neural network for performance, but rather on exploring a wide variety of random graph architectures in order to identify what features of a graph are related to good or bad performance of the associated neural network. Thus we need to study both strong and weak architectures in order to ascertain whether a given feature is, or is not *predictive* of good performance. We hope that our findings will help to develop new NAS search spaces.

Random Network Connectivity. There were already some prior approaches which focused on introducing randomness or irregularity into the network connectivity pattern. The work of [21] proposed stochastic connections between consecutive feed-forward layers, while in [13] entire blocks of layers were randomly dropped during training. However, the first paper which, to our knowledge, really investigated neural networks on random geometries was the pioneering work of [28]. This paper proposed a concrete construction of a neural network based on a set of underlying graphs (one for each resolution stage of the network). Several models based on classical random graph generators were evaluated on the ImageNet dataset, achieving competitive results to the models obtained by NAS or hand-engineered approaches. Using the same mapping, [20] investigated neural networks based on the connectomics of the mouse visual cortex and the biological neural network of *C.Elegans*, obtaining high accuracies on the MNIST and FashionMNIST datasets.

Although the works discussed above showed that deep learning models based on random or biologically inspired architectures can indeed be successfully trained without a loss in the predictive performance, they did not investigate what kind of graph properties characterize the best (and worst) performing topologies. The idea of analyzing the architecture of the network by investigating its graph structure has been raised in [30]. However, this work focused on exploring the properties of the introduced relational graph, which defined the communication pattern of a network layer. Such a pattern was then repeated sequentially to form a deep model. In addition, [16] have also analyzed machine learning models with the tools of network science, but their research was devoted to Restricted Boltzmann Machines.

The main goal of our work is to perform a detailed study of numerical graph characteristics in relation to the associated neural network performance. Contrary to [30] we are not concentrating on exploring the fine-grained architecture of a layer in a sequential network. Instead, we keep the low-level operation pattern fixed and encapsulated in the elementary computational node. We focus on the high-level connectivity of the network, by analyzing the graph characteristics of neural network architectures based on arbitrary directed acyclic graphs.

3   From a Graph to a Neural Network

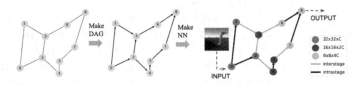

Fig. 1. The graph to neural network mapping. First, a graph is sampled from a predefined set of random graph generators. Next, the graph is transformed to a DAG by selecting a node ordering and enforcing the connections to be oriented accordingly to that ordering. Such DAG is treated as a blueprint for a neural network architecture. Nodes with different colors work on different resolutions of the feature maps. The beige (*interstage*) edges indicate the connections on which a resolution reduction is performed. The black edges (*intrastage*) link nodes that work within the same resolution. Best viewed in color.

In order to transform a graph into a neural network, we adopt the approach presented in [28]. In that paper, a graph is sampled from a predefined list of generators and transformed into a directed acyclic graph (DAG). Next, the DAG is mapped to a neural network architecture as follows:

The edges of the graph represent the flow of the information in the network and the nodes correspond to the operations performed on the data. The computation is performed accordingly to the topological order. In each node, the input from the ingoing edges is firstly aggregated using a weighted sum. Next, a ReLU - Conv2d - Batch-Norm block is applied. The result of this procedure is then propagated independently by each outgoing edge. When the computations

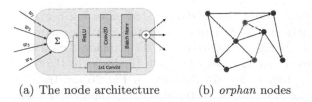

(a) The node architecture   (b) *orphan* nodes

Fig. 2. (a): The node is represented by the green-shaded area. The black arrows illustrate the graph edges labeled with the associated weights. The gray arrows indicate the ordering of the operations performed in the node as well as the residual connection. **(b)**: The gray nodes (orphan nodes) in the DAG either do not have an input from previous stages of processing or do not have an output. Hence we add the red edges from the immediately preceding node or to the immediately succeeding node.

leaves the last node, a global average pooling is performed, followed by a dense layer with the number of output neurons equal to the target dimension.

The network nodes are divided into three sets of equal size, referred to as *stages* (denoted by different colors in the figures). The first stage operates on the original input resolution, with the number of channels C being set in the first (input) node of the graph. The subsequent stages operate on a decreased input resolution and increased number of output channels by a factor of 2, with respect to the previous stage. In order to perform the downsampling, on every edge that crosses two stages the same block of operations as in a standard node is executed, but with the use of convolutions with stride 2 (when crossing subsequent stages) or stride 4 (when crossing from the first stage to the last). In the figures in the present paper, we represent such resolution-changing edges with beige color, and refer to them as *interstage*. See Fig. 1 for a visualization of the above described mapping. We introduce three modifications to this procedure:

Firstly, in [28] there were separate random graphs for each of the three stages of the neural network. This means that subsequent stages were connected only by one edge. In our case, we have a random graph for the *whole* network. Dimensionality reduction is performed on a graph edge when necessary, by a node with stride 2 or 4 convolution, as described above. In consequence, we do not bias the model to have a single bottleneck connection between the computations performed on different spatial resolutions. Moreover, we observe that the introduction of such a bottleneck generally deteriorates the network performance (we discuss this issue in Sect. 6.2).

Secondly, we introduce an additional residual connection from the aggregated signal to the output of the triplet block in the node. The residual connection always performs a projection (implemented by a 1×1-convolution, similar to ResNet C-type connections of [11] — see Fig. 2a). The residual skip connection shifts the responsibility of taking care of the vanishing gradient problem from edges to the nodes, allowing the global connectivity structure to focus on the information flow, with the low-level benefits of the residual structure already built-in.

Thirdly, we improve the method of transforming a graph into a DAG so that it automatically takes into account the graph structure. This is achieved by ordering the nodes accordingly to a 2D Kamada-Kawai embedding [14] and setting the directionality of an edge from the lower to the higher node number. Any arising orphan nodes like the ones in Fig. 2b are then fixed by adding a connection from the node with the preceding number or adding a connection to the node with the succeeding number. We observe that this approach leads to approximately 2x fewer orphan nodes than the random ordering, and circa 1.5x less than the original ordering returned by the generator, which was used in [28]. A detailed description of the DAG transformation process together with a comparison of various node orderings can be found in Appendix B and C.

4   The Space of Random Graphs and DAGs

We performed a massive empirical study of over 1000 neural network architectures based on 5 graph families and 2 auxiliary constructions. We summarize below their main characteristics.

- *Erdős-Rényi* (er) – In this model, given a parameter $p \in [0,1]$, each possible (undirected) edge arises independently of all the other edges with probability p [7]. Small p usually results in sparse graphs, while increasing p increases also the chance of obtaining a graph with dense connections.
- *Barabási-Albert* (ba) – Given a set of m initially connected nodes, new nodes are added to the graph iteratively. In each step, a new node is connected with at most m other nodes with probability proportional to the nodes' degrees. The Barabási-Albert model favors the formation of hubs, as the few nodes with a high degree are more likely to get even more connections in each iteration. Therefore graphs produced by this model are associated with scale-free networks [3].
- *Watts-Strogatz* (ws) – The Watts-Strogatz model starts with a regular ring of nodes, where each of the nodes is connected to k of its nearest neighbors. Then, iteratively, every edge (u, v) which was initially present in the graph is replaced with probability p by an edge (u, w), where the node w is sampled uniformly at random from all the other nodes. The graphs obtained by this method tend to have the small-world property [26].
- *Random-DAG* (rdag) – The models mentioned so far produce undirected graphs, which need to be later transformed to DAGs. We choose to also study models produced by an algorithm that directly constructs a random DAG. An advantage of this algorithm over other existing DAG-generating methods is that it allows to easily model neural networks with mostly short-range or mostly long-range connections, which was the main reason for implementing this construction. This procedure and its parameters are thoroughly explained in Sect. 4.1.
- *fMRI based* (fmri) – In addition to the above algorithmic generators we also introduce a family of graphs that are based on resting-state functional MRI

data. We use the network connectomes provided by the Human Connectome Project [24] obtained from the resting-state fMRI data of 1003 subjects [22]. As input for graph construction, we use the released (z-score transformed) partial correlation matrix for 50- and 100-component spatial group-ICA parcellation. We describe in detail the exact method of deriving DAGs from the fMRI partial correlation matrices in Appendix D. Apart from the number of nodes, this family has a single thresholding parameter.

Moreover, we considered two auxiliary types of graphs:

- *Bottleneck graphs* (`bottleneck`) – For some graphs from the above families, we introduced a bottleneck between the various resolution stages (see Sect. 6.2).
- *Composite graphs* (`composite`) – We obtained these graphs by maximizing in a Monte-Carlo simulation the expression $\left(\frac{log\_num\_paths}{num\_nodes} \right)^{\frac{1}{2}} - 2grc - avg\_clustering$ where grc is the global reaching centrality of the graph. This construction was motivated by a certain working hypothesis investigated at an early stage of this work which was later discontinued. Nevertheless, we kept the graphs for additional structural variety.

For each of the above families, we fix a set of representative parameters[2]. Then for every family-parameters pair, we sample 5 versions of the model by passing different random seeds to the generator. Using this procedure we create 475 networks with 30 nodes and 545 networks with 60 nodes. We train all networks for 100 epochs with the same settings on the CIFAR-10 dataset[3]. For each network, we set the number of initial channels C in order to obtain approximately the same number of parameters as in ResNet-56 (853k).

4.1 Direct Construction of Random DAGs

In order to study some specific questions, like the role of long-range versus short-range connectivity, we implement a procedure for directly constructing random DAGs that allows for more fine-grained control than the standard random graph generators and is flexible enough to generate various qualitatively different kinds of graph behaviors. As an additional benefit, we do not need to pass through the slightly artificial process of transforming an arbitrary undirected graph to DAG.

We present the method in Algorithm 1. We start with N nodes, with a prescribed ordering given by integers $0, \ldots, N-1$. For each node i, we fix the number of outgoing edges n_i^{out} (clearly $n_i^{out} < N-i$). Here we have various choices leading to qualitatively different graphs. For example, sampling n_i^{out} from a Gaussian and rounding to a positive integer (or setting n_i^{out} to a constant) would yield approximately homogeneous graphs. Taking a long-tailed distribution would yield some outgoing hubs. One could also select the large outgoing hubs by hand and place them in a background of constant and small n_i^{out}.

---

[2] Refer to Appendix I for a full list.
[3] We provide a full description of the training procedure in Appendix A.

Algorithm 1. Random DAG

1: **Input:** nodes $i = 0, \ldots, N - 1$,
     number of outgoing edges n_i^{out},
     size of a local neighbourhood B,
     real α, function $f(x)$
2: **for** $i = 0$ **to** $N - 2$ **do**
3:    **if** node $i + 1$ does not have an ingoing connection **then**
4:      make an edge $i \to i + 1$
5:    **end if**
6:    **while** not all n_i^{out} outgoing edges chosen **do**
7:      Make randomly the edge $i \to j$ with probability $p_j = \frac{w_{ij}}{\sum_{k>i} w_{ik}}$
8:      where the weight w_{ij} is given by
9:      $w_{ij} = (n_j^{out})^\alpha f\left(\lfloor \frac{j-i}{B} \rfloor\right)$
10:      provided $j > i$ and $i \to j$ does not exist so far
11:    **end while**
12: **end for**

For each node i we then randomly choose (with weight w_{ij} given in Algorithm 1) nodes $j > i$ to saturate the required n_i^{out} connections. The freedom in the choice of weight w_{ij} gives us the flexibility of preferential attachment through the parameter α, and the possibility of imposing local or semi-local structure through the choice of the weighting function $f\left(\lfloor \frac{j-i}{B} \rfloor\right)$.

Different choices of f lead to different connectivity structures of the DAG. An exponential $f(x) = \exp(-Cx)$ results in short-range connections and local connectivity. The power law scaling $f(x) = 1/x$ produces occasionally longer range connections, while $f(x) = 1$ does not imply any nontrivial spatial structure at all. In this work, we investigated all three of the above possibilities. Since we do not want the integer node labels i or j to be effectively a 1d coordinate, we define a local neighborhood size B so that differences of node labels of order B would not matter. This motivates the form of the argument of the weighting function $f(x) \equiv f\left(\lfloor \frac{j-i}{B} \rfloor\right)$, where $\lfloor a \rfloor$ denotes the floor of a. In the simulations we set $B = 5$ or $B = 10$.

Through the choice of the function $f(.)$, we can model graphs with varying proportions of short- to long-range connections with the parameter B defining the size of the local neighborhood. The choice of multiplicity distribution of n_i^{out} allows to model, within the same framework, a uniform graph, a graph with power law outgoing degree scaling, or a graph with a few hubs with very high multiplicity. Finally, the parameter α enables to control preferential attachment of the connections. Consequently, the algorithm allows to produce DAGs with diverse architectural characteristics well suited for neural network analysis.

Let us note that the presented procedure is somewhat similar to the *latent position random graph* model [1] with graph features $X_i = (i, n_i^{out})$ and kernel $\kappa(X_i, X_j) = (n_j^{out})^\alpha f\left(\lfloor \frac{j-i}{B} \rfloor\right)$. However, in such a case, the kernel would still need to be normalized by the weights of all nodes smaller than j, as in line 7 of

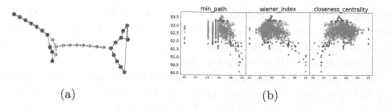

(a)   (b)

Fig. 3. (a): One of the worst networks with 30 nodes. The worst networks are typically characterized by sparse connections and long chains of operations. For more examples of the worst networks see Appendix J. **(b)**: The test accuracy versus selected network features. We indicate the best (equal or above 93.25%) models as red, the worst (below 92%) as blue, and the rest as gray. The first presented feature is the length of the shortest path between the input and the output node. The second one is the Wiener index [27], and the third is closeness centrality [8]. All features are rescaled using min-max scaling (For more details on data processing refer to Appendix F).

Algorithm 1. Moreover, such formulation could produce DAGs with disconnected components, whereas in our setting we ensure that every node $n > 0$ has an incoming connection (see lines 3–4).

5 Results

In this Section, we first exhibit the inadequacy of classical graph invariants to select the best performing networks and describe the generic features of the worst networks. Then we introduce a class of well-performing networks (which we call *quasi-1-dimensional* or Q1D) and provide their characterization in terms of a novel numerical graph invariant.

5.1 The Inadequacy of Classical Graph Characteristics

The key motivation for this work was to understand what features of the underlying graph are correlated with the test performance of the corresponding neural network. To this end, for the analysis, we use 54 graph features, mostly provided by the `networkx` library [10] as well as some simple natural ones, like the logarithm of the total number of paths between the input and output or the relative number of connections between stages with various resolutions. For a full list of the features see Appendix F.

It turns out that none of the classical features by itself is enough to isolate the best-performing networks. However, the worst networks form outliers for several of the tested graph properties and thus can be more or less identified (see Fig. 3b for a representative example and more plots in the Appendix H).

5.2 The Worst Networks

As mentioned before, several investigated network features seem to be able to discriminate the worst networks. In Fig. 3b features: the minimal length of a path

between the input and output node, the Wiener index [27], and the closeness
centrality [8] of the output node. The Wiener index is the sum of the lengths of
all-pair shortest paths and the closeness centrality of a node is the reciprocal of
the average shortest path distance to that node.

The above properties show that the worst networks are usually characterized
by long distances between any two nodes in the graph, resulting in long chains
of operations and sparse connections. An example of such a graph is presented
in Fig. 3a. In addition, we verified that purely sequential 1d chain graphs (node
i is connected only to node $i + 1$) gave indeed the worst performance.

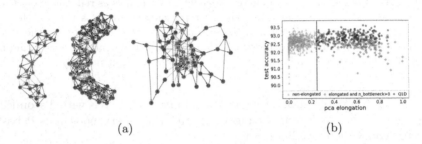

(a)   (b)

Fig. 4. (a): The best network with 30 nodes (left), with 60 nodes (center) and an
example of a highly ranked fMRI based network. For more examples of the best networks see Appendix J. **(b)**: The visualization of the Q1D criterion. The green triangles
indicate graphs without a global elongated structure and the gray diamonds are used
to represented the elongated graphs with bottlenecks. Networks with Q1D property are
drawn as red dots. The black vertical line illustrates the threshold $\tau = 0.25$. The Q1D
criterion for this threshold successfully selects the best networks from the elongated
group.

5.3   The Best Networks

Crucial to our results is the observation that the best networks belonged predominantly to the Random DAG category with short-range connections (i.e.
exponential $f(x)$). One generic visual feature of these graphs is that they have
a definite global ordering in the feed-forward processing sequence defining the
1d structure, yet locally there are lots of interconnections that most probably
implement rich expressiveness of intermediate feature representations (see the
first two graphs in Fig. 4a). We call such graphs *quasi-1-dimensional* (quasi-1d
for short). These models have a very large number of paths between the input and
the output. This is, however, not the feature responsible for good performance,
as maximally connected DAGs that have the maximal possible number of paths
do not fall into this category and give worse results (see Fig. 8 in Appendix). In
contrast, filament-like, almost sequential models such as some Watts-Strogatz
networks (recall Fig. 3a) have in fact significantly worse performance, so sequen-
tiality by itself also does not ensure good generalization.

We would like to formally characterize these graphs purely in terms of some
numerical graph features without recourse to their method of construction. This

is not *a priori* a trivial task. This is because one has to be sensitive to the globally elongated structure. However, the filament-like, almost sequential graphs are quite similar in this respect, yet they yield very bad performance. So numerical graph properties which are positively correlated with a stretched topology tend to have similar or even larger values for the very bad graphs. A condition that can eliminate the filament-like graphs is $n_{bottlenecks} = 0$, where a bottleneck edge is defined by the property that cutting that edge would split the graph into two separate components.

In order to numerically encode the elongated character of a network, we perform PCA on the set of node coordinates returned by the Kamada-Kawai embedding and require a sufficiently anisotropic explained variance ratio. Note that despite appearances this is a quite complex invariant of the original abstract graph, as the Kamada-Kawai embedding depends on the whole global adjacency structure through the spring energy minimization. Hence the nature of the embedding encodes nontrivial relevant information about the structure of the graph. We define then the elongation of the network as

$$pca\_elongation = 2 \cdot (variance\_ratio - 0.5), \tag{1}$$

where *variance_ratio* is the percentage of the variance explained by the component corresponding to the largest eigenvalue computed during the PCA decomposition. Networks with very large *pca_elongation* tend to have only one main computational path, while small *pca_elongation* is associated with graphs with many global inter-connections. For instance, almost purely sequential graphs have *pca_elongation* close to 1.0, while for fully connected DAGs this property is equal to 0. In order to use this continuous feature to define a discrete class of graphs with a visible hierarchical structure of the Kamada-Kawai embedding we need to specify a threshold τ and consider only graphs for which *pca_elongation* $> \tau$. Accordingly, we formally define the quasi-1d graphs (Q1D) as satisfying the condition:

$$pca\_elongation > \tau \quad \text{and} \quad n_{bottlenecks} = 0, \tag{2}$$

This condition is visualized in Fig. 4b. The first term of the Q1D definition accounts for networks that have a global one-dimensional order (like the two first networks in Fig. 4a). The second condition eliminates graphs containing bottlenecks which form the bulk of badly performing elongated graphs (denoted by gray dots in Fig. 4b). In our analysis, we set $\tau = 0.25$, which is a visual estimate motivated by Fig. 4b. This value is of course not set in stone and could just as well be a bit higher or lower. The rough choice of τ is also corroborated by the fact that the Q1D criterion for this threshold is strongly correlated with performance, as we discuss below.

We find that among the top 50 networks, 68% have the Q1D property. Moreover, out of the remaining 970 graphs, only 17% are Q1D. A breakdown of the top-50 and bottom-50 by specific graph families and the Q1D property is presented in Table 1. One may observe that Q1D successfully selects almost every of the best performing **rdags** and half of the **fmri** graphs (fourth column). Those

Table 1. For each graph family we report in percentage the number of all graphs having Q1D property, the number of graphs in top-50, the share of the given family in top-50 and the number of Q1D graphs within the ones present in top-50, followed by analogous statistics for the graphs in bottom-50. The Q1D criterion selects almost every best performing `rdag` and more than half `fmri` graphs (fourth column), which are the majority in top-50 (second column). None of the worst performing graph satisfies the Q1D criterion (last column).

MODEL	WITH Q1D PROPERTY	IN TOP-50	SHARE IN TOP-50	Q1D TOP-50	IN BOT-50	SHARE IN BOT-50	Q1D BOT-50
ba	0.00	4.00	4.00	0.00	0.00	0.00	0.00
bottleneck	0.00	0.67	2.00	0.00	6.67	20.00	0.00
composite	0.00	0.00	0.00	0.00	0.00	0.00	0.00
er	1.33	2.67	4.00	0.00	2.67	4.00	0.00
fmri	32.86	12.86	18.00	55.56	1.43	2.00	0.00
rdag	66.98	13.95	60.00	93.33	0.93	4.00	0.00
ws	7.05	1.36	12.00	16.67	7.95	70.00	0.00
all	19.50	4.90	–	**68.00**	4.90	–	**0.00**

two families are also the most representative among the top-50. Furthermore, *none* of the graphs in the bottom-50 has the Q1D property (last column).

The Q1D criterion is able to single out *one type* of the best performing networks, being at the same time *agnostic* about the details of the graph generation procedure. This is especially important considering the failure of classical graph features in this regard.

Finally, let us also mention that there are some qualitatively different networks (see for example the `fmri` network in Fig. 4a) in the `fmri` class as well as in the `ba` class, which achieve good performance. Those networks are often not elongated (as indicated by several green points with high test accuracy in Fig. 4b) and therefore do not satisfy the Q1D criterion. It seems, however, quite difficult to identify a numerical characterization that would pick out the best networks from this category.

6   Impact on Architecture Design

The key components of the Q1D graphs are elongated structure and lack of bottlenecks. In this section we further analyze the importance of those characteristics as guidelines in the design of neural network connectivity. We start with a study of the effect of short- vs. long-range connections in the `rdag` graphs and follow with a commentary about the role of many resolution-changing pathways. Finally, we also perform a comparison of the CIFAR-10 results with results on CIFAR-100 in order to ascertain the consistency of the identification of the best and worst-performing network families.

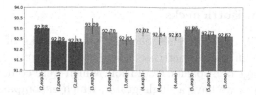

Fig. 5. The CIFAR-10 test accuracy averaged over different versions (random seeds) of random DAG models with 30 nodes and constant number (2–5) of output edges n_i^{out}. The symbol *exp3* stands for exponential weighting function $f(x)$, *pow1* for a power law and *one* for a constant. It may be observed that the networks with primarily local connections (*exp3* - the first bar in each set) have the best performance.

6.1   Long- vs. Short-range Connections

The algorithm for directly generating random DAGs allows for modifying, in a controllable way, the pattern of long- versus short-range connectivity. This is achieved by changing the function $f(x)$ from an exponential, leading to local connections, through a power law, which allows for occasional long-range connections, to a constant function, which does not impose any spatial order and allows connections at all scales. The results are presented in Fig. 5. We observe that within this class of networks the best performance comes from networks with primarily short-range connections and deteriorates with their increasing length.

This may at first glance seem counter-intuitive, as skip connections are typically considered beneficial. However, the effect of long-range connections which is associated with easier gradient propagation is already taken care of by the residual structure of each node in our neural networks (recall Sect. 3). One can understand the deterioration of the network performance with the introduction of long-term connections as coming from an inconsistency of the network with the natural hierarchical semantic structure of images. This result leads also to some caution in relation to physical intuition from critical systems where all kinds of power law properties abound. The dominance of short-range over long-range connections is also consistent with the good performance of *quasi-1-dimensional* networks as discussed in Sect. 5.3.

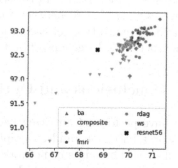

Fig. 6. The CIFAR-10 (y-axis) and CIFAR-100 (x-axis) test accuracies. Each datapoint contains results averaged over the random versions of the models. The results are strongly correlated, yielding Pearson correlation coefficient equal to 0.868.

6.2   Influence of Bottlenecks

As noted in Sect. 3, one difference between the networks of [28] and our construction was that in the former case, there were separate random graphs for each processing stage of a specific resolution, which were connected with a single gateway. In our case, we have a single graph, which encompasses all resolutions. Thus generally there are many independent resolution-reducing edges in the network instead of a single one. In order to verify whether such a single gateway between different resolutions is beneficial or not, for a selected set of graphs, we artificially introduced such a bottleneck by first erasing all inter-resolution edges. Next, we create a single edge from the last node in the preceding stage to the first node in the consequent stage and then fixing possible orphans as in Fig. 2b[4]. We found that, systematically, the introduction of a bottleneck deteriorates performance (see Fig. 8. in Appendix). Hence multiple resolution reduction pathways are beneficial. Let us note that this result is coherent with our findings from Sect. 5.3, where bottleneck edges (also within a single resolution stage) typically appear in badly performing networks.

6.3   CIFAR-10 Versus CIFAR-100 Consistency

In addition to the CIFAR-10 task, we trained all networks with 60 nodes (except for the bottleneck ablations) on the CIFAR-100 dataset. We used the same training procedure as the one for CIFAR-10. The motivation for this experiment was to verify whether the graph families which performed best in the first problem achieve also high results in the second. Indeed, we observe a significant correlation 0.868 (see Fig. 6) between the respective test accuracies (averaged over the 5 random realizations of each graph type). Especially noteworthy is the consistency between the groups of best and worst graphs for the two datasets.

7   Conclusions and Outlook

We have performed an extensive study of the performance of artificial neural networks based on random graphs of various types, keeping the training protocol fixed. One class of networks which had the best performance in our simulations were networks, which could be characterized as quasi-1-dimensional, having mostly local connections with a definite 1-dimensional hierarchy in data processing (one can dub this structure as *local chaos and global order*). These were predominantly networks in the rdag family. We also introduced a very compact numerical characterization of such graphs. It is worth noting, that some of the fMRI-based graphs were also among the best-performing ones (together with some ws and ba ones). We lack, however, a clear-cut numerical characterization of these "good" graphs as there exist graphs with apparently similar structure and numerical invariants but much worse performance.

---

[4] See a visualization of a bottleneck graph in Appendix I.

Among other structural observations made in this project, we noted that long-range connections were predominantly negatively impacting network performance. Similarly, artificially imposing a bottleneck between the processing stages of various resolutions also caused the results to deteriorate. Thus, a general guideline in devising neural network architectures which can be formed in consequence of our study is to prefer networks with rich local connections composed into an overall hierarchical computational flow, with multiple resolution-reducing pathways and no bottleneck edges. These characteristics seem to consistently lead to good performance among the vast panorama of connectivity patterns investigated in the present paper.

Acknowledgments. Work carried out within the research project *Bio-inspired artificial neural network* (grant no. POIR.04.04.00-00-14DE/18-00) within the Team-Net program of the Foundation for Polish Science co-financed by the European Union under the European Regional Development Fund. The fMRI data were provided by the Human Connectome Project, WU-Minn Consortium (PIs: David Van Essen and Kamil Ugurbil; 1U54MH091657) funded by 16 NIH Institutes and Centers that support the NIH Blueprint for Neuroscience Research; and by the McDonnell Center for Systems Neuroscience at Washington University.

References

1. Athreya, A., et al.: Statistical inference on random dot product graphs: a survey. J. Mach. Learn. Res. **18**(1), 8393–8484 (2017)
2. Baker, B., Gupta, O., Naik, N., Raskar, R.: Designing neural network architectures using reinforcement learning. In: International Conference on Learning Representations (2016)
3. Barabási, A.L., Albert, R.: Emergence of scaling in random networks. Science **286**(5439), 509–512 (1999)
4. Bullmore, E.T., Bassett, D.S.: Brain graphs: graphical models of the human brain connectome. Annu. Rev. Clin. Psychol. **7**, 113–140 (2011)
5. Dong, X., Yang, Y.: NAS-Bench-201: extending the scope of reproducible neural architecture search. In: International Conference on Learning Representations (2019)
6. Elsken, T., Metzen, J.H., Hutter, F.: Neural architecture search: a survey. J. Mach. Learn. Res. **20**(55), 1–21 (2019)
7. Erdős, P., Rényi, A.: On the evolution of random graphs. Publ. Math. Inst. Hung. Acad. Sci. **5**, 17–61 (1960)
8. Freeman, L.C.: Centrality in social networks conceptual clarification. Soc. Netw. **1**(3), 215–239 (1978)
9. Fruchterman, T.M., Reingold, E.M.: Graph drawing by force-directed placement. Softw. Pract. Exp. **21**(11), 1129–1164 (1991)
10. Hagberg, A., Swart, P., S Chult, D.: Exploring network structure, dynamics, and function using networkx. Technical report, Los Alamos National Lab. (LANL), Los Alamos, NM (United States) (2008)
11. He, K., Zhang, X., Ren, S., Sun, J.: Deep residual learning for image recognition. In: Proceedings of the IEEE Conference on Computer Vision and Pattern Recognition, pp. 770–778 (2016)

12. Huang, G., Liu, Z., Van Der Maaten, L., Weinberger, K.Q.: Densely connected convolutional networks. In: Proceedings of the IEEE Conference on Computer Vision and Pattern Recognition, pp. 4700–4708 (2017)
13. Huang, G., Sun, Yu., Liu, Z., Sedra, D., Weinberger, K.Q.: Deep networks with stochastic depth. In: Leibe, B., Matas, J., Sebe, N., Welling, M. (eds.) ECCV 2016. LNCS, vol. 9908, pp. 646–661. Springer, Cham (2016). https://doi.org/10. 1007/978-3-319-46493-0_39
14. Kamada, T., Kawai, S.: An algorithm for drawing general undirected graphs. Inf. Process. Lett. **31**(1), 7–15 (1989)
15. Liu, H., Simonyan, K., Yang, Y.: Darts: differentiable architecture search. In: International Conference on Learning Representations (2019)
16. Mocanu, D.C., Mocanu, E., Nguyen, P.H., Gibescu, M., Liotta, A.: A topological insight into restricted Boltzmann machines. Mach. Learn. **104**(2), 243–270 (2016)
17. Orsini, C., et al.: Quantifying randomness in real networks. Nat. Commun. **6**(1), 1–10 (2015)
18. Paszke, A., et al.: Pytorch: an imperative style, high-performance deep learning library. arXiv preprint arXiv:1912.01703 (2019)
19. Real, E., Aggarwal, A., Huang, Y., Le, Q.V.: Regularized evolution for image classifier architecture search. In: Proceedings of the AAAI Conference on Artificial Intelligence, vol. 33, pp. 4780–4789 (2019)
20. Roberts, N., Yap, D.A., Prabhu, V.U.: Deep connectomics networks: neural network architectures inspired by neuronal networks. arXiv preprint arXiv:1912.08986 (2019)
21. Shafiee, M.J., Siva, P., Wong, A.: StochasticNet: forming deep neural networks via stochastic connectivity. IEEE Access **4**, 1915–1924 (2016)
22. Smith, S.M., Beckmann, C.F., Andersson, J., Auerbach, E.J., et al.: Resting-state fMRI in the human connectome project. Neuroimage **80**, 144–168 (2013)
23. Szegedy, C., et al.: Going deeper with convolutions. In: Proceedings of the IEEE Conference on Computer Vision and Pattern Recognition, pp. 1–9 (2015)
24. Van Essen, D.C., Smith, S.M., Barch, D.M., Behrens, T.E., Yacoub, E., Ugurbil, K.: The WU-Minn human connectome project: an overview. Neuroimage **80**, 62–79 (2013)
25. Veit, A., Wilber, M., Belongie, S.: Residual networks behave like ensembles of relatively shallow networks. arXiv preprint arXiv:1605.06431 (2016)
26. Watts, D.J., Strogatz, S.H.: Collective dynamics of 'small-world' networks. Nature **393**(6684), 440 (1998)
27. Wiener, H.: Structural determination of paraffin boiling points. J. Am. Chem. Soc. **69**(1), 17–20 (1947)
28. Xie, S., Kirillov, A., Girshick, R., He, K.: Exploring randomly wired neural networks for image recognition. In: Proceedings of the IEEE International Conference on Computer Vision, pp. 1284–1293 (2019)
29. Ying, C., Klein, A., Christiansen, E., Real, E., Murphy, K., Hutter, F.: NAS-Bench-101: towards reproducible neural architecture search. In: International Conference on Machine Learning, pp. 7105–7114 (2019)
30. You, J., Leskovec, J., He, K., Xie, S.: Graph structure of neural networks. In: Proceedings of the International Conference on Machine Learning (2020)
31. Zoph, B., Le, Q.V.: Neural architecture search with reinforcement learning. In: International Conference on Learning Representations (2017)

Context Abstraction to Improve Decentralized Machine Learning in Structured Sensing Environments

Massinissa Hamidi and Aomar Osmani[✉]

LIPN-UMR CNRS 7030, Université Sorbonne Paris Nord, Paris, France
{hamidi,ao}@lipn.univ-paris13.fr

Abstract. In Internet of Things applications, data generated from devices with different characteristics and located at different positions are embedded into different contexts. This poses major challenges for decentralized machine learning as the data distribution across these devices and locations requires consideration for the invariants that characterize them, e.g., in activity recognition applications, the acceleration recorded by hand device must be corrected by the invariant related to the movement of the hand relative to the body. In this article, we propose a new approach that abstracts the exact context surrounding data generators and improves the reconciliation process for decentralized machine learning. Local learners are trained to decompose the learned representations into (i) universal components shared among devices and locations and (ii) local components that capture the specific context of device and location dependencies. The explicit representation of the relative geometry of devices through the special Euclidean Group $SE(3)$ imposes additional constraints that improve the decomposition process. Comprehensive experimental evaluations are carried out on sensor-based activity recognition datasets featuring multi-location and multi-device data collected in a structured sensing environment. Obtained results show the superiority of the proposed method compared with the advanced solutions.

Keywords: Meta-learning · Federated learning · Internet of things

1 Introduction

In Internet of Things (IoT) applications, data generated from different devices (or sensors) and locations are embodied with varying contexts. The devices offer specific perspectives on the problem of interest depending on their location. The movements of the area on which the devices are positioned generate data of two different but complementary natures. For instance, in Fig. 1, the data of the movement collected from the hand sensors combines data of the whole body intertwined with data related to the movement of the hand in relation to the body. In the case of human activity recognition (HAR), we notice, for example,

© The Author(s), under exclusive license to Springer Nature Switzerland AG 2023
M.-R. Amini et al. (Eds.): ECML PKDD 2022, LNAI 13715, pp. 647–663, 2023.
https://doi.org/10.1007/978-3-031-26409-2_39

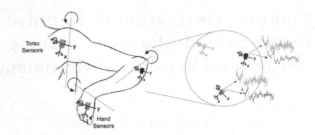

Fig. 1. Example of phenomena surrounded by a structured sensing environment. The hand sensor undergoes two types of movements. One is of the same nature as the torso and linked to the translational movement of the body. The other is linked to the movement of the hand locally relative to the body.

that the kinetics of the hand movements during a race can be decomposed into a circular movement (CM) of the hand relative to the shoulder and a translation movement (TM) associated with the whole body [23].

These characteristics pose significant challenges for decentralized machine learning as the data distribution across these devices and locations is skewed. Federated learning [16,22] is an appropriate framework that handles decentralized and distributed settings. In particular, the locally learned weights are aggregated into a central model during the conciliation phase. Decentralized machine learning suffers from objective inconsistency caused by the heterogeneity in local updates and by the interpretation of the locally collected data. Additional phenomena like the evolution of the local variables over time (concept drift) [15] or relativity of viewpoints (see Fig. 1) must also be considered.

Recent advances in machine learning literature, e.g., [31], seek the notions of invariance and symmetries within the phenomena of interest. Symmetry is one of the invariants that is leveraged for its powerful properties and its promising ability to drastically reduce the problem size [4,6,27] by requiring fewer training examples than standard approaches for achieving the same performance. Group theory provides a useful tool for reasoning about invariance and equivariance. For instance, in HAR [25,26], the acceleration recorded by the device held in hand must be corrected by the invariant related to the movement of the hand relative to the body so that the acceleration data related to the whole body is accurate. More generally, when the sensors are placed in a structured environment that exhibits regular dependencies between the locations of the sensors, it is possible to devise models of data transformations to reduce biases such as position biases. These models correspond to automatic changes in data representation to project them onto the same space while minimizing the impact of structure and location on the final data.

In this paper, we propose a novel approach that abstracts the exact context surrounding the data generators and hence improves decentralized machine learning. Local learners are trained to decompose the learned representations into (i) universal components shared across devices and locations and (ii) local components which capture the specific device- and location-dependent context.

We introduce the notion of relativity between data generators and model it via the special Euclidean group, denoted by $SE(3)$, which encompasses arbitrary combinations of translations and rotations. The relative contribution of a data generator in the description of the phenomena of interest is expressed using elements of this group and used to constrain the separation process. In particular, building on the symmetry-based disentanglement learning [12], the symmetry structure induced by the relative data generators is reflected in the latent space. This allows us to further leverage the notion of sharing which is reflected into the conciliation process of the decentralized learning setting by promising improvements. Comprehensive experimental evaluations are conducted to assess the effectiveness of the proposed approach. Obtained results demonstrate the superiority of the proposed method over more advanced solutions.

The main contributions of the paper are: (i) a novel approach that leverages additional knowledge in the terms of symmetries and invariants that emerge in these kinds of environments. These symmetries and invariants are explicitly represented in the form of group actions and incorporated into the learning process; (ii) a proposition of separation process of the data into universal and position-specific components improves collaboration across the decentralized devices materialized by the conciliation (or aggregation) process; (iii) extensive experiments on two large-scale real-world wearable benchmark datasets featuring structured sensing environments. Obtained results are promising noticeably in terms of the quality of the conciliation which open-up perspectives for the development of more efficient collaboration schemes in structured environments.

2   Background and Motivation

Here we provide a background on decentralized machine learning approaches and highlight their key principles. Then we review the impact of the various contexts surrounding the distributed data generators on the learning process in real-world IoT applications and *a priori* knowledge can be leveraged to deal with this challenge.

2.1   IoT Deployments

We consider settings where a collection $\mathcal{S}$ of M sensors (also called data sources), denoted $\{s_1, \ldots, s_M\}$, are positioned respectively at positions $\{p_1, \ldots, p_M\}$ on the object of interest, e.g., human body. Each sensor s_i generates a stream $\mathbf{x}^i = (x_1^i, x_2^i, \ldots)$ of observations of a certain modality like *acceleration*, *gravity*, or *video*, distributed according to an unknown generative process. Furthermore, each observation can be composed of channels, e.g. three axes of an accelerometer. The goal is to continuously recognize a set of human activity target concepts $\mathcal{Y}$ like *running* or *biking*. In the case of the SHL dataset, the data are generated from 4 smartphones, carried simultaneously at (*hand*, *torso*, *hips*, and *bag* body locations. Sensors distributed in various positions of the space provide rich perspectives and contribute in different ways to the learning process, and the decentralization of the sensors has the potential to offer better guarantees of the quality of the generalization.

2.2   Decentralized Machine Learning

In the decentralized machine learning setting, a set of M clients, each corresponding to a sensor of the above IoT deployment, aim to collectively solve the following optimization problem:

$$\min_{w \in \mathbb{R}^d} \left\{ F(w) := \sum_{p=1}^{M} \alpha_p \cdot f_p(w_p) \right\}, \tag{1}$$

where $f_p(w) = \frac{1}{n_p} \sum_{\zeta \in D_p} \ell_p(x; \zeta)$ is the local objective function at the p-th client, with ℓ_p the loss function and ζ a random data sample of size n_p drawn from local dataset D_p according to the distribution of position p. At each communication round r, each client runs independently τ_p iterations of the local solver, e.g., stochastic gradient descent, starting from the current global model (set of weights) $w_p^{(r,0)}$ until the step $w_p^{(r,\tau_p)}$ to optimize its own local objective. Then the updates of a subset of clients are sent to the central server where they are aggregated into a global model. Only parameter vectors are exchanged between the clients and the server during communication rounds while raw data are kept locally which complies with privacy-preserving constraints. Various algorithms were proposed for aggregating the locally learned parameter vectors into a global model, including [22] which updates the shared global model as follows:

$$w^{(r+1,0)} - w^{(r,0)} = -\sum_{p=1}^{M} \alpha_p \cdot \eta \sum_{k=0}^{\tau_p - 1} g_p(w_p^{(r,k)}) \tag{2}$$

where $w_p^{(r,k)}$ denotes the model of client p after the k-th local update in the r-th communication round. Also, η is the client learning rate and g_p represents the stochastic gradient computed over a mini-batch of samples.

2.3   IoT Deployments and Impact of the Context

Long lines of research studied the impact of the varying contexts on machine learning algorithms and showed their fragility to viewpoint variations [14]. For example, basic convolutional networks are found to fail when presented with out-of-distribution category-viewpoint combinations, i.e., combinations not seen during training. Similarly, in activity recognition, the diversity of users, their specific ways of performing activities, and the varying characteristics of the sensing devices have a substantial impact on performances [10,29]. In these cases, the conditional distributions may vary across clients even if the label distribution is shared [15]. In decentralized approaches, several theoretical analyses bound this drift by assuming bounded gradients [36], viewing it as additional noise [17], or assuming that the client optima are ϵ-close [19]. As a practical example, SCAFFOLD [16] tries to correct for this client-drift by estimating the update direction for the server model ($\mathbf{c}$) and the update direction for each client $\mathbf{c}_p$. Then, the difference ($\mathbf{c} - \mathbf{c}_p$) is used as the estimator of the client-drift which is used to correct the local update steps. The local models are, then, updated as $w_p^{(r+1,0)} - w_p^{(r,0)} = -\eta \cdot (g_p(w_p) + \mathbf{c} - \mathbf{c}_p)$.

The impact of varying contexts is not limited to a skewed distribution of labels but is rather predominantly related to the aspects of the phenomenon being captured by the sensing devices depending on their intrinsic characteristics and locations. Depending on their disposition w.r.t. to the phenomena of interest, the sensing devices generate different views of the same problem. The heterogeneity brought by these configurations in terms of views is beneficial but must be explicitly handled. Reconciling the various perspectives offered by these deployments using decentralized learning approaches requires several relaxations limiting their potential capabilities when the impact of the context on the data generation process is essential. Indeed, how to reconcile these different points of view which can potentially be redundant or even seemingly contradictory to each other? When additional knowledge is available about the structure of the sensing environment, these challenges can be handled efficiently.

2.4 Relativity of Viewpoints in Structured Sensing Environments

Very often, knowledge about the relative geometry of the sensing devices and domain models describing the dynamics of the phenomena is available and can be leveraged and incorporated into the learning process. For example, the spatial structure of the sensors deployment and the induced views, sensors capabilities and the perspectives (views) through which the data is collected (sensing model, range, coverage, position in space, position on the body, and type of captured modality) [1,11,33]. A long line of research work around activity recognition reviewed in, e.g. [9,34], has focused on the problem of optimal placement and combination of sensors on the body in order to improve *a priori* models' performance. Additionally, domain models derived from biomechanical studies like [3,23] are often used to describe body movements and the relative interactions between various body parts in a structured manner. Alternatively, considering the structure of the sensing devices explicitly during the learning process is more promising but challenging. An approach close to ours for the relativity of perspectives is that of [5] which describes the different perspectives by discrete subgroup of the rotation group.

Integrating these additional models into the learning process has promising implications noticeably on the conciliation process of decentralized machine learning algorithms: one can exhibit the relative contribution of the individual views to the bigger picture. The primary goal of this paper is to develop a robust approach that integrates knowledge about the structure of sensing devices in a principled way to achieve better collaboration.

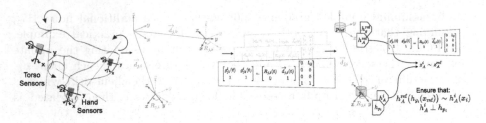

Fig. 2. Framework of the proposed approach. Explicitly representing the relative geometry of the decentralized devices and their symmetries using elements of the special Euclidean group $SE(3)$ and leveraging them to constrain the learning process with the goal of reducing the problem size and improving data efficiency.

3 FEDABSTRACT Algorithm

We propose an original approach based on local abstraction of the position-specific artifacts and aggregation of universal components in the data. We leverage knowledge about the structure of the sensing deployment by representing the relative geometry of the sensing devices with group transformations. At a given decentralized location, there are three different elements that are learned: (1) the universal (or group-invariant) and (2) position-specific representations (Sect. 3.1), and (3) the group of relative geometry representation (Sect. 3.2). The generalization capabilities of the universal representation are improved collaboratively across the decentralized sensing devices via the conciliation (or aggregation) process (Sect. 3.3). Figure 2 summarizes the proposed approach.

3.1 Learning Group-Invariant and Position-Specific Representations

The idea is to express the data generated from a decentralized device (e.g., hand sensors in the case of on-body sensor deployments) relative to the coordinate system of a referential (e.g., torso.) This way, the exact relative contribution of the sensing device is captured without the contextual artifacts. To do this, we have to capture the variations due to the relative location of the decentralized device w.r.t. a global coordinate system and capture invariant aspects that are shared across the devices. The latter aspects are universal components that are shared with the central model while the former ones are considered as specific components which add noise to the learning process, thus requiring to be discarded from it.

Invariance. A mapping $h(\cdot)$ is invariant to a set of transformations G if when we apply any transformation induced by g to the input of h, the output remains unchanged. A common example of invariance in deep learning is the translation invariance of convolutional layers. In the structured sensing environments considered here, the elements g of $SE(3)$ act on the spatial disposition of the data generators and ultimately the data they generate: if we translate the data representation learned at sensor position p_i to position p_j, the representation remains unchanged. Formally, if $h : A \rightarrow A$, and G is a set of transformations acting on A, h is said to be invariant to G if $\forall a \in A$, $\forall g \in G$, $h(ga) = h(a)$.

We construct at the level of each client i a representation that maps the observation space X to a latent space V with $h_A : X \rightarrow V$ (universal) and $h_{p_i} : X \rightarrow V$ (position-specific). The universal representation has to remain invariant to the relative location of the decentralized nodes. We also ensure during the learning process that the universal and location-specific transformations are orthogonal to each other ($h_A \perp h_{p_i}$). In other words, we want these two transformations to capture completely different factors of variations in the data. To do that, we enforce h_{p_i} to be insensitive to the factors of variations linked to the representation h_A using representation disentanglement techniques. We use in our approach, a family of models based on variational autoencoders (VAEs) [18] for their ability to deal with entangled representations.

Learning h_A and h_{p_i} Locally. The data x_i captured at a given location i are generated from two underlying factors: one reflecting the position-specific components and the other the position-invariant (or universal) components. The task here is to learn these factors of variation, commonly referred to as learning a disentangled representation. In other words, we want these two transformations to capture completely different factors of variations in the data. To do that, we enforce h_{p_i} to be insensitive to the factors of variations linked to the representation h_A using representation disentanglement techniques. It corresponds to finding a representation where each of its dimensions is sensitive to the variations of exactly one precise underlying factor and not the others. Note that the inputs to h_A in the local learners are the raw sensory data x_i generated locally.

At this point, we are left with two alternatives for jointly learning the universal transformation h_A and the position-specific transformation h_{p_i} at the local learner level: (1) using a separate VAE for each transformation and training each one of them jointly using the raw sensory data as inputs; (2) using a single VAE and train it to automatically factorize the learned representation so that each axis captures specific components. Recent advances in unsupervised disentangling based on VAEs demonstrated noticeable successes in many fields using the β-VAE, which leads to improved disentanglement [13]. It uses a unique representation vector and assigns an additional parameter ($\beta > 1$) to the VAE objective, precisely, on the Kullback-Leibler (KL) divergence between the variational posterior and the prior, which is intended to put implicit independence pressure on the learned posterior. The improved objective becomes:

$$\mathcal{L}(x; \theta, \varphi) = \mathbb{E}_{q_\varphi(z|x)}[\log p_\theta(x|z)] \text{ (autoencoder reconstruction term)}$$
$$- \beta D_{KL}(q_\varphi(z|x) \| p(z)) - \alpha D_{KL}(q_\varphi(z) \| p(z)),$$

where the term controlled by α allows to specify a much richer class of properties and more complex constraints on the dimensions of the learned representation other than independence. Indeed, the proposed conciliation step is challenging due to the dissimilarity of the data distributions across the local learners, leading to discrepancies between their respective learned representations.

One way to deal with this issue is by imposing sparsity on the latent representation in a way that only a few dimensions get activated depending on the learner and activities. We ensure the emergence of such sparse representations using the appropriate structure in the prior $p(z)$ such that the targeted underlying factors are captured by precise and homogeneous dimensions of the latent representation. We set the sparse prior as $p(z) = \prod_d (1 - \gamma)\mathcal{N}(z_d; 0, 1) + \gamma\mathcal{N}(z_d; 0, \sigma_0^2)$ with $\mathcal{N}$ is the Gaussian distribution. This distribution can be interpreted as a mixture of samples being either activated or not, whose proportion is controlled by the weight parameter γ [21].

Now, we have to represent the notion of data generators relativity and its induced symmetries in the form of group elements whose action on the data leaves the universal component of the learned representation invariant.

3.2 Relative Geometry for Data Generators

We model the relative geometry of sensors and the perspectives they provide via the special Euclidean group $SE(3)$. Let $\mathbf{x}^i$ and $\mathbf{x}^j$ be the stream of observations generated by the data sources s_i and s_j. At each time step t, the observations x_i and x_j generated by these data sources are related together via an element $g_j^i \in SE(3)$ of the group of symmetries, i.e., the observation x_i is obtained by applying g_j^i on x_j. Here, we want to learn a mapping h_{g_i} for each decentralized device, so that the biases that stem from the context (exact position) are corrected before its contribution is communicated to the global model.

Special Euclidean Group $SE(3)$. The special Euclidean group, denoted by $SE(3)$, encompasses arbitrary combinations of translations and rotations. The elements of this group are called rigid motions or Euclidean motions and correspond to the set of all 4 by 4 matrices of the form $P(R, \overrightarrow{d}) = \begin{pmatrix} R & \overrightarrow{d} \\ 0 & 1 \end{pmatrix}$, with $\overrightarrow{d} \in \mathbb{R}^3$ a translation vector, and $R \in \mathbb{R}^{3\times3}$ a rotation matrix. Members of $SE(3)$ act on points $z \in \mathbb{R}^3$ by rotating and translating them: $\begin{pmatrix} R & \overrightarrow{d} \\ 0 & 1 \end{pmatrix}\begin{pmatrix} z \\ 1 \end{pmatrix} = \begin{pmatrix} Rz + \overrightarrow{d} \\ 1 \end{pmatrix}$.

Relative Geometry Representation. Given a pair of sensing devices s_i and s_j located at positions p_i and p_j, each having its own local coordinate system attached to it. We represent the relative geometry of this pair by expressing each of the devices in the local coordinate system of the other (see Fig. 3). Similarly to [32], the local coordinate system attached to p_i is the result of a translation $\overrightarrow{d}_{j,i}$ and a rotation $R_{j,i}$, where the subscript j, i denotes the sense of the transformation being from p_j to p_i. While the translation corresponds to the alignment of the origins of the two coordinate systems, the rotation is obtained by rotating the global coordinate system such that the x-axis of the two coordinate systems coincide: $\begin{pmatrix} g_{j1}^i(t) & g_{j2}^i(t) \\ 1 & 1 \end{pmatrix} = \begin{pmatrix} R_{j,i}(t) & \overrightarrow{d}_{j,i}(t) \\ 0 & 1 \end{pmatrix}\begin{pmatrix} 0 & l_{ij} \\ 0 & 0 \\ 0 & 0 \\ 1 & 1 \end{pmatrix}$.

The relative geometry of the data generators is considered to be elements of $SE(3)$ and supposed to capture the transformations acting on the data generators. Without explicit information about the exact locations of the data generators, these transformations have to be learned. For this, we parameterize the transformation matrices used to represent the relative geometry of the data generators, with learnable weights. In particular, we parameterize as in [27] the n-dimensional representation of a rotation R as the product of $\frac{n(n-1)}{2}$ rotations, denoted $R^{v,w}$, each of which corresponds to the rotation in the v, w plane embedded in the n-dimensional representation. For example, a 3-dimensional representation has three learnable parameters, $g = g(\theta^{1,2}, \theta^{1,3}, \theta^{2,3})$, each parameterizing a single rotation, such as $R^{1,3}(\theta^{1,3}) = \begin{pmatrix} \cos \theta^{1,3} & 0 & \sin \theta^{1,3} \\ 0 & 1 & 0 \\ -\sin \theta^{1,3} & 0 & \cos \theta^{1,3} \end{pmatrix}$.

Learning h_A and h_g in the Central Server. The referential learner (or central server) happens also to be a learner similar to the local learners. The main difference is that the referential learner is located in a particular position of the sensors deployment, i.e., the referential coordinate system, which imposes it to perform additional processing. Let's denote the referential learner with subscript ref (the orange data source in Fig. 3). The referential learner maintains the specific h_g's corresponding to each individual position of the sensors deployment and ensures that:

$$h_A(h_{g_i}(x_{\text{ref}})) = h_A(x_i), \forall i \tag{3}$$

where h_{g_i} is the learned representation corresponding to the group action acting on the data x_i generated by the sensor located at position i and x_{ref} is the data generated by the sensor located at the referential point. The h_{g_i} transformation is learned by the referential learner using the raw data generated at the central server level. The constraint imposing the invariance, i.e., $h_A(h_{g_i}(x_{\text{ref}})) = h_A(x_i), \forall i$, is the pivotal element that makes it possible to effectively learn this transformation.

By drawing a parallel with the construction of manifolds in latent spaces, this transformation can be interpreted as an operator projecting the data, generated by the data source positioned on ref, towards a latent space shifted by the action of the group elements so that the universal components learned by the transformation h_A (at the referential) coincide with those transformations $(h_A^i, \forall i)$ learned by the local learners attached to the other positions. h_g must therefore act on different subgroups of the latent space. We ensure that the learned universal transformation h_A is invariant to the action of the group $SE(3)$, i.e., $h_A(gx) = h_A(x), g \in SE(3)$. For this we map the group $SE(3)$ to a linear representation GL on V, i.e., $\rho : SE(3) \rightarrow GL(V)$. Our goal is to map observations to a vector space V and interactions to elements of $GL(V)$ to obtain a disentangled representation of the relative geometry.

As there are many different group representations (one for each position of the deployment of the sensors) at the referential learner's level, we have to ensure that the learned representation h_g acts on specific subspaces of the latent space. At the central server, each client is considered to generate a subgroup of

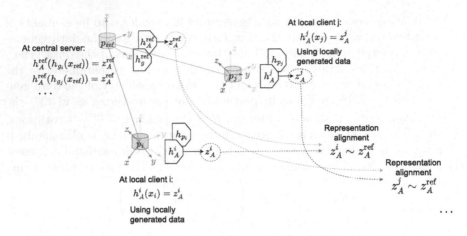

Fig. 3. Network architecture of FedAbstract. The local learners (red and blue) perform a set of updates on their proper version of the universal representation. The referential learner at position p_{ref} (in orange) maintains the specific h_g's corresponding to each individual position of the sensors deployment and ensures that: $h_A(h_{g_i}(x_{\text{ref}})) = h_A(x_i), \forall i$ where h_{g_i} is the learned representation corresponding to the group elements acting on the data x_i generated at position i and x_{ref} the data generated at the referential point. Notice that only gradient updates are shared to the central server and the data generated at a given location are processed exclusively by the local learner. (Color figure online)

relative geometry. During the learning process, each subgroup of the symmetry group is made to act on a specific subspace of the latent space. Formally, let $\cdot : G \times X \to X$ be a group action such that the group G decomposes as a direct product $G = G_1 \times G_2$. According to [12], the action is disentangled (w.r.t. the decomposition of G) if there is a decomposition $X = X_1 \times X_2$, and actions $\cdot_i : G_i \times X_i \to X_i, i \in \{1, 2\}$ such that: $(g_1, g_2) \cdot (v_1, v_2) = (g_1 \cdot_1 v_1, g_2 \cdot_2 v_2)$, where $\cdot$ denotes the action of the full group, and the actions of each subgroup as $\cdot_i$. An G_1 element is said to act on X_1 but leaves X_2 fixed, and vice versa. We end up here in the same situation as in the disentanglement of universal and position-specific components, i.e., either we use a separate VAE for each group transformation or a single one for all the groups with the additional constraint stating that the action of each subgroup act on specific regions of the latent space manifold and leave the other regions fixed. This can be achieved via clustering of the latent space using a Gaussian mixture prior [21] $p(z) = \sum_{c=1}^{C} \pi^c \prod_d \mathcal{N}(z_d | \mu_d^c, \sigma_d^d)$, with C the number of desired clusters and π^c the prior probability of the c-th Gaussian.

3.3   Conciliation Process

At the local learner's level, the proposed model is trained in an end-to-end fashion. The generalization capabilities of the representation h_A are improved via the conciliation process performed across the nodes of the deployment.

Algorithm 1: Multi-level abstraction of sensor position

Input : $\{\mathbf{x}^p\}_{p=1}^M$ streams of annotated observations

1   $w \leftarrow$ initWeights() ;   % *Initialize global learner's weights*
2   distributeWeights($w, \mathcal{S}$) ;   % *Weights distribution*
3   **while** *not converged* **do**
4   **foreach** *position p* **do**
5   **for** $t \in \tau_p$ steps **do**
6   Sample mini-batch $\{x_i^p\}_{i=1}^{n_p}$
7   Evaluate $\nabla_{w_p}\ell(w_p)$ w.r.t. the mini-batch
8   └ Subject to $J(z_A^p, z_A^{\mathrm{ref}})$ (e.g., correlation-based alignment [2])
9         $w_p^{(t)} \leftarrow w_p^{(t-1)} - \eta \nabla_{w_p}\ell(w_p)$
10   Ensure $h_A \perp h_{p_i}$ (see §3.1)
11   **end**
12   **Communicate** w_A (with $w_p = [w_A, w_{p_i}]$)
13   **end**
14     $w_A \leftarrow w_A + \sum_{p=1}^M \alpha_p \cdot \Delta w_A^p$;   % *Central updates*
15   Enforce group action disentanglement
16 **end**

Result: Globally shared universal representation h_A

Each local learner pursues its own version of the universal representation but has not to diverge from the referential universal representation h_A^{ref}, which constitutes a consensus among all local learners. After a predefined number of local update steps, we conduct a conciliation step (see the dotted arrows in Fig. 3). Each conciliation step t produces a new version of the referential learner $w_{\mathrm{ref}}^{(t)}$ and, a new version of the referential universal representation z_A^{ref}. The conciliation step has to be performed on the learned representations z_A^p via regularization, for example. In our approach, the conciliation step is performed via representation alignment, e.g., correlation-based alignment [2]. More formally, we instrument the objective function of the local learners with an additional term derived from the representation alignment [30]. The optimization problem (1) becomes:

$$\min_{w \in \mathbb{R}^d}\left\{ F(w) = \frac{1}{M}\sum_{p=1}^M \alpha_p(f_p(w_p) + \lambda J(z_A^p, z_A^{\mathrm{ref}})) \right\}, \tag{4}$$

where J is a regularization term responsible for aligning the locally learned universal components with the ones learned by the referential learner and $\lambda \in [0,1]$ is a regularization parameter that balances between the local objective and the regularization term. Algorithm 1 summarizes the process of the proposed approach and Fig. 3 illustrates its bigger picture.

4 Experiments and Results

We perform an empirical evaluation of the proposed approach, consisting of two major stages: (1) we verify the effectiveness of the proposed approach in the HAR task via a comparative analysis which includes representative related baselines (Sect. 4.1); (2) we also conduct extensive experiments and ablation analysis to demonstrate the effectiveness of the various components of our proposed approach (Sect. 4.2).

Experimental Setup. We evaluate our proposed approach on two large-scale real-world wearable benchmark datasets featuring structured sensing environments: SHL [8] and Fusion [28] datasets. We compare our approach with the following closely related baselines.

- **DeepConvLSTM** [24]: a model encompassing 4 convolutional layers responsible for extracting features from the sensory inputs and 2 long short-term memory (LSTM) cells used to capture their temporal dependence.
- **DeepSense** [35]: a variant of the DeepConvLSTM model combining convolutional and Gated Recurrent Units (GRU) in place of the LSTM cells.
- **AttnSense** [20]: features an additional attention mechanism on top of the DeepSense model forcing it to capture the most prominent sensory inputs both in the space and time domains to make the final predictions.
- **GILE** [26]: proposes to explicitly disentangle domain (or position)-specific and domain-agnostic features using two encoders. To constrain the disentanglement process, their proposed additional classifier is trained in a supervised manner with labels corresponding to the actual domain to which the learning examples belong. Here, we use the exact location of the data sources as domain labels.

To make these baselines comparable with our models, we make sure to get the same complexity, i.e., a comparable number of parameters. We use the f1-score in order to assess performances of the architectures. We compute this metric following the method recommended in [7] to alleviate bias that could stem from unbalanced class distribution. In addition, to alleviate the performance over-estimation problem due to neighborhood bias, we rely in our experiments on meta-segmented partitioning.

4.1 Performance Comparison

We conduct extensive experiments to evaluate the performance of the proposed algorithm in the following two settings: activity recognition (or classification) task and representation disentanglement. For the activity recognition setting, Table 1 summarizes the performance comparison of the baselines in terms of the f1-score obtained on the SHL and Fusion datasets. Here we assess the usefulness of the separated components per se by leveraging them in a traditional discriminative setting. In other words, we take the learned representation and add a

Table 1. Recognition performances (f1-score) of the baseline models on different representative related datasets. Evaluation based on the meta-segmented cross-validation.

Model	Fusion	SHL (Acc.)	SHL
DeepConvLSTM	68.5 ± .002	64.4 ± .0078	65.3 ± .0206
DeepSense	69.1 ± .0017	64.8 ± .0033	66.5 ± .006
AttnSense	70.3 ± .0027	69.6 ± .0072	68.4 ± .03
GILE	71.7 ± .014	71.1 ± .035	69.0 ± .001
FedAbstract	75.7 ± .047	75.7 ± .047	77.3 ± .017

simple dense layer on top of it. This additional layer is trained to minimize classification loss while the rest of the circuit is kept frozen. Experimental results show that the proposed approach exhibits superior performance compared to the baselines. The proposed method achieves promising improvements in terms of f1-score over the baseline methods. In particular, our proposed approach improves recognition performances by approximately 7–9% on Fusion and SHL, while the improvement of attention-based methods is only about 1–2%. Compared to GILE, our approach shows consistent improvement on the considered configurations. This demonstrates that leveraging knowledge about the structure of the deployment, instead of simply using domain labels corresponding to the exact location of the data sources, improves disentanglement and ultimately activity recognition.

In the representation disentanglement setting, we assess the separation quality between the universal and position-specific components as well as those related to the actions of each subgroup. For this, the average latent magnitude computed for each dimension of the learned representations constitutes an appropriate measure. Figure 4 illustrates the average latent magnitude computed for the group of relative geometry representation. It shows the activated latent dimensions depending on the subgroup of transformations (among Bag, Hand, and Hips) acting on the data sources. We can see in particular that specific dimensions are activated depending on the subgroup of transformations that are used to stimulate the learned representation. These dimensions are also independent of each other. Furthermore, in complementary experiments, one can observe the evolution of the dimensions of the central learner's latent representation where some of them are getting more activated than others, which is a sign of the emergence of the desired universal components shared across the learners.

4.2 Ablation Study

To demonstrate the generalization and effectiveness of each component of our proposed approach, we further design and perform ablation experiments on the SHL and Fusion datasets. We compare FedAbstract to FedAvg [22] and advanced solutions which try to correct for client-drift including SCAFFOLD [16]. FedAvg

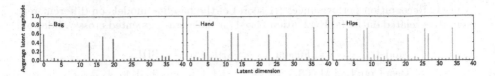

Fig. 4. Average latent encoding magnitude in the SHL dataset. It shows the repartition of the latent dimensions being activated between the different subgroups of transformations acting on the data sources (Bag, Hand, and Hips positions).

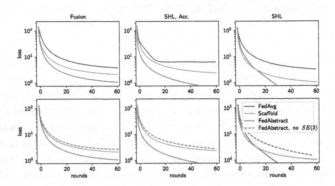

Fig. 5. Evolution of the loss during decentralized learning. (top) FedAbstract with both the relativity and decomposition constraints. (bottom) FedAbstract without the relativity representation constraints (FedAbstract, no $SE(3)$).

and SCAFFOLD do not perform explicit separation of the local data and thus constitute suitable baselines to assess the impact of each of FedAbstract's components. The experimental results illustrated in Fig. 5 (top) are obtained using Fed-Abstract with both the relativity and decomposition constraints. These results suggest that the evolution of the loss in the case of FedAvg gets slower as we increase the number of local steps, which corresponds to the common observation that client-drift increases proportionally to the number of local steps, hindering progress. At the same time, we observe that FedAbstract has excellent performance, slightly better than SCAFFOLD, suggesting a close connection between the estimate of the client-drift c_i and the position-specific components obtained via our proposed separation process.

Furthermore, we evaluate the effectiveness of explicitly representing the data generators' relativity via group actions while learning the universal and position-specific transformations. For this, we evaluate the performance of our proposed approach against a setting that does not specifically consider the relative geometry of the data generators. Basically, in this setting, the constraint imposing the relative geometry is not enforced during the learning process. Figure 5 (bottom) illustrates the obtained results in terms of the loss evolution on both SHL and Fusion datasets. We notice that compared to the basic setting, enforcement of the relative geometry consistently improves the convergence by 5% on SHL and 3% on Fusion. We see that these differences correspond to the gap between SCAFFOLD and our proposed approach. This demonstrates that the separation process constrained by the explicit representation of relativity ultimately leads to improving collaboration across the decentralized devices.

5   Conclusion and Future Work

In this work, we address the problem of decentralized learning in structured sensing environments. We propose a novel approach that leverages additional knowledge in terms of symmetries and invariants that emerge in these kinds of environments. These symmetries and invariants are explicitly represented in the form of group actions and incorporated into the learning process. Further, the proposed separation process of the data into universal and position-specific components improves collaboration across the decentralized devices materialized by the conciliation (or aggregation) process. Obtained results on activity recognition, an example of real-world structured sensing applications, are encouraging and open-up perspectives for studying more symmetries, invariants, and also equivariants that emerge in these environments. Future work also includes leveraging these symmetries and invariants from a theoretical perspective like Lie group and corresponding algebra, a special and large class of continuous groups that includes many valuable transformations like translations, rotations, and scalings and which also proposes a principled way for handling operations on the transformations such as composition, inversion, differentiation, and interpolation. The broader idea is that *universal* data is not directly accessible. On the other hand, it can be attained through various decentralized points of view. Collaboration is not a confrontation but rather the addition of relevant symmetries and complementary information from each viewpoint whose contribution can be determined precisely. The model we propose achieves this.

References

1. Aghajan, H., Cavallaro, A.: Multi-camera Networks: Principles and Applications. Academic Press (2009)
2. Andrew, G., Arora, R., Bilmes, J., Livescu, K.: Deep canonical correlation analysis. In: ICML, pp. 1247–1255. PMLR (2013)
3. Carollo, J., et al.: Relative phase measures of intersegmental coordination describe motor control impairments in children with cerebral palsy who exhibit stiff-knee gait. Clin. Biomech. **59**, 40–46 (2018)
4. Caselles-Dupré, et al.: Symmetry-based disentangled representation learning requires interaction with environments. In: NeurIPS, vol. 32, pp. 4606–4615 (2019)
5. Esteves, C., Xu, Y., Allen-Blanchette, C., Daniilidis, K.: Equivariant multi-view networks. In: IEEE/CVF ICCV, pp. 1568–1577 (2019)
6. Finzi, M., et al.: Generalizing convolutional neural networks for equivariance to lie groups on arbitrary continuous data. In: ICML, pp. 3165–3176 (2020)
7. Forman, G., Scholz, M.: Apples-to-apples in cross-validation studies. ACM SIGKDD **12**(1), 49–57 (2010)

8. Gjoreski, H., et al.: The university of Sussex-Huawei locomotion and transportation dataset for multimodal analytics with mobile devices. IEEE (2018)
9. Hamidi, M., Osmani, A.: Data generation process modeling for activity recognition. In: Dong, Y., Mladenić, D., Saunders, C. (eds.) ECML PKDD 2020. LNCS (LNAI), vol. 12460, pp. 374–390. Springer, Cham (2021). https://doi.org/10.1007/978-3-030-67667-4_23
10. Hamidi, M., Osmani, A.: Human activity recognition: a dynamic inductive bias selection perspective. Sensors **21**(21), 7278 (2021)
11. Hamidi, M., Osmani, A., Alizadeh, P.: A multi-view architecture for the SHL challenge. In: UbiComp-ISWC, pp. 317–322 (2020)
12. Higgins, I., Amos, D., Pfau, D., Racaniere, S., Matthey, L., et al.: Towards a definition of disentangled representations. arXiv:1812.02230 (2018)
13. Higgins, I., Matthey, L., Pal, A., Burgess, C., Glorot, X., et al.: Beta-VAE: learning basic visual concepts with a constrained variational framework. In: ICLR (2017)
14. Hsieh, K., Phanishayee, A., Mutlu, O., Gibbons, P.: The non-IID data quagmire of decentralized machine learning. In: ICML, pp. 4387–4398 (2020)
15. Kairouz, P., et al.: Advances and open problems in federated learning. arXiv:1912.04977 (2019)
16. Karimireddy, S.P., et al.: SCAFFOLD: stochastic controlled averaging for federated learning. In: ICML, pp. 5132–5143 (2020)
17. Khaled, A., Mishchenko, K., Richtárik, P.: Tighter theory for local SGD on identical and heterogeneous data. In: AISTATS, pp. 4519–4529 (2020)
18. Kingma, D., Welling, M.: Auto-encoding variational bayes. arXiv:1312.6114 (2013)
19. Li, T., Sahu, A.K., Zaheer, M., Sanjabi, M., Talwalkar, A., Smith, V.: Federated optimization in heterogeneous networks. MLSys **2**, 429–450 (2020)
20. Ma, H., Li, W., Zhang, X., Gao, S., Lu, S.: AttnSense: multi-level attention mechanism for multimodal human activity recognition. In: IJCAI, pp. 3109–3115 (2019)
21. Mathieu, E., Rainforth, T., Siddharth, N., Teh, Y.W.: Disentangling disentanglement in variational autoencoders. In: ICML, pp. 4402–4412 (2019)
22. McMahan, B., et al.: Communication-efficient learning of deep networks from decentralized data. In: AISTATS, pp. 1273–1282 (2017)
23. Melendez-Calderon, A., Shirota, C., Balasubramanian, S.: Estimating movement smoothness from inertial measurement units. bioRxiv (2020)
24. Ordóñez, F.J., Roggen, D.: Deep convolutional and LSTM recurrent neural networks for multimodal wearable activity recognition. Sensors **16**(1), 115 (2016)
25. Osmani, A., Hamidi, M.: Reduction of the position bias via multi-level learning for activity recognition. In: Gama, J., Li, T., Yu, Y., Chen, E., Zheng, Y., Teng, F. (eds.) PAKDD 2022. LNCS, pp. 289–302. Springer, Heidelberg (2022). https://doi.org/10.1007/978-3-031-05936-0_23
26. Qian, H., et al.: Latent independent excitation for generalizable sensor-based cross-person activity recognition. In: AAAI, vol. 35, pp. 11921–11929 (2021)
27. Quessard, R., Barrett, T., Clements, W.: Learning disentangled representations and group structure of dynamical environments. In: NeurIPS, vol. 33 (2020)
28. Shoaib, M., Bosch, S., Incel, O.D., et al.: Fusion of smartphone motion sensors for physical activity recognition. Sensors **14**(6), 10146–10176 (2014)
29. Stisen, A., et al.: Smart devices are different: assessing and mitigatingmobile sensing heterogeneities for activity recognition. In: ACM SenSys, pp. 127–140 (2015)
30. Dinh, C.T., Tran, N., Nguyen, T.D.: Personalized federated learning with Moreau envelopes. In: NeurIPS, vol. 33 (2020)
31. Vapnik, V., Izmailov, R.: Complete statistical theory of learning: learning using statistical invariants. In: COPA, pp. 4–40. PMLR (2020)

32. Vemulapalli, R., Arrate, F., Chellappa, R.: Human action recognition by representing 3D skeletons as points in a lie group. In: IEEE CVPR, pp. 588–595 (2014)
33. Wu, C., Khalili, A.H., Aghajan, H.: Multiview activity recognition in smart homes with spatio-temporal features. In: ACM/IEEE ICDSC, pp. 142–149 (2010)
34. Yang, J.Y., et al.: Using acceleration measurements for activity recognition. Pattern Recogn. Lett. **29**(16), 2213–2220 (2008)
35. Yao, S., et al.: DeepSense: a unified deep learning framework for time-series mobile sensing data processing. In: WWW, pp. 351–360 (2017)
36. Yu, H., et al.: Parallel restarted SGD with faster convergence and less communication. In: AAAI, pp. 5693–5700 (2019)

Efficient Automated Deep Learning
for Time Series Forecasting

Difan Deng[1(✉)], Florian Karl[2,3], Frank Hutter[4,5], Bernd Bischl[2,3],
and Marius Lindauer[1]

[1] Leibniz University Hannover, Hannover, Germany
d.deng@ai.uni-hannover.de
[2] Ludwig-Maximilian University, Munich, Germany
[3] Fraunhofer Institute for Integrated Circuits (IIS), Erlangen, Germany
[4] University of Freiburg, Freiburg, Germany
[5] Bosch Center for Artificial Intelligence, Renningen, Germany

Abstract. Recent years have witnessed tremendously improved efficiency of Automated Machine Learning (AutoML), especially Automated Deep Learning (AutoDL) systems, but recent work focuses on tabular, image, or NLP tasks. So far, little attention has been paid to general AutoDL frameworks for time series forecasting, despite the enormous success in applying different novel architectures to such tasks. In this paper, we propose an efficient approach for the joint optimization of neural architecture and hyperparameters of the entire data processing pipeline for time series forecasting. In contrast to common NAS search spaces, we designed a novel neural architecture search space covering various state-of-the-art architectures, allowing for an efficient macro-search over different DL approaches. To efficiently search in such a large configuration space, we use Bayesian optimization with multi-fidelity optimization. We empirically study several different budget types enabling efficient multi-fidelity optimization on different forecasting datasets. Furthermore, we compared our resulting system, dubbed `Auto-PyTorch-TS`, against several established baselines and show that it significantly outperforms all of them across several datasets.

Keywords: AutoML · Deep learning · Time series forecasting · Neural architecture search

1   Introduction

Time series (TS) forecasting plays a key role in many business and industrial problems, because an accurate forecasting model is a crucial part of a data-driven decision-making system. Previous forecasting approaches mainly consider each individual time series as one task and create a local model [3,7,26]. In recent years, with growing dataset size and the ascent of Deep Learning (DL), research interests have shifted to global forecasting models that are able to learn information across all time series in a dataset collected from similar sources [20,41].

Supplementary Information The online version contains supplementary material available at https://doi.org/10.1007/978-3-031-26409-2_40.

Given the strong ability of DL models to learn complex feature representations from a large amount of data, there is a growing trend of applying new DL models to forecasting tasks [38, 46, 50, 57].

Automated machine learning (AutoML) addresses the need of choosing the architecture and its hyperparameters depending on the task at hand to achieve peak predictive performance. The former is formalized as neural architecture search (NAS) [14] and the latter as hyperparameter optimization (HPO) [17]. Several techniques from the fields of NAS and HPO have been successfully applied to tabular and image benchmarks [15, 18, 33, 62]. Recent works have also shown that jointly optimizing both problems provides superior models that better capture the underlying structure of the target task [61, 62].

Although the principle idea of applying AutoML to time series forecasting models is very natural, there are only few prior approaches addressing this [32, 37, 43, 52]. In fact, combining state-of-the-art AutoML methods, such as Bayesian Optimization with multi-fidelity optimization [16, 30, 34, 36], with state-of-the-art time series forecasting models leads to several challenges we address in this paper. First, recent approaches for NAS mainly cover cell search spaces, allowing only for a very limited design space, that does not support different macro designs [12, 60]. Our goal is to search over a large variety of different architectures covering state-of-the-art ideas. Second, evaluating DL models for time series forecasting is fairly expensive and a machine learning practicioner may not be able to afford many model evaluations. Multi-fidelity optimization, e.g. [36], was proposed to alleviate this problem by only allocating a fraction of the resources to evaluated configurations and promoting the most promising configurations to give them additional resources. Third, as a consequence of applying multi-fidelity optimization, we have to choose how different fidelities are defined, i.e. what kind of budget is used. Examples for such *budget types* are number of epochs, dataset size or time series length. Depending on the correlation between lower and highest fidelity, multi-fidelity optimization can boost the efficiency of AutoML greatly or even slow it down in the worst case. Since we are the first to consider multi-fidelity optimization for AutoML on time series forecasting, we studied the efficiency of different budget types across many datasets. Fourth, all of these need to be put together; to that effect, we propose a new open-source package for Automated Deep Learning (AutoDL) for time series forecasting, dubbed `Auto-PyTorch-TS`.[1] Specifically, our contributions are as follows:

1. We propose the AutoDL framework `Auto-PyTorch-TS` that is able to jointly optimize the architecture and the corresponding hyperparameters for a given dataset for time series forecasting.
2. We present a unified architecture configuration space that contains several state-of-the-art forecasting architectures, allowing for a flexible and powerful macro-search.
3. We provide insights into the configuration space of `Auto-PyTorch-TS` by studying the most important design decisions and show that different architectures are reasonable for different datasets.

---

[1] The code is available under https://github.com/automl/Auto-PyTorch.

4. We show that `Auto-PyTorch-TS` is able to outperform a set of well-known traditional statistical models and modern deep learning models with an average relative error reduction of 19% against the best baseline across many forecasting datasets.

2   Related Work

We start by discussing the most closely related work in DL for time series forecasting, AutoDL, and AutoML for time series forecasting.

2.1   Deep Learning Based Forecasting

Early work on forecasting focused on building a local model for each individual series to predict future values, ignoring the correlation between different series. In contrast, global forecasting models are able to capture information of multiple time series in a dataset and use this at prediction time [31]. With growing dataset size and availability of multiple time series from similar sources, this becomes increasingly appealing over local models. We will in the following briefly introduce some popular forecasting DL models.

Simple feed-forward MLPs have been used for time series forecasting and extended to more complex models. For example, the N-BEATS framework [46] is composed of multiple stacks, each consisting of several blocks. This architectural choice aligns with the main principle of modern architecture design: Networks should be designed in a block-wise manner instead of layer-wise [63].

Additionally, RNNs [9,23] were proposed to process sequential data and thus they are directly applicable to time series forecasting [22,57]. A typical RNN-based model is the Seq2Seq network [9] that contains an RNN encoder and decoder. Wen et al. [57] further replaced the Seq2Seq's RNN decoder with a multi-head MLP. Flunkert et al. [50] proposed DeepAR that wraps an RNN encoder as an auto-regressive model and uses it to iteratively generate new sample points based on sampled trajectories from the last time step.

In contrast, CNNs can extract local, spatially-invariant relationships. Similarly, time series data may have time-invariant relationships, which makes CNN-based models suitable for time series tasks, e.g. WaveNet [6,45] and Temporal Convolution Networks (TCN) [4]. Similar to RNNs, CNNs could also be wrapped by an auto-regressive model to recursively forecast future targets [6,45].

Last but not least, attention mechanisms and transformers have shown superior performance over RNNs on natural language processing tasks [56] and over CNNs on computer vision tasks [13]. Transformers and RNNs can also be combined; e.g. Lim et al. [38] proposed temporal fusion transformers (TFT) that stack a transformer layer on top of an RNN to combine the best of two worlds.

2.2   Automated Deep Learning (AutoDL)

State-of-the-art AutoML approaches include Bayesian Optimization (BO) [18], Evolutionary Algorithms (EA) [44], reinforcement learning [63] or ensembles [15].

Most of them consider AutoML system as a black-box optimization problem that aims at finding the most promising machine learning models and their optimal corresponding hyperparameters. Neural Architecture Search (NAS), on the other hand, only contains one search space: its architecture. NAS aims at finding the optimal architecture for the given task with a fixed set of hyperparameters. Similar to the traditional approach, the architecture could be optimized with BO [33,62], EA [49] or Reinforcement Learning [63] among others, but there also exist many NAS-specific speedup techniques, such as one-shot models [59] and zero-cost proxies [1]. In this work we follow the state-of-the-art approach from Auto-PyTorch [62] and search for both the optimal architecture and its hyperparameters with BO.

Training a deep neural network requires lots of computational resources. Multi-fidelity optimization [16,30,36] is a common approach to accelerate AutoML and AutoDL. It prevents the optimizer from investing too many resources on the poorly performing configurations and allows for spending more on the most promising ones. However, the correlation between different fidelities might be weak [60] for DL models, in which case the result on a lower fidelity will provide little information for those on higher fidelities. Thus, it is an open question how to properly select the budget type for a given target task, and researchers often revert to application-specific decisions.

2.3 AutoML for Time Series Forecasting

While automatic forecasting has been of interest in the research community in the past [28], dedicated AutoML approaches for time series forecasting problems have only been explored recently [21,32,35,42,51]. Optimization methods such as random search [55], genetic algorithms [10], monte carlo tree search and algorithms akin to multi-fidelity optimization [51] have been used among others. Paldino et al. [47] showed that AutoML frameworks not intended for time series forecasting originally - in combination with feature engineering - were not able to significantly outperform simple forecasting strategies; a similar approach is presented in [10]. As part of a review of AutoML for forecasting pipelines, Meisenbacher et al. [42] concluded that there is a need for optimizing the entire pipeline as existing works tend to only focus on certain parts. We took all of these into account by proposing `Auto-PyTorch-TS` as a framework that is specifically designed to optimize over a flexible and powerful configuration space of forecasting pipelines.

3 AutoPyTorch Forecasting

For designing an AutoML system, we need to consider the following components: optimization targets, configuration space and optimization algorithm. The high-level workflow of our `Auto-PyTorch-TS` framework is shown in Fig. 1; in many ways it functions similar to existing state-of-the-art AutoML frameworks [17,62]. To better be able to explain unique design choice for time series forecasting, we

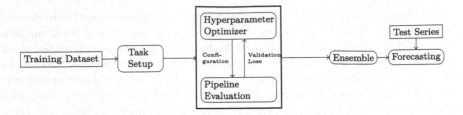

Fig. 1. An overview of `Auto-PyTorch-TS`. Given a dataset, `Auto-PyTorch-TS` automatically prepares the data to fit the requirement of a forecasting pipeline. The AutoML optimizer will then use the selected budget type to search for desirable neural architectures and hyperparameters from the pipeline configuration space. Finally, we create an ensemble out of the most promising pipelines to do the final forecasting on the test sets.

first present a formal statement of the forecasting problem and discuss challenges in evaluating forecasting pipelines before describing the components in detail.

3.1 Problem Definition

A multi-series forecasting task is defined as follows: given a dataset that contains N series: $\mathcal{D} = \{\mathcal{D}_i\}_{i=1}^N$ and $\mathcal{D}_i$ represents one series in the dataset: $\mathcal{D}_i = \{\mathbf{y}_{i,1:T_i}, \mathbf{x}_{i,1:T_i}^{(p)}, \mathbf{x}_{i,T_i+1:T_i+H}^{(f)}\}^2$, where T is the number of time steps until forecasting starts; H is the forecasting horizon that the model is required to predict; $\mathbf{y}_{1:T}$, $\mathbf{x}_{1:T}^{(p)}$ and $\mathbf{x}_{T_i+1:T_i+H}^{(f)}$ are the sets of observed past targets, past features and known future features values, respectively. The task of time series forecasting is to predict the possible future values with a model trained on $\mathcal{D}$:

$$\hat{\mathbf{y}}_{T+1:T+H} = f(\mathbf{y}_{1:T}, \mathbf{x}_{1:T+H}; \boldsymbol{\theta}) \tag{1}$$

where $\mathbf{x}_{1:T+H} := [\mathbf{x}_{1:T}^{(p)}, \mathbf{x}_{T+1:T+H}^{(f)}]$, $\boldsymbol{\theta}$ are the model parameters that are optimized with training losses $\mathcal{L}_{train}$, and $\hat{\mathbf{y}}_{T+1:T+H}$ are the predicted future target values. Depending on the model type, $\hat{\mathbf{y}}_{T+1:T+H}$ can be distributions [50] or scalar values [46]. Finally, the forecasting quality is measured by the discrepancy between the predicted targets $\hat{\mathbf{y}}_{T+1:T+H}$ and the ground truth future targets $\mathbf{y}_{T+1:T+H}$ according to a defined loss function $\mathcal{L}$. The most commonly applied metrics include mean absolute scaled error (MASE), mean absolute percentage error (MAPE), symmetric mean absolute percentage error (sMAPE) and mean absolute error (MAE) [19,29,46].

3.2 Evaluating Forecasting Pipelines

We split each sequence into three parts to obtain: a training set $\mathcal{D}_{\text{train}} = \{\mathbf{y}_{1:T-H}, \mathbf{x}_{1:T+H}\}$, a validation set $\mathcal{D}_{\text{val}} = \{\mathbf{y}_{T-H+1:T}, \mathbf{x}_{T-H+1:T}\}$ and a test

---

^{2} For the sake of brevity, we omit the sequence index i in the following part of this paper unless stated otherwise.

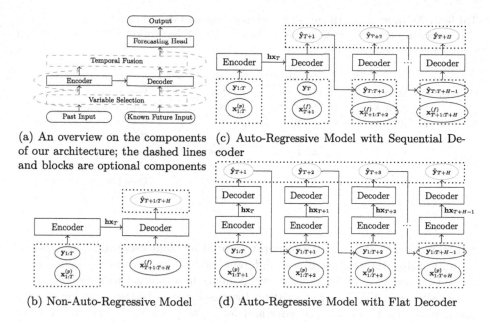

(a) An overview on the components of our architecture; the dashed lines and blocks are optional components

(c) Auto-Regressive Model with Sequential Decoder

(b) Non-Auto-Regressive Model

(d) Auto-Regressive Model with Flat Decoder

Fig. 2. Overview of the architectures that can be built by our framework. (a) shows the main components of our architecture space. (b)–(d) are specific instances of (a) and its data flow given different architecture properties.

set $\mathcal{D}_{test} = \{\mathbf{y}_{T+1:T+H}, \mathbf{x}_{T+1:T+H}\}$, i.e., the tails of each sequences are reserved as $\mathcal{D}_{val}$. At each iteration, our AutoML optimizer suggests a new hyperparameter and architecture configuration $\boldsymbol{\lambda}$, trains it on $\mathcal{D}_{train}$ and evaluates it on $\mathcal{D}_{val}$.

Both in AutoML frameworks [18,62] and in forecasting frameworks [46], ensembling of models is a common approach. We combine these two worlds in `Auto-PyTorch-TS` by using ensemble selection [8] to construct a weighted ensemble that is composed of the best k forecasting models from the previously evaluated configurations $\mathcal{D}_{hist}$. Finally, we retrain all ensemble members on $\mathcal{D}_{val} \cup \mathcal{D}_{train}$ before evaluating on $\mathcal{D}_{test}$.

3.3 Forecasting Pipeline Configuration Space

Existing DL packages for time series forecasting [2,5] follow the typical structure of traditional machine learning libraries: models are built individually with their own hyperparameters. Similar to other established AutoML tools [15,18,44], we designed the configuration space of `Auto-PyTorch-TS` as a combined algorithm selection and hyperparameter (CASH) problem [53], i.e., the optimizer first selects the most promising algorithms and then optimizes for their optimal hyperparameter configurations, with a hierarchy of design decisions. Deep neural networks, however, are built with stacked blocks [63] that can be disentangled to fit different requirements [58]. For instance, Seq2Seq [9], MQ-RNN [57] and DeepAR [50] all contain an RNN as their encoders. These models naturally share

Table 1. An overview of the possible combinations and design decisions of the models that exists in our configuration space. Only the TFT Network contains the optional components presented in Fig. 2a.

Encoder		Decoder	Auto-regressive	Architecture class
Flat encoder	MLP	MLP	No	Feed Forward Network
	N-BEATS	N-BEATS	No	N-BEATS [46]
Seq. encoder	RNN/Transformer	RNN/Transformer	Yes	Seq2Seq [9]
			No	TFT [38]
		MLP	Yes	DeepAR [50]
			No	MQ-RNN [57]
	TCN	MLP	Yes	DeepAR [50]/WaveNet [45]
			No	MQ-CNN [57]

common aspects and cannot be simply treated as completely different models. To fully utilize the relationships of different models, we propose a configuration space that includes all the possible components in a forecasting network.

As shown in Fig. 2a, most existing forecasting architectures can be decomposed into 3 parts: *encoder, decoder* and *forecasting heads*: the encoder receives the past target values and embeds them into the latent space. The latent embedding, together with the known future features (if applicable), are fed to the decoder network; the output of the decoder network is finally passed to the forecasting head to generate a sequence of scalar values or distributions, depending on the type of forecasting head. Additionally, the *variable selection, temporal fusion* and *skip connection layers* introduced by TFT [38] can be seamlessly integrated into our networks and are treated as optional components.

Table 1 lists all possible choices of encoders, decoders, and their corresponding architectures in our configuration space. Specifically, we define two types of network components: sequential encoder (Seq. Encoder) and flat encoder (Flat Encoder). The former (e.g., RNN, Transformer and TCN) directly processes sequential data and output a new sequence; the latter (e.g., MLP and N-BEATS) needs to flatten the sequential data into a 2D matrix to fuse the information from different time steps. Through this configuration space, `Auto-PyTorch-TS` is able to encompass the "convex hull" of several state-of-the-art global forecasting models and tune them.

As shown in Fig. 2, given the properties of encoders, decoders, and models themselves, we construct three types of architectures that forecast the future targets in different ways. Non-Auto-Regressive models (Fig. 2b), including MLP, MQ-RNN, MQ-CNN, N-BEATS and TFT, forecast the multi-horizontal predictions within one single step. In contrast, Auto-Regressive models do only one-step forecasting within each forward pass. The generated forecasting values are then iteratively fed to the network to forecast the value at the next time step. All the auto-regressive models are trained with teacher forcing [22]. Only sequential networks could serve as an encoder in auto-regressive models, however, we could select both sequential and flat decoders for auto-regressive models. Sequential decoders are capable of independently receiving the newly generated predictions. We consider this class of architectures as a Seq2Seq [9] model: we first feed the past input values to the encoder to generate its output **hx** and then pass **hx** to the decoder, as shown in Fig. 2c. Having acquired **hx**, the decoder then generates a sequence of predictions with the generated predictions and known future values by itself. Finally, Auto-Regressive Models with flat decoders are classified as the family of DeepAR models [50]. As the decoder could not collect more information as the number of generated samples increases, we need to feed the generated samples back to the encoder, as shown in Fig. 2d.

Besides its architectures, hyperparemeters also play an important role on the performance of a deep neural network [61], for the details of other hyperparameters in our configuration space, we refer to the Appendix.

3.4 Hyperparameter Optimization

We optimize the loss on the validation set $\mathcal{L}_{\mathcal{D}_{val}}$ with BO [17]. It is known for its sample efficiency, making it a good approach for expensive black-box optimization tasks, such as AutoDL for expensive global forecasting DL models. Specifically, we optimize the hyperparameters with SMAC [25][3] that constructs a random forest to model the loss distribution over the configuration space.

Similar to other AutoML tools [18,62] for supervised classification, we utilize multi-fidelity optimization to achieve better any-time performance. Multi-fidelity optimizers start with the lowest budget and gradually assign higher budgets to well-performing configurations. Thereby, the choice of what budget type to use is essential for the efficiency of a multi-fidelity optimizer. The most popular choices of budget type in DL tasks are the number of epochs and dataset size. For time series forecasting, we propose the following four different types of budget:

- Number of Epochs (*#Epochs*)
- Series Resolution (*Resolution*)
- Number of Series (*#Series*)
- Number of Samples in each Series (*#SMPs per Ser.*)

A higher *Resolution* indicates an extended sample interval. The sample interval is computed by the inverse of the fidelity value, e.g., a resolution fidelity of 0.1 indicates for each series we take every tenth point: we shrink the size of the sliding window accordingly to ensure that the lower fidelity optimizer does not receive more information than the higher fidelity optimizer. *#Series* means that we only sample a fraction of sequences to train our model. Finally, *#SMPs per Ser.* indicates that we decrease the expected value of the number of samples within each sequence; see Sect. 3.2 for sample-generation method. Next to these multi-fidelity variants, we also consider vanilla Bayesian optimization (*Vanilla BO*) using the maximum of all these fidelities.

---

[3] We used SMAC3 [39] from https://github.com/automl/SMAC3.

3.5   Proxy-Evaluation on Many Time Series

All trained models must query every series to evaluate $\mathcal{L}_{val}$. However, the number of series could be quite large. Additionally, many forecasting models (e.g., DeepAR) are cheap to be trained but expensive during inference time. As a result, rather than training time, inference time is more likely to become a bottleneck to optimize the hyperparameters on a large dataset (for instance, with 10k series or more), where configuration with lower fidelities would no longer provide the desirable speed-up when using the full validation set. Thereby, we consider a different evaluation strategy on large datasets (with more than 1k series) and lower budgets: we ask the model to only evaluate a fraction of the validation set (we call this fraction "proxy validation set") while the other series are predicted by a dummy forecaster (which simply repeats the last target value in the training series, i.e., $\mathbf{y}_T$, H times). The size of the proxy validation set is proportional to the budget allocated to the configuration: maximal budget indicates that the model needs to evaluate the entire validation set. We set the minimal number of series in the proxy set to be 1k to ensure that it contains enough information from the validation set. The proxy validation set is generated with a grid to ensure that all the configurations under the same fidelity are evaluated on the same proxy set.

4   Experiments

We evaluate `Auto-PyTorch-TS` on the established benchmarks of the Monash Time Series Forecasting Repository [20][4]. This repository contains various datasets that come from different domains, which allows us to assess the robustness of our framework against different data distributions. Additionally, it records the performance of several models, including local models [3,7,11,26,27], global traditional machine learning models [48,54], and global DL models [2,6,46,50,56] on $\mathcal{D}_{\text{test}}$, see [20] for details. For evaluating `Auto-PyTorch-TS`, we will follow the exact same protocol and dataset splits. We focus our comparison of `Auto-PyTorch-TS` against two types of baselines: (i) the overall single best baseline from [20], assuming a user would have the required expert knowledge and (ii) the best dataset-specific baseline. We note that the latter is a very strong baseline and a priori it is not known which baseline would be best for a given dataset; thus we call it the *theoretical oracle baseline*. Since the Monash Time Series Forecasting Repository does not record the standard deviation of each method, we reran those baselines on our cluster for 5 times. Compared to the repository, our configuration space includes one more strong class of algorithms, TFT [38], which we added to our set of baselines to ensure a fair and even harder comparison.

We set up our task following the method described in Sect. 3.2: HPO is only executed on $\mathcal{D}_{train/val}$ while H is given by the original repository. As described in Sect. 3.2, we create an ensemble with size 20 that collects multiple models

---

[4] https://forecastingdata.org/.

during the course of optimization. When the search finishes, we refit the ensemble to the union of $\mathcal{D}_{train/val}$ and evaluate the refitted model on $\mathcal{D}_{\text{test}}$. Both $\mathcal{L}_{val}$ and $\mathcal{L}_{test}$ are measured with the mean value of MASE [29] across all the series in the dataset. To leverage available expert knowledge, `Auto-PyTorch-TS` runs an initial design with the default configurations of each model in Table 1. Please note that this initial design will be evaluated on the smallest available fidelity. All multi-fidelity variants of `Auto-PyTorch-TS` start with the cheapest fidelity of 1/9, use then 1/3 and end with the highest fidelity (1.0). The runs of `Auto-PyTorch-TS` are repeated 5 times with different random seeds.

We ran all the datasets on a cluster node equipped with 8 Intel Xeon Gold 6254@ 3.10 GHz CPU cores and one NVIDIA GTX 2080TI GPU equipped with PyTorch 1.10 and Cuda 11.6. The hyperparameters were optimized with SMAC3 v1.0.1 for 10 h, and then we refit the ensemble on $\mathcal{D}_{train/val}$ and evaluate it on the test set. All the jobs were finished within 12 h.

4.1   Time Series Forecasting

Table 2 shows how different variants of `Auto-PyTorch-TS` perform against the two types of baselines across multiple datasets. Even using the theoretical oracle baseline for comparison, `Auto-PyTorch-TS` is able to outperform it on 18 out of 24 datasets. On the other 6 datasets, it achieved nearly the same performance as the baselines. On average, we were able to reduce the MASE by up to 5% against the oracle and by up to 19% against the single best baseline, establishing a new robust state-of-the-art overall.

Surprisingly, the forecasting-specific budget types did not perform significantly better than the number of epochs (the common budget type in classification). Nevertheless, the optimal choice of budget type varies across datasets, which aligns with our intuition that on a given dataset the correlation between lower and higher fidelities may be stronger for certain budget types than for other types. If we were to construct a theoretically optimal budget-type selector, which utilizes the best-performing budget type for a given dataset, we would reduce the relative error by 2% over the single best (i.e., # SMPs per Ser.).

4.2   Hyperparameter Importance

Although HPO is often considered as a black-box optimization problem [17], it is important to shed light on the importance of different hyperparameters to provide insights into the design choice of DL models and to indicate how to design the next generation of AutoDL systems.

Here we evaluate the importance of the hyperparameters with a global analysis based on fANOVA [24], which measures the importance of hyperparameters by the variance caused by changing one single hyperparameter while marginalizing over the effect of all other hyperparameters. Results on individual datasets can be found in appendix.

For each of the 10 most important hyperparameters in our configuration space (of more than 200 dimensions), Fig. 3 shows a box plot of the importance

Table 2. We compare variants of Auto-PyTorch-TS against the single best baseline (TBATS) and a theoretically optimal oracle of choosing the correct baseline for each dataset wrt mean MASE errors on the test sets. We show the mean and standard deviation for each dataset. The best results are highlighted in bold-face. We computed the relative improvement wrt the Oracle Baseline on each dataset and used the geometric average for aggregation over the datasets.

	Auto-PyTorch-TS					Best dataset-specific baseline	Overall single best baseline
	# Epochs	Resolution	# Series	# SMPs per Ser.	Vanilla BO		
M3 yearly	2.73(0.10)	2.66(0.05)	2.76(0.09)	**2.64**(0.09)	2.68(0.08)	2.77(0.00)	3.13(0.00)
M3 quarterly	**1.08**(0.01)	1.10(0.01)	1.10(0.01)	1.09(0.02)	1.12(0.03)	1.12(0.00)	1.26(0.00)
M3 monthly	**0.85**(0.01)	0.89(0.02)	0.86(0.01)	0.87(0.04)	0.86(0.02)	0.86(0.00)	0.86(0.00)
M3 Other	1.90(0.07)	1.82(0.03)	1.98(0.13)	1.92(0.05)	1.95(0.15)	**1.81**(0.00)	1.85(0.00)
M4 quarterly	1.15(0.01)	1.13(0.01)	**1.13**(0.01)	1.15(0.01)	1.15(0.02)	1.16(0.00)	1.19(0.00)
M4 monthly	0.93(0.02)	0.93(0.02)	0.93(0.02)	**0.93**(0.02)	0.96(0.02)	0.95(0.00)	1.05(0.00)
M4 weekly	0.44(0.01)	0.45(0.02)	**0.43**(0.02)	0.44(0.02)	0.45(0.01)	0.48(0.00)	0.50(0.00)
M4 daily	1.14(0.01)	1.18(0.07)	1.16(0.06)	1.14(0.04)	1.38(0.41)	**1.13**(0.02)	1.16(0.00)
M4 hourly	0.86(0.12)	0.95(0.11)	**0.78**(0.07)	0.85(0.07)	0.85(0.06)	1.66(0.00)	2.66(0.00)
M4 yearly	**3.05**(0.03)	3.08(0.04)	3.05(0.01)	3.09(0.04)	3.10(0.02)	3.38(0.00)	3.44(0.00)
Tourism quarterly	1.61(0.03)	1.57(0.05)	1.59(0.05)	1.59(0.02)	1.55(0.03)	**1.50**(0.01)	1.83(0.00)
Tourism monthly	**1.42**(0.03)	1.44(0.03)	1.45(0.04)	1.47(0.02)	1.42(0.02)	1.44(0.02)	1.75(0.00)
Dominick	0.51(0.04)	0.49(0.00)	**0.49**(0.01)	0.49(0.01)	0.49(0.01)	0.51(0.00)	0.72(0.00)
Kdd cup	1.20(0.02)	1.18(0.02)	1.18(0.03)	1.18(0.03)	1.20(0.03)	**1.17**(0.01)	1.39(0.00)
Weather	0.63(0.08)	0.58(0.04)	0.59(0.02)	0.59(0.06)	**0.57**(0.00)	0.64(0.01)	0.69(0.00)
NN5 daily	0.79(0.01)	0.80(0.01)	0.81(0.04)	**0.78**(0.01)	0.79(0.01)	0.86(0.00)	0.86(0.00)
NN5 weekly	0.76(0.01)	0.76(0.03)	0.76(0.01)	0.77(0.01)	**0.76**(0.01)	0.77(0.01)	0.87(0.00)
Hospital	0.76(0.01)	0.76(0.00)	0.76(0.00)	**0.75**(0.01)	0.75(0.01)	0.76(0.00)	0.77(0.00)
Traffic weekly	1.04(0.07)	1.10(0.03)	1.04(0.05)	1.08(0.09)	1.03(0.07)	**0.99**(0.03)	1.15(0.00)
Electricity weekly	0.78(0.04)	1.06(0.13)	0.80(0.04)	**0.74**(0.07)	0.85(0.11)	0.76(0.01)	0.79(0.00)
Electricity hourly	1.52(0.05)	1.54(0.00)	1.58(0.08)	1.54(0.06)	**1.51**(0.05)	1.60(0.02)	3.69(0.00)
Kaggle web traffic weekly	0.56(0.01)	0.56(0.00)	**0.55**(0.00)	0.57(0.01)	0.59(0.01)	0.61(0.02)	0.62(0.00)
Covid deaths	5.11(1.60)	4.54(0.05)	**4.43**(0.13)	4.58(0.30)	4.53(0.24)	5.16(0.04)	5.72(0.00)
Temperature rain	0.76(0.05)	0.75(0.01)	0.73(0.02)	0.73(0.03)	0.71(0.04)	**0.71**(0.03)	1.23(0.00)
Ø Rel. Impr Best	0.82	0.83	**0.81**	0.81	0.82	0.86	1.0
Ø Rel. Impr Oracle	0.96	0.97	**0.95**	0.95	0.96	1.0	1.17

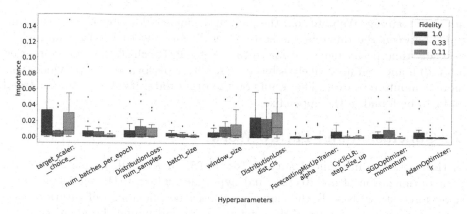

Fig. 3. Hyperparameter importance with fANOVA across all datasets of Table 2

across our datasets. The most important hyperparameters are closely associated with the training procedure: 3 of them control the optimizer of the neural network and its learning rate. Additionally, 4 hyperparameters (*window_size, num_batches_per_epoch, batch_size, target_scaler*) contribute to the sampler and data preprocessing, showing the importance of the data fed to the network. Finally, the fact that two hyperparameters controlling the data distribution are amongst the most important ones indicates that identifying the correct potential data distribution might be beneficial to the performance of the model.

4.3 Ablation Study

In Sect. 3.5, we propose to partially evaluate the validation set on larger datasets to further accelerate the optimization process. To study the efficiency gain of this approach, we compare evaluation on the full validation set vs the proxy-evaluation on parts of the validation set. We ran this ablation study on the largest dataset, namely "Dominick" (115 704 series).

Figure 4 shows the results. It takes much less time for our optimizer (blue) to finish the first configuration evaluations on the lowest fidelity, improving efficiency early on and showing the need of efficient validation and

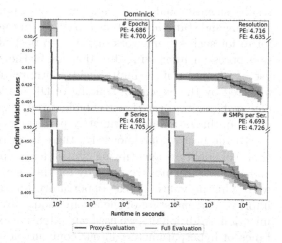

Fig. 4. Validation losses over time with different multi-fidelity approaches. We compute the area under the curve (AUC) of our approach (PE) and naive multi-fidelity optimizer (FE) and list them in the figures.

not only training. We note that the final performance does not change substantially between the different methods. Overall, `Auto-PyTorch-TS` achieves the best any-time performance. We note that `Auto-PyTorch-TS` has not converged after 10 h and will most likely achieve even better performance if provided with more compute resources. The results on the other datasets show a similar trend and can be found in the appendix.

5   Conclusion and Future Work

In this work, we introduced `Auto-PyTorch-TS`, an AutoDL framework for the joint optimization of architecture and hyperparameters of DL models for time series forecasting tasks. To this end, we propose a new flexible configuration space encompassing several state-of-the-art forecasting DL models by identifying key concepts in different model classes and combining them into a single framework.

Given the flexibility of our configuration space, new developers can easily adapt their architectures to our framework under the assumption that they can be formulated as an encoder-decoder-head architecture. Despite recent advances and competitive results, DL methods have until now not been considered the undisputed best approach in time series forecasting tasks: Traditional machine learning approaches and statistical methods have remained quite competitive [20,40]. By conducting a large benchmark, we demonstrated, that our proposed `Auto-PyTorch-TS` framework is able to outperform current state-of-the-art methods on a variety of forecasting datasets from different domains and even improves over a theoretically optimal oracle comprised of the best possible baseline model for each dataset.

While we were able to show superior performance over existing methods, our results suggest, that a combination of DL approaches with traditional machine learning and statistical methods could further improve performance. The optimal setup for such a framework and how to best utilize these model classes side by side poses an interesting direction for further research. Our framework makes use of BO and utilizes multi-fidelity optimization in order to alleviate the costs incurred by the expensive training of DL models. Our experiments empirically demonstrate, that the choice of budget type can have an influence on the quality of the optimization and ultimately performance.

To the best of our knowledge there is currently no research concerning the choice of fidelity when utilizing multi-fidelity optimization for architecture search and HPO of DL models; not only for time series forecasting, but other tasks as well. This provides a great opportunity for future research and could further improve current state-of-the-art methods already utilizing multi-fidelity optimization. Additionally, we used our extensive experiments to examine the importance of hyperparameters in our configuration space and were able to identify some of the critical choices for the configuration of DL architectures for time series forecasting. Finally, in contrast to previous AutoML systems, to the best of our knowledge, time series forecasting is the first task, where not only efficient training is important but also efficient validation. Although we showed empirical evidence for the problem and took a first step in the direction of efficient validation, it remains an open challenge for future work. `Auto-PyTorch-TS` can automatically optimize the hyperparameter configuration for a given task and can be viewed as a benchmark tool that isolates the influence of hyperparameter configurations of the model. This makes our framework an asset to the research community as it enables researchers to conveniently compare their methods to existing DL models.

Acknowledgements. Difan Deng and Marius Lindauer acknowledge financial support by the Federal Ministry for Economic Affairs and Energy of Germany in the project CoyPu under Grant No. 01MK21007L. Bernd Bischl acknowledges funding by the German Federal Ministry of Education and Research (BMBF) under Grant No. 01IS18036A. Florian Karl acknowledges support by the Bavarian Ministry of Economic Affairs, Regional Development and Energy through the Center for Analytics - Data - Applications (ADACenter) within the framework of BAYERN DIGITAL II (20-3410-2-9-8). Frank Hutter acknowledges support by European Research Council (ERC) Consolidator Grant "Deep Learning 2.0" (grant no. 101045765).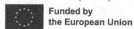

References

1. Abdelfattah, M.S., Mehrotra, A., Dudziak, Ł., Lane, N.D.: Zero-cost proxies for lightweight NAS. In: ICLR (2021)
2. Alexandrov, A., et al.: GluonTS: probabilistic and neural time series modeling in python. J. Mach. Learn. Res. **21**, 4629–4634 (2020)
3. Assimakopoulos, V., Nikolopoulos, K.: The theta model: a decomposition approach to forecasting. Int. J. Forecast. **16**(4), 521–530 (2000)
4. Bai, S., Kolter, J.Z., Koltun, V.: An empirical evaluation of generic convolutional and recurrent networks for sequence modeling. arXiv:1803.01271 (2018)
5. Beitner, J.: PyTorch forecasting: time series forecasting with PyTorch (2020)
6. Borovykh, A., Bohte, S., Oosterlee, C.W.: Conditional time series forecasting with convolutional neural networks. arXiv:1703.04691 (2017)
7. Box, G.E., Jenkins, G.M., Reinsel, G.C., Ljung, G.M.: Time Series Analysis: Forecasting and Control (2015)
8. Caruana, R., Niculescu-Mizil, A., Crew, G., Ksikes, A.: Ensemble selection from libraries of models. In: ICML (2004)
9. Cho, K., et al.: Learning phrase representations using RNN encoder-decoder for statistical machine translation (2014)
10. Dahl, S.M.J.: TSPO: an autoML approach to time series forecasting. Ph.D. thesis, Universidade NOVA de Lisboa (2020)
11. De Livera, A.M., Hyndman, R.J., Snyder, R.D.: Forecasting time series with complex seasonal patterns using exponential smoothing. J. Am. Stat. Assoc. **106**(496), 1513–1527 (2011)
12. Dong, X., Yang, Y.: NAS-Bench-201: extending the scope of reproducible neural architecture search. In: ICLR (2020)
13. Dosovitskiy, A., et al.: An image is worth 16×16 words: transformers for image recognition at scale. In: ICLR (2021)
14. Elsken, T., Metzen, J.H., Hutter, F.: Neural architecture search. In: Automatic Machine Learning: Methods, Systems, Challenges (2019)

15. Erickson, N., et al.: Autogluon-tabular: Robust and accurate automl for structured data. arXiv:2003.06505 (2020)
16. Falkner, S., Klein, A., Hutter, F.: BOHB: robust and efficient hyperparameter optimization at scale. In: ICML (2018)
17. Feurer, M., Hutter, F.: Hyperparameter optimization. In: Automatic Machine Learning: Methods, Systems, Challenges (2019)
18. Feurer, M., Klein, A., Eggensperger, K., Springenberg, J.T., Blum, M., Hutter, F.: Efficient and robust automated machine learning. In: NeurIPS (2015)
19. Flores, B.E.: A pragmatic view of accuracy measurement in forecasting. Omega **14**, 93–98 (1986)
20. Godahewa, R., Bergmeir, C., Webb, G.I., Hyndman, R.J., Montero-Manso, P.: Monash time series forecasting archive. In: NeurIPS Track on Datasets and Benchmarks (2021)
21. Halvari, T., Nurminen, J.K., Mikkonen, T.: Robustness of automl for time series forecasting in sensor networks. In: IFIP Networking Conference (2021)
22. Hewamalage, H., Bergmeir, C., Bandara, K.: Recurrent neural networks for time series forecasting: Current status and future directions. Int. J. Forecast. **37**, 388–427 (2021)
23. Hochreiter, S., Schmidhuber, J.: Long short-term memory. Neural Comput. **9**, 1735–1780 (1997)
24. Hutter, F., Hoos, H., Leyton-Brown, K.: An efficient approach for assessing hyperparameter importance. In: ICML (2014)
25. Hutter, F., Hoos, H.H., Leyton-Brown, K.: Sequential model-based optimization for general algorithm configuration. In: Coello, C.A.C. (ed.) LION 2011. LNCS, vol. 6683, pp. 507–523. Springer, Heidelberg (2011). https://doi.org/10.1007/978-3-642-25566-3_40
26. Hyndman, R., Koehler, A.B., Ord, J.K., Snyder, R.D.: Forecasting with Exponential Smoothing: The State Space Approach (2008)
27. Hyndman, R.J., Athanasopoulos, G.: Forecasting: Principles and Practice (2021)
28. Hyndman, R.J., Khandakar, Y.: Automatic time series forecasting: the forecast package for R. J. Stat. Softw. **27**, 1–22 (2008)
29. Hyndman, R.J., Koehler, A.B.: Another look at measures of forecast accuracy. Int. J. Forecast. **22**, 679–688 (2006)
30. Jamieson, K.G., Talwalkar, A.: Non-stochastic best arm identification and hyperparameter optimization. In: AISTA (2016)
31. Januschowski, T., et al.: Criteria for classifying forecasting methods. Int. J. Forecast. **36**(1), 167–177 (2020)
32. Javeri, I.Y., Toutiaee, M., Arpinar, I.B., Miller, J.A., Miller, T.W.: Improving neural networks for time-series forecasting using data augmentation and automl. In: BigDataService (2021)
33. Jin, H., Song, Q., Hu, X.: Auto-keras: an efficient neural architecture search system. In: SIGKDD (2019)
34. Klein, A., Tiao, L., Lienart, T., Archambeau, C., Seeger, M.: Model-based asynchronous hyperparameter and neural architecture search (2020)
35. Kurian, J.J., Dix, M., Amihai, I., Ceusters, G., Prabhune, A.: BOAT: a bayesian optimization automl time-series framework for industrial applications. In: BigDataService (2021)
36. Li, L., Jamieson, K.G., DeSalvo, G., Rostamizadeh, A., Talwalkar, A.: Hyperband: a novel bandit-based approach to hyperparameter optimization. J. Mach. Learn. Res. **18**, 6765–6816 (2017)

37. Li, T., Zhang, J., Bao, K., Liang, Y., Li, Y., Zheng, Y.: Autost: efficient neural architecture search for spatio-temporal prediction. In: SIGKDD (2020)
38. Lim, B., Arık, S.Ö., Loeff, N., Pfister, T.: Temporal fusion transformers for interpretable multi-horizon time series forecasting. Int. J. Forecast. **37**, 1748–1764 (2021)
39. Lindauer, M., et al.: SMAC3: a versatile Bayesian optimization package for hyperparameter optimization. J. Mach. Learn. Res. **23**, 1–9 (2022)
40. Makridakis, S., Spiliotis, E., Assimakopoulos, V.: The m4 competition: results, findings, conclusion and way forward. Int. J. Forecast. **34**, 802–808 (2018)
41. Makridakis, S., Spiliotis, E., Assimakopoulos, V.: The m4 competition: 100,000 time series and 61 forecasting methods. Int. J. Forecast. **36**, 54–74 (2020)
42. Meisenbacher, S., et al.: Review of automated time series forecasting pipelines. arXiv: 2202.01712 (2022)
43. Montero-Manso, P., Athanasopoulos, G., Hyndman, R.J., Talagala, T.S.: FFORMA: feature-based forecast model averaging. Int. J. Forecast. **36**(1), 86–92 (2020)
44. Olson, R.S., Bartley, N., Urbanowicz, R.J., Moore, J.H.: Evaluation of a tree-based pipeline optimization tool for automating data science. In: GECCO (2016)
45. van den Oord, A. et al.: WaveNet: a generative model for raw audio. In: ISCA Speech Synthesis Workshop (2016)
46. Oreshkin, B.N., Carpov, D., Chapados, N., Bengio, Y.: N-BEATS: neural basis expansion analysis for interpretable time series forecasting. In: ICLR (2020)
47. Paldino, G.M., De Stefani, J., De Caro, F., Bontempi, G.: Does automl outperform naive forecasting? In: Engineering Proceedings, vol. 5 (2021)
48. Prokhorenkova, L., Gusev, G., Vorobev, A., Dorogush, A.V., Gulin, A.: CatBoost: unbiased boosting with categorical features. In: NeurIPS (2018)
49. Real, E., Aggarwal, A., Huang, Y., Le, Q.V.: Regularized evolution for image classifier architecture search. In: AAAI (2019)
50. Salinas, D., Flunkert, V., Gasthaus, J., Januschowski, T.: DeepAR: probabilistic forecasting with autoregressive recurrent networks. Int. J. Forecast. **36**(3), 1181–1191 (2020)
51. Shah, S.Y., et al.: AutoAI-TS: Autoai for time series forecasting. In: SIGMOD (2021)
52. Talagala, T.S., Hyndman, R.J., Athanasopoulos, G., et al.: Meta-learning how to forecast time series. Monash Econometrics Bus. Stat. Working Pap. **6**, 16 (2018)
53. Thornton, C., Hutter, F., Hoos, H.H., Leyton-Brown, K.: Auto-WEKA: combined selection and hyperparameter optimization of classification algorithms. In: SIGKDD (2013)
54. Trapero, J.R., Kourentzes, N., Fildes, R.: On the identification of sales forecasting models in the presence of promotions. J. Oper. Res. Soc. **66**, 299–307 (2015)
55. van Kuppevelt, D., Meijer, C., Huber, F., van der Ploeg, A., Georgievska, S., van Hees, V.: Mcfly: automated deep learning on time series. SoftwareX **12**, 100548 (2020)
56. Vaswani, A., et al.: Attention is all you need. In: NeurIPS (2017)
57. Wen, R., Torkkola, K., Narayanaswamy, B., Madeka, D.: A multi-horizon quantile recurrent forecaster. In: 31st Conference on NeurIPS, Time Series Workshop (2017)
58. Wu, B., et al.: FBNetV5: Neural architecture search for multiple tasks in one run. arXiv:2111.10007 (2021)
59. Xiao, Y., Qiu, Y., Li, X.: A survey on one-shot neural architecture search. In: IOP Conference Series: Materials Science and Engineering, vol. 750. IOP Publishing (2020)

60. Ying, C., Klein, A., Christiansen, E., Real, E., Murphy, K., Hutter, F.: NAS-bench-101: towards reproducible neural architecture search. In: ICML (2019)
61. Zela, A., Klein, A., Falkner, S., Hutter, F.: Towards automated deep learning: efficient joint neural architecture and hyperparameter search. In: ICML 2018 AutoML Workshop (2018)
62. Zimmer, L., Lindauer, M., Hutter, F.: Auto-pytorch tabular: multi-fidelity metalearning for efficient and robust autodl. IEEE Trans. Pattern Anal. Mach. Intell. **43**(9), 3079–3090 (2021)
63. Zoph, B., Vasudevan, V., Shlens, J., Le, Q.V.: Learning transferable architectures for scalable image recognition. In: CVPR (2018)

Author Index

Printed in the United States
by Baker & Taylor Publisher Services